Essentials of Elementary & Interme

A Combined Course

Charles P. McKeague

SECOND PRINTING — December 2011
- Corrections made to odd answers section

THIRD PRINTING — January 2012
- Corrections made to selected chapter tests

FOURTH PRINTING — June 2012
- Miscellaneous changes made

FIFTH PRINTING — October 2012

Essentials of Elementary & Intermediate Algebra: A Combined Course

Charles P. McKeague

Publisher: XYZ Textbooks

Project Manager: Matthew Hoy

Editorial Assistants: Elizabeth Andrews, Stefanie Cohen, Graham Culbertson, Katherine Hofstra, Gordon Kirby, Aaron Salisbury, Katrina Smith, CJ Teuben

Composition: XYZ Textbooks

Sales: Amy Jacobs, Richard Jones, Bruce Spears, Rachael Hillman

Printed in the United States of America

ISBN-13: 978-1-936368-25-9 / ISBN-10: 1-936368-25-0

For product information and technology assistance, contact us at
XYZ Textbooks, 1-877-745-3499

For permission to use material from this text or product,
e-mail: **info@mathtv.com**

XYZ Textbooks
1339 Marsh Street
San Luis Obispo, CA 93401
USA

For your course and learning solutions, visit **www.xyztextbooks.com**

Brief Contents

1 Linear Equations and Inequalities 1

2 Linear Equations and Inequalities in Two Variables 85

3 Functions and Function Notation 149

4 Systems of Linear Equations 197

5 Exponents and Polynomials 237

6 Factoring 299

7 Quadratic Equations 353

8 Rational Expressions and Rational Functions 421

9 Rational Exponents and Roots 493

10 Exponential and Logarithmic Functions 555

Contents

1 Linear Equations and Inequalities 1

1.1 Simplifying Expressions 3
1.2 Addition Property of Equality 11
1.3 Multiplication Property of Equality 19
1.4 Solving Linear Equations 27
1.5 Formulas 35
1.6 Applications 47
1.7 More Applications 59
1.8 Linear Inequalities 67
1.9 Compound Inequalities 75
Chapter 1 Summary 81
Chapter 1 Test 84

2 Linear Equations and Inequalities in Two Variables 85

2.1 Paired Data and Graphing Ordered Pairs 87
2.2 Solutions to Linear Equations in Two Variables 95
2.3 Graphing Linear Equations in Two Variables 101
2.4 More on Graphing: Intercepts 111
2.5 The Slope of a Line 119
2.6 Finding the Equation of a Line 129
2.7 Linear Inequalities in Two Variables 139
Chapter 2 Summary 145
Chapter 2 Test 147

3 Functions and Function Notation 149

3.1 Introduction to Functions 151
3.2 Function Notation 165
3.3 Variation 175
3.4 Algebra and Composition with Functions 185
Chapter 3 Summary 193
Chapter 3 Test 195

4 Systems of Linear Equations 197

4.1 Solving Linear Systems by Graphing 199
4.2 The Elimination Method 207
4.3 The Substitution Method 217
4.4 Applications 225
Chapter 4 Summary 233
Chapter 4 Test 235

5 Exponents and Polynomials 237

5.1 Multiplication with Exponents 239
5.2 Division with Exponents 247
5.3 Operations with Monomials 259
5.4 Addition and Subtraction of Polynomials 267
5.5 Multiplication with Polynomials 273
5.6 Binomial Squares and Other Special Products 279
5.7 Dividing a Polynomial by a Monomial 285
5.8 Dividing a Polynomial by a Polynomial 289
Chapter 5 Summary 295
Chapter 5 Test 297

6 Factoring 299

6.1 The Greatest Common Factor and Factoring by Grouping 301
6.2 Factoring Trinomials 309
6.3 More Trinomials to Factor 313
6.4 The Difference of Two Squares 319
6.5 The Sum and Difference of Two Cubes 325
6.6 Factoring: A General Review 329
6.7 Solving Equations by Factoring 333
6.8 Applications 339
Chapter 6 Summary 349
Chapter 6 Test 351

7 Quadratic Equations 353

7.1 Completing the Square 355
7.2 The Quadratic Formula 367
7.3 Additional Items Involving Solutions to Equations 377
7.4 More Equations 385
7.5 Graphing Parabolas 393
7.6 Quadratic Inequalities 407
Chapter 7 Summary 417
Chapter 7 Test 419

8 Rational Expressions and Rational Functions 421

8.1 Basic Properties and Reducing to Lowest Terms 423
8.2 Multiplication and Division of Rational Expressions 433
8.3 Addition and Subtraction of Rational Expressions 441
8.4 Complex Fractions 451
8.5 Equations With Rational Expressions 457
8.6 Applications 467
8.7 Division of Polynomials 479
Chapter 8 Summary 489
Chapter 8 Test 492

9 Rational Exponents and Roots 493

9.1 Rational Exponents 495
9.2 Simplified Form for Radicals 507
9.3 Addition and Subtraction of Radical Expressions 519
9.4 Multiplication and Division of Radical Expressions 525
9.5 Equations Involving Radicals 533
9.6 Complex Numbers 543
Chapter 9 Summary 551
Chapter 9 Test 553

10 Exponential and Logarithmic Functions 555

10.1 Exponential Functions 557

10.2 The Inverse of a Function 567

10.3 Logarithms are Exponents 577

10.4 Properties of Logarithms 585

10.5 Common Logarithms and Natural Logarithms 593

10.6 Exponential Equations and Change of Base 603

Chapter 10 Summary 611

Chapter 10 Test 613

Preface to the Instructor

We have designed this book to help solve problems that you may encounter in the classroom.

Solutions to Your Problems

Problem: Some students may ask, "What are we going to use this for?"
Solution: Chapter and Section openings feature real-world examples, which show students how the material they are learning appears in the world around them.

Problem: Many students do not read the book.
Solution: At the end of each section, under the heading *Getting Ready for Class*, are four questions for students to answer from the reading. Even a minimal attempt to answer these questions enhances the students' in-class experience.

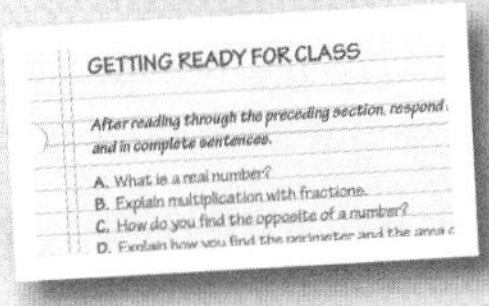

Problem: Some students may not see how the topics are connected.
Solution: At the conclusion of the problem set for each section are a series of problems under the heading *Getting Ready for the Next Section*. These problems are designed to bridge the gap between topics covered previously, and topics introduced in the next section. Students intuitively see how topics lead into, and out of, each other.

Problem: Some students lack good study skills, but may not know how to improve them.
Solution: Study skills and success skills appear throughout the book, as well as online at MathTV.com. Students learn the skills they need to become successful in this class, and in their other courses as well.

Problem: Students do well on homework, then forget everything a few days later.
Solution: We have designed this textbook so that no topic is covered and then discarded. Throughout the book, features such as *Getting Ready for the Next Section*, *Maintaining Your Skills*, the *Chapter Summary*, and the *Chapter Test* continually reinforce the skills students need to master. If students need still more practice, there are a variety of worksheets online at MathTV.com.

Problem: Some students just watch the videos at MathTV.com, but are not actively involved in learning.
Solution: The Matched Problems worksheets (available online at MathTV.com) contain problems similar to the video examples. Assigning the Matched Problems worksheets ensures that students will be actively involved with the videos.

Other Helpful Solutions

Blueprint for Problem Solving: Students can use these step-by-step methods for solving common application problems.

Facts from Geometry: Students see how topics from geometry are related to the algebra they are using.

Using Technology: Scattered throughout the book are optional exercises that demonstrate how students can use graphing calculators to enhance their understanding of the topics being covered.

Supplements for the Instructor

Please contact your sales representative.

MathTV.com With more than 6,000 videos, MathTV.com provides the instructor with a useful resource to help students learn the material. MathTV.com features videos of every example in the book, explained by the author and a variety of peer instructors. If a problem can be completed more than one way, the peer instructors often solve it by different methods. Instructors can also use the site's *Build a Playlist* feature to create a custom list of videos for posting on their class blog or website.

Online Homework XYZHomework.com provides powerful online instructional tools for faculty and students. Randomized questions provide unlimited practice and instant feedback with all the benefits of automatic grading. Tools for instructors include the following:

- Quick setup of your online class
- More than 1,500 randomized questions, similar to those in the textbook, for use in a variety of assessments, including online homework, quizzes and tests
- Text and videos designed to supplement your instruction
- Automated grading of online assignments
- Flexible gradebook
- Message boards and other communication tools, enhanced with calculator-style input for proper mathematics notation

Supplements for the Student

MathTV.com MathTV.com gives students access to math instruction 24 hours a day, seven days a week. Assistance with any problem or subject is never more than a few clicks away.

Online book This text is available online for both instructors and students. Tightly integrated with MathTV.com, students can read the book and watch videos of the author and peer instructors explaining each example. Access to the online book is available free with the purchase of a new book.

Additional worksheets A variety of worksheets are available to students online at MathTV.com's premium site. Worksheets include *Matched Problems, Multiple Choice, Find the Mistake*, and *Additional Problems.*

Online Homework XYZHomework.com provides powerful online instruction and homework practice for students. Benefits for the student include the following:

- Unlimited practice with problems similar to those in the text
- Online quizzes and tests for instant feedback on performance
- Online video examples
- Convenient tracking of class progress

Preface to the Student

I often find my students asking themselves the question "Why can't I understand this stuff the first time?" The answer is "You're not expected to." Learning a topic in mathematics isn't always accomplished the first time around. There are many instances when you will find yourself reading over new material a number of times before you can begin to work problems. That's just the way things are in mathematics. If you don't understand a topic the first time you see it, that doesn't mean there is something wrong with you. Understanding mathematics takes time. The process of understanding requires reading the book, studying the examples, working problems, and getting your questions answered.

How to Be Successful in Mathematics

1. **If you are in a lecture class, be sure to attend all class sessions on time.** You cannot know exactly what goes on in class unless you are there. Missing class and then expecting to find out what went on from someone else is not the same as being there yourself.
2. **Read the book.** It is best to read the section that will be covered in class beforehand. Reading in advance, even if you do not understand everything you read, is still better than going to class with no idea of what will be discussed.
3. **Work problems every day and check your answers.** The key to success in mathematics is working problems. The more problems you work, the better you will become at working them. The answers to the odd-numbered problems are given in the back of the book. When you have finished an assignment, be sure to compare your answers with those in the book. If you have made a mistake, find out what it is, and correct it.
4. **Do it on your own.** Don't be misled into thinking someone else's work is your own. Having someone else show you how to work a problem is not the same as working the same problem yourself. It is okay to get help when you are stuck. As a matter of fact, it is a good idea. Just be sure you do the work yourself.
5. **Review every day.** After you have finished the problems your instructor has assigned, take another 15 minutes and review a section you have already completed. The more you review, the longer you will retain the material you have learned.
6. **Don't expect to understand every new topic the first time you see it.** Sometimes you will understand everything you are doing, and sometimes you won't. That's just the way things are in mathematics. Expecting to understand each new topic the first time you see it can lead to disappointment and frustration. The process of understanding takes time. It requires that you read the book, work problems, and get your questions answered.
7. **Spend as much time as it takes for you to master the material.** No set formula exists for the exact amount of time you need to spend on mathematics to master it. You will find out as you go along what is or isn't enough time for you. If you end up spending 2 or more hours on each section in order to master the material there, then that's how much time it takes; trying to get by with less will not work.
8. **Relax.** It's probably not as difficult as you think.

Linear Equations and Inequalities

1

Chapter Outline

1.1 Simplifying Expressions
1.2 Addition Property of Equality
1.3 Multiplication Property of Equality
1.4 Solving Linear Equations
1.5 Formulas
1.6 Applications
1.7 More Applications
1.8 Linear Inequalities
1.9 Compound Inequalities

iStockphoto.com © Enge

Just before starting work on this edition of your text, I flew to Europe for vacation. From time to time the television screens on the plane displayed statistics about the flight. At one point during the flight the temperature outside the plane was −60°F. When I returned home, I did some research and found that the relationship between temperature T and altitude A can be described with the formula

$$T = -0.0035A + 70$$

when the temperature on the ground is 70°F. The table and the line graph also describe this relationship.

Air Temperature and Altitude

Altitude (feet)	Temperature (°F)
0	70
10,000	35
20,000	0
30,000	−35
40,000	−70

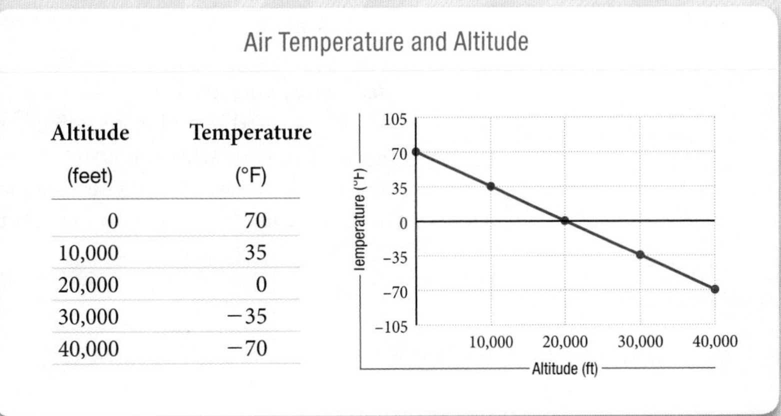

In this chapter we will start our work with formulas, and you will see how we use formulas to produce tables and line graphs like the ones above.

Study Skills

Some of the students enrolled in my college algebra classes develop difficulties early in the course. Their difficulties are not associated with their ability to learn mathematics; they all have the potential to pass the course. Students who get off to a poor start do so because they have not developed the study skills necessary to be successful in algebra. Here is a list of things you can do to begin to develop effective study skills.

1. **Put Yourself on a Schedule** The general rule is that you spend 2 hours on homework for every hour you are in class. Make a schedule for yourself in which you set aside 2 hours each day to work on algebra. Once you make the schedule, stick to it. Don't just complete your assignments and stop. Use all the time you have set aside. If you complete an assignment and have time left over, read the next section in the book, and then work more problems.

2. **Find Your Mistakes and Correct Them** There is more to studying algebra than just working problems. You must always check your answers with the answers in the back of the book. When you have made a mistake, find out what it is and correct it. Making mistakes is part of the process of learning mathematics. In the prologue to The Book of Squares, Leonardo Fibonacci (ca. 1170–ca. 1250) had this to say about the content of his book:

   > I have come to request indulgence if in any place it contains something more or less than right or necessary; for to remember everything and be mistaken in nothing is divine rather than human . . .

   Fibonacci knew, as you know, that human beings make mistakes. You cannot learn algebra without making mistakes.

3. **Gather Information on Available Resources** You need to anticipate that you will need extra help sometime during the course. One resource is your instructor; you need to know your instructor's office hours and where the office is located. Another resource is the math lab or study center, if they are available at your school. It also helps to have the phone numbers of other students in the class, in case you miss class. You want to anticipate that you will need these resources, so now is the time to gather them together.

Simplifying Expressions 1.1

If a cellular phone company charges \$35 per month plus \$0.25 for each minute, or fraction of a minute, that you use one of their cellular phones, then the amount of your monthly bill is given by the expression $35 + 0.25t$. To find the amount you will pay for using that phone 30 minutes in one month, you substitute 30 for t and simplify the resulting expression. This process is one of the topics we will study in this section.

As you will see in the next few sections, the first step in solving an equation is to simplify both sides as much as possible. In the first part of this section, we will practice simplifying expressions by combining what are called *similar* (or like) terms.

For our immediate purposes, a *term* is a number or a number and one or more variables multiplied together. For example, the number 5 is a term, as are the expressions $3x$, $-7y$, and $15xy$.

DEFINITION *similar terms*

Two or more terms with the same variable part are called ***similar*** (or ***like***) ***terms.***

The terms $3x$ and $4x$ are similar because their variable parts are identical. Likewise, the terms $18y$, $-10y$, and $6y$ are similar terms.

To simplify an algebraic expression, we simply reduce the number of terms in the expression. We accomplish this by applying the distributive property along with our knowledge of addition and subtraction of positive and negative real numbers. The following examples illustrate the procedure.

EXAMPLE 1 Simplify by combining similar terms.

a. $3x + 4x$ **b.** $7a - 10a$ **c.** $18y - 10y + 6y$

SOLUTION We combine similar terms by applying the distributive property.

a. $3x + 4x = (3 + 4)x$ Distributive property

$= 7x$ Addition of 3 and 4

b. $7a - 10a = (7 - 10)a$ Distributive property

$= -3a$ Addition of 7 and -10

c. $18y - 10y + 6y = (18 - 10 + 6)y$ Distributive property

$= 14y$ Addition of 18, -10, and 6

When the expression we intend to simplify is more complicated, we use the commutative and associative properties first.

EXAMPLE 2 Simplify each expression.

a. $3x + 5 + 2x - 3$ **b.** $4a - 7 - 2a + 3$ **c.** $5x + 8 - x - 6$

SOLUTION We combine similar terms by applying the distributive property.

a.
$$\begin{aligned} 3x + 5 + 2x - 3 &= 3x + 2x + 5 - 3 && \text{Commutative property} \\ &= (3x + 2x) + (5 - 3) && \text{Associative property} \\ &= (3 + 2)x + (5 - 3) && \text{Distributive property} \\ &= 5x + 2 && \text{Addition} \end{aligned}$$

b.
$$\begin{aligned} 4a - 7 - 2a + 3 &= (4a - 2a) + (-7 + 3) && \text{Commutative and associative properties} \\ &= (4 - 2)a + (-7 + 3) && \text{Distributive property} \\ &= 2a - 4 && \text{Addition} \end{aligned}$$

c.
$$\begin{aligned} 5x + 8 - x - 6 &= (5x - x) + (8 - 6) && \text{Commutative and associative properties} \\ &= (5 - 1)x + (8 - 6) && \text{Distributive property} \\ &= 4x + 2 && \text{Addition} \end{aligned}$$

Notice that in each case the result has fewer terms than the original expression. Because there are fewer terms, the resulting expression is said to be simpler than the original expression.

Simplifying Expressions Containing Parentheses

If an expression contains parentheses, it is often necessary to apply the distributive property to remove the parentheses before combining similar terms.

EXAMPLE 3 Simplify the expression $5(2x - 8) - 3$.

SOLUTION We begin by distributing the 5 across $2x - 8$. We then combine similar terms:

$$\begin{aligned} 5(2x - 8) - 3 &= 10x - 40 - 3 && \text{Distributive property} \\ &= 10x - 43 \end{aligned}$$

EXAMPLE 4 Simplify $7 - 3(2y + 1)$.

SOLUTION By the rule for order of operations, we must multiply before we add or subtract. For that reason, it would be incorrect to subtract 3 from 7 first. Instead, we multiply -3 and $2y + 1$ to remove the parentheses and then combine similar terms:

$$\begin{aligned} 7 - 3(2y + 1) &= 7 - 6y - 3 && \text{Distributive property} \\ &= -6y + 4 \end{aligned}$$

EXAMPLE 5 Simplify $5(x - 2) - (3x + 4)$.

SOLUTION We begin by applying the distributive property to remove the parentheses. The expression $-(3x + 4)$ can be thought of as $-1(3x + 4)$. Thinking of it in this way allows us to apply the distributive property:

$$-1(3x + 4) = -1(3x) + (-1)(4)$$
$$= -3x - 4$$

The complete solution looks like this:

$$5(x - 2) - (3x + 4) = 5x - 10 - 3x - 4 \quad \text{Distributive property}$$
$$= 2x - 14 \quad \text{Combine similar terms}$$

As you can see from the explanation in Example 5, we use the distributive property to simplify expressions in which parentheses are preceded by a negative sign. In general we can write

$$-(a + b) = -1(a + b)$$
$$= -a + (-b)$$
$$= -a - b$$

The negative sign outside the parentheses ends up changing the sign of each term within the parentheses. In words, we say "the opposite of a sum is the sum of the opposites."

The Value of an Expression

An expression like $3x + 2$ has a certain value depending on what number we assign to x. For instance, when x is 4, $3x + 2$ becomes $3(4) + 2$, or 14. When x is -8, $3x + 2$ becomes $3(-8) + 2$, or -22. The value of an expression is found by replacing the variable with a given number.

EXAMPLE 6 Find the value of the following expressions by replacing the variable with the given number.

Expression	The Variable	Value of the Expression
a. $3x - 1$	$x = 2$	$3(2) - 1 = 6 - 1 = 5$
b. $7a + 4$	$a = -3$	$7(-3) + 4 = -21 + 4 = -17$
c. $2x - 3 + 4x$	$x = -1$	$2(-1) - 3 + 4(-1) = -2 - 3 + (-4)$ $= -9$
d. $2x - 5 - 8x$	$x = 5$	$2(5) - 5 - 8(5) = 10 - 5 - 40$ $= -35$
e. $y^2 - 6y + 9$	$y = 4$	$4^2 - 6(4) + 9 = 16 - 24 + 9 = 1$

Simplifying an expression should not change its value; that is, if an expression has a certain value when x is 5, then it will always have that value no matter how much it has been simplified as long as x is 5. If we were to simplify the expression in Example 6d first, it would look like

$$2x - 5 - 8x = -6x - 5$$

When x is 5, the simplified expression $-6x - 5$ is

$$-6(5) - 5 = -30 - 5 = -35$$

It has the same value as the original expression when x is 5.

We also can find the value of an expression that contains two variables if we know the values for both variables.

EXAMPLE 7 Find the value of the expression $2x - 3y + 4$ when x is -5 and y is 6.

SOLUTION Substituting -5 for x and 6 for y, the expression becomes

$$\begin{aligned} 2(-5) - 3(6) + 4 &= -10 - 18 + 4 \\ &= -28 + 4 \\ &= -24 \end{aligned}$$

EXAMPLE 8 Find the value of the expression $x^2 - 2xy + y^2$ when x is 3 and y is -4.

SOLUTION Replacing each x in the expression with the number 3 and each y in the expression with the number -4 gives us

$$\begin{aligned} 3^2 - 2(3)(-4) + (-4)^2 &= 9 - 2(3)(-4) + 16 \\ &= 9 - (-24) + 16 \\ &= 33 + 16 \\ &= 49 \end{aligned}$$

More About Sequences

As the next example indicates, when we substitute the counting numbers, in order, into algebraic expressions, we form some of the sequences of numbers that we studied in Chapter 1. To review, recall that the sequence of counting numbers (also called the sequence of positive integers) is

$$\text{Counting numbers} = 1, 2, 3, \ldots$$

EXAMPLE 9 Substitute 1, 2, 3, and 4 for n in the expression $2n - 1$.

SOLUTION Substituting as indicated, we have

When $n = 1$, $2n - 1 = 2 \cdot 1 - 1 = 1$

When $n = 2$, $2n - 1 = 2 \cdot 2 - 1 = 3$

When $n = 3$, $2n - 1 = 2 \cdot 3 - 1 = 5$

When $n = 4$, $2n - 1 = 2 \cdot 4 - 1 = 7$

As you can see, substituting the first four counting numbers into the formula $2n - 1$ produces the first four numbers in the sequence of odd numbers.

The next example is similar to Example 9 but uses tables to display the information.

EXAMPLE 10 Fill in the tables below to find the sequences formed by substituting the first four counting numbers into the expressions $2n$ and n^2.

a.

n	1	2	3	4
$2n$				

b.

n	1	2	3	4
n^2				

SOLUTION Proceeding as we did in the previous example, we substitute the numbers 1, 2, 3, and 4 into the given expressions.

a. When $n = 1, 2n = 2 \cdot 1 = 2$
When $n = 2, 2n = 2 \cdot 2 = 4$
When $n = 3, 2n = 2 \cdot 3 = 6$
When $n = 4, 2n = 2 \cdot 4 = 8$

As you can see, the expression $2n$ produces the sequence of even numbers when n is replaced by the counting numbers. Placing these results into our first table gives us

n	1	2	3	4
$2n$	2	4	6	8

b. The expression n^2 produces the sequence of squares when n is replaced by 1, 2, 3, and 4. In table form we have

n	1	2	3	4
n^2	1	4	9	16

GETTING READY FOR CLASS

After reading through the preceding section, respond in your own words and in complete sentences.

A. What are similar terms?
B. Explain how the distributive property is used to combine similar terms.
C. What is wrong with writing $3x + 4x = 7x^2$?
D. Explain how you would find the value of $5x + 3$ when x is 6.

Problem Set 1.1

Simplify the following expressions.

1. $3x - 6x$ **2.** $7x - 5x$ **3.** $-2a + a$

4. $3a - a$ **5.** $7x + 3x + 2x$ **6.** $8x - 2x - x$

7. $3a - 2a + 5a$ **8.** $7a - a + 2a$ **9.** $4x - 3 + 2x$

10. $5x + 6 - 3x$ **11.** $3a + 4a + 5$ **12.** $6a + 7a + 8$

13. $2x - 3 + 3x - 2$ **14.** $6x + 5 - 2x + 3$ **15.** $3a - 1 + a + 3$

16. $-a + 2 + 8a - 7$ **17.** $-4x + 8 - 5x - 10$ **18.** $-9x - 1 + x - 4$

19. $7a + 3 + 2a + 3a$ **20.** $8a - 2 + a + 5a$ **21.** $5(2x - 1) + 4$

22. $2(4x - 3) + 2$ **23.** $7(3y + 2) - 8$ **24.** $6(4y + 2) - 7$

25. $-3(2x - 1) + 5$ **26.** $-4(3x - 2) - 6$ **27.** $5 - 2(a + 1)$

28. $7 - 8(2a + 3)$ **29.** $6 - 4(x - 5)$ **30.** $12 - 3(4x - 2)$

31. $-9 - 4(2 - y) + 1$ **32.** $-10 - 3(2 - y) + 3$ **33.** $-6 + 2(2 - 3x) + 1$

34. $-7 - 4(3 - x) + 1$ **35.** $(4x - 7) - (2x + 5)$ **36.** $(7x - 3) - (4x + 2)$

37. $8(2a + 4) - (6a - 1)$ **38.** $9(3a + 5) - (8a - 7)$ **39.** $3(x - 2) + (x - 3)$

40. $2(2x + 1) - (x + 4)$ **41.** $4(2y - 8) - (y + 7)$ **42.** $5(y - 3) - (y - 4)$

43. $-9(2x + 1) - (x + 5)$ **44.** $-3(3x - 2) - (2x + 3)$

Evaluate the following expressions when x is 2. (Find the value of the expressions if x is 2.)

45. $3x - 1$ **46.** $4x + 3$ **47.** $-2x - 5$ **48.** $-3x + 6$

49. $x^2 - 8x + 16$ **50.** $x^2 - 10x + 25$ **51.** $(x - 4)^2$ **52.** $(x - 5)^2$

Evaluate the following expressions when x is -5. Then simplify the expression, and check to see that it has the same value for $x = -5$.

53. $7x - 4 - x - 3$ **54.** $3x + 4 + 7x - 6$

55. $5(2x + 1) + 4$ **56.** $2(3x - 10) + 5$

Evaluate the following expressions when x is -3 and y is 5.

57. $x^2 - 2xy + y^2$ **58.** $x^2 + 2xy + y^2$ **59.** $(x - y)^2$

60. $(x + y)^2$ **61.** $x^2 + 6xy + 9y^2$ **62.** $x^2 + 10xy + 25y^2$

63. $(x + 3y)^2$ **64.** $(x + 5y)^2$

Find the value of $12x - 3$ for each of the following values of x.

65. $\frac{1}{2}$ **66.** $\frac{1}{3}$ **67.** $\frac{1}{4}$ **68.** $\frac{1}{6}$

69. $\frac{3}{2}$ **70.** $\frac{2}{3}$ **71.** $\frac{3}{4}$ **72.** $\frac{5}{6}$

73. Fill in the tables below to find the sequences formed by substituting the first four counting numbers into the expressions $3n$ and n^3.

a.

n	1	2	3	4
$3n$				

b.

n	1	2	3	4
n^3				

74. Fill in the tables below to find the sequences formed by substituting the first four counting numbers into the expressions $2n - 1$ and $2n + 1$.

a.

n	1	2	3	4
$2n - 1$				

b.

n	1	2	3	4
$2n + 1$				

Find the sequences formed by substituting the first four counting numbers, in order, into the following expressions.

75. $3n - 2$ **76.** $2n - 3$ **77.** $n^2 - 2n + 1$ **78.** $(n - 1)^2$

Here are some problems you will see later in the book. Simplify.

79. $7 - 3(2y + 1)$

80. $4(3x - 2) - (6x - 5)$

81. $0.08x + 0.09x$

82. $0.04x + 0.05x$

83. $(x + y) + (x - y)$

84. $(-12x - 20y) + (25x + 20y)$

85. $3x + 2(x - 2)$

86. $2(x - 2) + 3(5x)$

87. $4(x + 1) + 3(x - 3)$

88. $5(x + 1) + 3(x - 1)$

89. $x + (x + 3)(-3)$

90. $x - 2(x + 2)$

91. $3(4x - 2) - (5x - 8)$

92. $2(5x - 3) - (2x - 4)$

93. $-(3x + 1) - (4x - 7)$

94. $-(6x + 2) - (8x - 3)$

95. $(x + 3y) + 3(2x - y)$

96. $(2x - y) - 2(x + 3y)$

97. $3(2x + 3y) - 2(3x + 5y)$

98. $5(2x + 3y) - 3(3x + 5y)$

99. $-6\left(\frac{1}{2}x - \frac{1}{3}y\right) + 12\left(\frac{1}{4}x + \frac{2}{3}y\right)$

100. $6\left(\frac{1}{3}x + \frac{1}{2}y\right) - 4\left(x + \frac{3}{4}y\right)$

101. $0.08x + 0.09(x + 2{,}000)$

102. $0.06x + 0.04(x + 7{,}000)$

103. $0.10x + 0.12(x + 500)$

104. $0.08x + 0.06(x + 800)$

Find the value of $b^2 - 4ac$ for the given values of a, b, and c. (You will see these problems later in the book.)

105. $a = 1, b = -5, c = -6$

106. $a = 1, b = -6, c = 7$

107. $a = 2, b = 4, c = -3$

108. $a = 3, b = 4, c = -2$

Applying the Concepts

109. Temperature and Altitude If the temperature on the ground is 70°F, then the temperature at A feet above the ground can be found from the expression $-0.0035A + 70$. Find the temperature at the following altitudes.

a. 8,000 feet **b.** 12,000 feet **c.** 24,000 feet

110. Perimeter of a Rectangle The expression $2l + 2w$ gives the perimeter of a rectangle with length l and width w. Find the perimeter of the rectangles with the following lengths and widths.

a. Length = 8 meters
Width = 5 meters

b. Length = 10 feet
Width = 3 feet

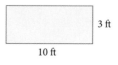

3 ft
10 ft

111. Cellular Phone Rates A cellular phone company charges \$35 per month plus \$0.25 for each minute, or fraction of a minute, that you use one of their cellular phones. The expression $35 + 0.25t$ gives the amount of money you will pay for using one of their phones for t minutes a month. Find the monthly bill for using one of their phones.

a. 10 minutes in a month **b.** 20 minutes in a month

c. 30 minutes in a month

112. Cost of Bottled Water A water bottling company charges \$7.00 per month for their water dispenser and \$1.10 for each gallon of water delivered. If you have g gallons of water delivered in a month, then the expression $7 + 1.1g$ gives the amount of your bill for that month. Find the monthly bill for each of the following deliveries.

a. 10 gallons **b.** 20 gallons **c.** 30 gallons

Getting Ready for the Next Section

These are problems that you must be able to work in order to understand the material in the next section. The problems below are exactly the type of problems you will see in the explanations and examples in the next section.

Simplify.

113. $17 - 5$

114. $12 + (-2)$

115. $2 - 5$

116. $25 - 20$

117. $-2.4 + (-7.3)$

118. $8.1 + 2.7$

119. $-\frac{1}{2} + \left(-\frac{3}{4}\right)$

120. $-\frac{1}{6} + \left(-\frac{2}{3}\right)$

121. $4(2 \cdot 9 - 3) - 7$

122. $5(3 \cdot 45 - 4) - 14 \cdot 45$

123. $4(2a - 3) - 7a$

124. $5(3a - 4) - 14a$

125. Find the value of $2x - 3$ when x is 5

126. Find the value of $3x + 4$ when x is -2

Addition Property of Equality

1.2

When light comes into contact with any object, it is reflected, absorbed, and transmitted, as shown below.

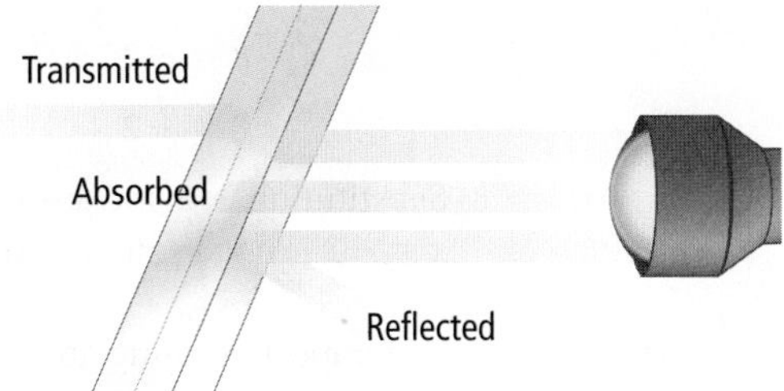

For a certain type of glass, 88% of the light hitting the glass is transmitted through to the other side, whereas 6% of the light is absorbed into the glass. To find the percent of light that is reflected by the glass, we can solve the equation

$$88 + R + 6 = 100$$

Solving equations of this type is what we study in this section. To solve an equation we must find all replacements for the variable that make the equation a true statement.

DEFINITION *solution set*

The ***solution set*** for an equation is the set of all numbers that when used in place of the variable make the equation a true statement

For example, the equation $x + 2 = 5$ has the solution set $\{3\}$ because when x is 3 the equation becomes the true statement $3 + 2 = 5$, or $5 = 5$.

EXAMPLE 1 Is 5 a solution to $2x - 3 = 7$?

SOLUTION We substitute 5 for x in the equation, and then simplify to see if a true statement results. A true statement means we have a solution; a false statement indicates the number we are using is not a solution.

When	$x = 5$	
the equation	$2x - 3 = 7$	
becomes	$2(5) - 3 \stackrel{?}{=} 7$	
	$10 - 3 \stackrel{?}{=} 7$	
	$7 = 7$	A true statement

Note: We can use a question mark over the equal signs to show that we don't know yet whether the two sides of the equation are equal.

Because $x = 5$ turns the equation into the true statement $7 = 7$, we know 5 is a solution to the equation.

EXAMPLE 2 Is -2 a solution to $8 = 3x + 4$?

SOLUTION Substituting -2 for x in the equation, we have

$$8 \stackrel{?}{=} 3(-2) + 4$$

$$8 \stackrel{?}{=} -6 + 4$$

$$8 = -2 \qquad \text{A false statement}$$

$$8 \neq -2$$

Substituting -2 for x in the equation produces a false statement. Therefore, $x = -2$ is not a solution to the equation.

The important thing about an equation is its solution set. We therefore make the following definition to classify together all equations with the same solution set.

DEFINITION *equivalent equation*

Two or more equations with the same solution set are said to be ***equivalent equations.***

Equivalent equations may look different but must have the same solution set.

EXAMPLE 3

a. $x + 2 = 5$ and $x = 3$ are equivalent equations because both have solution set $\{3\}$.

b. $a - 4 = 3$, $a - 2 = 5$, and $a = 7$ are equivalent equations because they all have solution set $\{7\}$.

c. $y + 3 = 4$, $y - 8 = -7$, and $y = 1$ are equivalent equations because they all have solution set $\{1\}$.

If two numbers are equal and we increase (or decrease) both of them by the same amount, the resulting quantities are also equal. We can apply this concept to equations. Adding the same amount to both sides of an equation always produces an equivalent equation—one with the same solution set. This fact about equations is called the *addition property of equality* and can be stated more formally as follows.

Note: We will use this property many times in the future. Be sure you understand it completely by the time you finish this section.

PROPERTY *Addition Property of Equality*

For any three algebraic expressions A, B, and C,

if $\quad A = B$

then $\quad A + C = B + C$

In words: Adding the same quantity to both sides of an equation will not change the solution set.

This property is just as simple as it seems. We can add any amount to both sides of an equation and always be sure we have not changed the solution set.

Consider the equation $x + 6 = 5$. We want to solve this equation for the value of x that makes it a true statement. We want to end up with x on one side of the equal sign and a number on the other side. Because we want x by itself, we will add -6 to both sides:

$$x + 6 + (-6) = 5 + (-6) \quad \text{Addition property of equality}$$

$$x + 0 = -1 \quad \text{Addition}$$

$$x = -1$$

All three equations say the same thing about x. They all say that x is -1. All three equations are equivalent. The last one is just easier to read.

Here are some further examples of how the addition property of equality can be used to solve equations.

EXAMPLE 4 Solve the equation $x - 5 = 12$ for x.

SOLUTION Because we want x alone on the left side, we choose to add 5 to both sides:

$$x - 5 + 5 = 12 + 5 \quad \text{Addition property of equality}$$

$$x + 0 = 17$$

$$x = 17$$

To check our solution to Example 4, we substitute 17 for x in the original equation:

When	$x = 17$	
the equation	$x - 5 = 12$	
becomes	$17 - 5 \stackrel{?}{=} 12$	
	$12 = 12$	A true statement

As you can see, our solution checks. The purpose for checking a solution to an equation is to catch any mistakes we may have made in the process of solving the equation.

EXAMPLE 5 Solve for a: $a + \frac{3}{4} = -\frac{1}{2}$.

SOLUTION Because we want a by itself on the left side of the equal sign, we add the opposite of $\frac{3}{4}$ to each side of the equation.

$$a + \frac{3}{4} + \left(-\frac{3}{4}\right) = -\frac{1}{2} + \left(-\frac{3}{4}\right) \quad \text{Addition property of equality}$$

$$a + 0 = -\frac{1}{2} \cdot \frac{2}{2} + \left(-\frac{3}{4}\right) \quad \text{LCD on the right side is 4}$$

$$a = -\frac{2}{4} + \left(-\frac{3}{4}\right) \quad \frac{2}{4} \text{ is equivalent to } \frac{1}{2}$$

$$a = -\frac{5}{4} \quad \text{Add fractions}$$

The solution is $a = -\frac{5}{4}$. To check our result, we replace a with $-\frac{5}{4}$ in the original equation. The left side then becomes $-\frac{5}{4} + \frac{3}{4}$, which reduces to $-\frac{1}{2}$, so our solution checks.

EXAMPLE 6 Solve for x: $7.3 + x = -2.4$.

SOLUTION Again, we want to isolate x, so we add the opposite of 7.3 to both sides:

$$7.3 + (-7.3) + x = -2.4 + (-7.3) \quad \text{Addition property of equality}$$
$$0 + x = -9.7$$
$$x = -9.7$$

Sometimes it is necessary to simplify each side of an equation before using the addition property of equality. The reason we simplify both sides first is that we want as few terms as possible on each side of the equation before we use the addition property of equality. The following examples illustrate this procedure.

EXAMPLE 7 Solve for x: $-x + 2 + 2x = 7 + 5$.

SOLUTION We begin by combining similar terms on each side of the equation. Then we use the addition property to solve the simplified equation.

$$x + 2 = 12 \quad \text{Simplify both sides first}$$
$$x + 2 + (-2) = 12 + (-2) \quad \text{Addition property of equality}$$
$$x + 0 = 10$$
$$x = 10$$

EXAMPLE 8 Solve $4(2a - 3) - 7a = 2 - 5$.

SOLUTION We must begin by applying the distributive property to separate terms on the left side of the equation. Following that, we combine similar terms and then apply the addition property of equality.

$$4(2a - 3) - 7a = 2 - 5 \quad \text{Original equation}$$
$$8a - 12 - 7a = 2 - 5 \quad \text{Distributive property}$$
$$a - 12 = -3 \quad \text{Simplify each side}$$
$$a - 12 + 12 = -3 + 12 \quad \text{Add 12 to each side}$$
$$a = 9 \quad \text{Addition}$$

To check our solution, we replace a with 9 in the original equation.

$$4(2 \cdot 9 - 3) - 7 \cdot 9 \stackrel{?}{=} 2 - 5$$
$$4(15) - 63 \stackrel{?}{=} -3$$
$$60 - 63 \stackrel{?}{=} -3$$
$$-3 = -3 \quad \text{A true statement}$$

Note: Again, we place a question mark over the equal sign because we don't know yet whether the expressions on the left and right side of the equal sign will be equal.

We can also add a term involving a variable to both sides of an equation.

EXAMPLE 9 Solve $3x - 5 = 2x + 7$.

SOLUTION We can solve this equation in two steps. First, we add $-2x$ to both sides of the equation. When this has been done, x appears on the left side only. Second, we add 5 to both sides:

$$3x + (-2x) - 5 = 2x + (-2x) + 7 \quad \text{Add } -2x \text{ to both sides}$$

$$x - 5 = 7 \quad \text{Simplify each side}$$

$$x - 5 + 5 = 7 + 5 \quad \text{Add 5 to both sides}$$

$$x = 12 \quad \text{Simplify each side}$$

Note: In my experience teaching algebra, I find that students make fewer mistakes if they think in terms of addition rather than subtraction. So, you are probably better off if you continue to use the addition property just the way we have used it in the examples in this section. But, if you are curious as to whether you can subtract the same number from both sides of an equation, the answer is yes.

PROPERTY ***A Note on Subtraction***

Although the addition property of equality is stated for addition only, we can subtract the same number from both sides of an equation as well. Because subtraction is defined as addition of the opposite, subtracting the same quantity from both sides of an equation does not change the solution.

$$x + 2 = 12 \quad \text{Original equation}$$

$$x + 2 - 2 = 12 - 2 \quad \text{Subtract 2 from each side}$$

$$x = 10 \quad \text{Subtraction}$$

GETTING READY FOR CLASS

After reading through the preceding section, respond in your own words and in complete sentences.

A. What is a solution to an equation?
B. What are equivalent equations?
C. Explain in words the addition property of equality.
D. How do you check a solution to an equation?

Problem Set 1.2

Solve the following equations.

1. $x - 3 = 8$
2. $x - 2 = 7$
3. $x + 2 = 6$
4. $x + 5 = 4$
5. $a + \frac{1}{2} = -\frac{1}{4}$
6. $a + \frac{1}{3} = -\frac{5}{6}$
7. $x + 2.3 = -3.5$
8. $x + 7.9 = 23.4$
9. $y + 11 = -6$
10. $y - 3 = -1$
11. $x - \frac{5}{8} = -\frac{3}{4}$
12. $x - \frac{2}{5} = -\frac{1}{10}$
13. $m - 6 = -10$
14. $m - 10 = -6$
15. $6.9 + x = 3.3$
16. $7.5 + x = 2.2$
17. $5 = a + 4$
18. $12 = a - 3$
19. $-\frac{5}{9} = x - \frac{2}{5}$
20. $-\frac{7}{8} = x - \frac{4}{5}$

Simplify both sides of the following equations as much as possible, and then solve.

21. $4x + 2 - 3x = 4 + 1$
22. $5x + 2 - 4x = 7 - 3$
23. $8a - \frac{1}{2} - 7a = \frac{3}{4} + \frac{1}{8}$
24. $9a - \frac{4}{5} - 8a = \frac{3}{10} - \frac{1}{5}$
25. $-3 - 4x + 5x = 18$
26. $10 - 3x + 4x = 20$
27. $-11x + 2 + 10x + 2x = 9$
28. $-10x + 5 - 4x + 15x = 0$
29. $-2.5 + 4.8 = 8x - 1.2 - 7x$
30. $-4.8 + 6.3 = 7x - 2.7 - 6x$
31. $2y - 10 + 3y - 4y = 18 - 6$
32. $15 - 21 = 8x + 3x - 10x$

The following equations contain parentheses. Apply the distributive property to remove the parentheses, then simplify each side before using the addition property of equality.

33. $2(x + 3) - x = 4$
34. $5(x + 1) - 4x = 2$
35. $-3(x - 4) + 4x = 3 - 7$
36. $-2(x - 5) + 3x = 4 - 9$
37. $5(2a + 1) - 9a = 8 - 6$
38. $4(2a - 1) - 7a = 9 - 5$
39. $-(x + 3) + 2x - 1 = 6$
40. $-(x - 7) + 2x - 8 = 4$
41. $4y - 3(y - 6) + 2 = 8$
42. $7y - 6(y - 1) + 3 = 9$
43. $-3(2m - 9) + 7(m - 4) = 12 - 9$
44. $-5(m - 3) + 2(3m + 1) = 15 - 8$

Solve the following equations by the method used in Example 9 in this section. Check each solution in the original equation.

45. $4x = 3x + 2$
46. $6x = 5x - 4$
47. $8a = 7a - 5$
48. $9a = 8a - 3$
49. $2x = 3x + 1$
50. $4x = 3x + 5$
51. $3y + 4 = 2y + 1$
52. $5y + 6 = 4y + 2$
53. $2m - 3 = m + 5$
54. $8m - 1 = 7m - 3$
55. $4x - 7 = 5x + 1$
56. $3x - 7 = 4x - 6$
57. $5x - \frac{2}{3} = 4x + \frac{4}{3}$
58. $3x - \frac{5}{4} = 2x + \frac{1}{4}$
59. $8a - 7.1 = 7a + 3.9$
60. $10a - 4.3 = 9a + 4.7$

61. $11y - 2.9 = 12y + 2.9$

62. $20y + 9.9 = 21y - 9.9$

Applying the Concepts

63. Light When light comes into contact with any object, it is reflected, absorbed, and transmitted, as shown in the following figure. If T represents the percent of light transmitted, R the percent of light reflected, and A the percent of light absorbed by a surface, then the equation $T + R + A = 100$ shows one way these quantities are related.

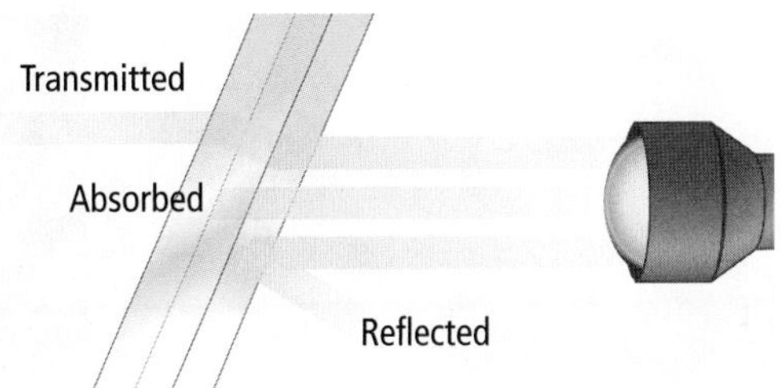

a. For glass, $T = 88$ and $A = 6$, meaning that 88% of the light hitting the glass is transmitted and 6% is absorbed. Substitute $T = 88$ and $A = 6$ into the equation $T + R + A = 100$ and solve for R to find the percent of light that is reflected.

b. For flat black paint, $A = 95$ and no light is transmitted, meaning that $T = 0$. What percent of light is reflected by flat black paint?

c. A pure white surface can reflect 98% of light, so $R = 98$. If no light is transmitted, what percent of light is absorbed by the pure white surface?

d. Typically, shiny gray metals reflect 70–80% of light. Suppose a thick sheet of aluminum absorbs 25% of light. What percent of light is reflected by this shiny gray metal? (Assume no light is transmitted.)

64. Geometry The three angles shown in the triangle at the front of the tent in the following figure add up to 180°. Use this fact to write an equation containing x, and then solve the equation to find the number of degrees in the angle at the top of the triangle.

Getting Ready for the Next Section

To understand all of the explanations and examples in the next section you must be able to work the problems below.

Simplify.

65. $\frac{3}{2}\left(\frac{2}{3}y\right)$ **66.** $\frac{5}{2}\left(-\frac{2}{5}y\right)$ **67.** $\frac{1}{5}(5x)$ **68.** $-\frac{1}{4}(-4a)$

69. $\frac{1}{5}(30)$ **70.** $-\frac{1}{4}(24)$ **71.** $\frac{3}{2}(4)$ **72.** $\frac{1}{26}(13)$

73. $12\left(-\frac{3}{4}\right)$ **74.** $12\left(\frac{1}{2}\right)$ **75.** $\frac{3}{2}\left(-\frac{5}{4}\right)$ **76.** $\frac{5}{3}\left(-\frac{6}{5}\right)$

77. $13 + (-5)$ **78.** $-13 + (-5)$ **79.** $-\frac{3}{4} + \left(-\frac{1}{2}\right)$ **80.** $-\frac{7}{10} + \left(-\frac{1}{2}\right)$

81. $7x + (-4x)$ **82.** $5x + (-2x)$

Multiplication Property of Equality 1.3

As we have mentioned before, we all have to pay taxes. According to Figure 1, people have been paying taxes for quite a long time.

FIGURE 1 *Collection of taxes, ca. 3000 b.c. Clerks and scribes appear at the right, with pen and papyrus, and officials and taxpayers appear at the left.*

Suppose 21% of your monthly pay is withheld for federal income taxes and another 8% is withheld for Social Security, state income tax, and other miscellaneous items, leaving you with \$987.50 a month in take-home pay. The amount you earned before the deductions were removed from your check, your gross income G, is given by the equation

$$G - 0.21G - 0.08G = 987.5$$

In this section we will learn how to solve equations of this type.

In the previous section, we found that adding the same number to both sides of an equation never changed the solution set. The same idea holds for multiplication by numbers other than zero. We can multiply both sides of an equation by the same nonzero number and always be sure we have not changed the solution set. (The reason we cannot multiply both sides by zero will become apparent later.) This fact about equations is called the *multiplication property of equality*, which can be stated formally as follows.

Note: This property is also used many times throughout the book. Make every effort to understand it completely.

[Δ≠Σ PROPERTY ***Multiplication Property of Equality***

For any three algebraic expressions A, B, and C, where $C \neq 0$,

if $A = B$
then $AC = BC$

In words: Multiplying both sides of an equation by the same nonzero number will not change the solution set.

Suppose we want to solve the equation $5x = 30$. We have $5x$ on the left side but would like to have just x. We choose to multiply both sides by $\frac{1}{5}$ because $\left(\frac{1}{5}\right)(5) = 1$. Here is the solution:

$$5x = 30$$

$$\frac{1}{5}(5x) = \frac{1}{5}(30) \quad \text{Multiplication property of equality}$$

$$\left(\frac{1}{5} \cdot 5\right)x = \frac{1}{5}(30) \quad \text{Associative property of multiplication}$$

$$1x = 6$$

$$x = 6$$

We chose to multiply by $\frac{1}{5}$ because it is the reciprocal of 5. We can see that multiplication by any number except zero will not change the solution set. If, however, we were to multiply both sides by zero, the result would always be $0 = 0$ because multiplication by zero always results in zero. Although the statement $0 = 0$ is true, we have lost our variable and cannot solve the equation. This is the only restriction of the multiplication property of equality. We are free to multiply both sides of an equation by any number except zero.

Here are some more examples that use the multiplication property of equality.

EXAMPLE 1 Solve for a: $-4a = 24$.

SOLUTION Because we want a alone on the left side, we choose to multiply both sides by $-\frac{1}{4}$:

$$-\frac{1}{4}(-4a) = -\frac{1}{4}(24) \quad \text{Multiplication property of equality}$$

$$\left[-\frac{1}{4}(-4)\right]a = -\frac{1}{4}(24) \quad \text{Associative property}$$

$$a = -6$$

EXAMPLE 2 Solve for t: $-\frac{t}{3} = 5$.

SOLUTION Because division by 3 is the same as multiplication by $\frac{1}{3}$, we can write $-\frac{t}{3}$ as $-\frac{1}{3}t$. To solve the equation, we multiply each side by the reciprocal of $-\frac{1}{3}$, which is -3.

$$-\frac{t}{3} = 5 \quad \text{Original equation}$$

$$-\frac{1}{3}t = 5 \quad \text{Dividing by 3 is equivalent to multiplying by } \frac{1}{3}$$

$$-3\left(-\frac{1}{3}t\right) = -3(5) \quad \text{Multiply each side by } -3$$

$$t = -15 \quad \text{Multiplication}$$

EXAMPLE 3 Solve $\frac{2}{3}y = 4$.

SOLUTION We can multiply both sides by $\frac{3}{2}$ and have $1y$ on the left side:

$$\frac{3}{2}\left(\frac{2}{3}y\right) = \frac{3}{2}(4) \quad \text{Multiplication property of equality}$$

$$\left(\frac{3}{2} \cdot \frac{2}{3}\right)y = \frac{3}{2}(4) \quad \text{Associative property}$$

$$y = 6 \quad \text{Simplify } \frac{3}{2}(4) = \frac{3}{2}\left(\frac{4}{1}\right) = \frac{12}{2} = 6$$

Note: Notice in Examples 1 through 3 that if the variable is being multiplied by a number like -4 or $\frac{2}{3}$, we always multiply by the number's reciprocal, $-\frac{1}{4}$ or $\frac{3}{2}$, to end up with just the variable on one side of the equation.

EXAMPLE 4 Solve $5 + 8 = 10x + 20x - 4x$.

SOLUTION Our first step will be to simplify each side of the equation:

$$13 = 26x \quad \text{Simplify both sides first}$$

$$\frac{1}{26}(13) = \frac{1}{26}(26x) \quad \text{Multiplication property of equality}$$

$$\frac{13}{26} = x \quad \text{Multiplication}$$

$$\frac{1}{2} = x \quad \text{Reduce to lowest terms}$$

In the next three examples, we will use both the addition property of equality and the multiplication property of equality.

EXAMPLE 5 Solve for $x: 6x + 5 = -13$.

SOLUTION We begin by adding -5 to both sides of the equation:

$$6x + 5 + (-5) = -13 + (-5) \quad \text{Add } -5 \text{ to both sides}$$

$$6x = -18 \quad \text{Simplify}$$

$$\frac{1}{6}(6x) = \frac{1}{6}(-18) \quad \text{Multiply both sides by } \frac{1}{6}$$

$$x = -3$$

EXAMPLE 6 Solve for $x: 5x = 2x + 12$.

SOLUTION We begin by adding $-2x$ to both sides of the equation:

$$5x + (-2x) = 2x + (-2x) + 12 \quad \text{Add } -2x \text{ to both sides}$$

$$3x = 12 \quad \text{Simplify}$$

$$\frac{1}{3}(3x) = \frac{1}{3}(12) \quad \text{Multiply both sides by } \frac{1}{3}$$

$$x = 4 \quad \text{Simplify}$$

Note: Notice that in Example 6 we used the addition property of equality first to combine all the terms containing x on the left side of the equation. Once this had been done, we used the multiplication property to isolate x on the left side.

EXAMPLE 7 Solve for $x: 3x - 4 = -2x + 6$.

SOLUTION We begin by adding $2x$ to both sides:

$$3x + 2x - 4 = -2x + 2x + 6 \quad \text{Add } 2x \text{ to both sides}$$

$$5x - 4 = 6 \quad \text{Simplify}$$

Now we add 4 to both sides:

$$5x - 4 + 4 = 6 + 4 \quad \text{Add 4 to both sides}$$

$$5x = 10 \quad \text{Simplify}$$

$$\frac{1}{5}(5x) = \frac{1}{5}(10) \quad \text{Multiply by } \frac{1}{5}$$

$$x = 2 \quad \text{Simplify}$$

The next example involves fractions. You will see that the properties we use to solve equations containing fractions are the same as the properties we used to solve the previous equations. Also, the LCD that we used previously to add fractions can be used with the multiplication property of equality to simplify equations containing fractions.

EXAMPLE 8 Solve $\frac{2}{3}x + \frac{1}{2} = -\frac{3}{4}$.

SOLUTION We can solve this equation by applying our properties and working with the fractions, or we can begin by eliminating the fractions.

Method 1 Working with the fractions.

$$\frac{2}{3}x + \frac{1}{2} + \left(-\frac{1}{2}\right) = -\frac{3}{4} + \left(-\frac{1}{2}\right) \quad \text{Add } -\tfrac{1}{2} \text{ to each side}$$

$$\frac{2}{3}x = -\frac{5}{4} \quad \text{Note that } -\tfrac{3}{4} + \left(-\tfrac{1}{2}\right) = -\tfrac{3}{4} + \left(-\tfrac{2}{4}\right)$$

$$\frac{3}{2}\left(\frac{2}{3}x\right) = \frac{3}{2}\left(-\frac{5}{4}\right) \quad \text{Multiply each side by } \tfrac{3}{2}$$

$$x = -\frac{15}{8}$$

Method 2 Eliminating the fractions in the beginning.

$$12\left(\frac{2}{3}x + \frac{1}{2}\right) = 12\left(-\frac{3}{4}\right) \quad \text{Multiply each side by the LCD 12}$$

$$12\left(\frac{2}{3}x\right) + 12\left(\frac{1}{2}\right) = 12\left(-\frac{3}{4}\right) \quad \text{Distributive property on the left side}$$

$$8x + 6 = -9 \quad \text{Multiply}$$

$$8x = -15 \quad \text{Add } -6 \text{ to each side}$$

$$x = -\frac{15}{8} \quad \text{Multiply each side by } \tfrac{1}{8}$$

Note: Our original equation has denominators of 3, 2, and 4. The LCD for these three denominators is 12, and it has the property that all three denominators will divide it evenly. Therefore, if we multiply both sides of our equation by 12, each denominator will divide into 12 and we will be left with an equation that does not contain any denominators other than 1.

As the third line in Method 2 indicates, multiplying each side of the equation by the LCD eliminates all the fractions from the equation.

As you can see, both methods yield the same solution.

PROPERTY *A Note on Division*

Because *division* is defined as multiplication by the reciprocal, multiplying both sides of an equation by the same number is equivalent to dividing both sides of the equation by the reciprocal of that number; that is, multiplying each side of an equation by $\frac{1}{3}$ and dividing each side of the equation by 3 are equivalent operations. If we were to solve the equation $3x = 18$ using division instead of multiplication, the steps would look like this:

$$3x = 18 \quad \text{Original equation}$$

$$\frac{3x}{3} = \frac{18}{3} \quad \text{Divide each side by 3}$$

$$x = 6 \quad \text{Division}$$

Using division instead of multiplication on a problem like this may save you some writing. However, with multiplication, it is easier to explain "why" we end up with just one x on the left side of the equation. (The "why" has to do with the associative property of multiplication.) My suggestion is that you continue to use multiplication to solve equations like this one until you understand the process completely. Then, if you find it more convenient, you can use division instead of multiplication.

GETTING READY FOR CLASS

After reading through the preceding section, respond in your own words and in complete sentences.

A. Explain in words the multiplication property of equality.
B. If an equation contains fractions, how do you use the multiplication property of equality to clear the equation of fractions?
C. Why is it okay to divide both sides of an equation by the same nonzero number?
D. Explain in words how you would solve the equation $3x = 7$ using the multiplication property of equality.

Problem Set 1.3

Solve the following equations. Be sure to show your work.

1. $5x = 10$ **2.** $6x = 12$ **3.** $7a = 28$ **4.** $4a = 36$

5. $-8x = 4$ **6.** $-6x = 2$ **7.** $8m = -16$ **8.** $5m = -25$

9. $-3x = -9$ **10.** $-9x = -36$ **11.** $-7y = -28$ **12.** $-15y = -30$

13. $2x = 0$ **14.** $7x = 0$ **15.** $-5x = 0$ **16.** $-3x = 0$

17. $\frac{x}{3} = 2$ **18.** $\frac{x}{4} = 3$ **19.** $-\frac{m}{5} = 10$ **20.** $-\frac{m}{7} = 1$

21. $-\frac{x}{2} = -\frac{3}{4}$ **22.** $-\frac{x}{3} = \frac{5}{6}$ **23.** $\frac{2}{3}a = 8$ **24.** $\frac{3}{4}a = 6$

25. $-\frac{3}{5}x = \frac{9}{5}$ **26.** $-\frac{2}{5}x = \frac{6}{15}$ **27.** $-\frac{5}{8}y = -20$ **28.** $-\frac{7}{2}y = -14$

Simplify both sides as much as possible, and then solve.

29. $-4x - 2x + 3x = 24$ **30.** $7x - 5x + 8x = 20$

31. $4x + 8x - 2x = 15 - 10$ **32.** $5x + 4x + 3x = 4 + 8$

33. $-3 - 5 = 3x + 5x - 10x$ **34.** $10 - 16 = 12x - 6x - 3x$

35. $18 - 13 = \frac{1}{2}a + \frac{3}{4}a - \frac{5}{8}a$ **36.** $20 - 14 = \frac{1}{3}a + \frac{5}{6}a - \frac{2}{3}a$

Solve the following equations by multiplying both sides by -1.

37. $-x = 4$ **38.** $-x = -3$ **39.** $-x = -4$ **40.** $-x = 3$

41. $15 = -a$ **42.** $-15 = -a$ **43.** $-y = \frac{1}{2}$ **44.** $-y = -\frac{3}{4}$

Solve each of the following equations using the method shown in Examples 5–8 in this section.

45. $3x - 2 = 7$ **46.** $2x - 3 = 9$ **47.** $2a + 1 = 3$ **48.** $5a - 3 = 7$

49. $\frac{1}{8} + \frac{1}{2}x = \frac{1}{4}$ **50.** $\frac{1}{3} + \frac{1}{7}x = -\frac{8}{21}$

51. $6x = 2x - 12$ **52.** $8x = 3x - 10$

53. $2y = -4y + 18$ **54.** $3y = -2y - 15$ **55.** $-7x = -3x - 8$

56. $-5x = -2x - 12$ **57.** $8x + 4 = 2x - 5$ **58.** $5x + 6 = 3x - 6$

59. $x + \frac{1}{2} = \frac{1}{4}x - \frac{5}{8}$ **60.** $\frac{1}{3}x + \frac{2}{5} = \frac{1}{5}x - \frac{2}{5}$

61. $6m - 3 = m + 2$ **62.** $6m - 5 = m + 5$

63. $\frac{1}{2}m - \frac{1}{4} = \frac{1}{12}m + \frac{1}{6}$ **64.** $\frac{1}{2}m - \frac{5}{12} = \frac{1}{12}m + \frac{5}{12}$

65. $9y + 2 = 6y - 4$ **66.** $6y + 14 = 2y - 2$

67. Solve each equation.

a. $2x = 3$ **b.** $2 + x = 3$

c. $2x + 3 = 0$ **d.** $2x + 3 = -5$

e. $2x + 3 = 7x - 5$

68. Solve each equation.

a. $5t = 10$

b. $5 + t = 10$

c. $5t + 10 = 0$

d. $5t + 10 = 12$

e. $5t + 10 = 8t + 12$

Applying the Concepts

69. Break-Even Point Movie theaters pay a certain price for the movies that you and I see. Suppose a theater pays \$1,500 for each showing of a popular movie. If they charge \$7.50 for each ticket they sell, then the equation $7.5x = 1{,}500$ gives the number of tickets they must sell to equal the \$1,500 cost of showing the movie. This number is called the break-even point. Solve the equation for x to find the break-even point.

70. Basketball Laura plays basketball for her community college. In one game she scored 13 points total, with a combination of free throws, field goals, and three-pointers. Each free throw is worth 1 point, each field goal is 2 points, and each three-pointer is worth 3 points. If she made 1 free throw and 3 field goals, then solving the equation

$$1 + 3(2) + 3x = 13$$

will give us the number of three-pointers she made. Solve the equation to find the number of three-point shots Laura made.

71. Taxes Suppose 21% of your monthly pay is withheld for federal income taxes and another 8% is withheld for Social Security, state income tax, and other miscellaneous items. If you are left with \$987.50 a month in take-home pay, then the amount you earned before the deductions were removed from your check is given by the equation

$$G - 0.21G - 0.08G = 987.5$$

Solve this equation to find your gross income.

72. Rhind Papyrus The *Rhind Papyrus* is an ancient document that contains mathematical riddles. One problem asks the reader to find a quantity such that when it is added to one-fourth of itself the sum is 15. The equation that describes this situation is

$$x + \frac{1}{4}x = 15$$

Solve this equation.

Getting Ready for the Next Section

To understand all of the explanations and examples in the next section you must be able to work the problems below.

Solve each equation.

73. $2x = 4$

74. $3x = 24$

75. $30 = 5x$

76. $0 = 5x$

77. $0.17x = 510$

78. $0.1x = 400$

Apply the distributive property and then simplify if possible.

79. $3(x - 5) + 4$ **80.** $5(x - 3) + 2$ **81.** $0.09(x + 2{,}000)$

82. $0.04(x + 7{,}000)$ **83.** $7 - 3(2y + 1)$ **84.** $4 - 2(3y + 1)$

85. $3(2x - 5) - (2x - 4)$ **86.** $4(3x - 2) - (6x - 5)$

Simplify.

87. $10x + (-5x)$ **88.** $12x + (-7x)$ **89.** $0.08x + 0.09x$ **90.** $0.06x + 0.04x$

Solving Linear Equations 1.4

We will now use the material we have developed in the first three sections of this chapter to build a method for solving any linear equation.

DEFINITION *linear equation*

A ***linear equation*** in one variable is any equation that can be put in the form $ax + b = 0$, where a and b are real numbers and a is not zero.

Each of the equations we will solve in this section is a linear equation in one variable. The steps we use to solve a linear equation in one variable are listed here.

Note: You may have some previous experience solving equations. Even so, you should solve the equations in this section using the method developed here. Your work should look like the examples in the text. If you have learned shortcuts or a different method of solving equations somewhere else, you can always go back to them later. What is important now is that you are able to solve equations by the methods shown here.

PROPERTY *Strategy for Solving Linear Equations in One Variable*

Step 1a: Use the distributive property to separate terms, if necessary.
1b: If fractions are present, consider multiplying both sides by the LCD to eliminate the fractions. If decimals are present, consider multiplying both sides by a power of 10 to clear the equation of decimals.
1c: Combine similar terms on each side of the equation.
Step 2: Use the addition property of equality to get all variable terms on one side of the equation and all constant terms on the other side. A variable term is a term that contains the variable (for example, $5x$). A constant term is a term that does not contain the variable (the number 3, for example).
Step 3: Use the multiplication property of equality to get x (that is, $1x$) by itself on one side of the equation.
Step 4: Check your solution in the original equation to be sure that you have not made a mistake in the solution process.

As you will see as you work through the examples in this section, it is not always necessary to use all four steps when solving equations. The number of steps used depends on the equation. In Example 1 there are no fractions or decimals in the original equation, so step 1b will not be used. Likewise, after applying the distributive property to the left side of the equation in Example 1, there are no similar terms to combine on either side of the equation, making step 1c also unnecessary.

EXAMPLE 1 Solve $2(x + 3) = 10$.

SOLUTION To begin, we apply the distributive property to the left side of the equation to separate terms:

Step 1a: $2x + 6 = 10$ — Distributive property

Step 2:
$2x + 6 + (-6) = 10 + (-6)$ — Addition property of equality
$2x = 4$

Step 3:
$\frac{1}{2}(2x) = \frac{1}{2}(4)$ — Multiply each side by $\frac{1}{2}$
$x = 2$ — The solution is 2

The solution to our equation is 2. We check our work (to be sure we have not made either a mistake in applying the properties or an arithmetic mistake) by substituting 2 into our original equation and simplifying each side of the result separately.

Check: When $x = 2$

the equation $2(x + 3) = 10$

Step 4: becomes $2(2 + 3) \stackrel{?}{=} 10$

$2(5) \stackrel{?}{=} 10$

$10 = 10$ A true statement

Our solution checks.

The general method of solving linear equations is actually very simple. It is based on the properties we developed in Chapter 1 and on two very simple new properties. We can add any number to both sides of the equation and multiply both sides by any nonzero number. The equation may change in form, but the solution set will not. If we look back to Example 1, each equation looks a little different from each preceding equation. What is interesting and useful is that each equation says the same thing about x. They all say x is 2. The last equation, of course, is the easiest to read, and that is why our goal is to end up with x by itself.

The examples that follow show a variety of equations and their solutions. When you have finished this section and worked the problems in the problem set, the steps in the solution process should be a description of how you operate when solving equations. That is, you want to work enough problems so that the Strategy for Solving Linear Equations is second nature to you.

EXAMPLE 2 Solve for $x: 3(x - 5) + 4 = 13$.

SOLUTION Our first step will be to apply the distributive property to the left side of the equation:

Step 1a: $3x - 15 + 4 = 13$ Distributive property

Step 1c: $3x - 11 = 13$ Simplify the left side

Step 2: $\begin{cases} 3x - 11 + 11 = 13 + 11 \\ 3x = 24 \end{cases}$ Add 11 to both sides

Step 3: $\begin{cases} \frac{1}{3}(3x) = \frac{1}{3}(24) \\ x = 8 \end{cases}$ Multiply both sides by $\frac{1}{3}$; The solution is 8

Check: When $x = 8$

the equation $3(x - 5) + 4 = 13$

becomes $3(8 - 5) + 4 \stackrel{?}{=} 13$

Step 4: $3(3) + 4 \stackrel{?}{=} 13$

$9 + 4 \stackrel{?}{=} 13$

$13 = 13$ A true statement

EXAMPLE 3 Solve $5(x - 3) + 2 = 5(2x - 8) - 3$.

SOLUTION In this case we apply the distributive property on each side of the equation:

Step 1a: $5x - 15 + 2 = 10x - 40 - 3$ — Distributive property

Step 1c: $5x - 13 = 10x - 43$ — Simplify each side

Step 2:

$5x + (-5x) - 13 = 10x + (-5x) - 43$ — Add $-5x$ to both sides

$-13 = 5x - 43$

$-13 + 43 = 5x - 43 + 43$ — Add 43 to both sides

$30 = 5x$

Step 3:

$\frac{1}{5}(30) = \frac{1}{5}(5x)$ — Multiply both sides by $\frac{1}{5}$

$6 = x$ — The solution is 6

Check: Replacing x with 6 in the original equation, we have

Step 4:

$5(6 - 3) + 2 \stackrel{?}{=} 5(2 \cdot 6 - 8) - 3$

$5(3) + 2 \stackrel{?}{=} 5(12 - 8) - 3$

$5(3) + 2 \stackrel{?}{=} 5(4) - 3$

$15 + 2 \stackrel{?}{=} 20 - 3$

$17 = 17$ — A true statement

Note: It makes no difference on which side of the equal sign x ends up. Most people prefer to have x on the left side because we read from left to right, and it seems to sound better to say x is 6 rather than 6 is x. Both expressions, however, have exactly the same meaning.

EXAMPLE 4 Solve the equation $0.08x + 0.09(x + 2{,}000) = 690$.

SOLUTION We can solve the equation in its original form by working with the decimals, or we can eliminate the decimals first by using the multiplication property of equality and solving the resulting equation. Both methods follow.

Method 1

Working with the decimals.

$0.08x + 0.09(x + 2{,}000) = 690$ — Original equation

Step 1a: $0.08x + 0.09x + 0.09(2{,}000) = 690$ — Distributive property

Step 1c: $0.17x + 180 = 690$ — Simplify the left side

Step 2:

$0.17x + 180 + (-180) = 690 + (-180)$ — Add -180 to each side

$0.17x = 510$

Step 3:

$\frac{0.17x}{0.17} = \frac{510}{0.17}$ — Divide each side by 0.17

$x = 3{,}000$

Note that we divided each side of the equation by 0.17 to obtain the solution. This is still an application of the multiplication property of equality because dividing by 0.17 is equivalent to multiplying by $\frac{1}{0.17}$.

Method 2

Eliminating the decimals in the beginning.

	$0.08x + 0.09(x + 2{,}000) = 690$	Original equation
Step 1a:	$0.08x + 0.09x + 180 = 690$	Distributive property
Step 1b:	$100(0.08x + 0.09x + 180) = 100(690)$	Multiply both sides by 100
	$8x + 9x + 18{,}000 = 69{,}000$	
Step 1c:	$17x + 18{,}000 = 69{,}000$	Simplify the left side
Step 2:	$17x = 51{,}000$	Add $-18{,}000$ to each side
Step 3:	$\frac{17x}{17} = \frac{51{,}000}{17}$	Divide each side by 17
	$x = 3{,}000$	

Substituting 3,000 for x in the original equation, we have

Step 4:	$0.08(3{,}000) + 0.09(3{,}000 + 2{,}000) \stackrel{?}{=} 690$	
	$0.08(3{,}000) + 0.09(5{,}000) \stackrel{?}{=} 690$	
	$240 + 450 \stackrel{?}{=} 690$	
	$690 = 690$	A true statement

EXAMPLE 5 Solve $7 - 3(2y + 1) = 16$.

SOLUTION We begin by multiplying -3 times the sum of $2y$ and 1:

Step 1a:	$7 - 6y - 3 = 16$	Distributive property
Step 1c:	$-6y + 4 = 16$	Simplify the left side
Step 2:	$-6y + 4 + (-4) = 16 + (-4)$	Add -4 to both sides
	$-6y = 12$	
Step 3:	$-\frac{1}{6}(-6y) = -\frac{1}{6}(12)$	Multiply both sides by $-\frac{1}{6}$
	$y = -2$	

There are two things to notice about the example that follows: first, the distributive property is used to remove parentheses that are preceded by a negative sign, and, second, the addition property and the multiplication property are not shown in as much detail as in the previous examples.

EXAMPLE 6 Solve $3(2x - 5) - (2x - 4) = 6 - (4x + 5)$.

SOLUTION When we apply the distributive property to remove the grouping symbols and separate terms, we have to be careful with the signs. Remember, we can think of $-(2x - 4)$ as $-1(2x - 4)$, so that

$$-(2x - 4) = -1(2x - 4) = -2x + 4$$

It is not uncommon for students to make a mistake with this type of simplification and write the result as $-2x - 4$, which is incorrect. Here is the complete solution to our equation:

$3(2x - 5) - (2x - 4) = 6 - (4x + 5)$	Original equation
$6x - 15 - 2x + 4 = 6 - 4x - 5$	Distributive property
$4x - 11 = -4x + 1$	Simplify each side
$8x - 11 = 1$	Add $4x$ to each side
$8x = 12$	Add 11 to each side
$x = \frac{12}{8}$	Multiply each side by $\frac{1}{8}$
$x = \frac{3}{2}$	Reduce to lowest terms

The solution, $\frac{3}{2}$, checks when replacing x in the original equation. ■

GETTING READY FOR CLASS

After reading through the preceding section, respond in your own words and in complete sentences.

A. What is the first step in solving a linear equation containing parentheses?
B. What is the last step in solving a linear equation?
C. Explain in words how you would solve the equation $2x - 3 = 8$.
D. If an equation contains decimals, what can you do to eliminate the decimals?

Problem Set 1.4

Solve each of the following equations using the four steps shown in this section.

1. $2(x + 3) = 12$ **2.** $3(x - 2) = 6$ **3.** $6(x - 1) = -18$

4. $4(x + 5) = 16$ **5.** $2(4a + 1) = -6$ **6.** $3(2a - 4) = 12$

7. $14 = 2(5x - 3)$ **8.** $-25 = 5(3x + 4)$ **9.** $-2(3y + 5) = 14$

10. $-3(2y - 4) = -6$ **11.** $-5(2a + 4) = 0$ **12.** $-3(3a - 6) = 0$

13. $1 = \frac{1}{2}(4x + 2)$ **14.** $1 = \frac{1}{3}(6x + 3)$ **15.** $3(t - 4) + 5 = -4$

16. $5(t - 1) + 6 = -9$

Solve each equation.

17. $4(2x + 1) - 7 = 1$ **18.** $6(3y + 2) - 8 = -2$

19. $\frac{1}{2}(x - 3) = \frac{1}{4}(x + 1)$ **20.** $\frac{1}{3}(x - 4) = \frac{1}{2}(x - 6)$

21. $-0.7(2x - 7) = 0.3(11 - 4x)$ **22.** $-0.3(2x - 5) = 0.7(3 - x)$

23. $-2(3y + 1) = 3(1 - 6y) - 9$ **24.** $-5(4y - 3) = 2(1 - 8y) + 11$

25. $\frac{3}{4}(8x - 4) + 3 = \frac{2}{5}(5x + 10) - 1$ **26.** $\frac{5}{6}(6x + 12) + 1 = \frac{2}{3}(9x - 3) + 5$

27. $0.06x + 0.08(100 - x) = 6.5$ **28.** $0.05x + 0.07(100 - x) = 6.2$

29. $6 - 5(2a - 3) = 1$ **30.** $-8 - 2(3 - a) = 0$

31. $0.2x - 0.5 = 0.5 - 0.2(2x - 13)$ **32.** $0.4x - 0.1 = 0.7 - 0.3(6 - 2x)$

33. $2(t - 3) + 3(t - 2) = 28$ **34.** $-3(t - 5) - 2(2t + 1) = -8$

35. $5(x - 2) - (3x + 4) = 3(6x - 8) + 10$

36. $3(x - 1) - (4x - 5) = 2(5x - 1) - 7$

37. $2(5x - 3) - (2x - 4) = 5 - (6x + 1)$

38. $3(4x - 2) - (5x - 8) = 8 - (2x + 3)$

39. $-(3x + 1) - (4x - 7) = 4 - (3x + 2)$

40. $-(6x + 2) - (8x - 3) = 8 - (5x + 1)$

41. $x + (2x - 1) = 2$ **42.** $x + (5x + 2) = 20$

43. $x - (3x + 5) = -3$ **44.** $x - (4x - 1) = 7$

45. $15 = 3(x - 1)$ **46.** $12 = 4(x - 5)$

47. $4x - (-4x + 1) = 5$ **48.** $-2x - (4x - 8) = -1$

49. $5x - 8(2x - 5) = 7$ **50.** $3x + 4(8x - 15) = 10$

51. $7(2y - 1) - 6y = -1$ **52.** $4(4y - 3) + 2y = 3$

53. $0.2x + 0.5(12 - x) = 3.6$ **54.** $0.3x + 0.6(25 - x) = 12$

55. $0.5x + 0.2(18 - x) = 5.4$ **56.** $0.1x + 0.5(40 - x) = 32$

57. $x + (x + 3)(-3) = x - 3$ **58.** $x - 2(x + 2) = x - 2$

59. $5(x + 2) + 3(x - 1) = -9$ **60.** $4(x + 1) + 3(x - 3) = 2$

61. $3(x - 3) + 2(2x) = 5$ **62.** $2(x - 2) + 3(5x) = 30$

63. $5(y + 2) = 4(y + 1)$ **64.** $3(y - 3) = 2(y - 2)$

65. $3x + 2(x - 2) = 6$

66. $5x - (x - 5) = 25$

67. $50(x - 5) = 30(x + 5)$

68. $34(x - 2) = 26(x + 2)$

69. $0.08x + 0.09(x + 2{,}000) = 860$

70. $0.11x + 0.12(x + 4{,}000) = 940$

71. $0.10x + 0.12(x + 500) = 214$

72. $0.08x + 0.06(x + 800) = 104$

73. $5x + 10(x + 8) = 245$

74. $5x + 10(x + 7) = 175$

75. $5x + 10(x + 3) + 25(x + 5) = 435$

76. $5(x + 3) + 10x + 25(x + 7) = 390$

The next two problems are intended to give you practice reading, and paying attention to, the instructions that accompany the problems you are working. Working these problems is an excellent way to get ready for a test or a quiz.

77. Work each problem according to the instructions given.

a. Solve: $4x - 5 = 0$

b. Solve: $4x - 5 = 25$

c. Add: $(4x - 5) + (2x + 25)$

d. Solve: $4x - 5 = 2x + 25$

e. Multiply: $4(x - 5)$

f. Solve: $4(x - 5) = 2x + 25$

78. Work each problem according to the instructions given.

a. Solve: $3x + 6 = 0$

b. Solve: $3x + 6 = 4$

c. Add: $(3x + 6) + (7x + 4)$

d. Solve: $3x + 6 = 7x + 4$

e. Multiply: $3(x + 6)$

f. Solve: $3(x + 6) = 7x + 4$

Getting Ready for the Next Section

To understand all of the explanations and examples in the next section you must be able to work the problems below.

Solve each equation.

79. $40 = 2x + 12$

80. $80 = 2x + 12$

81. $12 + 2y = 6$

82. $3x + 18 = 6$

83. $24x = 6$

84. $45 = 0.75x$

85. $70 = x \cdot 210$

86. $15 = x \cdot 80$

Apply the distributive property.

87. $\frac{1}{2}(-3x + 6)$

88. $-\frac{1}{4}(-5x + 20)$

Napa Valley College

You may think that all your mathematics instructors started their college math sequence with precalculus or calculus, but that is not always the case. Diane van Deusen, a full time mathematics instructor at Napa Valley College in Napa, California, started her career in mathematics in elementary algebar. Here is part of her story from her website:

Dear Student,

Welcome to elementary algebra! Since we will be spending a significant amount of time together this semester, I thought I should introduce myself to you, and tell you how I ended up with a career in education.

I was not encouraged to attend college after high school, and in fact, had no interest in "more school". Consequently, I didn't end up taking a college class until I was 31 years old! Before returning to and while attending college, I worked locally in the restaurant business as a waitress and bartender and in catering. In fact, I sometimes wait tables a few nights a week during my summer breaks.

When I first came back to school, at Napa Valley College (NVC), I thought I might like to enter the nursing program but soon found out nursing was not for me. As I started working on general education requirements, I took elementary algebra and was surprised to learn that I really loved mathematics, even though I had failed 8th grade algebra! As I continued to appreciate and value my own education, I decided to become a teacher so that I could support other people seeking education goals. After earning my AA degree from NVC, I transferred to Sonoma State where I earned my bachelors degree in mathematics with a concentration in statistics. Finally, I attended Cal State Hayward to earn my master's degree in applied statistics. It took me ten years in all to do this.

I feel that having been a returning student while a single, working parent, also an EOPS and Financial Aid recipient, I fully understand the complexity of the life of a community college student. If at any time you have questions about the college, the class or just need someone to talk to, my door is open.

I sincerely hope that my classroom will provide a positive and satisfying learning experience for you.

Diane Van Deusen

Algebra is a great place to start you journey into college mathematics. You can start here and go as far as you want in mathematics. Who knows, you may end up teaching mathematics one day, just like Diane Van Deusen.

Formulas 1.5

In this section we continue solving equations by working with formulas. To begin, here is the definition of a formula.

DEFINITION *formula*

In mathematics, a ***formula*** is an equation that contains more than one variable.

The equation $P = 2l + 2w$, which tells us how to find the perimeter of a rectangle, is an example of a formula.

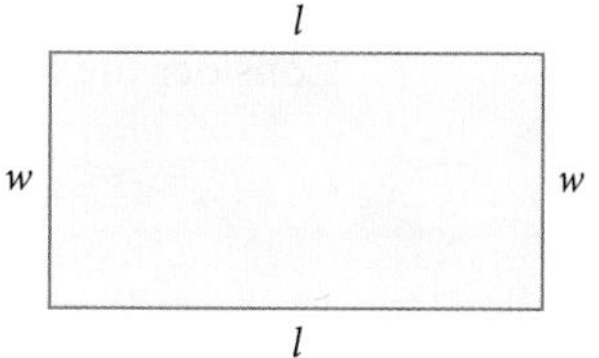

To begin our work with formulas, we will consider some examples in which we are given numerical replacements for all but one of the variables.

EXAMPLE 1 The perimeter P of a rectangular livestock pen is 40 feet. If the width w is 6 feet, find the length.

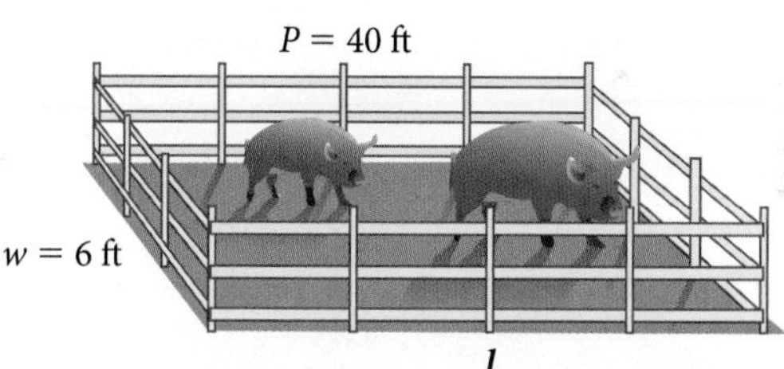

SOLUTION First we substitute 40 for P and 6 for w in the formula $P = 2l + 2w$. Then we solve for l:

When	$P = 40$ and $w = 6$	
the formula	$P = 2l + 2w$	
becomes	$40 = 2l + 2(6)$	
or	$40 = 2l + 12$	Multiply 2 and 6
	$28 = 2l$	Add -12 to each side
	$14 = l$	Multiply each side by $\frac{1}{2}$

To summarize our results, if a rectangular pen has a perimeter of 40 feet and a width of 6 feet, then the length must be 14 feet.

EXAMPLE 2 Find y when $x = 4$ in the formula $3x + 2y = 6$.

SOLUTION We substitute 4 for x in the formula and then solve for y:

When	$x = 4$	
the formula	$3x + 2y = 6$	
becomes	$3(4) + 2y = 6$	
or	$12 + 2y = 6$	Multiply 3 and 4
	$2y = -6$	Add -12 to each side
	$y = -3$	Multiply each side by $\frac{1}{2}$

In the next examples we will solve a formula for one of its variables without being given numerical replacements for the other variables.

Consider the formula for the area of a triangle:

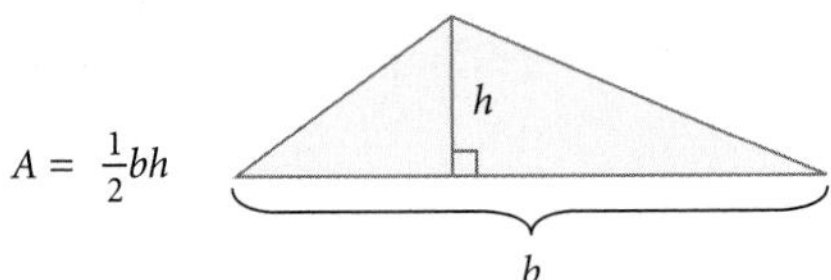

where A = area, b = length of the base, and h = height of the triangle.

Suppose we want to solve this formula for h. What we must do is isolate the variable h on one side of the equal sign. We begin by multiplying both sides by 2, because it is the reciprocal of $\frac{1}{2}$:

$$2 \cdot A = 2 \cdot \frac{1}{2}bh$$

$$2A = bh$$

Then we divide both sides by b:

$$\frac{2A}{b} = \frac{bh}{b}$$

$$h = \frac{2A}{b}$$

The original formula $A = \frac{1}{2}bh$ and the final formula $h = \frac{2A}{b}$ both give the same relationship among A, b, and h. The first one has been solved for A and the second one has been solved for h.

RULE

To solve a formula for one of its ***variables***, we must isolate that variable on either side of the equal sign. All other variables and constants will appear on the other side.

EXAMPLE 3 Solve $3x + 2y = 6$ for y.

SOLUTION To solve for y, we must isolate y on the left side of the equation. To begin, we use the addition property of equality to add $-3x$ to each side:

$$3x + 2y = 6 \quad \text{Original formula}$$

$$3x + (-3x) + 2y = (-3x) + 6 \quad \text{Add } -3x \text{ to each side}$$

$$2y = -3x + 6 \quad \text{Simplify the left side}$$

$$\frac{1}{2}(2y) = \frac{1}{2}(-3x + 6) \quad \text{Multiply each side by } \frac{1}{2}$$

$$y = -\frac{3}{2}x + 3 \quad \text{Multiplication}$$

EXAMPLE 4 Solve $h = vt - 16t^2$ for v.

SOLUTION Let's begin by interchanging the left and right sides of the equation. That way, the variable we are solving for, v, will be on the left side.

$$vt - 16t^2 = h \quad \text{Exchange sides}$$

$$vt - 16t^2 + 16t^2 = h + 16t^2 \quad \text{Add } 16t^2 \text{ to each side}$$

$$vt = h + 16t^2$$

$$\frac{vt}{t} = \frac{h + 16t^2}{t} \quad \text{Divide each side by } t$$

$$v = \frac{h + 16t^2}{t}$$

We know we are finished because we have isolated the variable we are solving for on the left side of the equation and it does not appear on the other side.

EXAMPLE 5 Solve for y: $\frac{y - 1}{x} = \frac{3}{2}$.

SOLUTION Although we will do more extensive work with formulas of this form later in the book, we need to know how to solve this particular formula for y in order to understand some things in the next chapter. We begin by multiplying each side of the formula by x. Doing so will simplify the left side of the equation, and make the rest of the solution process simple.

$$\frac{y - 1}{x} = \frac{3}{2} \quad \text{Original formula}$$

$$x \cdot \frac{y - 1}{x} = \frac{3}{2} \cdot x \quad \text{Multiply each side by } x$$

$$y - 1 = \frac{3}{2}x \quad \text{Simplify each side}$$

$$y = \frac{3}{2}x + 1 \quad \text{Add 1 to each side}$$

This is our solution. If we look back to the first step, we can justify our result on the left side of the equation this way: Dividing by x is equivalent to multiplying by its reciprocal $\frac{1}{x}$. Here is what it looks like when written out completely:

$$x \cdot \frac{y - 1}{x} = x\,\frac{1}{x}\,(y - 1) = 1(y - 1) = (y - 1)$$

FACTS FROM GEOMETRY ***More on Complementary and Supplementary Angles***

In Chapter 1 we defined complementary angles as angles that add to 90°; that is, if x and y are complementary angles, then

$$x + y = 90°$$

If we solve this formula for y, we obtain a formula equivalent to our original formula:

$$y = 90° - x$$

Because y is the complement of x, we can generalize by saying that the complement of angle x is the angle $90° - x$. By a similar reasoning process, we can say that the supplement of angle x is the angle $180° - x$. To summarize, if x is an angle, then

The complement of x is $90° - x$, and

The supplement of x is $180° - x$

If you go on to take a trigonometry class, you will see this formula again.

EXAMPLE 6 Find the complement and the supplement of 25°.

SOLUTION We can use the formulas $90° - x$ and $180° - x$.

The complement of 25° is $90° - 25° = 65°$.

The supplement of 25° is $180° - 25° = 155°$.

Basic Percent Problems

The next examples in this section show how basic percent problems can be translated directly into equations. To understand these examples, you must recall that *percent* means "per hundred" that is, 75% is the same as $\frac{75}{100}$, 0.75, and, in reduced fraction form, $\frac{3}{4}$. Likewise, the decimal 0.25 is equivalent to 25%. To change a decimal to a percent, we move the decimal point two places to the right and write the % symbol. To change from a percent to a decimal, we drop the % symbol and move the decimal point two places to the left. The table that follows gives some of the most commonly used fractions and decimals and their equivalent percents.

Fraction	Decimal	Percent
$\frac{1}{2}$	0.5	50%
$\frac{1}{4}$	0.25	25%
$\frac{3}{4}$	0.75	75%
$\frac{1}{3}$	$0.33\frac{1}{3}$	$33\frac{1}{3}\%$
$\frac{2}{3}$	$0.66\frac{2}{3}$	$66\frac{2}{3}\%$
$\frac{1}{5}$	0.2	20%
$\frac{2}{5}$	0.4	40%

EXAMPLE 7 What number is 25% of 60?

SOLUTION To solve a problem like this, we let $x =$ the number in question (that is, the number we are looking for). Then, we translate the sentence directly into an equation by using an equal sign for the word "is" and multiplication for the word "of." Here is how it is done:

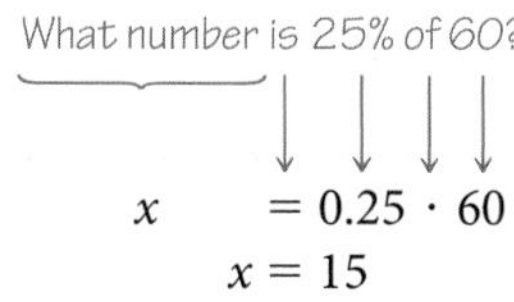

Notice that we must write 25% as a decimal in order to do the arithmetic in the problem.

The number 15 is 25% of 60.

EXAMPLE 8 What percent of 24 is 6?

SOLUTION Translating this sentence into an equation, as we did in Example 7, we have:

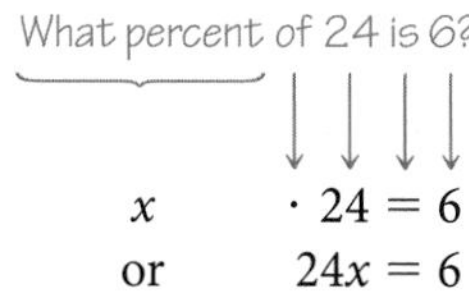

Next, we multiply each side by $\frac{1}{24}$. (This is the same as dividing each side by 24.)

$$\frac{1}{24}(24x) = \frac{1}{24}(6)$$

$$x = \frac{6}{24}$$

$$= \frac{1}{4}$$

$$= 0.25, \text{ or } 25\%$$

25% of 24 is 6, or in other words, the number 6 is 25% of 24.

EXAMPLE 9 45 is 75% of what number?

SOLUTION Again, we translate the sentence directly:

45 is 75% of what number?

$$45 = 0.75 \cdot x$$

Next, we multiply each side by $\frac{1}{0.75}$ (which is the same as dividing each side by 0.75):

$$\frac{1}{0.75}(45) = \frac{1}{0.75}(0.75x)$$

$$\frac{45}{0.75} = x$$

$$60 = x$$

The number 45 is 75% of 60.

EXAMPLE 10 The American Dietetic Association (ADA) recommends eating foods in which the calories from fat are less than 30% of the total calories. The nutrition labels from two kinds of granola bars are shown in Figure 1. For each bar, what percent of the total calories come from fat?

BAR I

Nutrition Facts
Serving Size 2 bars (47g)
Servings Per Container 6

Amount Per Serving	
Calories	210
Calories from Fat	**70**
	% Daily Value*
Total Fat 8g	**12%**
Saturated Fat 1g	**5%**
Cholesterol 0mg	**0%**
Sodium 150mg	**6%**
Total Carbohydrate 32g	**11%**
Dietary Fiber 2g	**10%**
Sugars 12g	
Protein 4g	

* Percent Daily Values are based on a 2,000 calorie diet. Your daily values may be higher or lower depending on your calorie needs.

BAR II

Nutrition Facts
Serving Size 1 bar (21g)
Servings Per Container 8

Amount Per Serving	
Calories	80
Calories from Fat	**15**
	% Daily Value*
Total Fat 1.5g	**2%**
Saturated Fat 0g	**0%**
Cholesterol 0mg	**0%**
Sodium 60mg	**3%**
Total Carbohydrate 16g	**5%**
Dietary Fiber 1g	**4%**
Sugars 5g	
Protein 2g	

* Percent Daily Values are based on a 2,000 calorie diet. Your daily values may be higher or lower depending on your calorie needs.

FIGURE 1

SOLUTION The information needed to solve this problem is located towards the top of each label. Each serving of Bar I contains 210 calories, of which 70 calories come from fat. To find the percent of total calories that come from fat, we must answer this question:

70 is what percent of 210?

For Bar II, one serving contains 80 calories, of which 15 calories come from fat. To find the percent of total calories that come from fat, we must answer this question:

15 is what percent of 80?

Translating each equation into symbols, we have

70 is what percent of 210	15 is what percent of 80
$70 = x \cdot 210$	$15 = x \cdot 80$
$x = \dfrac{70}{210}$	$x = \dfrac{15}{80}$
$x = 0.33$ to the nearest hundredth	$x = 0.19$ to the nearest hundredth
$x = 33\%$	$x = 19\%$

Comparing the two bars, 33% of the calories in Bar I are fat calories, whereas 19% of the calories in Bar II are fat calories. According to the ADA, Bar II is the healthier choice.

Applying the Concepts

As we mentioned in Chapter 1, in the U.S. system, temperature is measured on the Fahrenheit scale. In the metric system, temperature is measured on the Celsius scale. On the Celsius scale, water boils at 100 degrees and freezes at 0 degrees. To denote a temperature of 100 degrees on the Celsius scale, we write

100°C, which is read "100 degrees Celsius"

Table 1 is intended to give you an intuitive idea of the relationship between the two temperature scales. Table 2 gives the formulas, in both symbols and words, that are used to convert between the two scales.

Table 1

	Temperature	
Situation	**Fahrenheit**	**Celsius**
Water freezes	32°F	0°C
Room temperature	68°F	20°C
Normal body temperature	98.6°F	37°C
Water boils	212°F	100°C

Table 2

To Convert from	Formula in Symbols	Formula in Words
Fahrenheit to Celsius	$C = \frac{5}{9}(F - 32)$	Subtract 32, multiply by 5, then divide by 9
Celsius to Fahrenheit	$F = \frac{9}{5}C + 32$	Multiply by $\frac{9}{5}$, then add 32

EXAMPLE 11 Mr. McKeague traveled to Buenos Aires with a group of friends. It was a hot day when they arrived. One of the bank kiosks indicated the temperature was 25°C. Someone asked what that would be on the Fahrenheit scale (the scale they were familiar with), and Budd, one of his friends said, "just multiply by 2 and add 30."

©Nikada/iStockPhoto.com

a. What was the temperature in °F according to Budd's approximation?

b. What is the actual temperature in °F?

c. Why does Budd's estimate work?

d. Write a formula for Budd's estimate.

SOLUTION

a. According to Budd, we multiply by 2 and add 30, so

$$2 \cdot 25 + 30 = 50 + 30 = 80°F$$

b. Using the formula $F = \frac{9}{5}C + 32$, with C = 25, we have

$$F = \frac{9}{5}(25) + 32 = 45 + 32 = 77°F$$

c. Budd's estimate works because $\frac{9}{5}$ is approximately 2 and 30 is close to 32.

d. In symbols, Budd's estimate is $F = 2 \cdot C + 30$.

GETTING READY FOR CLASS

After reading through the preceding section, respond in your own words and in complete sentences.

A. What is a formula?
B. How do you solve a formula for one of its variables?
C. What are complementary angles?
D. What does percent mean?

Use the formula $P = 2l + 2w$ to find the length l of a rectangular lot if

1. The width w is 50 feet and the perimeter P is 300 feet.
2. The width w is 75 feet and the perimeter P is 300 feet.

Use the formula $2x + 3y = 6$ to find y when

3. x is 3
4. x is -2
5. x is 0
6. x is -3

Use the formula $2x - 5y = 20$ to find x when

7. y is 2
8. y is -4
9. y is 0
10. y is -6

Use the equation $y = (x + 1)^2 - 3$ to find the value of y when

11. $x = -2$
12. $x = -1$
13. $x = 1$
14. $x = 2$

15. Use the formula $y = \frac{20}{x}$ to find y when
    - a. $x = 10$
    - b. $x = 5$
16. Use the formula $y = 2x^2$ to find y when
    - a. $x = 5$
    - b. $x = -6$
17. Use the formula $y = Kx$ to find K when
    - a. $y = 15$ and $x = 3$
    - b. $y = 72$ and $x = 4$
18. Use the formula $y = Kx^2$ to find K when
    - a. $y = 32$ and $x = 4$
    - b. $y = 45$ and $x = 3$

Solve each of the following for the indicated variable.

19. $A = lw$ for l
20. $d = rt$ for r
21. $V = lwh$ for h
22. $PV = nRT$ for P
23. $P = a + b + c$ for a
24. $P = a + b + c$ for b
25. $x - 3y = -1$ for x
26. $x + 3y = 2$ for x
27. $-3x + y = 6$ for y
28. $2x + y = -17$ for y
29. $2x + 3y = 6$ for y
30. $4x + 5y = 20$ for y
31. $y - 3 = -2(x + 4)$ for y
32. $y + 5 = 2(x + 2)$ for y
33. $y - 3 = -\frac{2}{3}(x + 3)$ for y
34. $y - 1 = -\frac{1}{2}(x + 4)$ for y
35. $P = 2l + 2w$ for w
36. $P = 2l + 2w$ for l
37. $h = vt + 16t^2$ for v
38. $h = vt - 16t^2$ for v
39. $A = \pi r^2 + 2\pi rh$ for h
40. $A = 2\pi r^2 + 2\pi rh$ for h

41. Solve for y.
    - a. $\frac{y-1}{x} = \frac{3}{5}$
    - b. $\frac{y-2}{x} = \frac{1}{2}$
    - c. $\frac{y-3}{x} = 4$
42. Solve for y.
    - a. $\frac{y+1}{x} = -\frac{3}{5}$
    - b. $\frac{y+2}{x} = -\frac{1}{2}$
    - c. $\frac{y+3}{x} = -4$

Solve each formula for y.

43. $\frac{x}{7} - \frac{y}{3} = 1$

44. $\frac{x}{5} - \frac{y}{9} = 1$

45. $-\frac{1}{4}x + \frac{1}{8}y = 1$

46. $-\frac{1}{9}x + \frac{1}{3}y = 1$

Find the complement and the supplement of each angle.

47. 30°

48. 60°

49. 45°

50. 15°

Translate each of the following into an equation, and then solve that equation.

51. What number is 25% of 40?

52. What number is 75% of 40?

53. What number is 12% of 2,000?

54. What number is 9% of 3,000?

55. What percent of 28 is 7?

56. What percent of 28 is 21?

57. What percent of 40 is 14?

58. What percent of 20 is 14?

59. 32 is 50% of what number?

60. 16 is 50% of what number?

61. 240 is 12% of what number?

62. 360 is 12% of what number?

63. Let $F = 212$ in the formula $C = \frac{5}{9}(F - 32)$, and solve for C. Does the value of C agree with the information in Table 1?

64. Let $C = 100$ in the formula $F = \frac{9}{5}C + 32$, and solve for F. Does the value of F agree with the information in Table 1?

65. Let $F = 68$ in the formula $C = \frac{5}{9}(F - 32)$, and solve for C. Does the value of C agree with the information in Table 1?

66. Let $C = 37$ in the formula $F = \frac{9}{5}C + 32$, and solve for F. Does the value of F agree with the information in Table 1?

67. Solve the formula $F = \frac{9}{5}C + 32$ for C.

68. Solve the formula $C = \frac{5}{9}(F - 32)$ for F.

69. How far off is Budd's estimate when the temperature is 30°C? (See Example 11)

70. How far off is Budd's estimate when the temperature is 0°C? (See Example 11)

Circumference The circumference of a circle is given by the formula $C = 2\pi r$. Find r if

71. The circumference C is 44 meters and π is $\frac{22}{7}$

72. The circumference C is 176 meters and π is $\frac{22}{7}$

73. The circumference is 9.42 inches and π is 3.14

74. The circumference is 12.56 inches and π is 3.14

Volume The volume of a cylinder is given by the formula $V = \pi r^2 h$. Find the height h if

75. The volume V is 42 cubic feet, the radius is $\frac{7}{22}$ feet, and π is $\frac{22}{7}$

76. The volume V is 84 cubic inches, the radius is $\frac{7}{11}$ inches, and π is $\frac{22}{7}$

77. The volume is 6.28 cubic centimeters, the radius is 3 centimeters, and π is 3.14.

78. The volume is 12.56 cubic centimeters, the radius is 2 centimeters, and π is 3.14.

Nutrition Labels The nutrition label in Figure 2 is from a quart of vanilla ice cream. The label in Figure 3 is from a pint of vanilla frozen yogurt. Use the information on these labels for problems 79–82. Round your answers to the nearest tenth of a percent.

Nutrition Facts
Serving Size 1/2 cup (65g)
Servings 8

Amount/Serving	
Calories 150	Calories from Fat 90
	% Daily Value*
Total Fat 10g	**16%**
Saturated Fat 6g	**32%**
Cholesterol 35mg	**12%**
Sodium 30mg	**1%**
Total Carbohydrate 14g	**5%**
Dietary Fiber 0g	**0%**
Sugars 11g	
Protein 2g	
Vitamin A 6% •	Vitamin C 0%
Calcium 6% •	Iron 0%

* Percent Daily Values are based on a 2,000 calorie diet.

FIGURE 2 *Vanilla ice cream*

Nutrition Facts
Serving Size 1/2 cup (98g)
Servings Per Container 4

Amount Per Serving	
Calories 160	Calories from Fat 25
	% Daily Value*
Total Fat 2.5g	**4%**
Saturated Fat 1.5g	**7%**
Cholesterol 45mg	**15%**
Sodium 55mg	**2%**
Total Carbohydrate 26g	**9%**
Dietary Fiber 0g	**0%**
Sugars 19g	
Protein 8g	
Vitamin A 0% •	Vitamin C 0%
Calcium 25% •	Iron 0%

* Percent Daily Values are based on a 2,000 calorie diet.

FIGURE 3 *Vanilla frozen yogurt*

79. What percent of the calories in one serving of the vanilla ice cream are fat calories?

80. What percent of the calories in one serving of the frozen yogurt are fat calories?

81. One serving of frozen yogurt is 98 grams, of which 26 grams are carbohydrates. What percent of one serving are carbohydrates?

82. One serving of vanilla ice cream is 65 grams. What percent of one serving is sugar?

Getting Read for the Next Section

To understand all of the explanations and examples in the next section you must be able to work the problems below.

Write an equivalent expression in English. Include the words *sum* and *difference* when possible.

83. $4 + 1 = 5$ **84.** $7 + 3 = 10$ **85.** $6 - 2 = 4$ **86.** $8 - 1 = 7$

87. $x - 15 = -12$ **88.** $2x + 3 = 7$

89. $x + 3 = 4(x - 3)$ **90.** $2(2x - 5) = 2x - 34$

For each of the following expressions, write an equivalent equation.

91. Twice the sum of 6 and 3 is 18.

92. Four added to the product of 5 and -1 is -1.

93. The sum of twice 5 and 3 is 13.

94. Twice the difference of 8 and 2 is 12.

95. The sum of a number and five is thirteen.

96. The difference of ten and a number is negative eight.

97. Five times the sum of a number and seven is thirty.

98. Five times the difference of twice a number and six is negative twenty.

Applications 1.6

As you begin reading through the examples in this section, you may find yourself asking why some of these problems seem so contrived. The title of the section is "Applications," but many of the problems here don't seem to have much to do with "real life." You are right about that. Example 3 is what we refer to as an "age problem." But imagine a conversation in which you ask someone how old her children are and she replies, "Bill is 6 years older than Tom. Three years ago the sum of their ages was 21. You figure it out." Although many of the "application" problems in this section are contrived, they are also good for practicing the strategy we will use to solve all application problems.

To begin this section, we list the steps used in solving application problems. We call this strategy the *Blueprint for Problem Solving*. It is an outline that will overlay the solution process we use on all application problems.

BLUEPRINT FOR PROBLEM SOLVING

Step 1: ***Read*** the problem, and then mentally ***list*** the items that are known and the items that are unknown.

Step 2: ***Assign a variable*** to one of the unknown items. (In most cases this will amount to letting x = the item that is asked for in the problem.) Then ***translate*** the other ***information*** in the problem to expressions involving the variable.

Step 3: ***Reread*** the problem, and then ***write an equation***, using the items and variables listed in steps 1 and 2, that describes the situation.

Step 4: ***Solve the equation*** found in step 3.

Step 5: ***Write*** your ***answer*** using a complete sentence.

Step 6: ***Reread*** the problem, and ***check*** your solution with the original words in the problem.

There are a number of substeps within each of the steps in our blueprint. For instance, with steps 1 and 2 it is always a good idea to draw a diagram or picture if it helps visualize the relationship between the items in the problem. In other cases a table helps organize the information. As you gain more experience using the blueprint to solve application problems, you will find additional techniques that expand the blueprint.

To help with problems of the type shown next in Example 1, here are some common English words and phrases and their mathematical translations.

English	Algebra
The sum of a and b	$a + b$
The difference of a and b	$a - b$
The product of a and b	$a \cdot b$
The quotient of a and b	$\frac{a}{b}$
of	$\cdot$ (multiply)
is	$=$ (equals)
A number	x
4 more than x	$x + 4$
4 times x	$4x$
4 less than x	$x - 4$

Number Problems

EXAMPLE 1 The sum of twice a number and three is seven. Find the number.

SOLUTION Using the Blueprint for Problem Solving as an outline, we solve the problem as follows:

Step 1: ***Read*** the problem, and then mentally ***list*** the items that are known and the items that are unknown.

Known items: The numbers 3 and 7

Unknown items: The number in question

Step 2: ***Assign a variable*** to one of the unknown items. Then ***translate*** the other ***information*** in the problem to expressions involving the variable.

Let $x =$ the number asked for in the problem, then "The sum of twice a number and three" translates to $2x + 3$.

Step 3: ***Reread*** the problem, and then ***write an equation,*** using the items and variables listed in steps 1 and 2, that describes the situation.
With all word problems, the word *is* translates to $=$.

$$\underbrace{\text{The sum of twice } x \text{ and } 3}_{2x + 3} \ \underset{=}{\text{is}} \ \underset{7}{7}$$

Step 4: ***Solve the equation*** found in step 3.

$$2x + 3 = 7$$
$$2x + 3 + (-3) = 7 + (-3)$$
$$2x = 4$$
$$\frac{1}{2}(2x) = \frac{1}{2}(4)$$
$$x = 2$$

Step 5: ***Write*** your ***answer*** using a complete sentence.

The number is 2.

Step 6: ***Reread*** the problem, and ***check*** your solution with the original words in the problem.

The sum of twice 2 and 3 is 7; a true statement.

You may find some examples and problems in this section that you can solve without using algebra or our blueprint. It is very important that you solve these problems using the methods we are showing here. The purpose behind these problems is to give you experience using the blueprint as a guide to solving problems written in words. Your answers are much less important than the work that you show to obtain your answer. You will be able to condense the steps in the blueprint later in the course. For now, though, you need to show your work in the same detail that we are showing in the examples in this section.

EXAMPLE 2 One number is three more than twice another; their sum is eighteen. Find the numbers.

SOLUTION

Step 1: **Read and list.**

Known items: Two numbers that add to 18. One is 3 more than twice the other.

Unknown items: The numbers in question.

Step 2: **Assign a variable, and translate information.**

Let $x =$ the first number. The other is $2x + 3$.

Step 3: **Reread, and write an equation.**

Their sum is 18

$$x + (2x + 3) = 18$$

Step 4: **Solve the equation.**

$$x + (2x + 3) = 18$$
$$3x + 3 = 18$$
$$3x + 3 + (-3) = 18 + (-3)$$
$$3x = 15$$
$$x = 5$$

Step 5: **Write the answer.**

The first number is 5. The other is $2 \cdot 5 + 3 = 13$.

Step 6: **Reread, and check.**

The sum of 5 and 13 is 18, and 13 is 3 more than twice 5.

Age Problem

Remember as you read through the steps in the solutions to the examples in this section that step 1 is done mentally. Read the problem, and then mentally list the items that you know and the items that you don't know. The purpose of step 1 is to give you direction as you begin to work application problems. Finding the solution to an application problem is a process; it doesn't happen all at once. The first step is to read the problem with a purpose in mind. That purpose is to mentally note the items that are known and the items that are unknown.

EXAMPLE 3 Bill is 6 years older than Tom. Three years ago Bill's age was four times Tom's age. Find the age of each boy now.

SOLUTION Applying the Blueprint for Problem Solving, we have

Step 1: **Read and list.**

Known items: Bill is 6 years older than Tom. Three years ago Bill's age was four times Tom's age.

Unknown items: Bill's age and Tom's age

Step 2: **Assign a variable, and translate information.**

Let $x =$ Tom's age now. That makes Bill $x + 6$ years old now. A table like the one shown here can help organize the information in an age problem. Notice how we placed the x in the box that corresponds to Tom's age now.

	Three Years Ago	Now
Bill		$x + 6$
Tom		x

If Tom is x years old now, 3 years ago he was $x - 3$ years old. If Bill is $x + 6$ years old now, 3 years ago he was $x + 6 - 3 = x + 3$ years old. We use this information to fill in the remaining squares in the table.

	Three Years Ago	Now
Bill	$x + 3$	$x + 6$
Tom	$x - 3$	x

Step 3: **Reread, and write an equation.**
Reading the problem again, we see that 3 years ago Bill's age was four times Tom's age. Writing this as an equation, we have Bill's age 3 years ago $= 4 \cdot$ (Tom's age 3 years ago):

$$x + 3 = 4(x - 3)$$

Step 4: **Solve the equation.**

$$\begin{aligned} x + 3 &= 4(x - 3) \\ x + 3 &= 4x - 12 \\ x + (-x) + 3 &= 4x + (-x) - 12 \\ 3 &= 3x - 12 \\ 3 + 12 &= 3x - 12 + 12 \\ 15 &= 3x \\ x &= 5 \end{aligned}$$

Step 5: **Write the answer.**
Tom is 5 years old. Bill is 11 years old.

Step 6: **Reread, and check.**
If Tom is 5 and Bill is 11, then Bill is 6 years older than Tom. Three years ago Tom was 2 and Bill was 8. At that time, Bill's age was four times Tom's age. As you can see, the answers check with the original problem. ■

Geometry Problem

To understand Example 4 completely, you need to recall from Chapter 1 that the perimeter of a rectangle is the sum of the lengths of the sides. The formula for the perimeter is $P = 2l + 2w$.

EXAMPLE 4 The length of a rectangle is 5 inches more than twice the width. The perimeter is 34 inches. Find the length and width.

SOLUTION When working problems that involve geometric figures, a sketch of the figure helps organize and visualize the problem.

Step 1: **Read and list.**

Known items: The figure is a rectangle. The length is 5 inches more than twice the width. The perimeter is 34 inches.

Unknown items: The length and the width

Step 2: **Assign a variable, and translate information.**

Because the length is given in terms of the width (the length is 5 more than twice the width), we let $x =$ the width of the rectangle. The length is 5 more than twice the width, so it must be $2x + 5$. The diagram below is a visual description of the relationships we have listed so far.

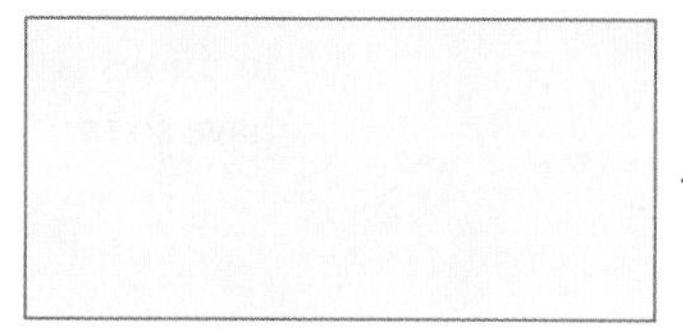

Step 3: **Reread, and write an equation.**

The equation that describes the situation is

Twice the length + twice the width is the perimeter

$$2(2x + 5) \quad + \quad 2x \quad = \quad 34$$

Step 4: **Solve the equation.**

$$2(2x + 5) + 2x = 34 \qquad \text{Original equation}$$
$$4x + 10 + 2x = 34 \qquad \text{Distributive property}$$
$$6x + 10 = 34 \qquad \text{Add } 4x \text{ and } 2x$$
$$6x = 24 \qquad \text{Add } -10 \text{ to each side}$$
$$x = 4 \qquad \text{Divide each side by } 6$$

Step 5: **Write the answer.**

The width x is 4 inches. The length is $2x + 5 = 2(4) + 5 = 13$ inches.

Step 6: **Reread, and check.**

If the length is 13 and the width is 4, then the perimeter must be $2(13) + 2(4) = 26 + 8 = 34$, which checks with the original problem.

Coin Problem

EXAMPLE 5 Jennifer has \$2.45 in dimes and nickels. If she has 8 more dimes than nickels, how many of each coin does she have?

SOLUTION

Step 1: **Read and list.**

Known items: The type of coins, the total value of the coins, and that there are 8 more dimes than nickels.

Unknown items: The number of nickels and the number of dimes

Step 2: **Assign a variable, and translate information.**

If we let $x =$ the number of nickels, then $x + 8 =$ the number of dimes. Because the value of each nickel is 5 cents, the amount of money in nickels is $5x$. Similarly, because each dime is worth 10 cents, the amount of money in dimes is $10(x + 8)$. Here is a table that summarizes the information we have so far:

	Nickels	Dimes
Number	x	$x + 8$
Value (in cents)	$5x$	$10(x + 8)$

Step 3: **Reread, and write an equation.**

Because the total value of all the coins is 245 cents, the equation that describes this situation is

Amount of money in nickels	+	Amount of money in dimes	=	Total amount of money
$5x$	+	$10(x + 8)$	=	245

Step 4: **Solve the equation.**

To solve the equation, we apply the distributive property first.

$$5x + 10x + 80 = 245 \quad \text{Distributive property}$$

$$15x + 80 = 245 \quad \text{Add } 5x \text{ and } 10x$$

$$15x = 165 \quad \text{Add } -80 \text{ to each side}$$

$$x = 11 \quad \text{Divide each side by } 15$$

Step 5: **Write the answer.**

The number of nickels is $x = 11$.

The number of dimes is $x + 8 = 11 + 8 = 19$.

Step 6: **Reread, and check.**

To check our results

11 nickels are worth $5(11) =$ 55 cents

19 dimes are worth $10(19) =$ 190 cents

The total value is 245 cents = \$2.45

When you begin working the problems in the problem set that follows, there are a few things to remember. The first is that you may have to read the problems a number of times before you begin to see how to solve them. The second thing to remember is that word problems are not always solved correctly the first time you try them. Sometimes it takes a few attempts and some wrong answers before you can set up and solve these problems correctly.

GETTING READY FOR CLASS

After reading through the preceding section, respond in your own words and in complete sentences.

A. What is the first step in the Blueprint for Problem Solving?
B. What is the last thing you do when solving an application problem?
C. What good does it do you to solve application problems even when they don't have much to do with real life?
D. Write an application problem whose solution depends on solving the equation $2x + 3 = 7$.

SPOTLIGHT ON SUCCESS *Student Instructor Cynthia*

Each time we face our fear, we gain strength, courage, and confidence in the doing.
—Unknown

I must admit, when it comes to math, it takes me longer to learn the material compared to other students. Because of that, I was afraid to ask questions, especially when it seemed like everyone else understood what was going on. Because I wasn't getting my questions answered, my quiz and exam scores were only getting worse. I realized that I was already paying a lot to go to college and that I couldn't afford to keep doing poorly on my exams. I learned how to overcome my fear of asking questions by studying the material before class, and working on extra problem sets until I was confident enough that at least I understood the main concepts. By preparing myself beforehand, I would often end up answering the question myself. Even when that wasn't the case, the professor knew that I tried to answer the question on my own. If you want to be successful, but you are afraid to ask a question, try putting in a little extra time working on problems before you ask your instructor for help. I think you will find, like I did, that it's not as bad as you imagined it, and you will have overcome an obstacle that was in the way of your success.

Problem Set 1.6

Solve the following word problems. Follow the steps given in the Blueprint for Problem Solving.

Number Problems

1. The sum of a number and five is thirteen. Find the number.
2. The difference of ten and a number is negative eight. Find the number.
3. The sum of twice a number and four is fourteen. Find the number.
4. The difference of four times a number and eight is sixteen. Find the number.
5. Five times the sum of a number and seven is thirty. Find the number.
6. Five times the difference of twice a number and six is negative twenty. Find the number.
7. One number is two more than another. Their sum is eight. Find both numbers.
8. One number is three less than another. Their sum is fifteen. Find the numbers.
9. One number is four less than three times another. If their sum is increased by five, the result is twenty-five. Find the numbers.
10. One number is five more than twice another. If their sum is decreased by ten, the result is twenty-two. Find the numbers.

Age Problems

11. Shelly is 3 years older than Michele. Four years ago the sum of their ages was 67. Find the age of each person now.

	Four Years Ago	Now
Shelly	$x - 1$	$x + 3$
Michele	$x - 4$	x

12. Cary is 9 years older than Dan. In 7 years the sum of their ages will be 93. Find the age of each man now. (Begin by filling in the table.)

	Now	In Seven Years
Cary	$x + 9$	
Dan	x	$x + 7$

13. Cody is twice as old as Evan. Three years ago the sum of their ages was 27. Find the age of each boy now.

	Three Years Ago	Now
Cody		
Evan	$x - 3$	x

14. Justin is 2 years older than Ethan. In 9 years the sum of their ages will be 30. Find the age of each boy now.

	Now	In Nine Years
Justin		
Ethan	x	

15. Fred is 4 years older than Barney. Five years ago the sum of their ages was 48. How old are they now?

	Five Years Ago	Now
Fred		
Barney		x

16. Tim is 5 years older than JoAnn. Six years from now the sum of their ages will be 79. How old are they now?

	Now	Six Years From Now
Tim		
JoAnn	x	

17. Jack is twice as old as Lacy. In 3 years the sum of their ages will be 54. How old are they now?

18. John is 4 times as old as Martha. Five years ago the sum of their ages was 50. How old are they now?

19. Pat is 20 years older than his son Patrick. In 2 years Pat will be twice as old as Patrick. How old are they now?

20. Diane is 23 years older than her daughter Amy. In 6 years Diane will be twice as old as Amy. How old are they now?

Geometry Problems

21. The perimeter of a square is 36 inches. Find the length of one side.

22. The perimeter of a square is 44 centimeters. Find the length of one side.

23. The perimeter of a square is 60 feet. Find the length of one side.

24. The perimeter of a square is 84 meters. Find the length of one side.

25. One side of a triangle is three times the shortest side. The third side is 7 feet more than the shortest side. The perimeter is 62 feet. Find all three sides.

26. One side of a triangle is half the longest side. The third side is 10 meters less than the longest side. The perimeter is 45 meters. Find all three sides.

27. One side of a triangle is half the longest side. The third side is 12 feet less than the longest side. The perimeter is 53 feet. Find all three sides.

28. One side of a triangle is 6 meters more than twice the shortest side. The third side is 9 meters more than the shortest side. The perimeter is 75 meters. Find all three sides.

29. The length of a rectangle is 5 inches more than the width. The perimeter is 34 inches. Find the length and width.

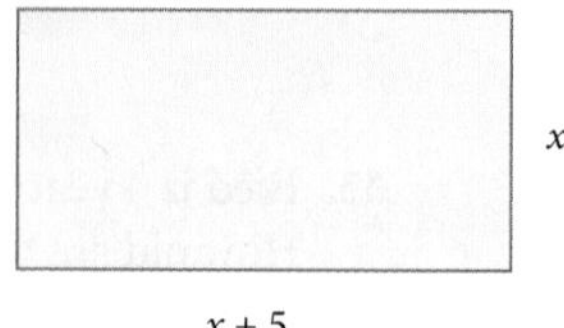

30. The width of a rectangle is 3 feet less than the length. The perimeter is 10 feet. Find the length and width.

31. The length of a rectangle is 7 inches more than twice the width. The perimeter is 68 inches. Find the length and width.

32. The length of a rectangle is 4 inches more than three times the width. The perimeter is 72 inches. Find the length and width.

33. The length of a rectangle is 6 feet more than three times the width. The perimeter is 36 feet. Find the length and width.

34. The length of a rectangle is 3 feet less than twice the width. The perimeter is 54 feet. Find the length and width.

Coin Problems

35. Marissa has $4.40 in quarters and dimes. If she has 5 more quarters than dimes, how many of each coin does she have?

	Dimes	Quarters
Number	x	$x + 5$
Value (in cents)	$10(x)$	$25(x + 5)$

36. Kendra has $2.75 in dimes and nickels. If she has twice as many dimes as nickels, how many of each coin does she have?

	Nickels	Dimes
Number	x	$2x$
Value (in cents)	$5(x)$	

37. Tanner has $4.35 in nickels and quarters. If he has 15 more nickels than quarters, how many of each coin does he have?

	Nickels	Quarters
Number	$x + 15$	x
Value (in cents)		

38. Connor has \$9.00 in dimes and quarters. If he has twice as many quarters as dimes, how many of each coin does he have?

	Dimes	Quarters
Number	x	$2x$
Value (in cents)		

39. Sue has \$2.10 in dimes and nickels. If she has 9 more dimes than nickels, how many of each coin does she have? (Completing the table may help you get started.)

40. Mike has \$1.55 in dimes and nickels. If he has 7 more nickels than dimes, how many of each coin does he have?

41. Katie has a collection of nickels, dimes, and quarters with a total value of \$4.35. There are 3 more dimes than nickels and 5 more quarters than nickels. How many of each coin is in her collection? (*Hint:* Let $x =$ the number of nickels.)

	Nickels	Dimes	Quarters
Number	x		
Value			

42. Mary Jo has \$3.90 worth of nickels, dimes, and quarters. The number of nickels is 3 more than the number of dimes. The number of quarters is 7 more than the number of dimes. How many of each coin does she have? (*Hint:* Let $x =$ the number of dimes.)

	Nickels	Dimes	Quarters
Number			
Value			

43. Cory has a collection of nickels, dimes, and quarters with a total value of \$2.55. There are 6 more dimes than nickels and twice as many quarters as nickels. How many of each coin is in her collection?

	Nickels	Dimes	Quarters
Number	x		
Value			

44. Kelly has a collection of nickels, dimes, and quarters with a total value of \$7.40. There are four more nickels than dimes and twice as many quarters as nickels. How many of each coin is in her collection?

	Nickels	Dimes	Quarters
Number			
Value			

Getting Ready for the Next Section

To understand all of the explanations and examples in the next section you must be able to work the problems below.

Simplify the following expressions.

45. $x + 2x + 2x$ **46.** $x + 2x + 3x$ **47.** $x + 0.075x$ **48.** $x + 0.065x$

49. $0.09(x + 2{,}000)$ **50.** $0.06(x + 1{,}500)$

51. $0.02x + 0.06(x + 1{,}500) = 570$ **52.** $0.08x + 0.09(x + 2{,}000) = 690$

53. $x + 2x + 3x = 180$ **54.** $2x + 3x + 5x = 180$

More Applications

1.7

Now that you have worked through a number of application problems using our blueprint, you probably have noticed that step 3, in which we write an equation that describes the situation, is the key step. Anyone with experience solving application problems will tell you that there will be times when your first attempt at step 3 results in the wrong equation. Remember, mistakes are part of the process of learning to do things correctly. Many times the correct equation will become obvious after you have written an equation that is partially wrong. In any case it is better to write an equation that is partially wrong and be actively involved with the problem than to write nothing at all. Application problems, like other problems in algebra, are not always solved correctly the first time.

Consecutive Integers

Our first example involves consecutive integers. When we ask for consecutive integers, we mean integers that are next to each other on the number line, like 5 and 6, or 13 and 14, or -4 and -3. In the dictionary, consecutive is defined as following one another in uninterrupted order. If we ask for consecutive odd integers, then we mean odd integers that follow one another on the number line. For example, 3 and 5, 11 and 13, and -9 and -7 are consecutive odd integers. As you can see, to get from one odd integer to the next consecutive odd integer we add 2.

If we are asked to find two consecutive integers and we let x equal the first integer, the next one must be $x + 1$, because consecutive integers always differ by 1. Likewise, if we are asked to find two consecutive odd or even integers, and we let x equal the first integer, then the next one will be $x + 2$ because consecutive even or odd integers always differ by 2. Here is a table that summarizes this information.

In Words	Using Algebra	Example
Two consecutive integers	$x, x + 1$	The sum of two consecutive integers is 15. $x + (x + 1) = 15$ or $7 + 8 = 15$
Three consecutive integers	$x, x + 1, x + 2$	The sum of three consecutive integers is 24. $x + (x + 1) + (x + 2) = 24$ or $7 + 8 + 9 = 24$
Two consecutive odd integers	$x, x + 2$	The sum of two consecutive odd integers is 16. $x + (x + 2) = 16$ or $7 + 9 = 16$
Two consecutive even integers	$x, x + 2$	The sum of two consecutive even integers is 18. $x + (x + 2) = 18$ or $8 + 10 = 18$

EXAMPLE 1 The sum of two consecutive odd integers is 28. Find the two integers.

SOLUTION

Step 1: **Read and list.**

Known items: Two consecutive odd integers. Their sum is equal to 28.
Unknown items: The numbers in question.

Step 2: **Assign a variable, and translate information.**
If we let $x =$ the first of the two consecutive odd integers, then $x + 2$ is the next consecutive one.

Step 3: **Reread, and write an equation.**
Their sum is 28.

$$x + (x + 2) = 28$$

Step 4: **Solve the equation.**

$$2x + 2 = 28 \quad \text{Simplify the left side}$$

$$2x = 26 \quad \text{Add } -2 \text{ to each side}$$

$$x = 13 \quad \text{Multiply each side by } \frac{1}{2}$$

Step 5: **Write the answer.**
The first of the two integers is 13. The second of the two integers will be two more than the first, which is 15.

Step 6: **Reread, and check.**
Suppose the first integer is 13. The next consecutive odd integer is 15. The sum of 15 and 13 is 28.

Interest

EXAMPLE 2 Suppose you invest a certain amount of money in an account that earns 8% in annual interest. At the same time, you invest $2,000 more than that in an account that pays 9% in annual interest. If the total interest from both accounts at the end of the year is $690, how much is invested in each account?

SOLUTION

Step 1: **Read and list.**
Known items: The interest rates, the total interest earned, and how much more is invested at 9%
Unknown items: The amounts invested in each account

Step 2: **Assign a variable, and translate information.**
Let $x =$ the amount of money invested at 8%. From this, $x + 2{,}000 =$ the amount of money invested at 9%. The interest earned on x dollars invested at 8% is $0.08x$. The interest earned on $x + 2{,}000$ dollars invested at 9% is $0.09(x + 2{,}000)$.

Here is a table that summarizes this information:

	Dollars Invested at 8%	Dollars Invested at 9%
Number of	x	$x + 2{,}000$
Interest on	$0.08x$	$0.09(x + 2{,}000)$

Step 3: **Reread, and write an equation.**
Because the total amount of interest earned from both accounts is $690, the equation that describes the situation is

Interest earned at 8%	+	Interest earned at 9%	=	Total interest earned
$0.08x$	+	$0.09(x + 2{,}000)$	=	690

Step 4: **Solve the equation.**

$$0.08x + 0.09(x + 2{,}000) = 690$$

$$0.08x + 0.09x + 180 = 690 \quad \text{Distributive property}$$

$$0.17x + 180 = 690 \quad \text{Add } 0.08x \text{ and } 0.09x$$

$$0.17x = 510 \quad \text{Add } -180 \text{ to each side}$$

$$x = 3{,}000 \quad \text{Divide each side by } 0.17$$

Step 5: **Write the answer:**
The amount of money invested at 8% is \$3,000, whereas the amount of money invested at 9% is $x + 2{,}000 = 3{,}000 + 2{,}000 =$ \$5,000.

Step 6: **Reread, and check.**

The interest at 8% is 8% of 3,000 = 0.08(3,000) = \$240
The interest at 9% is 9% of 5,000 = 0.09(5,000) = \$450
The total interest is \$690

FACTS FROM GEOMETRY *Labeling Triangles and the Sum of the Angles in a Triangle*

One way to label the important parts of a triangle is to label the vertices with capital letters and the sides with small letters, as shown in Figure 1.

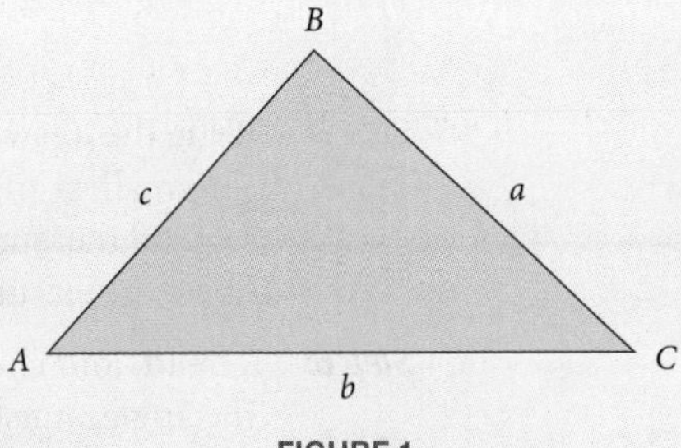

FIGURE 1

In Figure 1, notice that side a is opposite vertex A, side b is opposite vertex B, and side c is opposite vertex C. Also, because each vertex is the vertex of one of the angles of the triangle, we refer to the three interior angles as A, B, and C.

In any triangle, the sum of the interior angles is 180°. For the triangle shown in Figure 1, the relationship is written

$$A + B + C = 180°$$

EXAMPLE 3 The angles in a triangle are such that one angle is twice the smallest angle, whereas the third angle is three times as large as the smallest angle. Find the measure of all three angles.

SOLUTION

Step 1: **Read and list.**
Known items: The sum of all three angles is 180°, one angle is twice the smallest angle, the largest angle is three times the smallest angle.
Unknown items: The measure of each angle

Step 2: **Assign a variable, and translate information.**
Let x be the smallest angle, then $2x$ will be the measure of another angle and $3x$ will be the measure of the largest angle.

Step 3: **Reread, and write an equation.**
When working with geometric objects, drawing a generic diagram sometimes will help us visualize what it is that we are asked to find. In Figure 2, we draw a triangle with angles A, B, and C.

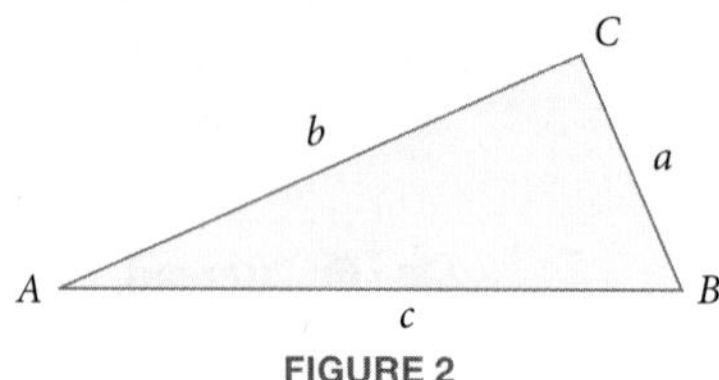

FIGURE 2

We can let the value of $A = x$, the value of $B = 2x$, and the value of $C = 3x$. We know that the sum of angles A, B, and C will be 180°, so our equation becomes

$$x + 2x + 3x = 180°$$

Step 4: **Solve the equation.**

$$x + 2x + 3x = 180°$$
$$6x = 180°$$
$$x = 30°$$

Step 5: **Write the answer.**
The smallest angle A measures 30°
Angle B measures $2x$, or $2(30°) = 60°$
Angle C measures $3x$, or $3(30°) = 90°$

Step 6: **Reread, and check.**
The angles must add to 180°:

$$A + B + C = 180°$$
$$30° + 60° + 90° \stackrel{?}{=} 180°$$
$$180° = 180° \quad \text{Our answers check}$$

GETTING READY FOR CLASS

After reading through the preceding section, respond in your own words and in complete sentences.

A. How do we label triangles?
B. What rule is always true about the three angles in a triangle?
C. Write an application problem whose solution depends on solving the equation $x + 0.075x = 500$.
D. Write an application problem whose solution depends on solving the equation $0.05x + 0.06(x + 200) = 67$.

Consecutive Integer Problems

1. The sum of two consecutive integers is 11. Find the numbers.
2. The sum of two consecutive integers is 15. Find the numbers.
3. The sum of two consecutive integers is -9. Find the numbers.
4. The sum of two consecutive integers is -21. Find the numbers.
5. The sum of two consecutive odd integers is 28. Find the numbers.
6. The sum of two consecutive odd integers is 44. Find the numbers.
7. The sum of two consecutive even integers is 106. Find the numbers.
8. The sum of two consecutive even integers is 66. Find the numbers.
9. The sum of two consecutive even integers is -30. Find the numbers.
10. The sum of two consecutive odd integers is -76. Find the numbers.
11. The sum of three consecutive odd integers is 57. Find the numbers.
12. The sum of three consecutive odd integers is -51. Find the numbers.
13. The sum of three consecutive even integers is 132. Find the numbers.
14. The sum of three consecutive even integers is -108. Find the numbers.

Interest Problems

15. Suppose you invest money in two accounts. One of the accounts pays 8% annual interest, whereas the other pays 9% annual interest. If you have \$2,000 more invested at 9% than you have invested at 8%, how much do you have invested in each account if the total amount of interest you earn in a year is \$860? (Begin by completing the following table.)

	Dollars Invested at 8%	Dollars Invested at 9%
Number of	x	
Interest on		

16. Suppose you invest a certain amount of money in an account that pays 11% interest annually, and \$4,000 more than that in an account that pays 12% annually. How much money do you have in each account if the total interest for a year is \$940?

	Dollars Invested at 11%	Dollars Invested at 12%
Number of	x	
Interest on		

17. Tyler has two savings accounts that his grandparents opened for him. The two accounts pay 10% and 12% in annual interest; there is \$500 more in the account that pays 12% than there is in the other account. If the total interest for a year is \$214, how much money does he have in each account?

18. Travis has a savings account that his parents opened for him. It pays 6% annual interest. His uncle also opened an account for him, but it pays 8% annual interest. If there is \$800 more in the account that pays 6%, and the total interest from both accounts is \$104, how much money is in each of the accounts?

19. A stockbroker has money in three accounts. The interest rates on the three accounts are 8%, 9%, and 10%. If she has twice as much money invested at 9% as she has invested at 8%, three times as much at 10% as she has at 8%, and the total interest for the year is \$280, how much is invested at each rate? (*Hint:* Let $x =$ the amount invested at 8%.)

20. An accountant has money in three accounts that pay 9%, 10%, and 11% in annual interest. He has twice as much invested at 9% as he does at 10% and three times as much invested at 11% as he does at 10%. If the total interest from the three accounts is \$610 for the year, how much is invested at each rate? (*Hint:* Let $x =$ the amount invested at 10%.)

Triangle Problems

21. Two angles in a triangle are equal and their sum is equal to the third angle in the triangle. What are the measures of each of the three interior angles?

22. One angle in a triangle measures twice the smallest angle, whereas the largest angle is six times the smallest angle. Find the measures of all three angles.

23. The smallest angle in a triangle is $\frac{1}{5}$ as large as the largest angle. The third angle is twice the smallest angle. Find the three angles.

24. One angle in a triangle is half the largest angle but three times the smallest. Find all three angles.

25. A right triangle has one 37° angle. Find the other two angles.

26. In a right triangle, one of the acute angles is twice as large as the other acute angle. Find the measure of the two acute angles.

27. One angle of a triangle measures 20° more than the smallest, while a third angle is twice the smallest. Find the measure of each angle.

28. One angle of a triangle measures 50° more than the smallest, while a third angle is three times the smallest. Find the measure of each angle.

Miscellaneous Problems

29. Ticket Prices Miguel is selling tickets to a barbecue. Adult tickets cost \$6.00 and children's tickets cost \$4.00. He sells six more children's tickets than adult tickets. The total amount of money he collects is \$184. How many adult tickets and how many children's tickets did he sell?

	Adult	Child
Number	x	$x + 6$
Income	$6(x)$	$4(x + 6)$

30. **Working Two Jobs** Maggie has a job working in an office for \$10 an hour and another job driving a tractor for \$12 an hour. One week she works in the office twice as long as she drives the tractor. Her total income for that week is \$416. How many hours did she spend at each job?

Job	Office	Tractor
Hours Worked	$2x$	x
Wages Earned	$10(2x)$	$12x$

31. **Phone Bill** The cost of a long-distance phone call is \$0.41 for the first minute and \$0.32 for each additional minute. If the total charge for a long-distance call is \$5.21, how many minutes was the call?

32. **Phone Bill** Danny, who is 1 year old, is playing with the telephone when he accidentally presses one of the buttons his mother has programmed to dial her friend Sue's number. Sue answers the phone and realizes Danny is on the other end. She talks to Danny, trying to get him to hang up. The cost for a call is \$0.23 for the first minute and \$0.14 for every minute after that. If the total charge for the call is \$3.73, how long did it take Sue to convince Danny to hang up the phone?

33. **Hourly Wages** JoAnn works in the publicity office at the state university. She is paid \$12 an hour for the first 35 hours she works each week and \$18 an hour for every hour after that. If she makes \$492 one week, how many hours did she work?

34. **Hourly Wages** Diane has a part-time job that pays her \$6.50 an hour. During one week she works 26 hours and is paid \$178.10. She realizes when she sees her check that she has been given a raise. How much per hour is that raise?

35. **Office Numbers** Professors Wong and Gil have offices in the mathematics building at Miami Dade College. Their office numbers are consecutive odd integers with a sum of 14,660. What are the office numbers of these two professors?

36. **Cell Phone Numbers** Diana and Tom buy two cell phones. The phone numbers assigned to each are consecutive integers with a sum of 11,109,295. If the smaller number is Diana's, what are their phone numbers?

37. **Age** Marissa and Kendra are 2 years apart in age. Their ages are two consecutive even integers. Kendra is the younger of the two. If Marissa's age is added to twice Kendra's age, the result is 26. How old is each girl?

38. **Age** Justin's and Ethan's ages form two consecutive odd integers. What is the difference of their ages?

39. **Arrival Time** Jeff and Carla Cole are driving separately from San Luis Obispo, California, to the north shore of Lake Tahoe, a distance of 425 miles. Jeff leaves San Luis Obispo at 11:00 AM and averages 55 miles per hour on the drive, Carla leaves later, at 1:00 PM but averages 65 miles per hour. Which person arrives in Lake Tahoe first?

40. **Piano Lessons** Tyler is taking piano lessons. Because he doesn't practice as often as his parents would like him to, he has to pay for part of the lessons himself. His parents pay him \$0.50 to do the laundry and \$1.25 to mow the lawn. In one month, he does the laundry 6 more times than he mows the lawn. If his parents pay him \$13.50 that month, how many times did he mow the lawn?

At one time, the Texas Junior College Teachers Association annual conference was held in Austin. At that time a taxi ride in Austin was \$1.25 for the first $\frac{1}{5}$ of a mile and \$0.25 for each additional $\frac{1}{5}$ of a mile. Use this information for Problems 41 and 42.

41. Cost of a Taxi Ride If the distance from one of the convention hotels to the airport is 7.5 miles, how much will it cost to take a taxi from that hotel to the airport?

42. Cost of a Taxi Ride Suppose the distance from one of the hotels to one of the western dance clubs in Austin is 12.4 miles. If the fare meter in the taxi gives the charge for that trip as \$16.50, is the meter working correctly?

43. Geometry The length and width of a rectangle are consecutive even integers. The perimeter is 44 meters. Find the length and width.

44. Geometry The length and width of a rectangle are consecutive odd integers. The perimeter is 128 meters. Find the length and width.

45. Geometry The angles of a triangle are three consecutive integers. Find the measure of each angle.

46. Geometry The angles of a triangle are three consecutive even integers. Find the measure of each angle.

Ike and Nancy Lara give western dance lessons at the Elk's Lodge on Sunday nights. The lessons cost \$3.00 for members of the lodge and \$5.00 for nonmembers. Half of the money collected for the lesson is paid to Ike and Nancy. The Elk's Lodge keeps the other half. One Sunday night Ike counts 36 people in the dance lesson. Use this information to work Problems 47 through 50.

47. Dance Lessons What is the least amount of money Ike and Nancy will make?

48. Dance Lessons What is the largest amount of money Ike and Nancy will make?

49. Dance Lessons At the end of the evening, the Elk's Lodge gives Ike and Nancy a check for \$80 to cover half of the receipts. Can this amount be correct?

50. Dance Lessons Besides the number of people in the dance lesson, what additional information does Ike need to know to always be sure he is being paid the correct amount?

Getting Ready for the Next Section

To understand all the explanations and examples in the next section you must be able to work the problems below.

Solve the following equations.

51. **a.** $x - 3 = 6$ **b.** $x + 3 = 6$ **c.** $-x - 3 = 6$ **d.** $-x + 3 = 6$

52. **a.** $x - 7 = 16$ **b.** $x + 7 = 16$ **c.** $-x - 7 = 16$ **d.** $-x + 7 = 16$

53. **a.** $\frac{x}{4} = -2$ **b.** $-\frac{x}{4} = -2$ **c.** $\frac{x}{4} = 2$ **d.** $-\frac{x}{4} = 2$

54. **a.** $3a = 15$ **b.** $3a = -15$ **c.** $-3a = 15$ **d.** $-3a = -15$

55. $2.5x - 3.48 = 4.9x + 2.07$

56. $2(1 - 3x) + 4 = 4x - 14$

57. $3(x - 4) = -2$

58. Solve for y: $2x - 3y = 6$

Linear Inequalities 1.8

Linear inequalities are solved by a method similar to the one used in solving linear equations. The only real differences between the methods are in the multiplication property for inequalities and in graphing the solution set.

An inequality differs from an equation only with respect to the comparison symbol between the two quantities being compared. In place of the equal sign, we use $<$ (less than), $\leq$ (less than or equal to), $>$ (greater than), or $\geq$ (greater than or equal to). The addition property for inequalities is almost identical to the addition property for equality.

PROPERTY *Addition Property for Inequalities*

For any three algebraic expressions A, B, and C,

if $\quad A < B$

then $\quad A + C < B + C$

In words: Adding the same quantity to both sides of an inequality will not change the solution set.

It makes no difference which inequality symbol we use to state the property. Adding the same amount to both sides always produces an inequality equivalent to the original inequality. Also, because subtraction can be thought of as addition of the opposite, this property holds for subtraction as well as addition.

EXAMPLE 1 Solve the inequality $x + 5 < 7$.

SOLUTION To isolate x, we add -5 to both sides of the inequality:

$$x + 5 < 7$$

$$x + 5 + (-5) < 7 + (-5) \qquad \text{Addition property for inequalities}$$

$$x < 2$$

We can go one step further here and graph the solution set. The solution set is all real numbers less than 2. To graph this set, we simply draw a straight line and label the center 0 (zero) for reference. Then we label the 2 on the right side of zero and extend an arrow beginning at 2 and pointing to the left. We use an open circle at 2 because it is not included in the solution set. Here is the graph.

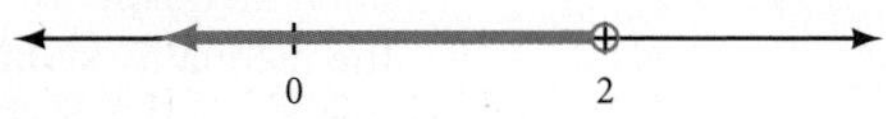

EXAMPLE 2 Solve $x - 6 \leq -3$.

SOLUTION Adding 6 to each side will isolate x on the left side:

$$x - 6 \leq -3$$

$$x - 6 + 6 \leq -3 + 6 \qquad \text{Add 6 to both sides}$$

$$x \leq 3$$

The graph of the solution set is

Notice that the dot at the 3 is darkened because 3 is included in the solution set. We always will use open circles on the graphs of solution sets with $<$ or $>$ and closed (darkened) circles on the graphs of solution sets with $\leq$ or $\geq$.

To see the idea behind the multiplication property for inequalities, we will consider three true inequality statements and explore what happens when we multiply both sides by a positive number and then what happens when we multiply by a negative number.

Consider the following three true statements:

$$3 < 5 \qquad -3 < 5 \qquad -5 < -3$$

Now multiply both sides by the positive number 4:

$$4(3) < 4(5) \qquad 4(-3) < 4(5) \qquad 4(-5) < 4(-3)$$

$$12 < 20 \qquad -12 < 20 \qquad -20 < -12$$

Note: This discussion is intended to show why the multiplication property for inequalities is written the way it is. You may want to look ahead to the property itself and then come back to this discussion if you are having trouble making sense out of it.

In each case, the inequality symbol in the result points in the same direction it did in the original inequality. We say the "sense" of the inequality doesn't change when we multiply both sides by a positive quantity.

Notice what happens when we go through the same process but multiply both sides by -4 instead of 4:

$$3 < 5 \qquad -3 < 5 \qquad -5 < -3$$

$$-4(3) > -4(5) \qquad -4(-3) > -4(5) \qquad -4(-5) > -4(-3)$$

$$-12 > -20 \qquad 12 > -20 \qquad 20 > 12$$

In each case, we have to change the direction in which the inequality symbol points to keep each statement true. Multiplying both sides of an inequality by a negative quantity always reverses the sense of the inequality. Our results are summarized in the multiplication property for inequalities.

Note: Because division is defined in terms of multiplication, this property is also true for division. We can divide both sides of an inequality by any nonzero number we choose. If that number happens to be negative, we must also reverse the direction of the inequality symbol.

[Δ≠Σ] PROPERTY ***Multiplication Property for Inequalities***

For any three algebraic expressions A, B, and C,

if	$A < B$	
then	$AC < BC$	when C is positive
and	$AC > BC$	when C is negative

In words: Multiplying both sides of an inequality by a positive number does not change the solution set. When multiplying both sides of an inequality by a negative number, it is necessary to reverse the inequality symbol to produce an equivalent inequality.

We can multiply both sides of an inequality by any nonzero number we choose. If that number happens to be negative, we must also reverse the sense of the inequality.

EXAMPLE 3 Solve $3a < 15$ and graph the solution.

SOLUTION We begin by multiplying each side by $\frac{1}{3}$. Because $\frac{1}{3}$ is a positive number, we do not reverse the direction of the inequality symbol:

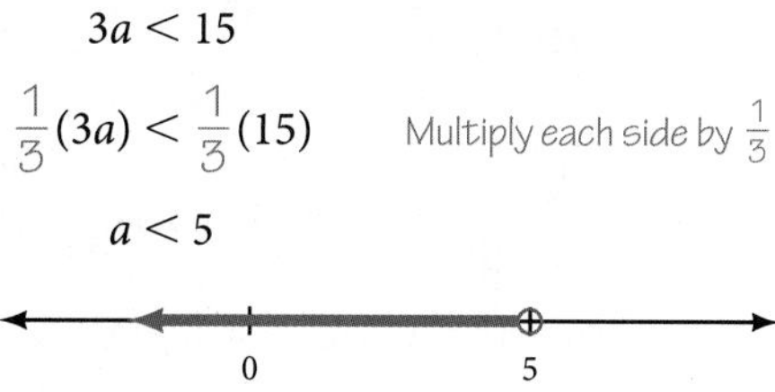

$$3a < 15$$

$$\frac{1}{3}(3a) < \frac{1}{3}(15) \quad \text{Multiply each side by } \frac{1}{3}$$

$$a < 5$$

EXAMPLE 4 Solve $-3a \le 18$, and graph the solution.

SOLUTION We begin by multiplying both sides by $-\frac{1}{3}$. Because $-\frac{1}{3}$ is a negative number, we must reverse the direction of the inequality symbol at the same time that we multiply by $-\frac{1}{3}$.

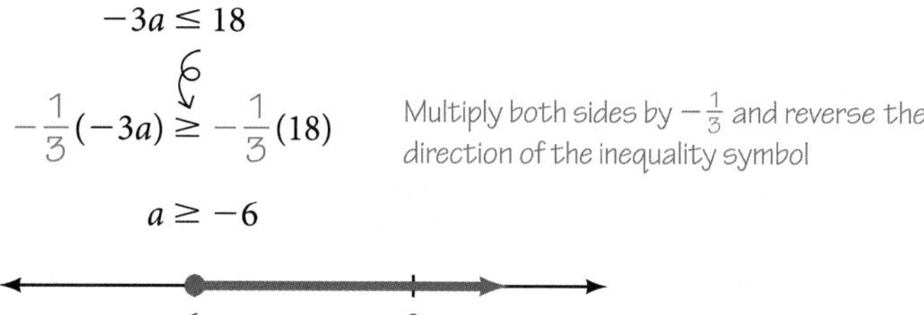

$$-3a \le 18$$

$$-\frac{1}{3}(-3a) \ge -\frac{1}{3}(18) \quad \text{Multiply both sides by } -\frac{1}{3} \text{ and reverse the direction of the inequality symbol}$$

$$a \ge -6$$

EXAMPLE 5 Solve $-\frac{x}{4} > 2$ and graph the solution.

SOLUTION To isolate x, we multiply each side by -4. Because -4 is a negative number, we also must reverse the direction of the inequality symbol:

$$-\frac{x}{4} > 2$$

$$-4\left(-\frac{x}{4}\right) < -4(2) \quad \text{Multiply each side by } -4\text{, and reverse the direction of the inequality symbol}$$

$$x < -8$$

-8 0

To solve more complicated inequalities, we use the following steps.

HOW TO ***Solving Linear Inequalities in One Variable***

Step 1a: Use the distributive property to separate terms, if necessary.

1b: If fractions are present, consider multiplying both sides by the LCD to eliminate the fractions. If decimals are present, consider multiplying both sides by a power of 10 to clear the inequality of decimals.

1c: Combine similar terms on each side of the inequality.

Step 2: Use the addition property for inequalities to get all variable terms on one side of the inequality and all constant terms on the other side.

Step 3: Use the multiplication property for inequalities to get x by itself on one side of the inequality.

Step 4: Graph the solution set.

EXAMPLE 6 Solve $2.5x - 3.48 < -4.9x + 2.07$.

SOLUTION We have two methods we can use to solve this inequality. We can simply apply our properties to the inequality the way it is currently written and work with the decimal numbers, or we can eliminate the decimals to begin with and solve the resulting inequality.

Method 1 Working with the decimals.

$$2.5x - 3.48 < -4.9x + 2.07 \qquad \text{Original inequality}$$

$$2.5x + 4.9x - 3.48 < -4.9x + 4.9x + 2.07 \qquad \text{Add 4.9x to each side}$$

$$7.4x - 3.48 < 2.07$$

$$7.4x - 3.48 + 3.48 < 2.07 + 3.48 \qquad \text{Add 3.48 to each side}$$

$$7.4x < 5.55$$

$$\frac{7.4x}{7.4} < \frac{5.55}{7.4} \qquad \text{Divide each side by 7.4}$$

$$x < 0.75$$

Method 2 Eliminating the decimals in the beginning.

Because the greatest number of places to the right of the decimal point in any of the numbers is 2, we can multiply each side of the inequality by 100 and we will be left with an equivalent inequality that contains only whole numbers.

$$2.5x - 3.48 < -4.9x + 2.07 \qquad \text{Original inequality}$$

$$100(2.5x - 3.48) < 100(-4.9x + 2.07) \qquad \text{Multiply each side by 100}$$

$$100(2.5x) - 100(3.48) < 100(-4.9x) + 100(2.07) \qquad \text{Distributive property}$$

$$250x - 348 < -490x + 207 \qquad \text{Multiplication}$$

$$740x - 348 < 207 \qquad \text{Add 490x to each side}$$

$$740x < 555 \qquad \text{Add 348 to each side}$$

$$\frac{740x}{740} < \frac{555}{740} \qquad \text{Divide each side by 740}$$

$$x < 0.75$$

The solution by either method is $x < 0.75$. Here is the graph:

0 0.75

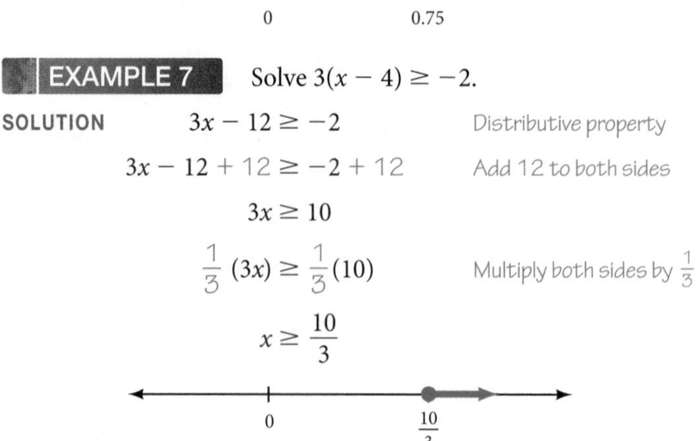

EXAMPLE 7 Solve $3(x - 4) \geq -2$.

SOLUTION

$$3x - 12 \geq -2 \qquad \text{Distributive property}$$

$$3x - 12 + 12 \geq -2 + 12 \qquad \text{Add 12 to both sides}$$

$$3x \geq 10$$

$$\frac{1}{3}(3x) \geq \frac{1}{3}(10) \qquad \text{Multiply both sides by } \frac{1}{3}$$

$$x \geq \frac{10}{3}$$

0 $\frac{10}{3}$

EXAMPLE 8 Solve and graph $2(1 - 3x) + 4 < 4x - 14$.

SOLUTION

$2 - 6x + 4 < 4x - 14$	Distributive property
$-6x + 6 < 4x - 14$	Simplify
$-6x + 6 + (-6) < 4x - 14 + (-6)$	Add -6 to both sides
$-6x < 4x - 20$	
$-6x + (-4x) < 4x + (-4x) - 20$	Add $-4x$ to both sides
$-10x < -20$	
$\left(-\frac{1}{10}\right)(-10x) > \left(-\frac{1}{10}\right)(-20)$	Multiply by $-\frac{1}{10}$, reverse the direction of the inequality
$x > 2$	

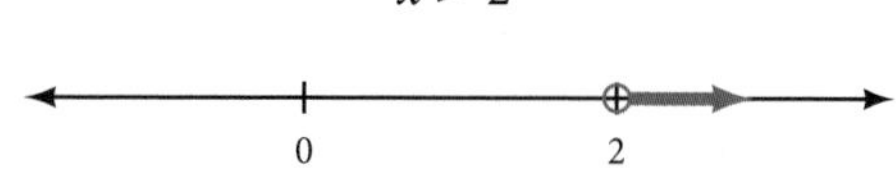

EXAMPLE 9 Solve $2x - 3y < 6$ for y.

SOLUTION We can solve this formula for y by first adding $-2x$ to each side and then multiplying each side by $-\frac{1}{3}$. When we multiply by $-\frac{1}{3}$ we must reverse the direction of the inequality symbol. Because this is a formula, we will not graph the solution.

$2x - 3y < 6$	Original formula
$2x + (-2x) - 3y < (-2x) + 6$	Add $-2x$ to each side
$-3y < -2x + 6$	
$-\frac{1}{3}(-3y) > -\frac{1}{3}(-2x + 6)$	Multiply each side by $-\frac{1}{3}$
$y > \frac{2}{3}x - 2$	Distributive property

When working application problems that involve inequalities, the phrases "at least" and "at most" translate as follows:

In Words	*In Symbols*
x is at least 30	$x \geq 30$
x is at most 20	$x \leq 20$

Our next example is similar to an example done earlier in this chapter. This time it involves an inequality instead of an equation.

We can modify our Blueprint for Problem Solving to solve application problems whose solutions depend on writing and then solving inequalities.

EXAMPLE 10 The sum of two consecutive odd integers is at most 28. What are the possibilities for the first of the two integers?

SOLUTION When we use the phrase "their sum is at most 28," we mean that their sum is less than or equal to 28.

Step 1: **Read and list.**
Known items: Two consecutive odd integers. Their sum is less than or equal to 28.
Unknown items: The numbers in question.

Step 2: **Assign a variable, and translate information.**
If we let $x =$ the first of the two consecutive odd integers, then $x + 2$ is the next consecutive one.

Step 3: **Reread, and write an inequality.**
Their sum is at most 28.

$$x + (x + 2) \leq 28$$

Step 4: **Solve the inequality.**

$$2x + 2 \leq 28 \quad \text{Simplify the left side}$$

$$2x \leq 26 \quad \text{Add } -2 \text{ to each side}$$

$$x \leq 13 \quad \text{Multiply each side by } \tfrac{1}{2}.$$

Step 5: **Write the answer.**
The first of the two integers must be an odd integer that is less than or equal to 13. The second of the two integers will be two more than whatever the first one is.

Step 6: **Reread, and check.**
Suppose the first integer is 13. The next consecutive odd integer is 15. The sum of 15 and 13 is 28. If the first odd integer is less than 13, the sum of it and the next consecutive odd integer will be less than 28.

GETTING READY FOR CLASS

After reading through the preceding section, respond in your own words and in complete sentences.

A. State the addition property for inequalities.
B. How is the multiplication property for inequalities different from the multiplication property of equality?
C. When do we reverse the direction of an inequality symbol?
D. Under what conditions do we not change the direction of the inequality symbol when we multiply both sides of an inequality by a number?

Solve the following inequalities using the addition property of inequalities. Graph each solution set.

1. $x - 5 < 7$ **2.** $x + 3 < -5$ **3.** $a - 4 \le 8$ **4.** $a + 3 \le 10$

5. $x - 4.3 > 8.7$ **6.** $x - 2.6 > 10.4$ **7.** $y + 6 \ge 10$ **8.** $y + 3 \ge 12$

9. $2 < x - 7$ **10.** $3 < x + 8$

Solve the following inequalities using the multiplication property of inequalities. If you multiply both sides by a negative number, be sure to reverse the direction of the inequality symbol. Graph the solution set.

11. $3x < 6$ **12.** $2x < 14$ **13.** $5a \le 25$ **14.** $4a \le 16$

15. $\frac{x}{3} > 5$ **16.** $\frac{x}{7} > 1$ **17.** $-2x > 6$ **18.** $-3x \ge 9$

19. $-3x \ge -18$ **20.** $-8x \ge -24$ **21.** $-\frac{x}{5} \le 10$ **22.** $-\frac{x}{9} \ge -1$

23. $-\frac{2}{3}y > 4$ **24.** $-\frac{3}{4}y > 6$

Solve the following inequalities. Graph the solution set in each case.

25. $2x - 3 < 9$ **26.** $3x - 4 < 17$ **27.** $-\frac{1}{5}y - \frac{1}{3} \le \frac{2}{3}$

28. $-\frac{1}{6}y - \frac{1}{2} \le \frac{2}{3}$ **29.** $-7.2x + 1.8 > -19.8$ **30.** $-7.8x - 1.3 > 22.1$

31. $\frac{2}{3}x - 5 \le 7$ **32.** $\frac{3}{4}x - 8 \le 1$ **33.** $-\frac{2}{5}a - 3 > 5$

34. $-\frac{4}{5}a - 2 > 10$ **35.** $5 - \frac{3}{5}y > -10$ **36.** $4 - \frac{5}{6}y > -11$

37. $0.3(a + 1) \le 1.2$ **38.** $0.4(a - 2) \le 0.4$ **39.** $2(5 - 2x) \le -20$

40. $7(8 - 2x) > 28$ **41.** $3x - 5 > 8x$ **42.** $8x - 4 > 6x$

43. $\frac{1}{3}y - \frac{1}{2} \le \frac{5}{6}y + \frac{1}{2}$ **44.** $\frac{7}{6}y + \frac{4}{3} \le \frac{11}{6}y - \frac{7}{6}$

45. $-2.8x + 8.4 < -14x - 2.8$ **46.** $-7.2x - 2.4 < -2.4x + 12$

47. $3(m - 2) - 4 \ge 7m + 14$ **48.** $2(3m - 1) + 5 \ge 8m - 7$

49. $3 - 4(x - 2) \le -5x + 6$ **50.** $8 - 6(x - 3) \le -4x + 12$

Solve each of the following formulas for y.

51. $3x + 2y < 6$ **52.** $-3x + 2y < 6$ **53.** $2x - 5y > 10$

54. $-2x - 5y > 5$ **55.** $-3x + 7y \le 21$ **56.** $-7x + 3y \le 21$

57. $2x - 4y \ge -4$ **58.** $4x - 2y \ge -8$

The next two problems are intended to give you practice reading, and paying attention to, the instructions that accompany the problems you are working.

59. Work each problem according to the instructions given.

a. Evaluate when $x = 0$: $-5x + 3$ **b.** Solve: $-5x + 3 = -7$

c. Is 0 a solution to $-5x + 3 < -7$ **d.** Solve: $-5x + 3 < -7$

60. Work each problem according to the instructions given.

a. Evaluate when $x = 0$: $-2x - 5$ **b.** Solve: $-2x - 5 = 1$

c. Is 0 a solution to $-2x - 5 > 1$ **d.** Solve: $-2x - 5 > 1$

For each graph below, write an inequality whose solution is the graph.

61. 

62.

63.

64.

Applying the Concepts

65. Consecutive Integers The sum of two consecutive integers is at least 583. What are the possibilities for the first of the two integers?

66. Consecutive Integers The sum of two consecutive integers is at most 583. What are the possibilities for the first of the two integers?

67. Number Problems The sum of twice a number and six is less than ten. Find all solutions.

68. Number Problems Twice the difference of a number and three is greater than or equal to the number increased by five. Find all solutions.

69. Number Problems The product of a number and four is greater than the number minus eight. Find the solution set.

70. Number Problems The quotient of a number and five is less than the sum of seven and two. Find the solution set.

71. Geometry Problems The length of a rectangle is 3 times the width. If the perimeter is to be at least 48 meters, what are the possible values for the width? (If the perimeter is at least 48 meters, then it is greater than or equal to 48 meters.)

72. Geometry Problems The length of a rectangle is 3 more than twice the width. If the perimeter is to be at least 51 meters, what are the possible values for the width? (If the perimeter is at least 51 meters, then it is greater than or equal to 51 meters.)

73. Geometry Problems The numerical values of the three sides of a triangle are given by three consecutive even integers. If the perimeter is greater than 24 inches, what are the possibilities for the shortest side?

74. Geometry Problems The numerical values of the three sides of a triangle are given by three consecutive odd integers. If the perimeter is greater than 27 inches, what are the possibilities for the shortest side?

Getting Ready for the Next Section

Solve each inequality. Do not graph.

75. $2x - 1 \geq 3$ **76.** $3x + 1 \geq 7$ **77.** $-2x > -8$ **78.** $-3x > -12$

79. $-3 > 4x + 1$ **80.** $4x + 1 \leq 9$

Compound Inequalities 1.9

The instrument panel on most cars includes a temperature gauge. The one shown below indicates that the normal operating temperature for the engine is from 50°F to 270°F.

We can represent the same situation with an inequality by writing $50 \le F \le 270$, where F is the temperature in degrees Fahrenheit. This inequality is a *compound inequality*. In this section we present the notation and definitions associated with compound inequalities.

The *union* of two sets A and B is the set of all elements that are in A or in B. The word *or* is the key word in the definition. The *intersection* of two sets A and B is the set of elements contained in both A and B. The key word in this definition is *and*. We can put the words *and* and *or* together with our methods of graphing inequalities to find the solution sets for compound inequalities.

DEFINITION *compound inequality*

A ***compound inequality*** is two or more inequalities connected by the word *and* or *or*.

EXAMPLE 1 Graph the solution set for the compound inequality

$$x < -1 \quad \text{or} \quad x \ge 3$$

SOLUTION Graphing each inequality separately, we have

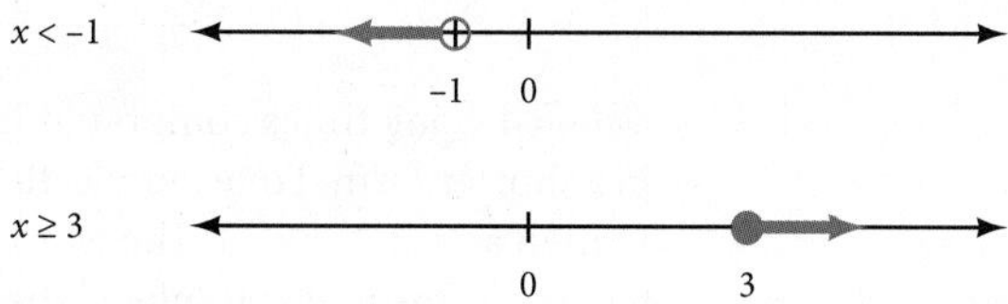

Because the two inequalities are connected by *or*, we want to graph their union; that is, we graph all points that are on either the first graph or the second graph. Essentially, we put the two graphs together on the same number line.

$$x < -1 \quad \text{or} \quad x \ge 3$$

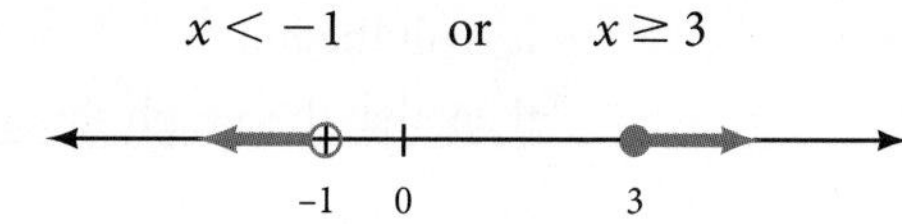

EXAMPLE 2 Graph the solution set for the compound inequality

$$x > -2 \qquad \text{and} \qquad x < 3$$

SOLUTION Graphing each inequality separately, we have

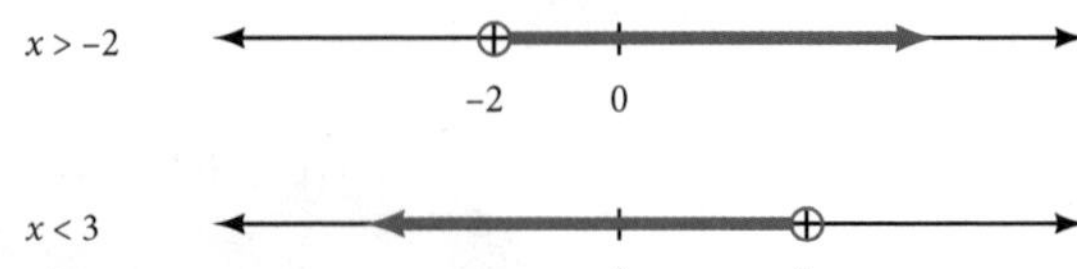

Because the two inequalities are connected by the word *and,* we will graph their intersection, which consists of all points that are common to both graphs; that is, we graph the region where the two graphs overlap.

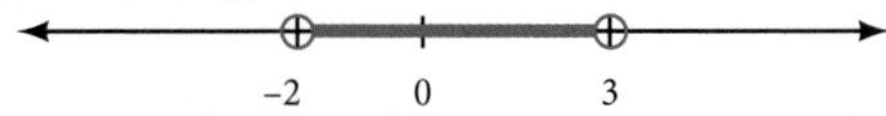

EXAMPLE 3 Solve and graph the solution set for

$$2x - 1 \geq 3 \qquad \text{and} \qquad -3x > -12$$

SOLUTION Solving the two inequalities separately, we have

$$2x - 1 \geq 3 \qquad \text{and} \qquad -3x > -12$$

$$2x \geq 4 \qquad\qquad -\frac{1}{3}(-3x) < -\frac{1}{3}(-12)$$

$$x \geq 2 \qquad \text{and} \qquad x < 4$$

Because the word *and* connects the two graphs, we will graph their intersection—the points they have in common:

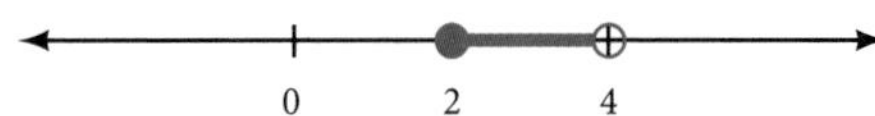

Notation Sometimes compound inequalities that use the word *and* can be written in a shorter form. For example, the compound inequality $-2 < x$ and $x < 3$ can be written as $-2 < x < 3$. The word *and* does not appear when an inequality is written in this form; it is implied. The solution set for $-2 < x$ and $x < 3$ is

It is all the numbers between -2 and 3 on the number line. It seems reasonable then, that this graph should be the graph of

$$-2 < x < 3$$

In both the graph and the inequality, x is said to be between -2 and 3.

EXAMPLE 4 Solve and graph $-3 \le 2x - 1 \le 9$.

SOLUTION To solve for x, we must add 1 to the center expression and then divide the result by 2. Whatever we do to the center expression, we also must do to the two expressions on the ends. In this way we can be sure we are producing equivalent inequalities. The solution set will not be affected.

$$-3 \le 2x - 1 \le 9$$

$$-2 \le 2x \le 10 \quad \text{Add 1 to each expression}$$

$$-1 \le x \le 5 \quad \text{Multiply each expression by } \frac{1}{2}$$

-1 5

GETTING READY FOR CLASS

After reading through the preceding section, respond in your own words and in complete sentences.

A. What is a compound inequality?

B. Explain the shorthand notation that can be used to write two inequalities connected by the word *and*.

C. Write two inequalities connected by the word *and* that together are equivalent to $-1 < x < 2$.

D. Explain in words how you would graph the compound inequality $x < 2$ or $x > -3$.

Problem Set 1.9

Graph the following compound inequalities.

1. $x < -1$ or $x > 5$
2. $x \le -2$ or $x \ge -1$
3. $x < -3$ or $x \ge 0$
4. $x < 5$ and $x > 1$
5. $x \le 6$ and $x > -1$
6. $x \le 7$ and $x > 0$
7. $x > 2$ and $x < 4$
8. $x < 2$ or $x > 4$
9. $x \ge -2$ and $x \le 4$
10. $x \le 2$ or $x \ge 4$
11. $x < 5$ and $x > -1$
12. $x > 5$ or $x < -1$
13. $-1 < x < 3$
14. $-1 \le x \le 3$
15. $-3 < x \le -2$
16. $-5 \le x \le 0$

Solve the following compound inequalities. Graph the solution set in each case.

17. $3x - 1 < 5$ or $5x - 5 > 10$
18. $x + 1 < -3$ or $x - 2 > 6$
19. $x - 2 > -5$ and $x + 7 < 13$
20. $3x + 2 \le 11$ and $2x + 2 \ge 0$
21. $11x < 22$ or $12x > 36$
22. $-5x < 25$ and $-2x \ge -12$
23. $3x - 5 < 10$ and $2x + 1 > -5$
24. $5x + 8 < -7$ or $3x - 8 > 10$
25. $2x - 3 < 8$ and $3x + 1 > -10$
26. $11x - 8 > 3$ or $12x + 7 < -5$
27. $2x - 1 < 3$ and $3x - 2 > 1$
28. $3x + 9 < 7$ or $2x - 7 > 11$
29. $-1 \le x - 5 \le 2$
30. $0 \le x + 2 \le 3$
31. $-4 \le 2x \le 6$
32. $-5 < 5x < 10$
33. $-3 < 2x + 1 < 5$
34. $-7 \le 2x - 3 \le 7$
35. $0 \le 3x + 2 \le 7$
36. $2 \le 5x - 3 \le 12$
37. $-7 < 2x + 3 < 11$
38. $-5 < 6x - 2 < 8$
39. $-1 \le 4x + 5 \le 9$
40. $-8 \le 7x - 1 \le 13$

For each graph below, write an inequality whose solution is the graph.

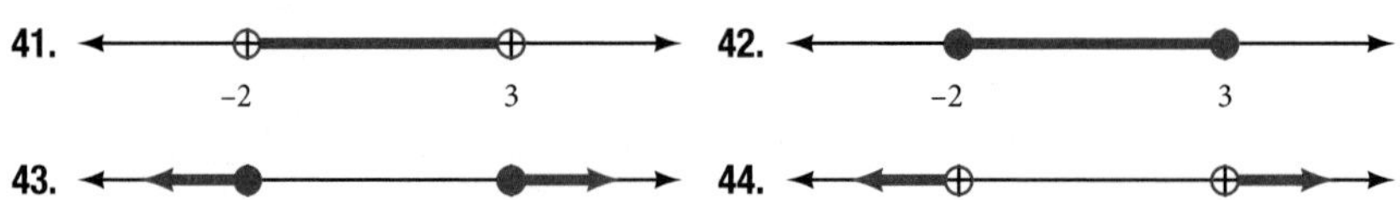

Applying the Concepts

Triangle Inequality The triangle inequality states that the sum of any two sides of a triangle must be greater than the third side.

45. The following triangle *RST* has sides of length x, $2x$, and 10 as shown.

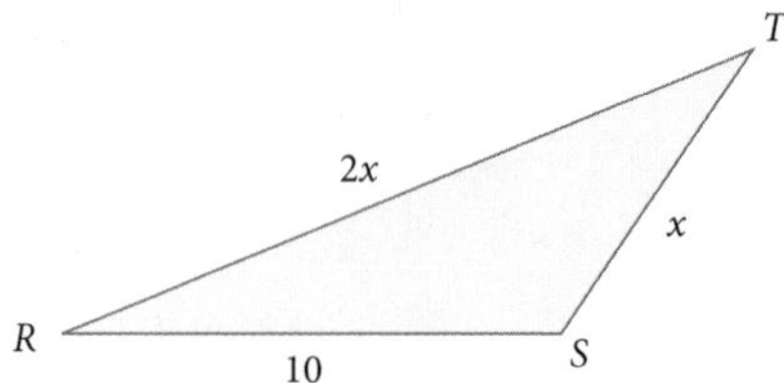

a. Find the three inequalities, which must be true based on the sides of the triangle.

b. Write a compound inequality based on your results above.

46. The following triangle ABC has sides of length x, $3x$, and 16 as shown.

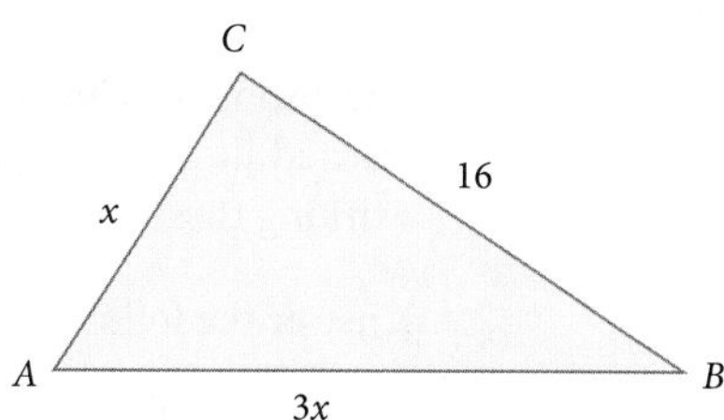

a. Find the three inequalities, which must be true based on the sides of the triangle.

b. Write a compound inequality based on your results above.

47. Engine Temperature The engine in a car gives off a lot of heat due to the combustion in the cylinders. The water used to cool the engine keeps the temperature within the range $50 \le F \le 266$ where F is in degrees Fahrenheit. Graph this inequality on the number line.

48. Engine Temperature To find the engine temperature range from Problem 47 in degrees Celsius, we use the fact that $F = \frac{9}{5}C + 32$ to rewrite the inequality as

$$50 \le \frac{9}{5}C + 32 \le 266$$

Solve this inequality and graph the solution set.

49. Number Problem The difference of twice a number and 3 is between 5 and 7. Find the number.

50. Number Problem The sum of twice a number and 5 is between 7 and 13. Find the number.

51. Perimeter The length of a rectangle is 4 inches longer than the width. The perimeter is between 20 inches and 30 inches.

a. Write the perimeter as a compound inequality. $\_\_\_\_ < P < \_\_\_\_$

b. Write the width as a compound inequality. $\_\_\_\_ < w < \_\_\_\_$

c. Write the length as a compound inequality. $\_\_\_\_ < l < \_\_\_\_$

52. Perimeter The length of a rectangle is 6 feet longer than the width. The perimeter is between 24 feet and 36 feet.

a. Write the perimeter as a compound inequality. $\_\_\_\_ < P < \_\_\_\_$

b. Write the width as a compound inequality. $\_\_\_\_ < w < \_\_\_\_$

c. Write the length as a compound inequality. $\_\_\_\_ < l < \_\_\_\_$

Maintaining Your Skills

The problems that follow review some of the more important skills you have learned in previous sections and chapters. You can consider the time you spend working these problems as time spent studying for exams.

Answer the following percent problems.

53. What number is 25% of 32?

54. What number is 15% of 75?

55. What number is 20% of 120?

56. What number is 125% of 300?

57. What percent of 36 is 9?

58. What percent of 16 is 9?

59. What percent of 50 is 5?

60. What percent of 140 is 35?

61. 16 is 20% of what number?

62. 6 is 3% of what number?

63. 8 is 2% of what number?

64. 70 is 175% of what number?

Simplify each expression.

65. $-|-5|$

66. $\left(-\frac{2}{3}\right)^3$

67. $-3 - 4(-2)$

68. $2^4 + 3^3 \div 9 - 4^2$

69. $5|3 - 8| - 6|2 - 5|$

70. $7 - 3(2 - 6)$

71. $5 - 2[-3(5 - 7) - 8]$

72. $\frac{5 + 3(7 - 2)}{2(-3) - 4}$

73. Find the difference of -3 and -9.

74. If you add -4 to the product of -3 and 5, what number results?

75. Apply the distributive property to $\frac{1}{2}(4x - 6)$.

76. Use the associative property to simplify $-6\left(\frac{1}{3}x\right)$.

For the set $\left\{-3, -\frac{4}{5}, 0, \frac{5}{8}, 2, \sqrt{5}\right\}$, which numbers are

77. Integers

78. Rational numbers

Chapter 1 Summary

EXAMPLES

Similar Terms [1.1]

1. The terms $2x$, $5x$, and $-7x$ are all similar because their variable parts are the same.

A *term* is a number or a number and one or more variables multiplied together. *Similar terms* are terms with the same variable part.

Simplifying Expressions [1.1]

2. Simplify $3x + 4x$.

$$3x + 4x = (3 + 4)x$$
$$= 7x$$

In this chapter we simplified expressions that contained variables by using the distributive property to combine similar terms.

Solution Set [1.2]

3. The solution set for the equation $x + 2 = 5$ is $\{3\}$ because when x is 3 the equation is $3 + 2 = 5$, or $5 = 5$.

The *solution set* for an equation (or inequality) is all the numbers that, when used in place of the variable, make the equation (or inequality) a true statement.

Equivalent Equations [1.2]

4. The equation $a - 4 = 3$ and $a - 2 = 5$ are equivalent because both have solution set $\{7\}$.

Two equations are called *equivalent* if they have the same solution set.

Addition Property of Equality [1.2]

5. Solve $x - 5 = 12$.

$$x - 5\ (+\ \mathbf{5}) = 12\ (+\ \mathbf{5})$$
$$x + 0 = 17$$
$$x = 17$$

When the same quantity is added to both sides of an equation, the solution set for the equation is unchanged. Adding the same amount to both sides of an equation produces an equivalent equation.

Multiplication Property of Equality [1.3]

6. Solve $3x = 18$.

$$\tfrac{1}{3}(3x) = \tfrac{1}{3}(18)$$
$$x = 6$$

If both sides of an equation are multiplied by the same nonzero number, the solution set is unchanged. Multiplying both sides of an equation by a nonzero quantity produces an equivalent equation.

Strategy for Solving Linear Equations in One Variable [1.4]

7. Solve $2(x + 3) = 10$.

$$2x + 6 = 10$$
$$2x + 6 + (\mathbf{-6}) = 10 + (\mathbf{-6})$$
$$2x = 4$$
$$\tfrac{1}{2}(2x) = \tfrac{1}{2}(4)$$
$$x = 2$$

Step 1a: Use the distributive property to separate terms, if necessary.

1b: If fractions are present, consider multiplying both sides by the LCD to eliminate the fractions. If decimals are present, consider multiplying both sides by a power of 10 to clear the equation of decimals.

1c: Combine similar terms on each side of the equation.

Step 2: Use the addition property of equality to get all variable terms on one side of the equation and all constant terms on the other side. A variable term is a term that contains the variable (for example, $5x$). A constant term is a term that does not contain the variable (the number 3, for example).

Step 3: Use the multiplication property of equality to get x (that is, $1x$) by itself on one side of the equation.

Step 4: Check your solution in the original equation to be sure that you have not made a mistake in the solution process.

Formulas [1.5]

8. Solving $P = 2l + 2w$ for l, we have

$$P - 2w = 2l$$

$$\frac{P - 2w}{2} = l$$

A formula is an equation with more than one variable. To solve a formula for one of its variables, we use the addition and multiplication properties of equality to move everything except the variable in question to one side of the equal sign so the variable in question is alone on the other side.

Blueprint for Problem Solving [1.6, 1.7]

Step 1: ***Read*** the problem, and then mentally ***list*** the items that are known and the items that are unknown.

Step 2: ***Assign a variable*** to one of the unknown items. (In most cases this will amount to letting $x =$ the item that is asked for in the problem.) Then ***translate*** the other ***information*** in the problem to expressions involving the variable.

Step 3: ***Reread*** the problem, and then ***write an equation,*** using the items and variables listed in steps 1 and 2, that describes the situation.

Step 4: ***Solve the equation*** found in step 3.

Step 5: ***Write*** your ***answer*** using a complete sentence.

Step 6: ***Reread*** the problem, and ***check*** your solution with the original words in the problem.

Addition Property for Inequalities [1.8]

9. Solve $x + 5 < 7$.

$$x + 5 + (-5) < 7 + (-5)$$

$$x < 2$$

Adding the same quantity to both sides of an inequality produces an equivalent inequality, one with the same solution set.

Multiplication Property for Inequalities [1.8]

10. Solve $-3a \leq 18$.

$$-\frac{1}{3}(-3a) \geq -\frac{1}{3}(18)$$

$$a \geq -6$$

Multiplying both sides of an inequality by a positive number never changes the solution set. If both sides are multiplied by a negative number, the sign of the inequality must be reversed to produce an equivalent inequality.

Strategy for Solving Linear Inequalities in One Variable [1.8]

Step 1a: Use the distributive property to separate terms, if necessary.

Step 1b: If fractions are present, consider multiplying both sides by the LCD to eliminate the fractions. If decimals are present, consider multiplying both sides by a power of 10 to clear the inequality of decimals.

Step 1c: Combine similar terms on each side of the inequality.

Step 2: Use the addition property for inequalities to get all variable terms on one side of the inequality and all constant terms on the other side.

Step 3: Use the multiplication property for inequalities to get x by itself on one side of the inequality.

Step 4: Graph the solution set.

11. Solve $3(x - 4) \geq -2$.

$$3x - 12 \geq -2$$
$$3x - 12 + \mathbf{12} \geq -2 + \mathbf{12}$$
$$3x \geq 10$$
$$\tfrac{1}{3}(3x) \geq \tfrac{1}{3}(10)$$
$$x \geq \frac{10}{3}$$

Compound Inequalities [1.9]

Two inequalities connected by the word *and* or *or* form a compound inequality. If the connecting word is *or*, we graph all points that are on either graph. If the connecting word is *and*, we graph only those points that are common to both graphs. The inequality $-2 \leq x \leq 3$ is equivalent to the compound inequality $-2 \leq x$ and $x \leq 3$.

12. $x < -3$ or $x > 1$

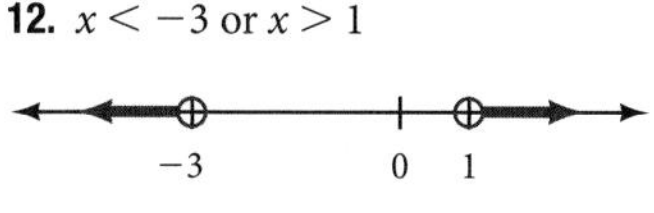

$-2 \leq x \leq 3$

Chapter 1 Test

Simplify each of the following expressions. [1.1]

1. $5y - 3 - 6y + 4$

2. $3x - 4 + x + 3$

3. $4 - 2(y - 3) - 6$

4. $3(3x - 4) - 2(4x + 5)$

5. Find the value of $3x + 12 + 2x$ when $x = -3$. [1.1]

6. Find the value of $x^2 - 3xy + y^2$ when $x = -2$ and $y = -4$. [1.1]

7. Fill in the tables below to find the sequences formed by substituting the first four counting numbers into the expressions $(n + 2)^2$ and $n^2 + 2$. [1.1]

a.

n	$(n + 2)^2$
1	
2	
3	
4	

b.

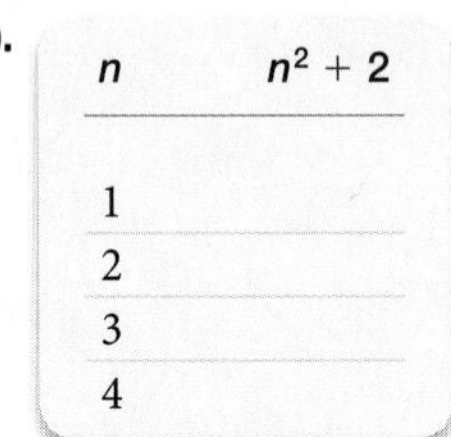

n	$n^2 + 2$
1	
2	
3	
4	

Solve the following equations. [1.2, 1.3, 1.4]

8. $3x - 2 = 7$

9. $4y + 15 = y$

10. $\frac{1}{4}x - \frac{1}{12} = \frac{1}{3}x - \frac{1}{6}$

11. $-3(3 - 2x) - 7 = 8$

12. $3x - 9 = -6$

13. $0.05 + 0.07(100 - x) = 3.2$

14. $4(t - 3) + 2(t + 4) = 2t - 16$

15. $4x - 2(3x - 1) = 2x - 8$

For each of the following expressions, write an equivalent equation. [1.5]

16. What number is 40% of 56?

17. 720 is 24% of what number?

Solve each formula for the appropriate variable. [1.5]

18. If $3x - 4y = 16$, find y when $x = 4$.

19. If $3x - 4y = 16$, find x when $y = 2$.

20. Solve $2x + 6y = 12$ for y.

21. Solve $x^2 = v^2 + 2ad$ for a.

Solve each word problem. [1.6, 1.7]

22. Age Problem Paul is twice as old as Becca. Five years ago the sum of their ages was 44. How old are they now?

23. Geometry The length of a rectangle is 5 less than 3 times the width. The perimeter is 150 centimeters. What are the length and width?

24. Coin Problem A man has a collection of dimes and nickels with a total value of \$1.70. If he has 8 more dimes than nickels, how many of each coin does he have?

25. Investing A woman has money in two accounts. One account pays 6% annual interest, whereas the other pays 12% annual interest. If she has \$500 more invested at 12% than she does at 6% and her total interest for a year is \$186, how much does she have in each account?

Solve each inequality, and graph the solution. [1.8]

26. $\frac{1}{2}x - 2 > 3$

27. $-6y \le 24$

28. $0.3 - 0.2x < 1.1$

29. $3 - 2(n - 1) \ge 9$

Solve each inequality, and graph the solution. [1.9]

30. $5x - 3 < 2x$ or $2x > 6$

31. $-3 \le 2x - 7 \le 9$

Linear Equations and Inequalities in Two Variables

2

Chapter Outline

2.1 Paired Data and Graphing Ordered Pairs

2.2 Solutions to Linear Equations in Two Variables

2.3 Graphing Linear Equations in Two Variables

2.4 More on Graphing: Intercepts

2.5 The Slope of a Line

2.6 Finding the Equation of a Line

2.7 Linear Inequalities in Two Variables

iStockphoto.com © AtollPhotography

When light comes into contact with a surface that does not transmit light, then all the light that contacts the surface is either reflected off the surface or absorbed into the surface. If we let R represent the percentage of light reflected and A represent the percentage of light absorbed, then the relationship between these two variables can be written as

$$R + A = 100$$

which is a linear equation in two variables. The following table and graph show the same relationship as that described by the equation. The table is a numerical description; the graph is a visual description.

Reflected and Absorbed Light

Percent Reflected	Percent Absorbed
0	100
20	80
40	60
60	40
80	20
100	0

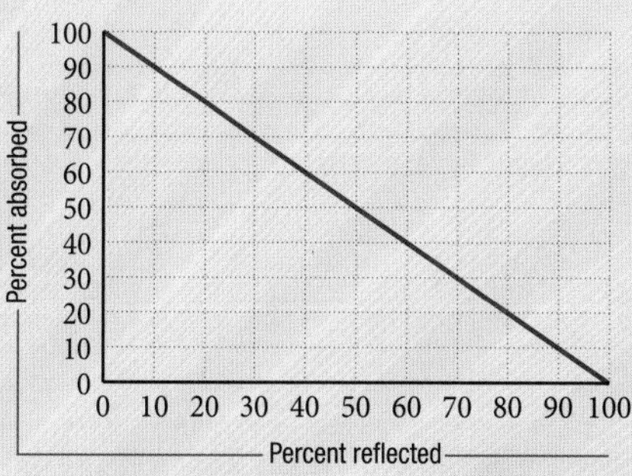

In this chapter we learn how to build tables and draw graphs from linear equations in two variables.

Study Skills

If you have successfully completed Chapter 1, then you have made a good start at developing the study skills necessary to succeed in all math classes. Some of the study skills for this chapter are a continuation of the skills from Chapter 1, while others are new to this chapter.

1. **Continue to Set and Keep a Schedule** Sometimes I find students do well in Chapter 1 and then become overconfident. They will begin to put in less time with their homework. Don't do it. Keep to the same schedule.

2. **Increase Effectiveness** You want to become more and more effective with the time you spend on your homework. Increase those activities that are the most beneficial and decrease those that have not given you the results you want.

3. **List Difficult Problems** Begin to make lists of problems that give you the most difficulty. These are the problems in which you are repeatedly making mistakes.

4. **Begin to Develop Confidence With Word Problems** It seems that the main difference between people who are good at working word problems and those who are not is confidence. People with confidence know that no matter how long it takes them, they will eventually be able to solve the problem. Those without confidence begin by saying to themselves, "I'll never be able to work this problem." If you are in this second category, then instead of telling yourself that you can't do word problems, decide to do whatever it takes to master them. The more word problems you work, the better you will become at them.

   Many of my students keep a notebook that contains everything that they need for the course: class notes, homework, quizzes, tests, and research projects. A three-ring binder with tabs is ideal. Organize your notebook so that you can easily get to any item you want to look at.

Paired Data and Graphing Ordered Pairs 2.1

This table and figure show the relationship between the table of values for the speed of a race car and the corresponding bar chart. In Figure 1, the horizontal line that shows the elapsed time in seconds is called the *horizontal axis*, and the vertical line that shows the speed in miles per hour is called the *vertical axis*.

The data in the table are called *paired data* because the information is organized so that each number in the first column is paired with a specific number in the second column. Each pair of numbers is associated with one of the solid bars in Figure 1. For example, the third bar in the bar chart is associated with the pair of numbers 3 seconds and 162.8 miles per hour. The first number, 3 seconds, is associated with the horizontal axis, and the second number, 162.8 miles per hour, is associated with the vertical axis.

Speed of a Race Car

Time in Seconds	Speed in Miles per Hour
0	0
1	72.7
2	129.9
3	162.8
4	192.2
5	212.4
6	228.1

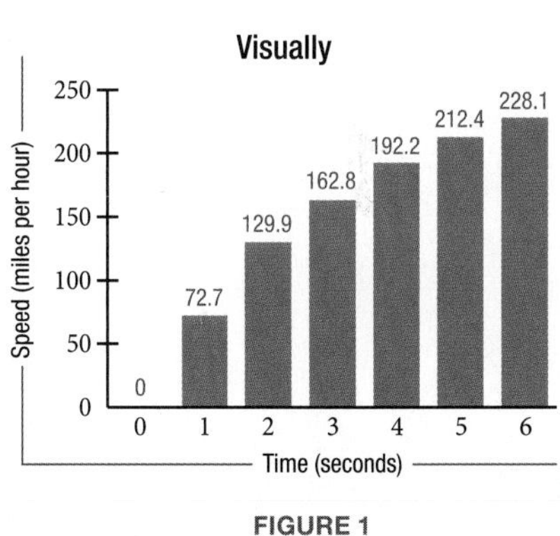

FIGURE 1

Scatter Diagrams and Line Graphs

The information in the table can be visualized with a *scatter diagram* and *line graph* as well. Figure 2 is a scatter diagram of the information in Table 1. We use dots instead of the bars shown in Figure 1 to show the speed of the race car at each second during the race. Figure 3 is called a *line graph*. It is constructed by taking the dots in Figure 2 and connecting each one to the next with a straight line. Notice that we have labeled the axes in these two figures a little differently than we did with the bar chart by making the axes intersect at the number 0.

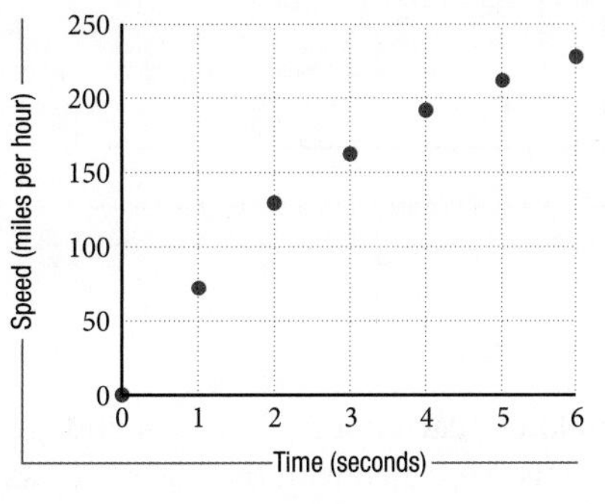

FIGURE 2

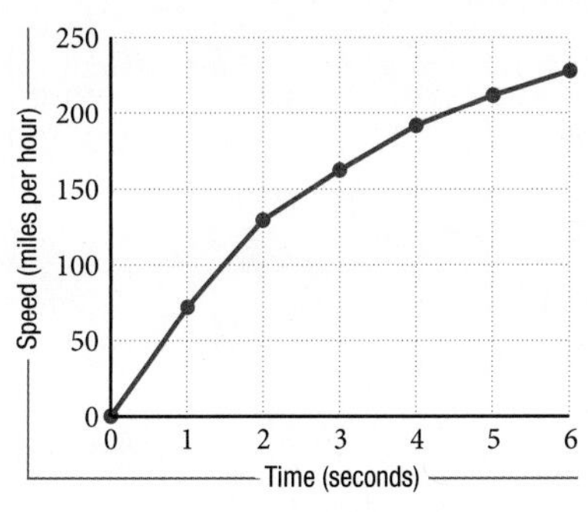

FIGURE 3

The number sequences we have worked with in the past can also be written as paired data by associating each number in the sequence with its position in the sequence. For instance, in the sequence of odd numbers

$$1, 3, 5, 7, 9, \ldots$$

the number 7 is the fourth number in the sequence. Its position is 4, and its value is 7. Here is the sequence of odd numbers written so that the position of each term is noted:

Position	1, 2, 3, 4, 5, . . .
Value	1, 3, 5, 7, 9, . . .

EXAMPLE 1 The tables below give the first five terms of the sequence of odd numbers and the sequence of squares as paired data. In each case construct a scatter diagram.

Odd Numbers

Position	Value
1	1
2	3
3	5
4	7
5	9

Squares

Position	Value
1	1
2	4
3	9
4	16
5	25

SOLUTION The two scatter diagrams are based on the data from these tables shown here. Notice how the dots in Figure 4 seem to line up in a straight line, whereas the dots in Figure 5 give the impression of a curve. We say the points in Figure 4 suggest a linear relationship between the two sets of data, whereas the points in Figure 5 suggest a nonlinear relationship.

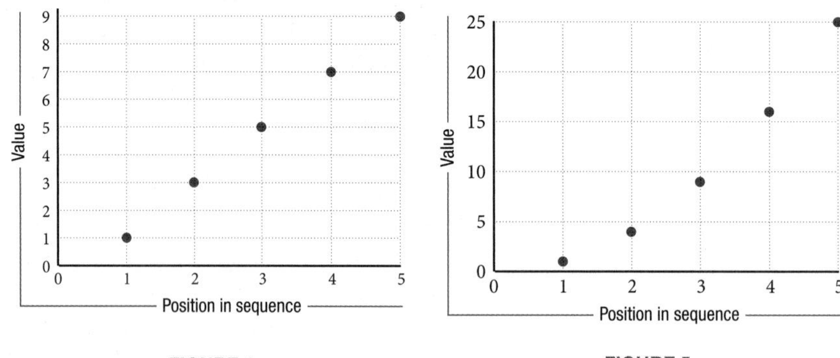

FIGURE 4

FIGURE 5

As you know, each dot in Figures 4 and 5 corresponds to a pair of numbers, one of which is associated with the horizontal axis and the other with the vertical axis. Paired data play a very important role in the equations we will solve in the next section. To prepare ourselves for those equations, we need to expand the concept of paired data to include negative numbers. At the same time, we want to standardize the position of the axes in the diagrams that we use to visualize paired data.

DEFINITION *x-coordinate, y-coordinate*

A pair of numbers enclosed in parentheses and separated by a comma, such as $(-2, 1)$, is called an ordered pair of numbers. The first number in the pair is called the x-coordinate of the ordered pair; the second number is called the y-coordinate. For the ordered pair $(-2, 1)$, the x-coordinate is -2 and the y-coordinate is 1.

Ordered pairs of numbers are important in the study of mathematics because they give us a way to visualize solutions to equations. To see the visual component of ordered pairs, we need the diagram shown in Figure 6. It is called the *rectangular coordinate system*.

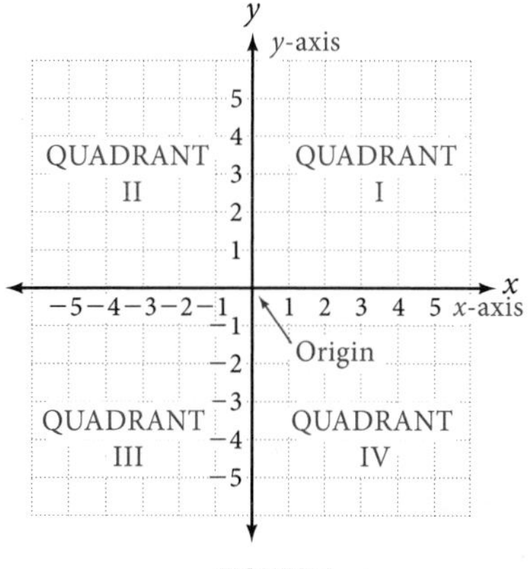

FIGURE 6

The rectangular coordinate system is built from two number lines oriented perpendicular to each other. The horizontal number line is exactly the same as our real number line and is called the *x-axis*. The vertical number line is also the same as our real number line with the positive direction up and the negative direction down. It is called the *y-axis*. The point where the two axes intersect is called the *origin*. As you can see from Figure 6, the axes divide the plane into four *quadrants*, which are numbered I through IV in a counterclockwise direction.

Graphing Ordered Pairs

To graph the ordered pair (a, b), we start at the origin and move a units forward or back (forward if a is positive and back if a is negative). Then we move b units up or down (up if b is positive, down if b is negative). The point where we end up is the graph of the ordered pair (a, b). To graph the ordered pair $(5, 2)$, we start at the origin and move 5 units to the right. Then, from that position, we move 2 units up.

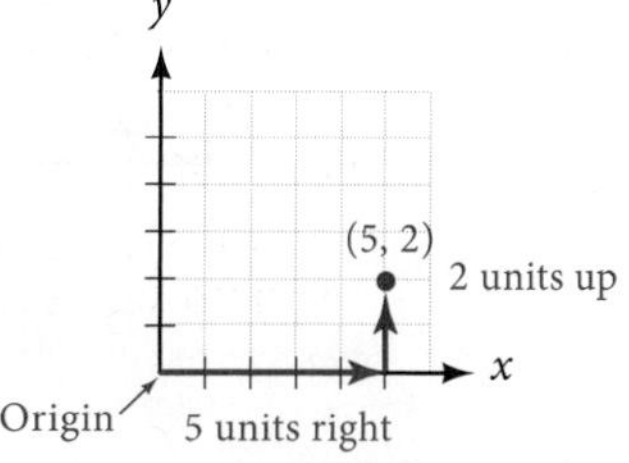

EXAMPLE 2 Graph the ordered pairs (3, 4), (3, −4), (−3, 4), and (−3, −4).

SOLUTION

Note: It is very important that you graph ordered pairs quickly and accurately. Remember, the first coordinate goes with the horizontal axis and the second coordinate goes with the vertical axis.

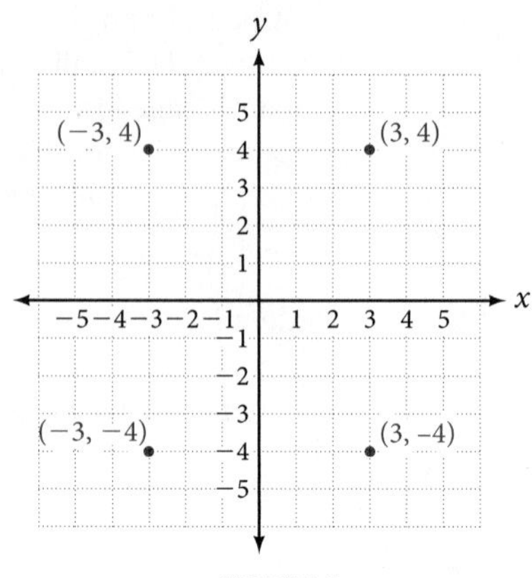

FIGURE 7

We can see in Figure 7 that when we graph ordered pairs, the x-coordinate corresponds to movement parallel to the x-axis (horizontal) and the y-coordinate corresponds to movement parallel to the y-axis (vertical).

EXAMPLE 3 Graph the ordered pairs (−1, 3), (2, 5), (0, 0), (0, −3), and (4, 0).

SOLUTION See Figure 8.

Note: If we do not label the axes of a coordinate system, we assume that each square is one unit long and one unit wide.

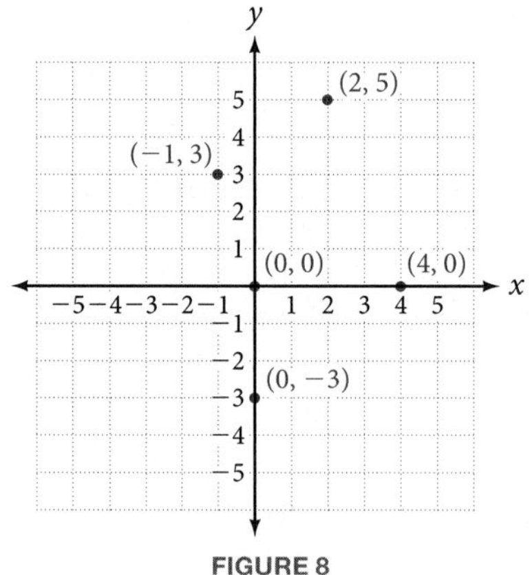

FIGURE 8

GETTING READY FOR CLASS

After reading through the preceding section, respond in your own words and in complete sentences.

A. What is an ordered pair of numbers?
B. Explain in words how you would graph the ordered pair (3, 4).
C. How do you construct a rectangular coordinate system?
D. Where is the origin on a rectangular coordinate system?

Graph the following ordered pairs.

1. $(3, 2)$ **2.** $(3, -2)$ **3.** $(-3, 2)$ **4.** $(-3, -2)$

5. $(5, 1)$ **6.** $(5, -1)$ **7.** $(1, 5)$ **8.** $(1, -5)$

9. $(-1, 5)$ **10.** $(-1, -5)$ **11.** $\left(2, \frac{1}{2}\right)$ **12.** $\left(3, \frac{3}{2}\right)$

13. $\left(-4, -\frac{5}{2}\right)$ **14.** $\left(-5, -\frac{3}{2}\right)$ **15.** $(3, 0)$ **16.** $(-2, 0)$

17. $(0, 5)$ **18.** $(0, 0)$

Give the coordinates of each numbered point in the figure.

19–28.

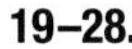

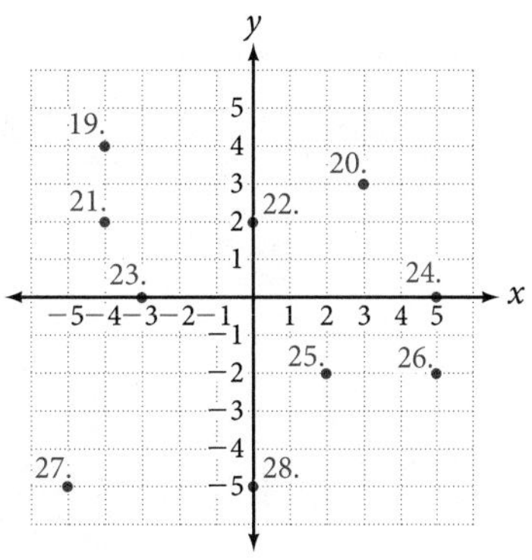

Graph the points (4, 3) and (−4, −1), and draw a straight line that passes through both of them. Then answer the following questions.

29. Does the graph of (2, 2) lie on the line?

30. Does the graph of (−2, 0) lie on the line?

31. Does the graph of (0, −2) lie on the line?

32. Does the graph of (−6, 2) lie on the line?

Graph the points (−2, 4) and (2, −4), and draw a straight line that passes through both of them. Then answer the following questions.

33. Does the graph of (0, 0) lie on the line?

34. Does the graph of (−1, 2) lie on the line?

35. Does the graph of (2, −1) lie on the line?

36. Does the graph of (1, −2) lie on the line?

Draw a straight line that passes through the points (3, 4) and (3, −4). Then answer the following questions.

37. Is the graph of (3, 0) on this line?

38. Is the graph of (0, 3) on this line?

39. Is there any point on this line with an x-coordinate other than 3?

40. If you extended the line, would it pass through a point with a y-coordinate of 10?

Draw a straight line that passes through the points (3, 4) and (−3, 4). Then answer the following questions.

41. Is the graph of (4, 0) on this line?

42. Is the graph of (0, 4) on this line?

43. Is there any point on this line with a y-coordinate other than 4?

44. If you extended the line, would it pass through a point with an x-coordinate of 10?

Applying the Concepts

45. Hourly Wages Jane takes a job at the local Marcy's department store. Her job pays $8.00 per hour. The graph shows how much Jane earns for working from 0 to 40 hours in a week.

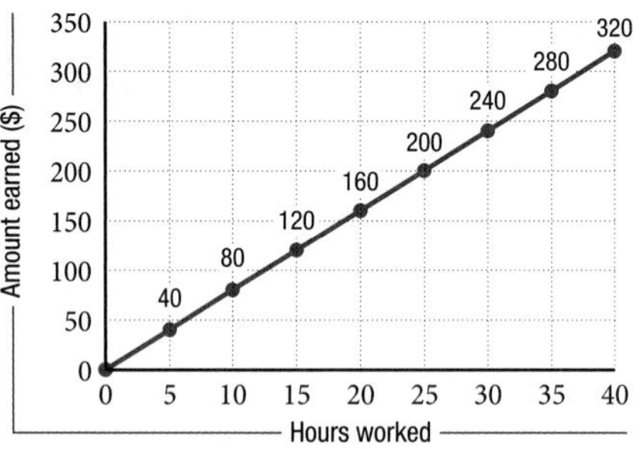

a. List three ordered pairs that lie on the line graph.

b. How much will she earn for working 40 hours?

c. If her check for one week is $240, how many hours did she work?

d. She works 35 hours one week, but her paycheck before deductions are subtracted out is for $260. Is this correct? Explain.

46. Hourly Wages Judy takes a job at Gigi's boutique. Her job pays $6.00 per hour plus $50 per week in commission. The graph shows how much Judy earns for working from 0 to 40 hours in a week.

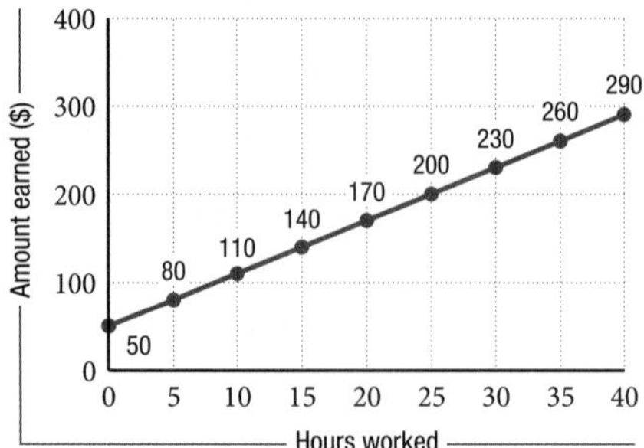

a. List three ordered pairs that lie on the line graph.

b. How much will she earn for working 40 hours?

c. If her check for one week is $230, how many hours did she work?

d. She works 35 hours one week, but her paycheck before deductions are subtracted out is for $260. Is this correct? Explain.

47. Non-Camera Phone Sales The table and bar chart shown here show what are the projected sales of non-camera phones for the years 2006–2010. Use the information from the table and chart to construct a line graph.

Year	Sales (in Millions)
2006	300
2007	250
2008	175
2009	150
2010	125

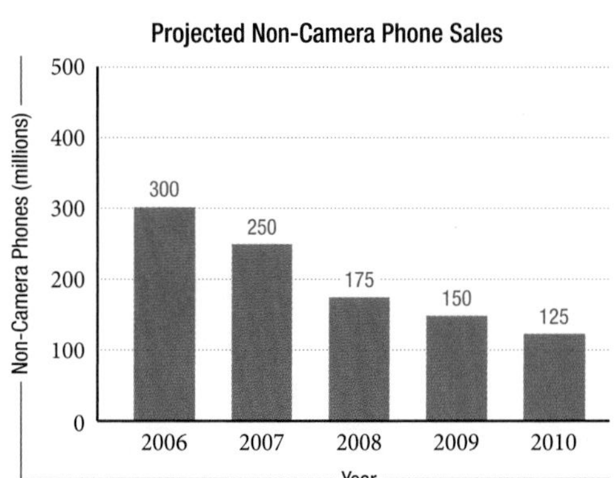

48. Camera Phone Sales The table and bar chart shown here show the projected sales of camera phones from 2006 to 2010. Use the information from the table and chart to construct a line graph.

Year	Sales (in Millions)
2006	500
2007	650
2008	750
2009	875
2010	900

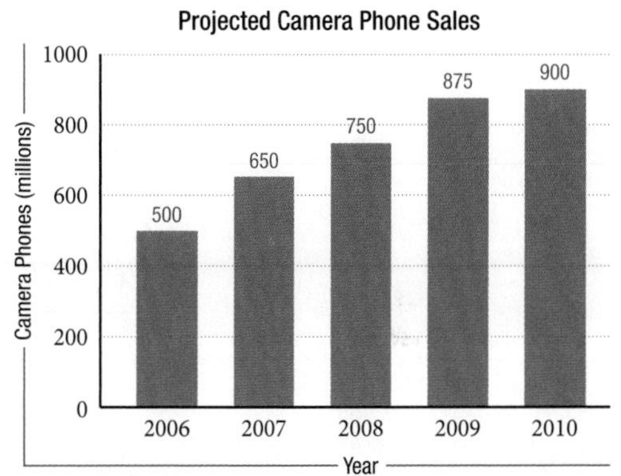

49. Kentucky Derby The line graph gives the monetary bets placed at the Kentucky Derby for specific years. If x represents the year in question and y represents the total wagering for that year, write five ordered pairs that describe the information in the table.

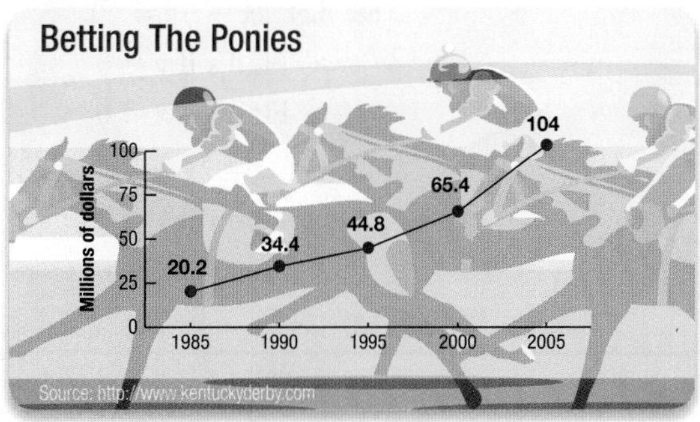

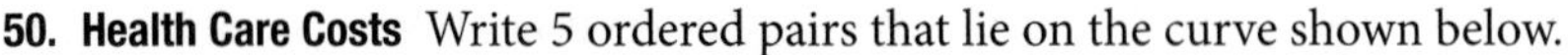

50. Health Care Costs Write 5 ordered pairs that lie on the curve shown below.

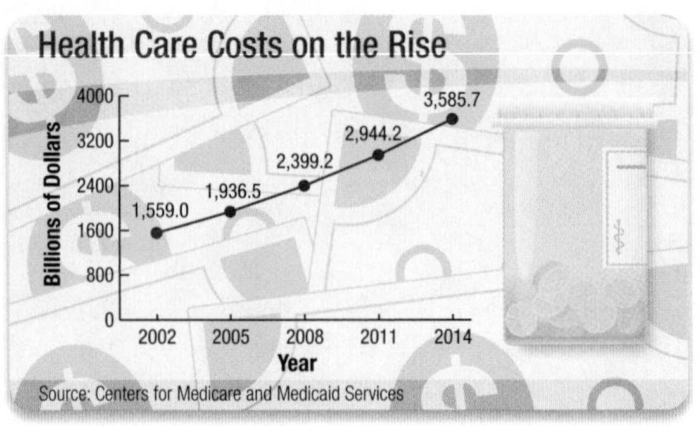

51. Right triangle *ABC* (Figure 9) has legs of length 5. Point *C* is the ordered pair (6, 2). Find the coordinates of *A* and *B*.

52. Right triangle *ABC* (Figure 10) has legs of length 7. Point *C* is the ordered pair $(-8, -3)$. Find the coordinates of *A* and *B*.

53. Rectangle *ABCD* (Figure 11) has a length of 5 and a width of 3. Point *D* is the ordered pair (7, 2). Find points *A*, *B*, and *C*.

54. Rectangle *ABCD* (Figure 12) has a length of 5 and a width of 3. Point *D* is the ordered pair $(-1, 1)$. Find points *A*, *B*, and *C*.

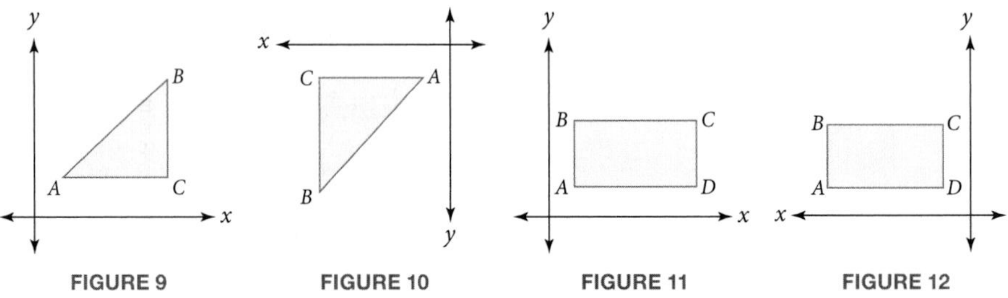

FIGURE 9 FIGURE 10 FIGURE 11 FIGURE 12

Getting Ready for the Next Section

55. Let $2x + 3y = 6$

a. Find x if $y = 4$

b. Find x if $y = -2$

c. Find y if $x = 3$

d. Find y if $x = 9$

56. Let $2x - 5y = 20$

a. Find x if $y = 0$

b. Find x if $y = -6$

c. Find y if $x = 0$

d. Find y if $x = 5$

57. Let $y = 2x - 1$

a. Find x if $y = 7$

b. Find x if $y = 3$

c. Find y if $x = 0$

d. Find y if $x = 5$

58. Let $y = 3x - 2$

a. Find x if $y = 4$

b. Find x if $y = 3$

c. Find y if $x = 2$

d. Find y if $x = -3$

Solutions to Linear Equations in Two Variables 2.2

In this section we will begin to investigate equations in two variables. As you will see, equations in two variables have pairs of numbers for solutions. Because we know how to use paired data to construct tables, histograms, and other charts, we can take our work with paired data further by using equations in two variables to construct tables of paired data. Let's begin this section by reviewing the relationship between equations in one variable and their solutions.

If we solve the equation $3x - 2 = 10$, the solution is $x = 4$. If we graph this solution, we simply draw the real number line and place a dot at the point whose coordinate is 4. The relationship between linear equations in one variable, their solutions, and the graphs of those solutions look like this:

Equation	*Solution*	*Graph of Solution Set*
$3x - 2 = 10$	$x = 4$	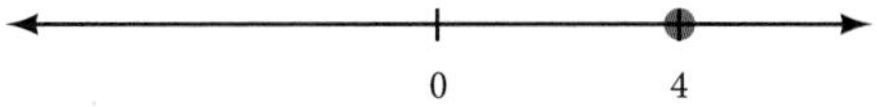
$x + 5 = 7$	$x = 2$	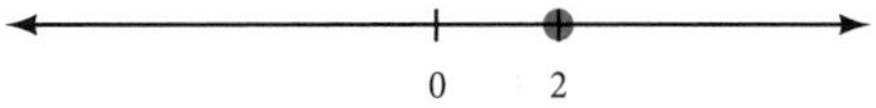
$2x = -6$	$x = -3$	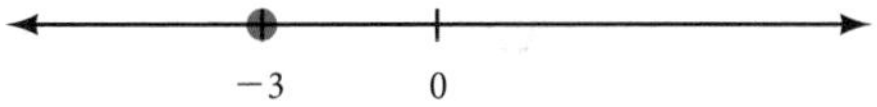

Note: If this discussion seems a little long and confusing, you may want to look over some of the examples first and then come back and read this. Remember, it isn't always easy to read material in mathematics. What is important is that you understand what you are doing when you work problems. The reading is intended to assist you in understanding what you are doing. It is important to read everything in the book, but you don't always have to read it in the order it is written.

When the equation has one variable, the solution is a single number whose graph is a point on a line.

Now, consider the equation $2x + y = 3$. The first thing we notice is that there are two variables instead of one. Therefore, a solution to the equation $2x + y = 3$ will be not a single number but a pair of numbers, one for x and one for y, that makes the equation a true statement. One pair of numbers that works is $x = 2$, $y = -1$ because when we substitute them for x and y in the equation, we get a true statement.

$$2(2) + (-1) \stackrel{?}{=} 3$$
$$4 - 1 = 3$$
$$3 = 3 \quad \text{A true statement}$$

The pair of numbers $x = 2, y = -1$ is written as $(2, -1)$. As you know from Section 3.1, $(2, -1)$ is called an *ordered pair* because it is a pair of numbers written in a specific order. The first number is always associated with the variable x, and the second number is always associated with the variable y. We call the first number in the ordered pair the *x-coordinate* (or x component) and the second number the *y-coordinate* (or y component) of the ordered pair.

Let's look back to the equation $2x + y = 3$. The ordered pair $(2, -1)$ is not the only solution. Another solution is $(0, 3)$ because when we substitute 0 for x and 3 for y we get

$$2(0) + 3 \stackrel{?}{=} 3$$
$$0 + 3 = 3$$
$$3 = 3 \quad \text{A true statement}$$

Still another solution is the ordered pair $(5, -7)$ because

$$2(5) + (-7) \stackrel{?}{=} 3$$
$$10 - 7 = 3$$
$$3 = 3 \quad \text{A true statement}$$

As a matter of fact, for any number we want to use for x, there is another number we can use for y that will make the equation a true statement. There is an infinite number of ordered pairs that satisfy (are solutions to) the equation $2x + y = 3$; we have listed just a few of them.

EXAMPLE 1 Given the equation $2x + 3y = 6$, complete the following ordered pairs so they will be solutions to the equation: (0,), (, 1), (3,).

SOLUTION To complete the ordered pair (0,), we substitute 0 for x in the equation and then solve for y:

$$2(0) + 3y = 6$$
$$3y = 6$$
$$y = 2$$

The ordered pair is (0, 2).

To complete the ordered pair (, 1), we substitute 1 for y in the equation and solve for x:

$$2x + 3(1) = 6$$
$$2x + 3 = 6$$
$$2x = 3$$
$$x = \frac{3}{2}$$

The ordered pair is $\left(\frac{3}{2}, 1\right)$.

To complete the ordered pair (3,), we substitute 3 for x in the equation and solve for y:

$$2(3) + 3y = 6$$
$$6 + 3y = 6$$
$$3y = 0$$
$$y = 0$$

The ordered pair is (3, 0).

Notice in each case that once we have used a number in place of one of the variables, the equation becomes a linear equation in one variable. We then use the method explained in Chapter 2 to solve for that variable.

EXAMPLE 2 Complete the following table for the equation $2x - 5y = 20$.

x	y
0	
	2
	0
−5	

SOLUTION Filling in the table is equivalent to completing the following ordered pairs: (0,), (, 2), (, 0), (−5,). So we proceed as in Example 1.

When $x = 0$, we have

$$2(0) - 5y = 20$$
$$0 - 5y = 20$$
$$-5y = 20$$
$$y = -4$$

When $y = 2$, we have

$$2x - 5(2) = 20$$
$$2x - 10 = 20$$
$$2x = 30$$
$$x = 15$$

When $y = 0$, we have

$$2x - 5(0) = 20$$
$$2x - 0 = 20$$
$$2x = 20$$
$$x = 10$$

When $x = -5$, we have

$$2(-5) - 5y = 20$$
$$-10 - 5y = 20$$
$$-5y = 30$$
$$y = -6$$

The completed table looks like this:

x	y
0	−4
15	2
10	0
−5	−6

which is equivalent to the ordered pairs (0, −4), (15, 2), (10, 0), and (−5, −6).

EXAMPLE 3 Complete the following table for the equation $y = 2x - 1$.

x	y
0	
5	
	7
	3

SOLUTION When $x = 0$, we have

$$y = 2(0) - 1$$
$$y = 0 - 1$$
$$y = -1$$

When $x = 5$, we have

$$y = 2(5) - 1$$
$$y = 10 - 1$$
$$y = 9$$

When $y = 7$, we have

$$7 = 2x - 1$$
$$8 = 2x$$
$$4 = x$$

When $y = 3$, we have

$$3 = 2x - 1$$
$$4 = 2x$$
$$2 = x$$

The completed table is

x	y
0	−1
5	9
4	7
2	3

which means the ordered pairs $(0, -1)$, $(5, 9)$, $(4, 7)$, and $(2, 3)$ are among the solutions to the equation $y = 2x - 1$.

EXAMPLE 4 Which of the ordered pairs $(2, 3)$, $(1, 5)$, and $(-2, -4)$ are solutions to the equation $y = 3x + 2$?

SOLUTION If an ordered pair is a solution to the equation, then it must satisfy the equation; that is, when the coordinates are used in place of the variables in the equation, the equation becomes a true statement.

Try $(2, 3)$ in $y = 3x + 2$:

$$3 \stackrel{?}{=} 3(2) + 2$$
$$3 = 6 + 2$$
$$3 = 8 \qquad \text{A false statement}$$

Try $(1, 5)$ in $y = 3x + 2$:

$$5 \stackrel{?}{=} 3(1) + 2$$
$$5 = 3 + 2$$
$$5 = 5 \qquad \text{A true statement}$$

Try $(-2, -4)$ in $y = 3x + 2$:

$$-4 \stackrel{?}{=} 3(-2) + 2$$
$$-4 = -6 + 2$$
$$-4 = -4 \qquad \text{A true statement}$$

The ordered pairs $(1, 5)$ and $(-2, -4)$ are solutions to the equation $y = 3x + 2$, and $(2, 3)$ is not.

GETTING READY FOR CLASS

After reading through the preceding section, respond in your own words and in complete sentences.

A. How can you tell if an ordered pair is a solution to an equation?
B. How would you find a solution to $y = 3x - 5$?
C. Why is $(3, 2)$ not a solution to $y = 3x - 5$?
D. How many solutions are there to an equation that contains two variables?

For each equation, complete the given ordered pairs.

1. $2x + y = 6$ $(0, \ ), (\ , 0), (\ , -6)$

2. $3x - y = 5$ $(0, \ ), (1, \ ), (\ , 5)$

3. $3x + 4y = 12$ $(0, \ ), (\ , 0), (-4, \ )$

4. $5x - 5y = 20$ $(0, \ ), (\ , -2), (1, \ )$

5. $y = 4x - 3$ $(1, \ ), (\ , 0), (5, \ )$

6. $y = 3x - 5$ $(\ , 13), (0, \ ), (-2, \ )$

7. $y = 7x - 1$ $(2, \ ), (\ , 6), (0, \ )$

8. $y = 8x + 2$ $(3, \ ), (\ , 0), (\ , -6)$

9. $x = -5$ $(\ , 4), (\ , -3), (\ , 0)$

10. $y = 2$ $(5, \ ), (-8, \ ), \left(\frac{1}{2}, \ \right)$

For each of the following equations, complete the given table.

11. $y = 3x$

x	y
1	3
−3	
	12
	18

12. $y = -2x$

x	y
−4	
0	
	10
	12

13. $y = 4x$

x	y
0	
	−2
−3	
	12

14. $y = -5x$

x	y
3	
	0
−2	
	−20

15. $x + y = 5$

x	y
2	
3	
	0
	−4

16. $x - y = 8$

x	y
0	
4	
	−3
	−2

17. $2x - y = 4$

x	y
	0
	2
1	
−3	

18. $3x - y = 9$

x	y
	0
	−9
5	
−4	

19. $y = 6x - 1$

x	y
0	
	−7
−3	
	8

20. $y = 5x + 7$

x	y
0	
−2	
−4	
	−8

For the following equations, tell which of the given ordered pairs are solutions.

21. $2x - 5y = 10$ $(2, 3), (0, -2), \left(\frac{5}{2}, 1\right)$

22. $3x + 7y = 21$ $(0, 3), (7, 0), (1, 2)$

23. $y = 7x - 2$ $(1, 5), (0, -2), (-2, -16)$

24. $y = 8x - 3$ $(0, 3), (5, 16), (1, 5)$

25. $y = 6x$ $(1, 6), (-2, 12), (0, 0)$

26. $y = -4x$ $(0, 0), (2, 4), (-3, 12)$

27. $x + y = 0$ $(1, 1), (2, -2), (3, 3)$

28. $x - y = 1$ $(0, 1), (0, -1), (1, 2)$

29. $x = 3$ $(3, 0), (3, -3), (5, 3)$

30. $y = -4$ $(3, -4), (-4, 4), (0, -4)$

Applying the Concepts

31. Perimeter If the perimeter of a rectangle is 30 inches, then the relationship between the length l and the width w is given by the equation

$$2l + 2w = 30$$

What is the length when the width is 3 inches?

32. Perimeter The relationship between the perimeter P of a square and the length of its side s is given by the formula $P = 4s$. If each side of a square is 5 inches, what is the perimeter? If the perimeter of a square is 28 inches, how long is a side?

33. Janai earns \$12 per hour working as a math tutor. We can express the amount she earns each week, y, for working x hours with the equation $y = 12x$. Indicate with a yes or no, which of the following could be one of Janai's paychecks. If you answer no, explain your answer.

a. \$60 for working five hours. **b.** \$100 for working nine hours
c. \$80 for working seven hours. **d.** \$168 for working 14 hours

34. Erin earns \$15 per hour working as a graphic designer. We can express the amount she earns each week, y, for working x hours with the equation $y = 15x$. Indicate with a yes or no which of the following could be one of Erin's paychecks. If you answer no, explain your answer.

a. \$75 for working five hours. **b.** \$125 for working nine hours
c. \$90 for working six hours. **d.** \$500 for working 35 hours

35. The equation $V = -45{,}000t + 600{,}000$, can be used to find the value, V, of a small crane at the end of t years.

a. What is the value of the crane at the end of five years?
b. When is the crane worth \$330,000?
c. Is it true that the crane with be worth \$150,000 after nine years?
d. How much did the crane cost?

36. The equation $P = -400t + 2{,}500$, can be used to find the price, P, of a notebook computer at the end of t years.

a. What is the value of the notebook computer at the end of four years?
b. When is the notebook computer worth \$1,700?
c. Is it true that the notebook computer with be worth \$100 after five years?
d. How much did the notebook computer cost?

Getting Ready for the Next Section

37. Find y when x is 4 in the formula $3x + 2y = 6$.

38. Find y when x is 0 in the formula $3x + 2y = 6$.

39. Find y when x is 0 in $y = -\frac{1}{3}x + 2$. **40.** Find y when x is 3 in $y = -\frac{1}{3}x + 2$.

41. Find y when x is 2 in $y = \frac{3}{2}x - 3$. **42.** Find y when x is 4 in $y = \frac{3}{2}x - 3$.

43. Solve $5x + y = 4$ for y. **44.** Solve $-3x + y = 5$ for y.

45. Solve $3x - 2y = 6$ for y. **46.** Solve $2x - 3y = 6$ for y.

Graphing Linear Equations in Two Variables

2.3

In this section we will use the rectangular coordinate system introduced in Section 3.1 to obtain a visual picture of *all* solutions to a linear equation in two variables. The process we use to obtain a visual picture of all solutions to an equation is called *graphing*. The picture itself is called the *graph* of the equation.

EXAMPLE 1 Graph the solution set for $x + y = 5$.

SOLUTION We know from the previous section that an infinite number of ordered pairs are solutions to the equation $x + y = 5$. We can't possibly list them all. What we can do is list a few of them and see if there is any pattern to their graphs.

Some ordered pairs that are solutions to $x + y = 5$ are (0, 5), (2, 3), (3, 2), (5, 0). The graph of each is shown in Figure 1.

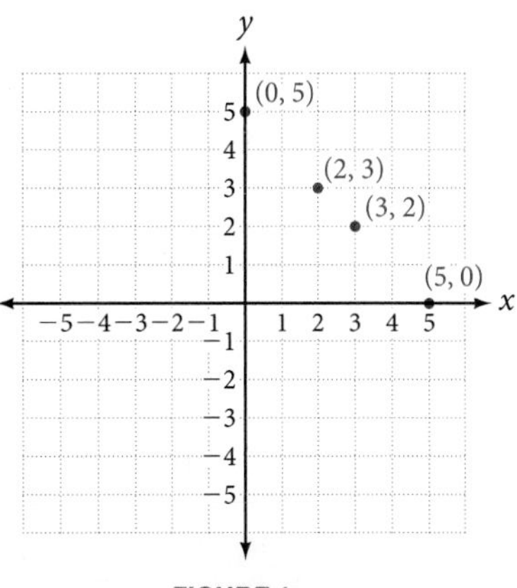

FIGURE 1

Now, by passing a straight line through these points we can graph the solution set for the equation $x + y = 5$. Linear equations in two variables always have graphs that are straight lines. The graph of the solution set for $x + y = 5$ is shown in Figure 2.

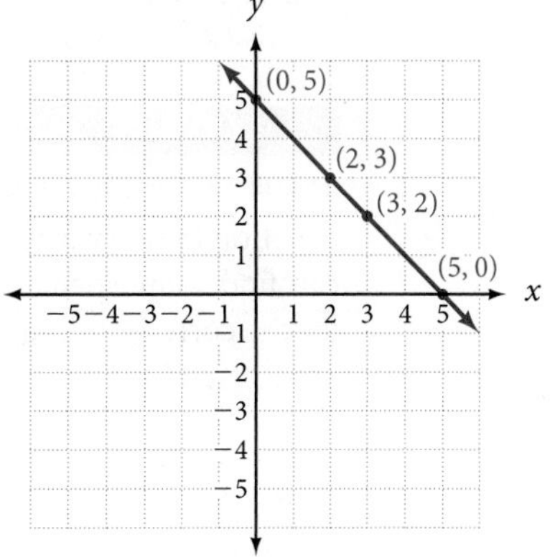

FIGURE 2

Every ordered pair that satisfies $x + y = 5$ has its graph on the line, and any point on the line has coordinates that satisfy the equation. So, there is a one-to-one correspondence between points on the line and solutions to the equation.

Our ability to graph an equation as we have done in Example 1 is due to the invention of the rectangular coordinate system. The French philosopher René Descartes (1595–1650) is the person usually credited with the invention of the rectangular coordinate system. As a philosopher, Descartes is responsible for the statement "I think, therefore I am." Until Descartes invented his coordinate system in 1637, algebra and geometry were treated as separate subjects. The rectangular coordinate system allows us to connect algebra and geometry by associating geometric shapes with algebraic equations.

Here is the precise definition for a linear equation in two variables.

DEFINITION *Linear Equation in Two Variables, Standard Form*

Any equation that can be put in the form $ax + by = c$, where a, b, and c are real numbers and a and b are not both 0, is called a ***linear equation in two variables.*** The graph of any equation of this form is a straight line (that is why these equations are called "linear"). The form $ax + by = c$ is called ***standard form.***

To graph a linear equation in two variables, we simply graph its solution set; that is, we draw a line through all the points whose coordinates satisfy the equation. Here are the steps to follow.

PROPERTY *To Graph a Linear Equation in Two Variables*

Step 1: Find any three ordered pairs that satisfy the equation. This can be done by using a convenient number for one variable and solving for the other variable.

Step 2: Graph the three ordered pairs found in step 1. Actually, we need only two points to graph a straight line. The third point serves as a check. If all three points do not line up, there is a mistake in our work.

Step 3: Draw a straight line through the three points graphed in step 2.

Note: The meaning of the convenient numbers referred to in step 1 will become clear as you read the next two examples.

EXAMPLE 2 Graph the equation $y = 3x - 1$.

SOLUTION Because $y = 3x - 1$ can be put in the form $ax + by = c$, it is a linear equation in two variables. Hence, the graph of its solution set is a straight line. We can find some specific solutions by substituting numbers for x and then solving for the corresponding values of y. We are free to choose any numbers for x, so let's use 0, 2, and -1.

Note: It may seem that we have simply picked the numbers 0, 2, and -1 out of the air and used them for x. In fact we have done just that. Could we have used numbers other than these? The answer is yes, we can substitute any number for x; there will always be a value of y to go with it.

Let $x = 0$: $y = 3(0) - 1$

$y = 0 - 1$

$y = -1$

The ordered pair $(0, -1)$ is one solution.

Let $x = 2$: $y = 3(2) - 1$

$y = 6 - 1$

$y = 5$

The ordered pair $(2, 5)$ is a second solution.

Let $x = -1$: $y = 3(-1) - 1$

$y = -3 - 1$

$y = -4$

The ordered pair $(-1, -4)$ is a third solution.

In table form

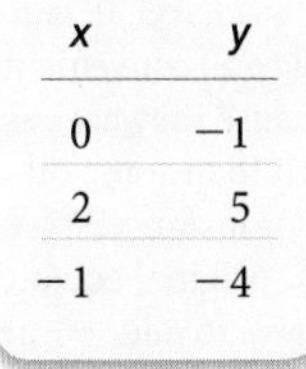

x	y
0	-1
2	5
-1	-4

Next, we graph the ordered pairs $(0, -1)$, $(2, 5)$, $(-1, -4)$ and draw a straight line through them.

The line we have drawn in Figure 3 is the graph of $y = 3x - 1$.

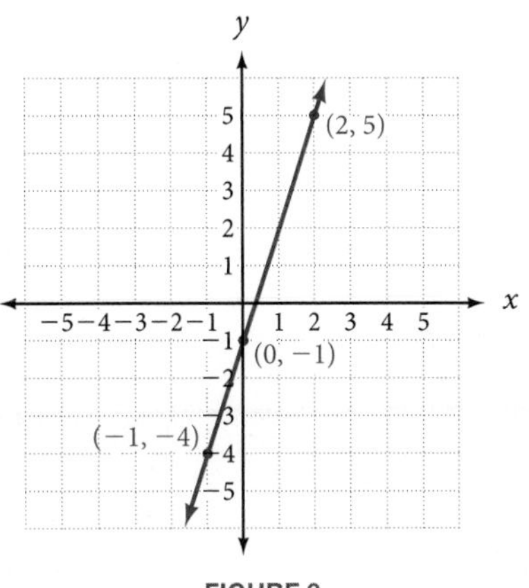

FIGURE 3

Example 2 again illustrates the connection between algebra and geometry that we mentioned previously. Descartes' rectangular coordinate system allows us to associate the equation $y = 3x - 1$ (an algebraic concept) with a specific straight line (a geometric concept). The study of the relationship between equations in algebra and their associated geometric figures is called *analytic geometry*. The rectangular coordinate system often is referred to as the *Cartesian coordinate system* in honor of Descartes.

EXAMPLE 3 Graph the equation $y = -\frac{1}{3}x + 2$.

SOLUTION We need to find three ordered pairs that satisfy the equation. To do so, we can let x equal any numbers we choose and find corresponding values of y. But, every value of x we substitute into the equation is going to be multiplied by $-\frac{1}{3}$. Let's use numbers for x that are divisible by 3, like -3, 0, and 3. That way, when we multiply them by $-\frac{1}{3}$, the result will be an integer.

> *Note:* In Example 3 the values of x we used, -3, 0, and 3, are referred to as convenient values of x because they are easier to work with than some other numbers. For instance, if we let $x = 2$ in our original equation, we would have to add $-\frac{2}{3}$ and 2 to find the corresponding value of y. Not only would the arithmetic be more difficult but also the ordered pair we obtained would have a fraction for its y-coordinate, making it more difficult to graph accurately.

Let $x = -3$: $\quad y = -\frac{1}{3}(-3) + 2$

$$y = 1 + 2$$

$$y = 3$$

The ordered pair $(-3, 3)$ is one solution.

Let $x = 0$: $\quad y = -\frac{1}{3}(0) + 2$

$$y = 0 + 2$$

$$y = 2$$

The ordered pair $(0, 2)$ is a second solution.

Let $x = 3$: $\quad y = -\frac{1}{3}(3) + 2$

$$y = -1 + 2$$

$$y = 1$$

The ordered pair $(3, 1)$ is a third solution.

In table form

x	y
-3	3
0	2
3	1

Graphing the ordered pairs $(-3, 3)$, $(0, 2)$, and $(3, 1)$ and drawing a straight line through their graphs, we have the graph of the equation $y = -\frac{1}{3}x + 2$, as shown in Figure 4.

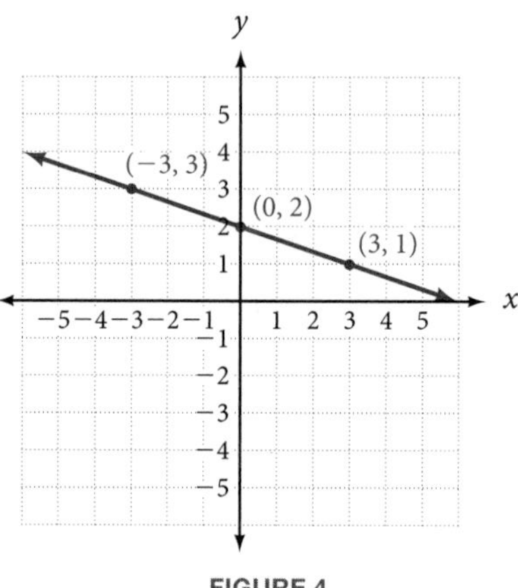

FIGURE 4

EXAMPLE 4 Graph the solution set for $3x - 2y = 6$.

SOLUTION It will be easier to find convenient values of x to use in the equation if we first solve the equation for y. To do so, we add $-3x$ to each side, and then we multiply each side by $-\frac{1}{2}$.

$3x - 2y = 6$	Original equation
$-2y = -3x + 6$	Add $-3x$ to each side
$-\frac{1}{2}(-2y) = -\frac{1}{2}(-3x + 6)$	Multiply each side by $-\frac{1}{2}$
$y = \frac{3}{2}x - 3$	Simplify each side

Now, because each value of x will be multiplied by $\frac{3}{2}$, it will be to our advantage to choose values of x that are divisible by 2. That way, we will obtain values of y that do not contain fractions. This time, let's use 0, 2, and 4 for x.

When $x = 0$: $\quad y = \frac{3}{2}(0) - 3$

$$y = 0 - 3$$

$$y = -3$$

The ordered pair (0, −3) is one solution.

When $x = 2$: $\quad y = \frac{3}{2}(2) - 3$

$$y = 3 - 3$$

$$y = 0$$

The ordered pair (2, 0) is a second solution.

When $x = 4$: $\quad y = \frac{3}{2}(4) - 3$

$$y = 6 - 3$$

$$y = 3$$

The ordered pair (4, 3) is a third solution

Graphing the ordered pairs (0, −3), (2, 0), and (4, 3) and drawing a line through them, we have the graph shown in Figure 5.

Note: After reading through Example 4, many students ask why we didn't use -2 for x when we were finding ordered pairs that were solutions to the original equation. The answer is, we could have. If we were to let $x = -2$, the corresponding value of y would have been -6. As you can see by looking at the graph in Figure 5, the ordered pair $(-2, -6)$ is on the graph.

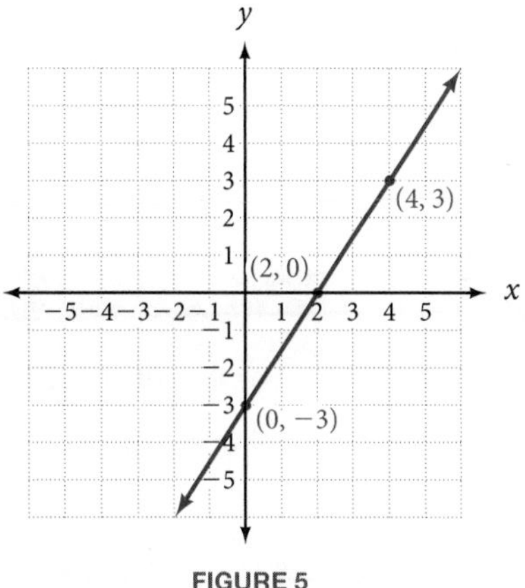

FIGURE 5

EXAMPLE 5 Graph each of the following lines.

a. $y = \frac{1}{2}x$ **b.** $x = 3$ **c.** $y = -2$

SOLUTION

a. The line $y = \frac{1}{2}x$ passes through the origin because (0, 0) satisfies the equation. To sketch the graph we need at least one more point on the line. When x is 2, we obtain the point (2, 1), and when x is -4, we obtain the point $(-4, -2)$. The graph of $y = \frac{1}{2}x$ is shown in Figure 6A.

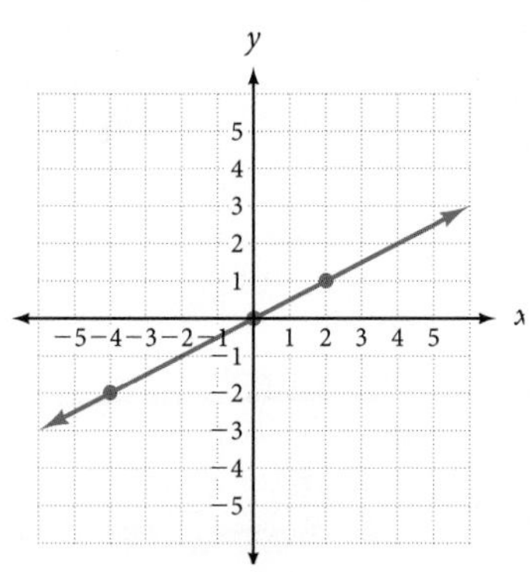

FIGURE 6A

b. The line $x = 3$ is the set of all points whose does not appear in the equation, so the y-coordinate can be any number. Note that we can write our equation as a linear equation in two variables by writing it as $x + 0y = 3$. Because the product of 0 and y will always be 0, y can be any number. The graph of $x = 3$ is the vertical line shown in Figure 6B.

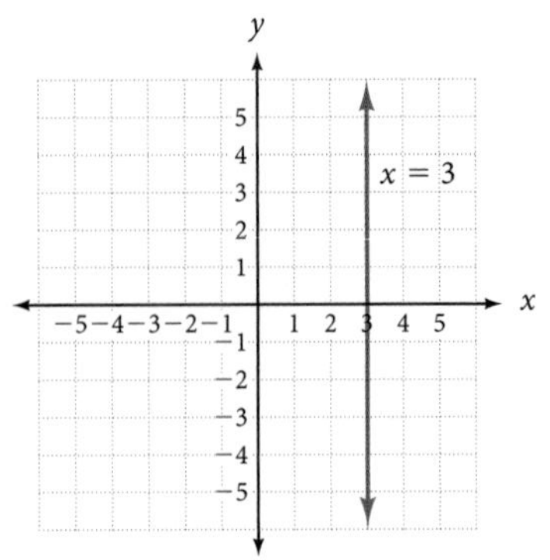

FIGURE 6B

c. The line $y = -2$ is the set of all points whose y-coordinate is -2. The variable x does not appear in the equation, so the x-coordinate can be any number. Again, we can write our equation as a linear equation in two variables by writing it as $0x + y = -2$. Because the product of 0 and x will always be 0, x can be any number. The graph of $y = -2$ is the horizontal line shown in Figure 6C.

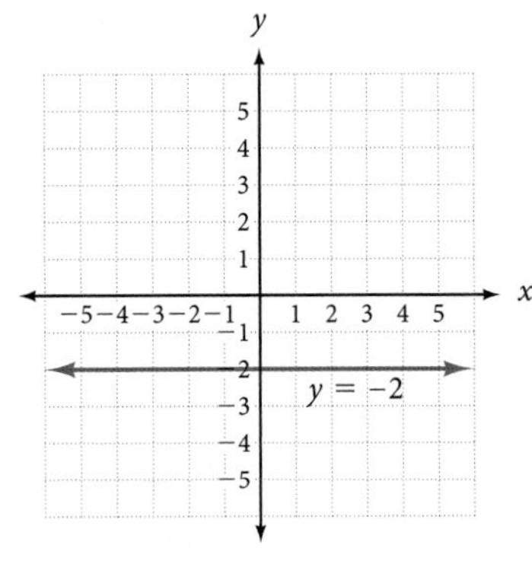

FIGURE 6C

FACTS FROM GEOMETRY *Special Equations and Their Graphs*

For the equations below, m, a, and b are real numbers.

Through the Origin

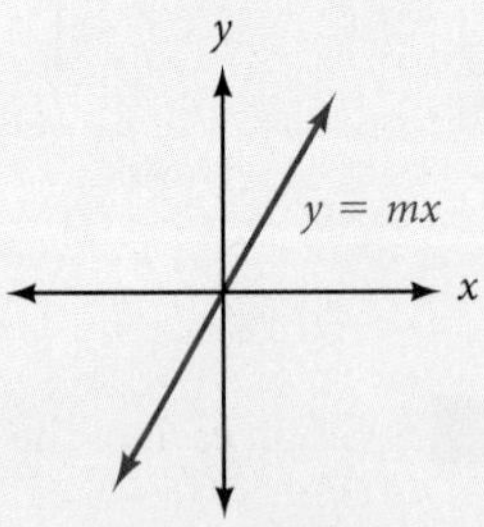

$y = mx$

FIGURE 7A Any equation of the form $y = mx$ has a graph that passes through the origin.

Vertical Line

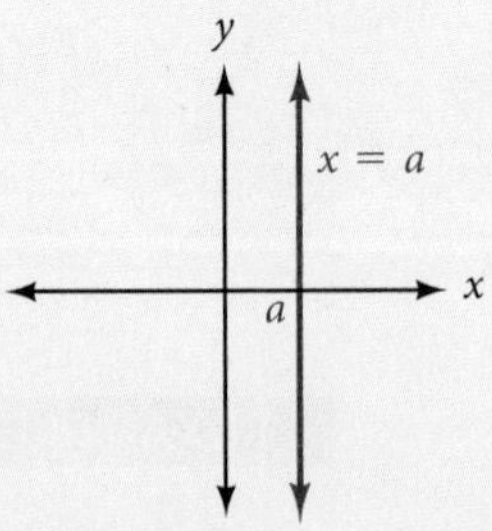

$x = a$

FIGURE 7B Any equation of the form $x = a$ has a vertical line for its graph.

Horizontal Line

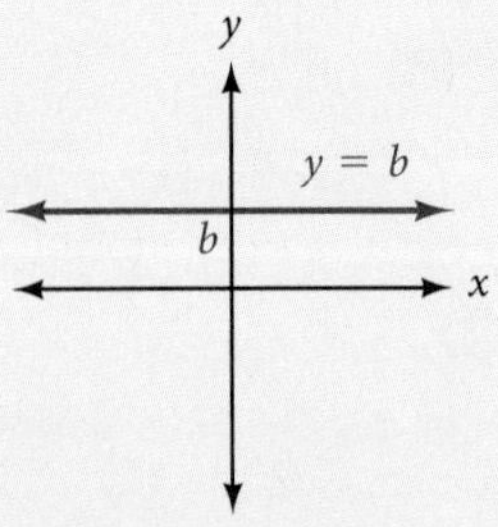

$y = b$

FIGURE 7C Any equation of the form $y = b$ has a horizontal line for its graph.

GETTING READY FOR CLASS

After reading through the preceding section, respond in your own words and in complete sentences.

A. Explain how you would go about graphing the line $x + y = 5$.

B. When graphing straight lines, why is it a good idea to find three points, when every straight line is determined by only two points?

C. What kind of equations have vertical lines for graphs?

D. What kind of equations have horizontal lines for graphs?

SPOTLIGHT ON SUCCESS *University of North Alabama*

Pride is a personal commitment.
It is an attitude which separates excellence from mediocrity.
—William Blake

The University of Northern Alabama places its Pride Rock, a 60-pound granite stone engraved with a lion's paw print, behind the north end zone at all home football games. The rock reminds current Lion players of the proud athletic traditions that has been established at the school, and to take pride in their efforts on the field.

Photo courtesy UNA

The same idea holds true for your work in your math class. Take pride in it. When you turn in an assignment, it should be accurate and easy for the instructor to read. It shows that you care about your progress in the course and that you take pride in your work. The work that you turn in to your instructor is a reflection of you. As the quote from William Blake indicates, pride is a personal commitment; a decision that you make, yourself. And once you make that commitment to take pride in the work you do in your math class, you have directed yourself toward excellence, and away from mediocrity.

Problem Set 2.3

For the following equations, complete the given ordered pairs, and use the results to graph the solution set for the equation.

1. $x + y = 4 \quad (0,\ ), (2,\ ), (\ , 0)$
2. $x - y = 3 \quad (0,\ ), (2,\ ), (\ , 0)$
3. $x + y = 3 \quad (0,\ ), (2,\ ), (\ , -1)$
4. $x - y = 4 \quad (1,\ ), (-1,\ ), (\ , 0)$
5. $y = 2x \quad (0,\ ), (-2,\ ), (2,\ )$
6. $y = \frac{1}{2}x \quad (0,\ ), (-2,\ ), (2,\ )$
7. $y = \frac{1}{3}x \quad (-3,\ ), (0,\ ), (3,\ )$
8. $y = 3x \quad (-2,\ ), (0,\ ), (2,\ )$
9. $y = 2x + 1 \quad (0,\ ), (-1,\ ), (1,\ )$
10. $y = -2x + 1 \quad (0,\ ), (-1,\ ), (1,\ )$
11. $y = 4 \quad (0,\ ), (-1,\ ), (2,\ )$
12. $x = 3 \quad (\ , -2), (\ , 0), (\ , 5)$
13. $y = \frac{1}{2}x + 3 \quad (-2,\ ), (0,\ ), (2,\ )$
14. $y = \frac{1}{2}x - 3 \quad (-2,\ ), (0,\ ), (2,\ )$
15. $y = -\frac{2}{3}x + 1 \quad (-3,\ ), (0,\ ), (3,\ )$
16. $y = -\frac{2}{3}x - 1 \quad (-3,\ ), (0,\ ), (3,\ )$

Solve each equation for *y*. Then, complete the given ordered pairs, and use them to draw the graph.

17. $2x + y = 3 \quad (-1,\ ), (0,\ ), (1,\ )$
18. $3x + y = 2 \quad (-1,\ ), (0,\ ), (1,\ )$
19. $3x + 2y = 6 \quad (0,\ ), (2,\ ), (4,\ )$
20. $2x + 3y = 6 \quad (0,\ ), (3,\ ), (6,\ )$
21. $-x + 2y = 6 \quad (-2,\ ), (0,\ ), (2,\ )$
22. $-x + 3y = 6 \quad (-3,\ ), (0,\ ), (3,\ )$

Find three solutions to each of the following equations, and then graph the solution set.

23. $y = -\frac{1}{2}x$
24. $y = -2x$
25. $y = 3x - 1$
26. $y = -3x - 1$
27. $-2x + y = 1$
28. $-3x + y = 1$
29. $3x + 4y = 8$
30. $3x - 4y = 8$
31. $x = -2$
32. $y = 3$
33. $y = 2$
34. $x = -3$

Graph each equation.

35. $y = \frac{3}{4}x + 1$
36. $y = \frac{2}{3}x + 1$
37. $y = \frac{1}{3}x + \frac{2}{3}$
38. $y = \frac{1}{2}x + \frac{1}{2}$
39. $y = \frac{2}{3}x + \frac{2}{3}$
40. $y = -\frac{3}{4}x + \frac{3}{2}$

For each equation in each table below, indicate whether the graph is horizontal (H), or vertical (V), or whether it passes through the origin (O).

41.

Equation	H, V, and/or O
$x = 3$	
$y = 3$	
$y = 3x$	
$y = 0$	

42.

Equation	H, V, and/or O
$x = \frac{1}{2}$	
$y = \frac{1}{2}$	
$y = \frac{1}{2}x$	
$x = 0$	

43.

Equation	H, V, and/or O
$x = -\frac{3}{5}$	
$y = -\frac{3}{5}$	
$y = -\frac{3}{5}x$	
$x = 0$	

44.

Equation	H, V, and/or O
$x = -4$	
$y = -4$	
$y = -4x$	
$y = 0$	

45. Use the graph at the right to complete the table.

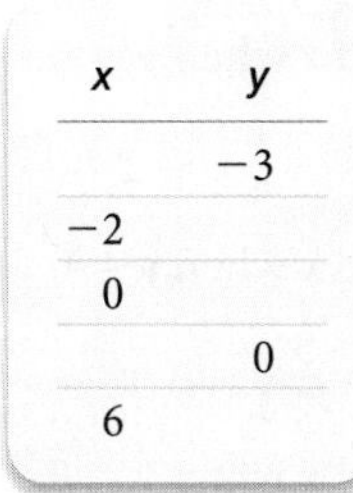

x	y
	−3
−2	
0	
	0
6	

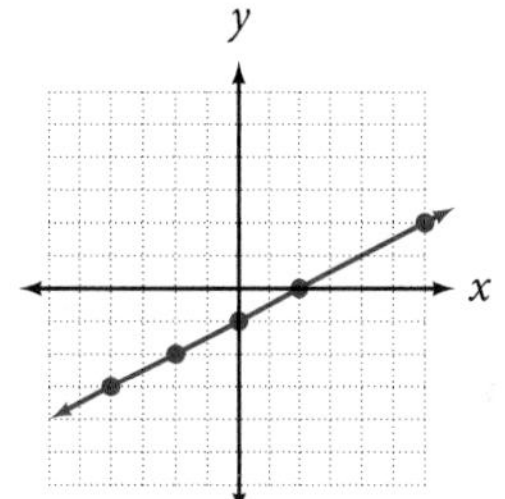

46. Use the graph at the right to complete the table. (*Hint:* Some parts have two answers.)

x	y
−3	6
	4
0	3
	1
6	

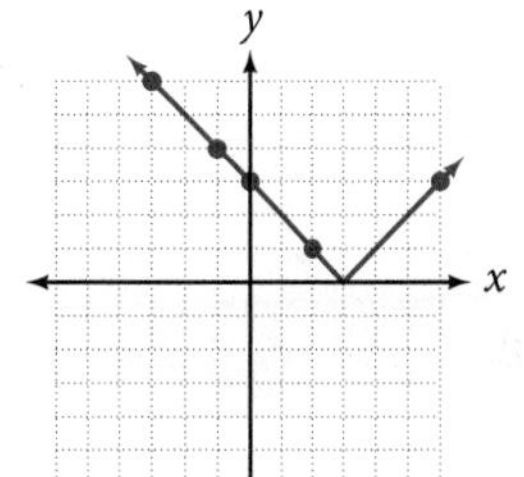

The next two problems are intended to give you practice reading, and paying attention to, the instructions that accompany the problems you are working. Working these problems is an excellent way to get ready for a test or a quiz.

47. Work each problem according to the instructions given.

a. Solve: $2x + 5 = 10$
b. Find x when y is 0: $2x + 5y = 10$
c. Find y when x is 0: $2x + 5y = 10$
d. Graph: $2x + 5y = 10$
e. Solve for y: $2x + 5y = 10$

48. Work each problem according to the instructions given.

a. Solve: $x - 2 = 6$
b. Find x when y is 0: $x - 2y = 6$
c. Find y when x is 0: $x - 2y = 6$
d. Graph: $x - 2y = 6$
e. Solve for y: $x - 2y = 6$

Getting Ready for the Next Section

49. Let $3x + 2y = 6$

a. Find x when $y = 0$ **b.** Find y when $x = 0$

50. Let $2x - 5y = 10$

a. Find x when $y = 0$ **b.** Find y when $x = 0$

51. Let $-x + 2y = 4$

a. Find x when $y = 0$ **b.** Find y when $x = 0$

52. Let $3x - y = 6$

a. Find x when $y = 0$ **b.** Find y when $x = 0$

53. Let $y = -\frac{1}{3}x + 2$

a. Find x when $y = 0$ **b.** Find y when $x = 0$

54. Let $y = \frac{3}{2}x - 3$

a. Find x when $y = 0$ **b.** Find y when $x = 0$

More on Graphing: Intercepts 2.4

In this section we continue our work with graphing lines by finding the points where a line crosses the axes of our coordinate system. To do so, we use the fact that any point on the x-axis has a y-coordinate of 0 and any point on the y-axis has an x-coordinate of 0. We begin with the following definition.

DEFINITION *x-intercept, y-intercept*

The ***x-intercept*** of a straight line is the x-coordinate of the point where the graph crosses the x-axis. The ***y-intercept*** is defined similarly. It is the y-coordinate of the point where the graph crosses the y-axis.

If the x-intercept is a, then the point $(a, 0)$ lies on the graph. (This is true because any point on the x-axis has a y-coordinate of 0.)

If the y-intercept is b, then the point $(0, b)$ lies on the graph. (This is true because any point on the y-axis has an x-coordinate of 0.)

Graphically, the relationship is shown in Figure 1.

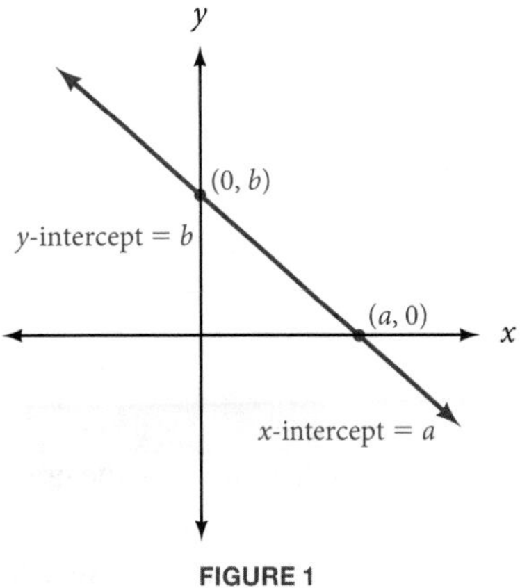

FIGURE 1

EXAMPLE 1 Find the x- and y-intercepts for $3x - 2y = 6$, and then use them to draw the graph.

SOLUTION To find where the graph crosses the x-axis, we let $y = 0$. (The y-coordinate of any point on the x-axis is 0.)

***x*-intercept:**

When	$y = 0$
the equation	$3x - 2y = 6$
becomes	$3x - 2(0) = 6$
	$3x - 0 = 6$
	$x = 2$ Multiply each side by $\frac{1}{3}$

The graph crosses the x-axis at $(2, 0)$, which means the x-intercept is 2.

***y*-intercept:**

When	$x = 0$
the equation	$3x - 2y = 6$
becomes	$3(0) - 2y = 6$
	$0 - 2y = 6$
	$-2y = 6$
	$y = -3$ Multiply each side by $-\frac{1}{2}$

The graph crosses the *y*-axis at $(0, -3)$, which means the *y*-intercept is -3.

Plotting the *x*- and *y*-intercepts and then drawing a line through them, we have the graph of $3x - 2y = 6$, as shown in Figure 2.

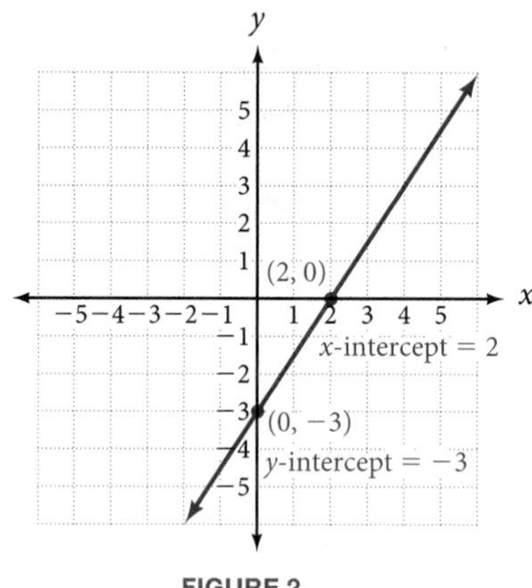

FIGURE 2

EXAMPLE 2 Graph $-x + 2y = 4$ by finding the intercepts and using them to draw the graph.

SOLUTION Again, we find the *x*-intercept by letting $y = 0$ in the equation and solving for *x*. Similarly, we find the *y*-intercept by letting $x = 0$ and solving for *y*.

***x*-intercept:**

When	$y = 0$
the equation	$-x + 2y = 4$
becomes	$-x + 2(0) = 4$
	$-x + 0 = 4$
	$-x = 4$
	$x = -4$ Multiply each side by -1

The *x*-intercept is -4, indicating that the point $(-4, 0)$, is on the graph of $-x + 2y = 4$.

***y*-intercept:**

When	$x = 0$
the equation	$-x + 2y = 4$
becomes	$-0 + 2y = 4$
	$2y = 4$
	$y = 2$ Multiply each side by $\frac{1}{2}$

The y-intercept is 2, indicating that the point (0, 2) is on the graph of $-x + 2y = 4$.

Plotting the intercepts and drawing a line through them, we have the graph of $-x + 2y = 4$, as shown in Figure 3.

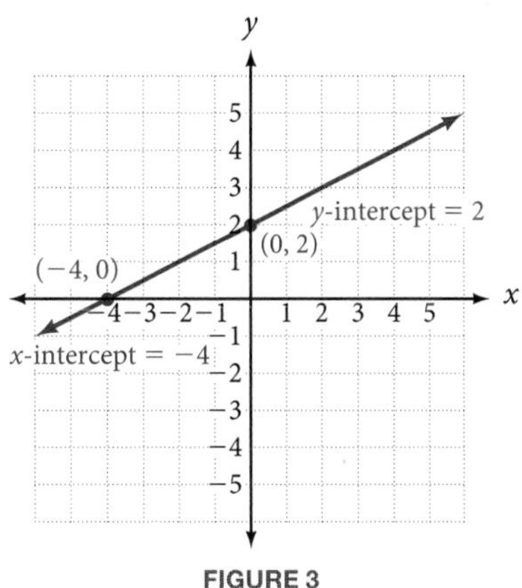

FIGURE 3

Graphing a line by finding the intercepts, as we have done in Examples 1 and 2, is an easy method of graphing if the equation has the form $ax + by = c$ and both the numbers a and b divide the number c evenly.

In our next example we use the intercepts to graph a line in which y is given in terms of x.

EXAMPLE 3 Use the intercepts for $y = -\frac{1}{3}x + 2$ to draw its graph.

SOLUTION We graphed this line previously in Example 3 of Section 3.3 by substituting three different values of x into the equation and solving for y. This time we will graph the line by finding the intercepts.

***x*-intercept:**

When	$y = 0$	
the equation	$y = -\frac{1}{3}x + 2$	
becomes	$0 = -\frac{1}{3}x + 2$	
	$-2 = -\frac{1}{3}x$	Add -2 to each side
	$6 = x$	Multiply each side by -3

The x-intercept is 6, which means the graph passes through the point (6, 0).

***y*-intercept:**

When	$x = 0$
the equation	$y = -\frac{1}{3}x + 2$
becomes	$y = -\frac{1}{3}(0) + 2$
	$y = 2$

The y-intercept is 2, which means the graph passes through the point (0, 2).

The graph of $y = -\frac{1}{3}x + 2$ is shown in Figure 4. Compare this graph, and the method used to obtain it, with Example 3 in Section 3.3.

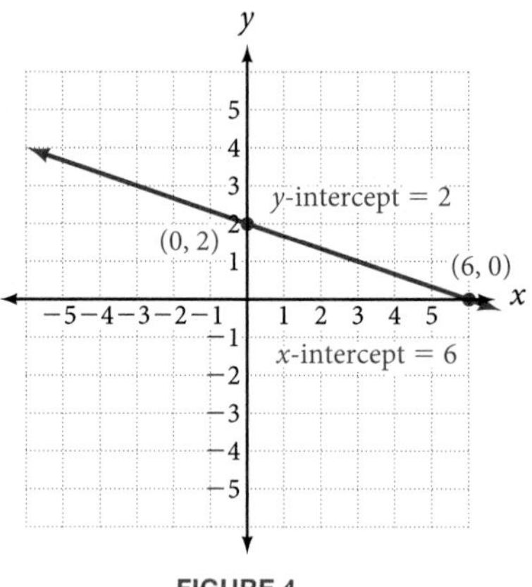

FIGURE 4

GETTING READY FOR CLASS

After reading through the preceding section, respond in your own words and in complete sentences.

A. What is the x-intercept for a graph?

B. What is the y-intercept for a graph?

C. How do we find the y-intercept for a line from the equation?

D. How do we graph a line using its intercepts?

Find the x- and y-intercepts for the following equations. Then use the intercepts to graph each equation.

1. $2x + y = 4$ **2.** $2x + y = 2$ **3.** $-x + y = 3$ **4.** $-x + y = 4$

5. $-x + 2y = 2$ **6.** $-x + 2y = 4$ **7.** $5x + 2y = 10$ **8.** $2x + 5y = 10$

9. $4x - 2y = 8$ **10.** $2x - 4y = 8$ **11.** $-4x + 5y = 20$ **12.** $-5x + 4y = 20$

13. $y = 2x - 6$ **14.** $y = 2x + 6$ **15.** $y = 2x + 2$ **16.** $y = -2x + 2$

17. $y = 2x - 1$ **18.** $y = -2x - 1$ **19.** $y = \frac{1}{2}x + 3$ **20.** $y = \frac{1}{2}x - 3$

21. $y = -\frac{1}{3}x - 2$ **22.** $y = -\frac{1}{3}x + 2$

For each of the following lines the x-intercept and the y-intercept are both 0, which means the graph of each will go through the origin, (0, 0). Graph each line by finding a point on each, other than the origin, and then drawing a line through that point and the origin.

23. $y = -2x$ **24.** $y = \frac{1}{2}x$ **25.** $y = -\frac{1}{3}x$ **26.** $y = -3x$

27. $y = \frac{2}{3}x$ **28.** $y = \frac{3}{2}x$

Complete each table.

29.

Equation	x-intercept	y-intercept
$3x + 4y = 12$		
$3x + 4y = 4$		
$3x + 4y = 3$		
$3x + 4y = 2$		

30.

Equation	x-intercept	y-intercept
$-2x + 3y = 6$		
$-2x + 3y = 3$		
$-2x + 3y = 2$		
$-2x + 3y = 1$		

31.

Equation	x-intercept	y-intercept
$x - 3y = 2$		
$y = \frac{1}{3}x - \frac{2}{3}$		
$x - 3y = 0$		
$y = \frac{1}{3}x$		

32.

Equation	x-intercept	y-intercept
$x - 2y = 1$		
$y = \frac{1}{2}x - \frac{1}{2}$		
$x - 2y = 0$		
$y = \frac{1}{2}x$		

The next two problems are intended to give you practice reading, and paying attention to, the instructions that accompany the problems you are working. Working these problems is an excellent way to get ready for a test or a quiz.

33. Work each problem according to the instructions given.

a. Solve: $2x - 3 = -3$ **b.** Find the x-intercept: $2x - 3y = -3$
c. Find y when x is 0: $2x - 3y = -3$ **d.** Graph: $2x - 3y = -3$
e. Solve for y: $2x - 3y = -3$

34. Work each problem according to the instructions given.

a. Solve: $3x - 4 = -4$ **b.** Find the y-intercept: $3x - 4y = -4$
c. Find x when y is 0: $3x - 4y = -4$ **d.** Graph: $3x - 4y = -4$
e. Solve for y: $3x - 4y = -4$

From the graphs below, find the x- and y-intercepts for each line.

35.

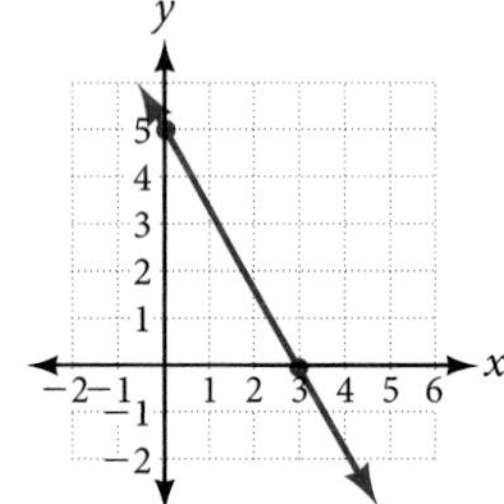

36.

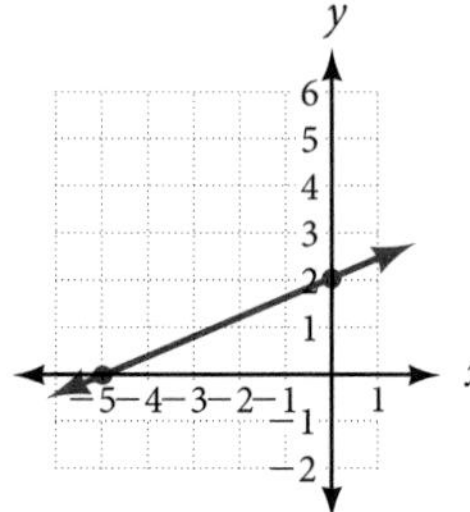

37.

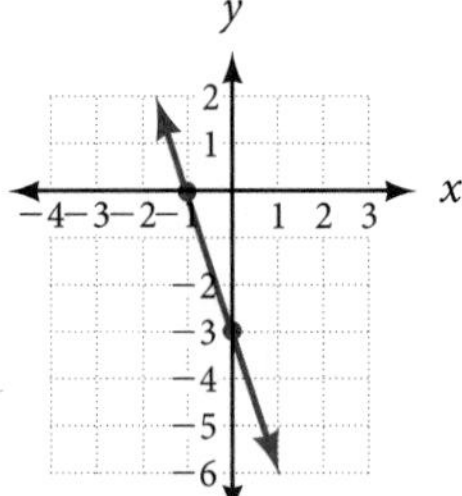

38.

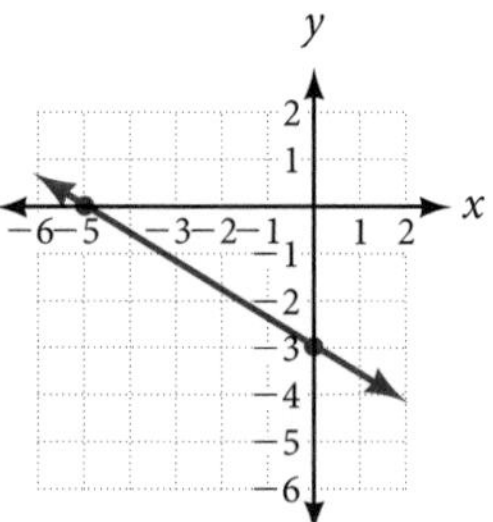

39. Graph the line that passes through the point $(-4, 4)$ and has an x-intercept of -2. What is the y-intercept of this line?

40. Graph the line that passes through the point $(-3, 4)$ and has a y-intercept of 3. What is the x-intercept of this line?

41. A line passes through the point $(1, 4)$ and has a y-intercept of 3. Graph the line and name its x-intercept.

42. A line passes through the point $(3, 4)$ and has an x-intercept of 1. Graph the line and name its y-intercept.

43. Graph the line that passes through the points $(-2, 5)$ and $(5, -2)$. What are the x- and y-intercepts for this line?

44. Graph the line that passes through the points $(5, 3)$ and $(-3, -5)$. What are the x- and y-intercepts for this line?

45. Use the graph at the right to complete the following table.

x	y
-2	
0	
	0
	-2

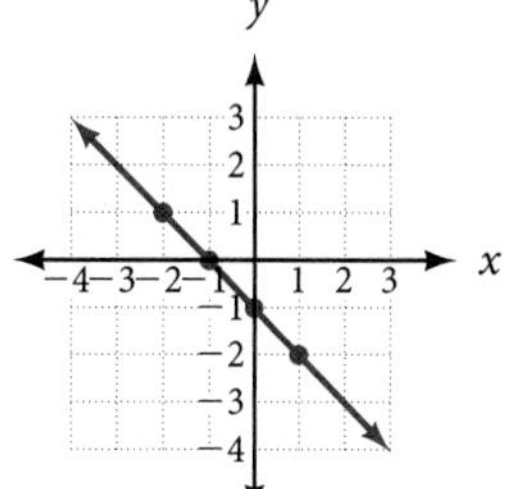

46. Use the graph at the right to complete the following table.

x	y
-2	
	0
	6

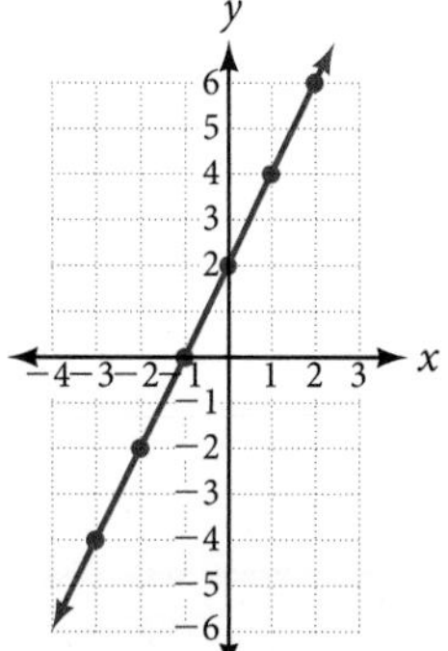

47. The vertical line $x = 3$ has only one intercept. Graph $x = 3$, and name its intercept. [Remember, ordered pairs (x, y) that are solutions to the equation $x = 3$ are ordered pairs with an x-coordinate of 3 and any y-coordinate.]

48. Graph the vertical line $x = -2$. Then name its intercept.

49. The horizontal line $y = 4$ has only one intercept. Graph $y = 4$, and name its intercept. [Ordered pairs (x, y) that are solutions to the equation $y = 4$ are ordered pairs with a y-coordinate of 4 and any x-coordinate.]

50. Graph the horizontal line $y = -3$. Then name its intercept.

Applying the Concepts

51. Complementary Angles The following diagram shows sunlight hitting the ground. Angle α (*alpha*) is called the angle of inclination, and angle θ (*theta*) is called the angle of incidence. As the sun moves across the sky, the values of these angles change. Assume that $\alpha + \theta = 90$, where both α and θ are in degrees measure. Graph this equation on a coordinate system where the horizontal axis is the α-axis and the vertical axis is the θ-axis. Find the intercepts first, and limit your graph to the first quadrant only.

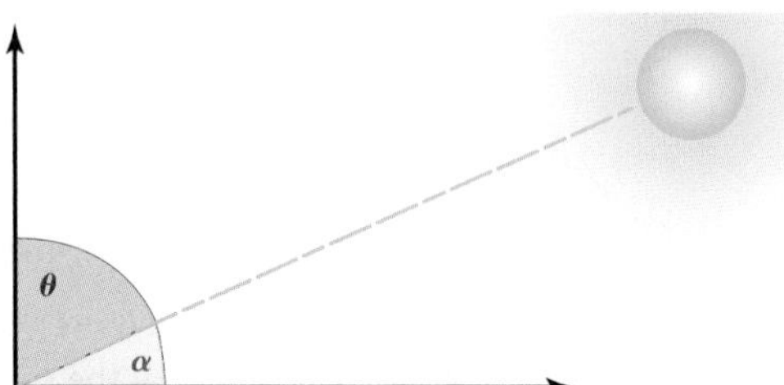

52. Light When light comes into contact with an impenetrable object, such as a thick piece of wood or metal, it is reflected or absorbed, but not transmitted, as shown in the following diagram. If we let R represent the percentage of light reflected and A the percentage of light absorbed by a surface, then the relationship between R and A is $R + A = 100$. Graph this equation on a coordinate system where the horizontal axis is the A-axis and the vertical axis is the R-axis. Find the intercepts first, and limit your graph to the first quadrant.

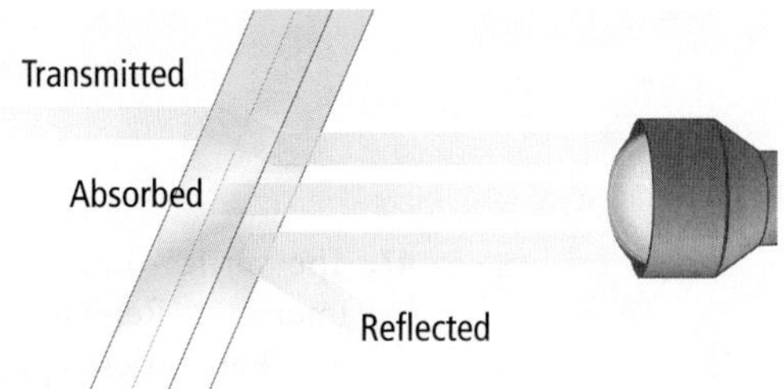

Getting Ready for the Next Section

53. Evaluate

a. $\dfrac{5 - 2}{3 - 1}$ **b.** $\dfrac{2 - 5}{1 - 3}$

54. Evaluate

a. $\dfrac{-4 - 1}{5 - (-2)}$ **b.** $\dfrac{1 + 4}{-2 - 5}$

55. Evaluate the following expressions when $x = 3$, and $y = 5$.

a. $\dfrac{y - 2}{x - 1}$ **b.** $\dfrac{2 - y}{1 - x}$

56. Evaluate the following expressions when $x = 4$, and $y = -1$.

a. $\dfrac{-4 - y}{5 - x}$ **b.** $\dfrac{y + 4}{x - 5}$

The Slope of a Line 2.5

In defining the slope of a straight line, we are looking for a number to associate with a straight line that does two things. First of all, we want the slope of a line to measure the "steepness" of the line; that is, in comparing two lines, the slope of the steeper line should have the larger numerical value. Second, we want a line that *rises* going from left to right to have a *positive* slope. We want a line that *falls* going from left to right to have a *negative* slope. (A line that neither rises nor falls going from left to right must, therefore, have 0 slope.) These are illustrated in Figure 1.

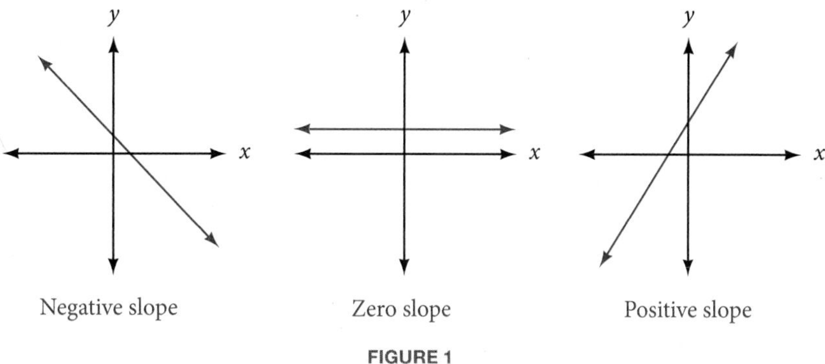

FIGURE 1

Suppose we know the coordinates of two points on a line. Because we are trying to develop a general formula for the slope of a line, we will use general points—call the two points $P_1(x_1, y_1)$ and $P_2(x_2, y_2)$. They represent the coordinates of any two different points on our line. We define the *slope* of our line to be the ratio of the vertical change to the horizontal change as we move from point (x_1, y_1) to point (x_2, y_2) on the line. (See Figure 2.)

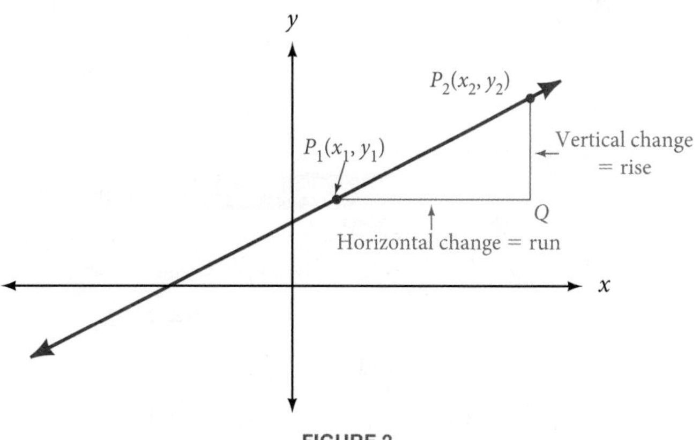

FIGURE 2

We call the vertical change the *rise* in the graph and the horizontal change the *run* in the graph. The slope, then, is

$$\text{Slope} = \frac{\text{vertical change}}{\text{horizontal change}} = \frac{\text{rise}}{\text{run}}$$

We would like to have a numerical value to associate with the rise in the graph and a numerical value to associate with the run in the graph. A quick study of Figure 2 shows that the coordinates of point Q must be (x_2, y_1), because Q is directly below point P_2 and right across from point P_1. We can draw our diagram again in the manner shown in Figure 3. It is apparent from this graph that the rise can be expressed as $(y_2 - y_1)$ and the run as $(x_2 - x_1)$. We usually denote the slope of a line by the letter m. The complete definition of slope follows along with a diagram (Figure 3) that illustrates the definition.

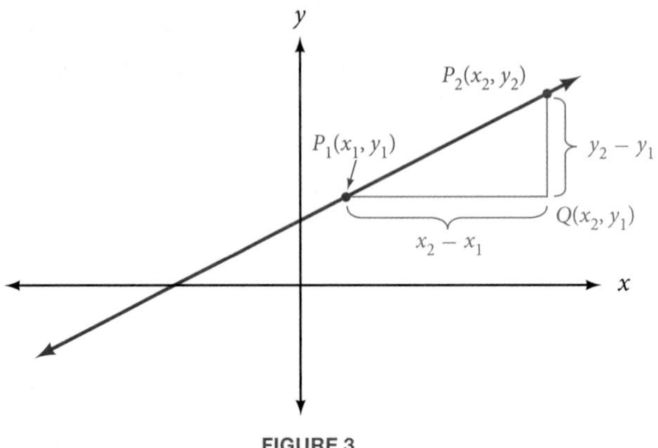

FIGURE 3

DEFINITION *slope*

If points (x_1, y_1) and (x_2, y_2) are any two different points, then the ***slope*** of the line on which they lie is

$$\text{Slope} = m = \frac{\text{rise}}{\text{run}} = \frac{y_2 - y_1}{x_2 - x_1}$$

This definition of the *slope* of a line does just what we want it to do. If the line rises going from left to right, the slope will be positive. If the line falls from left to right, the slope will be negative. Also, the steeper the line, the larger numerical value the slope will have.

EXAMPLE 1 Find the slope of the line between the points (1, 2) and (3, 5).

SOLUTION We can let

$$(x_1, y_1) = (1, 2)$$

and

$$(x_2, y_2) = (3, 5)$$

then

$$m = \frac{y_2 - y_1}{x_2 - x_1} = \frac{5 - 2}{3 - 1} = \frac{3}{2}$$

The slope is $\frac{3}{2}$. For every vertical change of 3 units, there will be a corresponding horizontal change of 2 units. (See Figure 4.)

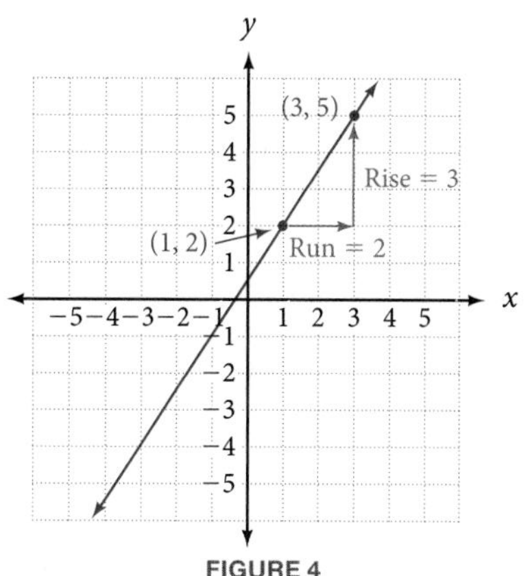

FIGURE 4

EXAMPLE 2 Find the slope of the line through $(-2, 1)$ and $(5, -4)$.

SOLUTION It makes no difference which ordered pair we call (x_1, y_1) and which we call (x_2, y_2).

$$\text{Slope} = m = \frac{y_2 - y_1}{x_2 - x_1} = \frac{-4 - 1}{5 - (-2)} = -\frac{5}{7}$$

The slope is $-\frac{5}{7}$. Every vertical change of -5 units (down 5 units) is accompanied by a horizontal change of 7 units (to the right 7 units). (See Figure 5.)

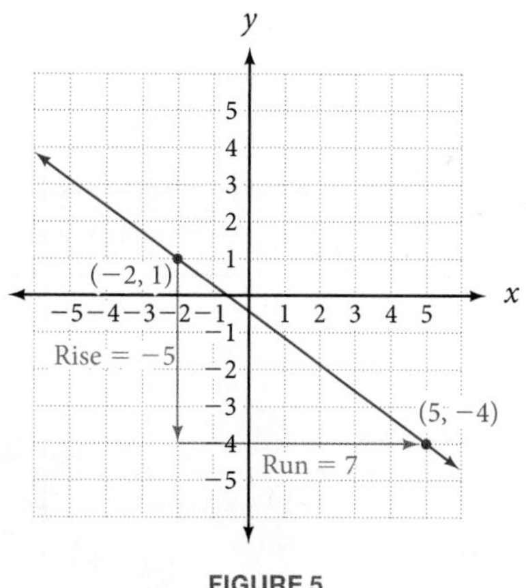

FIGURE 5

EXAMPLE 3 Graph the line with slope $\frac{3}{2}$ and y-intercept 1.

SOLUTION Because the y-intercept is 1, we know that one point on the line is $(0, 1)$. So, we begin by plotting the point $(0, 1)$, as shown in Figure 6.

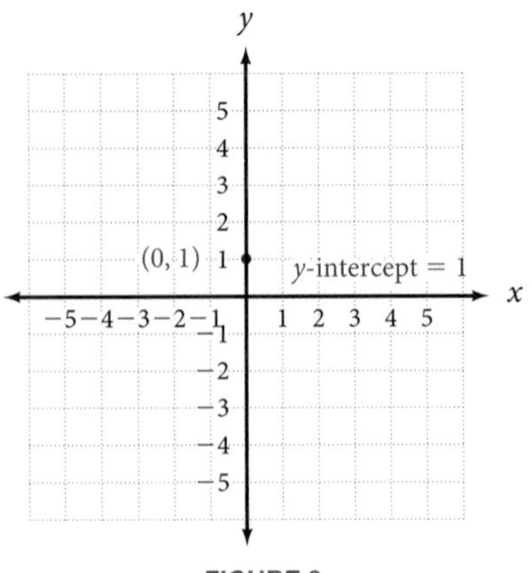

FIGURE 6

There are many lines that pass through the point shown in Figure 6, but only one of those lines has a slope of $\frac{3}{2}$. The slope, $\frac{3}{2}$, can be thought of as the rise in the graph divided by the run in the graph. Therefore, if we start at the point (0, 1) and move 3 units up (that's a rise of 3) and then 2 units to the right (a run of 2), we will be at another point on the graph. Figure 7 shows that the point we reach by doing so is the point (2, 4).

$$\text{Slope} = m = \frac{\text{rise}}{\text{run}} = \frac{3}{2}$$

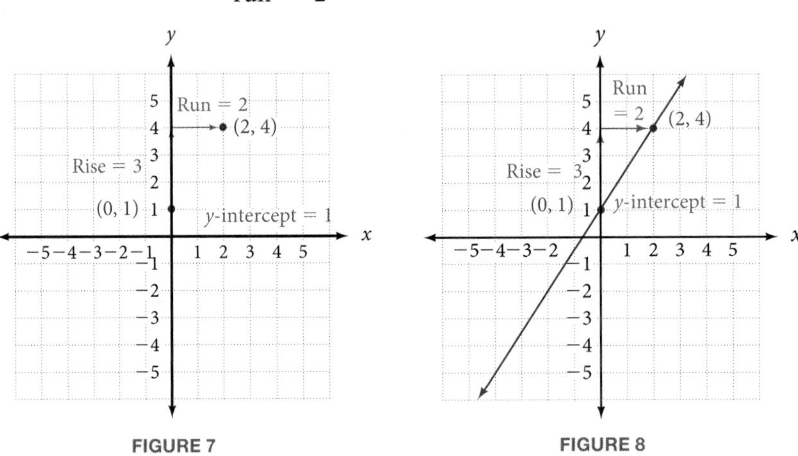

FIGURE 7 FIGURE 8

To graph the line with slope $\frac{3}{2}$ and y-intercept 1, we simply draw a line through the two points in Figure 7 to obtain the graph shown in Figure 8.

EXAMPLE 4 Find the slope of the line containing (3, −1) and (3, 4).

SOLUTION Using the definition for slope, we have

$$m = \frac{y_2 - y_1}{x_2 - x_1} = \frac{4 - (-1)}{3 - 3} = \frac{5}{0}$$

The expression $\frac{5}{0}$ is undefined; that is, there is no real number to associate with it. In this case, we say the line *has no slope*.

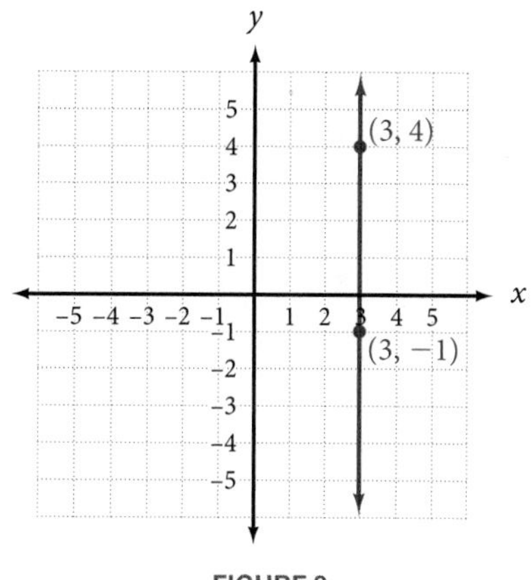

FIGURE 9

The graph of our line is shown in Figure 9. Our line with no slope is a vertical line. All vertical lines have no slope. (And all horizontal lines, as we mentioned earlier, have 0 slope.)

As a final note, the summary reminds us that all horizontal lines have equations of the form $y = b$ and slopes of 0. Because they cross the y-axis at b, the y-intercept is b; there is no x-intercept. Vertical lines have no slope and equations of the form $x = a$. Each will have an x-intercept at a and no y-intercept. Finally, equations of the form $y = mx$ have graphs that pass through the origin. The slope is always m and both the x-intercept and the y-intercept are 0.

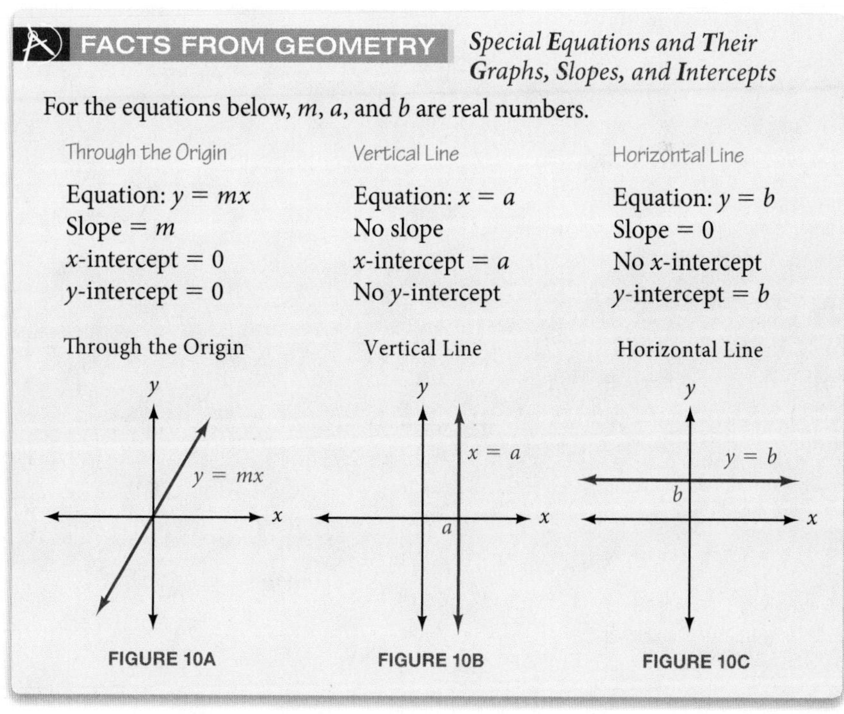

FACTS FROM GEOMETRY *Special Equations and Their Graphs, Slopes, and Intercepts*

For the equations below, m, a, and b are real numbers.

Through the Origin	Vertical Line	Horizontal Line
Equation: $y = mx$	Equation: $x = a$	Equation: $y = b$
Slope $= m$	No slope	Slope $= 0$
x-intercept $= 0$	x-intercept $= a$	No x-intercept
y-intercept $= 0$	No y-intercept	y-intercept $= b$

FIGURE 10A FIGURE 10B FIGURE 10C

GETTING READY FOR CLASS

After reading through the preceding section, respond in your own words and in complete sentences.

A. What is the slope of a line?

B. Would you rather climb a hill with a slope of 1 or a slope of 3? Explain why.

C. Describe how to obtain the slope of a line if you know the coordinates of two points on the line.

D. Describe how you would graph a line from its slope and y-intercept.

Find the slope of the line through the following pairs of points. Then plot each pair of points, draw a line through them, and indicate the rise and run in the graph in the same manner shown in Examples 1 and 2.

1. (2, 1), (4, 4) **2.** (3, 1), (5, 4) **3.** (1, 4), (5, 2)

4. (1, 3), (5, 2) **5.** (1, −3), (4, 2) **6.** (2, −3), (5, 2)

7. (−3, −2), (1, 3) **8.** (−3, −1), (1, 4) **9.** (−3, 2), (3, −2)

10. (−3, 3), (3, −1) **11.** (2, −5), (3, −2) **12.** (2, −4), (3, −1)

In each of the following problems, graph the line with the given slope and y-intercept b.

13. $m = \frac{2}{3}, b = 1$ **14.** $m = \frac{3}{4}, b = -2$ **15.** $m = \frac{3}{2}, b = -3$

16. $m = \frac{4}{3}, b = 2$ **17.** $m = -\frac{4}{3}, b = 5$ **18.** $m = -\frac{3}{5}, b = 4$

19. $m = 2, b = 1$ **20.** $m = -2, b = 4$ **21.** $m = 3, b = -1$

22. $m = 3, b = -2$

Find the slope and y-intercept for each line.

23.

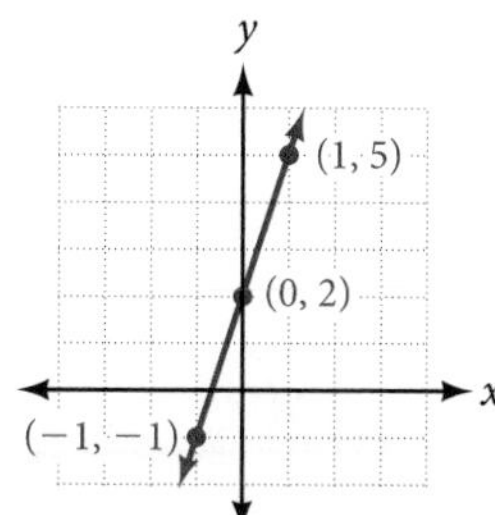

24.

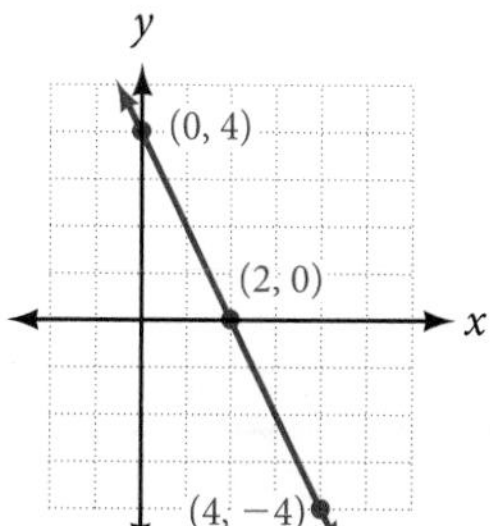

25.

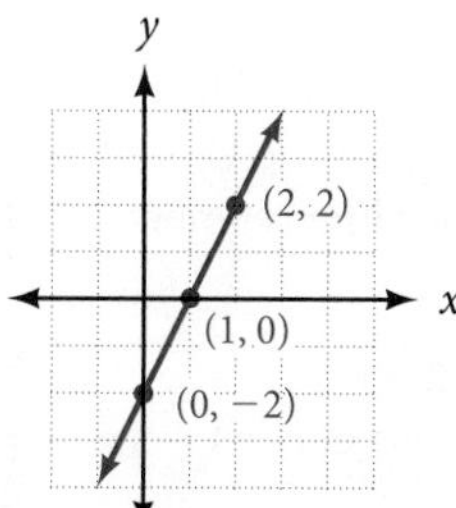

26.

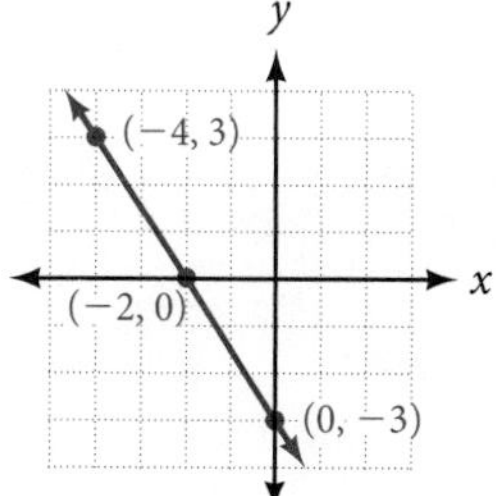

27. Graph the line that has an x-intercept of 3 and a y-intercept of −2. What is the slope of this line?

28. Graph the line that has an x-intercept of 2 and a y-intercept of −3. What is the slope of this line?

29. Graph the line with x-intercept 4 and y-intercept 2. What is the slope of this line?

30. Graph the line with x-intercept −4 and y-intercept −2. What is the slope of this line?

31. Graph the line $y = 2x - 3$, then name the slope and y-intercept by looking at the graph.

32. Graph the line $y = -2x + 3$, then name the slope and y-intercept by looking at the graph.

33. Graph the line $y = \frac{1}{2}x + 1$, then name the slope and y-intercept by looking at the graph.

34. Graph the line $y = -\frac{1}{2}x - 2$, then name the slope and y-intercept by looking at the graph.

35. Find y if the line through $(4, 2)$ and $(6, y)$ has a slope of 2.

36. Find y if the line through $(1, y)$ and $(7, 3)$ has a slope of 6.

For each equation in each table, give the slope of the graph.

37.

Equation	Slope
$x = 3$	
$y = 3$	
$y = 3x$	

38.

Equation	Slope
$y = \frac{3}{2}$	
$x = \frac{3}{2}$	
$y = \frac{3}{2}x$	

39.

Equation	Slope
$y = -\frac{2}{3}$	
$x = -\frac{2}{3}$	
$y = -\frac{2}{3}x$	

40.

Equation	Slope
$x = -2$	
$y = -2$	
$y = -2x$	

Applying the Concepts

41. Garbage Production The table and completed line graph gives the annual production of garbage in the United States for some specific years. Find the slope of each of the four line segments, A, B, C, and D.

Year	Garbage (millions of tons)
1960	88
1970	121
1980	152
1990	205
2000	224

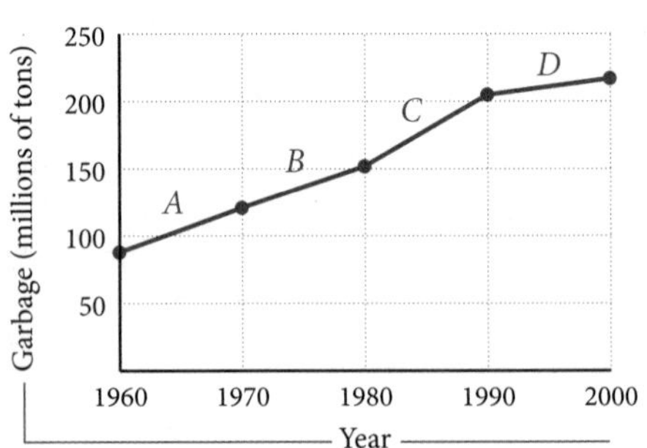

42. Grass Height The table and completed line graph gives the growth of a certain plant species over time. Find the slopes of the line segments labeled *A*, *B*, and *C*.

Day	Plant Height
0	0
2	1
4	3
6	6
8	13
10	23

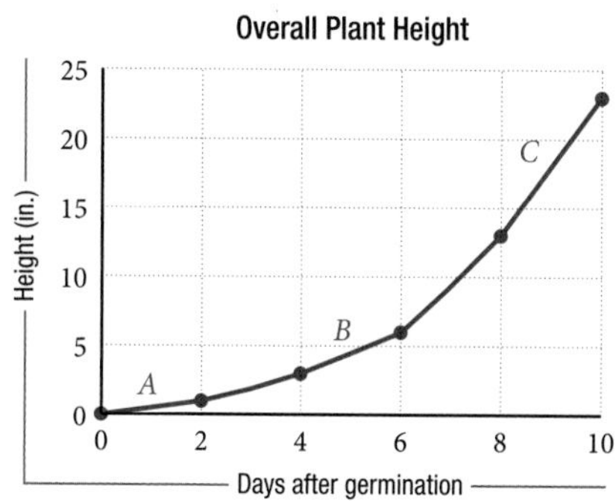

43. Non-Camera Phone Sales The table and line graph here each show the projected non-camera phone sales each year through 2010. Find the slope of each of the three line segments, *A*, *B*, and *C*.

Year	Sales (in millions)
2006	300
2007	250
2008	175
2009	150
2010	125

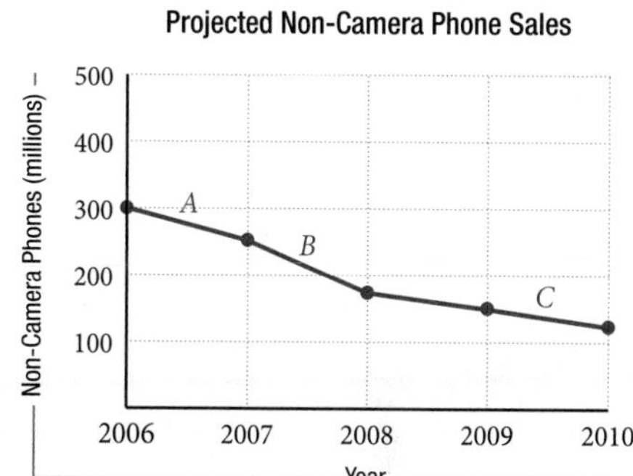

44. Camera Phone Sales The table from Problem 48 in Problem Set 3.1 and a line graph are shown here. Each shows the projected sales of camera phones from 2006 to 2010. Find the slopes of line segments *A*, *B*, and *C*.

Year	Sales (in millions)
2006	500
2007	650
2008	750
2009	875
2010	900

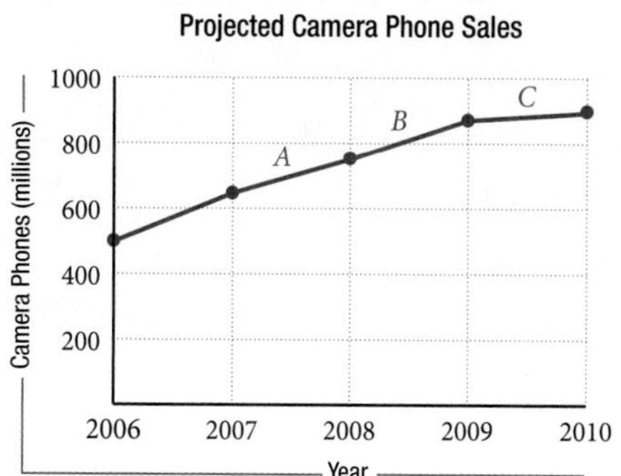

Getting Ready for the Next Section

Solve each equation for y.

45. $-2x + y = 4$

46. $-4x + y = -2$

47. $2x + y = 3$

48. $3x + 2y = 6$

49. $4x - 5y = 20$

50. $-2x - 5y = 10$

51. $-y - 3 = -2(x + 4)$

52. $-y + 5 = 2(x + 2)$

53. $-y - 3 = -\frac{2}{3}(x + 3)$

54. $-y - 1 = -\frac{1}{2}(x + 4)$

55. $-\frac{y - 1}{x} = \frac{3}{2}$

56. $-\frac{y + 1}{x} = \frac{3}{2}$

Finding the Equation of a Line 2.6

To this point in the chapter, most of the problems we have worked have used the equation of a line to find different types of information about the line. For instance, given the equation of a line, we can find points on the line, the graph of the line, the intercepts, and the slope of the line. In this section we reverse things somewhat and move in the other direction; we will use information about a line, such as its slope and y-intercept, to find the equation of a line.

There are three main types of problems to solve in this section.

1. Find the equation of a line from the slope and y-intercept.
2. Find the equation of a line given one point on the line and the slope of the line.
3. Find the equation of a line given two points on the line.

Examples 1 and 2 illustrate the first type of problem. Example 5 solves the second type of problem. The third type of problem is solved in Example 6.

The Slope-Intercept Form of an Equation of a Straight Line

EXAMPLE 1 Find the equation of the line with slope $\frac{3}{2}$ and y-intercept 1.

SOLUTION We graphed the line with slope $\frac{3}{2}$ and y-intercept 1 in Example 3 of the previous section. Figure 1 shows that graph.

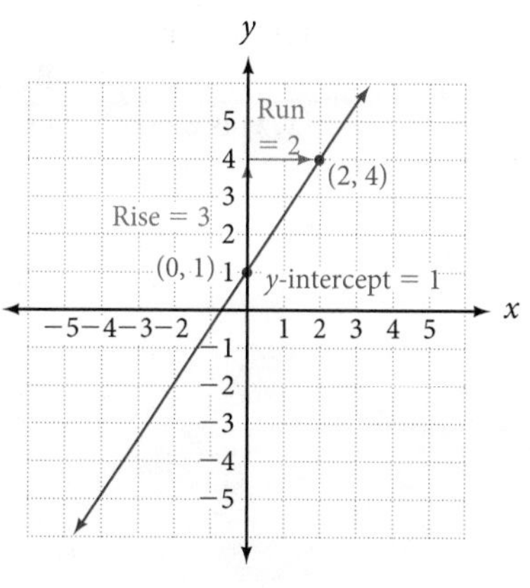

FIGURE 1

What we want to do now is find the equation of the line shown in Figure 1. To do so, we take any other point (x, y) on the line and apply our slope formula to that point and the point $(0, 1)$. We set that result equal to $\frac{3}{2}$, because $\frac{3}{2}$ is the slope of our line and a diagram of the situation follows.

$$\frac{y-1}{x-0} = \frac{3}{2} \qquad \text{Slope} = \frac{\text{vertical change}}{\text{horizontal change}}$$

$$\frac{y-1}{x} = \frac{3}{2} \qquad x - 0 = x$$

$$y - 1 = \frac{3}{2}x \qquad \text{Multiply each side by } x$$

$$y = \frac{3}{2}x + 1 \qquad \text{Add 1 to each side}$$

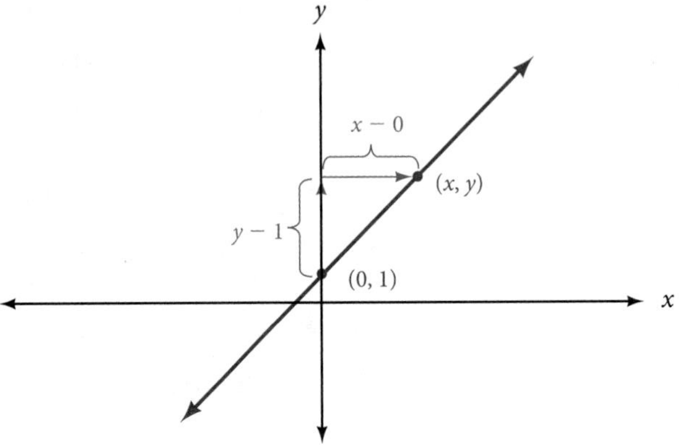

What is interesting and useful about the equation we have just found is that the number in front of x is the slope of the line and the constant term is the y-intercept. It is no coincidence that it turned out this way. Whenever an equation has the form $y = mx + b$, the graph is always a straight line with slope m and y-intercept b. To see that this is true in general, suppose we want the equation of a line with slope m and y-intercept b. Because the y-intercept is b, then the point $(0, b)$ is on the line. If (x, y) is any other point on the line, then we apply our slope formula to get

$$\frac{y-b}{x-0} = m \qquad \text{Slope} = \frac{\text{vertical change}}{\text{horizontal change}}$$

$$\frac{y-b}{x} = m \qquad x - 0 = x$$

$$y - b = mx \qquad \text{Multiply each side by } x$$

$$y = mx + b \qquad \text{Add } b \text{ to each side}$$

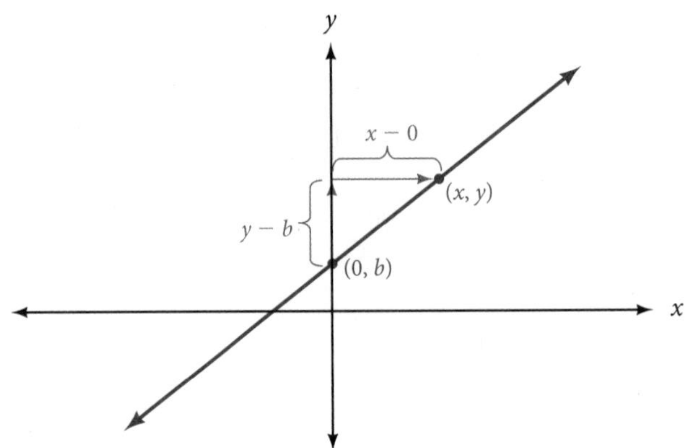

Here is a summary of what we have just found.

PROPERTY *Slope-Intercept Form of the Equation of a Line*

The equation of the line with slope m and y-intercept b is always given by

$$y = mx + b$$

EXAMPLE 2 Find the equation of the line with slope $-\frac{4}{3}$ and y-intercept 5. Then, graph the line.

SOLUTION Substituting $m = -\frac{4}{3}$ and $b = 5$ into the equation $y = mx + b$, we have

$$y = -\frac{4}{3}x + 5$$

Finding the equation from the slope and y-intercept is just that easy. If the slope is m and the y-intercept is b, then the equation is always $y = mx + b$.

Because the y-intercept is 5, the graph goes through the point (0, 5). To find a second point on the graph, we start at (0, 5) and move 4 units down (that's a rise of -4) and 3 units to the right (a run of 3). The point we reach is (3, 1). Drawing a line that passes through (0, 5) and (3, 1), we have the graph of our equation. (Note that we could also let the rise = 4 and the run = -3 and obtain the same graph.) The graph is shown in Figure 2.

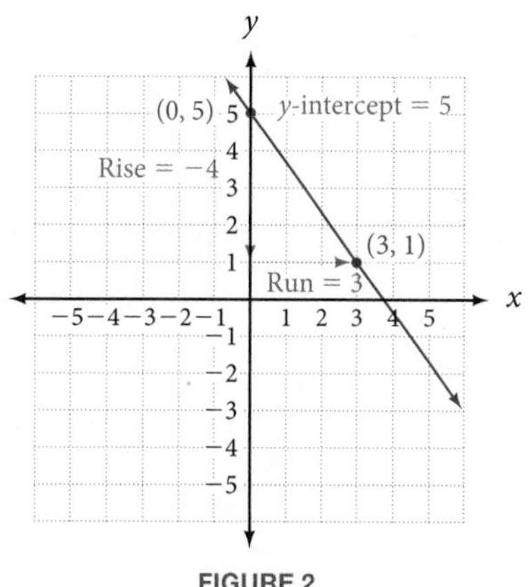

FIGURE 2

EXAMPLE 3 Find the slope and y-intercept for $-2x + y = -4$. Then, use them to draw the graph.

SOLUTION To identify the slope and y-intercept from the equation, the equation must be in the form $y = mx + b$ (slope-intercept form). To write our equation in this form, we must solve the equation for y. To do so, we simply add $2x$ to each side of the equation.

$$-2x + y = -4 \qquad \text{Original equation}$$

$$y = 2x - 4 \qquad \text{Add } 2x \text{ to each side}$$

The equation is now in slope-intercept form, so the slope must be 2 and the y-intercept must be -4. The graph, therefore, crosses the y-axis at $(0, -4)$. Because the slope is 2, we can let the rise $= 2$ and the run $= 1$ and find a second point on the graph. The graph is shown in Figure 3.

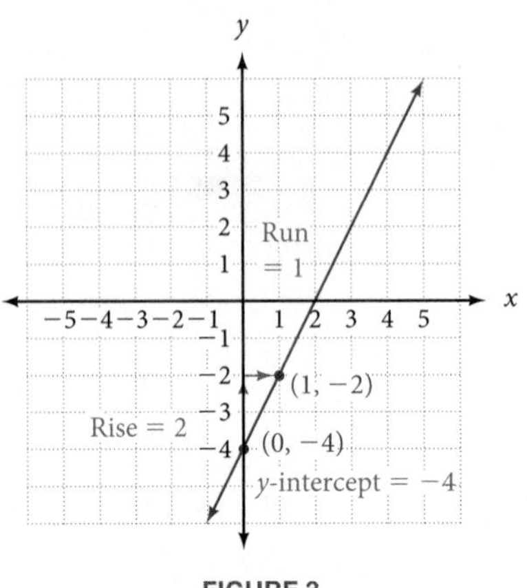

FIGURE 3

EXAMPLE 4 Find the slope and y-intercept for $3x - 2y = 6$.

SOLUTION To find the slope and y-intercept from the equation, we must write the equation in the form $y = mx + b$. This means we must solve the equation $3x - 2y = 6$ for y.

$$3x - 2y = 6 \qquad \text{Original equation}$$

$$-2y = -3x + 6 \qquad \text{Add } -3x \text{ to each side}$$

$$-\frac{1}{2}(-2y) = -\frac{1}{2}(-3x + 6) \qquad \text{Multiply each side by } -\frac{1}{2}$$

$$y = \frac{3}{2}x - 3 \qquad \text{Simplify each side}$$

Now that the equation is written in slope-intercept form, we can identify the slope as $\frac{3}{2}$ and the y-intercept as -3. The graph is shown in Figure 4.

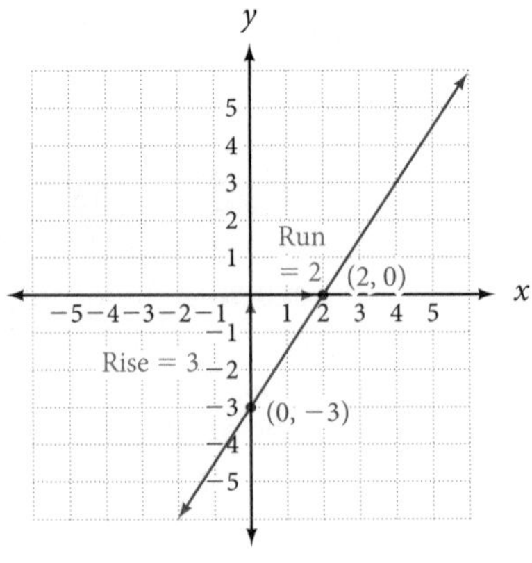

FIGURE 4

The Point-Slope Form of an Equation of a Straight Line

A second useful form of the equation of a straight line is the point-slope form.

Let line l contain the point (x_1, y_1) and have slope m. If (x, y) is any other point on l, then by the definition of slope we have

$$\frac{y - y_1}{x - x_1} = m$$

Multiplying both sides by $(x - x_1)$ gives us

$$(x - x_1) \cdot \frac{y - y_1}{x - x_1} = m(x - x_1)$$

$$y - y_1 = m(x - x_1)$$

This last equation is known as the *point-slope form* of the equation of a straight line.

PROPERTY *Point-Slope Form of the Equation of a Line*

The equation of the line through (x_1, y_1) with slope m is given by

$$y - y_1 = m(x - x_1)$$

This form is used to find the equation of a line, either given one point on the line and the slope, or given two points on the line.

EXAMPLE 5 Find the equation of the line with slope -2 that contains the point $(-4, 3)$. Write the answer in slope-intercept form.

SOLUTION Using $(x_1, y_1) = (-4, 3)$ and $m = -2$

in $y - y_1 = m(x - x_1)$ — Point-slope form

gives us $y - 3 = -2(x + 4)$ — Note: $x - (-4) = x + 4$

$y - 3 = -2x - 8$ — Multiply out right side

$y = -2x - 5$ — Add 3 to each side

Figure 5 is the graph of the line that contains $(-4, 3)$ and has a slope of -2. Notice that the y-intercept on the graph matches that of the equation we found.

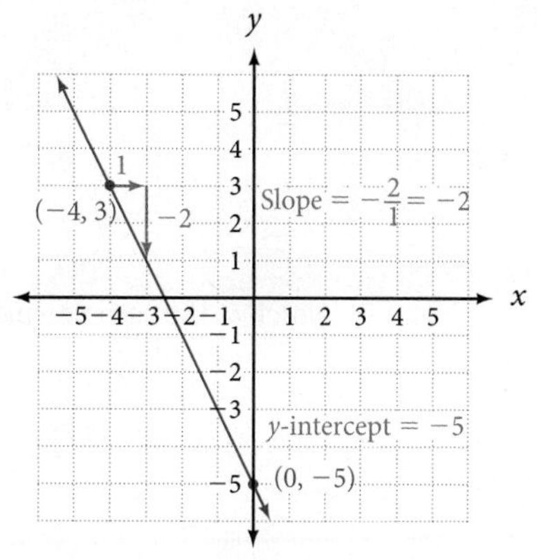

FIGURE 5

EXAMPLE 6 Find the equation of the line that passes through the points $(-3, 3)$ and $(3, -1)$.

SOLUTION We begin by finding the slope of the line:

$$m = \frac{3 - (-1)}{-3 - 3} = \frac{4}{-6} = -\frac{2}{3}$$

Using $(x_1, y_1) = (3, -1)$ and $m = -\frac{2}{3}$ in $y - y_1 = m(x - x_1)$ yields

$$y + 1 = -\frac{2}{3}(x - 3)$$

$$y + 1 = -\frac{2}{3}x + 2 \quad \text{Multiply out right side}$$

$$y = -\frac{2}{3}x + 1 \quad \text{Add } -1 \text{ to each side}$$

Figure 6 shows the graph of the line that passes through the points $(-3, 3)$ and $(3, -1)$. As you can see, the slope and y-intercept are $-\frac{2}{3}$ and 1, respectively.

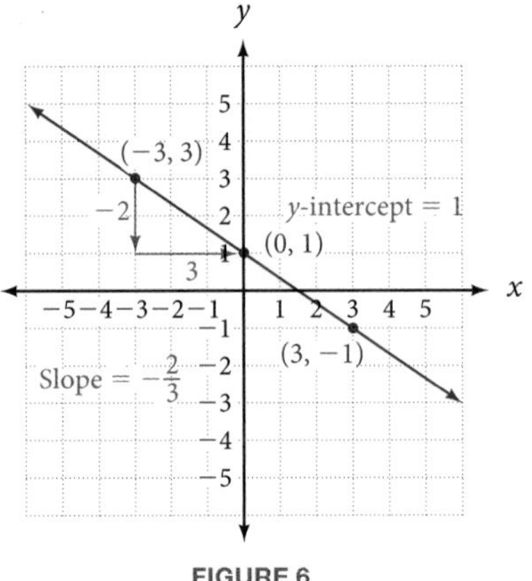

FIGURE 6

Note In Example 6 we could have used the point $(-3, 3)$ instead of $(3, -1)$ and obtained the same equation; that is, using $(x_1, y_1) = (-3, 3)$ and $m = -\frac{2}{3}$ in $y - y_1 = m(x - x_1)$ gives us

$$y - 3 = -\frac{2}{3}(x + 3)$$

$$y - 3 = -\frac{2}{3}x - 2$$

$$y = -\frac{2}{3}x + 1$$

which is the same result we obtained using $(3, -1)$.

Methods of Graphing Lines

1. Substitute convenient values of x into the equation, and find the corresponding values of y. We used this method first for equations like $y = 2x - 3$. To use this method for equations that looked like $2x - 3y = 6$, we first solved them for y.
2. Find the x- and y-intercepts. This method works best for equations of the form $3x + 2y = 6$ where the numbers in front of x and y divide the constant term evenly.
3. Find the slope and y-intercept. This method works best when the equation has the form $y = mx + b$ and b is an integer.

GETTING READY FOR CLASS

After reading through the preceding section, respond in your own words and in complete sentences.

A. What are m and b in the equation $y = mx + b$?
B. How would you find the slope and y-intercept for the line $3x - 2y = 6$?
C. What is the point-slope form of the equation of a line?
D. How would you find the equation of a line from two points on the line?

Problem Set 2.6

In each of the following problems, give the equation of the line with the given slope and y-intercept.

1. $m = \frac{2}{3}, b = 1$ **2.** $m = \frac{3}{4}, b = -2$ **3.** $m = \frac{3}{2}, b = -1$ **4.** $m = \frac{4}{3}, b = 2$

5. $m = -\frac{2}{3}, b = 3$ **6.** $m = -\frac{3}{5}, b = 4$ **7.** $m = 2, b = -4$ **8.** $m = -2, b = 4$

Find the slope and y-intercept for each of the following equations by writing them in the form $y = mx + b$. Then, graph each equation.

9. $-2x + y = 4$ **10.** $-2x + y = 2$ **11.** $3x + y = 3$ **12.** $3x + y = 6$

13. $3x + 2y = 6$ **14.** $2x + 3y = 6$ **15.** $4x - 5y = 20$ **16.** $2x - 5y = 10$

17. $-2x - 5y = 10$ **18.** $-4x + 5y = 20$

For each of the following problems, the slope and one point on a line are given. In each case use the point-slope form to find the equation of that line. (Write your answers in slope-intercept form.)

19. $(-2, -5), m = 2$ **20.** $(-1, -5), m = 2$ **21.** $(-4, 1), m = -\frac{1}{2}$

22. $(-2, 1), m = -\frac{1}{2}$ **23.** $(2, -3), m = \frac{3}{2}$ **24.** $(3, -4), m = \frac{4}{3}$

25. $(-1, 4), m = -3$ **26.** $(-2, 5), m = -3$

Find the equation of the line that passes through each pair of points. Write your answers in slope-intercept form.

27. $(-2, -4), (1, -1)$ **28.** $(2, 4), (-3, -1)$ **29.** $(-1, -5), (2, 1)$

30. $(-1, 6), (1, 2)$ **31.** $(-3, -2), (3, 6)$ **32.** $(-3, 6), (3, -2)$

33. $(-3, -1), (3, -5)$ **34.** $(-3, -5), (3, 1)$

Find the slope and y-intercept for each line. Then write the equation of each line in slope-intercept form.

35.

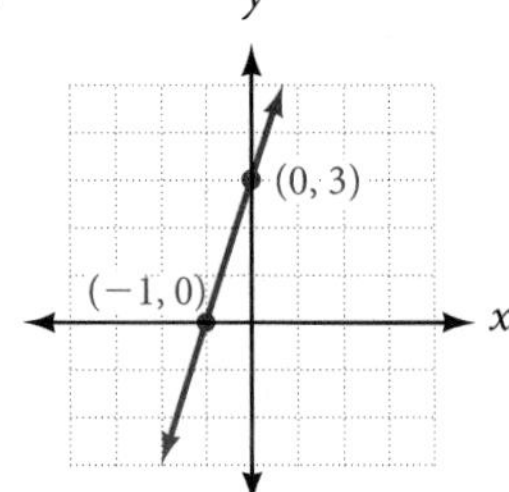

36.

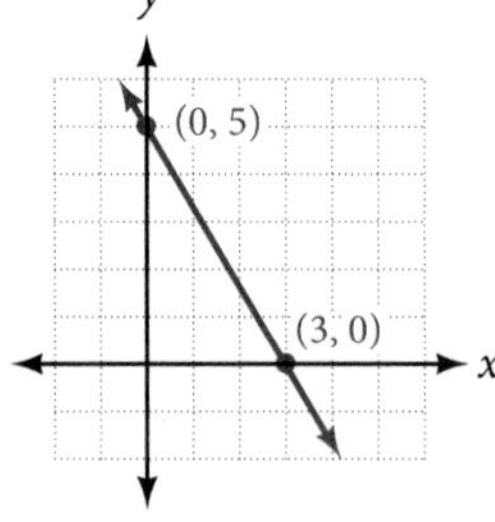

37.

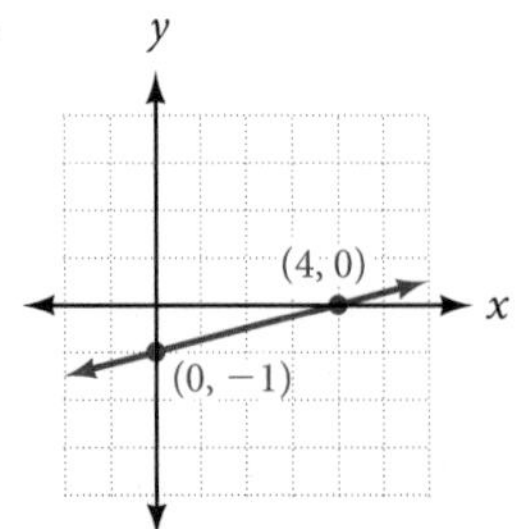

38.

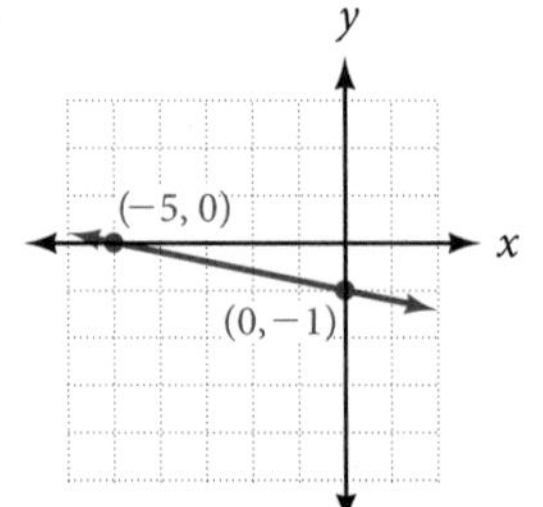

The next two problems are intended to give you practice reading, and paying attention to, the instructions that accompany the problems you are working. Working these problems is an excellent way to get ready for a test or a quiz.

39. Work each problem according to the instructions given.

a. Solve: $-2x + 1 = 6$

b. Write in slope-intercept form: $-2x + y = 6$

c. Find the y-intercept: $-2x + y = 6$

d. Find the slope: $-2x + y = 6$

e. Graph: $-2x + y = 6$

40. Work each problem according to the instructions given.

a. Solve: $x + 3 = -6$

b. Write in slope-intercept form: $x + 3y = -6$

c. Find the y-intercept: $x + 3y = -6$

d. Find the slope: $x + 3y = -6$

e. Graph: $x + 3y = -6$

41. Find the equation of the line with x-intercept 3 and y-intercept 2.

42. Find the equation of the line with x-intercept 2 and y-intercept 3.

43. Find the equation of the line with x-intercept -2 and y-intercept -5.

44. Find the equation of the line with x-intercept -3 and y-intercept -5.

45. The equation of the vertical line that passes through the points $(3, -2)$ and $(3, 4)$ is either $x = 3$ or $y = 3$. Which one is it?

46. The equation of the horizontal line that passes through the points $(2, 3)$ and $(-1, 3)$ is either $x = 3$ or $y = 3$. Which one is it?

Applying the Concepts

47. Value of a Copy Machine Cassandra buys a new color copier for her small business. It will cost $21,000 and will decrease in value each year. The graph below shows the value of the copier after the first 5 years of ownership.

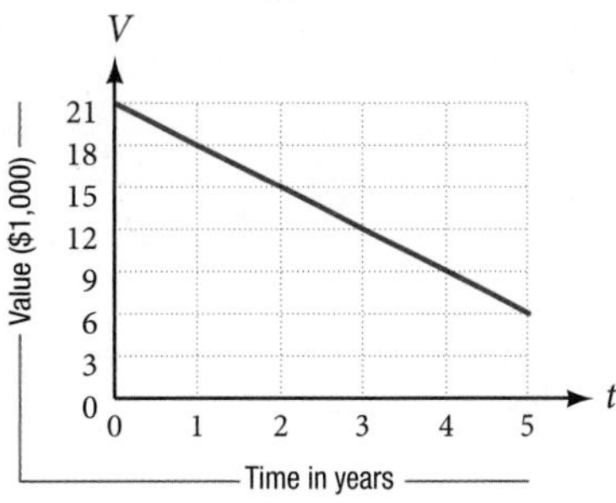

a. How much is the copier worth after 5 years?

b. After how many years is the copier worth $12,000?

c. Find the slope of this line.

d. By how many dollars per year is the copier decreasing in value?

e. Find the equation of this line where V is the value after t years.

48. Salesperson's Income Kevin starts a new job in sales next month. He will earn $1,000 per month plus a certain amount for each shirt he sells. The graph below shows the amount Kevin will earn per month based on how many shirts he sells.

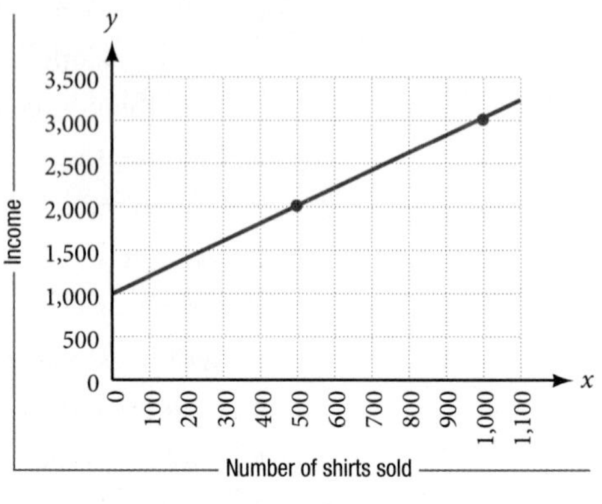

FIGURE 2

a. How much will he earn for selling 1,000 shirts?

b. How many shirts must he sell to earn $2,000 for a month?

c. Find the slope of this line.

d. How much money does Kevin earn for each shirt he sells?

e. Find the equation of this line where y is the amount he earns for selling x number of shirts.

Getting Ready for the Next Section

Graph each of the following lines.

49. $x + y = 4$ **50.** $x - y = -2$ **51.** $y = 2x - 3$ **52.** $y = 2x + 3$

53. $y = 2x$ **54.** $y = -2x$

Linear Inequalities in Two Variables

2.7

A linear inequality in two variables is any expression that can be put in the form

$$ax + by < c$$

where a, b, and c are real numbers (a and b not both 0). The inequality symbol can be any of the following four: $<$, $\leq$, $>$, $\geq$.

Some examples of linear inequalities are

$$2x + 3y < 6 \qquad y \geq 2x + 1 \qquad x - y \leq 0$$

Although not all of these inequalities have the form $ax + by < c$, each one can be put in that form.

The solution set for a linear inequality is a section of the coordinate plane. The boundary for the section is found by replacing the inequality symbol with an equal sign and graphing the resulting equation. The boundary is included in the solution set (and represented with a solid line) if the inequality symbol used originally is $\leq$ or $\geq$. The boundary is not included (and is represented with a broken line) if the original symbol is $<$ or $>$.

Let's look at some examples.

EXAMPLE 1 Graph the solution set for $x + y \leq 4$.

SOLUTION The boundary for the graph is the graph of $x + y = 4$. The boundary is included in the solution set because the inequality symbol is $\leq$.

The graph of the boundary is shown in Figure 1.

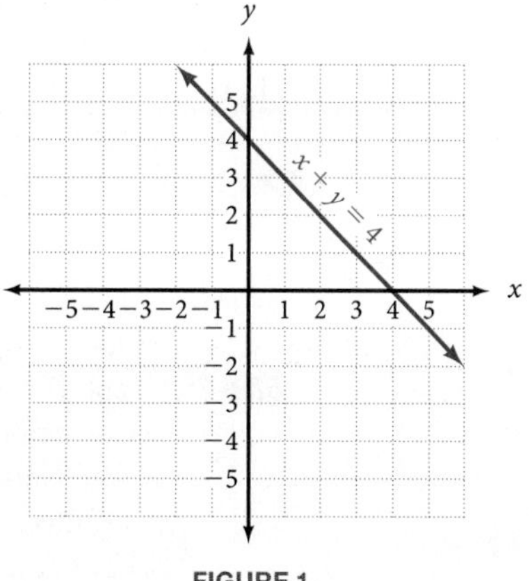

FIGURE 1

The boundary separates the coordinate plane into two sections, or regions: the region above the boundary and the region below the boundary. The solution set for $x + y \leq 4$ is one of these two regions along with the boundary. To find the correct region, we simply choose any convenient point that is *not* on the boundary. We then substitute the coordinates of the point into the original inequality $x + y \leq 4$. If the point we choose satisfies the inequality, then it is a member of the solution set, and we can assume that all points on the same side of the boundary as the chosen point are also in the solution set. If the coordinates of our point do not satisfy the original inequality, then the solution set lies on the other side of the boundary.

In this example a convenient point not on the boundary is the origin. Substituting (0, 0) into $x + y \le 4$ gives us

$$0 + 0 \overset{?}{\le} 4$$

$$0 \le 4 \quad \text{A true statement}$$

Because the origin is a solution to the inequality $x + y \le 4$, and the origin is below the boundary, all other points below the boundary are also solutions.

The graph of $x + y \le 4$ is shown in Figure 2.

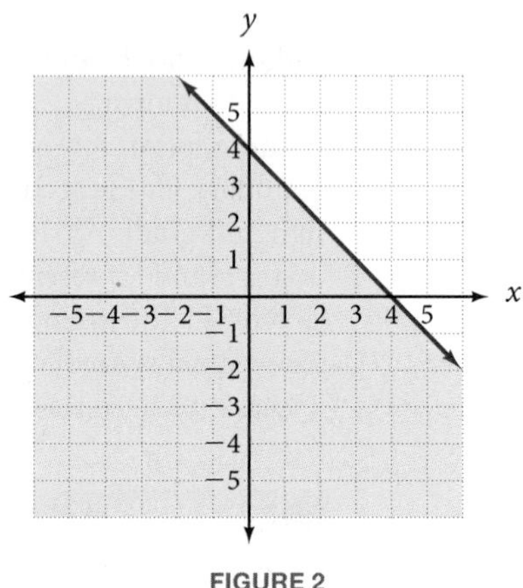

FIGURE 2

The region above the boundary is described by the inequality $x + y > 4$.

Here is a list of steps to follow when graphing the solution set for linear inequalities in two variables.

PROPERTY *To Graph the Solution Set for Linear Inequalities in Two Variables*

Step 1: Replace the inequality symbol with an equal sign. The resulting equation represents the boundary for the solution set.

Step 2: Graph the boundary found in step 1 using a *solid line* if the boundary is included in the solution set (that is, if the original inequality symbol was either $\le$ or $\ge$). Use a *broken line* to graph the boundary if it is *not* included in the solution set. (It is not included if the original inequality was either $<$ or $>$).

Step 3: Choose any convenient point not on the boundary and substitute the coordinates into the *original* inequality. If the resulting statement is *true*, the graph lies on the *same* side of the boundary as the chosen point. If the resulting statement is *false*, the solution set lies on the *opposite* side of the boundary.

EXAMPLE 2 Graph the solution set for $y < 2x - 3$.

SOLUTION The boundary is the graph of $y = 2x - 3$. The boundary is not included because the original inequality symbol is $<$. We therefore use a broken line to represent the boundary, as shown in Figure 3.

A convenient test point is again the origin. Using (0, 0) in $y < 2x - 3$, we have

$$0 \overset{?}{<} 2(0) - 3$$

$$0 < -3 \qquad \text{A false statement}$$

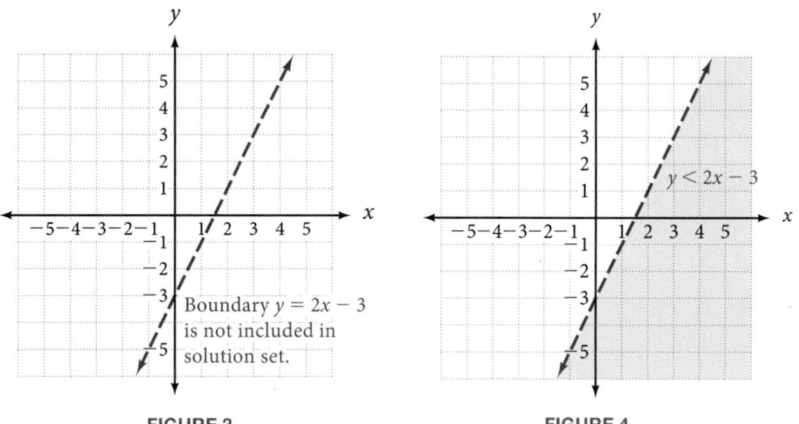

FIGURE 3 FIGURE 4

Because our test point gives us a false statement and it lies above the boundary, the solution set must lie on the other side of the boundary, as shown in Figure 4.

EXAMPLE 3 Graph the inequality $2x + 3y \leq 6$.

SOLUTION We begin by graphing the boundary $2x + 3y = 6$. The boundary is included in the solution because the inequality symbol is $\leq$.

If we use (0, 0) as our test point, we see that it yields a true statement when its coordinates are substituted into $2x + 3y \leq 6$. The graph, therefore, lies below the boundary, as shown in Figure 5.

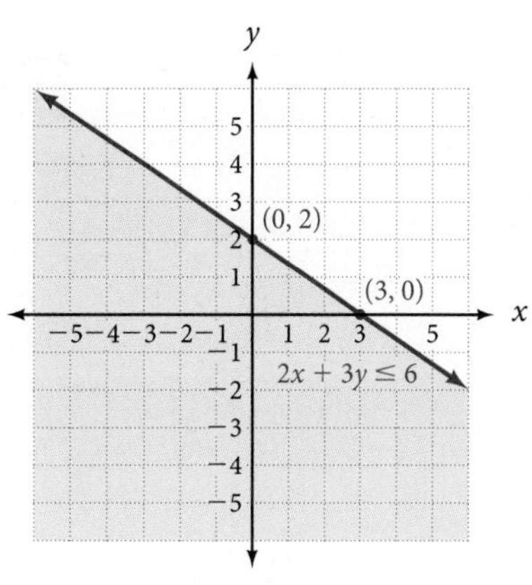

FIGURE 5

The ordered pair (0, 0) is a solution to $2x + 3y \leq 6$; all points on the same side of the boundary as (0, 0) also must be solutions to the inequality $2x + 3y \leq 6$.

EXAMPLE 4 Graph the solution set for $x \le 5$.

SOLUTION The boundary is $x = 5$, which is a vertical line. All points to the left have x-coordinates less than 5, and all points to the right have x-coordinates greater than 5, as shown in Figure 6.

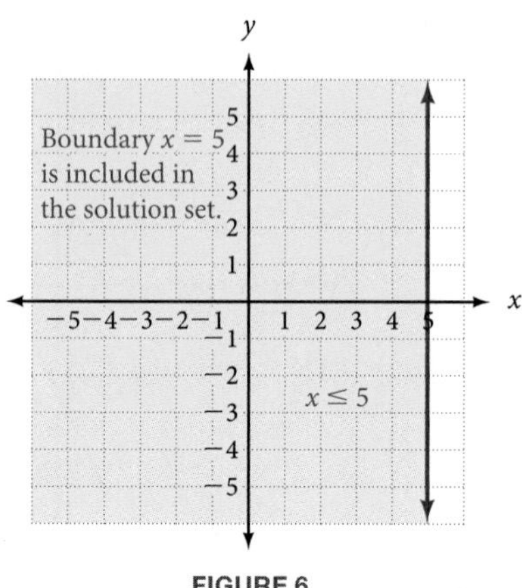

FIGURE 6

GETTING READY FOR CLASS

After reading through the preceding section, respond in your own words and in complete sentences.

A. When graphing a linear inequality in two variables, how do you find the equation of the boundary line?

B. What is the significance of a broken line in the graph of an inequality?

C. When graphing a linear inequality in two variables, how do you know which side of the boundary line to shade?

D. Describe the set of ordered pairs that are solutions to $x + y < 6$.

Problem Set 2.7

Graph the following linear inequalities.

1. $2x - 3y < 6$ **2.** $3x + 2y \geq 6$ **3.** $x - 2y \leq 4$ **4.** $2x + y > 4$

5. $x - y \leq 2$ **6.** $x - y \leq 1$ **7.** $3x - 4y \geq 12$ **8.** $4x + 3y < 12$

9. $5x - y \leq 5$ **10.** $4x + y > 4$ **11.** $2x + 6y \leq 12$ **12.** $x - 5y > 5$

13. $x \geq 1$ **14.** $x < 5$ **15.** $x \geq -3$ **16.** $y \leq -4$

17. $y < 2$ **18.** $3x - y > 1$ **19.** $2x + y > 3$ **20.** $5x + 2y < 2$

21. $y \leq 3x - 1$ **22.** $y \geq 3x + 2$ **23.** $y \leq -\frac{1}{2}x + 2$ **24.** $y < \frac{1}{3}x + 3$

The next two problems are intended to give you practice reading, and paying attention to, the instructions that accompany the problems you are working.

25. Work each problem according to the instructions given.

a. Solve: $4 + 3y < 12$ **b.** Solve: $4 - 3y < 12$

c. Solve for y: $4x + 3y = 12$ **d.** Graph: $y < -\frac{4}{3}x + 4$

26. Work each problem according to the instructions given.

a. Solve: $3x + 2 \geq 6$ **b.** Solve: $-3x + 2 \geq 6$

c. Solve for y: $3x + 2y = 6$ **d.** Graph: $y \geq -\frac{3}{2}x + 3$

27. Find the equation of the line in part a, then use this information to find the inequalities for the graphs on parts b and c.

a. **b.**

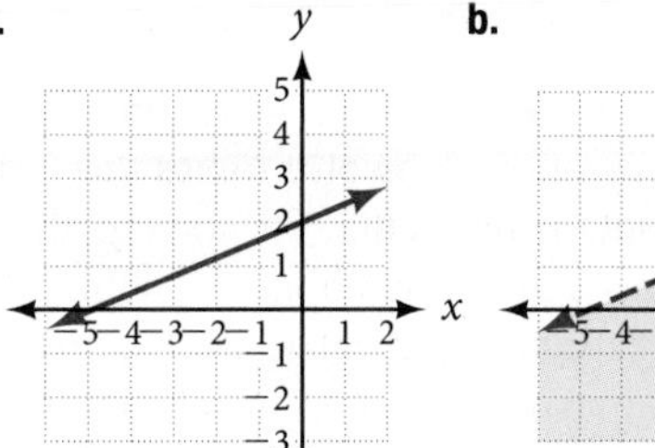

c.

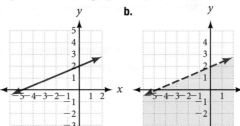

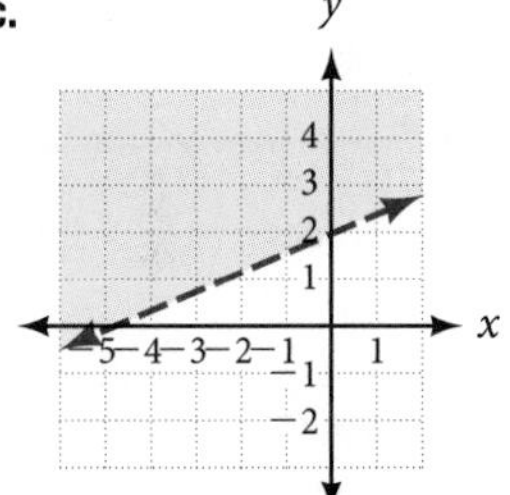

28. Find the equation of the line in part a, then use this information to find the inequalities for the graphs on parts b and c.

a.

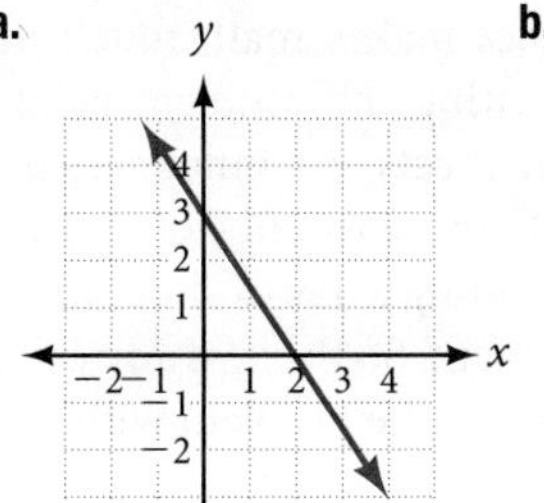

b.

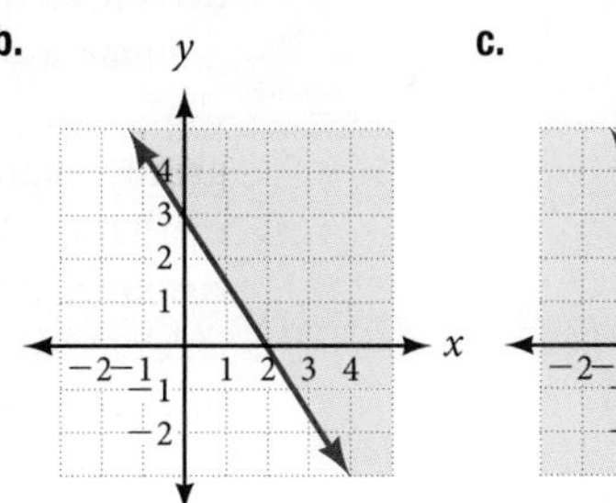

c.

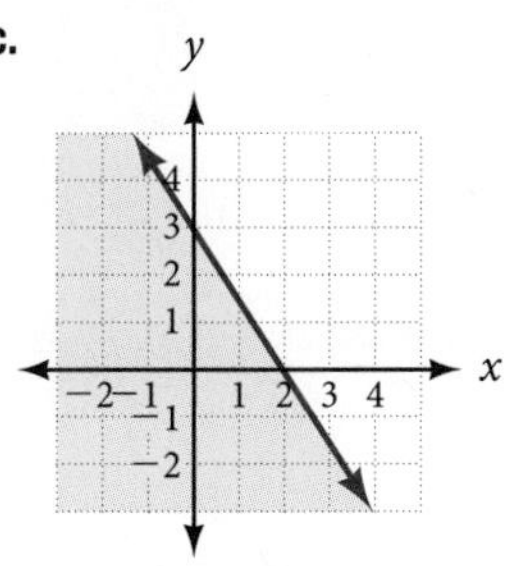

Maintaining Your Skills

29. Simplify the expression $7 - 3(2x - 4) - 8$.

30. Find the value of $x^2 - 2xy + y^2$ when $x = 3$ and $y = -4$.

Solve each equation.

31. $-\frac{3}{2}x = 12$

32. $2x - 4 = 5x + 2$

33. $8 - 2(x + 7) = 2$

34. $3(2x - 5) - (2x - 4) = 6 - (4x + 5)$

35. Solve the formula $P = 2l + 2w$ for w.

Solve each inequality, and graph the solution.

36. $-4x < 20$

37. $3 - 2x > 5$

38. $3 - 4(x - 2) \geq -5x + 6$

39. Solve the formula $3x - 2y \leq 12$ for y.

40. What number is 12% of 2,000?

41. Geometry The length of a rectangle is 5 inches more than 3 times the width. If the perimeter is 26 inches, find the length and width.

SPOTLIGHT ON SUCCESS *Student Instructor Lauren*

There are a lot of word problems in algebra and many of them involve topics that I don't know much about. I am better off solving these problems if I know something about the subject. So, I try to find something I can relate to. For instance, an example may involve the amount of fuel used by a pilot in a jet airplane engine. In my mind, I'd change the subject to something more familiar, like the mileage I'd be getting in my car and the amount spent on fuel, driving from my hometown to my college. Changing these problems to more familiar topics makes math much more interesting and gives me a better chance of getting the problem right. It also helps me to understand how greatly math affects and influences me in my everyday life. We really do use math more than we would like to admit—budgeting our income, purchasing gasoline, planning a day of shopping with friends—almost everything we do is related to math. So the best advice I can give with word problems is to learn how to associate the problem with something familiar to you.

You should know that I have always enjoyed math. I like working out problems and love the challenges of solving equations like individual puzzles. Although there are more interesting subjects to me, and I don't plan on pursuing a career in math or teaching, I do think it's an important subject that will help you in any profession.

Chapter 2 Summary

Linear Equation in Two Variables [2.3]

A linear equation in two variables is any equation that can be put in the form $ax + by = c$. The graph of every linear equation is a straight line.

EXAMPLES

1. The equation $3x + 2y = 6$ is an example of a linear equation in two variables.

Strategy for Graphing Linear Equations in Two Variables [2.3]

Step 1 Find any three ordered pairs that satisfy the equation. This can be done by using a convenient number for one variable and solving for the other variable.

Step 2: Graph the three ordered pairs found in step 1. Actually, we need only two points to graph a straight line. The third point serves as a check. If all three points do not line up, there is a mistake in our work.

Step 3: Draw a straight line through the three points graphed in step 2.

2. The graph of $y = -\frac{2}{3}x - 1$ is shown below.

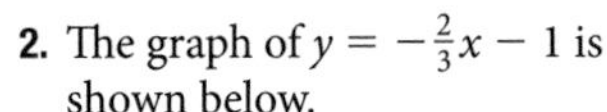

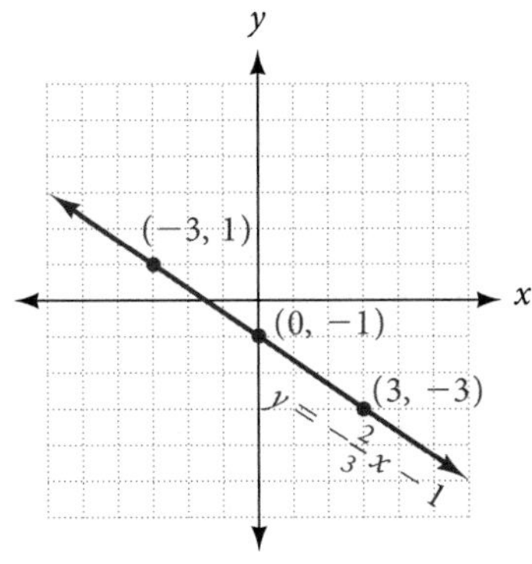

Intercepts [2.4]

The *x-intercept* of an equation is the *x-coordinate* of the point where the graph crosses the *x-axis*. The *y-intercept* is the *y-coordinate* of the point where the graph crosses the *y-axis*. We find the *y-intercept* by substituting $x = 0$ into the equation and solving for y. The *x-intercept* is found by letting $y = 0$ and solving for x.

3. To find the x-intercept for $3x + 2y = 6$, we let $y = 0$ and get

$$3x = 6$$
$$x = 2$$

In this case the x-intercept is 2, and the graph crosses the x-axis at $(2, 0)$.

Slope of a Line [2.5]

The *slope* of the line containing the points (x_1, y_1) and (x_2, y_2) is given by

$$\text{Slope} = m = \frac{y_2 - y_1}{x_2 - x_1} = \frac{\text{rise}}{\text{run}}$$

4. The slope of the line through $(3, -5)$ and $(-2, 1)$ is

$$m = \frac{-5 - 1}{3 - (-2)} = \frac{-6}{5} = -\frac{6}{5}$$

Slope-Intercept Form of a Straight Line [2.6]

The equation of the line with a slope of m and a y-intercept of b is

$$y = mx + b$$

5. The equation of the line with a slope of 2 and a y-intercept of 5 is

$$y = 2x + 5$$

Point-Slope Form of a Straight Line [2.6]

6. The equation of the line through (1, 2) with a slope of 3 is

$$y - 2 = 3(x - 1)$$
$$y - 2 = 3x - 3$$
$$y = 3x - 1$$

If a line has a slope of m and contains the point (x_1, y_1), the equation can be written as

$$y - y_1 = m(x - x_1)$$

To Graph a Linear Inequality in Two Variables [2.7]

7. Graph $x - y \geq 3$.

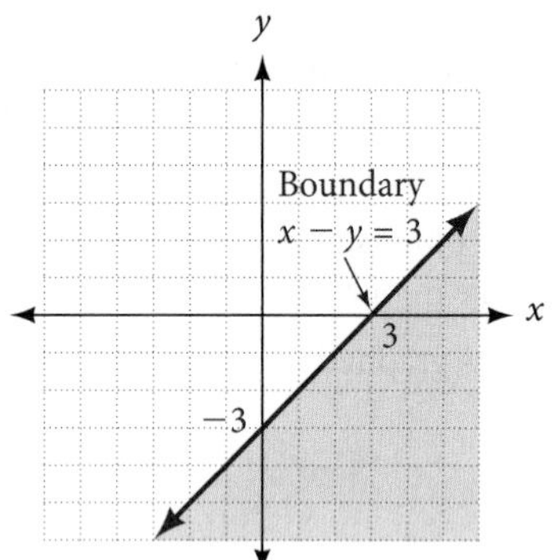

Step 1: Replace the inequality symbol with an equal sign. The resulting equation represents the boundary for the solution set.

Step 2: Graph the boundary found in step 1, using a *solid line* if the original inequality symbol was either ≤, or ≥. Use a *broken line* otherwise.

Step 3: Choose any convenient point not on the boundary and substitute the coordinates into the *original* inequality. If the resulting statement is *true*, the graph lies on the *same* side of the boundary as the chosen point. If the resulting statement is *false*, the solution set lies on the *opposite* side of the boundary.

Graph the ordered pairs. [2.1]

1. (2, −1) **2.** (−4, 3) **3.** (−3, −2) **4.** (0, −4)

5. Fill in the following ordered pairs for the equation $3x - 2y = 6$. [2.2]

(0,) (, 0) (4,) (, −6)

6. Which of the following ordered pairs are solutions to $y = -3x + 7$? [2.2]

(0, 7) (2, −1) (4, −5) (−5, −3)

Graph each line. [2.3]

7. $y = -\frac{1}{2}x + 4$ **8.** $x = -3$

Find the *x*- and *y*-intercepts. [2.4]

9. $8x - 4y = 16$ **10.** $y = \frac{3}{2}x + 6$

11. $y = 3$

Find the slope of the line through each pair of points. [2.5]

12. (3, 2), (−5, 6) **13.** (0, 9), (7, 1)

Find the slope of each line. [2.5]

14.

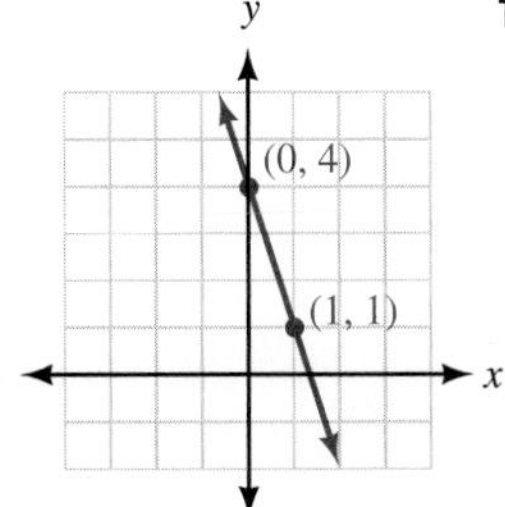

15.

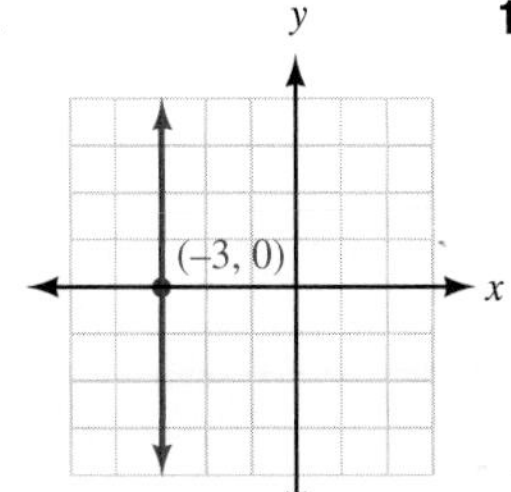

16.

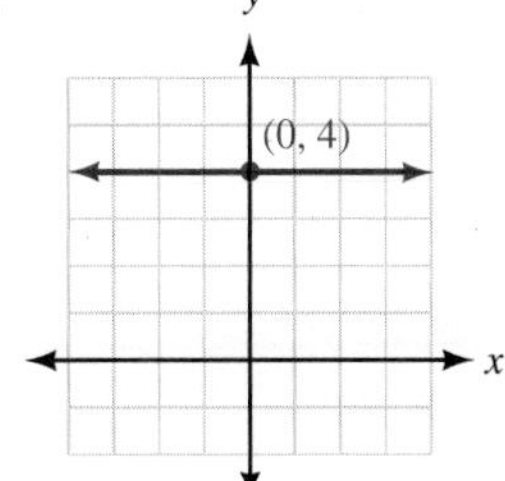

17. Find the equation of the line through (4, 1) with a slope of $-\frac{1}{2}$. [2.6]

18. Find the equation of the line with a slope of 3 and *y*-intercept −5. [2.6]

19. Find the equation of the line passing through the points (3, −4) and (−6, 2) [2.6]

20. A straight line has an *x*-intercept 3 and contains the point (−2, 6). Find its equation. [2.6]

Graph each linear inequality in two variables. [2.7]

21. $y > x - 6$ **22.** $6x - 9y \leq 18$

Functions and Function Notation

3

Chapter Outline

3.1 Introduction to Functions

3.2 Function Notation

3.3 Variation

3.4 Algebra and Composition with Functions

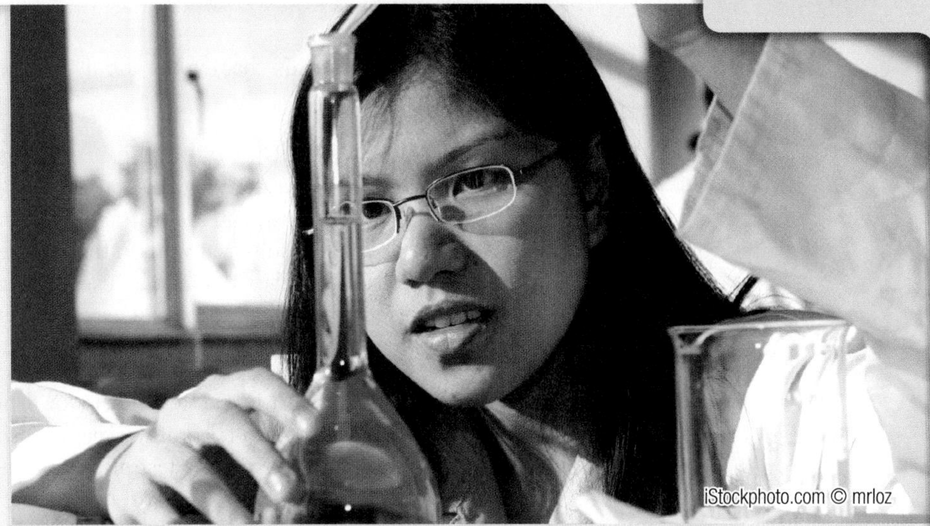

iStockphoto.com © mrloz

A student is heating water in a chemistry lab. As the water heats, she records the temperature readings from two thermometers, one giving temperature in degrees Fahrenheit and the other in degrees Celsius. The table below shows some of the data she collects. The scatter diagram that gives a visual representation of the data in the table.

Corresponding Temperatures

Degrees Fahrenheit	Degrees Celsius
77	25
95	35
167	75
212	100

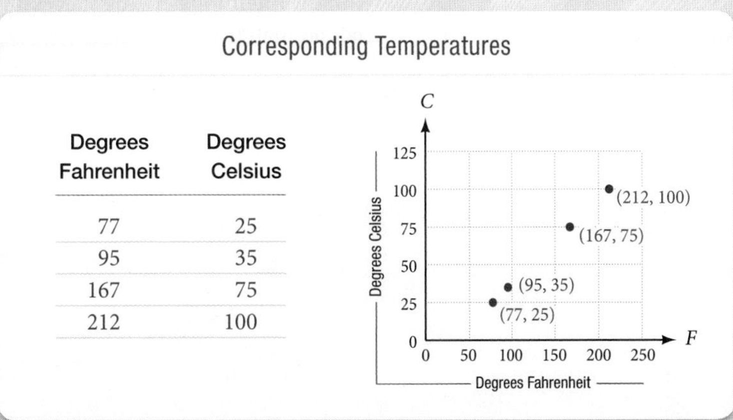

The exact relationship between the Fahrenheit and Celsius temperature scales is given by the formula

$$C = \frac{5}{9}(F - 32)$$

We have three ways to describe the relationship between the two temperature scales: a table, a graph, and an equation. But, most important to us, we don't need to accept this formula on faith. Later, you will derive the formula from the data in the table above.

Study Skills

The study skills for this chapter are about attitude. They are points of view that point toward success.

1. **Be Focused, Not Distracted** I have students who begin their assignments by asking themselves, "Why am I taking this class?" If you are asking yourself similar questions, you are distracting yourself from doing the things that will produce the results you want in this course. Don't dwell on questions and evaluations of the class that can be used as excuses for not doing well. If you want to succeed in this course, focus your energy and efforts toward success, rather than distracting yourself from your goals.

2. **Be Resilient** Don't let setbacks keep you from your goals. You want to put yourself on the road to becoming a person who can succeed in this class, or any class in college. Failing a test or quiz, or having a difficult time on some topics, is normal. No one goes through college without some setbacks. Don't let a temporary disappointment keep you from succeeding in this course. A low grade on a test or quiz is simply a signal that you need to reevaluate your study habits.

3. **Intend to Succeed** I have a few students who simply go through the motions of studying without intending to master the material. It is more important to them to look like they are studying than to actually study. You need to study with the intention of being successful in the course. Intend to master the material, no matter what it takes.

Introduction to Functions

3.1

The ad shown here appeared in the Help Wanted section of the local newspaper the day I was writing this section of the book. If you held the job described in the ad, you would earn \$7.50 for every hour you worked. The amount of money you make in one week depends on the number of hours you work that week. In mathematics, we say that your weekly earnings are a *function* of the number of hours you work.

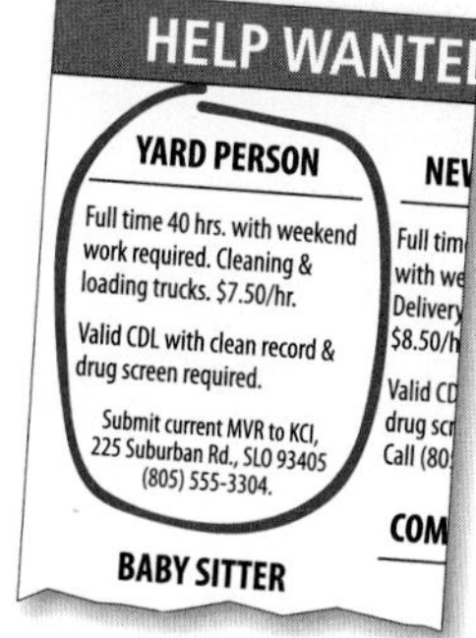

An Informal Look at Functions

Suppose you have a job that pays \$7.50 per hour and that you work anywhere from 0 to 40 hours per week. If we let the variable x represent hours and the variable y represent the money you make, then the relationship between x and y can be written as

$$y = 7.5x \quad \text{for} \quad 0 \le x \le 40$$

EXAMPLE 1 Construct a table and graph for the function

$$y = 7.5x \quad \text{for} \quad 0 \le x \le 40$$

SOLUTION Table 1 gives some of the paired data that satisfy the equation $y = 7.5x$. Figure 1 is the graph of the equation with the restriction $0 \le x \le 40$.

TABLE 1 Weekly Wages

Hours Worked	Rule	Pay
x	$y = 7.5x$	y
0	$y = 7.5(0)$	0
10	$y = 7.5(10)$	75
20	$y = 7.5(20)$	150
30	$y = 7.5(30)$	225
40	$y = 7.5(40)$	300

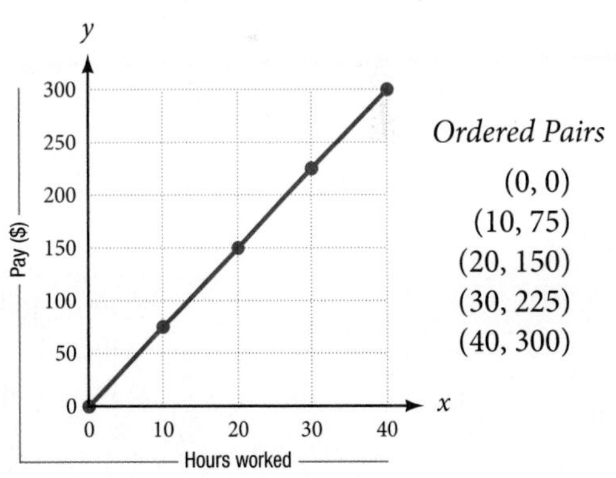

FIGURE 1 *Weekly wages at \$7.50 per hour*

The equation $y = 7.5x$ with the restriction $0 \le x \le 40$, Table 1, and Figure 1 are three ways to describe the same relationship between the number of hours you work in one week and your gross pay for that week. In all three, we *input* values of x, and then use the function rule to *output* values of y.

Domain and Range of a Function

We began this discussion by saying that the number of hours worked during the week was from 0 to 40, so these are the values that x can assume. From the line graph in Figure 1, we see that the values of y range from 0 to 300. We call the complete set of values that x can assume the *domain* of the function. The values that are assigned to y are called the *range* of the function.

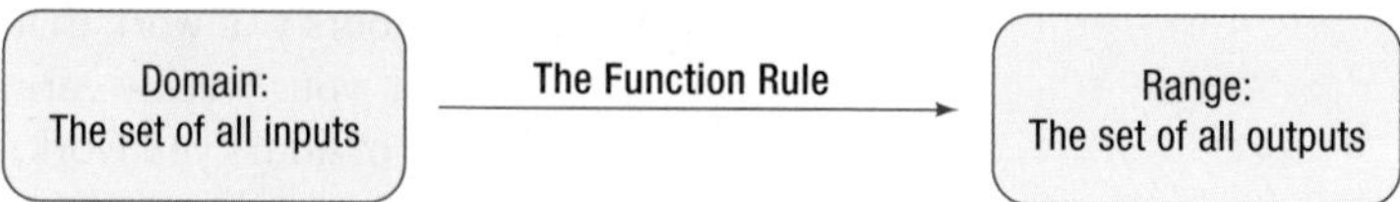

EXAMPLE 2 State the domain and range for the function

$$y = 7.5x, \quad 0 \le x \le 40$$

SOLUTION From the previous discussion, we have

Domain $= \{x \mid 0 \le x \le 40\}$

Range $= \{y \mid 0 \le y \le 300\}$

Function Maps

Another way to visualize the relationship between x and y is with the diagram in Figure 2, which we call a *function map*.

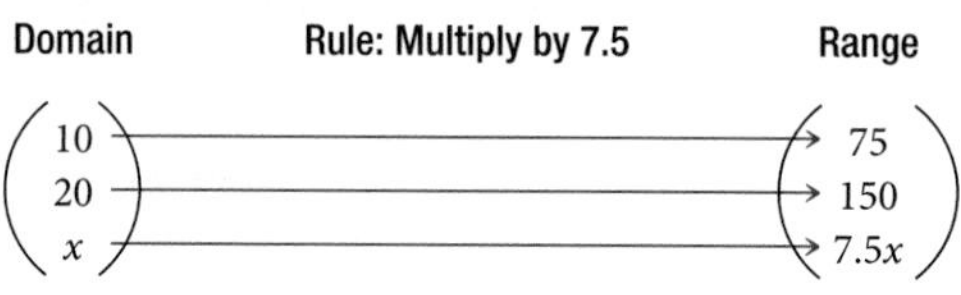

FIGURE 2 *A function map*

Although the diagram in Figure 2 does not show all the values that x and y can assume, it does give us a visual description of how x and y are related. It shows that values of y in the range come from values of x in the domain according to a specific rule (multiply by 7.5 each time).

A Formal Look at Functions

We are now ready for the formal definition of a function.

DEFINITION *function*

A ***function*** is a rule that pairs each element in one set, called the ***domain,*** with exactly one element from a second set, called the ***range.***

In other words, a function is a rule for which each input is paired with exactly one output.

EXAMPLE 3 Kendra tosses a softball into the air with an underhand motion. The distance of the ball above her hand is given by the function

$$h = 32t - 16t^2 \quad \text{for} \quad 0 \leq t \leq 2$$

where h is the height of the ball in feet and t is the time in seconds. Construct a table that gives the height of the ball at quarter-second intervals, starting with $t = 0$ and ending with $t = 2$, then graph the function.

SOLUTION We construct Table 2 using the following values of t: $0, \frac{1}{4}, \frac{1}{2}, \frac{3}{4}, 1, \frac{5}{4}, \frac{3}{2}, \frac{7}{4}, 2$. Then we construct the graph in Figure 3 from the table. The graph appears only in the first quadrant because neither t nor h can be negative.

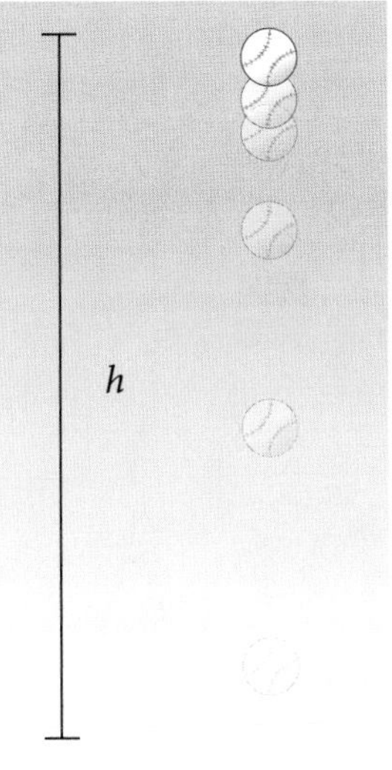

TABLE 2 Tossing a Softball into the Air

Input		Output
Time (sec) t	Function Rule $h = 32t - 16t^2$	Distance (ft) h
0	$h = 32(0) - 16(0)^2 = 0 - 0 = 0$	0
$\frac{1}{4}$	$h = 32\left(\frac{1}{4}\right) - 16\left(\frac{1}{4}\right)^2 = 8 - 1 = 7$	7
$\frac{1}{2}$	$h = 32\left(\frac{1}{2}\right) - 16\left(\frac{1}{2}\right)^2 = 16 - 4 = 12$	12
$\frac{3}{4}$	$h = 32\left(\frac{3}{4}\right) - 16\left(\frac{3}{4}\right)^2 = 24 - 9 = 15$	15
1	$h = 32(1) - 16(1)^2 = 32 - 16 = 16$	16
$\frac{5}{4}$	$h = 32\left(\frac{5}{4}\right) - 16\left(\frac{5}{4}\right)^2 = 40 - 25 = 15$	15
$\frac{3}{2}$	$h = 32\left(\frac{3}{2}\right) - 16\left(\frac{3}{2}\right)^2 = 48 - 36 = 12$	12
$\frac{7}{4}$	$h = 32\left(\frac{7}{4}\right) - 16\left(\frac{7}{4}\right)^2 = 56 - 49 = 7$	7
2	$h = 32(2) - 16(2)^2 = 64 - 64 = 0$	0

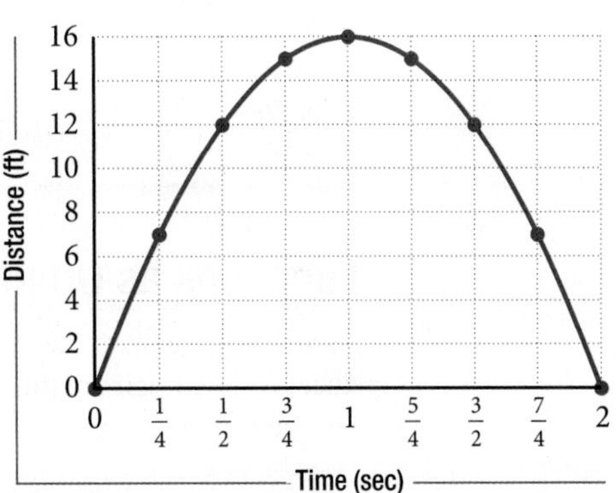

FIGURE 3

Here is a summary of what we know about functions as it applies to this example: We input values of t and output values of h according to the function rule

$$h = 32t - 16t^2 \quad \text{for} \quad 0 \leq t \leq 2$$

The domain is given by the inequality that follows the equation; it is

$$\text{Domain} = \{t \mid 0 \leq t \leq 2\}$$

The range is the set of all outputs that are possible by substituting the values of t from the domain into the equation. From our table and graph, it seems that the range is

$$Range = \{h \mid 0 \le h \le 16\}$$

USING TECHNOLOGY *More About Example 3*

Most graphing calculators can easily produce the information in Table 2. Simply set Y_1 equal to $32X - 16X^2$. Then set up the table so it starts at 0 and increases by an increment of 0.25 each time. (On a TI-82/83, use the TBLSET key to set up the table.)

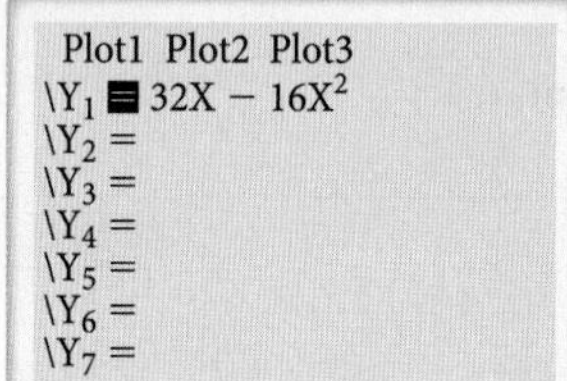

TABLE SETUP
TblStart = 0
ΔTbl = .25
Indpnt: Auto Ask
Depend: Auto Ask

The table will look like this:

X	Y_1
0	0
.25	7
.5	12
.75	15
1	16
1.25	15
1.5	12

Graph each equation and build a table as indicated.

1. $y = 64t - 16t^2$ TblStart = 0 ΔTbl = 1

2. $y = \frac{1}{2}x - 4$ TblStart = −5 ΔTbl = 1

3. $y = \frac{12}{x}$ TblStart = 0.5 ΔTbl = 0.5

Functions as Ordered Pairs

As you can see from the examples we have done to this point, the function rule produces ordered pairs of numbers. We use this result to write an alternative definition for a function.

ALTERNATE DEFINITION *function*

A ***function*** is a set of ordered pairs in which no two different ordered pairs have the same first coordinate. The set of all first coordinates is called the ***domain*** of the function. The set of all second coordinates is called the ***range*** of the function.

The restriction on first coordinates in the alternative definition keeps us from assigning a number in the domain to more than one number in the range.

A Relationship That is Not a Function

You may be wondering if any sets of paired data fail to qualify as functions. The answer is yes, as the next example reveals.

EXAMPLE 4 Table 3 shows the prices of used Ford Mustangs that were listed in the local newspaper. The diagram in Figure 4 is called a *scatter diagram*. It gives a visual representation of the data in Table 3. Why is this data not a function?

TABLE 3 Used Mustang Prices

Year x	Price (\$) y
1997	13,925
1997	11,850
1997	9,995
1996	10,200
1996	9,600
1995	9,525
1994	8,675
1994	7,900
1993	6,975

Used Mustang Prices

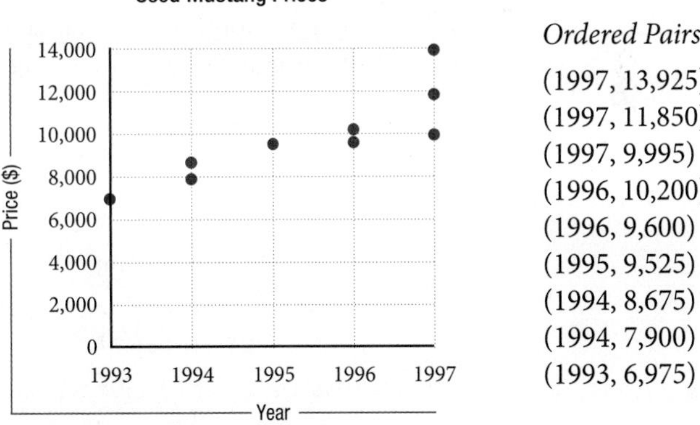

FIGURE 4 *Scatter diagram of data in Table 3*

SOLUTION In Table 3, the year 1997 is paired with three different prices: \$13,925, \$11,850, and \$9,995. That is enough to disqualify the data from belonging to a function. For a set of paired data to be considered a function, each number in the domain must be paired with exactly one number in the range.

Still, there is a relationship between the first coordinates and second coordinates in the used car data. It is not a function relationship, but it is a relationship. To classify all relationships specified by ordered pairs, whether they are functions or not, we include the following two definitions.

DEFINITION *relation*

A ***relation*** is a rule that pairs each element in one set, called the domain, with ***one or more elements*** from a second set, called the ***range.***

ALTERNATE DEFINITION *relation*

A ***relation*** is a set of ordered pairs. The set of all first coordinates is the ***domain*** of the relation. The set of all second coordinates is the ***range*** of the relation.

Here are some facts that will help clarify the distinction between relations and functions:

1. Any rule that assigns numbers from one set to numbers in another set is a relation. If that rule makes the assignment so no input has more than one output, then it is also a function.
2. Any set of ordered pairs is a relation. If none of the first coordinates of those ordered pairs is repeated, the set of ordered pairs is also a function.
3. Every function is a relation.
4. Not every relation is a function.

EXAMPLE 5 Sketch the graph of $x = y^2$.

SOLUTION Without going into much detail, we graph the equation $x = y^2$ by finding a number of ordered pairs that satisfy the equation, plotting these points, then drawing a smooth curve that connects them. A table of values for x and y that satisfy the equation follows, along with the graph of $x = y^2$ shown in Figure 5.

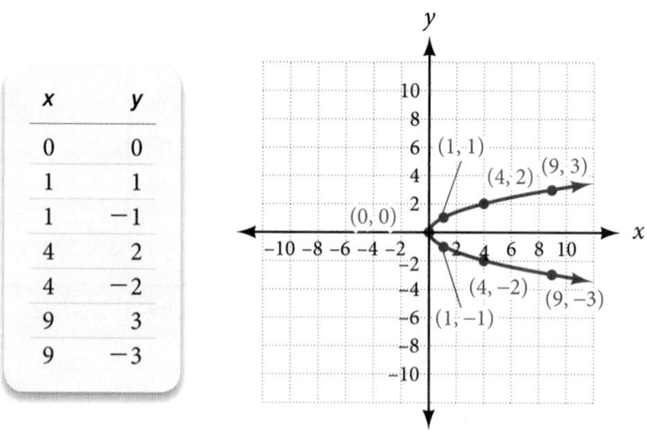

x	y
0	0
1	1
1	−1
4	2
4	−2
9	3
9	−3

FIGURE 5

As you can see from looking at the table and the graph in Figure 5, several ordered pairs whose graphs lie on the curve have repeated first coordinates, for instance (1, 1) and (1, −1), (4, 2) and (4, −2), as well as (9, 3) and (9, −3). The graph is therefore not the graph of a function.

Vertical Line Test

Look back at the scatter diagram for used Mustang prices shown in Figure 4. Notice that some of the points on the diagram lie above and below each other along vertical lines. This is an indication that the data do not constitute a function. Two data points that lie on the same vertical line must have come from two ordered pairs with the same first coordinates.

Now, look at the graph shown in Figure 5. The reason this graph is the graph of a relation, but not of a function, is that some points on the graph have the same first coordinates, for example, the points (4, 2) and (4, -2). Furthermore, any time two points on a graph have the same first coordinates, those points must lie on a vertical line. [To convince yourself, connect the points (4, 2) and (4, -2) with a straight line. You will see that it must be a vertical line.] This allows us to write the following test that uses the graph to determine whether a relation is also a function.

RULE *Vertical Line Test*

If a vertical line crosses the graph of a relation in more than one place, the relation cannot be a function. If no vertical line can be found that crosses a graph in more than one place, then the graph represents a function.

If we look back to the graph of $h = 32t - 16t^2$ as shown in Figure 3, we see that no vertical line can be found that crosses this graph in more than one place. The graph shown in Figure 3 is therefore the graph of a function.

EXAMPLE 6 Match each relation with its graph, then indicate which relations are functions

a. $y = |x| - 4$ **b.** $y = x^2 - 4$ **c.** $y = 2x + 2$

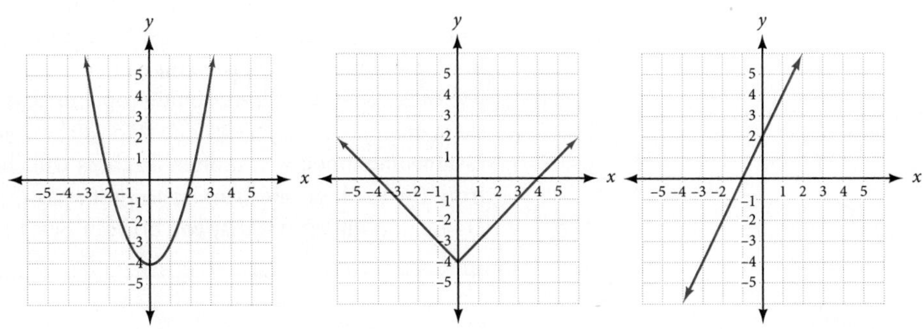

FIGURE 6 FIGURE 7 FIGURE 8

SOLUTION Using the basic graphs for a guide along with our knowledge of translations, we have the following:

a. Figure 7 **b.** Figure 6 **c.** Figure 8

And, since all graphs pass the vertical line test, all are functions.

GETTING READY FOR CLASS

After reading through the preceding section, respond in your own words and in complete sentences.

A. What is a function?
B. What is the vertical line test?
C. Is every line the graph of a function? Explain.
D. Which variable is usually associated with the domain of a function?

SPOTLIGHT ON SUCCESS *Student Instructor CJ*

We are what we repeatedly do.
Excellence, then, is not an act, but a habit.
—Aristotle

Something that has worked for me in college, in addition to completing the assigned homework, is working on some extra problems from each section. Working on these extra problems is a great habit to get into because it helps further your understanding of the material, and you see the many different types of problems that can arise. If you have completed every problem that your book offers, and you still don't feel confident that you have a full grasp of the material, look for more problems. Many problems can be found online or in other books. Your professors may even have some problems that they would suggest doing for extra practice. The biggest benefit to working all the problems in the course's assigned textbook is that often teachers will choose problems either straight from the book or ones similar to problems that were not assigned for tests. Doing this will ensure that you do your best in all your classes.

For each of the following relations, give the domain and range, and indicate which are also functions.

1. (1, 2), (3, 4), (5, 6), (7, 8)

2. (2, 1), (4, 3), (6, 5), (8, 7)

3. (2, 5), (3, 4), (1, 4), (0, 6)

4. (0, 4), (1, 6), (2, 4), (1, 5)

5. $(a, 3), (b, 4), (c, 3), (d, 5)$

6. $(a, 5), (b, 5), (c, 4), (d, 5)$

7. $(a, 1), (a, 2), (a, 3), (a, 4)$

8. $(a, 1), (b, 1), (c, 1), (d, 1)$

State whether each of the following graphs represents a function.

9.

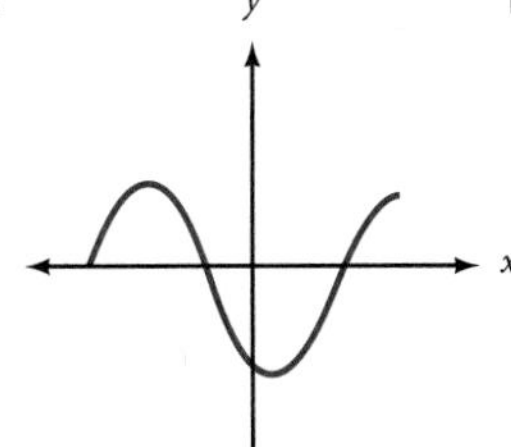

10.

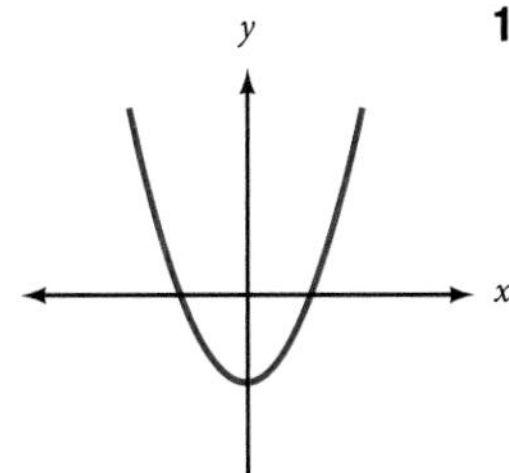

11.

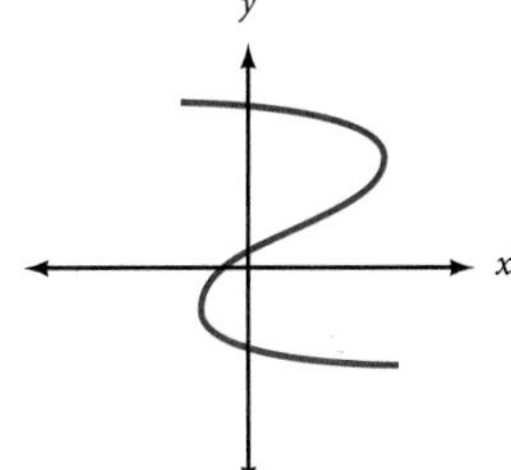

12.

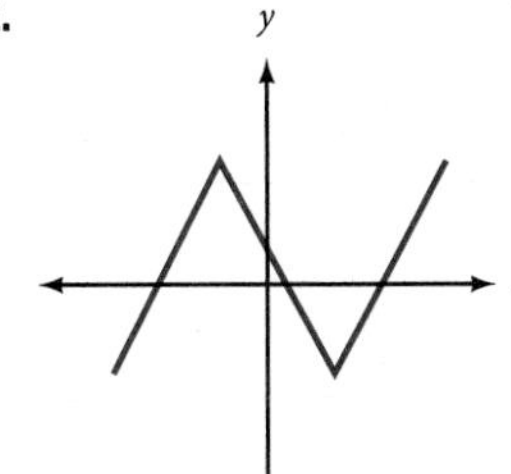

13.

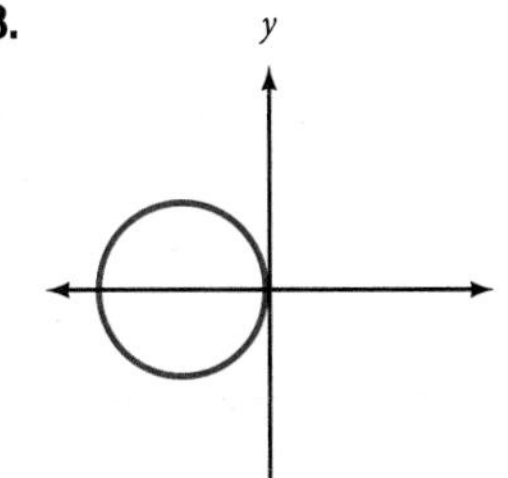

14.

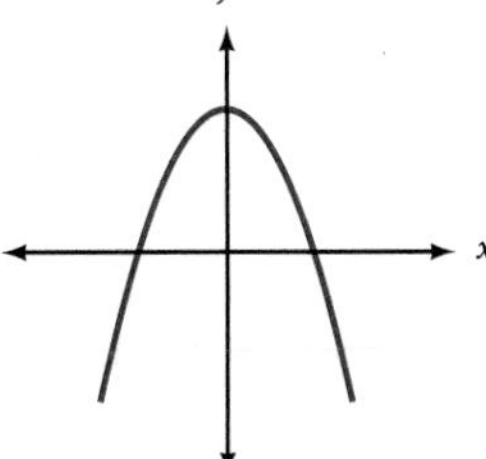

15.

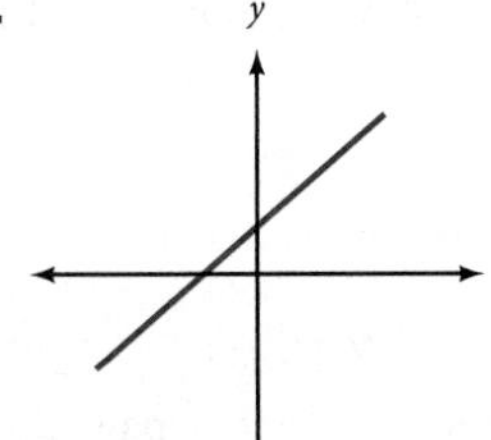

16.

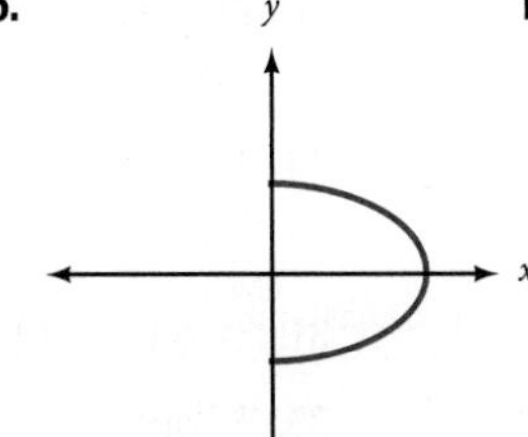

17.

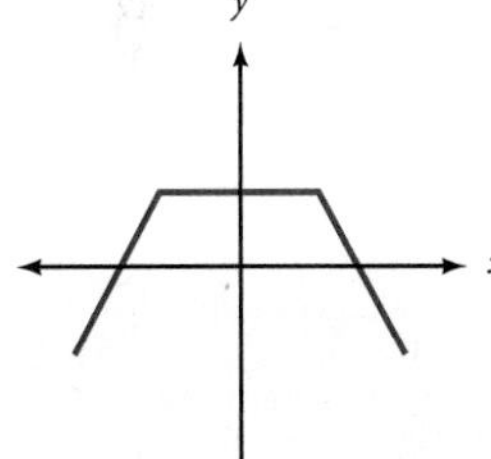

18.

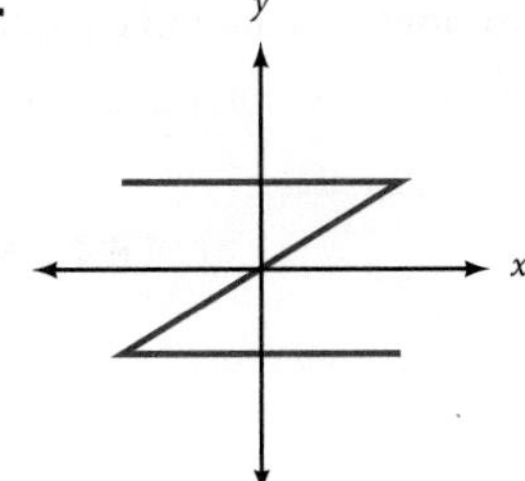

Determine the domain and range of the following functions. Assume the *entire* function is shown.

19.

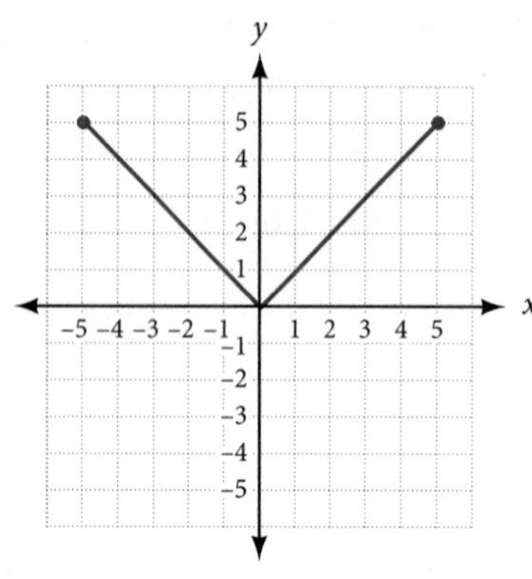

20.

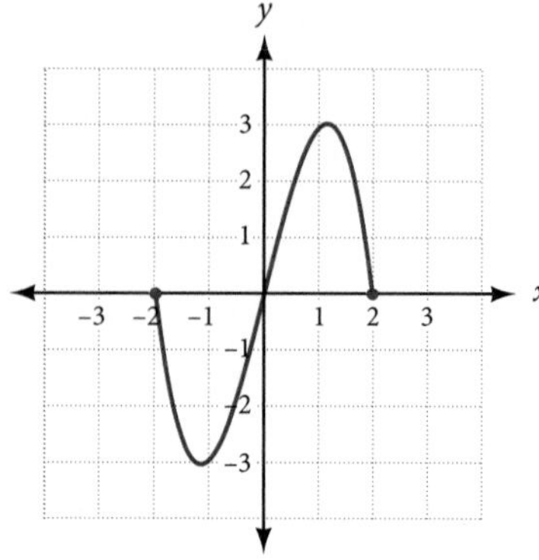

21.

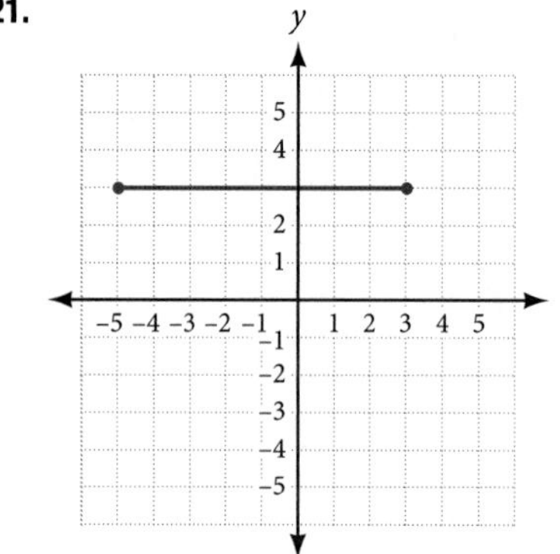

22.

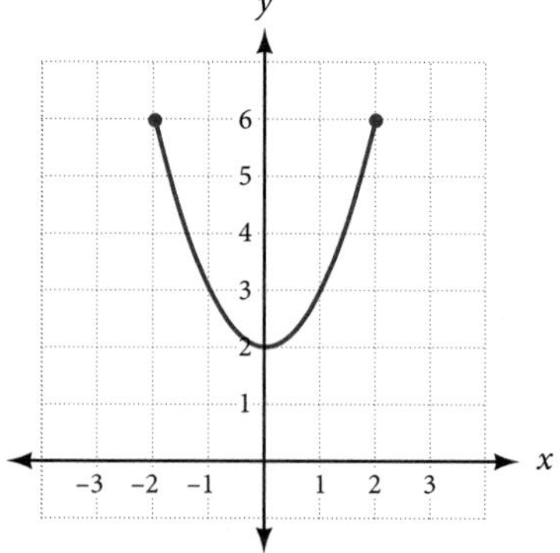

Graph each of the following relations. In each case, use the graph to find the domain and range, and indicate whether the graph is the graph of a function.

23. $y = x^2 - 1$ **24.** $y = x^2 + 1$ **25.** $y = x^2 + 4$ **26.** $y = x^2 - 9$

27. $x = y^2 - 1$ **28.** $x = y^2 + 1$ **29.** $y = (x + 2)^2$ **30.** $y = (x - 3)^2$

31. $x = (y + 1)^2$ **32.** $x = 3 - y^2$

33. Suppose you have a job that pays \$8.50 per hour and you work anywhere from 10 to 40 hours per week.

a. Write an equation, with a restriction on the variable x, that gives the amount of money, y, you will earn for working x hours in one week.

b. Use the function rule you have written in part **a.** to complete Table 4.

TABLE 4 Weekly Wages

Hours Worked	Function Rule	Gross Pay ($)
x		y
10		
20		
30		
40		

c. Construct a line graph from the information in Table 4.

d. State the domain and range of this function.

e. What is the minimum amount you can earn in a week with this job? What is the maximum amount?

34. The ad shown here was in the local newspaper. Suppose you are hired for the job described in the ad.

312 HELP WANTED

ESPRESSO BAR OPERATOR

Must be dependable, honest, serivce-oriented. Coffee exp desired. 15-30 hrs per wk. $5.25/hour. Start 5/31. Apply in person:

Espresso Yourself
Central Coast Mall

Deadline 5/23

a. If x is the number of hours you work per week and y is your weekly gross pay, write the equation for y. (Be sure to include any restrictions on the variable x that are given in the ad.)

b. Use the function rule you have written in part **a.** to complete Table 5.

TABLE 5 Weekly Wages

Hours Worked	Function Rule	Gross Pay ($)
x		y
15		
20		
25		
30		

c. Construct a line graph from the information in Table 5.

d. State the domain and range of this function.

e. What is the minimum amount you can earn in a week with this job? What is the maximum amount?

35. Camera Phones The chart shows the estimated number of camera phones and non-camera phones sold from 2004 to 2010. Using the chart, list all the values in the domain and range for the total phones sales.

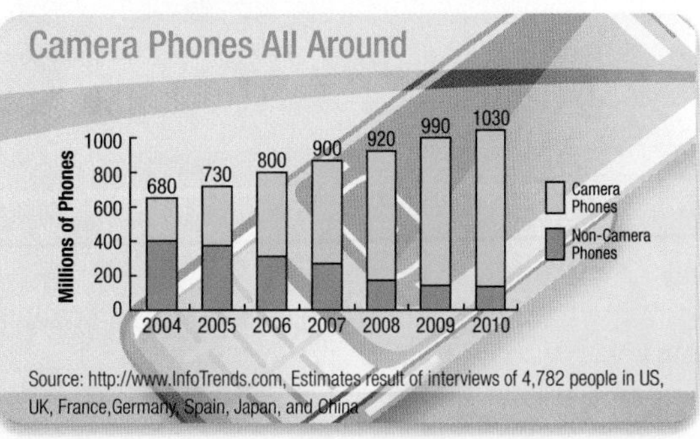

36. Light Bulbs The chart shows a comparison of power usage between incandescent and energy efficient light bulbs. Use the chart to state the domain and range of the function for an energy efficient bulb.

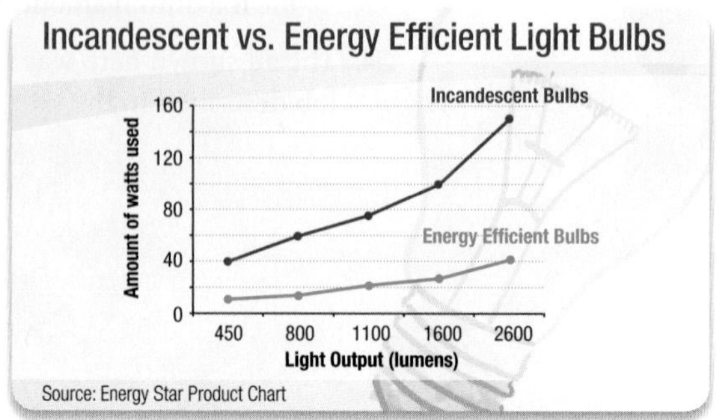

37. Profits Match each of the following statements to the appropriate graph indicated by labels I–IV.

a. Sarah works 25 hours to earn $250.

b. Justin works 35 hours to earn $560.

c. Rosemary works 30 hours to earn $360.

d. Marcus works 40 hours to earn $320.

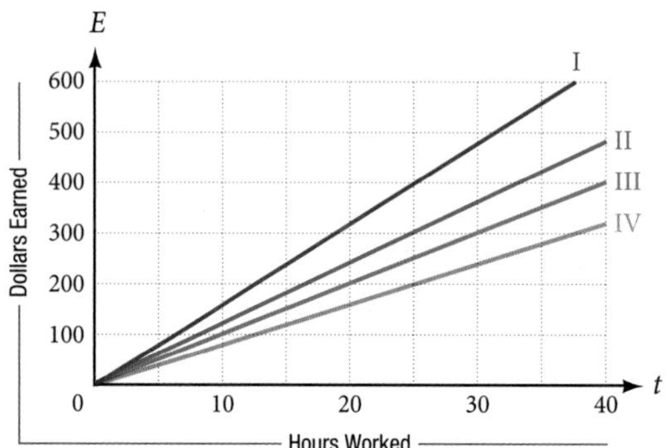

38. Find an equation for each of the functions shown in the graph. Show dollars earned, E, as a function of hours worked, t. Then, indicate the domain and range of each function.

a. Graph I: $E =$ Domain $= \{t \mid \quad\}$ Range $= \{E \mid \quad\}$

b. Graph II: $E =$ Domain $= \{t \mid \quad\}$ Range $= \{E \mid \quad\}$

c. Graph III: $E =$ Domain $= \{t \mid \quad\}$ Range $= \{E \mid \quad\}$

d. Graph IV: $E =$ Domain $= \{t \mid \quad\}$ Range $= \{E \mid \quad\}$

Getting Ready for the Next Section

Simplify. Round to the nearest whole number if necessary.

39. $4(3.14)(9)$

40. $\frac{4}{3}(3.14) \cdot 3^3$

41. $4(-2) - 1$

42. $3(3)^2 + 2(3) - 1$

43. If $s = \frac{60}{t}$, find s when

a. $t = 10$

b. $t = 8$

44. If $y = 3x^2 + 2x - 1$, find y when

a. $x = 0$

b. $x = -2$

45. Find the value of $x^2 + 2$ for

a. $x = 5$

b. $x = -2$

46. Find the value of $125 \cdot 2^t$ for

a. $t = 0$

b. $t = 1$

For the equation $y = x^2 - 3$:

47. Find y if x is 2.

48. Find y if x is -2.

49. Find y if x is 0.

50. Find y if x is -4.

The problems that follow review some of the more important skills you have learned in previous sections and chapters.

51. If $x - 2y = 4$, and $x = \frac{8}{5}$ find y.

52. If $\frac{x^2}{25} + \frac{y^2}{9} = 1$, find y when x is -4.

53. Let $x = 0$ and $y = 0$ in $y = a(x - 8)^2 + 70$ and solve for a.

54. Find R if $p = 2.5$ and $R = (900 - 300p)p$.

3.2 Function Notation

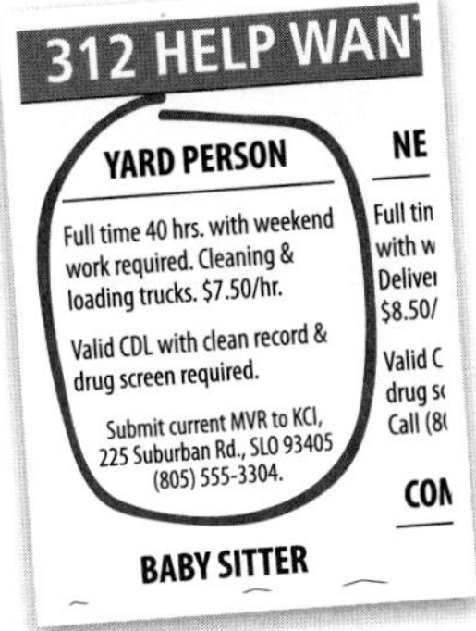

Let's return to the discussion that introduced us to functions. If a job pays \$7.50 per hour for working from 0 to 40 hours a week, then the amount of money y earned in one week is a function of the number of hours worked x. The exact relationship between x and y is written

$$y = 7.5x \quad \text{for} \quad 0 \le x \le 40$$

Because the amount of money earned y depends on the number of hours worked x, we call y the *dependent variable* and x the *independent variable*. Furthermore, if we let f represent all the ordered pairs produced by the equation, then we can write

$$f = \{(x, y) \mid y = 7.5x \quad \text{and} \quad 0 \le x \le 40\}$$

Once we have named a function with a letter, we can use an alternative notation to represent the dependent variable y. The alternative notation for y is $f(x)$. It is read "f of x" and can be used instead of the variable y when working with functions. The notation y and the notation $f(x)$ are equivalent. That is,

$$y = 7.5x \Leftrightarrow f(x) = 7.5x$$

When we use the notation $f(x)$ we are using *function notation*. The benefit of using function notation is that we can write more information with fewer symbols than we can by using just the variable y. For example, asking how much money a person will make for working 20 hours is simply a matter of asking for $f(20)$. Without function notation, we would have to say, "Find the value of y that corresponds to a value of $x = 20$." To illustrate further, using the variable y, we can say "y is 150 when x is 20." Using the notation $f(x)$, we simply say "$f(20) = 150$." Each expression indicates that you will earn \$150 for working 20 hours.

EXAMPLE 1 If $f(x) = 7.5x$, find $f(0)$, $f(10)$, and $f(20)$.

SOLUTION To find $f(0)$, we substitute 0 for x in the expression $7.5x$ and simplify. We find $f(10)$ and $f(20)$ in a similar manner—by substitution.

If $\quad f(x) = 7.5x$

then $\quad f(0) = 7.5(0) = 0$

$f(10) = 7.5(10) = 75$

$f(20) = 7.5(20) = 150$

Note Some students like to think of functions as machines. Values of x are put into the machine, which transforms them into values of $f(x)$, which are then output by the machine.

If we changed the example in the discussion that opened this section so the hourly wage was \$6.50 per hour, we would have a new equation to work with, namely,

$$y = 6.5x \qquad \text{for} \qquad 0 \le x \le 40$$

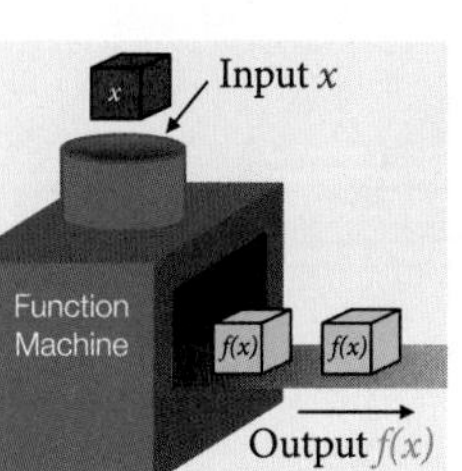

Suppose we name this new function with the letter g. Then

$$g = \{(x, y) \mid y = 6.5x \quad \text{and} \quad 0 \le x \le 40\}$$

and

$$g(x) = 6.5x$$

If we want to talk about both functions in the same discussion, having two different letters, f and g, makes it easy to distinguish between them. For example, since

$f(x) = 7.5x$ and $g(x) = 6.5x$, asking how much money a person makes for working 20 hours is simply a matter of asking for $f(20)$ or $g(20)$, avoiding any confusion over which hourly wage we are talking about.

The diagrams shown in Figure 1 further illustrate the similarities and differences between the two functions we have been discussing.

Note The symbol $\in$ means "is a member of".

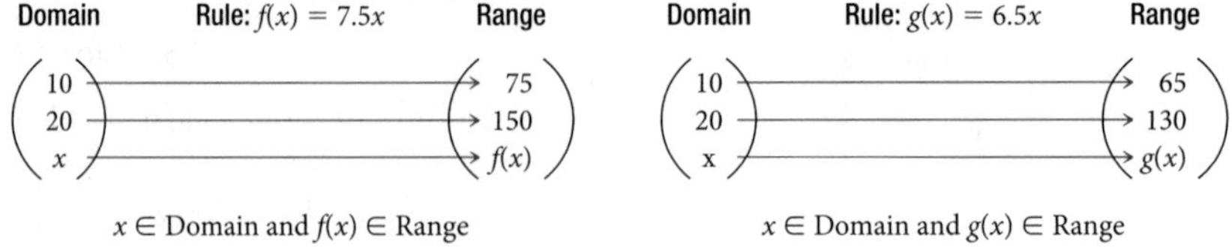

FIGURE 1 *Function maps*

Function Notation and Graphs

We can visualize the relationship between x and $f(x)$ on the graph of the function. Figure 2 shows the graph of $f(x) = 7.5x$ along with two additional line segments. The horizontal line segment corresponds to $x = 20$, and the vertical line segment corresponds to $f(20)$. (Note that the domain is restricted to $0 \leq x \leq 40$.)

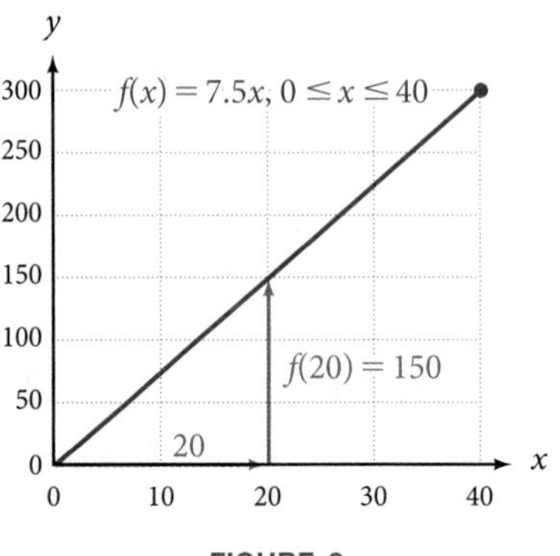

FIGURE 2

We can use functions and function notation to talk about numbers in the chart on gasoline prices. Let's let x represent one of the years in the chart.

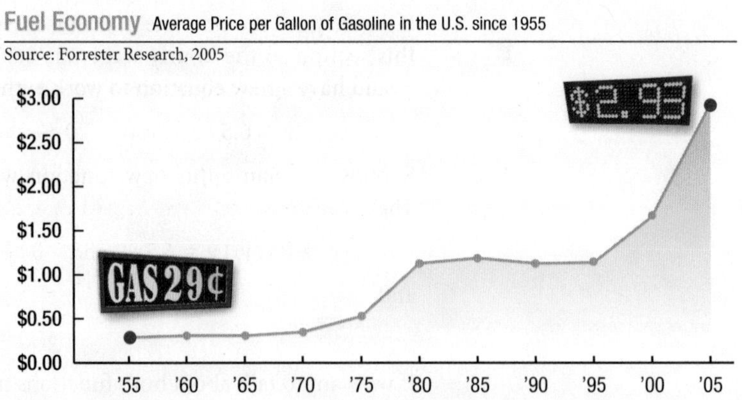

If the function f pairs each year in the chart with the average price of regular gasoline for that year, then each statement below is true:

$$f(1955) = \$0.29$$

The domain of f =
{1955, 1960, 1965, 1970, 1975, 1980, 1985, 1990, 1995, 2000, 2005}

In general, when we refer to the function f we are referring to the domain, the range, and the rule that takes elements in the domain and outputs elements in the range. When we talk about $f(x)$ we are talking about the rule itself, or an element in the range, or the variable y.

The function f

Domain of f	$y = f(x)$	Range of f
Inputs	*Rule*	*Outputs*

Using Function Notation

The remaining examples in this section show a variety of ways to use and interpret function notation.

EXAMPLE 2 If it takes Lorena t minutes to run a mile, then her average speed s, in miles per hour, is given by the formula

$$s(t) = \frac{60}{t} \qquad \text{for} \qquad t > 0$$

Find $s(10)$ and $s(8)$, and then explain what they mean.

SOLUTION To find $s(10)$, we substitute 10 for t in the equation and simplify:

$$s(10) = \frac{60}{10} = 6$$

In words: When Lorena runs a mile in 10 minutes, her average speed is 6 miles per hour.

We calculate $s(8)$ by substituting 8 for t in the equation. Doing so gives us

$$s(8) = \frac{60}{8} = 7.5$$

In words: Running a mile in 8 minutes is running at a rate of 7.5 miles per hour.

EXAMPLE 3 A painting is purchased as an investment for \$125. If its value increases continuously so that it doubles every 5 years, then its value is given by the function

$$V(t) = 125 \cdot 2^{t/5} \qquad \text{for} \qquad t \geq 0$$

where t is the number of years since the painting was purchased, and V is its value (in dollars) at time t. Find $V(5)$ and $V(10)$, and explain what they mean.

SOLUTION The expression $V(5)$ is the value of the painting when $t = 5$ (5 years after it is purchased). We calculate $V(5)$ by substituting 5 for t in the equation $V(t) = 125 \cdot 2^{t/5}$. Here is our work:

$$V(5) = 125 \cdot 2^{5/5} = 125 \cdot 2^1 = 125 \cdot 2 = 250$$

In words: After 5 years, the painting is worth \$250.

The expression $V(10)$ is the value of the painting after 10 years. To find this number, we substitute 10 for t in the equation:

$$V(10) = 125 \cdot 2^{10/5} = 125 \cdot 2^2 = 125 \cdot 4 = 500$$

In words: The value of the painting 10 years after it is purchased is \$500.

EXAMPLE 4 A balloon has the shape of a sphere with a radius of 3 inches. Use the following formulas to find the volume and surface area of the balloon.

$$V(r) = \frac{4}{3}\pi r^3 \qquad S(r) = 4\pi r^2$$

SOLUTION As you can see, we have used function notation to write the formulas for volume and surface area, because each quantity is a function of the radius. To find these quantities when the radius is 3 inches, we evaluate $V(3)$ and $S(3)$:

$$\begin{aligned} V(3) &= \frac{4}{3}\pi \cdot 3^3 = \frac{4}{3}\pi \cdot 27 \\ &= 36\pi \text{ cubic inches, or 113 cubic inches} \\ &\quad \text{(to the nearest whole number)} \\ S(3) &= 4\pi \cdot 3^2 \\ &= 36\pi \text{ square inches, or 113 square inches} \\ &\quad \text{(to the nearest whole number)} \end{aligned}$$

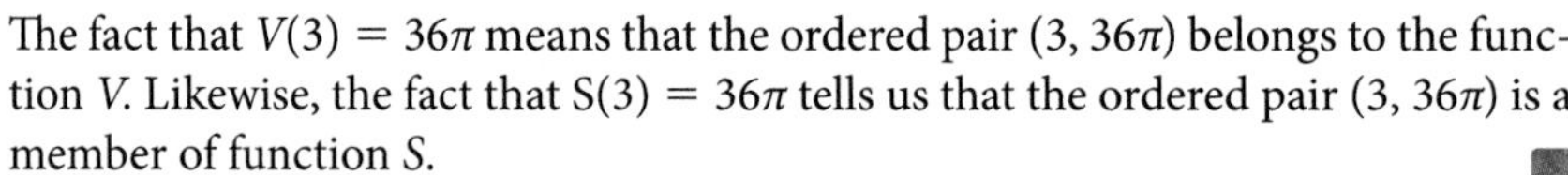

The fact that $V(3) = 36\pi$ means that the ordered pair $(3, 36\pi)$ belongs to the function V. Likewise, the fact that $S(3) = 36\pi$ tells us that the ordered pair $(3, 36\pi)$ is a member of function S.

We can generalize the discussion at the end of Example 4 this way:

$$(a, b) \in f \qquad \text{if and only if} \qquad f(a) = b$$

USING TECHNOLOGY *More About Example 4*

If we look at Example 4, we see that when the radius of a sphere is 3, the numerical values of the volume and surface area are equal. How unusual is this? Are there other values of r for which $V(r)$ and $S(r)$ are equal? We can answer this question by looking at the graphs of both V and S.

To graph the function $V(r) = \frac{4}{3}\pi r^3$, set $Y_1 = 4\pi X^3/3$. To graph $S(r) = 4\pi r^2$, set $Y_2 = 4\pi X^2$. Graph the two functions in each of the following windows:

Window 1: X from -4 to 4, Y from -2 to 10

Window 2: X from 0 to 4, Y from 0 to 50

Window 3: X from 0 to 4, Y from 0 to 150

Then use the Trace and Zoom features of your calculator to locate the point in the first quadrant where the two graphs intersect. How do the coordinates of this point compare with the results in Example 4?

EXAMPLE 5 If $f(x) = 3x^2 + 2x - 1$, find $f(0)$, $f(3)$, and $f(-2)$.

SOLUTION Since $f(x) = 3x^2 + 2x - 1$, we have

$$f(0) = 3(0)^2 + 2(0) - 1 = 0 - 1 = -1$$
$$f(3) = 3(3)^2 + 2(3) - 1 = 27 + 6 - 1 = 32$$
$$f(-2) = 3(-2)^2 + 2(-2) - 1 = 12 - 4 - 1 = 7$$

In Example 5, the function f is defined by the equation $f(x) = 3x^2 + 2x - 1$. We could just as easily have said $y = 3x^2 + 2x - 1$. That is, $y = f(x)$. Saying $f(-2) = 7$ is exactly the same as saying y is 7 when x is -2.

EXAMPLE 6 If $f(x) = 4x - 1$ and $g(x) = x^2 + 2$, then

$$f(5) = 4(5) - 1 = 19 \quad \text{and} \quad g(5) = 5^2 + 2 = 27$$
$$f(-2) = 4(-2) - 1 = -9 \quad \text{and} \quad g(-2) = (-2)^2 + 2 = 6$$
$$f(0) = 4(0) - 1 = -1 \quad \text{and} \quad g(0) = 0^2 + 2 = 2$$
$$f(z) = 4z - 1 \quad \text{and} \quad g(z) = z^2 + 2$$
$$f(a) = 4a - 1 \quad \text{and} \quad g(a) = a^2 + 2$$

$$\begin{aligned} f(a + 3) &= 4(a + 3) - 1 \\ &= 4a + 12 - 1 \\ &= 4a + 11 \end{aligned} \qquad \begin{aligned} g(a + 3) &= (a + 3)^2 + 2 \\ &= (a^2 + 6a + 9) + 2 \\ &= a^2 + 6a + 11 \end{aligned}$$

USING TECHNOLOGY *More About Example 6*

Most graphing calculators can use tables to evaluate functions. To work Example 6 using a graphing calculator table, set Y_1 equal to $4X - 1$ and Y_2 equal to $X^2 + 2$. Then set the independent variable in the table to Ask instead of Auto. Go to your table and input 5, -2, and 0. Under Y_1 in the table, you will find $f(5)$, $f(-2)$, and $f(0)$. Under Y_2, you will find $g(5)$, $g(-2)$, and $g(0)$.

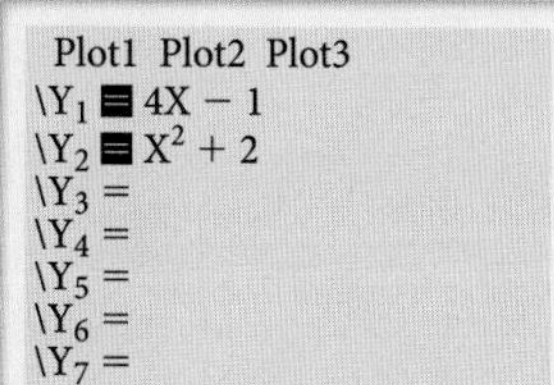

TABLE SETUP
TblStart = 0
ΔTbl = 1
Indpnt: Auto **Ask**
Depend: Auto **Ask**

The table will look like this:

X	Y_1	Y_2
5	19	27
−2	−9	6
0	−1	2

Although the calculator asks us for a table increment, the increment doesn't matter because we are inputting the X values ourselves.

EXAMPLE 7 If the function f is given by

$$f = \{(-2, 0), (3, -1), (2, 4), (7, 5)\}$$

then $f(-2) = 0, f(3) = -1, f(2) = 4$, and $f(7) = 5$.

EXAMPLE 8 If $f(x) = 2x^2$ and $g(x) = 3x - 1$, find

a. $f[g(2)]$ **b.** $g[f(2)]$

SOLUTION The expression $f[g(2)]$ is read "f of g of 2."

a. Because $g(2) = 3(2) - 1 = 5$,

$$f[g(2)] = f(5) = 2(5)^2 = 50$$

b. Because $f(2) = 2(2)^2 = 8$,

$$g[f(2)] = g(8) = 3(8) - 1 = 23$$

GETTING READY FOR CLASS

After reading through the preceding section, respond in your own words and in complete sentences.

A. Explain what you are calculating when you find $f(2)$ for a given function f.

B. If $s(t) = \frac{60}{t}$ how do you find $s(10)$?

C. If $f(2) = 3$ for a function f, what is the relationship between the numbers 2 and 3 and the graph of f?

D. If $f(6) = 0$ for a particular function f, then you can immediately graph one of the intercepts. Explain.

Problem Set 3.2

Let $f(x) = 2x - 5$ and $g(x) = x^2 + 3x + 4$. Evaluate the following.

1. $f(2)$ **2.** $f(3)$ **3.** $f(-3)$ **4.** $g(-2)$

5. $g(-1)$ **6.** $f(-4)$ **7.** $g(-3)$ **8.** $g(2)$

9. $g(a)$ **10.** $f(a)$ **11.** $f(a + 6)$ **12.** $g(a + 6)$

Let $f(x) = 3x^2 - 4x + 1$ and $g(x) = 2x - 1$. Evaluate the following.

13. $f(0)$ **14.** $g(0)$ **15.** $g(-4)$ **16.** $f(1)$

17. $f(-1)$ **18.** $g(-1)$ **19.** $g\left(\frac{1}{2}\right)$ **20.** $g\left(\frac{1}{4}\right)$

21. $f(a)$ **22.** $g(a)$ **23.** $f(a + 2)$ **24.** $g(a + 2)$

If $f = \{(1, 4), (-2, 0), \left(3, \frac{1}{2}\right), (\pi, 0)\}$ and $g = \{(1, 1), (-2, 2), \left(\frac{1}{2}, 0\right)\}$, find each of the following values of f and g.

25. $f(1)$ **26.** $g(1)$ **27.** $g\left(\frac{1}{2}\right)$ **28.** $f(3)$

29. $g(-2)$ **30.** $f(\pi)$

Let $f(x) = x^2 - 2x$ and $g(x) = 5x - 4$. Evaluate the following.

31. $f(-4)$ **32.** $g(-3)$ **33.** $f(-2) + g(-1)$

34. $f(-1) + g(-2)$ **35.** $2f(x) - 3g(x)$ **36.** $f(x) - g(x^2)$

37. $f[g(3)]$ **38.** $g[f(3)]$

Let $f(x) = \frac{1}{x + 3}$ and $g(x) = \frac{1}{x} + 1$. Evaluate the following.

39. $f\left(\frac{1}{3}\right)$ **40.** $g\left(\frac{1}{3}\right)$ **41.** $f\left(-\frac{1}{2}\right)$ **42.** $g\left(-\frac{1}{2}\right)$

43. $f(-3)$ **44.** $g(0)$

45. For the function $f(x) = x^2 - 4$, evaluate each of the following expressions.

a. $f(a) - 3$ **b.** $f(a - 3)$ **c.** $f(x) + 2$

d. $f(x + 2)$ **e.** $f(a + b)$ **f.** $f(x + h)$

46. For the function $f(x) = 3x^2$, evaluate each of the following expressions.

a. $f(a) - 2$ **b.** $f(a - 2)$ **c.** $f(x) + 5$

d. $f(x + 5)$ **e.** $f(a + b)$ **f.** $f(x + h)$

47. Graph the function $f(x) = \frac{1}{2}x + 2$. Then draw and label the line segments that represent $x = 4$ and $f(4)$.

48. Graph the function $f(x) = -\frac{1}{2}x + 6$. Then draw and label the line segments that represent $x = 4$ and $f(4)$.

49. For the function $f(x) = \frac{1}{2}x + 2$, find the value of x for which $f(x) = x$.

50. For the function $f(x) = -\frac{1}{2}x + 6$, find the value of x for which $f(x) = x$.

51. Graph the function $f(x) = x^2$. Then draw and label the line segments that represent $x = 1$ and $f(1)$, $x = 2$ and $f(2)$ and, finally, $x = 3$ and $f(3)$.

52. Graph the function $f(x) = x^2 - 2$. Then draw and label the line segments that represent $x = 2$ and $f(2)$ and the line segments corresponding to $x = 3$ and $f(3)$.

Applying the Concepts

53. Investing in Art A painting is purchased as an investment for \$150. If its value increases continuously so that it doubles every 3 years, then its value is given by the function

$$V(t) = 150 \cdot 2^{t/3} \quad \text{for} \quad t \geq 0$$

where t is the number of years since the painting was purchased, and $V(t)$ is its value (in dollars) at time t. Find $V(3)$ and $V(6)$, and then explain what they mean.

54. Average Speed If it takes Minke t minutes to run a mile, then her average speed $s(t)$, in miles per hour, is given by the formula

$$s(t) = \frac{60}{t} \quad \text{for} \quad t > 0$$

Find $s(4)$ and $s(5)$, and then explain what they mean.

55. Antidepressant Sales Suppose x represents one of the years in the chart. Suppose further that we have three functions f, g, and h that do the following:

f pairs each year with the total sales of Zoloft in billions of dollars for that year.
g pairs each year with the total sales of Effexor in billions of dollars for that year.
h pairs each year with the total sales of Wellbutrin in billions of dollars for that year.

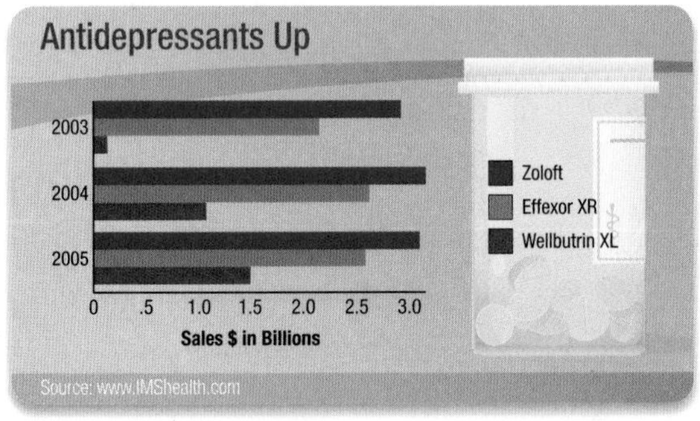

For each statement below, indicate whether the statement is true or false.

a. The domain of g is $\{2003, 2004, 2005\}$

b. The domain of g is $\{x \mid 2003 \leq x \leq 2005\}$

c. $f(2004) > g(2004)$

d. $h(2005) > 1.5$

e. $h(2005) > h(2004) > h(2003)$

56. Mobile Phone Sales Suppose x represents one of the years in the chart. Suppose further that we have three functions f, g, and h that do the following:

f pairs each year with the number of camera phones sold that year.
g pairs each year with the number of non-camera phones sold that year.
h is such that $h(x) = f(x) + g(x)$.

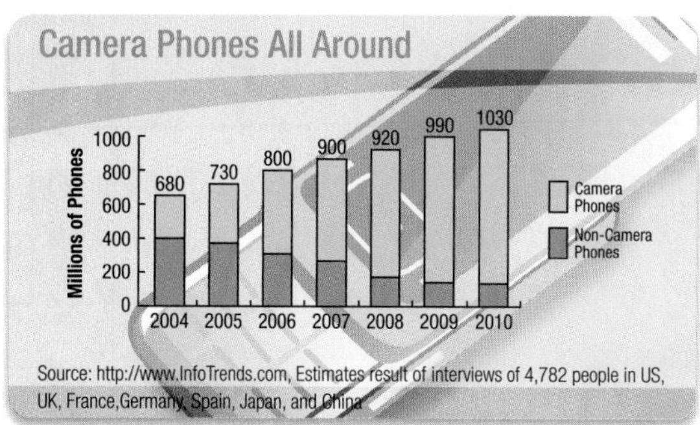

For each statement below, indicate whether the statement is true or false.

a. The domain of f is {2004, 2005, 2006, 2007, 2008, 2009, 2010}

b. $h(2005) = 741{,}000{,}000$

c. $f(2009) > g(2009)$

d. $f(2004) < f(2005)$

e. $h(2010) > h(2007) > h(2004)$

Straight-Line Depreciation Straight-line depreciation is an accounting method used to help spread the cost of new equipment over a number of years. It takes into account both the cost when new and the salvage value, which is the value of the equipment at the time it gets replaced.

57. Value of a Copy Machine The function $V(t) = -3{,}300t + 18{,}000$, where V is value and t is time in years, can be used to find the value of a large copy machine during the first 5 years of use.

a. What is the value of the copier after 3 years and 9 months?

b. What is the salvage value of this copier if it is replaced after 5 years?

c. State the domain of this function.

d. Sketch the graph of this function.

e. What is the range of this function?

f. After how many years will the copier be worth only $10,000?

58. Step Function Figure 3 shows the graph of the step function C that was used to calculate the first-class postage on a letter weighing x ounces in 2006. Use this graph to answer questions **a.** through **d.**

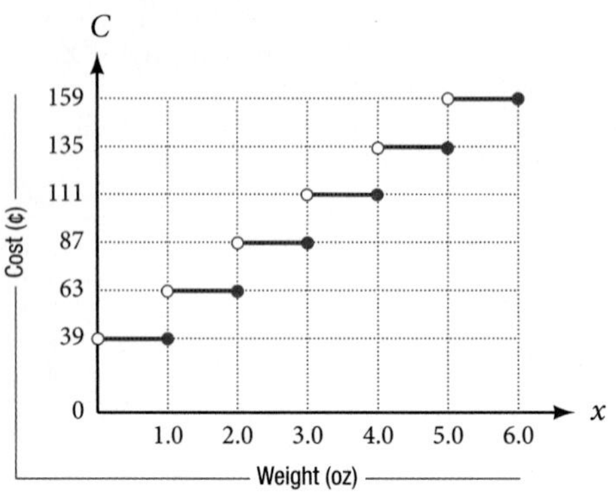

FIGURE 3 *The graph of C(x)*

a. Fill in the following table:

Weight (ounces)	0.6	1.0	1.1	2.5	3.0	4.8	5.0	5.3
Cost (cents)								

b. If a letter cost 87 cents to mail, how much does it weigh? State your answer in words. State your answer as an inequality.

c. If the entire function is shown in Figure 3, state the domain.

d. State the range of the function shown in Figure 3.

Getting Ready for the Next Section

Simplify.

59. $16(3.5)^2$

60. $\frac{2{,}400}{100}$

61. $\frac{180}{45}$

62. $4(2)(4)^2$

63. $\frac{0.0005(200)}{(0.25)^2}$

64. $\frac{0.2(0.5)^2}{100}$

65. If $y = Kx$, find K if $x = 5$ and $y = 15$.

66. If $d = Kt^2$, find K if $t = 2$ and $d = 64$.

67. If $P = \frac{K}{V}$, find K if $P = 48$ and $V = 50$.

68. If $y = Kxz^2$, find K if $x = 5$, $z = 3$, and $y = 180$.

Variation

3.3

If you are a runner and you average t minutes for every mile you run during one of your workouts, then your speed s in miles per hour is given by the equation and graph shown here. The graph (Figure 1) is shown in the first quadrant only because both t and s are positive.

$$s = \frac{60}{t}$$

Input	Output
t	s
4	15
6	10
8	7.5
10	6
12	5
14	4.3

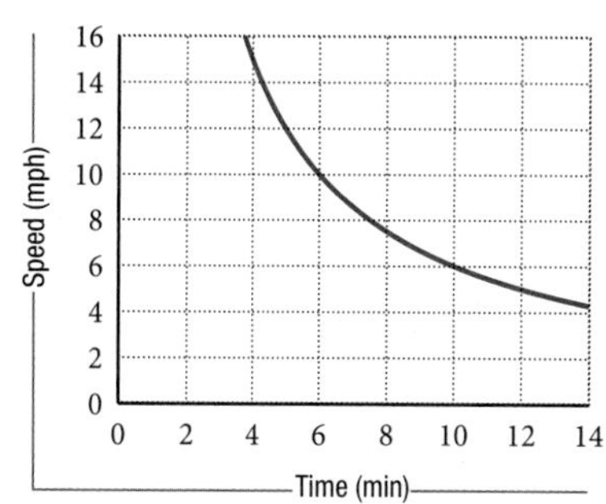

FIGURE 1

You know intuitively that as your average time per mile t increases, your speed s decreases. Likewise, lowering your time per mile will increase your speed. The equation and Figure 1 also show this to be true: Increasing t decreases s, and decreasing t increases s. Quantities that are connected in this way are said to *vary inversely* with each other. Inverse variation is one of the topics we will study in this section.

There are two main types of variation: *direct variation* and *inverse variation*. Variation problems are most common in the sciences, particularly in chemistry and physics.

Direct Variation

When we say the variable y *varies directly* with the variable x, we mean that the relationship can be written in symbols as $y = Kx$, where K is a nonzero constant called the *constant of variation* (or *proportionality constant*).

Another way of saying y varies directly with x is to say y is *directly proportional* to x.

Study the following list. It gives the mathematical equivalent of some direct variation statements.

Verbal Phrase	Algebraic Equation
y varies directly with x.	$y = Kx$
s varies directly with the square of t.	$s = Kt^2$
y is directly proportional to the cube of z.	$y = Kz^3$
u is directly proportional to the square root of v.	$u = K\sqrt{v}$

EXAMPLE 1 y varies directly with x. If y is 15 when x is 5, find y when x is 7.

SOLUTION The first sentence gives us the general relationship between x and y. The equation equivalent to the statement "y varies directly with x" is

$$y = Kx$$

The first part of the second sentence in our example gives us the information necessary to evaluate the constant K:

When	$y = 15$
and	$x = 5$
the equation	$y = Kx$
becomes	$15 = K \cdot 5$
or	$K = 3$

The equation can now be written specifically as

$$y = 3x$$

Letting $x = 7$, we have

$$y = 3 \cdot 7$$

$$y = 21$$

EXAMPLE 2 A skydiver jumps from a plane. Like any object that falls toward earth, the distance the skydiver falls is directly proportional to the square of the time he has been falling, until he reaches his terminal velocity. If the skydiver falls 64 feet in the first 2 seconds of the jump, then

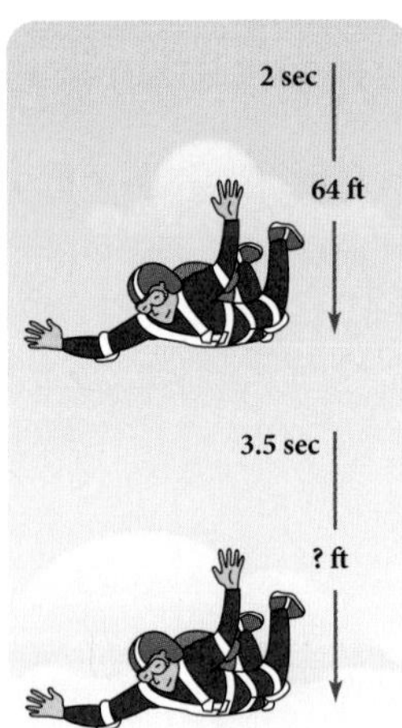

a. How far will he have fallen after 3.5 seconds?
b. Graph the relationship between distance and time.
c. How long will it take him to fall 256 feet?

SOLUTION We let t represent the time the skydiver has been falling, then we can let $d(t)$ represent the distance he has fallen.

a. Since $d(t)$ is directly proportional to the square of t, we have the general function that describes this situation:

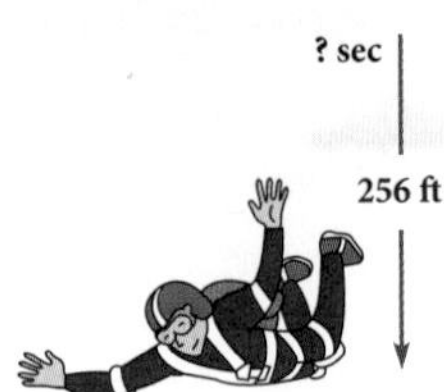

$$d(t) = Kt^2$$

Next, we use the fact that $d(2) = 64$ to find K.

$$64 = K(2)^2$$

$$K = 16$$

The specific equation that describes this situation is

$$d(t) = 16t^2$$

To find how far a skydiver will fall after 3.5 seconds, we find $d(3.5)$,

$$d(3.5) = 16(3.5)^2$$

$$d(3.5) = 196$$

A skydiver will fall 196 feet after 3.5 seconds.

b. To graph this equation, we use a table:

Input	Output
t	$d(t)$
0	0
1	16
2	64
3	144
4	256
5	400

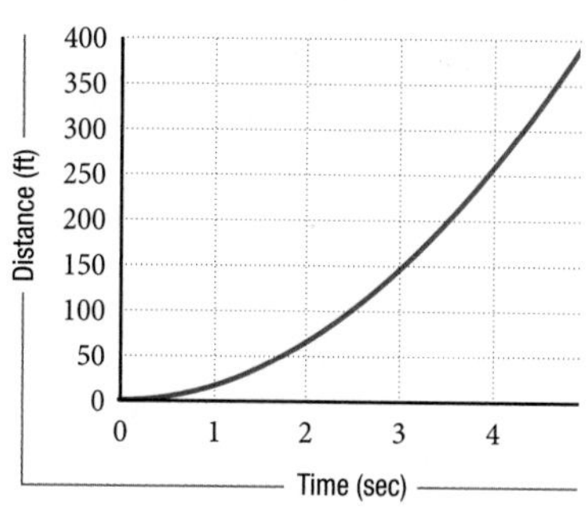

FIGURE 2

c. From the table or the graph (Figure 2), we see that it will take 4 seconds for the skydiver to fall 256 feet.

Inverse Variation

Running

From the introduction to this section, we know that the relationship between the number of minutes t it takes a person to run a mile and his or her average speed in miles per hour s can be described with the following equation and table, and with Figure 3.

$$s = \frac{60}{t}$$

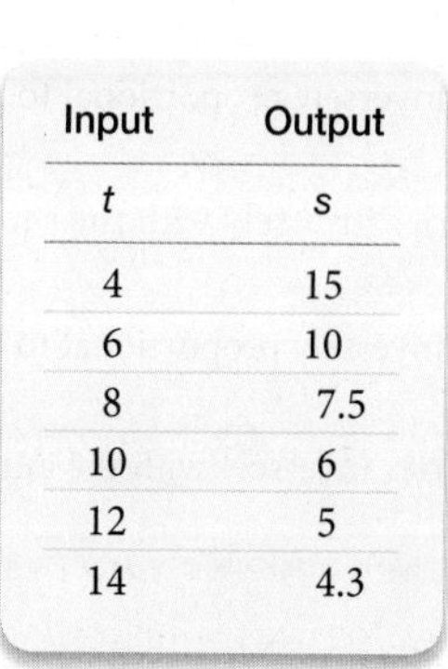

Input	Output
t	s
4	15
6	10
8	7.5
10	6
12	5
14	4.3

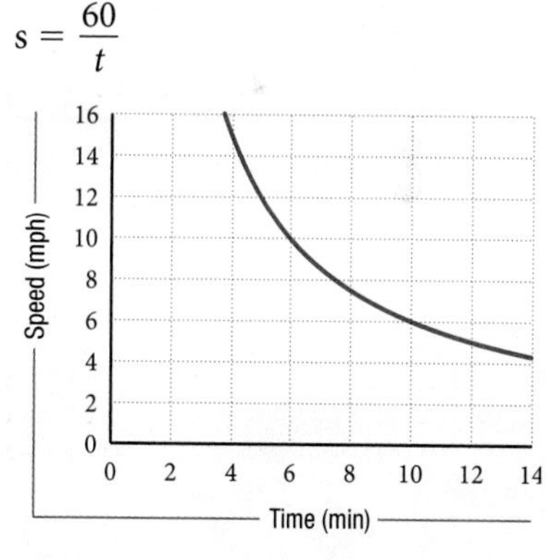

FIGURE 3

If t decreases, then s will increase, and if t increases, then s will decrease. The variable s is *inversely proportional* to the variable t. In this case, the *constant of proportionality* is 60.

Photography

If you are familiar with the terminology and mechanics associated with photography, you know that the *f*-stop for a particular lens will increase as the aperture (the maximum diameter of the opening of the lens) decreases. In mathematics, we say that *f*-stop and aperture vary inversely with each other. The following diagram illustrates this relationship.

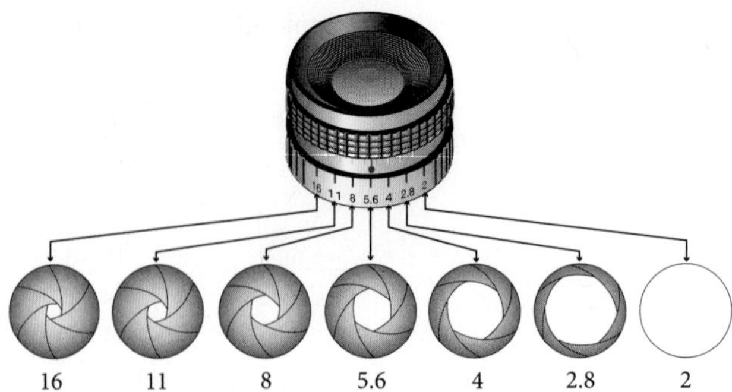

If f is the f-stop and d is the aperture, then their relationship can be written

$$f = \frac{K}{d}$$

In this case, K is the constant of proportionality. (Those of you familiar with photography know that K is also the focal length of the camera lens.)

In General

We generalize this discussion of inverse variation as follows: If y varies inversely with x, then

$$y = K\frac{1}{x} \quad \text{or} \quad y = \frac{K}{x}$$

We can also say y is inversely proportional to x. The constant K is again called the constant of variation or proportionality constant.

Verbal Phrase	Algebraic Equation
y is inversely proportional to x.	$y = \frac{K}{x}$
s varies inversely with the square of t.	$s = \frac{K}{t^2}$
y is inversely proportional to x^4.	$y = \frac{K}{x^4}$
z varies inversely with the cube root of t.	$z = \frac{K}{\sqrt[3]{t}}$

EXAMPLE 3 The volume of a gas is inversely proportional to the pressure of the gas on its container. If a pressure of 48 pounds per square inch corresponds to a volume of 50 cubic feet, what pressure is needed to produce a volume of 100 cubic feet?

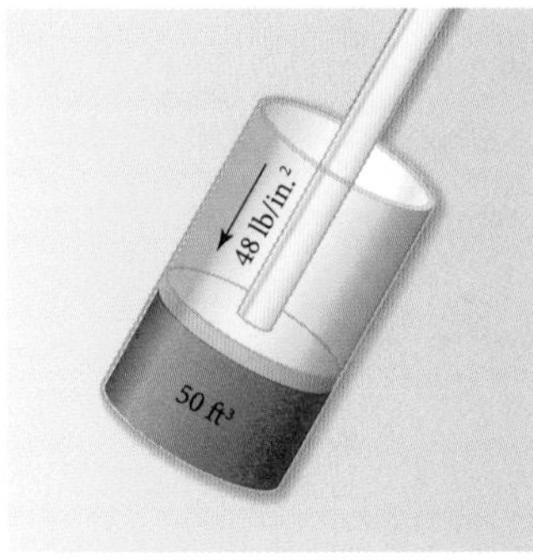

SOLUTION We can represent volume with V and pressure with P:

$$V = \frac{K}{P}$$

Using $P = 48$ and $V = 50$, we have

$$50 = \frac{K}{48}$$

$$K = 50(48)$$

$$K = 2{,}400$$

The equation that describes the relationship between P and V is

$$V = \frac{2{,}400}{P}$$

Here is a graph of this relationship.

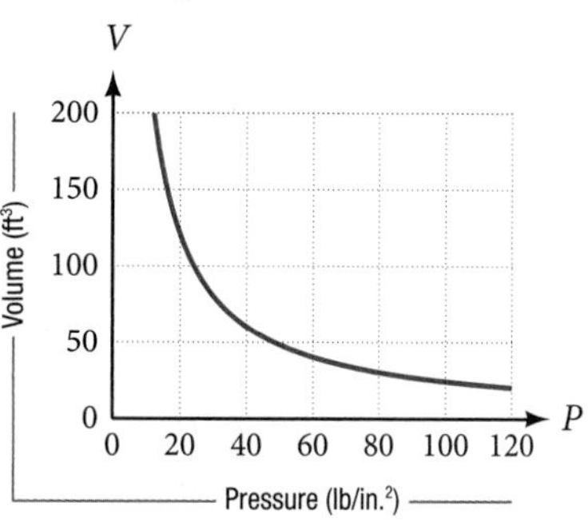

FIGURE 4

Note The relationship between pressure and volume as given in this example is known as Boyle's law and applies to situations such as those encountered in a piston-cylinder arrangement. It was Robert Boyle (1627–1691) who, in 1662, published the results of some of his experiments that showed, among other things, that the volume of a gas decreases as the pressure increases. This is an example of inverse variation.

Substituting $V = 100$ into our last equation, we get

$$100 = \frac{2{,}400}{P}$$

$$100P = 2{,}400$$

$$P = \frac{2{,}400}{100}$$

$$P = 24$$

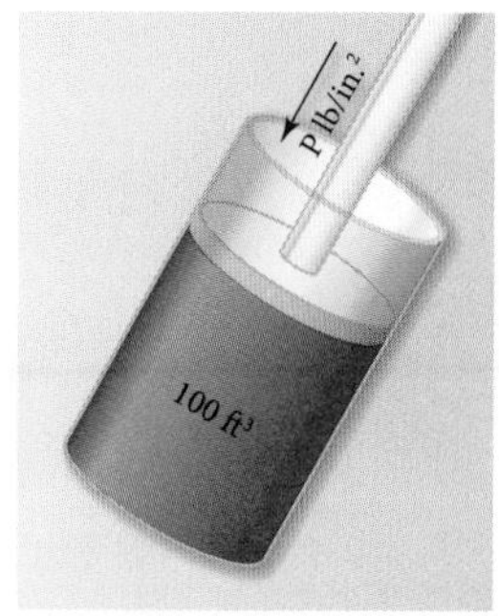

A volume of 100 cubic feet is produced by a pressure of 24 pounds per square inch.

Joint Variation and Other Variation Combinations

Many times relationships among different quantities are described in terms of more than two variables. If the variable y varies directly with *two* other variables, say x and z, then we say y varies *jointly* with x and z. In addition to *joint variation*, there are many other combinations of direct and inverse variation involving more than two variables. The following table is a list of some variation statements and their equivalent mathematical forms:

Verbal Phrase	Algebraic Equation
y varies jointly with x and z.	$y = Kxz$
z varies jointly with r and the square of s.	$z = Krs^2$
V is directly proportional to T and inversely proportional to P.	$V = \frac{KT}{P}$
F varies jointly with m_1 and m_2 and inversely with the square of r.	$F = \frac{Km_1m_2}{r^2}$

EXAMPLE 4 y varies jointly with x and the square of z. When x is 5 and z is 3, y is 180. Find y when x is 2 and z is 4.

SOLUTION The general equation is given by

$$y = Kxz^2$$

Substituting $x = 5$, $z = 3$, and $y = 180$, we have

$$180 = K(5)(3)^2$$
$$180 = 45K$$
$$K = 4$$

The specific equation is

$$y = 4xz^2$$

When $x = 2$ and $z = 4$, the last equation becomes

$$y = 4(2)(4)^2$$
$$y = 128$$

EXAMPLE 5 In electricity, the resistance of a cable is directly proportional to its length and inversely proportional to the square of the diameter. If a 100-foot cable 0.5 inch in diameter has a resistance of 0.2 ohm, what will be the resistance of a cable made from the same material if it is 200 feet long with a diameter of 0.25 inch?

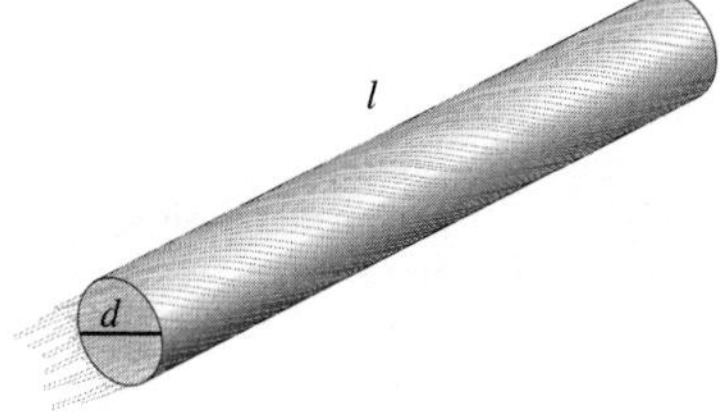

SOLUTION Let R = resistance, l = length, and d = diameter. The equation is

$$R = \frac{Kl}{d^2}$$

When $R = 0.2$, $l = 100$, and $d = 0.5$, the equation becomes

$$0.2 = \frac{K(100)}{(0.5)^2}$$

or

$$K = 0.0005$$

Using this value of K in our original equation, the result is

$$R = \frac{0.0005l}{d^2}$$

When $l = 200$ and $d = 0.25$, the equation becomes

$$R = \frac{0.0005(200)}{(0.25)^2}$$
$$R = 1.6 \text{ ohms}$$

GETTING READY FOR CLASS

After reading through the preceding section, respond in your own words and in complete sentences.

A. Give an example of a direct variation statement, and then translate it into symbols.

B. Translate the equation $y = \frac{k}{x}$ into words.

C. For the inverse variation equation $y = \frac{3}{x}$ what happens to the values of y as x gets larger?

D. How are direct variation statements and linear equations in two variables related?

Problem Set 3.3

For the following problems, y varies directly with x.

1. If y is 10 when x is 2, find y when x is 6.
2. If y is -32 when x is 4, find x when y is -40.

For the following problems, r is inversely proportional to s.

3. If r is -3 when s is 4, find r when s is 2.
4. If r is 8 when s is 3, find s when r is 48.

For the following problems, d varies directly with the square of r.

5. If $d = 10$ when $r = 5$, find d when $r = 10$.
6. If $d = 12$ when $r = 6$, find d when $r = 9$.

For the following problems, y varies inversely with the square of x.

7. If $y = 45$ when $x = 3$, find y when x is 5.
8. If $y = 12$ when $x = 2$, find y when x is 6.

For the following problems, z varies jointly with x and the square of y.

9. If z is 54 when x and y are 3, find z when $x = 2$ and $y = 4$.
10. If z is 27 when $x = 6$ and $y = 3$, find x when $z = 50$ and $y = 4$.

For the following problems, I varies inversely with the cube of w.

11. If $I = 32$ when $w = \frac{1}{2}$, find I when $w = \frac{1}{3}$.
12. If $I = \frac{1}{25}$ when $w = 5$, find I when $w = 10$.

For the following problems, z varies jointly with y and the square of x.

13. If $z = 72$ when $x = 3$ and $y = 2$, find z when $x = 5$ and $y = 3$.
14. If $z = 240$ when $x = 4$ and $y = 5$, find z when $x = 6$ and $y = 3$.
15. If $x = 1$ when $z = 25$ and $y = 5$, find x when $z = 160$ and $y = 8$.
16. If $x = 4$ when $z = 96$ and $y = 2$, find x when $z = 108$ and $y = 1$.

For the following problems, F varies directly with m and inversely with the square of d.

17. If $F = 150$ when $m = 240$ and $d = 8$, find F when $m = 360$ and $d = 3$.
18. If $F = 72$ when $m = 50$ and $d = 5$, find F when $m = 80$ and $d = 6$.
19. If $d = 5$ when $F = 24$ and $m = 20$, find d when $F = 18.75$ and $m = 40$.
20. If $d = 4$ when $F = 75$ and $m = 20$, find d when $F = 200$ and $m = 120$.

Applying the Concepts

21. **Length of a Spring** The length a spring stretches is directly proportional to the force applied. If a force of 5 pounds stretches a spring 3 inches, how much force is necessary to stretch the same spring 10 inches?

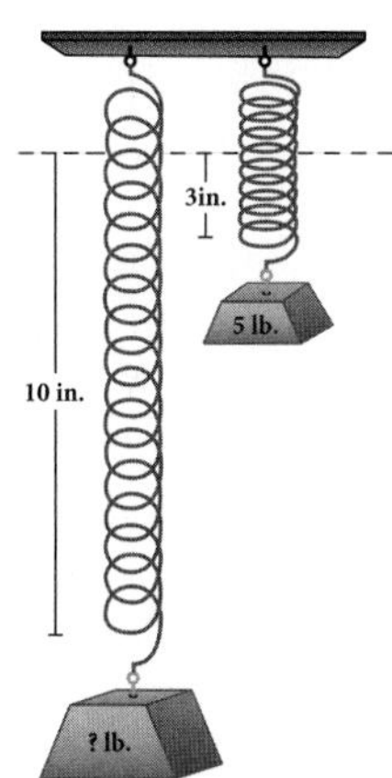

22. **Weight and Surface Area** The weight of a certain material varies directly with the surface area of that material. If 8 square feet weighs half a pound, how much will 10 square feet weigh?

23. **Pressure and Temperature** The temperature of a gas varies directly with its pressure. A temperature of 200 K produces a pressure of 50 pounds per square inch.
    a. Find the equation that relates pressure and temperature.
    b. Graph the equation from part **a.** in the first quadrant only.
    c. What pressure will the gas have at 280 K?

24. **Circumference and Diameter** The circumference of a wheel is directly proportional to its diameter. A wheel has a circumference of 8.5 feet and a diameter of 2.7 feet.
    a. Find the equation that relates circumference and diameter.
    b. Graph the equation from part **a.** in the first quadrant only.
    c. What is the circumference of a wheel that has a diameter of 11.3 feet?

25. **Volume and Pressure** The volume of a gas is inversely proportional to the pressure. If a pressure of 36 pounds per square inch corresponds to a volume of 25 cubic feet, what pressure is needed to produce a volume of 75 cubic feet?

26. **Wave Frequency** The frequency of an electromagnetic wave varies inversely with the wavelength. If a wavelength of 200 meters has a frequency of 800 kilocycles per second, what frequency will be associated with a wavelength of 500 meters?

27. **f-Stop and Aperture Diameter** The relative aperture, or *f*-stop, for a camera lens is inversely proportional to the diameter of the aperture. An *f*-stop of 2 corresponds to an aperture diameter of 40 millimeters for the lens on an automatic camera.
    a. Find the equation that relates *f*-stop and diameter.
    b. Graph the equation from part **a.** in the first quadrant only.
    c. What is the *f*-stop of this camera when the aperture diameter is 10 millimeters?

28. f-Stop and Aperture Diameter The relative aperture, or *f*-stop, for a camera lens is inversely proportional to the diameter of the aperture. An *f*-stop of 2.8 corresponds to an aperture diameter of 75 millimeters for a certain telephoto lens.

a. Find the equation that relates *f*-stop and diameter.

b. Graph the equation from part a. in the first quadrant only.

c. What aperture diameter corresponds to an *f*-stop of 5.6?

29. Surface Area of a Cylinder The surface area of a hollow cylinder varies jointly with the height and radius of the cylinder. If a cylinder with radius 3 inches and height 5 inches has a surface area of 94 square inches, what is the surface area of a cylinder with radius 2 inches and height 8 inches?

30. Capacity of a Cylinder The capacity of a cylinder varies jointly with its height and the square of its radius. If a cylinder with a radius of 3 centimeters and a height of 6 centimeters has a capacity of 169.56 cubic centimeters, what will be the capacity of a cylinder with radius 4 centimeters and height 9 centimeters?

31. Electrical Resistance The resistance of a wire varies directly with its length and inversely with the square of its diameter. If 100 feet of wire with diameter 0.01 inch has a resistance of 10 ohms, what is the resistance of 60 feet of the same type of wire if its diameter is 0.02 inch?

32. Volume and Temperature The volume of a gas varies directly with its temperature and inversely with the pressure. If the volume of a certain gas is 30 cubic feet at a temperature of 300 K and a pressure of 20 pounds per square inch, what is the volume of the same gas at 340 K when the pressure is 30 pounds per square inch?

33. Period of a Pendulum The time it takes for a pendulum to complete one period varies directly with the square root of the length of the pendulum. A 100-centimeter pendulum takes 2.1 seconds to complete one period.

a. Find the equation that relates period and pendulum length.

b. Graph the equation from part **a.** in quadrant I only.

c. How long does it take to complete one period if the pendulum hangs 225 centimeters?

Getting Ready for the Next Section

Multiply.

34. $x(35 - 0.1x)$

35. $0.6(M - 70)$

36. $(4x - 3)(x - 1)$

37. $(4x - 3)(4x^2 - 7x + 3)$

Simplify.

38. $(35x - 0.1x^2) - (8x + 500)$

39. $(4x - 3) + (4x^2 - 7x + 3)$

40. $(4x^2 + 3x + 2) - (2x^2 - 5x - 6)$

41. $(4x^2 + 3x + 2) + (2x^2 - 5x - 6)$

42. $4(2)^2 - 3(2)$

43. $4(-1)^2 - 7(-1)$

Algebra and Composition with Functions

3.4

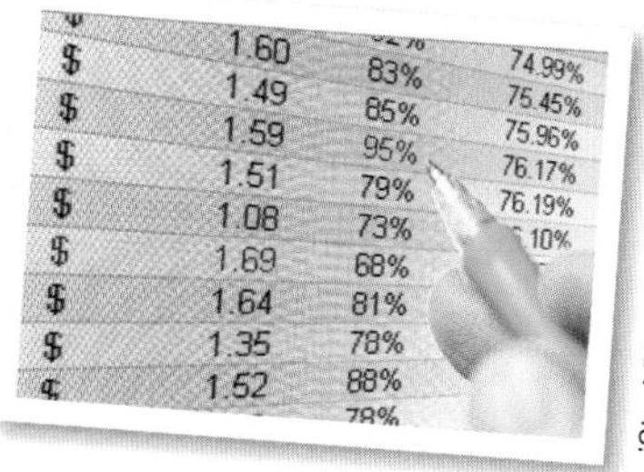

A company produces and sells copies of an accounting program for home computers. The price they charge for the program is related to the number of copies sold by the demand function

$$p(x) = 35 - 0.1x$$

We find the revenue for this business by multiplying the number of items sold by the price per item. When we do so, we are forming a new function by combining two existing functions. That is, if $n(x) = x$ is the number of items sold and $p(x) = 35 - 0.1x$ is the price per item, then revenue is

$$R(x) = n(x) \cdot p(x) = x(35 - 0.1x) = 35x - 0.1x^2$$

In this case, the revenue function is the product of two functions. When we combine functions in this manner, we are applying our rules for algebra to functions.

To carry this situation further, we know the profit function is the difference between two functions. If the cost function for producing x copies of the accounting program is $C(x) = 8x + 500$, then the profit function is

$$P(x) = R(x) - C(x) = (35x - 0.1x^2) - (8x + 500) = -500 + 27x - 0.1x^2$$

The relationship between these last three functions is represented visually in Figure 1.

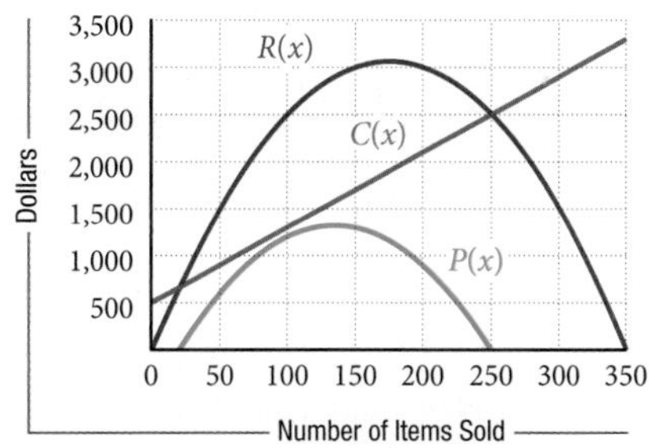

FIGURE 1

Algebra with Functions

Again, when we combine functions in the manner shown, we are applying our rules for algebra to functions. To begin this section, we take a formal look at addition, subtraction, multiplication, and division with functions.

If we are given two functions f and g with a common domain, we can define four other functions as follows.

DEFINITION

$(f+g)(x) = f(x) + g(x)$ The function $f+g$ is the sum of the functions f and g.

$(f-g)(x) = f(x) - g(x)$ The function $f-g$ is the difference of the functions f and g.

$(fg)(x) = f(x)g(x)$ The function fg is the product of the functions f and g.

$\left(\frac{f}{g}\right)(x) = \frac{f(x)}{g(x)}$ The function $\frac{f}{g}$ is the quotient of the functions f and g, where $g(x) \neq 0$.

EXAMPLE 1 If $f(x) = 4x^2 + 3x + 2$ and $g(x) = 2x^2 - 5x - 6$, write the formulas for the functions $f+g$, $f-g$, fg, and f/g.

SOLUTION The function $f+g$ is defined by

$$\begin{aligned}(f+g)(x) &= f(x) + g(x)\\ &= (4x^2 + 3x + 2) + (2x^2 - 5x - 6)\\ &= 6x^2 - 2x - 4\end{aligned}$$

The function $f-g$ is defined by

$$\begin{aligned}(f-g)(x) &= f(x) - g(x)\\ &= (4x^2 + 3x + 2) - (2x^2 - 5x - 6)\\ &= 4x^2 + 3x + 2 - 2x^2 + 5x + 6\\ &= 2x^2 + 8x + 8\end{aligned}$$

The function fg is defined by

$$\begin{aligned}(fg)(x) &= f(x)g(x)\\ &= (4x^2 + 3x + 2)(2x^2 - 5x - 6)\\ &= 8x^4 - 20x^3 - 24x^2 + 6x^3 - 15x^2 - 18x + 4x^2 - 10x - 12\\ &= 8x^4 - 14x^3 - 35x^2 - 28x - 12\end{aligned}$$

The function f/g is defined by

$$\begin{aligned}\left(\frac{f}{g}\right)(x) &= \frac{f(x)}{g(x)}\\ &= \frac{4x^2 + 3x + 2}{2x^2 - 5x - 6}\end{aligned}$$

EXAMPLE 2 Let $f(x) = 4x - 3$, $g(x) = 4x^2 - 7x + 3$, and $h(x) = x - 1$. Find $f+g$, fh, fg and $\frac{g}{f}$.

SOLUTION The function $f+g$, the sum of functions f and g, is defined by

$$\begin{aligned}(f+g)(x) &= f(x) + g(x)\\ &= (4x - 3) + (4x^2 - 7x + 3)\\ &= 4x^2 - 3x\end{aligned}$$

The function fh, the product of functions f and h, is defined by

$$\begin{aligned}(fh)(x) &= f(x)h(x)\\ &= (4x-3)(x-1)\\ &= 4x^2-7x+3\\ &= g(x)\end{aligned}$$

The function fg, the product of the functions f and g, is defined by

$$\begin{aligned}(fg)(x) &= f(x)g(x)\\ &= (4x-3)(4x^2-7x+3)\\ &= 16x^3-28x^2+12x-12x^2+21x-9\\ &= 16x^3-40x^2+33x-9\end{aligned}$$

The function $\frac{g}{f}$, the quotient of the functions g and f, is defined by

$$\begin{aligned}\left(\frac{g}{f}\right)(x) &= \frac{g(x)}{f(x)}\\ &= \frac{4x^2-7x+3}{4x-3}\end{aligned}$$

Factoring the numerator, we can reduce to lowest terms:

$$\begin{aligned}\left(\frac{g}{f}\right)(x) &= \frac{(4x-3)(x-1)}{4x-3}\\ &= x-1\\ &= h(x)\end{aligned}$$

EXAMPLE 3 If f, g, and h are the same functions defined in Example 2, evaluate $(f+g)(2)$, $(fh)(-1)$, $(fg)(0)$, and $\left(\frac{g}{f}\right)(5)$.

SOLUTION We use the formulas for $f+g$, fh, fg and $\frac{g}{f}$ found in Example 2:

$$\begin{aligned}(f+g)(2) &= 4(2)^2-3(2)\\ &= 16-6\\ &= 10\end{aligned}$$

$$\begin{aligned}(fh)(-1) &= 4(-1)^2-7(-1)+3\\ &= 4+7+3\\ &= 14\end{aligned}$$

$$\begin{aligned}(fg)(0) &= 16(0)^3-40(0)^2+33(0)-9\\ &= 0-0+0-9\\ &= -9\end{aligned}$$

$$\begin{aligned}\left(\frac{g}{f}\right)(5) &= 5-1\\ &= 4\end{aligned}$$

Composition of Functions

In addition to the four operations used to combine functions shown so far in this section, there is a fifth way to combine two functions to obtain a new function. It is called *composition of functions*. To illustrate the concept, recall from Chapter 2 the definition of training heart rate: training heart rate, in beats per minute, is resting heart rate plus 60% of the difference between maximum heart rate and resting heart rate. If your resting heart rate is 70 beats per minute, then your training heart rate is a function of your maximum heart rate M.

$$T(M) = 70 + 0.6(M - 70) = 70 + 0.6M - 42 = 28 + 0.6M$$

But your maximum heart rate is found by subtracting your age in years from 220. So, if x represents your age in years, then your maximum heart rate is

$$M(x) = 220 - x$$

Therefore, if your resting heart rate is 70 beats per minute and your age in years is x, then your training heart rate can be written as a function of x.

$$T(x) = 28 + 0.6(220 - x)$$

This last line is the composition of functions T and M. We input x into function M, which outputs $M(x)$. Then, we input $M(x)$ into function T, which outputs $T(M(x))$, which is the training heart rate as a function of age x. Here is a diagram, called a function map, of the situation:

Age		Maximum heart rate		Training heart rate
x	$\xrightarrow{M}$	$M(x)$	$\xrightarrow{T}$	$T(M(x))$

FIGURE 2

Now let's generalize the preceding ideas into a formal development of composition of functions. To find the composition of two functions f and g, we first require that the range of g have numbers in common with the domain of f. Then the composition of f with g, is defined this way:

$$(f \circ g)(x) = f(g(x))$$

To understand this new function, we begin with a number x, and we operate on it with g, giving us $g(x)$. Then we take $g(x)$ and operate on it with f, giving us $f(g(x))$. The only numbers we can use for the domain of the composition of f with g are numbers x in the domain of g, for which $g(x)$ is in the domain of f. The diagrams in Figure 3 illustrate the composition of f with g.

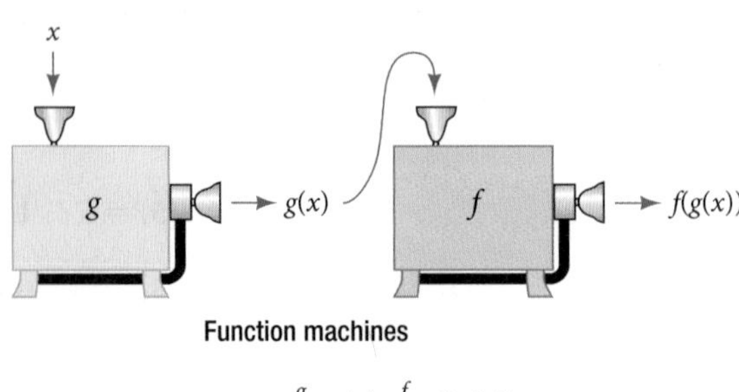

Function machines

$$x \xrightarrow{g} g(x) \xrightarrow{f} f(g(x))$$

FIGURE 3

Composition of functions is not commutative. The composition of f with g, $f \circ g$, may therefore be different from the composition of g with f, $g \circ f$.

$$(g \circ f)(x) = g(f(x))$$

Again, the only numbers we can use for the domain of the composition of g with f are numbers in the domain of f, for which $f(x)$ is in the domain of g. The diagrams in Figure 4 illustrate the composition of g with f.

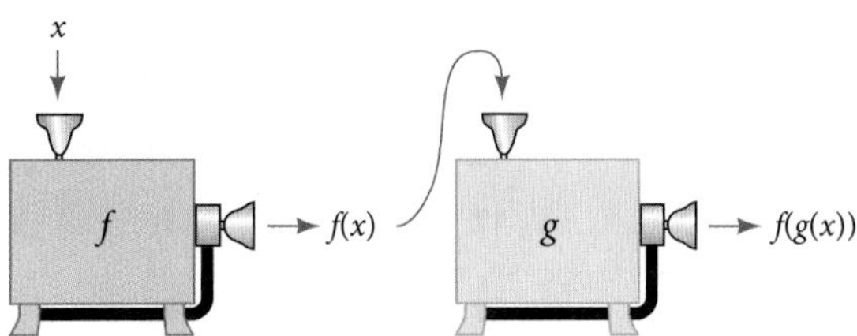

Function machines

$$x \xrightarrow{f} f(x) \xrightarrow{g} g(f(x))$$

FIGURE 4

EXAMPLE 4 If $f(x) = x + 5$ and $g(x) = x^2 - 2x$, find $(f \circ g)(x)$ and $(g \circ f)(x)$.

SOLUTION The composition of f with g is

$$\begin{aligned}(f \circ g)(x) &= f(g(x)) \\ &= f(x^2 - 2x) \\ &= (x^2 - 2x) + 5 \\ &= x^2 - 2x + 5\end{aligned}$$

The composition of g with f is

$$\begin{aligned}(g \circ f)(x) &= g(f(x)) \\ &= g(x + 5) \\ &= (x + 5)^2 - 2(x + 5) \\ &= (x^2 + 10x + 25) - 2x - 10 \\ &= x^2 + 8x + 15\end{aligned}$$

GETTING READY FOR CLASS

Respond in your own words and in complete sentences.

A. How are profit, revenue, and cost related?

B. How do you find maximum heart rate?

C. For functions f and g, how do you find the composition of f with g?

D. For functions f and g, how do you find the composition of g with f?

Problem Set 3.4

Let $f(x) = 4x - 3$ and $g(x) = 2x + 5$. Write a formula for each of the following functions.

1. $f + g$ **2.** $f - g$ **3.** $g - f$ **4.** $g + f$

5. fg **6.** $\frac{f}{g}$ **7.** $\frac{g}{f}$ **8.** ff

If the functions f, g, and h are defined by $f(x) = 3x - 5$, $g(x) = x - 2$ and $h(x) = 3x^2 - 11x + 10$, write a formula for each of the following functions.

9. $g + f$ **10.** $f + h$ **11.** $g + h$ **12.** $f - g$

13. $g - f$ **14.** $h - g$ **15.** fg **16.** gf

17. fh **18.** gh **19.** $\frac{h}{f}$ **20.** $\frac{h}{g}$

21. $\frac{f}{h}$ **22.** $\frac{g}{h}$ **23.** $f + g + h$ **24.** $h - g + f$

25. $h + fg$ **26.** $h - fg$

Let $f(x) = 2x + 1$, $g(x) = 4x + 2$, and $h(x) = 4x^2 + 4x + 1$, and find the following.

27. $(f + g)(2)$ **28.** $(f - g)(-1)$ **29.** $(fg)(3)$ **30.** $(f/g)(-3)$

31. $(h/g)(1)$ **32.** $(hg)(1)$ **33.** $(fh)(0)$ **34.** $(h - g)(-4)$

35. $(f + g + h)(2)$ **36.** $(h - f + g)(0)$ **37.** $(h + fg)(3)$ **38.** $(h - fg)(5)$

39. Let $f(x) = x^2$ and $g(x) = x + 4$, and find

a. $(f \circ g)(5)$ **b.** $(g \circ f)(5)$ **c.** $(f \circ g)(x)$ **d.** $(g \circ f)(x)$

40. Let $f(x) = 3 - x$ and $g(x) = x^3 - 1$, and find

a. $(f \circ g)(0)$ **b.** $(g \circ f)(0)$ **c.** $(f \circ g)(x)$ **d.** $(g \circ f)(x)$

41. Let $f(x) = x^2 + 3x$ and $g(x) = 4x - 1$, and find

a. $(f \circ g)(0)$ **b.** $(g \circ f)(0)$ **c.** $(f \circ g)(x)$ **d.** $(g \circ f)(x)$

42. Let $f(x) = (x - 2)^2$ and $g(x) = x + 1$, and find the following

a. $(f \circ g)(-1)$ **b.** $(g \circ f)(-1)$ **c.** $(f \circ g)(x)$ **d.** $(g \circ f)(x)$

For each of the following pairs of functions f and g, show that $(f \circ g)(x) = (g \circ f)(x) = x$.

43. $f(x) = 5x - 4$ and $g(x) = \frac{x + 4}{5}$ **44.** $f(x) = \frac{x}{6} - 2$ and $g(x) = 6x + 12$

Applying the Concepts

45. Profit, Revenue, and Cost A company manufactures and sells DVD's. Here are the equations they use in connection with their business.

Number of DVD's sold each day: $n(x) = x$
Selling price for each DVD's: $p(x) = 11.5 - 0.05x$
Daily fixed costs: $f(x) = 200$
Daily variable costs: $v(x) = 2x$
Find the following functions.

a. Revenue $= R(x) =$ the product of the number of DVD's sold each day and the selling price of each DVD's.

b. Cost $= C(x) =$ the sum of the fixed costs and the variable costs.

c. Profit $= P(x) =$ the difference between revenue and cost.

d. Average cost $= \overline{C}(x) =$ the quotient of cost and the number of tapes sold each day.

46. Profit, Revenue, and Cost A company manufactures and sells CD's for home computers. Here are the equations they use in connection with their business.

Number of CD's sold each day: $n(x) = x$

Selling price for each CD's: $p(x) = 3 - \frac{1}{300}x$
Daily fixed costs: $f(x) = 200$
Daily variable costs: $v(x) = 2x$
Find the following functions.

a. Revenue $= R(x) =$ the product of the number of CD's sold each day and the selling price of each diskette.

b. Cost $= C(x) =$ the sum of the fixed costs and the variable costs.

c. Profit $= P(x) =$ the difference between revenue and cost.

d. Average cost $= \overline{C}(x) =$ the quotient of cost and the number of CD's sold each day.

47. Training Heart Rate Find the training heart rate function, $T(M)$ for a person with a resting heart rate of 62 beats per minute, then find the following.

a. Find the maximum heart rate function, $M(x)$, for a person x years of age.

b. What is the maximum heart rate for a 24-year-old person?

c. What is the training heart rate for a 24-year-old person with a resting heart rate of 62 beats per minute?

d. What is the training heart rate for a 36-year-old person with a resting heart rate of 62 beats per minute?

e. What is the training heart rate for a 48-year-old person with a resting heart rate of 62 beats per minute?

48. Training Heart Rate Find the training heart rate function, $T(M)$ for a person with a resting heart rate of 72 beats per minute, then find the following to the nearest whole number.

a. Find the maximum heart rate function, $M(x)$, for a person x years of age.

b. What is the maximum heart rate for a 20-year-old person?

c. What is the training heart rate for a 20-year-old person with a resting heart rate of 72 beats per minute?

d. What is the training heart rate for a 30-year-old person with a resting heart rate of 72 beats per minute?

e. What is the training heart rate for a 40-year-old person with a resting heart rate of 72 beats per minute?

Maintaining Your Skills

The problems that follow review some of the more important skills you have learned in previous sections and chapters.

Solve the following equations.

49. $x - 5 = 7$ **50.** $3y = -4$ **51.** $5 - \frac{4}{7}a = -11$

52. $\frac{1}{5}x - \frac{1}{2} - \frac{1}{10}x + \frac{2}{5} = \frac{3}{10}x + \frac{1}{2}$

53. $5(x - 1) - 2(2x + 3) = 5x - 4$

54. $0.07 - 0.02(3x + 1) = -0.04x + 0.01$

Solve for the indicated variable.

55. $P = 2l + 2w$ for w

56. $A = \frac{1}{2}h(b + B)$ for B

Solve the following inequalities. Write the solution set using interval notation, then graph the solution set.

57. $-5t \leq 30$

58. $5 - \frac{3}{2}x > -1$

59. $1.6x - 2 < 0.8x + 2.8$

60. $3(2y + 4) \geq 5(y - 8)$

Solve the following equations.

61. $\left|\frac{1}{4}x - 1\right| = \frac{1}{2}$

62. $\left|\frac{2}{3}a + 4\right| = 6$

63. $|3 - 2x| + 5 = 2$

64. $5 = |3y + 6| - 4$

Chapter 3 Summary

Relations and Functions [3.1]

EXAMPLES

1. The relation

$$\{(8, 1), (6, 1), (-3, 0)\}$$

is also a function because no ordered pairs have the same first coordinates. The domain is $\{8, 6, -3\}$ and the range is $\{1, 0\}$.

A *function* is a rule that pairs each element in one set, called the *domain*, with exactly one element from a second set, called the *range*.

A *relation* is any set of ordered pairs. The set of all first coordinates is called the *domain* of the relation, and the set of all second coordinates is the *range* of the relation. A function is a relation in which no two different ordered pairs have the same first coordinates.

Vertical Line Test [3.1]

2. The graph of $x = y^2$ shown in Figure 5 in Section 3.5 fails the vertical line test. It is not the graph of a function.

If a vertical line crosses the graph of a relation in more than one place, the relation cannot be a function. If no vertical line can be found that crosses the graph in more than one place, the relation must be a function.

Function Notation [3.2]

3. If $f(x) = 5x - 3$ then

$$\begin{aligned} f(0) &= 5(0) - 3 \\ &= -3 \\ f(1) &= 5(1) - 3 \\ &= 2 \\ f(-2) &= 5(-2) - 3 \\ &= -13 \\ f(a) &= 5a - 3 \end{aligned}$$

The alternative notation for y is $f(x)$. It is read "f of x" and can be used instead of the variable y when working with functions. The notation y and the notation $f(x)$ are equivalent; that is, $y = f(x)$.

Variation [3.3]

If y *varies directly* with x (y is directly proportional to x), then

$$y = Kx$$

If y *varies inversely* with x (y is inversely proportional to x), then

$$y = \frac{K}{x}$$

If z *varies jointly* with x and y (z is directly proportional to both x and y), then

$$z = Kxy$$

In each case, K is called the *constant of variation.*

Algebra with Functions [3.4]

If f and g are any two functions with a common domain, then:

$(f+g)(x) = f(x) + g(x)$ The function $f+g$ is the sum of the functions f and g.

$(f-g)(x) = f(x) - g(x)$ The function $f-g$ is the difference of the functions f and g.

$(fg)(x) = f(x)g(x)$ The function fg is the product of the functions f and g.

$\frac{f}{g}(x) = \frac{f(x)}{g(x)}$ The function $\frac{f}{g}$ is the quotient of the functions f and g, where $g(x) \neq 0$

Composition of Functions [3.4]

If f and g are two functions for which the range of each has numbers in common with the domain of the other, then we have the following definitions:

The composition of f with g: $(f \circ g)(x) = f[g(x)]$

The composition of g with f: $(g \circ f)(x) = g[f(x)]$

COMMON MISTAKE

1. When graphing ordered pairs, the most common mistake is to associate the first coordinate with the y-axis and the second with the x-axis. If you make this mistake you would graph (3, 1) by going up 3 and to the right 1, which is just the reverse of what you should do. Remember, the first coordinate is always associated with the horizontal axis, and the second coordinate is always associated with the vertical axis.
2. The two most common mistakes students make when first working with the formula for the slope of a line are the following:
   a. Putting the difference of the x-coordinates over the difference of the y-coordinates.
   b. Subtracting in one order in the numerator and then subtracting in the opposite order in the denominator.
3. When graphing linear inequalities in two variables, remember to graph the boundary with a broken line when the inequality symbol is $<$ or $>$. The only time you use a solid line for the boundary is when the inequality symbol is $\leq$ or $\geq$.

State the domain and range for the following relations, and indicate which relations are also functions. [3.1]

1. $\{(-2, 0), (-3, 0), (-2, 1)\}$

2. $y = x^2 - 9$

Let $f(x) = x - 2$, $g(x) = 3x + 4$ and $h(x) = 3x^2 - 2x - 8$, and find the following. [3.2, 3.3]

3. $f(3) + g(2)$

4. $h(0) + g(0)$

5. $(f \circ g)(2)$

6. $(g \circ f)(2)$

Solve the following variation problems. [3.4]

7. Direct Variation Quantity y varies directly with the square of x. If y is 50 when x is 5, find y when x is 3.

8. Joint Variation Quantity z varies jointly with x and the cube of y. If z is 15 when x is 5 and y is 2, find z when x is 2 and y is 3.

9. Maximum Load The maximum load (L) a horizontal beam can safely hold varies jointly with the width (w) and the square of the depth (d) and inversely with the length (l). If a 10-foot beam with width 3 feet and depth 4 feet will safely hold up to 800 pounds, how many pounds will a 12-foot beam with width 3 feet and depth 4 feet hold?

Systems of Linear Equations

4

Chapter Outline

4.1 Solving Linear Systems by Graphing

4.2 The Elimination Method

4.3 The Substitution Method

4.4 Applications

Two companies offer Internet access to their customers. Company A charges \$10 a month plus \$3 for every hour of Internet connection. Company B charges \$18 a month plus \$1 for every hour of Internet connection. To compare the monthly charges of the two companies we form what is called a system of equations. Here is that system.

$$y = 3x + 10$$
$$y = x + 18$$

The top equation gives us information on company A; the bottom equation gives information on company B. Tables 1 and 2 and the graphs in Figure 1 give us additional information about this system of equations.

TABLE 1

Company A

Hours	Cost
0	\$10
1	\$13
2	\$16
3	\$19
4	\$22
5	\$25
6	\$28
7	\$31
8	\$34
9	\$37
10	\$40

TABLE 2

Company B

Hours	Cost
0	\$18
1	\$19
2	\$20
3	\$21
4	\$22
5	\$23
6	\$24
7	\$25
8	\$26
9	\$27
10	\$28

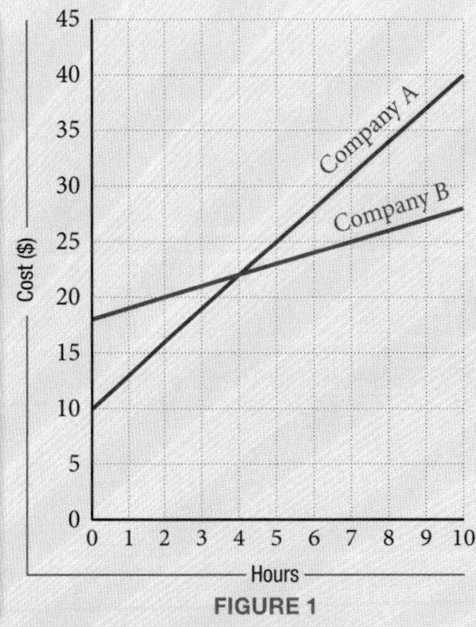

FIGURE 1

As you can see from looking at the tables and at the graphs in Figure 1, the monthly charges for the two companies will be equal if Internet use is exactly 4 hours. In this chapter we work with systems of linear equations.

Study Skills

The study skills for this chapter are concerned with getting ready to take an exam.

1. **Getting Ready to Take an Exam** Try to arrange your daily study habits so you have little studying to do the night before your next exam. The next two goals will help you achieve goal number 1.

2. **Review With the Exam in Mind** You should review material that will be covered on the next exam every day. Your review should consist of working problems. Preferably, the problems you work should be problems from your list of difficult problems.

3. **Continue to List Difficult Problems** You should continue to list and rework the problems that give you the most difficulty. It is this list that you will use to study for the next exam. Your goal is to go into the next exam knowing you can successfully work any problem from your list of hard problems.

4. **Pay Attention to Instructions** Taking a test is different from doing homework. When you take a test, the problems will be mixed up. When you do your homework, you usually work a number of similar problems. Sometimes students who do well on their homework become confused when they see the same problems on a test, because they have not paid attention to the instructions on their homework. For example, suppose you see the equation $y = 3x - 2$ on your next test. By itself, the equation is simply a statement. There isn't anything to do unless the equation is accompanied by instructions. Each of the following is a valid instruction with respect to the equation $y = 3x - 2$ and the result of applying the instructions will be different in each case:

   Find x when y is 10.
   Solve for x.
   Graph the equation.
   Find the intercepts.
   Find the slope.

   There are many things to do with the equation If you train yourself to pay attention to the instructions that accompany a problem as you work through the assigned problems, you will not find yourself confused about what to do with a problem when you see it on a test.

Solving Linear Systems by Graphing

4.1

Two linear equations considered at the same time make up what is called a *system of linear equations*. Both equations contain two variables and, of course, have graphs that are straight lines. The following are systems of linear equations:

$x + y = 3$	$y = 2x + 1$	$2x - y = 1$
$3x + 4y = 2$	$y = 3x + 2$	$3x - 2y = 6$

The solution set for a system of linear equations is all ordered pairs that are solutions to both equations. Because each linear equation has a graph that is a straight line, we can expect the intersection of the graphs to be a point whose coordinates are solutions to the system; that is, if we graph both equations on the same coordinate system, we can read the coordinates of the point of intersection and have the solution to our system. Here is an example.

EXAMPLE 1 Solve the following system by graphing.

$$x + y = 4$$
$$x - y = -2$$

SOLUTION On the same set of coordinate axes we graph each equation separately. Figure 1 shows both graphs, without showing the work necessary to get them. We can see from the graphs that they intersect at the point (1, 3). The point (1, 3) therefore must be the solution to our system because it is the only ordered pair whose graph lies on both lines. Its coordinates satisfy both equations.

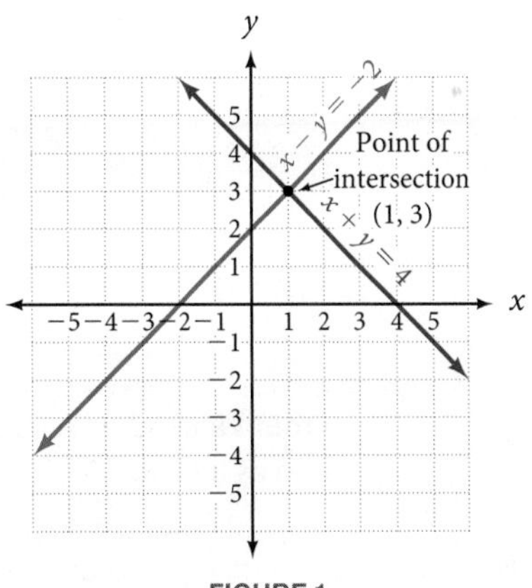

FIGURE 1

We can check our results by substituting the coordinates $x = 1$, $y = 3$ into both equations to see if they work.

When	$x = 1$	When	$x = 1$
and	$y = 3$	and	$y = 3$
the equation	$x + y = 4$	the equation	$x - y = -2$
becomes	$1 + 3 \stackrel{?}{=} 4$	becomes	$1 - 3 \stackrel{?}{=} -2$
	$4 = 4$		$-2 = -2$

The point (1, 3) satisfies both equations.

Here are some steps to follow in solving linear systems by graphing.

HOW TO ***Solving a Linear System by Graphing***

Step 1: Graph the first equation by the methods described in Section 3.3 or 3.4.
Step 2: Graph the second equation on the same set of axes used for the first equation.
Step 3: Read the coordinates of the point of intersection of the two graphs.
Step 4: Check the solution in both equations.

EXAMPLE 2 Solve the following system by graphing.

$$x + 2y = 8$$
$$2x - 3y = 2$$

SOLUTION Graphing each equation on the same coordinate system, we have the lines shown in Figure 2.

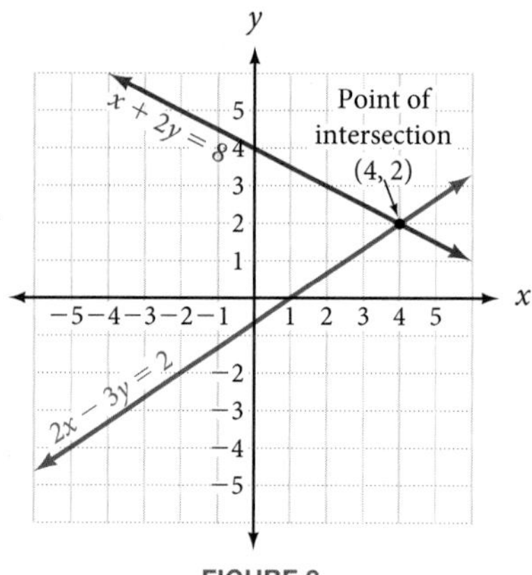

FIGURE 2

From Figure 2, we can see the solution for our system is (4, 2). We check this solution as follows.

When	$x = 4$	When	$x = 4$
and	$y = 2$	and	$y = 2$
the equation	$x + 2y = 8$	the equation	$2x - 3y = 2$
becomes	$4 + 2(2) \stackrel{?}{=} 8$	becomes	$2(4) - 3(2) \stackrel{?}{=} 2$
	$4 + 4 = 8$		$8 - 6 = 2$
	$8 = 8$		$2 = 2$

The point (4, 2) satisfies both equations and, therefore, must be the solution to our system.

EXAMPLE 3 Solve this system by graphing.

$$y = 2x - 3$$
$$x = 3$$

SOLUTION Graphing both equations on the same set of axes, we have Figure 3.

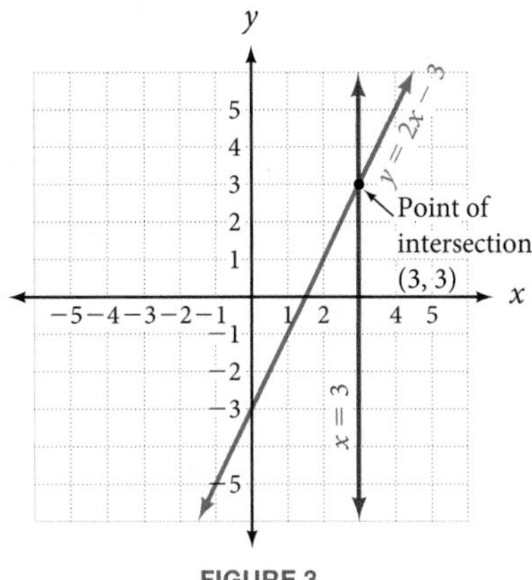

FIGURE 3

The solution to the system is the point (3, 3).

EXAMPLE 4 Solve by graphing.

$$y = x - 2$$
$$y = x + 1$$

SOLUTION Graphing both equations produces the lines shown in Figure 4. We can see in Figure 4 that the lines are parallel and therefore do not intersect. Our system has no ordered pair as a solution because there is no ordered pair that satisfies both equations. We say the solution set is the empty set and write $\varnothing$.

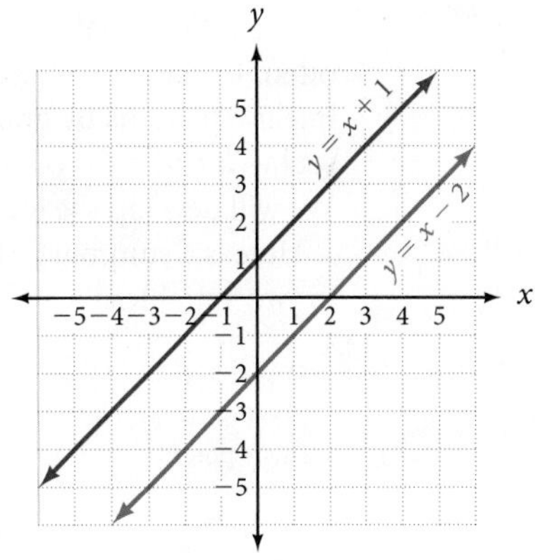

FIGURE 4

EXAMPLE 5 Graph the system.

$$2x + y = 4$$
$$4x + 2y = 8$$

SOLUTION Both graphs are shown in Figure 5. The two graphs coincide. The reason becomes apparent when we multiply both sides of the first equation by 2:

$$2x + y = 4$$
$$2(2x + y) = 2(4) \quad \text{Multiply both sides by 2}$$
$$4x + 2y = 8$$

The equations have the same solution set. Any ordered pair that is a solution to one is a solution to the system. The system has an infinite number of solutions. (Any point on the line is a solution to the system.)

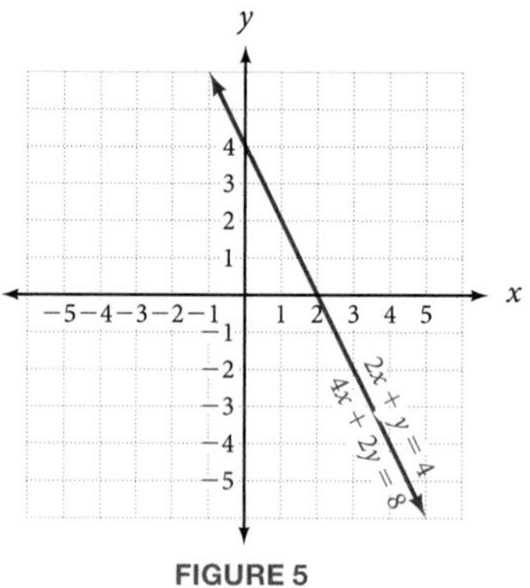

FIGURE 5

We sometimes use special vocabulary to describe the special cases shown in Examples 4 and 5. When a system of equations has no solution because the lines are parallel (as in Example 4), we say the system is *inconsistent*. When the lines coincide (as in Example 5), we say the equations are *dependent*.

The two special cases illustrated in the previous two examples do not happen often. Usually, a system has a single ordered pair as a solution. Solving a system of linear equations by graphing is useful only when the ordered pair in the solution set has integers for coordinates. Two other solution methods work well in all cases. We will develop the other two methods in the next two sections.

Here is a summary of three possible types of solutions to a system of equations in two variables.

One Solution	No Solution	Infinite Solutions
Lines intersect at a single point	Lines are parallel and never cross	Lines coincide
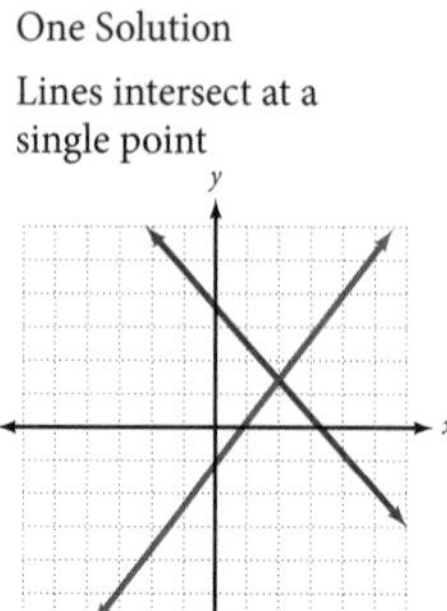	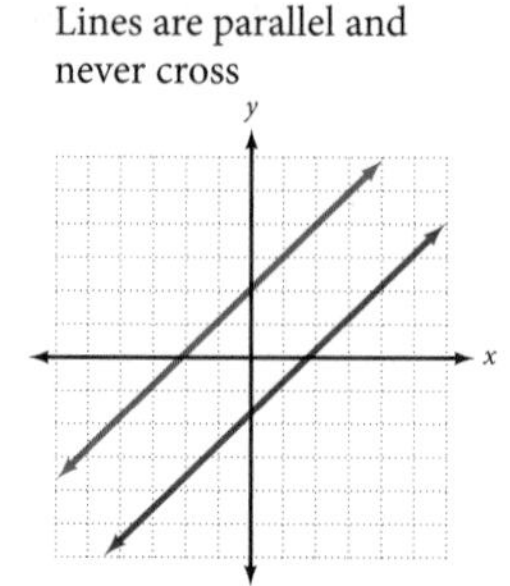	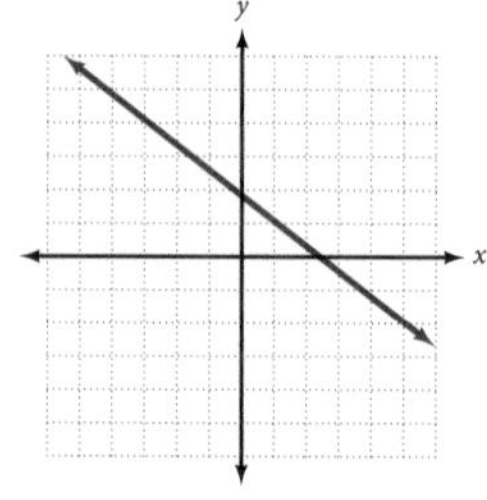
	System is inconsistent	Equations are dependent

GETTING READY FOR CLASS

After reading through the preceding section, respond in your own words and in complete sentences.

A. What is a system of two linear equations in two variables?
B. What is a solution to a system of linear equations?
C. How do we solve a system of linear equations by graphing?
D. Under what conditions will a system of linear equations not have a solution?

SPOTLIGHT ON SUCCESS *Student Instructor Gordon*

Math takes time. This fact holds true in the smallest of math problems as much as it does in the most math intensive careers. I see proof in each video I make. My videos get progressively better with each take, though I still make mistakes and find aspects I can improve on with each new video. In order to keep trying to improve in spite of any failures or lack of improvement, something else is needed. For me it is the sense of a specific goal in sight, to help me maintain the desire to put in continued time and effort.

When I decided on the number one university I wanted to attend, I wrote the name of that school in bold block letters on my door, written to remind myself daily of my ultimate goal. Stuck in the back of my head, this end result pushed me little by little to succeed and meet all of the requirements for the university I had in mind. And now I can say I'm at my dream school bringing with me that skill.

I recognize that others may have much more difficult circumstances than my own to endure, with the goal of improving or escaping those circumstances, and I deeply respect that. But that fact demonstrates to me how easy but effective it is, in comparison, to "stay with the problems longer" with a goal in mind of something much more easily realized, like a good grade on a test. I've learned to set goals, small or big, and to stick with them until they are realized.

Problem Set 4.1

Solve the following systems of linear equations by graphing.

1. $x + y = 3$
$x - y = 1$

2. $x + y = 2$
$x - y = 4$

3. $x + y = 1$
$-x + y = 3$

4. $x + y = 1$
$x - y = -5$

5. $x + y = 8$
$-x + y = 2$

6. $x + y = 6$
$-x + y = -2$

7. $3x - 2y = 6$
$x - y = 1$

8. $5x - 2y = 10$
$x - y = -1$

9. $6x - 2y = 12$
$3x + y = -6$

10. $4x - 2y = 8$
$2x + y = -4$

11. $4x + y = 4$
$3x - y = 3$

12. $5x - y = 10$
$2x + y = 4$

13. $x + 2y = 0$
$2x - y = 0$

14. $3x + y = 0$
$5x - y = 0$

15. $3x - 5y = 15$
$-2x + y = 4$

16. $2x - 4y = 8$
$2x - y = -1$

17. $y = 2x + 1$
$y = -2x - 3$

18. $y = 3x - 4$
$y = -2x + 1$

19. $x + 3y = 3$
$y = x + 5$

20. $2x + y = -2$
$y = x + 4$

21. $x + y = 2$
$x = -3$

22. $x + y = 6$
$y = 2$

23. $x = -4$
$y = 6$

24. $x = 5$
$y = -1$

25. $x + y = 4$
$2x + 2y = -6$

26. $x - y = 3$
$2x - 2y = 6$

27. $4x - 2y = 8$
$2x - y = 4$

28. $3x - 6y = 6$
$x - 2y = 4$

29. As you probably have guessed by now, it can be difficult to solve a system of equations by graphing if the solution to the system contains a fraction. The solution to the following system is $\left(\frac{1}{2}, 1\right)$. Solve the system by graphing.

$$y = -2x + 2$$
$$y = 4x - 1$$

30. The solution to the following system is $\left(\frac{1}{3}, -2\right)$. Solve the system by graphing.

$$y = 3x - 3$$
$$y = -3x - 1$$

31. A second difficulty can arise in solving a system of equations by graphing if one or both of the equations is difficult to graph. The solution to the following system is (2, 1). Solve the system by graphing.

$$3x - 8y = -2$$
$$x - y = 1$$

32. The solution to the following system is (−3, 2). Solve the system by graphing.

$$2x + 5y = 4$$
$$x - y = -5$$

Applying the Concepts

33. Job Comparison Jane is deciding between two sales positions. She can work for Marcy's and receive $8.00 per hour, or she can work for Gigi's, where she earns $6.00 per hour but also receives a $50 commission per week. The two lines in the following figure represent the money Jane will make for working at each of the jobs.

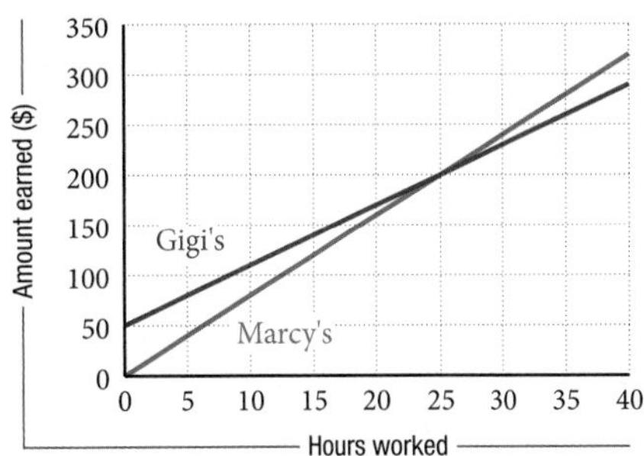

a. From the figure, how many hours would Jane have to work to earn the same amount at each of the positions?

b. If Jane expects to work less than 20 hours a week, which job should she choose?

c. If Jane expects to work more than 30 hours a week, which job should she choose?

34. Truck Rental You need to rent a moving truck for two days. Rider Moving Trucks charges $50 per day and $0.50 per mile. UMove Trucks charges $45 per day and $0.75 per mile. The following figure represents the cost of renting each of the trucks for two days.

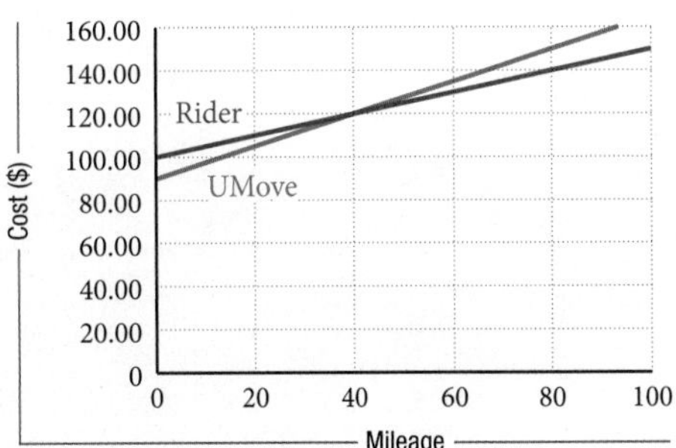

a. From the figure, after how many miles would the trucks cost the same?

b. Which company will give you a better deal if you drive less than 30 miles?

c. Which company will give you a better deal if you drive more than 60 miles?

Getting Ready For the Next Section

Simplify each of the following.

35. $(x + y) + (x - y)$

36. $(x + 2y) + (-x + y)$

37. $3(2x - y) + (x + 3y)$

38. $3(2x + 3y) - 2(3x + 5y)$

39. $-4(3x + 5y) + 5(5x + 4y)$

40. $(3x + 8y) - (3x - 2y)$

41. $6\left(\frac{1}{2}x - \frac{1}{3}y\right)$

42. $12\left(\frac{1}{4}x + \frac{2}{3}y\right)$

43. Let $x + y = 4$. If $x = 3$, find y.

44. Let $x + 2y = 4$. If $x = 3$, find y.

45. Let $x + 3y = 3$. If $x = 3$, find y.

46. Let $2x + 3y = -1$. If $y = -1$, find x.

47. Let $3x + 5y = -7$. If $x = 6$, find y.

48. Let $3x - 2y = 12$. If $y = 6$, find x.

The Elimination Method 4.2

The addition property states that if equal quantities are added to both sides of an equation, the solution set is unchanged. In the past we have used this property to help solve equations in one variable. We will now use it to solve systems of linear equations. Here is another way to state the addition property of equality.

Let A, B, C, and D represent algebraic expressions.

$$\begin{array}{ll} \text{If} & A = B \\ \text{and} & C = D \\ \text{then} & A + C = B + D \end{array}$$

Because C and D are equal (that is, they represent the same number), what we have done is added the same amount to both sides of the equation $A = B$. Let's see how we can use this form of the addition property of equality to solve a system of linear equations.

EXAMPLE 1 Solve the following system.

$$\begin{aligned} x + y &= 4 \\ x - y &= 2 \end{aligned}$$

Note: The graphs shown in our first three examples are not part of the solution shown in each example. The graphs are there simply to show you that the results we obtain by the elimination method are consistent with the results we would obtain by graphing.

SOLUTION The system is written in the form of the addition property of equality as written in this section. It looks like this:

$$\begin{aligned} A &= B \\ C &= D \end{aligned}$$

where A is $x + y$, B is 4, C is $x - y$, and D is 2.

We use the addition property of equality to add the left sides together and the right sides together.

$$\begin{array}{r} x + y = 4 \\ \underline{x - y = 2} \\ 2x + 0 = 6 \end{array}$$

We now solve the resulting equation for x.

$$\begin{aligned} 2x + 0 &= 6 \\ 2x &= 6 \\ x &= 3 \end{aligned}$$

The value we get for x is the value of the x-coordinate of the point of intersection of the two lines $x + y = 4$ and $x - y = 2$. To find the y-coordinate, we simply substitute $x = 3$ into either of the two original equations. Using the first equation, we get

$$\begin{aligned} 3 + y &= 4 \\ y &= 1 \end{aligned}$$

The solution to our system is the ordered pair (3, 1). It satisfies both equations.

When	$x = 3$	When	$x = 3$
and	$y = 1$	and	$y = 1$
the equation	$x + y = 4$	the equation	$x - y = 2$
becomes	$3 + 1 \stackrel{?}{=} 4$	becomes	$3 - 1 \stackrel{?}{=} 2$
	$4 = 4$		$2 = 2$

Figure 1 is visual evidence that the solution to our system is (3, 1).

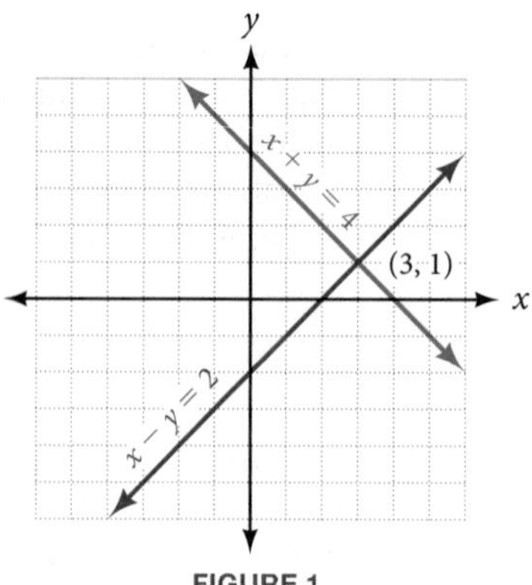

FIGURE 1

The most important part of this method of solving linear systems is eliminating one of the variables when we add the left and right sides together. In our first example, the equations were written so that the y variable was eliminated when we added the left and right sides together. If the equations are not set up this way to begin with, we have to work on one or both of them separately before we can add them together to eliminate one variable.

EXAMPLE 2 Solve the following system.

$$x + 2y = 4$$
$$x - y = -5$$

SOLUTION Notice that if we were to add the equations together as they are, the resulting equation would have terms in both x and y. Let's eliminate the variable x by multiplying both sides of the second equation by -1 before we add the equations together. (As you will see, we can choose to eliminate either the x or the y variable.) Multiplying both sides of the second equation by -1 will not change its solution, so we do not need to be concerned that we have altered the system.

$$x + 2y = 4 \xrightarrow{\text{No change}} x + 2y = 4$$
$$x - y = -5 \xrightarrow[\text{Multiply by } -1]{} \underline{-x + y = 5}$$
$$0 + 3y = 9 \quad \text{Add left and right sides to get}$$
$$3y = 9$$
$$y = 3 \quad \left\{ \begin{array}{l} y\text{-Coordinate of the} \\ \text{point of intersection} \end{array} \right.$$

Substituting $y = 3$ into either of the two original equations, we get $x = -2$. The solution to the system is $(-2, 3)$. It satisfies both equations. Figure 2 shows the solution to the system as the point where the two lines cross.

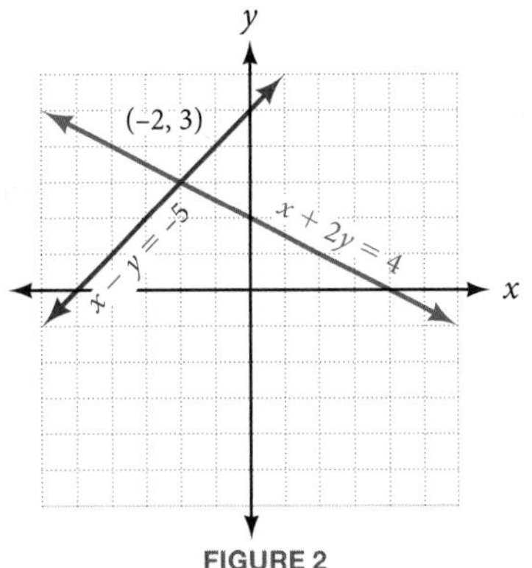

FIGURE 2

EXAMPLE 3 Solve the following system.

$$2x - y = 6$$
$$x + 3y = 3$$

SOLUTION Let's eliminate the y variable from the two equations. We can do this by multiplying the first equation by 3 and leaving the second equation unchanged.

$$2x - y = 6 \xrightarrow{\text{3 times both sides}} 6x - 3y = 18$$
$$x + 3y = 3 \xrightarrow[\text{No change}]{} x + 3y = 3$$

The important thing about our system now is that the coefficients (the numbers in front) of the y variables are opposites. When we add the terms on each side of the equal sign, then the terms in y will add to zero and be eliminated.

$$\begin{aligned} 6x - 3y &= 18 \\ x + 3y &= 3 \\ \hline 7x \qquad &= 21 \end{aligned}$$ *Add corresponding terms*

This gives us $x = 3$. Using this value of x in the second equation of our original system, we have

$$\begin{aligned} 3 + 3y &= 3 \\ 3y &= 0 \\ y &= 0 \end{aligned}$$

We could substitute $x = 3$ into any of the equations with both x and y variables and also get $y = 0$. The solution to our system is the ordered pair (3, 0). Figure 3 is a picture of the system of equations showing the solution (3, 0).

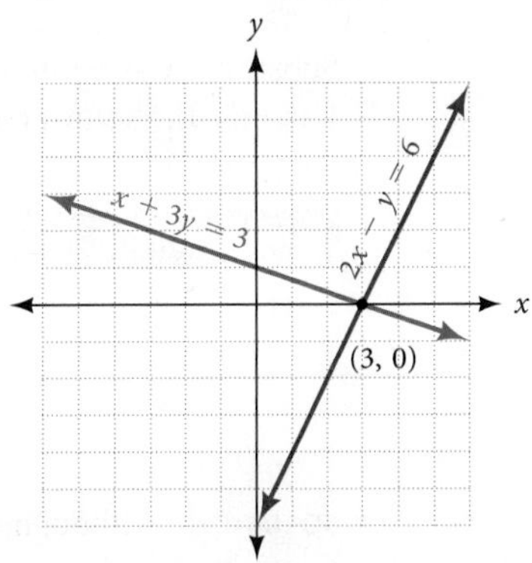

FIGURE 3

EXAMPLE 4 Solve the system.

$$2x + 3y = -1$$
$$3x + 5y = -2$$

SOLUTION Let's eliminate x from the two equations. If we multiply the first equation by 3 and the second by -2, the coefficients of x will be 6 and -6, respectively. The x terms in the two equations will then add to zero.

$$2x + 3y = -1 \xrightarrow{\text{Multiply by } 3} 6x + 9y = -3$$
$$3x + 5y = -2 \xrightarrow[\text{Multiply by } -2]{} -6x - 10y = 4$$

Note: If you are having trouble understanding this method of solution, it is probably because you can't see why we chose to multiply by 3 and -2 in the first step of Example 4. Look at the result of doing so: the $6x$ and $-6x$ will add to 0. We chose to multiply by 3 and -2 because they produce $6x$ and $-6x$, which will add to 0.

We now add the left and right sides of our new system together.

$$\begin{aligned} 6x + 9y &= -3 \\ -6x - 10y &= 4 \\ \hline -y &= 1 \\ y &= -1 \end{aligned}$$

Substituting $y = -1$ into the first equation in our original system, we have

$$\begin{aligned} 2x + 3(-1) &= -1 \\ 2x - 3 &= -1 \\ 2x &= 2 \\ x &= 1 \end{aligned}$$

The solution to our system is $(1, -1)$. It is the only ordered pair that satisfies both equations.

EXAMPLE 5 Solve the system.

$$3x + 5y = -7$$
$$5x + 4y = 10$$

SOLUTION Let's eliminate y by multiplying the first equation by -4 and the second equation by 5.

$$3x + 5y = -7 \xrightarrow{\text{Multiply by } -4} -12x - 20y = 28$$
$$5x + 4y = 10 \xrightarrow[\text{Multiply by } 5]{} 25x + 20y = 50$$

$$\begin{aligned} 13x &= 78 \\ x &= 6 \end{aligned}$$

Substitute $x = 6$ into either equation in our original system, and the result will be $y = -5$. The solution is therefore $(6, -5)$.

EXAMPLE 6 Solve the system.

$$\frac{1}{2}x - \frac{1}{3}y = 2$$
$$\frac{1}{4}x + \frac{2}{3}y = 6$$

SOLUTION Although we could solve this system without clearing the equations of fractions, there is probably less chance for error if we have only integer coefficients to work with. So let's begin by multiplying both sides of the top equation by 6 and both sides of the bottom equation by 12, to clear each equation of fractions.

$$\frac{1}{2}x - \frac{1}{3}y = 2 \xrightarrow{\text{Multiply by 6}} 3x - 2y = 12$$

$$\frac{1}{4}x + \frac{2}{3}y = 6 \xrightarrow[\text{Multiply by 12}]{} 3x + 8y = 72$$

Now we can eliminate x by multiplying the top equation by -1 and leaving the bottom equation unchanged.

$$\begin{aligned} 3x - 2y = 12 &\xrightarrow{\text{Multiply by } -1} & -3x + 2y &= -12 \\ 3x + 8y = 72 &\xrightarrow[\text{No change}]{} & 3x + 8y &= 72 \\ & & \hline 10y &= 60 \\ & & y &= 6 \end{aligned}$$

We can substitute $y = 6$ into any equation that contains both x and y. Let's use $3x - 2y = 12$.

$$\begin{aligned} 3x - 2(6) &= 12 \\ 3x - 12 &= 12 \\ 3x &= 24 \\ x &= 8 \end{aligned}$$

The solution to the system is (8, 6).

Our next two examples will show what happens when we apply the elimination method to a system of equations consisting of parallel lines and to a system in which the lines coincide.

EXAMPLE 7 Solve the system.

$$\begin{aligned} 2x - y &= 2 \\ 4x - 2y &= 12 \end{aligned}$$

SOLUTION Let us choose to eliminate y from the system. We can do this by multiplying the first equation by -2 and leaving the second equation unchanged.

$$\begin{aligned} 2x - y = 2 &\xrightarrow{\text{Multiply by } -2} -4x + 2y = -4 \\ 4x - 2y = 12 &\xrightarrow[\text{No change}]{} 4x - 2y = 12 \end{aligned}$$

If we add both sides of the resulting system, we have

$$\begin{aligned} -4x + 2y &= -4 \\ 4x - 2y &= 12 \\ \hline 0 + 0 &= 8 \\ 0 &= 8 \quad \text{A false statement} \end{aligned}$$

Both variables have been eliminated and we end up with the false statement $0 = 8$. We have tried to solve a system that consists of two parallel lines. There is no solution, and that is the reason we end up with a false statement. Figure 4 is a visual representation of the situation and is conclusive evidence that there is no solution to our system.

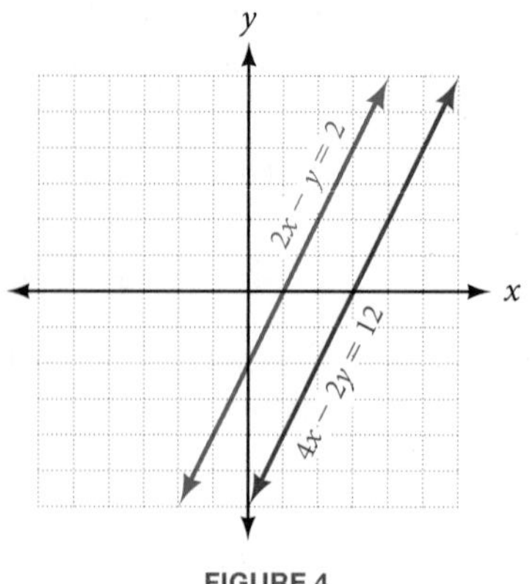

FIGURE 4

EXAMPLE 8 Solve the system.

$$4x - 3y = 2$$
$$8x - 6y = 4$$

SOLUTION Multiplying the top equation by -2 and adding, we can eliminate the variable x.

$$4x - 3y = 2 \xrightarrow{\text{Multiply by } -2} -8x + 6y = -4$$
$$8x - 6y = 4 \xrightarrow[\text{No change}]{} \underline{8x - 6y = 4}$$
$$0 = 0$$

Both variables have been eliminated, and the resulting statement $0 = 0$ is true. In this case the lines coincide because the equations are equivalent. The solution set consists of all ordered pairs that satisfy either equation.

The preceding two examples illustrate the two special cases in which the graphs of the equations in the system either coincide or are parallel.

Here is a summary of our results from these two examples:

Both variables are eliminated and the resulting statement is false.	$\leftrightarrow$	The lines are parallel and there is no solution to the system.
Both variables are eliminated and the resulting statement is true.	$\leftrightarrow$	The lines coincide and there is an infinite number of solutions to the system.

The main idea in solving a system of linear equations by the elimination method is to use the multiplication property of equality on one or both of the original equations, if necessary, to make the coefficients of either variable opposites. The following box shows some steps to follow when solving a system of linear equations by the elimination method.

HOW TO *Solving a System of Linear Equations by the Elimination Method*

Step 1: Decide which variable to eliminate. (In some cases one variable will be easier to eliminate than the other. With some practice you will notice which one it is.)

Step 2: Use the multiplication property of equality on each equation separately to make the coefficients of the variable that is to be eliminated opposites.

Step 3: Add the respective left and right sides of the system together.

Step 4: Solve for the variable remaining.

Step 5: Substitute the value of the variable from step 4 into an equation containing both variables and solve for the other variable.

Step 6: Check your solution in both equations, if necessary.

GETTING READY FOR CLASS

After reading through the preceding section, respond in your own words and in complete sentences.

A. How is the addition property of equality used in the elimination method of solving a system of linear equations?

B. What happens when we use the elimination method to solve a system of linear equations consisting of two parallel lines?

C. What does it mean when we solve a system of linear equations by the elimination method and we end up with the statement $0 = 8$?

D. What is the first step in solving a system of linear equations that contains fractions?

Problem Set 4.2

Solve the following systems of linear equations by elimination.

1. $x + y = 3$
   $x - y = 1$
2. $x + y = -2$
   $x - y = 6$
3. $x + y = 10$
   $-x + y = 4$
4. $x - y = 1$
   $-x - y = -7$
5. $x - y = 7$
   $-x - y = 3$
6. $x - y = 4$
   $2x + y = 8$
7. $x + y = -1$
   $3x - y = -3$
8. $2x - y = -2$
   $-2x - y = 2$
9. $3x + 2y = 1$
   $-3x - 2y = -1$
10. $-2x - 4y = 1$
    $2x + 4y = -1$

Solve each of the following systems by eliminating the *y* variable.

11. $3x - y = 4$
    $2x + 2y = 24$
12. $2x + y = 3$
    $3x + 2y = 1$
13. $5x - 3y = -2$
    $10x - y = 1$
14. $4x - y = -1$
    $2x + 4y = 13$
15. $11x - 4y = 11$
    $5x + y = 5$
16. $3x - y = 7$
    $10x - 5y = 25$

Solve each of the following systems by eliminating the *x* variable.

17. $3x - 5y = 7$
    $-x + y = -1$
18. $4x + 2y = 32$
    $x + y = -2$
19. $-x - 8y = -1$
    $-2x + 4y = 13$
20. $-x + 10y = 1$
    $-5x + 15y = -9$
21. $-3x - y = 7$
    $6x + 7y = 11$
22. $-5x + 2y = -6$
    $10x + 7y = 34$

Solve each of the following systems of linear equations by the elimination method.

23. $6x - y = -8$
    $2x + y = -16$
24. $5x - 3y = -3$
    $3x + 3y = -21$
25. $x + 3y = 9$
    $2x - y = 4$
26. $x + 2y = 0$
    $2x - y = 0$
27. $x - 6y = 3$
    $4x + 3y = 21$
28. $8x + y = -1$
    $4x - 5y = 16$
29. $2x + 9y = 2$
    $5x + 3y = -8$
30. $5x + 2y = 11$
    $7x + 8y = 7$
31. $\frac{1}{3}x + \frac{1}{4}y = \frac{7}{6}$
    $\frac{3}{2}x - \frac{1}{3}y = \frac{7}{3}$
32. $\frac{7}{12}x - \frac{1}{2}y = \frac{1}{6}$
    $\frac{2}{5}x - \frac{1}{3}y = \frac{11}{15}$
33. $3x + 2y = -1$
    $6x + 4y = 0$
34. $8x - 2y = 2$
    $4x - y = 2$
35. $11x + 6y = 17$
    $5x - 4y = 1$
36. $3x - 8y = 7$
    $10x - 5y = 45$
37. $\frac{1}{2}x + \frac{1}{6}y = \frac{1}{3}$
    $-x - \frac{1}{3}y = -\frac{1}{6}$
38. $-\frac{1}{3}x - \frac{1}{2}y = -\frac{2}{3}$
    $-\frac{2}{3}x - y = -\frac{4}{3}$

39. Multiply both sides of the second equation in the following system by 100, and then solve as usual.

$$x + y = 22$$
$$0.05x + 0.10y = 1.70$$

40. Multiply both sides of the second equation in the following system by 100, and then solve as usual.

$$x + y = 15{,}000$$
$$0.06x + 0.07y = 980$$

Getting Ready for the Next Section

Solve.

41. $x + (2x - 1) = 2$

42. $2x - 3(2x - 8) = 12$

43. $2(3y - 1) - 3y = 4$

44. $-2x + 4(3x + 6) = 14$

45. $4x + 2(-2x + 4) = 8$

46. $1.5x + 15 = 0.75x + 24.95$

Solve each equation for the indicated variable.

47. $x - 3y = -1$ for x

48. $-3x + y = 6$ for y

49. Let $y = 2x - 1$. If $x = 1$, find y.

50. Let $y = 2x - 8$. If $x = 5$, find y.

51. Let $x = 3y - 1$. If $y = 2$, find x.

52. Let $y = 3x + 6$. If $y = -6$, find x.

Let $y = 1.5x + 15$

53. If $x = 13$, find y.

54. If $x = 14$, find y.

Let $y = 0.75x + 24.95$

55. If $x = 12$, find y.

56. If $x = 16$, find y.

SPOTLIGHT ON SUCCESS *Howard College*

HOWARD HC COLLEGE

The motto at Howard College in Texas is "Education, for learning, for earning, for life."

The school's website, www.earnmydegree.com poses the question: "Does a college degree pay off?"

You can make much more money by earning a college degree.

The data shows that a college degree correlates directly to your salary range—and the relationship between compensation and education level is becoming even more prominent.

Employers have increasingly use diplomas and degrees as a way to screen applicants. And once you've landed the job you want, your salary will reflect your credentials. On average, a person with a master's degree earns $31,900 more per year than a high school graduate—a difference of as much as 105%!

Average Annual Earnings by Education Level

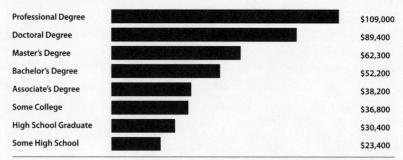

Average Annual Earnings — Different Levels of Education
Source: U.S. Census Bureau, Current Population Surveys, March 1998, 1999, and 2000

The Substitution Method 4.3

There is a third method of solving systems of equations. It is the substitution method, and, like the elimination method, it can be used on any system of linear equations. Some systems, however, lend themselves more to the substitution method than others do.

EXAMPLE 1 Solve the following system.

$$x + y = 2$$
$$y = 2x - 1$$

SOLUTION If we were to solve this system by the methods used in the previous section, we would have to rearrange the terms of the second equation so that similar terms would be in the same column. There is no need to do this, however, because the second equation tells us that y is $2x - 1$. We can replace the y variable in the first equation with the expression $2x - 1$ from the second equation; that is, we *substitute* $2x - 1$ from the second equation for y in the first equation. Here is what it looks like:

$$x + (2x - 1) = 2$$

The equation we end up with contains only the variable x. The y variable has been eliminated by substitution.

Note: Sometimes this method of solving systems of equations is confusing the first time you see it. If you are confused, you may want to read through this first example more than once.

Solving the resulting equation, we have

$$x + (2x - 1) = 2$$
$$3x - 1 = 2$$
$$3x = 3$$
$$x = 1$$

This is the x-coordinate of the solution to our system. To find the y-coordinate, we substitute $x = 1$ into the second equation of our system. (We could substitute $x = 1$ into the first equation also and have the same result.)

$$y = 2(1) - 1$$
$$y = 2 - 1$$
$$y = 1$$

The solution to our system is the ordered pair (1, 1). It satisfies both of the original equations. Figure 1 provides visual evidence that the substitution method yields the correct solution.

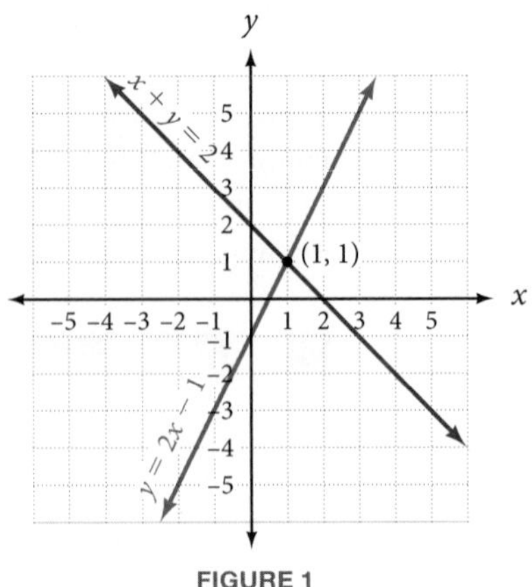

FIGURE 1

EXAMPLE 2 Solve the following system by the substitution method.

$$2x - 3y = 12$$
$$y = 2x - 8$$

SOLUTION Again, the second equation says y is $2x - 8$. Because we are looking for the ordered pair that satisfies both equations, the y in the first equation must also be $2x - 8$. Substituting $2x - 8$ from the second equation for y in the first equation, we have

$$2x - 3(2x - 8) = 12$$

This equation can still be read as $2x - 3y = 12$ because $2x - 8$ is the same as y. Solving the equation, we have

$$2x - 3(2x - 8) = 12$$
$$2x - 6x + 24 = 12$$
$$-4x + 24 = 12$$
$$-4x = -12$$
$$x = 3$$

To find the y-coordinate of our solution, we substitute $x = 3$ into the second equation in the original system.

When	$x = 3$
the equation	$y = 2x - 8$
becomes	$y = 2(3) - 8$
	$y = 6 - 8 = -2$

The solution to our system is $(3, -2)$.

EXAMPLE 3 Solve the following system by solving the first equation for x and then using the substitution method:

$$x - 3y = -1$$
$$2x - 3y = 4$$

SOLUTION We solve the first equation for x by adding $3y$ to both sides to get

$$x = 3y - 1$$

Using this value of x in the second equation, we have

$$\begin{aligned} 2(3y - 1) - 3y &= 4 \\ 6y - 2 - 3y &= 4 \\ 3y - 2 &= 4 \\ 3y &= 6 \\ y &= 2 \end{aligned}$$

Next, we find x.

When	$y = 2$
the equation	$x = 3y - 1$
becomes	$x = 3(2) - 1$
	$x = 6 - 1$
	$x = 5$

The solution to our system is (5, 2)

Here are the steps to use in solving a system of equations by the substitution method.

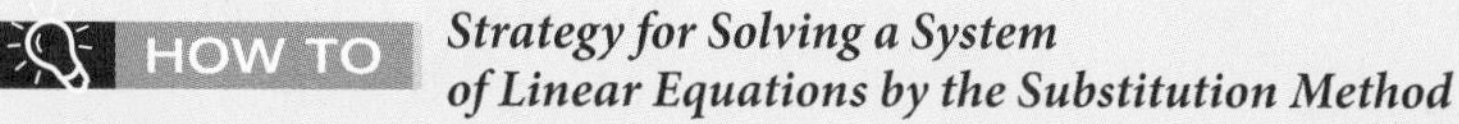

HOW TO ***Strategy for Solving a System of Linear Equations by the Substitution Method***

Step 1: Solve either one of the equations for x or y. (This step is not necessary if one of the equations is already in the correct form, as in Examples 1 and 2.)

Step 2: Substitute the expression for the variable obtained in step 1 into the other equation and solve it.

Step 3: Substitute the solution from step 2 into any equation in the system that contains both variables and solve it.

Step 4: Check your results, if necessary.

EXAMPLE 4 Solve by substitution.

$$\begin{aligned} -2x + 4y &= 14 \\ -3x + y &= 6 \end{aligned}$$

SOLUTION We can solve either equation for either variable. If we look at the system closely, it becomes apparent that solving the second equation for y is the easiest way to go. If we add $3x$ to both sides of the second equation, we have

$$y = 3x + 6$$

Substituting the expression $3x + 6$ back into the first equation in place of y yields the following result.

$$\begin{aligned} -2x + 4(3x + 6) &= 14 \\ -2x + 12x + 24 &= 14 \\ 10x + 24 &= 14 \\ 10x &= -10 \\ x &= -1 \end{aligned}$$

Substituting $x = -1$ into the equation $y = 3x + 6$ leaves us with

$$\begin{aligned} y &= 3(-1) + 6 \\ y &= -3 + 6 \\ y &= 3 \end{aligned}$$

The solution to our system is $(-1, 3)$.

EXAMPLE 5 Solve by substitution.

$$\begin{aligned} 4x + 2y &= 8 \\ y &= -2x + 4 \end{aligned}$$

SOLUTION Substituting the expression $-2x + 4$ for y from the second equation into the first equation, we have

$$\begin{aligned} 4x + 2(-2x + 4) &= 8 \\ 4x - 4x + 8 &= 8 \\ 8 &= 8 \qquad \text{A true statement} \end{aligned}$$

Both variables have been eliminated, and we are left with a true statement. Recall from the last section that a true statement in this situation tells us the lines coincide; that is, the equations $4x + 2y = 8$ and $y = -2x + 4$ have exactly the same graph. Any point on that graph has coordinates that satisfy both equations and is a solution to the system.

EXAMPLE 6 The following table shows two contract rates charged by GTE Wireless for cellular phone use. At how many minutes will the two rates cost the same amount?

	Flat Rate	Plus	Per Minute Charge
Plan 1	\$15		\$1.50
Plan 2	\$24.95		\$0.75

SOLUTION If we let $y =$ the monthly charge for x minutes of phone use, then the equations for each plan are

$$\text{Plan 1:} \quad y = 1.5x + 15$$

$$\text{Plan 2:} \quad y = 0.75x + 24.95$$

We can solve this system by substitution by replacing the variable y in Plan 2 with the expression $1.5x + 15$ from Plan 1. If we do so, we have

$$1.5x + 15 = 0.75x + 24.95$$

$$0.75x + 15 = 24.95$$

$$0.75x = 9.95$$

$$x = 13.27 \quad \text{to the nearest hundredth}$$

The monthly bill is based on the number of minutes you use the phone, with any fraction of a minute moving you up to the next minute. If you talk for a total of 13 minutes, you are billed for 13 minutes. If you talk for 13 minutes, 10 seconds, you are billed for 14 minutes. The number of minutes on your bill always will be a whole number. So, to calculate the cost for talking 13.27 minutes, we would replace x with 14 and find y. Let's compare the two plans at $x = 13$ minutes and at $x = 14$ minutes.

Plan 1:	$y = 1.5x + 15$	Plan 2:	$y = 0.75x + 24.95$
When	$x = 13, y = \$34.50$	When	$x = 13, y = \$34.70$
When	$x = 14, y = \$36.00$	When	$x = 14, y = \$35.45$

The two plans will never give the same cost for talking x minutes. If you talk 13 or less minutes, Plan 1 will cost less. If you talk for more than 13 minutes, you will be billed for 14 minutes, and Plan 2 will cost less than Plan 1.

GETTING READY FOR CLASS

After reading through the preceding section, respond in your own words and in complete sentences.

A. What is the first step in solving a system of linear equations by substitution?

B. When would substitution be more efficient than the elimination method in solving two linear equations?

C. What does it mean when we solve a system of linear equations by the substitution method and we end up with the statement $8 = 8$?

D. How would you begin solving the following system using the substitution method?

$$x + y = 2$$
$$y = 2x - 1$$

Problem Set 4.3

Solve the following systems by substitution. Substitute the expression in the second equation into the first equation and solve.

1. $x + y = 11$
$y = 2x - 1$

2. $x - y = -3$
$y = 3x + 5$

3. $x + y = 20$
$y = 5x + 2$

4. $3x - y = -1$
$x = 2y - 7$

5. $-2x + y = -1$
$y = -4x + 8$

6. $4x - y = 5$
$y = -4x + 1$

7. $3x - 2y = -2$
$x = -y + 6$

8. $2x - 3y = 17$
$x = -y + 6$

9. $5x - 4y = -16$
$y = 4$

10. $6x + 2y = 18$
$x = 3$

11. $5x + 4y = 7$
$y = -3x$

12. $10x + 2y = -6$
$y = -5x$

Solve the following systems by solving one of the equations for x or y and then using the substitution method.

13. $x + 3y = 4$
$x - 2y = -1$

14. $x - y = 5$
$x + 2y = -1$

15. $2x + y = 1$
$x - 5y = 17$

16. $2x - 2y = 2$
$x - 3y = -7$

17. $3x + 5y = -3$
$x - 5y = -5$

18. $2x - 4y = -4$
$x + 2y = 8$

19. $5x + 3y = 0$
$x - 3y = -18$

20. $x - 3y = -5$
$x - 2y = 0$

21. $-3x - 9y = 7$
$x + 3y = 12$

22. $2x + 6y = -18$
$x + 3y = -9$

Solve the following systems using the substitution method.

23. $5x - 8y = 7$
$y = 2x - 5$

24. $3x + 4y = 10$
$y = 8x - 15$

25. $7x - 6y = -1$
$x = 2y - 1$

26. $4x + 2y = 3$
$x = 4y - 3$

27. $-3x + 2y = 6$
$y = 3x$

28. $-2x - y = -3$
$y = -3x$

29. $5x - 6y = -4$
$x = y$

30. $2x - 4y = 0$
$y = x$

31. $3x + 3y = 9$
$y = 2x - 12$

32. $7x + 6y = -9$
$y = -2x + 1$

33. $7x - 11y = 16$
$y = 10$

34. $9x - 7y = -14$
$x = 7$

35. $-4x + 4y = -8$
$y = x - 2$

36. $-4x + 2y = -10$
$y = 2x - 5$

Solve each system by substitution. You can eliminate the decimals if you like, but you don't have to. The solution will be the same in either case.

37. $0.05x + 0.10y = 1.70$
$y = 22 - x$

38. $0.20x + 0.50y = 3.60$
$y = 12 - x$

Applying the Concepts

39. Gas Mileage Daniel is trying to decide whether to buy a car or a truck. The truck he is considering will cost him \$150 a month in loan payments, and it gets 20 miles per gallon in gas mileage. The car will cost \$180 a month in loan payments, but it gets 35 miles per gallon in gas mileage. Daniel estimates that he will pay \$1.40 per gallon for gas. This means that the monthly cost to drive the truck x miles will be $y = \frac{1.40}{20}x + 150$. The total monthly cost to drive the car x miles will be $y = \frac{1.40}{35}x + 180$. The following figure shows the graph of each equation.

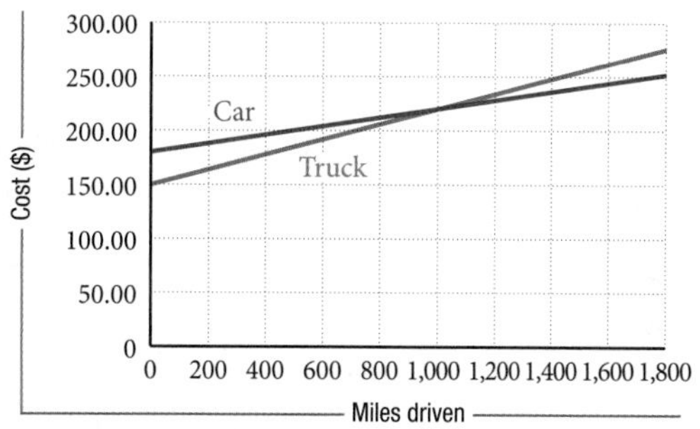

a. At how many miles do the car and the truck cost the same to operate?

b. If Daniel drives more than 1,200 miles, which will be cheaper?

c. If Daniel drives fewer than 800 miles, which will be cheaper?

d. Why do the graphs appear in the first quadrant only?

40. Video Production Pat runs a small company that duplicates videotapes. The daily cost and daily revenue for a company duplicating videos are shown in the following figure. The daily cost for duplicating x videos is $y = \frac{6}{5}x + 20$; the daily revenue (the amount of money he brings in each day) for duplicating x videos is $y = 1.7x$. The graphs of the two lines are shown in the following figure.

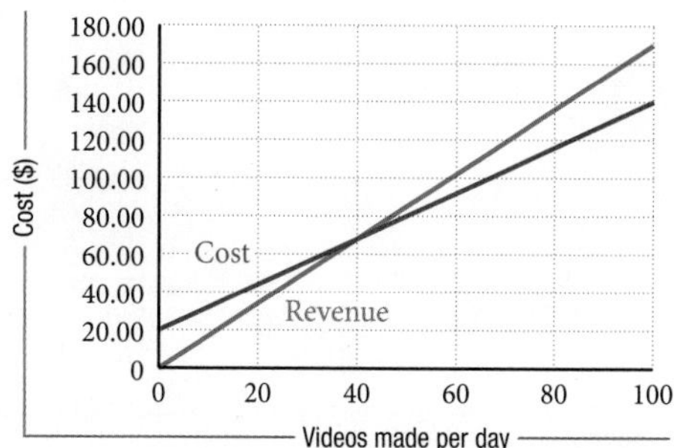

a. Pat will "break even" when his cost and his revenue are equal. How many videos does he need to duplicate to break even?

b. Pat will incur a loss when his revenue is less than his cost. If he duplicates 30 videos in one day, will he incur a loss?

c. Pat will make a profit when his revenue is larger than his costs. For what values of x will Pat make a profit?

d. Why does the graph appear in the first quadrant only?

Getting Ready for the Next Section

41. One number is eight more than five times another; their sum is 26. Find the numbers.

42. One number is three less than four times another; their sum is 27. Find the numbers.

43. The difference of two positive numbers is nine. The larger number is six less than twice the smaller number. Find the numbers.

44. The difference of two positive numbers is 17. The larger number is one more than twice the smaller number. Find the numbers.

45. The length of a rectangle is five inches more than three times the width. The perimeter is 58 inches. Find the length and width.

46. The length of a rectangle is three inches less than twice the width. The perimeter is 36 inches. Find the length and width.

47. John has $1.70 in nickels and dimes in his pocket. He has four more nickels than he does dimes. How many of each does he have?

48. Jamie has $2.65 in dimes and quarters in his pocket. He has two more dimes than she does quarters. How many of each does she have?

Applications 4.4

I often have heard students remark about the word problems in beginning algebra: "What does this have to do with real life?" Most of the word problems we will encounter don't have much to do with "real life." We are actually just practicing. Ultimately, all problems requiring the use of algebra are word problems; that is, they are stated in words first, then translated to symbols. The problem then is solved by some system of mathematics, like algebra. Most real applications involve calculus or higher levels of mathematics. So, if the problems we solve are upsetting or frustrating to you, then you are probably taking them too seriously.

The word problems in this section have two unknown quantities. We will write two equations in two variables (each of which represents one of the unknown quantities), which of course is a system of equations. We then solve the system by one of the methods developed in the previous sections of this chapter. Here are the steps to follow in solving these word problems.

BLUEPRINT FOR PROBLEM SOLVING

Using a System of Equations

Step 1: ***Read*** the problem, and then mentally ***list*** the items that are known and the items that are unknown.

Step 2: ***Assign variables*** to each of the unknown items; that is, let $x =$ one of the unknown items and $y =$ the other unknown item. Then ***translate*** the other ***information*** in the problem to expressions involving the two variables.

Step 3: ***Reread*** the problem, and then ***write a system of equations,*** using the items and variables listed in steps 1 and 2, that describes the situation.

Step 4: ***Solve the system*** found in step 3.

Step 5: ***Write*** your ***answers*** using complete sentences.

Step 6: ***Reread*** the problem, and ***check*** your solution with the original words in the problem.

Remember, the more problems you work, the more problems you will be able to work. If you have trouble getting started on the problem set, come back to the examples and work through them yourself. The examples are similar to the problems found in the problem set.

Number Problem

EXAMPLE 1 One number is 2 more than 5 times another number. Their sum is 20. Find the two numbers.

SOLUTION Applying the steps in our blueprint, we have

Step 1: We know that the two numbers have a sum of 20 and that one of them is 2 more than 5 times the other. We don't know what the numbers themselves are.

Step 2: Let x represent one of the numbers and y represent the other. "One number is 2 more than 5 times another" translates to

$$y = 5x + 2$$

"Their sum is 20" translates to

$$x + y = 20$$

Step 3: The system that describes the situation must be

$$\begin{aligned} x + y &= 20 \\ y &= 5x + 2 \end{aligned}$$

Step 4: We can solve this system by substituting the expression $5x + 2$ in the second equation for y in the first equation:

$$\begin{aligned} x + (5x + 2) &= 20 \\ 6x + 2 &= 20 \\ 6x &= 18 \\ x &= 3 \end{aligned}$$

Using $x = 3$ in either of the first two equations and then solving for y, we get $y = 17$.

Step 5: So 17 and 3 are the numbers we are looking for.

Step 6: The number 17 is 2 more than 5 times 3, and the sum of 17 and 3 is 20.

Interest Problem

EXAMPLE 2 Mr. Hicks had \$15,000 to invest. He invested some at 6% and the rest at 7%. If he earns \$980 in interest, how much did he invest at each rate?

SOLUTION Remember, step 1 is done mentally.

Step 1: We do not know the specific amounts invested in the two accounts. We do know that their sum is \$15,000 and that the interest rates on the two accounts are 6% and 7%.

Step 2: Let $x =$ the amount invested at 6% and $y =$ the amount invested at 7%. Because Mr. Hicks invested a total of \$15,000, we have

$$x + y = 15{,}000$$

The interest he earns comes from 6% of the amount invested at 6% and 7% of the amount invested at 7%. To find 6% of x, we multiply x by 0.06, which gives us $0.06x$. To find 7% of y, we multiply 0.07 times y and get $0.07y$.

$$\begin{matrix} \text{Interest} & + & \text{interest} & = & \text{total} \\ \text{at 6\%} & & \text{at 7\%} & & \text{interest} \end{matrix}$$

$$0.06x + 0.07y = 980$$

Step 3: The system is

$$\begin{aligned} x + y &= 15{,}000 \\ 0.06x + 0.07y &= 980 \end{aligned}$$

Step 4: We multiply the first equation by -6 and the second by 100 to eliminate x:

$$\begin{array}{lcr} x + \quad y = 15{,}000 & \xrightarrow{\text{Multiply by } -6} & -6x - 6y = -90{,}000 \\ 0.06x + 0.07y = 980 & \xrightarrow[\text{Multiply by } 100]{} & 6x + 7y = 98{,}000 \\ \hline & & y = 8{,}000 \end{array}$$

Substituting $y = 8{,}000$ into the first equation and solving for x, we get $x = 7{,}000$.

Step 5: He invested \$7,000 at 6% and \$8,000 at 7%.

Step 6: Checking our solutions in the original problem, we have: The sum of \$7,000 and \$8,000 is \$15,000, the total amount he invested. To complete our check, we find the total interest earned from the two accounts:

The interest on \$7,000 at 6% is $0.06(7{,}000) = \$420$
The interest on \$8,000 at 7% is $0.07(8{,}000) = \$560$
The total interest is \$980

Coin Problem

EXAMPLE 3 John has \$1.70 all in dimes and nickels. He has a total of 22 coins. How many of each kind does he have?

SOLUTION

Step 1: We know that John has 22 coins that are dimes and nickels. We know that a dime is worth 10 cents and a nickel is worth 5 cents. We do not know the specific number of dimes and nickels he has.

Step 2: Let x = the number of nickels and y = the number of dimes. The total number of coins is 22, so

$$x + y = 22$$

The total amount of money he has is \$1.70, which comes from nickels and dimes:

$$\begin{array}{ccccc} \text{Amount of money in nickels} & + & \text{amount of money in dimes} & = & \text{total amount of money} \\ 0.05x & + & 0.10y & = & 1.70 \end{array}$$

Step 3: The system that represents the situation is

$$\begin{array}{rl} x + y = 22 & \text{The number of coins} \\ 0.05x + 0.10y = 1.70 & \text{The value of the coins} \end{array}$$

Step 4: We multiply the first equation by -5 and the second by 100 to eliminate the variable x:

$$\begin{array}{lcr} x + \quad y = \quad 22 & \xrightarrow{\text{Multiply by } -5} & -5x - 5y = -110 \\ 0.05x + 0.10y = 1.70 & \xrightarrow[\text{Multiply by } 100]{} & 5x + 10y = 170 \\ \hline & & 5y = 60 \\ & & y = 12 \end{array}$$

Substituting $y = 12$ into our first equation, we get $x = 10$.

Step 5: John has 12 dimes and 10 nickels.

Step 6: Twelve dimes and 10 nickels total 22 coins.

$$\begin{aligned} \text{12 dimes are worth } 12(0.10) &= 1.20 \\ \text{10 nickels are worth } 10(0.05) &= 0.50 \\ \hline \text{The total value is } &\ \ \$1.70 \end{aligned}$$

Mixture Problem

EXAMPLE 4 How much of a 20% alcohol solution and 50% alcohol solution must be mixed to get 12 gallons of 30% alcohol solution?

SOLUTION To solve this problem we must first understand that a 20% alcohol solution is 20% alcohol and 80% water.

Step 1: We know there are two solutions that together must total 12 gallons. 20% of one of the solutions is alcohol and the rest is water, whereas the other solution is 50% alcohol and 50% water. We do not know how many gallons of each individual solution we need.

Step 2: Let $x =$ the number of gallons of 20% alcohol solution needed and $y =$ the number of gallons of 50% alcohol solution needed. Because the total number of gallons we will end up with is 12, and this 12 gallons must come from the two solutions we are mixing, our first equation is

$$x + y = 12$$

To obtain our second equation, we look at the amount of alcohol in our two original solutions and our final solution. The amount of alcohol in the x gallons of 20% solution is $0.20x$, and the amount of alcohol in y gallons of 50% solution is $0.50y$. The amount of alcohol in the 12 gallons of 30% solution is $0.30(12)$. Because the amount of alcohol we start with must equal the amount of alcohol we end up with, our second equation is

$$0.20x + 0.50y = 0.30(12)$$

The information we have so far can also be summarized with a table. Sometimes by looking at a table like the one that follows it is easier to see where the equations come from.

	20% Solution	50% Solution	Final Solution
Number of Gallons	x	y	12
Gallons of Alcohol	$0.20x$	$0.50y$	$0.30(12)$

Step 3: Our system of equations is

$$\begin{aligned} x + y &= 12 \\ 0.20x + 0.50y &= 0.30(12) \end{aligned}$$

Step 4: We can solve this system by substitution. Solving the first equation for y and substituting the result into the second equation, we have

$$0.20x + 0.50(12 - x) = 0.30(12)$$

Multiplying each side by 10 gives us an equivalent equation that is a little easier to work with.

$$2x + 5(12 - x) = 3(12)$$

$$2x + 60 - 5x = 36$$

$$-3x + 60 = 36$$

$$-3x = -24$$

$$x = 8$$

If x is 8, then y must be 4 because $x + y = 12$.

Step 5: It takes 8 gallons of 20% alcohol solution and 4 gallons of 50% alcohol solution to produce 12 gallons of 30% alcohol solution.

Step 6: Try it and see.

GETTING READY FOR CLASS

After reading through the preceding section, respond in your own words and in complete sentences.

A. If you were to apply the Blueprint for Problem Solving from Section 2.6 to the examples in this section, what would be the first step?

B. If you were to apply the Blueprint for Problem Solving from Section 2.6 to the examples in this section, what would be the last step?

C. Which method of solving these systems do you prefer? Why?

D. Write an application problem for which the solution depends on solving a system of equations.

Problem Set 4.4

Solve the following word problems. Be sure to show the equations used.

Number Problems

1. Two numbers have a sum of 25. One number is 5 more than the other. Find the numbers.
2. The difference of two numbers is 6. Their sum is 30. Find the two numbers.
3. The sum of two numbers is 15. One number is 4 times the other. Find the numbers.
4. The difference of two positive numbers is 28. One number is 3 times the other. Find the two numbers.
5. Two positive numbers have a difference of 5. The larger number is one more than twice the smaller. Find the two numbers.
6. One number is 2 more than 3 times another. Their sum is 26. Find the two numbers.
7. One number is 5 more than 4 times another. Their sum is 35. Find the two numbers.
8. The difference of two positive numbers is 8. The larger is twice the smaller decreased by 7. Find the two numbers.

Interest Problems

9. Mr. Wilson invested money in two accounts. His total investment was $20,000. If one account pays 6% in interest and the other pays 8% in interest, how much does he have in each account if he earned a total of $1,380 in interest in 1 year?
10. A total of $11,000 was invested. Part of the $11,000 was invested at 4%, and the rest was invested at 7%. If the investments earn $680 per year, how much was invested at each rate?
11. A woman invested 4 times as much at 5% as she did at 6%. The total amount of interest she earns in 1 year from both accounts is $520. How much did she invest at each rate?
12. Ms. Hagan invested twice as much money in an account that pays 7% interest as she did in an account that pays 6% in interest. Her total investment pays her $1,000 a year in interest. How much did she invest at each rate?

Coin Problems

13. Ron has 14 coins with a total value of $2.30. The coins are nickels and quarters. How many of each coin does he have?
14. Diane has $0.95 in dimes and nickels. She has a total of 11 coins. How many of each kind does she have?
15. Suppose Tom has 21 coins totaling $3.45. If he has only dimes and quarters, how many of each type does he have?
16. A coin collector has 31 dimes and nickels with a total face value of $2.40. (They are actually worth a lot more.) How many of each coin does she have?

Mixture Problems

17. How many liters of 50% alcohol solution and 20% alcohol solution must be mixed to obtain 18 liters of 30% alcohol solution?

	50% Solution	20% Solution	Final Solution
Number of Liters	x	y	18
Liters of Alcohol			

18. How many liters of 10% alcohol solution and 5% alcohol solution must be mixed to obtain 40 liters of 8% alcohol solution?

	10% Solution	5% Solution	Final Solution
Number of Liters	x	y	40
Liters of Alcohol			

19. A mixture of 8% disinfectant solution is to be made from 10% and 7% disinfectant solutions. How much of each solution should be used if 30 gallons of 8% solution are needed?

20. How much 50% antifreeze solution and 40% antifreeze solution should be combined to give 50 gallons of 46% antifreeze solution?

Miscellaneous Problems

21. For a Saturday matinee, adult tickets cost \$5.50 and kids under 12 pay only \$4.00. If 70 tickets are sold for a total of \$310, how many of the tickets were adult tickets and how many were sold to kids under 12?

22. The Bishop's Peak 4 – H club is having its annual fundraising dinner. Adults pay \$15 apiece and children pay \$10 apiece. If the number of adult tickets sold is twice the number of children's tickets sold, and the total income for the dinner is \$1,600, how many of each kind of ticket did the 4 – H club sell?

23. A farmer has 96 feet of fence with which to make a corral. If he arranges it into a rectangle that is twice as long as it is wide, what are the dimensions?

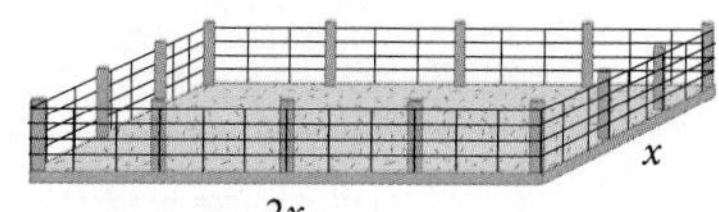

24. If a 22-inch rope is to be cut into two pieces so that one piece is 3 inches longer than twice the other, how long is each piece?

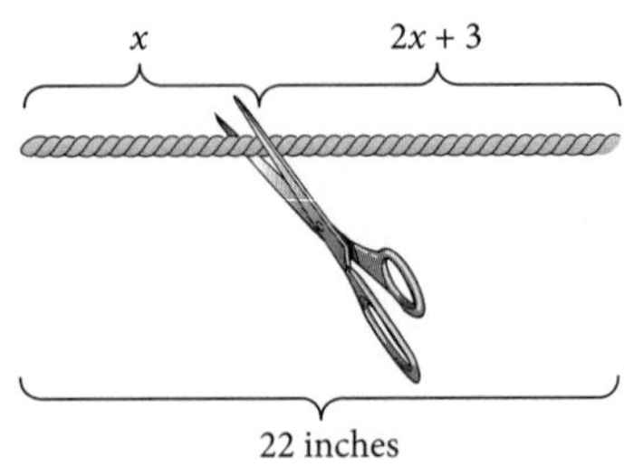

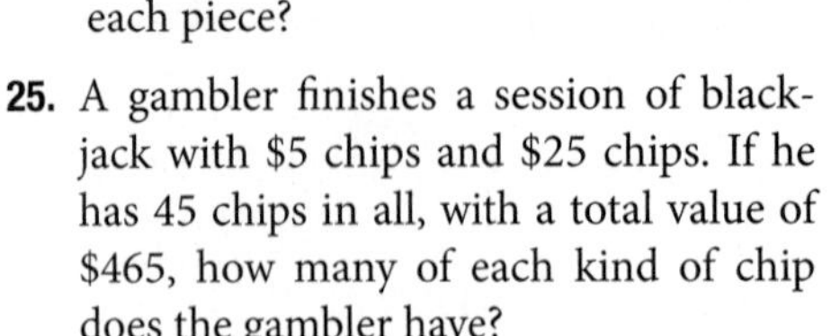

25. A gambler finishes a session of blackjack with \$5 chips and \$25 chips. If he has 45 chips in all, with a total value of \$465, how many of each kind of chip does the gambler have?

26. Tyler has been saving his winning lottery tickets. He has 23 tickets that are worth a total of \$175. If each ticket is worth either \$5 or \$10, how many of each does he have?

27. Mary Jo spends \$2,550 to buy stock in two companies. She pays \$11 a share to one of the companies and \$20 a share to the other. If she ends up with a total of 150 shares, how many shares did she buy at \$11 a share and how many did she buy at \$20 a share?

28. Kelly sells 62 shares of stock she owns for a total of \$433. If the stock was in two different companies, one selling at \$6.50 a share and the other at \$7.25 a share, how many of each did she sell?

Maintaining Your Skill

29. Reduce to lowest terms $\frac{x^2 - x - 6}{x^2 - 9}$

30. Divide using long division $\frac{x^2 - 2x + 6}{x - 4}$.

Perform the indicated operations.

31. $\frac{x^2 - 25}{x + 4} \cdot \frac{2x + 8}{x^2 - 9x + 20}$

32. $\frac{3x + 6}{x^2 - 4x + 3} \div \frac{x^2 + x - 2}{x^2 + 2x - 3}$

33. $\frac{x}{x^2 - 16} + \frac{4}{x^2 - 16}$

34. $\frac{2}{x^2 - 1} - \frac{5}{x^2 + 3x - 4}$

35. $\dfrac{1 - \frac{25}{x^2}}{1 - \frac{8}{x} + \frac{15}{x^2}}$

Solve each equation.

36. $\frac{x}{2} - \frac{5}{x} = -\frac{3}{2}$

37. $\frac{x}{x^2 - 9} - \frac{3}{x - 3} = \frac{1}{x + 3}$

38. Speed of a Boat A boat travels 30 miles up a river in the same amount of time it takes to travel 50 miles down the same river. If the current is 5 miles per hour, what is the speed of the boat in still water?

39. Filling a Pool A pool can be filled by an inlet pipe in 8 hours. The drain will empty the pool in 12 hours. How long will it take to fill the pool if both the inlet pipe and the drain are open?

40. Mixture Problem If 30 liters of a certain solution contains 2 liters of alcohol, how much alcohol is in 45 liters of the same solution?

41. y varies directly with x. If $y = 8$ when x is 12, find y when x is 36.

Chapter 4 Summary

EXAMPLES

1. The solution to the system

$$x + 2y = 4$$
$$x - y = 1$$

is the ordered pair (2, 1). It is the only ordered pair that satisfies both equations.

Definitions [4.1]

1. A *system of linear equations*, as the term is used in this book, is two linear equations that each contain the same two variables.
2. The *solution set* for a system of equations is the set of all ordered pairs that satisfy *both* equations. The solution set to a system of linear equations will contain:

Case I — One ordered pair when the graphs of the two equations intersect at only one point (this is the most common situation)

Case II — No ordered pairs when the graphs of the two equations are parallel lines

Case III — An infinite number of ordered pairs when the graphs of the two equations coincide (are the same line)

2. Solving the system in Example 1 by graphing looks like

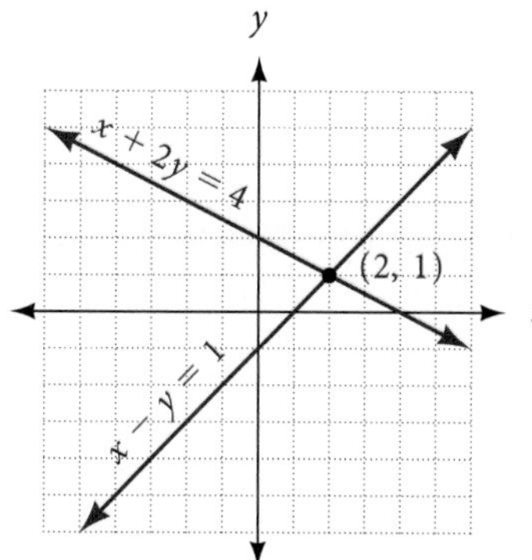

Strategy for Solving a System by Graphing [4.1]

Step 1: Graph the first equation.

Step 2: Graph the second equation on the same set of axes.

Step 3: Read the coordinates of the point where the graphs cross each other (the coordinates of the point of intersection).

Step 4: Check the solution to see that it satisfies *both* equations.

3. We can eliminate the y variable from the system in Example 1 by multiplying both sides of the second equation by 2 and adding the result to the first equation

$$\begin{aligned} x + 2y &= 4 & \qquad x + 2y &= 4 \\ x - y &= 1 \xrightarrow{\text{Multiply by 2}} & 2x - 2y &= 2 \\ & & 3x &= 6 \\ & & x &= 2 \end{aligned}$$

Substituting $x = 2$ into either of the original two equations gives $y = 1$. The solution is (2, 1).

Strategy for Solving a System by the Elimination Method [4.2]

Step 1: Look the system over to decide which variable will be easier to eliminate.

Step 2: Use the multiplication property of equality on each equation separately to ensure that the coefficients of the variable to be eliminated are opposites.

Step 3: Add the left and right sides of the system produced in step 2, and solve the resulting equation.

Step 4: Substitute the solution from step 3 back into any equation with both x and y variables, and solve.

Step 5: Check your solution in both equations, if necessary.

Strategy for Solving a System by the Substitution Method [4.3]

4. We can apply the substitution method to the system in Example 1 by first solving the second equation for x to get $x = y + 1$. Substituting this expression for x into the first equation, we have

$$\begin{aligned}(y + 1) + 2y &= 4\\ 3y + 1 &= 4\\ 3y &= 3\\ y &= 1\end{aligned}$$

Using $y = 1$ in either of the original equations gives $x = 2$.

Step 1: Solve either of the equations for one of the variables (this step is not necessary if one of the equations has the correct form already).

Step 2: Substitute the results of step 1 into the other equation, and solve.

Step 3: Substitute the results of step 2 into an equation with both x and y variables, and solve. (The equation produced in step 1 is usually a good one to use.)

Step 4: Check your solution, if necessary.

Special Cases [4.1, 4.2, 4.3]

In some cases, using the elimination or substitution method eliminates both variables. The situation is interpreted as follows.

1. If the resulting statement is *false*, then the lines are parallel and there is no solution to the system.
2. If the resulting statement is *true*, then the equations represent the same line (the lines coincide). In this case any ordered pair that satisfies either equation is a solution to the system.

COMMON MISTAKE

The most common mistake encountered in solving linear systems is the failure to complete the problem. Here is an example.

$$\begin{aligned}x + y &= 8\\ x - y &= 4\\ 2x &= 12\\ x &= 6\end{aligned}$$

This is only half the solution. To find the other half, we must substitute the 6 back into one of the original equations and then solve for y.

Remember, solutions to systems of linear equations always consist of ordered pairs. We need an x-coordinate and a y-coordinate; $x = 6$ can never be a solution to a system of linear equations.

1. Write the solution to the system which is graphed below. [4.1]

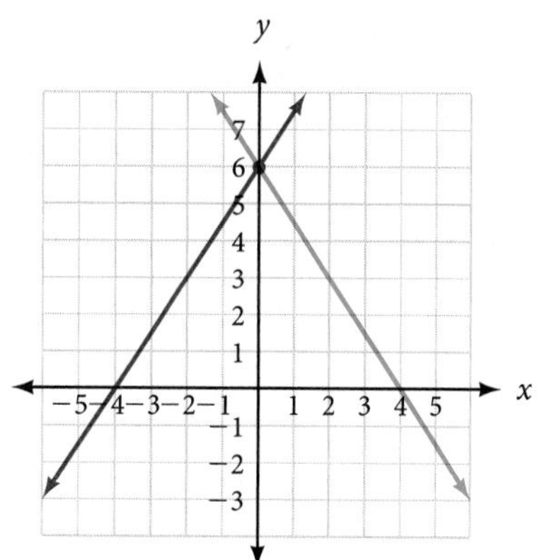

Solve each system by graphing. [4.1]

2. $4x - 2y = 8$
$y = \frac{2}{3}x$

3. $3x - 2y = 13$
$y = 4$

4. $2x - 2y = -12$
$-3x - y = 2$

Solve each system by the elimination method. [4.2]

5. $x - y = -9$
$2x + 3y = 7$

6. $3x - y = 1$
$5x - y = 3$

7. $2x + 3y = -3$
$x + 6y = 12$

8. $2x + 3y = 4$
$4x + 6y = 8$

Solve each system by the substitution method. [4.3]

9. $3x - y = 12$
$y = 2x - 8$

10. $3x + 6y = 3$
$x = 4y - 17$

11. $2x - 3y = -18$
$3x + y = -5$

12. $2x - 3y = 13$
$x - 4y = -1$

Solve the following word problems. In each case, be sure to show the system of equations that describes the situation. [4.4]

13. **Number Problem** The sum of two numbers is 15. Their difference is 11. Find the numbers.

14. **Number Problem** The sum of two numbers is 18. One number is 2 more than 3 times the other. Find the two numbers.

15. **Investing** Dave has $2,000 to invest. He would like to earn $135.20 per year in interest. How much should he invest at 6% if the rest is to be invested at 7%?

16. **Coin Problem** Maria has 19 coins that total $1.35. If the coins are all nickels and dimes, how many of each type does she have?

17. **Fencing Problem** A rancher wants to build a rectangular corral using 198 feet of fence. If the length of the corral is to be 15 feet longer than twice the width, find the dimensions of the corral.

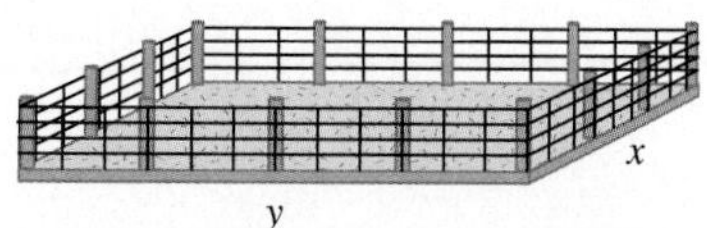

5 Exponents and Polynomials

Chapter Outline

5.1 Multiplication with Exponents

5.2 Division with Exponents

5.3 Operations with Monomials

5.4 Addition and Subtraction of Polynomials

5.5 Multiplication with Polynomials

5.6 Binomial Squares and Other Special Products

5.7 Dividing a Polynomial by a Monomial

5.8 Dividing a Polynomial by a Polynomial

If you were given a penny on the first day of September, and then each day after that you were given twice the amount of money you received the day before, how much money would you receive on September 30th? To begin, Table 1 and Figure 1 show the amount of money you would receive on each of the first 10 days of the month. As you can see, on the tenth day of the month you would receive \$5.12.

TABLE 1

Money That Doubles Each Day

Day	Money (in cents)
1	$1 = 2^0$
2	$2 = 2^1$
3	$4 = 2^2$
4	$8 = 2^3$
5	$16 = 2^4$
6	$32 = 2^5$
7	$64 = 2^6$
8	$128 = 2^7$
9	$256 = 2^8$
10	$512 = 2^9$

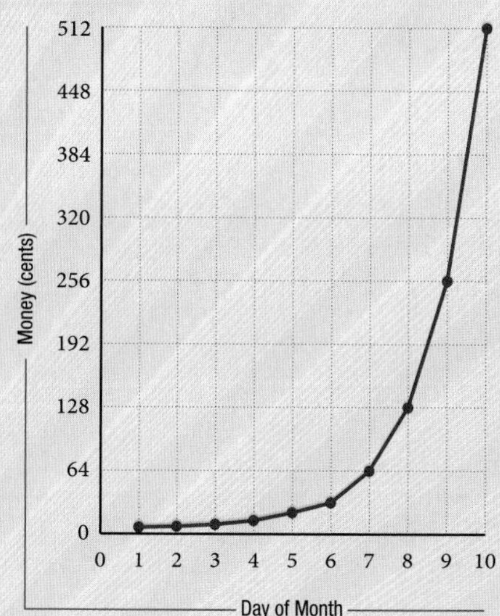

To find the amount of money on day 30, we could continue to double the amount on each of the next 20 days. Or, we could notice the pattern of exponents in the second column of the table and reason that the amount of money on day 30 would be 2^{29} cents, which is a very large number. In fact, 2^{29} cents is \$5,368,709.12—a little less than \$5.4 million. When you are finished with this chapter, you will have a good working knowledge of exponents.

Study Skills

The study skills for this chapter cover the way you approach new situations in mathematics. The first study skill is a point of view you hold about your natural instincts for what does and doesn't work in mathematics. The second study skill gives you a way of testing your instincts.

1. **Don't Let Your Intuition Fool You** As you become more experienced and more successful in mathematics you will be able to trust your mathematical intuition. For now, though, it can get in the way of your success. For example, if you ask some students to "subtract 3 from -5" they will answer -2 or 2. Both answers are incorrect, even though they may seem intuitively true. Likewise, some students will expand $(a + b)^2$ and arrive at $a^2 + b^2$, which is incorrect. In both cases, intuition leads directly to the wrong answer.

2. **Test Properties of Which You are Unsure** From time to time, you will be in a situation where you would like to apply a property or rule, but you are not sure it is true. You can always test a property or statement by substituting numbers for variables. For instance, I always have students that rewrite $(x + 3)^2$ as $x^2 + 9$, thinking that the two expressions are equivalent. The fact that the two expressions are not equivalent becomes obvious when we substitute 10 for x in each one.

   When $x = 10$, the expression $(x + 3)^2$ is $(10 + 3)^2 = 13^2 = 169$

   When $x = 10$, the expression $x^2 + 9 = 10^2 + 9 = 100 + 9 = 109$

   When you test the equivalence of expressions by substituting numbers for the variable, make it easy on yourself by choosing numbers that are easy to work with, such as 10. Don't try to verify the equivalence of expressions by substituting 0, 1, or 2 for the variable, as using these numbers will occasionally give you false results.

It is not good practice to trust your intuition or instincts in every new situation in algebra. If you have any doubt about the generalizations you are making, test them by replacing variables with numbers and simplifying.

Multiplication with Exponents 5.1

Recall that an *exponent* is a number written just above and to the right of another number, which is called the *base*. In the expression 5^2, for example, the exponent is 2 and the base is 5. The expression 5^2 is read "5 to the second power" or "5 squared." The meaning of the expression is

$$5^2 = 5 \cdot 5 = 25$$

In the expression 5^3, the exponent is 3 and the base is 5. The expression 5^3 is read "5 to the third power" or "5 cubed." The meaning of the expression is

$$5^3 = 5 \cdot 5 \cdot 5 = 125$$

Here are some further examples.

EXAMPLE 1 Write each expression as a single number.

a. 4^3 **b.** -3^4 **c.** $(-2)^5$ **d.** $\left(-\frac{3}{4}\right)^2$

SOLUTION

a. $4^3 = 4 \cdot 4 \cdot 4 = 16 \cdot 4 = 64$ Exponent 3, base 4

b. $-3^4 = -3 \cdot 3 \cdot 3 \cdot 3 = -81$ Exponent 4, base 3

c. $(-2)^5 = (-2)(-2)(-2)(-2)(-2) = -32$ Exponent 5, base -2

d. $\left(-\frac{3}{4}\right)^2 = \left(-\frac{3}{4}\right)\left(-\frac{3}{4}\right) = \frac{9}{16}$ Exponent 2, base $-\frac{3}{4}$

Question: In what way are $(-5)^2$ and -5^2 different?

Answer: In the first case, the base is -5. In the second case, the base is 5. The answer to the first is 25. The answer to the second is -25. Can you tell why? Would there be a difference in the answers if the exponent in each case were changed to 3?

We can simplify our work with exponents by developing some properties of exponents. We want to list the things we know are true about exponents and then use these properties to simplify expressions that contain exponents.

The first property of exponents applies to products with the same base. We can use the definition of exponents, as indicating repeated multiplication, to simplify expressions like $7^4 \cdot 7^2$.

$$\begin{aligned} 7^4 \cdot 7^2 &= (7 \cdot 7 \cdot 7 \cdot 7)(7 \cdot 7) \\ &= (7 \cdot 7 \cdot 7 \cdot 7 \cdot 7 \cdot 7) \\ &= 7^6 \end{aligned}$$

Notice: $4 + 2 = 6$

As you can see, multiplication with the same base resulted in addition of exponents. We can summarize this result with the following property.

PROPERTY *Property 1 for Exponents*

If a is any real number and r and s are integers, then

$$a^r \cdot a^s = a^{r+s}$$

In words: To multiply two expressions with the same base, add exponents and use the common base.

Here is an example using Property 1.

EXAMPLE 2 Use Property 1 to simplify the following expressions. Leave your answers in terms of exponents:

a. $5^3 \cdot 5^6$ **b.** $x^7 \cdot x^8$ **c.** $3^4 \cdot 3^8 \cdot 3^5$

SOLUTION

a. $5^3 \cdot 5^6 = 5^{3+6} = 5^9$

b. $x^7 \cdot x^8 = x^{7+8} = x^{15}$

c. $3^4 \cdot 3^8 \cdot 3^5 = 3^{4+8+5} = 3^{17}$

Note: In Example 2, notice that in each case the base in the original problem is the same base that appears in the answer and that it is written only once in the answer. A very common mistake that people make when they first begin to use Property 1 is to write a 2 in front of the base in the answer. For example, people making this mistake would get $2x^{15}$ or $(2x)^{15}$ as the result in Example 2. To avoid this mistake, you must be sure you understand the meaning of Property 1 exactly as it is written.

Another common type of expression involving exponents is one in which an expression containing an exponent is raised to another power. The expression $(5^3)^2$ is an example:

$$(5^3)^2 = (5^3)(5^3)$$
$$= 5^{3+3}$$
$$= 5^6 \qquad \text{Notice: } 3 \cdot 2 = 6$$

This result offers justification for the second property of exponents.

PROPERTY *Property 2 for Exponents*

If a is any real number and r and s are integers, then

$$(a^r)^s = a^{r \cdot s}$$

In words: A power raised to another power is the base raised to the product of the powers.

EXAMPLE 3 Simplify the following expressions:

a. $(4^5)^6$ **b.** $(x^3)^5$

SOLUTION

a. $(4^5)^6 = 4^{5 \cdot 6} = 4^{30}$

b. $(x^3)^5 = x^{3 \cdot 5} = x^{15}$

The third property of exponents applies to expressions in which the product of two or more numbers or variables is raised to a power. Let's look at how the expression $(2x)^3$ can be simplified:

$$(2x)^3 = (2x)(2x)(2x)$$
$$= (2 \cdot 2 \cdot 2)(x \cdot x \cdot x)$$
$$= 2^3 \cdot x^3 \qquad \text{Notice: The exponent 3 distributes over the product } 2x$$
$$= 8x^3$$

We can generalize this result into a third property of exponents.

PROPERTY *Property 3 for Exponents*

If a and b are any two real numbers and r is an integer, then

$$(ab)^r = a^r b^r$$

In words: The power of a product is the product of the powers.

Here are some examples using Property 3 to simplify expressions.

EXAMPLE 4 Simplify the following expressions:

a. $\left(-\frac{1}{4}x^2y^3\right)^2$ **b.** $(x^4)^3(x^2)^5$ **c.** $(2y)^3(3y^2)$ **d.** $(2x^2y^5)^3(3x^4y)^2$

SOLUTION

a. $\left(-\frac{1}{4}x^2y^3\right)^2 = \left(-\frac{1}{4}\right)^2(x^2)^2(y^3)^2$ Property 3

$= \frac{1}{16}x^4y^6$ Property 2

b. $(x^4)^3(x^2)^5 = x^{12} \cdot x^{10}$ Property 2

$= x^{22}$ Property 1

c. $(2y)^3(3y^2) = 2^3y^3(3y^2)$ Property 3

$= 8 \cdot 3(y^3 \cdot y^2)$ Commutative and associative properties

$= 24y^5$ Property 1

d. $(2x^2y^5)^3(3x^4y)^2 = 2^3(x^2)^3(y^5)^3 \cdot 3^2(x^4)^2y^2$ Property 3

$= 8x^6y^{15} \cdot 9x^8y^2$ Property 2

$= (8 \cdot 9)(x^6x^8)(y^{15}y^2)$ Commutative and associative properties

$= 72x^{14}y^{17}$ Property 1

FACTS FROM GEOMETRY ***Volume of a Rectangular Solid***

It is easy to see why the phrase "five squared" is associated with the expression 5^2. Simply find the area of the square shown in Figure 1 with a side of 5.

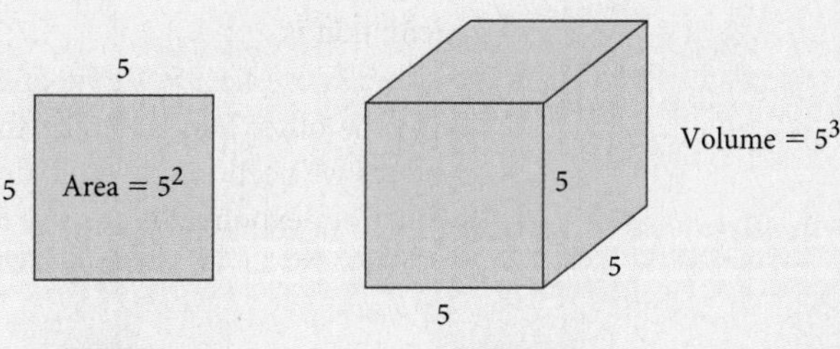

FIGURE 1 **FIGURE 2**

To see why the phrase "five cubed" is associated with the expression 5^3, we have to find the *volume* of a cube for which all three dimensions are 5 units long. The volume of a cube is a measure of the space occupied by the cube. To calculate the volume of the cube shown in Figure 2, we multiply the three dimensions together to get $5 \cdot 5 \cdot 5 = 5^3$.

The cube shown in Figure 2 is a special case of a general category of three dimensional geometric figures called *rectangular solids*. Rectangular solids have rectangles for sides, and all connecting sides meet at right angles. The three dimensions are length, width, and height. To find the volume of a rectangular solid, we find the product of the three dimensions.

Note: If we include units with the dimensions of the diagrams, then the units for the area will be square units and the units for volume will be cubic units. More specifically,

If a square has a side 5 inches long, then its area will be

$A = (5 \text{ inches})^2 = 25 \text{ inches}^2$

where the unit inches^2 stands for square inches.

If a cube has a single side 5 inches long, then its volume will be

$V = (5 \text{ inches})^3 = 125 \text{ inches}^3$

where the unit inches^3 stands for cubic inches.

If a rectangular solid has a length of 5 inches, a width of 4 inches, and a height of 3 inches, then its volume is

$V = (5 \text{ in.})(4 \text{ in.})(3 \text{ in.})$

$= 60 \text{ inches}^3$

Scientific Notation

Many branches of science require working with very large numbers. In astronomy, for example, distances commonly are given in light-years. A light-year is the distance light travels in a year. It is approximately

$$5{,}880{,}000{,}000{,}000 \text{ miles}$$

This number is difficult to use in calculations because of the number of zeros it contains. Scientific notation provides a way of writing very large numbers in a more manageable form.

DEFINITION *scientific notation*

A number is in ***scientific notation*** when it is written as the product of a number between 1 and 10 and an integer power of 10. A number written in scientific notation has the form

$$n \times 10^r$$

where $1 \le n < 10$ and $r =$ an integer.

EXAMPLE 5 Write 376,000 in scientific notation.

SOLUTION We must rewrite 376,000 as the product of a number between 1 and 10 and a power of 10. To do so, we move the decimal point 5 places to the left so that it appears between the 3 and the 7. Then we multiply this number by 10^5. The number that results has the same value as our original number and is written in scientific notation:

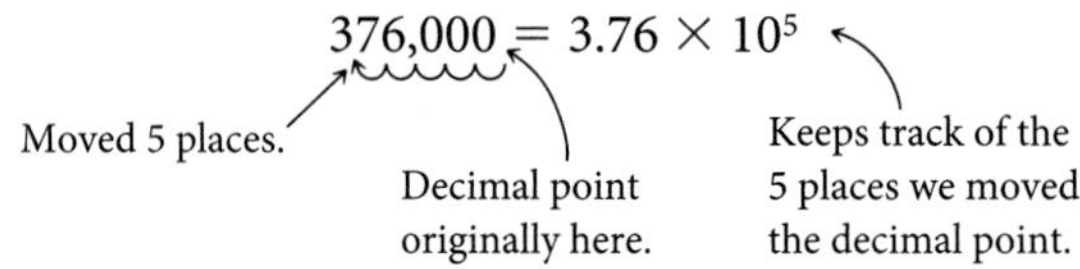

EXAMPLE 6 Write 4.52×10^3 in expanded form.

SOLUTION Since 10^3 is 1,000, we can think of this as simply a multiplication problem; that is,

$$4.52 \times 10^3 = 4.52 \times 1{,}000 = 4{,}520$$

On the other hand, we can think of the exponent 3 as indicating the number of places we need to move the decimal point to write our number in expanded form. Since our exponent is positive 3, we move the decimal point three places to the right:

$$4.52 \times 10^3 = 4{,}520$$

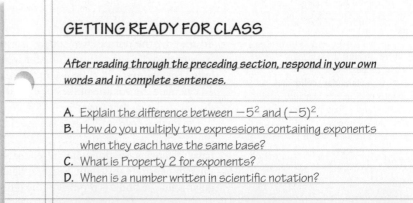

GETTING READY FOR CLASS

After reading through the preceding section, respond in your own words and in complete sentences.

A. Explain the difference between -5^2 and $(-5)^2$.
B. How do you multiply two expressions containing exponents when they each have the same base?
C. What is Property 2 for exponents?
D. When is a number written in scientific notation?

1. 4, 2, 16
2. 6, 2, 36
3. 0.3, 2, 0.09
4. 0.03, 2, 0.0009
5. 4, 3, 64
6. 10, 3, 1,000
7. −5, 2, 25
8. 5, 2, −25
9. 2, 3, −8
10. −2, 3, −8
11. 3, 4, 81
12. −3, 4, 81
13. $\frac{2}{3}, 2, \frac{4}{9}$
14. $\frac{2}{3}, 3, \frac{8}{27}$
15. $\frac{1}{2}, 4, \frac{1}{16}$
16. $\frac{4}{5}, 2, \frac{16}{25}$
19. x^9
20. x^{10}
21. y^{30}
22. y^{60}
23. 2^{12}
24. 4^9
25. x^{28}
26. x^{68}
27. x^{10}
28. x^{10}
29. 5^{12}
30. 5^{12}
31. y^9
32. y^4
33. 2^{50}
34. 10^{10}
35. a^{3x}
36. a^{5x}
37. b^{xy}
38. b^{rs}
39. $16x^2$
40. $16x^4$
41. $32y^5$
42. $25y^2$
43. $81x^4$
44. $-27x^3$
45. $0.25a^2b^2$
46. $0.16a^2b^2$
47. $64x^3y^3z^3$
48. $125x^3y^3z^3$
49. $8x^{12}$
50. $9x^{10}$
51. $16a^6$
52. $25a^4$
53. x^{14}
54. x^{25}
55. a^{11}
56. a^7

Name the base and exponent in each of the following expressions. Then use the definition of exponents as repeated multiplication to simplify.

1. 4^2 **2.** 6^2 **3.** $(0.3)^2$ **4.** $(0.03)^2$ **5.** 4^3 **6.** 10^3

7. $(-5)^2$ **8.** -5^2 **9.** -2^3 **10.** $(-2)^3$ **11.** 3^4 **12.** $(-3)^4$

13. $\left(\frac{2}{3}\right)^2$ **14.** $\left(\frac{2}{3}\right)^3$ **15.** $\left(\frac{1}{2}\right)^4$ **16.** $\left(\frac{4}{5}\right)^2$

17. a. Complete the following table.

Number x	1	2	3	4	5	6	7
Square x^2	1	4	9	16	25	36	49

b. Using the results of part **a**, fill in the blank in the following statement: For numbers larger than 1, the square of the number is larger than the number.

18. a. Complete the following table.

Number x	$\frac{1}{2}$	$\frac{1}{3}$	$\frac{1}{4}$	$\frac{1}{5}$	$\frac{1}{6}$	$\frac{1}{7}$	$\frac{1}{8}$
Square x^2	$\frac{1}{4}$	$\frac{1}{9}$	$\frac{1}{16}$	$\frac{1}{25}$	$\frac{1}{36}$	$\frac{1}{49}$	$\frac{1}{64}$

b. Using the results of part **a**, fill in the blank in the following statement: For numbers between 0 and 1, the square of the number is smaller than the number.

Use Property 1 to simplify the following expressions.

19. $x^4 \cdot x^5$ **20.** $x^7 \cdot x^3$ **21.** $y^{10} \cdot y^{20}$

22. $y^{30} \cdot y^{30}$ **23.** $2^5 \cdot 2^4 \cdot 2^3$ **24.** $4^2 \cdot 4^3 \cdot 4^4$

25. $x^4 \cdot x^6 \cdot x^8 \cdot x^{10}$ **26.** $x^{20} \cdot x^{18} \cdot x^{16} \cdot x^{14}$

Use Property 2 for exponents to write each of the following problems with a single exponent. (Assume all variables are positive numbers.)

27. $(x^2)^5$ **28.** $(x^5)^2$ **29.** $(5^4)^3$ **30.** $(5^3)^4$ **31.** $(y^3)^3$ **32.** $(y^2)^2$

33. $(2^5)^{10}$ **34.** $(10^5)^2$ **35.** $(a^3)^x$ **36.** $(a^5)^x$ **37.** $(b^x)^y$ **38.** $(b^r)^s$

Use Property 3 for exponents to simplify each of the following expressions.

39. $(4x)^2$ **40.** $(2x)^4$ **41.** $(2y)^5$ **42.** $(5y)^2$

43. $(-3x)^4$ **44.** $(-3x)^3$ **45.** $(0.5ab)^2$ **46.** $(0.4ab)^2$

47. $(4xyz)^3$ **48.** $(5xyz)^3$

Simplify the following expressions by using the properties of exponents.

49. $(2x^4)^3$ **50.** $(3x^5)^2$ **51.** $(4a^3)^2$ **52.** $(5a^2)^2$

53. $(x^2)^3(x^4)^2$ **54.** $(x^5)^2(x^3)^5$ **55.** $(a^3)^1(a^2)^4$ **56.** $(a^4)^1(a^1)^3$

57. $(2x)^3(2x)^4$ **58.** $(3x)^2(3x)^3$ **59.** $(3x^2)^3(2x)^4$ **60.** $(3x)^3(2x^3)^2$

57. $128x^7$
58. $243x^5$
59. $432x^{10}$
60. $108x^9$
61. $16x^4y^6$
62. $81x^6y^{10}$
63. $\frac{8}{27}a^{12}b^{15}$
64. $\frac{27}{64}a^3b^{21}$
69. 4.32×10^4
70. 4.32×10^5
71. 5.7×10^2
72. 5.7×10^3
73. 2.38×10^5
74. 2.38×10^6
75. 2,490
76. 24,900
77. 352
78. 352,000
79. 28,000
80. 2,800

61. $(4x^2y^3)^2$ **62.** $(9x^3y^5)^2$ **63.** $\left(\frac{2}{3}a^4b^5\right)^3$ **64.** $\left(\frac{3}{4}ab^7\right)^3$

65. Complete the following table, and then construct a line graph of the information in the table

Number x	-3	-2	-1	0	1	2	3
Square x^2	9	4	1	0	1	4	9

66. Complete the table, and then construct a line graph of the information in the table.

Number x	-3	-2	-1	0	1	2	3
Cube x^3	-27	-8	-1	0	1	8	27

67. Complete the table. When you are finished, notice how the points in this table could be used to refine the line graph you created in Problem 65.

Number x	-2.5	-1.5	-0.5	0	0.5	1.5	2.5
Square x^2	6.25	2.25	0.25	0	0.25	2.25	6.25

68. Complete the following table. When you are finished, notice that this table contains exactly the same entries as the table from Problem 67. This table uses fractions, whereas the table from Problem 67 uses decimals.

Number x	$-\frac{5}{2}$	$-\frac{3}{2}$	$-\frac{1}{2}$	0	$\frac{1}{2}$	$\frac{3}{2}$	$\frac{5}{2}$
Square x^2	$\frac{25}{4}$	$\frac{9}{4}$	$\frac{1}{4}$	0	$\frac{1}{4}$	$\frac{9}{4}$	$\frac{25}{4}$

Write each number in scientific notation.

69. 43,200 **70.** 432,000 **71.** 570

72. 5,700 **73.** 238,000 **74.** 2,380,000

Write each number in expanded form.

75. 2.49×10^3 **76.** 2.49×10^4 **77.** 3.52×10^2

78. 3.52×10^5 **79.** 2.8×10^4 **80.** 2.8×10^3

81. 27 inches^3
82. 27 feet^3
83. 15.6 inches^3
84. 5,832 inches^3
85. 36 inches^3
86. Around 100 inches^3
87. answers will vary
88. answers will vary
89. 6.5×10^8 seconds
90. 1.3×10^8 feet
91. \$740,000
92. 37,800,000 times
93. \$180,000
94. \$3,270

Applying the Concepts

81. Volume of a Cube Find the volume of a cube if each side is 3 inches long.

82. Volume of a Cube Find the volume of a cube if each side is 3 feet long.

83. Volume of a Cube A bottle of perfume is packaged in a box that is in the shape of a cube. Find the volume of the box if each side is 2.5 inches long. Round to the nearest tenth.

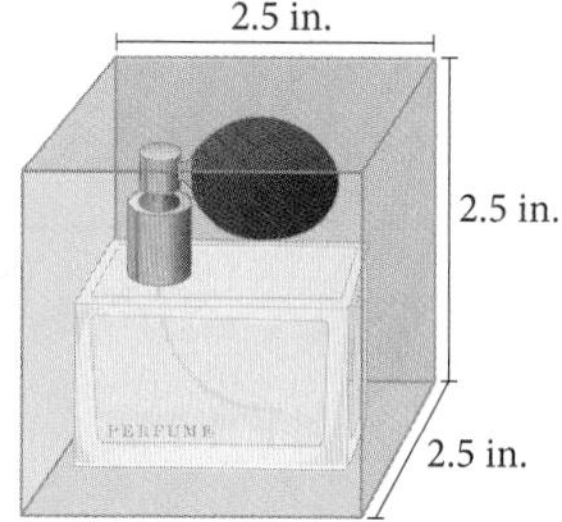

84. Volume of a Cube A television set is packaged in a box that is in the shape of a cube. Find the volume of the box if each side is 18 inches long.

85. Volume of a Box A rented videotape is in a plastic container that has the shape of a rectangular solid. Find the volume of the container if the length is 8 inches, the width is 4.5 inches, and the height is 1 inch.

86. Volume of a Box Your textbook is in the shape of a rectangular solid. Find the volume in cubic inches.

87. Volume of a Box If a box has a volume of 42 cubic feet, is it possible for you to fit inside the box? Explain your answer.

88. Volume of a Box A box has a volume of 45 cubic inches. Will a can of soup fit inside the box? Explain your answer.

89. Age in seconds If you are 21 years old, you have been alive for more than 650,000,000 seconds. Write this last number in scientific notation.

90. Distance Around the Earth The distance around the Earth at the equator is more than 130,000,000 feet. Write this number in scientific notation.

91. Lifetime Earnings If you earn at least \$12 an hour and work full-time for 30 years, you will make at least 7.4×10^5 dollars. Write this last number in expanded form.

92. Heart Beats per Year If your pulse is 72, then in one year your heart will beat at least 3.78×10^7 times. Write this last number in expanded form.

93. Investing If you put \$1,000 into a savings account every year from the time you are 25 years old until you are 55 years old, you will have more than 1.8×10^5 dollars in the account when you reach 55 years of age (assuming 10% annual interest). Write 1.8×10^5 in expanded form.

94. Investing If you put \$20 into a savings account every month from the time you are 20 years old until you are 30 years old, you will have more than 3.27×10^3 dollars in the account when you reach 30 years of age (assuming 6% annual interest compounded monthly). Write 3.27×10^3 in expanded form.

95. 219 inches^3
96. 253 inches^3
97. 182 inches^3
98. 220 inches^3
99. -3
100. -11
101. 11
102. 3
103. -5
104. 35
105. 5
106. -35
107. 2
108. 5
109. 6
110. -14
111. 4
112. 14
113. 3
114. 16

Displacement The displacement, in cubic inches, of a car engine is given by the formula

$$d = \pi \cdot s \cdot c \cdot \left(\frac{1}{2} \cdot b\right)^2$$

where s is the stroke and b is the bore, as shown in the figure, and c is the number of cylinders.

Calculate the engine displacement for each of the following cars. Use 3.14 to approximate π.

95. Ferrari Modena 8 cylinders, 3.35 inches of bore, 3.11 inches of stroke

96. Audi A8 8 cylinders, 3.32 inches of bore, 3.66 inches of stroke

97. Mitsubishi Eclipse 6 cylinders, 3.59 inches of bore, 2.99 inches of stroke

98. Porsche 911 GT3 6 cylinders, 3.94 inches of bore, 3.01 inches of stroke

Getting Ready for the Next Section

Subtract.

99. $4 - 7$

100. $-4 - 7$

101. $4 - (-7)$

102. $-4 - (-7)$

103. $15 - 20$

104. $15 - (-20)$

105. $-15 - (-20)$

106. $-15 - 20$

107. $2(3) - 4$

108. $5(3) - 10$

109. $4(3) - 3(2)$

110. $-8 - 2(3)$

111. $2(5 - 3)$

112. $2(3) - 4 - 3(-4)$

113. $5 + 4(-2) - 2(-3)$

114. $2(3) + 4(5) - 5(2)$

Division with Exponents

5.2

In Section 4.1 we found that multiplication with the same base results in addition of exponents; that is, $a^r \cdot a^s = a^{r+s}$. Since division is the inverse operation of multiplication, we can expect division with the same base to result in subtraction of exponents.

To develop the properties for exponents under division, we again apply the definition of exponents:

$$\begin{aligned}\frac{x^5}{x^3} &= \frac{x \cdot x \cdot x \cdot x \cdot x}{x \cdot x \cdot x} \\ &= \frac{x \cdot x \cdot x}{x \cdot x \cdot x}(x \cdot x) \\ &= 1(x \cdot x) \\ &= x^2 \quad \text{Notice: } 5 - 3 = 2\end{aligned}$$

$$\begin{aligned}\frac{2^4}{2^7} &= \frac{2 \cdot 2 \cdot 2 \cdot 2}{2 \cdot 2 \cdot 2 \cdot 2 \cdot 2 \cdot 2 \cdot 2} \\ &= \frac{2 \cdot 2 \cdot 2 \cdot 2}{2 \cdot 2 \cdot 2 \cdot 2} \cdot \frac{1}{2 \cdot 2 \cdot 2} \\ &= \frac{1}{2 \cdot 2 \cdot 2} \\ &= \frac{1}{2^3} \quad \text{Notice: } 7 - 4 = 3\end{aligned}$$

In both cases division with the same base resulted in subtraction of the smaller exponent from the larger. The problem is deciding whether the answer is a fraction. The problem is resolved easily by the following definition.

(def **DEFINITION** *Negative Exponents*

If r is a positive integer, then $a^{-r} = \frac{1}{a^r} = \left(\frac{1}{a}\right)^r \qquad (a \neq 0)$

The following examples illustrate how we use this definition to simplify expressions that contain negative exponents.

EXAMPLE 1 Write each expression with a positive exponent and then simplify:

a. 2^{-3} **b.** 5^{-2} **c.** $3x^{-6}$

SOLUTION

a. $2^{-3} = \frac{1}{2^3} = \frac{1}{8}$ Notice: Negative exponents do not indicate negative numbers. They indicate reciprocals

b. $5^{-2} = \frac{1}{5^2} = \frac{1}{25}$

c. $3x^{-6} = 3 \cdot \frac{1}{x^6} = \frac{3}{x^6}$ ■

Now let us look back to our original problem and try to work it again with the help of a negative exponent. We know that $\frac{2^4}{2^7} = \frac{1}{2^3}$. Let us decide now that with division of the same base, we will always subtract the exponent in the denominator from the exponent in the numerator and see if this conflicts with what we know is true.

$$\begin{aligned}\frac{2^4}{2^7} &= 2^{4-7} && \text{Subtracting the bottom exponent from the top exponent} \\ &= 2^{-3} && \text{Subtraction} \\ &= \frac{1}{2^3} && \text{Definition of negative exponents}\end{aligned}$$

Subtracting the exponent in the denominator from the exponent in the numerator and then using the definition of negative exponents gives us the same result we obtained previously. We can now continue the list of properties of exponents we started in Section 5.1.

[Δ≠Σ PROPERTY *Property 4 for Exponents*

If a is any real number and r and s are integers, then

$$\frac{a^r}{a^s} = a^{r-s} \qquad (a \neq 0)$$

In words: To divide with the same base, subtract the exponent in the denominator from the exponent in the numerator and raise the base to the exponent that results.

The following examples show how we use Property 4 and the definition for negative exponents to simplify expressions involving division.

EXAMPLE 2 Simplify the following expressions:

a. $\frac{x^9}{x^6}$ **b.** $\frac{x^4}{x^{10}}$ **c.** $\frac{2^{15}}{2^{20}}$

SOLUTION

a. $\frac{x^9}{x^6} = x^{9-6} = x^3$

b. $\frac{x^4}{x^{10}} = x^{4-10} = x^{-6} = \frac{1}{x^6}$

c. $\frac{2^{15}}{2^{20}} = 2^{15-20} = 2^{-5} = \frac{1}{2^5} = \frac{1}{32}$

Our final property of exponents is similar to Property 3 from Section 5.1, but it involves division instead of multiplication. After we have stated the property, we will give a proof of it. The proof shows why this property is true.

[Δ≠Σ PROPERTY *Property 5 for Exponents*

If a and b are any two real numbers ($b \neq 0$) and r is an integer, then

$$\left(\frac{a}{b}\right)^r = \frac{a^r}{b^r}$$

In words: A quotient raised to a power is the quotient of the powers.

Proof

$$\left(\frac{a}{b}\right)^r = \left(a \cdot \frac{1}{b}\right)^r \qquad \text{By the definition of division}$$

$$= a^r \cdot \left(\frac{1}{b}\right)^r \qquad \text{By Property 3}$$

$$= a^r \cdot b^{-r} \qquad \text{By the definition of negative exponents}$$

$$= a^r \cdot \frac{1}{b^r} \qquad \text{By the definition of negative exponents}$$

$$= \frac{a^r}{b^r} \qquad \text{By the definition of division}$$

EXAMPLE 3 Simplify the following expressions.

a. $\left(\frac{x}{2}\right)^3$ **b.** $\left(\frac{5}{y}\right)^2$ **c.** $\left(\frac{2}{3}\right)^4$

SOLUTION

a. $\left(\frac{x}{2}\right)^3 = \frac{x^3}{2^3} = \frac{x^3}{8}$

b. $\left(\frac{5}{y}\right)^2 = \frac{5^2}{y^2} = \frac{25}{y^2}$

c. $\left(\frac{2}{3}\right)^4 = \frac{2^4}{3^4} = \frac{16}{81}$

Zero and One as Exponents

We have two special exponents left to deal with before our rules for exponents are complete: 0 and 1. To obtain an expression for x^1, we will solve a problem two different ways:

$$\left.\begin{aligned} \frac{x^3}{x^2} &= \frac{x \cdot x \cdot x}{x \cdot x} = x \\ \frac{x^3}{x^2} &= x^{3-2} = x^1 \end{aligned}\right\} \quad \text{Hence } x^1 = x$$

Stated generally, this rule says that $a^1 = a$. This seems reasonable and we will use it since it is consistent with our property of division using the same base.

We use the same procedure to obtain an expression for x^0:

$$\left.\begin{aligned} \frac{5^2}{5^2} &= \frac{25}{25} = 1 \\ \frac{5^2}{5^2} &= 5^{2-2} = 5^0 \end{aligned}\right\} \quad \text{Hence } 5^0 = 1$$

It seems, therefore, that the best definition of x^0 is 1 for all x except $x = 0$. In the case of $x = 0$, we have 0^0, which we will not define. This definition will probably seem awkward at first. Most people would like to define x^0 as 0 when they first encounter it. Remember, the zero in this expression is an exponent, so x^0 does not mean to multiply by zero. Thus, we can make the general statement that $a^0 = 1$ for all real numbers except $a = 0$.

Here are some examples involving the exponents 0 and 1.

EXAMPLE 4 Simplify the following expressions:

a. 8^0 **b.** 8^1 **c.** $4^0 + 4^1$ **d.** $(2x^2y)^0$

SOLUTION

a. $8^0 = 1$

b. $8^1 = 8$

c. $4^0 + 4^1 = 1 + 4 = 5$

d. $(2x^2y)^0 = 1$

Here is a summary of the definitions and properties of exponents we have developed so far. For each definition or property in the list, a and b are real numbers, and r and s are integers.

Definitions		Properties	
$a^{-r} = \frac{1}{a^r} = \left(\frac{1}{a}\right)^r$	$a \neq 0$	**1.** $a^r \cdot a^s = a^{r+s}$	
$a^1 = a$		**2.** $(a^r)^s = a^{rs}$	
$a^0 = 1$	$a \neq 0$	**3.** $(ab)^r = a^r b^r$	
		4. $\frac{a^r}{a^s} = a^{r-s}$	$a \neq 0$
		5. $\left(\frac{a}{b}\right)^r = \frac{a^r}{b^r}$	$b \neq 0$

Here are some additional examples. These examples use a combination of the preceding properties and definitions.

EXAMPLES Simplify each expression. Write all answers with positive exponents only:

5. $\frac{(5x^3)^2}{x^4} = \frac{25x^6}{x^4}$ Properties 2 and 3

$= 25x^2$ Property 4

6. $\frac{x^{-8}}{(x^2)^3} = \frac{x^{-8}}{x^6}$ Property 2

$= x^{-8-6}$ Property 4

$= x^{-14}$ Subtraction

$= \frac{1}{x^{14}}$ Definition of negative exponents

7. $\left(\frac{y^5}{y^3}\right)^2 = \frac{(y^5)^2}{(y^3)^2}$ Property 5

$= \frac{y^{10}}{y^6}$ Property 2

$= y^4$ Property 4

Notice in Example 7 that we could have simplified inside the parentheses first and then raised the result to the second power:

$$\left(\frac{y^5}{y^3}\right)^2 = (y^2)^2 = y^4$$

8. $(3x^5)^{-2} = \frac{1}{(3x^5)^2}$ Definition of negative exponents

$= \frac{1}{9x^{10}}$ Properties 2 and 3

9. $x^{-8} \cdot x^5 = x^{-8+5}$ Property 1

$= x^{-3}$ Addition

$= \frac{1}{x^3}$ Definition of negative exponents

10. $\frac{(a^3)^2 a^{-4}}{(a^{-4})^3} = \frac{a^6 a^{-4}}{a^{-12}}$ Property 2

$= \frac{a^2}{a^{-12}}$ Property 1

$= a^{14}$ Property 4

In the next two examples we use division to compare the area and volume of geometric figures.

EXAMPLE 11 Suppose you have two squares, one of which is larger than the other. If the length of a side of the larger square is 3 times as long as the length of a side of the smaller square, how many of the smaller squares will it take to cover up the larger square?

SOLUTION If we let x represent the length of a side of the smaller square, then the length of a side of the larger square is $3x$. The area of each square, along with a diagram of the situation, is given in Figure 1.

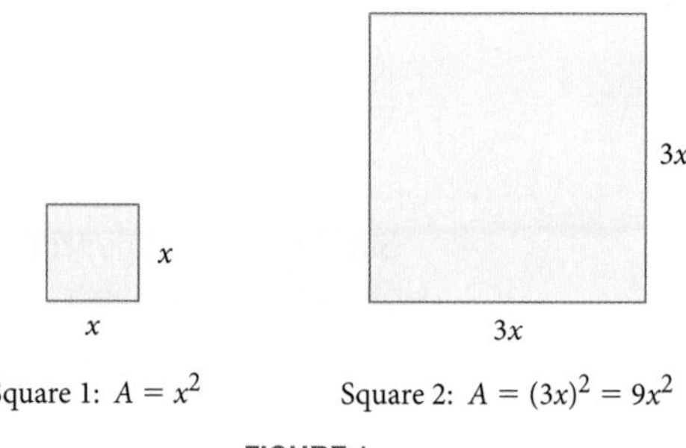

FIGURE 1

To find out how many smaller squares it will take to cover up the larger square, we divide the area of the larger square by the area of the smaller square.

$$\frac{\text{Area of square 2}}{\text{Area of square 1}} = \frac{9x^2}{x^2} = 9$$

It will take 9 of the smaller squares to cover the larger square.

EXAMPLE 12 Suppose you have two boxes, each of which is a cube. If the length of a side in the second box is 3 times as long as the length of a side of the first box, how many of the smaller boxes will fit inside the larger box?

SOLUTION If we let x represent the length of a side of the smaller box, then the length of a side of the larger box is $3x$. The volume of each box, along with a diagram of the situation, is given in Figure 2.

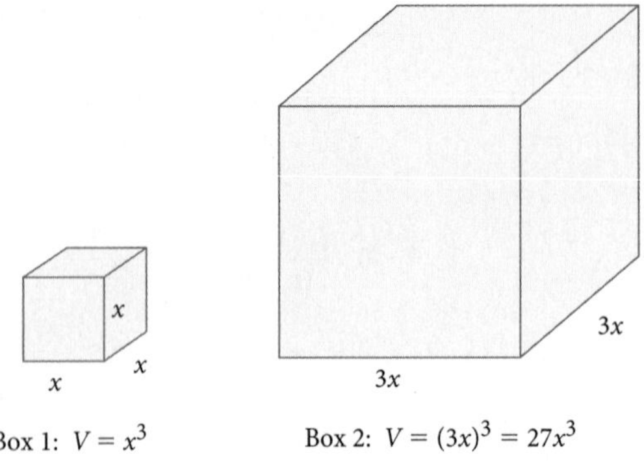

FIGURE 2

To find out how many smaller boxes will fit inside the larger box, we divide the volume of the larger box by the volume of the smaller box.

$$\frac{\text{Volume of box 2}}{\text{Volume of box 1}} = \frac{27x^3}{x^3} = 27$$

We can fit 27 of the smaller boxes inside the larger box.

More on Scientific Notation

Now that we have completed our list of definitions and properties of exponents, we can expand the work we did previously with scientific notation.

Recall that a number is in scientific notation when it is written in the form

$$n \times 10^r$$

where $1 \leq n < 10$ and r is an integer.

Since negative exponents give us reciprocals, we can use negative exponents to write very small numbers in scientific notation. For example, the number 0.00057, when written in scientific notation, is equivalent to 5.7×10^{-4}. Here's why:

$$5.7 \times 10^{-4} = 5.7 \times \frac{1}{10^4} = 5.7 \times \frac{1}{10{,}000} = \frac{5.7}{10{,}000} = 0.00057$$

The table below lists some other numbers in both scientific notation and expanded form.

Number Written the Long Way		Number Written Again in Scientific Notation
376,000	=	3.76×10^5
49,500	=	4.95×10^4
3,200	=	3.2×10^3
591	=	5.91×10^2
46	=	4.6×10^1
8	=	8×10^0
0.47	=	4.7×10^{-1}
0.093	=	9.3×10^{-2}
0.00688	=	6.88×10^{-3}
0.0002	=	2×10^{-4}
0.000098	=	9.8×10^{-5}

Notice that in each case, when the number is written in scientific notation, the decimal point in the first number is placed so that the number is between 1 and 10. The exponent on 10 in the second number keeps track of the number of places we moved the decimal point in the original number to get a number between 1 and 10:

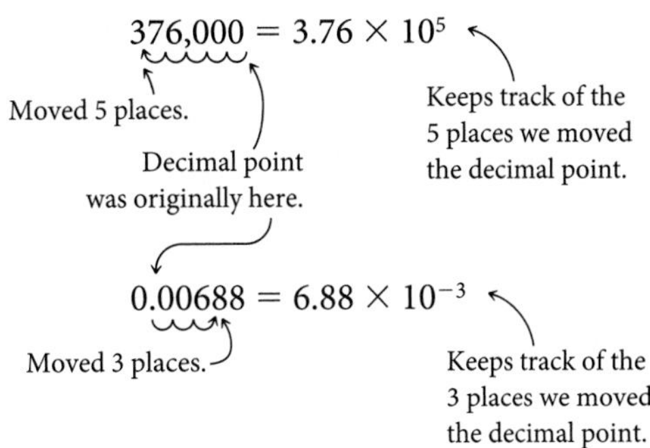

376,000 = 3.76 × 10^5
Moved 5 places.
Decimal point was originally here.
Keeps track of the 5 places we moved the decimal point.
0.00688 = 6.88 × 10^−3
Moved 3 places.
Keeps track of the 3 places we moved the decimal point.

GETTING READY FOR CLASS

After reading through the preceding section, respond in your own words and in complete sentences.

A. How do you divide two expressions containing exponents when they each have the same base?

B. Explain the difference between 3^2 and 3^{-2}.

C. If a positive base is raised to a negative exponent, can the result be a negative number?

D. Explain what happens when we use 0 as an exponent.

Problem Set 5.2

1. $\frac{1}{9}$
2. $\frac{1}{27}$
3. $\frac{1}{36}$
4. $\frac{1}{64}$
5. $\frac{1}{64}$
6. $\frac{1}{81}$
7. $\frac{1}{125}$
8. $\frac{1}{81}$
9. $\frac{2}{x^3}$
10. $\frac{5}{x}$
11. $\frac{1}{8x^3}$
12. $\frac{1}{5x}$
13. $\frac{1}{25y^2}$
14. $\frac{5}{y^2}$
15. $\frac{1}{100}$
16. $\frac{1}{1{,}000}$
19. $\frac{1}{25}$
20. $\frac{1}{49}$
21. x^6
22. $\frac{1}{x^6}$
23. 64
24. $\frac{1}{64}$
25. $8x^3$
26. $\frac{1}{8x^3}$
27. 6^{10}
28. 8^6
29. $\frac{1}{6^{10}}$
30. $\frac{1}{8^6}$
31. $\frac{1}{2^8}$
32. $\frac{1}{4}$
33. 2^8
34. 4
35. $27x^3$
36. $32x^5$
37. $81x^4y^4$
38. $64x^3y^3$
39. 1
40. 10
41. $2a^2b$
42. 1
43. $\frac{1}{49y^6}$
44. $\frac{1}{25y^8}$
45. $\frac{1}{x^8}$
46. x^2
47. $\frac{1}{y^3}$

Simplify each expression.

1. 3^{-2} **2.** 3^{-3} **3.** 6^{-2} **4.** 2^{-6} **5.** 8^{-2} **6.** 3^{-4}

7. 5^{-3} **8.** 9^{-2} **9.** $2x^{-3}$ **10.** $5x^{-1}$ **11.** $(2x)^{-3}$ **12.** $(5x)^{-1}$

13. $(5y)^{-2}$ **14.** $5y^{-2}$ **15.** 10^{-2} **16.** 10^{-3}

17. Complete the following table.

Number x	Square x^2	Power of 2 2^x
-3	9	$\frac{1}{8}$
-2	4	$\frac{1}{4}$
-1	1	$\frac{1}{2}$
0	0	1
1	1	2
2	4	4
3	9	8

18. Complete the following table.

Number x	Cube x^3	Power of 3 3^x
-3	-27	$\frac{1}{27}$
-2	-8	$\frac{1}{9}$
-1	-1	$\frac{1}{3}$
0	0	1
1	1	3
2	8	9
3	27	27

Use Property 4 to simplify each of the following expressions. Write all answers that contain exponents with positive exponents only.

19. $\frac{5^1}{5^3}$ **20.** $\frac{7^6}{7^8}$ **21.** $\frac{x^{10}}{x^4}$ **22.** $\frac{x^4}{x^{10}}$ **23.** $\frac{4^3}{4^0}$ **24.** $\frac{4^0}{4^3}$

25. $\frac{(2x)^7}{(2x)^4}$ **26.** $\frac{(2x)^4}{(2x)^7}$ **27.** $\frac{6^{11}}{6}$ **28.** $\frac{8^7}{8}$ **29.** $\frac{6}{6^{11}}$ **30.** $\frac{8}{8^7}$

31. $\frac{2^{-5}}{2^3}$ **32.** $\frac{2^{-5}}{2^{-3}}$ **33.** $\frac{2^5}{2^{-3}}$ **34.** $\frac{2^{-3}}{2^{-5}}$ **35.** $\frac{(3x)^{-5}}{(3x)^{-8}}$ **36.** $\frac{(2x)^{-10}}{(2x)^{-15}}$

Simplify the following expressions. Any answers that contain exponents should contain positive exponents only.

37. $(3xy)^4$ **38.** $(4xy)^3$ **39.** 10^0 **40.** 10^1

41. $(2a^2b)^1$ **42.** $(2a^2b)^0$ **43.** $(7y^3)^{-2}$ **44.** $(5y^4)^{-2}$

45. $x^{-3}x^{-5}$ **46.** $x^{-6} \cdot x^8$ **47.** $y^7 \cdot y^{-10}$ **48.** $y^{-4} \cdot y^{-6}$

48. $\frac{1}{y^{10}}$
49. x^2
50. x^5
51. a^6
52. a^5
53. $\frac{1}{y^9}$
54. $\frac{1}{y^{10}}$
55. y^{40}
56. $\frac{1}{y^4}$
57. $\frac{1}{x}$
58. $\frac{1}{x}$
59. x^9
60. $\frac{1}{x^{21}}$
61. a^{16}
62. a^{30}
63. $\frac{1}{a^4}$
64. a^{23}
67. 4.8×10^{-3}
68. 4.8×10^{-5}
69. 2.5×10^1
70. 3.5×10^1
71. 9×10^{-6}
72. 9×10^{-4}

49. $\frac{(x^2)^3}{x^4}$ **50.** $\frac{(x^5)^3}{x^{10}}$ **51.** $\frac{(a^4)^3}{(a^3)^2}$ **52.** $\frac{(a^5)^3}{(a^5)^2}$

53. $\frac{y^7}{(y^2)^8}$ **54.** $\frac{y^2}{(y^3)^4}$ **55.** $\left(\frac{y^7}{y^2}\right)^8$ **56.** $\left(\frac{y^2}{y^3}\right)^4$

57. $\frac{(x^{-2})^3}{x^{-5}}$ **58.** $\frac{(x^2)^{-3}}{x^{-5}}$ **59.** $\left(\frac{x^{-2}}{x^{-5}}\right)^3$ **60.** $\left(\frac{x^2}{x^{-5}}\right)^{-3}$

61. $\frac{(a^3)^2(a^4)^5}{(a^5)^2}$ **62.** $\frac{(a^4)^8(a^2)^5}{(a^3)^4}$ **63.** $\frac{(a^{-2})^3(a^4)^2}{(a^{-3})^{-2}}$ **64.** $\frac{(a^{-5})^{-3}(a^7)^{-1}}{(a^{-3})^5}$

65. Complete the following table, and then construct a line graph of the information in the table.

Number x	-3	-2	-1	0	1	2	3
Power of 2 2^x	$\frac{1}{8}$	$\frac{1}{4}$	$\frac{1}{2}$	1	2	4	8

66. Complete the following table, and then construct a line graph of the information in the table.

Number x	-3	-2	-1	0	1	2	3
Power of 3 3^x	$\frac{1}{27}$	$\frac{1}{9}$	$\frac{1}{3}$	1	3	9	27

Write each of the following numbers in scientific notation.

67. 0.0048 **68.** 0.000048

69. 25 **70.** 35

71. 0.000009 **72.** 0.0009

73. Complete the following table.

Expanded Form	Scientific Notation $n \times 10^r$
0.000357	3.57×10^{-4}
0.00357	3.57×10^{-3}
0.0357	3.57×10^{-2}
0.357	3.57×10^{-1}
3.57	3.57×10^{0}
35.7	3.57×10^{1}
357	3.57×10^{2}
3,570	3.57×10^{3}
35,700	3.57×10^{4}

75. 0.00423
76. 4,230
77. 0.00008
78. 800,000
79. 4.2
80. 42
81. 0.002
82. 0.00003
83. Craven/Bush 2×10^{-3}
Earnhardt/Irvan 5×10^{-3}
Harvick/Gordon 6×10^{-3}
Kahne/Kenseth 1×10^{-2}
Kenseth/Kahne 1×10^{-2}
84. 1.67×10^{-7}
85. 2.5×10^{4}
86. 2.5×10^{2}

74. Complete the following table.

Expanded Form	Scientific Notation $n \times 10^r$
0.000123	1.23×10^{-4}
0.00123	1.23×10^{-3}
0.0123	1.23×10^{-2}
0.123	1.23×10^{-1}
1.23	1.23×10^{0}
12.3	1.23×10^{1}
123	1.23×10^{2}
1,230	1.23×10^{3}
12,300	1.23×10^{4}

Write each of the following numbers in expanded form.

75. 4.23×10^{-3} **76.** 4.23×10^{3} **77.** 8×10^{-5}

78. 8×10^{5} **79.** 4.2×10^{0} **80.** 4.2×10^{1}

Applying the Concepts

Scientific Notation Problems

81. Some home computers can do a calculation in 2×10^{-3} seconds. Write this number in expanded form.

82. Some of the cells in the human body have a radius of 3×10^{-5} inches. Write this number in expanded form.

83. Margin of Victory Since 1993, the Nascar races with the smallest margin of victory are shown here.

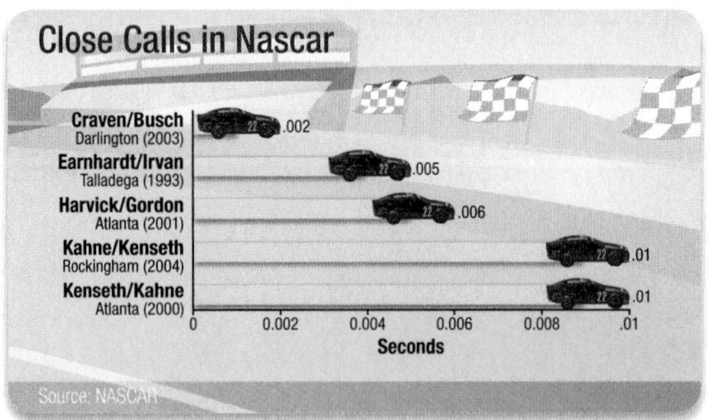

Write each number in scientific notation.

84. Some cameras used in scientific research can take one picture every 0.000000167 second. Write this number in scientific notation.

85. The number 25×10^{3} is not in scientific notation because 25 is larger than 10. Write 25×10^{3} in scientific notation.

86. The number 0.25×10^{3} is not in scientific notation because 0.25 is less than 1. Write 0.25×10^{3} in scientific notation.

87. 2.35×10^5
88. 3.75×10^5
89. 8.2×10^{-4}
90. 9.3×10^{-3}
91. 100 inches^2, 400 inches^2; 4
92. 4
93. x^2; $4x^2$; 4
94. 4
95. 216 inches^3; 1,728 inches^3; 8
96. 8
97. x^3; $8x^3$; 8
98. 8

87. The number 23.5×10^4 is not in scientific notation because 23.5 is not between 1 and 10. Rewrite 23.5×10^4 in scientific notation.

88. The number 375×10^3 is not in scientific notation because 375 is not between 1 and 10. Rewrite 375×10^3 in scientific notation.

89. The number 0.82×10^{-3} is not in scientific notation because 0.82 is not between 1 and 10. Rewrite 0.82×10^{-3} in scientific notation.

90. The number 0.93×10^{-2} is not in scientific notation because 0.93 is not between 1 and 10. Rewrite 0.93×10^{-2} in scientific notation.

Comparing Areas Suppose you have two squares, one of which is larger than the other. Suppose further that the side of the larger square is twice as long as the side of the smaller square.

91. If the length of the side of the smaller square is 10 inches, give the area of each square. Then find the number of smaller squares it will take to cover the larger square.

92. How many smaller squares will it take to cover the larger square if the length of the side of the smaller square is 1 foot?

93. If the length of the side of the smaller square is x, find the area of each square. Then find the number of smaller squares it will take to cover the larger square.

94. Suppose the length of the side of the larger square is 1 foot. How many smaller squares will it take to cover the larger square?

Comparing Volumes Suppose you have two boxes, each of which is a cube. Suppose further that the length of a side of the second box is twice as long as the length of a side of the first box.

95. If the length of a side of the first box is 6 inches, give the volume of each box. Then find the number of smaller boxes that will fit inside the larger box.

96. How many smaller boxes can be placed inside the larger box if the length of a side of the second box is 1 foot?

97. If the length of a side of the first box is x, find the volume of each box. Then find the number of smaller boxes that will fit inside the larger box.

98. Suppose the length of a side of the larger box is 12 inches. How many smaller boxes will fit inside the larger box?

99. 13.5
100. $\frac{5}{14}$
101. 8
102. 3.2
103. 26.52
104. −11
105. 12
106. x^3
107. x^8
108. y^3
109. x
110. x
111. $\frac{1}{y^2}$
112. $\frac{1}{x^3}$
113. 340
114. 0.0006

Getting Ready for the Next Section

Simplify.

99. $3(4.5)$

100. $\frac{1}{2} \cdot \frac{5}{7}$

101. $\frac{4}{5}(10)$

102. $\frac{9.6}{3}$

103. $6.8(3.9)$

104. $9 - 20$

105. $-3 + 15$

106. $2x \cdot x \cdot \frac{1}{2}x$

107. $x^5 \cdot x^3$

108. $y^2 \cdot y$

109. $\frac{x^3}{(x^2)}$

110. $\frac{x^2}{x}$

111. $\frac{y^3}{y^5}$

112. $\frac{x^2}{x^5}$

Write in expanded form.

113. 3.4×10^2

114. 6.0×10^{-4}

Operations with Monomials 5.3

We have developed all the tools necessary to perform the four basic operations on the simplest of polynomials: monomials.

(dĕf DEFINITION *monomial*

A ***monomial*** is a one-term expression that is either a constant (number) or the product of a constant and one or more variables raised to whole number exponents.

The following are examples of monomials:

$$-3 \qquad 15x \qquad -23x^2y \qquad 49x^4y^2z^4 \qquad \frac{3}{4}a^2b^3$$

The numerical part of each monomial is called the *numerical coefficient*, or just *coefficient*. Monomials are also called *terms*.

Multiplication and Division of Monomials

There are two basic steps involved in the multiplication of monomials. First, we rewrite the products using the commutative and associative properties. Then, we simplify by multiplying coefficients and adding exponents of like bases.

EXAMPLE 1 Multiply:

a. $(-3x^2)(4x^3)$ **b.** $\left(\frac{4}{5}x^5 \cdot y^2\right)(10x^3 \cdot y)$

SOLUTION

a. $(-3x^2)(4x^3) = (-3 \cdot 4)(x^2 \cdot x^3)$ Commutative and associative properties

$= -12x^5$ Multiply coefficients, add exponents

b. $\left(\frac{4}{5}x^5 \cdot y^2\right)(10x^3 \cdot y) = \left(\frac{4}{5} \cdot 10\right)(x^5 \cdot x^3)(y^2 \cdot y)$ Commutative and associative properties

$= 8x^8y^3$ Multiply coefficients, add exponents

You can see that in each case the work was the same—multiply coefficients and add exponents of the same base. We can expect division of monomials to proceed in a similar way. Since our properties are consistent, division of monomials will result in division of coefficients and subtraction of exponents of like bases.

EXAMPLE 2 Divide:

a. $\dfrac{15x^3}{3x^2}$ **b.** $\dfrac{39x^2y^3}{3xy^5}$

SOLUTION

a. $\dfrac{15x^3}{3x^2} = \dfrac{15}{3} \cdot \dfrac{x^3}{x^2}$ Write as separate fractions

$= 5x$ Divide coefficients, subtract exponents

b. $\dfrac{39x^2y^3}{3xy^5} = \dfrac{39}{3} \cdot \dfrac{x^2}{x} \cdot \dfrac{y^3}{y^5}$ Write as separate fractions

$= 13x \cdot \dfrac{1}{y^2}$ Divide coefficients, subtract exponents

$= \dfrac{13x}{y^2}$ Write answer as a single fraction

In Example 2b, the expression $\frac{y^3}{y^5}$ simplifies to $\frac{1}{y^2}$ because of Property 4 for exponents and the definition of negative exponents. If we were to show all the work in this simplification process, it would look like this:

$\dfrac{y^3}{y^5} = y^{3-5}$ Property 4 for exponents

$= y^{-2}$ Subtraction

$= \dfrac{1}{y^2}$ Definition of negative exponents

The point of this explanation is this: Even though we may not show all the steps when simplifying an expression involving exponents, the result we obtain still can be justified using the properties of exponents. We have not introduced any new properties in Example 2; we have just not shown the details of each simplification.

EXAMPLE 3 Divide $25a^5b^3$ by $50a^2b^7$.

SOLUTION

$\dfrac{25a^5b^3}{50a^2b^7} = \dfrac{25}{50} \cdot \dfrac{a^5}{a^2} \cdot \dfrac{b^3}{b^7}$ Write as separate fractions

$= \dfrac{1}{2} \cdot a^3 \cdot \dfrac{1}{b^4}$ Divide coefficients, subtract exponents

$= \dfrac{a^3}{2b^4}$ Write answer as a single fraction

Notice in Example 3 that dividing 25 by 50 results in $\frac{1}{2}$. This is the same result we would obtain if we reduced the fraction $\frac{25}{50}$ to lowest terms, and there is no harm in thinking of it that way. Also, notice that the expression $\frac{b^3}{b^7}$ simplifies to $\frac{1}{b^4}$ by Property 4 for exponents and the definition of negative exponents, even though we have not shown the steps involved in doing so.

Multiplication and Division of Numbers Written in Scientific Notation

We multiply and divide numbers written in scientific notation using the same steps we used to multiply and divide monomials.

EXAMPLE 4 Multiply $(4 \times 10^7)(2 \times 10^{-4})$.

SOLUTION Since multiplication is commutative and associative, we can rearrange the order of these numbers and group them as follows:

$$(4 \times 10^7)(2 \times 10^{-4}) = (4 \times 2)(10^7 \times 10^{-4})$$

$$= 8 \times 10^3$$

Notice that we add exponents, $7 + (-4) = 3$, when we multiply with the same base.

EXAMPLE 5 Divide $\dfrac{9.6 \times 10^{12}}{3 \times 10^4}$.

SOLUTION We group the numbers between 1 and 10 separately from the powers of 10 and proceed as we did in Example 4:

$$\frac{9.6 \times 10^{12}}{3 \times 10^4} = \frac{9.6}{3} \times \frac{10^{12}}{10^4}$$
$$= 3.2 \times 10^8$$

Notice that the procedure we used in both of these examples is very similar to multiplication and division of monomials, for which we multiplied or divided coefficients and added or subtracted exponents.

Addition and Subtraction of Monomials

Addition and subtraction of monomials will be almost identical since subtraction is defined as addition of the opposite. With multiplication and division of monomials, the key was rearranging the numbers and variables using the commutative and associative properties. With addition, the key is application of the distributive property. We sometimes use the phrase *combine monomials* to describe addition and subtraction of monomials.

DEFINITION *similar terms*

Two terms (monomials) with the same variable part (same variables raised to the same powers) are called ***similar*** (or like) ***terms.***

You can add only similar terms. This is because the distributive property (which is the key to addition of monomials) cannot be applied to terms that are not similar.

EXAMPLE 6 Combine the following monomials.

a. $-3x^2 + 15x^2$ **b.** $9x^2y - 20x^2y$ **c.** $5x^2 + 8y^2$

SOLUTION

a. $-3x^2 + 15x^2 = (-3 + 15)x^2$ Distributive property

$= 12x^2$ Add coefficients

b. $9x^2y - 20x^2y = (9 - 20)x^2y$ Distributive property

$= -11x^2y$ Add coefficients

c. $5x^2 + 8y^2$ In this case we cannot apply the distributive property, so we cannot add the monomials

The next examples show how we simplify expressions containing monomials when more than one operation is involved.

EXAMPLE 7 Apply the distributive property.

a. $x^2\left(1 - \dfrac{6}{x}\right)$ **b.** $ab\left(\dfrac{1}{b} - \dfrac{1}{a}\right)$

SOLUTION

a. $x^2\left(1 - \dfrac{6}{x}\right) = x^2 \cdot 1 - x^2 \cdot \dfrac{6}{x} = x^2 - \dfrac{6x^2}{x} = x^2 - 6x$

b. $ab\left(\dfrac{1}{b} - \dfrac{1}{a}\right) = ab \cdot \dfrac{1}{b} - ab \cdot \dfrac{1}{a} = \dfrac{ab}{b} - \dfrac{ab}{a} = a - b$

EXAMPLE 8 Simplify $\frac{(6x^4y)(3x^7y^5)}{9x^5y^2}$.

SOLUTION We begin by multiplying the two monomials in the numerator:

$$\frac{(6x^4y)(3x^7y^5)}{9x^5y^2} = \frac{18x^{11}y^6}{9x^5y^2} \quad \text{Simplify numerator}$$
$$= 2x^6y^4 \quad \text{Divide}$$

EXAMPLE 9 Simplify $\frac{(6.8 \times 10^5)(3.9 \times 10^{-7})}{7.8 \times 10^{-4}}$.

SOLUTION We group the numbers between 1 and 10 separately from the powers of 10:

$$\frac{(6.8)(3.9)}{7.8} \times \frac{(10^5)(10^{-7})}{10^{-4}} = 3.4 \times 10^{5+(-7)-(-4)}$$
$$= 3.4 \times 10^2$$

EXAMPLE 10 Simplify $\frac{14x^5}{2x^2} + \frac{15x^8}{3x^5}$.

SOLUTION Simplifying each expression separately and then combining similar terms gives

$$\frac{14x^5}{2x^2} + \frac{15x^8}{3x^5} = 7x^3 + 5x^3 \quad \text{Divide}$$
$$= 12x^3 \quad \text{Add}$$

EXAMPLE 11 A rectangular solid is twice as long as it is wide and one-half as high as it is wide. Write an expression for the volume.

SOLUTION We begin by making a diagram of the object (Figure 1) with the dimensions labeled as given in the problem.

FIGURE 1

The volume is the product of the three dimensions:

$$V = 2x \cdot x \cdot \frac{1}{2}x = x^3$$

The box has the same volume as a cube with side x, as shown in Figure 2.

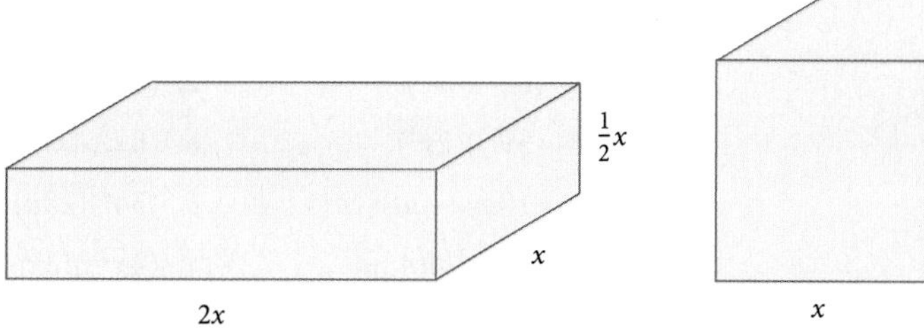

Equal Volumes

FIGURE 2

GETTING READY FOR CLASS

After reading through the preceding section, respond in your own words and in complete sentences.

A. What is a monomial?

B. Describe how you would multiply $3x^2$ and $5x^2$.

C. Describe how you would add $3x^2$ and $5x^2$.

D. Describe how you would multiply two numbers written in scientific notation.

Problem Set 5.3

1. $12x^7$
2. $-12x^7$
3. $-16y^{11}$
4. $10y^{15}$
5. $32x^2$
6. $35x^2$
7. $200a^6$
8. $500a^9$
9. $-24a^3b^3$
10. $-20a^4b^5$
11. $24x^6y^8$
12. $100x^9y^{10}$
13. $3x$
14. $5x$
15. $\frac{6}{y^3}$
16. $\frac{3}{y^3}$
17. $\frac{1}{2a}$
18. $\frac{1}{3a}$
19. $-\frac{3a}{b^2}$
20. $4a^4b$
21. $\frac{x^2}{9z^2}$
22. $\frac{x^2y^3}{6z}$
25. 6×10^8
26. 4×10^{14}
27. 1.75×10^{-1}
28. 1.42×10^{-2}
29. 1.21×10^{-6}
30. 1.53×10^{-7}
31. 4.2×10^3
32. 3.2×10^{14}
33. 3×10^{10}
34. 2×10^{15}
35. 5×10^{-3}
36. 5×10^{-5}

Multiply.

1. $(3x^4)(4x^3)$ **2.** $(6x^5)(-2x^2)$ **3.** $(-2y^4)(8y^7)$

4. $(5y^{10})(2y^5)$ **5.** $(8x)(4x)$ **6.** $(7x)(5x)$

7. $(10a^3)(10a)(2a^2)$ **8.** $(5a^4)(10a)(10a^4)$ **9.** $(6ab^2)(-4a^2b)$

10. $(-5a^3b)(4ab^4)$ **11.** $(4x^2y)(3x^3y^3)(2xy^4)$ **12.** $(5x^6)(-10xy^4)(-2x^2y^6)$

Divide. Write all answers with positive exponents only.

13. $\frac{15x^3}{5x^2}$ **14.** $\frac{25x^5}{5x^4}$ **15.** $\frac{18y^9}{3y^{12}}$ **16.** $\frac{24y^4}{8y^7}$

17. $\frac{32a^3}{64a^4}$ **18.** $\frac{25a^5}{75a^6}$ **19.** $\frac{21a^2b^3}{-7ab^5}$ **20.** $\frac{32a^5b^6}{8ab^5}$

21. $\frac{3x^3y^2z}{27xy^2z^3}$ **22.** $\frac{5x^5y^4z}{30x^3yz^2}$

23. Fill in the table.

a	b	ab	$\frac{a}{b}$	$\frac{b}{a}$
10	$5x$	$50x$	$\frac{2}{x}$	$\frac{x}{2}$
$20x^3$	$6x^2$	$120x^5$	$\frac{10x}{3}$	$\frac{3}{10x}$
$25x^5$	$5x^4$	$125x^9$	$5x$	$\frac{1}{5x}$
$3x^{-2}$	$3x^2$	9	$\frac{1}{x^4}$	x^4
$-2y^4$	$8y^7$	$-16y^{11}$	$-\frac{1}{4y^3}$	$-4y^3$

24. Fill in the table.

a	b	ab	$\frac{a}{b}$	$\frac{b}{a}$
$10y$	$2y^2$	$20y^3$	$\frac{5}{y}$	$\frac{y}{5}$
$10y^2$	$2y$	$20y^3$	$5y$	$\frac{1}{5y}$
$5y^3$	15	$75y^3$	$\frac{y^3}{3}$	$\frac{3}{y^3}$
5	$15y^3$	$75y^3$	$\frac{1}{3y^3}$	$3y^3$
$4y^{-3}$	$4y^3$	16	$\frac{1}{y^6}$	y^6

Find each product. Write all answers in scientific notation.

25. $(3 \times 10^3)(2 \times 10^5)$ **26.** $(4 \times 10^8)(1 \times 10^6)$

27. $(3.5 \times 10^4)(5 \times 10^{-6})$ **28.** $(7.1 \times 10^5)(2 \times 10^{-8})$

29. $(5.5 \times 10^{-3})(2.2 \times 10^{-4})$ **30.** $(3.4 \times 10^{-2})(4.5 \times 10^{-6})$

Find each quotient. Write all answers in scientific notation.

31. $\frac{8.4 \times 10^5}{2 \times 10^2}$ **32.** $\frac{9.6 \times 10^{20}}{3 \times 10^6}$ **33.** $\frac{6 \times 10^8}{2 \times 10^{-2}}$

34. $\frac{8 \times 10^{12}}{4 \times 10^{-3}}$ **35.** $\frac{2.5 \times 10^{-6}}{5 \times 10^{-4}}$ **36.** $\frac{4.5 \times 10^{-8}}{9 \times 10^{-4}}$

37. $8x^2$
38. $12x^3$
39. $-11x^5$
40. $25x^6$
41. 0
42. 0
43. $4x^3$
44. $3x^5$
45. $31ab^2$
46. $8a^3b^2$
49. $4x^3$
50. $3x^6$
51. $\frac{1}{b^2}$
52. $3a^{12}b^4$
53. $\frac{6y^{10}}{x^4}$
54. $\frac{x^5}{y^5}$
55. $x^2y + x$
56. $xy^2 + y$
57. $x + y$
58. $y - x$
59. $x^2 - 4$
60. $x^2 - 9$
61. $x^2 - x - 6$
62. $x^2 - 5x + 6$
63. $x^2 - 5x$
64. $x^2 - 3x$
65. $x^2 - 8x$
66. $x^2 - 6x$

Combine by adding or subtracting as indicated.

37. $3x^2 + 5x^2$

38. $4x^3 + 8x^3$

39. $8x^5 - 19x^5$

40. $75x^6 - 50x^6$

41. $2a + a - 3a$

42. $5a + a - 6a$

43. $10x^3 - 8x^3 + 2x^3$

44. $7x^5 + 8x^5 - 12x^5$

45. $20ab^2 - 19ab^2 + 30ab^2$

46. $18a^3b^2 - 20a^3b^2 + 10a^3b^2$

47. Fill in the table.

a	b	ab	a + b
$5x$	$3x$	$15x^2$	$8x$
$4x^2$	$2x^2$	$8x^4$	$6x^2$
$3x^3$	$6x^3$	$18x^6$	$9x^3$
$2x^4$	$-3x^4$	$-6x^8$	$-x^4$
x^5	$7x^5$	$7x^{10}$	$8x^5$

48. Fill in the table.

a	b	ab	a − b
$2y$	$3y$	$6y^2$	$-y$
$-2y$	$3y$	$-6y^2$	$-5y$
$4y^2$	$5y^2$	$20y^4$	$-y^2$
y^3	$-3y^3$	$-3y^6$	$4y^3$
$5y^4$	$7y^4$	$35y^8$	$-2y^4$

Simplify. Write all answers with positive exponents only.

49. $\frac{(3x^2)(8x^5)}{6x^4}$

50. $\frac{(7x^3)(6x^8)}{14x^5}$

51. $\frac{(9a^2b)(2a^3b^4)}{18a^5b^7}$

52. $\frac{(21a^5b)(2a^8b^4)}{14ab}$

53. $\frac{(4x^3y^2)(9x^4y^{10})}{(3x^5y)(2x^6y)}$

54. $\frac{(5x^4y^4)(10x^3y^3)}{(25xy^5)(2xy^7)}$

Apply the distributive property.

55. $xy\left(x + \frac{1}{y}\right)$

56. $xy\left(y + \frac{1}{x}\right)$

57. $xy\left(\frac{1}{y} + \frac{1}{x}\right)$

58. $xy\left(\frac{1}{x} - \frac{1}{y}\right)$

59. $x^2\left(1 - \frac{4}{x^2}\right)$

60. $x^2\left(1 - \frac{9}{x^2}\right)$

61. $x^2\left(1 - \frac{1}{x} - \frac{6}{x^2}\right)$

62. $x^2\left(1 - \frac{5}{x} + \frac{6}{x^2}\right)$

63. $x^2\left(1 - \frac{5}{x}\right)$

64. $x^2\left(1 - \frac{3}{x}\right)$

65. $x^2\left(1 - \frac{8}{x}\right)$

66. $x^2\left(1 - \frac{6}{x}\right)$

67. 2×10^6
68. 2×10^8
69. 1×10^1
70. 2×10^6
71. 4.2×10^{-6}
72. 4.2×10^{-13}
73. $9x^3$
74. $8x^6$
75. $-20a^2$
76. $-3a^8$
77. $6x^5y^2$
78. $7x^8y^5$
79. -5
80. 2
81. 6
82. -7
83. 76
84. 26
85. $6x^2$
86. $2x^2$
87. $2x$
88. $-x$
89. $-2x - 9$
90. $-4x^2 + 2x + 6$
91. 11
92. 81

Simplify each expression, and write all answers in scientific notation.

67. $\dfrac{(6 \times 10^8)(3 \times 10^5)}{9 \times 10^7}$

68. $\dfrac{(8 \times 10^4)(5 \times 10^{10})}{2 \times 10^7}$

69. $\dfrac{(5 \times 10^3)(4 \times 10^{-5})}{2 \times 10^{-2}}$

70. $\dfrac{(7 \times 10^6)(4 \times 10^{-4})}{1.4 \times 10^{-3}}$

71. $\dfrac{(2.8 \times 10^{-7})(3.6 \times 10^4)}{2.4 \times 10^3}$

72. $\dfrac{(5.4 \times 10^2)(3.5 \times 10^{-9})}{4.5 \times 10^6}$

Simplify.

73. $\dfrac{18x^4}{3x} + \dfrac{21x^7}{7x^4}$

74. $\dfrac{24x^{10}}{6x^4} + \dfrac{32x^7}{8x}$

75. $\dfrac{45a^6}{9a^4} - \dfrac{50a^8}{2a^6}$

76. $\dfrac{16a^9}{4a} - \dfrac{28a^{12}}{4a^4}$

77. $\dfrac{6x^7y^4}{3x^2y^2} + \dfrac{8x^5y^8}{2y^6}$

78. $\dfrac{40x^{10}y^{10}}{8x^2y^5} + \dfrac{10x^8y^8}{5y^3}$

Getting Ready for the Next Section

Simplify.

79. $3 - 8$

80. $-5 + 7$

81. $-1 + 7$

82. $1 - 8$

83. $3(5)^2 + 1$

84. $3(-2)^2 - 5(-2) + 4$

85. $2x^2 + 4x^2$

86. $3x^2 - x^2$

87. $-5x + 7x$

88. $x - 2x$

89. $-(2x + 9)$

90. $-(4x^2 - 2x - 6)$

91. Find the value of $2x + 3$ when $x = 4$

92. Find the value of $(3x)^2$ when $x = 3$

Addition and Subtraction of Polynomials

5.4

In this section we will extend what we learned in Section 5.3 to expressions called polynomials. We begin this section with the definition of a polynomial.

DEFINITION *polynomial*

A ***polynomial*** is a finite sum of monomials (terms).

Here are some examples of polynomials:

$$3x^2 + 2x + 1 \qquad 15x^2y + 21xy^2 - y^2 \qquad 3a - 2b + 4c - 5d$$

Polynomials can be further classified by the number of terms they contain. A polynomial with two terms is called a binomial. If it has three terms, it is a trinomial. As stated before, a monomial has only one term.

DEFINITION *degree*

The ***degree*** of a polynomial in one variable is the highest power to which the variable is raised.

Various degrees of polynomials:

$3x^5 + 2x^3 + 1$	A trinomial of degree 5
$2x + 1$	A binomial of degree 1
$3x^2 + 2x + 1$	A trinomial of degree 2
$3x^5$	A monomial of degree 5
-9	A monomial of degree 0

There are no new rules for adding one or more polynomials. We rely only on our previous knowledge. Here are some examples.

EXAMPLE 1 Add $(2x^2 - 5x + 3) + (4x^2 + 7x - 8)$.

SOLUTION We use the commutative and associative properties to group similar terms together and then apply the distributive property to add

$(2x^2 - 5x + 3) + (4x^2 + 7x - 8)$

$= (2x^2 + 4x^2) + (-5x + 7x) + (3 - 8)$ Commutative and associative properties

$= (2 + 4)x^2 + (-5 + 7)x + (3 - 8)$ Distributive property

$= 6x^2 + 2x - 5$ Addition

The results here indicate that to add two polynomials, we add coefficients of similar terms

EXAMPLE 2 Add $x^2 + 3x + 2x + 6$.

SOLUTION The only similar terms here are the two middle terms. We combine them as usual to get

$$x^2 + 3x + 2x + 6 = x^2 + 5x + 6$$

You will recall from Chapter 1 the definition of subtraction: $a - b = a + (-b)$. To subtract one expression from another, we simply add its opposite. The letters a and b in the definition can each represent polynomials. The opposite of a polynomial is the opposite of each of its terms. When you subtract one polynomial from another you subtract each of its terms.

EXAMPLE 3 Subtract $(3x^2 + x + 4) - (x^2 + 2x + 3)$.

SOLUTION To subtract $x^2 + 2x + 3$, we change the sign of each of its terms and add. If you are having trouble remembering why we do this, remember that we can think of $-(x^2 + 2x + 3)$ as $-1(x^2 + 2x + 3)$. If we distribute the -1 across $x^2 + 2x + 3$, we get $-x^2 - 2x - 3$:

$$(3x^2 + x + 4) - (x^2 + 2x + 3)$$

$$= 3x^2 + x + 4 - x^2 - 2x - 3$$ Take the opposite of each term in the second polynomial

$$= (3x^2 - x^2) + (x - 2x) + (4 - 3)$$

$$= 2x^2 - x + 1$$

EXAMPLE 4 Subtract $-4x^2 + 5x - 7$ from $x^2 - x - 1$.

SOLUTION The polynomial $x^2 - x - 1$ comes first, then the subtraction sign, and finally the polynomial $-4x^2 + 5x - 7$ in parentheses.

$$(x^2 - x - 1) - (-4x^2 + 5x - 7)$$

$$= x^2 - x - 1 + 4x^2 - 5x + 7$$ Take the opposite of each term in the second polynomial

$$= (x^2 + 4x^2) + (-x - 5x) + (-1 + 7)$$

$$= 5x^2 - 6x + 6$$

The last topic we want to consider in this section is finding the value of a polynomial for a given value of the variable.

To find the value of the polynomial $3x^2 + 1$ when x is 5, we replace x with 5 and simplify the result:

When	$x = 5$
the polynomial	$3x^2 + 1$
becomes	$3(5)^2 + 1 = 3(25) + 1$
	$= 75 + 1$
	$= 76$

EXAMPLE 5 Find the value of $3x^2 - 5x + 4$ when $x = -2$.

SOLUTION

When $x = -2$

the polynomial $3x^2 - 5x + 4$

becomes

$$3(-2)^2 - 5(-2) + 4 = 3(4) + 10 + 4$$
$$= 12 + 10 + 4$$
$$= 26$$

GETTING READY FOR CLASS

After reading through the preceding section, respond in your own words and in complete sentences.

A. What are similar terms?
B. What is the degree of a polynomial?
C. Describe how you would subtract one polynomial from another.
D. How you would find the value of $3x^2 - 5x + 4$ when x is -2?

Problem Set 5.4

1. Trinomial, 3
2. Trinomial, 2
3. Trinomial, 3
4. Trinomial, 4
5. Binomial, 1
6. Binomial, 1
7. Binomial, 2
8. Binomial, 3
9. Monomial, 2
10. Monomial, 1
11. Monomial, 0
12. Monomial, 0
13. $5x^2 + 5x + 9$
14. $2x^2 + 8x + 10$
15. $5a^2 - 9a + 7$
16. $9a^2 - 5a + 9$
17. $x^2 + 6x + 8$
18. $x^2 + 2x - 15$
19. $6x^2 - 13x + 5$
20. $10x^2 + 28x - 6$
21. $x^2 - 9$
22. $x^2 - 25$
23. $3y^2 - 11y + 10$
24. $y^2 - 16y - 12$
25. $6x^3 + 5x^2 - 4x + 3$
26. $5x^3 + 4x^2 + 8x + 1$
27. $2x^2 - x + 1$
28. $x^3 - x^2 + 1$
29. $2a^2 - 2a - 2$
30. $8a^2 + a - 10$
31. $-\frac{1}{9}x^3 - \frac{2}{3}x^2 - \frac{5}{2}x + \frac{7}{4}$
32. $-\frac{1}{2}x^3 - \frac{13}{20}x^2 + \frac{11}{8}x - \frac{11}{6}$
33. $-4y^2 + 15y - 22$
34. $-15y + 3$
35. $x^2 - 33x + 63$
36. $2x^2 - 2x + 5$
37. $8y^2 + 4y + 26$
38. $-12y^2 + 5y + 4$
39. $75x^2 - 150x - 75$
40. $-x^2 - 6x - 2$
41. $12x + 2$
42. $11x + 6$
43. 4
44. 4

Identify each of the following polynomials as a trinomial, binomial, or monomial, and give the degree in each case.

1. $2x^3 - 3x^2 + 1$ **2.** $4x^2 - 4x + 1$ **3.** $5 + 8a - 9a^3$

4. $6 + 12x^3 + x^4$ **5.** $2x - 1$ **6.** $4 + 7x$

7. $45x^2 - 1$ **8.** $3a^3 + 8$ **9.** $7a^2$

10. $90x$ **11.** -4 **12.** 56

Perform the following additions and subtractions.

13. $(2x^2 + 3x + 4) + (3x^2 + 2x + 5)$ **14.** $(x^2 + 5x + 6) + (x^2 + 3x + 4)$

15. $(3a^2 - 4a + 1) + (2a^2 - 5a + 6)$ **16.** $(5a^2 - 2a + 7) + (4a^2 - 3a + 2)$

17. $x^2 + 4x + 2x + 8$ **18.** $x^2 + 5x - 3x - 15$

19. $6x^2 - 3x - 10x + 5$ **20.** $10x^2 + 30x - 2x - 6$

21. $x^2 - 3x + 3x - 9$ **22.** $x^2 - 5x + 5x - 25$

23. $3y^2 - 5y - 6y + 10$ **24.** $y^2 - 18y + 2y - 12$

25. $(6x^3 - 4x^2 + 2x) + (9x^2 - 6x + 3)$

26. $(5x^3 + 2x^2 + 3x) + (2x^2 + 5x + 1)$

27. $\left(\frac{2}{3}x^2 - \frac{1}{5}x - \frac{3}{4}\right) + \left(\frac{4}{3}x^2 - \frac{4}{5}x + \frac{7}{4}\right)$

28. $\left(\frac{3}{8}x^3 - \frac{5}{7}x^2 - \frac{2}{5}\right) + \left(\frac{5}{8}x^3 - \frac{2}{7}x^2 + \frac{7}{5}\right)$

29. $(a^2 - a - 1) - (-a^2 + a + 1)$

30. $(5a^2 - a - 6) - (-3a^2 - 2a + 4)$

31. $\left(\frac{5}{9}x^3 + \frac{1}{3}x^2 - 2x + 1\right) - \left(\frac{2}{3}x^3 + x^2 + \frac{1}{2}x - \frac{3}{4}\right)$

32. $\left(4x^3 - \frac{2}{5}x^2 + \frac{3}{8}x - 1\right) - \left(\frac{9}{2}x^3 + \frac{1}{4}x^2 - x + \frac{5}{6}\right)$

33. $(4y^2 - 3y + 2) + (5y^2 + 12y - 4) - (13y^2 - 6y + 20)$

34. $(2y^2 - 7y - 8) - (6y^2 + 6y - 8) + (4y^2 - 2y + 3)$

35. Subtract $10x^2 + 23x - 50$ from $11x^2 - 10x + 13$.

36. Subtract $2x^2 - 3x + 5$ from $4x^2 - 5x + 10$.

37. Subtract $3y^2 + 7y - 15$ from $11y^2 + 11y + 11$.

38. Subtract $15y^2 - 8y - 2$ from $3y^2 - 3y + 2$.

39. Add $50x^2 - 100x - 150$ to $25x^2 - 50x + 75$.

40. Add $7x^2 - 8x + 10$ to $-8x^2 + 2x - 12$.

41. Subtract $2x + 1$ from the sum of $3x - 2$ and $11x + 5$.

42. Subtract $3x - 5$ from the sum of $5x + 2$ and $9x - 1$.

43. Find the value of the polynomial $x^2 - 2x + 1$ when x is 3.

44. Find the value of the polynomial $(x - 1)^2$ when x is 3.

45. 56.52 inches^3
46. 238.64 inches^3
47. 5
48. -12
49. -6
50. -56
51. $-20x^2$
52. $6x^2$
53. $-21x$
54. $-3x$
55. $2x$
56. $-13x$
57. $6x - 18$
58. $-4x^2 - 20x$

Applying the Concepts

45. Packaging A crystal ball with a diameter of 6 inches is being packaged for shipment. If the crystal ball is placed inside a circular cylinder with radius 3 inches and height 6 inches, how much volume will need to be filled with padding? (The volume of a sphere with radius r is $\frac{4}{3}\pi r^3$, and the volume of a right circular cylinder with radius r and height h is $\pi r^2 h$.) Use 3.14 to approximate π.

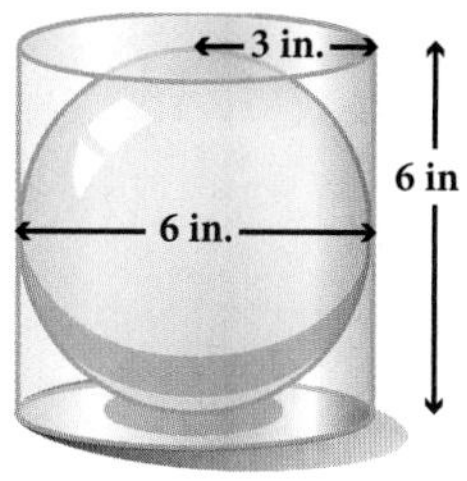

46. Packaging Suppose the circular cylinder of Problem 45 has a radius of 4 inches and a height of 7 inches. How much volume will need to be filled with padding?

Getting Ready for the Next Section

Simplify.

47. $(-5)(-1)$

48. $3(-4)$

49. $(-1)(6)$

50. $(-7) \cdot 8$

51. $(5x)(-4x)$

52. $(3x)(2x)$

53. $3x(-7)$

54. $3x(-1)$

55. $5x + (-3x)$

56. $-3x - 10x$

57. $3(2x - 6)$

58. $-4x(x + 5)$

SPOTLIGHT ON SUCCESS *Instructor Edwin*

You never fail until you stop trying.
—Albert Einstein

Coming to the United States at the age of 10 and not knowing how to speak English was a very difficult hurdle to overcome. However, with hard work and dedication I was able to rise above those obstacles. When I came to the U.S. our school did not have a strong English development program as it was known at that time, English as a Second Language (ESL). The approach back then was "sink or swim." When my self-esteem was low, my mom and my three older sisters were always there for me and they would always encourage me to do well. My mom was a single parent, and her number one priority was that we would receive a good education. My mother's perseverance is what has made me the person I am today. At a young age I was able to see that she had overcome more than what my situation was, and I would always tell myself, "if Mom can do it, I could also do it." Not only did she not have an education, but she also saved us from a civil war that was happening in my home country of El Salvador.

When things in school got hard, I would always reflect on all the hard work, sacrifice and effort of my mother. I would just tell myself that I should not have any excuses and that I needed to keep going. If my mother, who worked as a housekeeper, could send all four of her kids to college doesn't motivate you, I don't know what does. It definitely motivated me. The day everything began to change for me was when I was in eighth grade. I was sitting in my biology class not paying attention to the teacher because I was really focusing on a piece of paper on the wall. It said, "You never fail until you stop trying." I read it over and over, trying to digest what the quote meant. With my limited English I was doing my best to translate what it meant in my native language. It finally clicked! I was able to figure out what those seven words meant. I memorized the quote and began to apply it to my academics and to real-life situations. I began to really focus in my studies. I wanted to do well in school, and most important I wanted to improve my English. To this day I always reflect to that quote when I feel I can't do something.

I was able to finish junior high successfully. Going to high school was a lot easier and I ended up with very good grades and eventually I was accepted to an excellent college. I was never the smartest student on campus, but I always did well because I never quit. I earned my college degree and now I teach at a dual immersion elementary school. I have that same quote in my classroom and I constantly remind my students to never stop trying.

Multiplication with Polynomials 5.5

We begin our discussion of multiplication of polynomials by finding the product of a monomial and a trinomial.

EXAMPLE 1 Multiply $3x^2(2x^2 + 4x + 5)$.

SOLUTION Applying the distributive property gives us

$$3x^2(2x^2 + 4x + 5) = 3x^2(2x^2) + 3x^2(4x) + 3x^2(5) \quad \text{Distributive property}$$
$$= 6x^4 + 12x^3 + 15x^2 \quad \text{Multiplication}$$

The distributive property is the key to multiplication of polynomials. We can use it to find the product of any two polynomials. There are some shortcuts we can use in certain situations, however. Let's look at an example that involves the product of two binomials.

EXAMPLE 2 Multiply $(3x - 5)(2x - 1)$.

SOLUTION $(3x - 5)(2x - 1) = 3x(2x - 1) - 5(2x - 1)$

$$= 3x(2x) + 3x(-1) + (-5)(2x) + (-5)(-1)$$
$$= 6x^2 - 3x - 10x + 5$$
$$= 6x^2 - 13x + 5$$

If we look closely at the second and third lines of work in this example, we can see that the terms in the answer come from all possible products of terms in the first binomial with terms in the second binomial. This result is generalized as follows.

RULE

To multiply any two polynomials, multiply each term in the first with each term in the second.

There are two ways we can put this rule to work.

FOIL Method

If we look at the original problem in Example 2 and then to the answer, we see that the first term in the answer came from multiplying the first terms in each binomial:

$$3x \cdot 2x = 6x^2 \quad \text{First}$$

The middle term in the answer came from adding the products of the two outside terms with the two inside terms in each binomial:

$$3x(-1) = -3x \quad \text{Outside}$$
$$-5(2x) = -10x \quad \text{Inside}$$
$$= -13x$$

The last term in the answer came from multiplying the two last terms:

$$-5(-1) = 5 \quad \text{Last}$$

To summarize the FOIL method, we will multiply another two binomials.

EXAMPLE 3 Multiply $(2x + 3)(5x - 4)$.

SOLUTION $(2x + 3)(5x - 4) = \underbrace{2x(5x)}_{\text{First}} + \underbrace{2x(-4)}_{\text{Outside}} + \underbrace{3(5x)}_{\text{Inside}} + \underbrace{3(-4)}_{\text{Last}}$

$$= 10x^2 - 8x + 15x - 12$$

$$= 10x^2 + 7x - 12$$

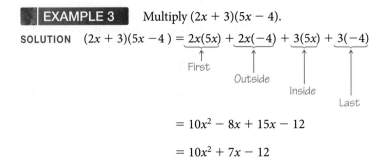

With practice $-8x + 15x = 7x$ can be done mentally.

COLUMN Method

The FOIL method can be applied only when multiplying two binomials. To find products of polynomials with more than two terms, we use what is called the COLUMN method.

The COLUMN method of multiplying two polynomials is very similar to long multiplication with whole numbers. It is just another way of finding all possible products of terms in one polynomial with terms in another polynomial.

EXAMPLE 4 Multiply $(2x + 3)(3x^2 - 2x + 1)$.

SOLUTION

$$\begin{array}{rl} 3x^2 - 2x + 1 & \\ 2x + 3 & \\ \hline 6x^3 - 4x^2 + 2x \phantom{{}+3} & \leftarrow 2x(3x^2 - 2x + 1) \\ 9x^2 - 6x + 3 & \leftarrow 3(3x^2 - 2x + 1) \\ \hline 6x^3 + 5x^2 - 4x + 3 & \leftarrow \text{Add similar terms} \end{array}$$

It will be to your advantage to become very fast and accurate at multiplying polynomials. You should be comfortable using either method. The following examples illustrate two types of multiplication.

EXAMPLE 5 Multiply:

a. $4a^2(2a^2 - 3a + 5) = 4a^2(2a^2) + 4a^2(-3a) + 4a^2(5)$

$$= 8a^4 - 12a^3 + 20a^2$$

b. $(x - 2)(y + 3) = \underset{\text{F}}{x(y)} + \underset{\text{O}}{x(3)} + \underset{\text{I}}{(-2)(y)} + \underset{\text{L}}{(-2)(3)}$

$$= xy + 3x - 2y - 6$$

c. $(x + y)(a - b) = \underset{\text{F}}{x(a)} + \underset{\text{O}}{x(-b)} + \underset{\text{I}}{y(a)} + \underset{\text{L}}{y(-b)}$

$$= xa - xb + ya - yb$$

d. $(5x - 1)(2x + 6) = \underset{\text{F}}{5x(2x)} + \underset{\text{O}}{5x(6)} + \underset{\text{I}}{(-1)(2x)} + \underset{\text{L}}{(-1)(6)}$

$$= 10x^2 + 30x + (-2x) + (-6)$$

$$= 10x^2 + 28x - 6$$

EXAMPLE 6 The length of a rectangle is 3 more than twice the width. Write an expression for the area of the rectangle.

SOLUTION We begin by drawing a rectangle and labeling the width with x. Since the length is 3 more than twice the width, we label the length with $2x + 3$.

$2x + 3$

x

Since the area A of a rectangle is the product of the length and width, we write our formula for the area of this rectangle as

$$A = x(2x + 3)$$

$$A = 2x^2 + 3x \qquad \text{Multiply}$$

Revenue

Suppose that a store sells x items at p dollars per item. The total amount of money obtained by selling the items is called the *revenue*. It can be found by multiplying the number of items sold, x, by the price per item, p. For example, if 100 items are sold for \$6 each, the revenue is $100(6) = \$600$. Similarly, if 500 items are sold for \$8 each, the total revenue is $500(8) = \$4{,}000$. If we denote the revenue with the letter R, then the formula that relates R, x, and p is

$$\text{Revenue} = (\text{number of items sold})(\text{price of each item})$$

In symbols: $R = xp$.

EXAMPLE 7 A store selling diskettes for home computers knows from past experience that it can sell x diskettes each day at a price of p dollars per diskette, according to the equation $x = 800 - 100p$. Write a formula for the daily revenue that involves only the variables R and p.

SOLUTION From our previous discussion we know that the revenue R is given by the formula

$$R = xp$$

But, since $x = 800 - 100p$, we can substitute $800 - 100p$ for x in the revenue equation to obtain

$$R = (800 - 100p)p$$

$$R = 800p - 100p^2$$

This last formula gives the revenue, R, in terms of the price, p.

GETTING READY FOR CLASS

After reading through the preceding section, respond in your own words and in complete sentences.

A. How do we multiply two polynomials?
B. Describe how the distributive property is used to multiply a monomial and a polynomial.
C. Describe how you would use the FOIL method to multiply two binomials.
D. Show how the product of two binomials can be a trinomial.

1. $6x^2 + 2x$
2. $8x^2 - 12x$
3. $6x^4 - 4x^3 + 2x^2$
4. $20x^4 - 25x^3 + 5x^2$
5. $2a^3b - 2a^2b^2 + 2ab$
6. $3a^5b + 3a^4b^3 + 3a^2b^4$
7. $3y^4 + 9y^3 + 12y^2$
8. $10y^3 - 15y^2 + 25y$
9. $8x^5y^2 + 12x^4y^3 + 32x^2y^4$
10. $12x^3y^3 + 30x^2y^4 + 72xy^5$
11. $x^2 + 7x + 12$
12. $x^2 + 7x + 10$
13. $x^2 + 7x + 6$
14. $x^2 + 5x + 4$
15. $x^2 + 2x + \frac{3}{4}$
16. $x^2 + x + \frac{6}{25}$
17. $a^2 + 2a - 15$
18. $a^2 - 6a - 16$
19. $xy + bx - ay - ab$
20. $xy - xb + ay - ab$
21. $x^2 - 36$
22. $x^2 - 9$
23. $y^2 - \frac{25}{36}$
24. $y^2 - \frac{16}{49}$
25. $2x^2 - 11x + 12$
26. $3x^2 - 11x + 10$
27. $2a^2 + 3a - 2$
28. $3a^2 - 16a - 12$
29. $6x^2 - 19x + 10$
30. $6x^2 + 9x - 6$
31. $2ax + 8x + 3a + 12$
32. $2ax - 8x - 3a + 12$
33. $25x^2 - 16$
34. $36x^2 - 25$
35. $2x^2 + \frac{5}{2}x - \frac{3}{4}$
36. $4x^2 + \frac{1}{2}x - \frac{3}{4}$
37. $3 - 10a + 8a^2$
38. $3 - 7a - 6a^2$
39. $(x + 2)(x + 3)$
    $= x^2 + 2x + 3x + 6$
    $= x^2 + 5x + 6$
40. $(x + 4)(x + 5)$
    $= x^2 + 4x + 5x + 20$
    $= x^2 + 9x + 20$
41. $(x + 1)(2x + 2) = 2x^2 + 4x + 2$
42. $(2x + 1)(2x + 2)$
    $= 4x^2 + 6x + 2$

Multiply the following by applying the distributive property.

1. $2x(3x + 1)$
2. $4x(2x - 3)$
3. $2x^2(3x^2 - 2x + 1)$
4. $5x(4x^3 - 5x^2 + x)$
5. $2ab(a^2 - ab + 1)$
6. $3a^2b(a^3 + a^2b^2 + b^3)$
7. $y^2(3y^2 + 9y + 12)$
8. $5y(2y^2 - 3y + 5)$
9. $4x^2y(2x^3y + 3x^2y^2 + 8y^3)$
10. $6xy^3(2x^2 + 5xy + 12y^2)$

Multiply the following binomials. You should do about half the problems using the FOIL method and the other half using the COLUMN method. Remember, you want to be comfortable using both methods.

11. $(x + 3)(x + 4)$
12. $(x + 2)(x + 5)$
13. $(x + 6)(x + 1)$
14. $(x + 1)(x + 4)$
15. $\left(x + \frac{1}{2}\right)\left(x + \frac{3}{2}\right)$
16. $\left(x + \frac{3}{5}\right)\left(x + \frac{2}{5}\right)$
17. $(a + 5)(a - 3)$
18. $(a - 8)(a + 2)$
19. $(x - a)(y + b)$
20. $(x + a)(y - b)$
21. $(x + 6)(x - 6)$
22. $(x + 3)(x - 3)$
23. $\left(y + \frac{5}{6}\right)\left(y - \frac{5}{6}\right)$
24. $\left(y - \frac{4}{7}\right)\left(y + \frac{4}{7}\right)$
25. $(2x - 3)(x - 4)$
26. $(3x - 5)(x - 2)$
27. $(a + 2)(2a - 1)$
28. $(a - 6)(3a + 2)$
29. $(2x - 5)(3x - 2)$
30. $(3x + 6)(2x - 1)$
31. $(2x + 3)(a + 4)$
32. $(2x - 3)(a - 4)$
33. $(5x - 4)(5x + 4)$
34. $(6x + 5)(6x - 5)$
35. $\left(2x - \frac{1}{2}\right)\left(x + \frac{3}{2}\right)$
36. $\left(4x - \frac{3}{2}\right)\left(x + \frac{1}{2}\right)$
37. $(1 - 2a)(3 - 4a)$
38. $(1 - 3a)(3 + 2a)$

For each of the following problems, fill in the area of each small rectangle and square, and then add the results together to find the indicated product.

39. $(x + 2)(x + 3)$

	x	3
x	x^2	$3x$
2	$2x$	6

40. $(x + 4)(x + 5)$

	x	5
x	x^2	$5x$
4	$4x$	20

41. $(x + 1)(2x + 2)$

	x	x	2
x	x^2	x^2	$2x$
1	x	x	2

42. $(2x + 1)(2x + 2)$

	x	x	2
x	x^2	x^2	$2x$
x	x^2	x^2	$2x$
1	x	x	2

43. $a^3 - 6a^2 + 11a - 6$
44. $a^3 + 7a^2 + 13a + 15$
45. $x^3 + 8$
46. $x^3 + 27$
47. $2x^3 + 17x^2 + 26x + 9$
48. $3x^3 - 23x^2 + 38x - 16$
49. $5x^4 - 13x^3 + 20x^2 + 7x + 5$
50. $2x^4 - 7x^3 + 3x^2 - x + 3$
51. $2x^4 + x^2 - 15$
52. $20x^6 - 24x^3 - 32$
53. $6a^6 + 15a^4 + 4a^2 + 10$
54. $28a^7 - 42a^4 - 32a^3 + 48$
55. $x^3 + 12x^2 + 47x + 60$
56. $x^3 - 12x^2 + 47x - 60$
57. $x^2 - 5x + 8$
58. $6x^2 - 11x - 14$
59. $8x^2 - 6x - 5$
60. $15x^2 + 19x - 4$
61. $x^2 - x - 30$
62. $x^2 - x - 20$
63. $x^2 + 4x - 6$
64. $6x + 19$
65. $x^2 + 13x$
66. $x^2 + 4x$
67. $x^2 + 2x - 3$
68. $2x^2 - 8x + 6$
69. $a^2 - 3a + 6$
70. $a^2 - 4a + 8$
71. $A = x(2x + 5) = 2x^2 + 5x$
72. $A = x(3x + 2) = 3x^2 + 2x$
73. $A = x(x + 1) = x^2 + x$
74. $A = x(x + 2) = x^2 + 2x$
75. 169
76. $9x^2$
77. $-10x$
78. $-12x$
79. 0
80. 0
81. 0
82. 0
83. $-12x + 16$
84. $-4x^2 - 14x$
85. $x^2 + x - 2$
86. $x^2 - x - 30$
87. $x^2 + 6x + 9$
88. $9x^2 - 12x + 4$

Multiply the following.

43. $(a - 3)(a^2 - 3a + 2)$

44. $(a + 5)(a^2 + 2a + 3)$

45. $(x + 2)(x^2 - 2x + 4)$

46. $(x + 3)(x^2 - 3x + 9)$

47. $(2x + 1)(x^2 + 8x + 9)$

48. $(3x - 2)(x^2 - 7x + 8)$

49. $(5x^2 + 2x + 1)(x^2 - 3x + 5)$

50. $(2x^2 + x + 1)(x^2 - 4x + 3)$

51. $(x^2 + 3)(2x^2 - 5)$

52. $(4x^3 - 8)(5x^3 + 4)$

53. $(3a^4 + 2)(2a^2 + 5)$

54. $(7a^4 - 8)(4a^3 - 6)$

55. $(x + 3)(x + 4)(x + 5)$

56. $(x - 3)(x - 4)(x - 5)$

Simplify.

57. $(x - 3)(x - 2) + 2$

58. $(2x - 5)(3x + 2) - 4$

59. $(2x - 3)(4x + 3) + 4$

60. $(3x + 8)(5x - 7) + 52$

61. $(x + 4)(x - 5) + (-5)(2)$

62. $(x + 3)(x - 4) + (-4)(2)$

63. $2(x - 3) + x(x + 2)$

64. $5(x + 3) + 1(x + 4)$

65. $3x(x + 1) - 2x(x - 5)$

66. $4x(x - 2) - 3x(x - 4)$

67. $x(x + 2) - 3$

68. $2x(x - 4) + 6$

69. $a(a - 3) + 6$

70. $a(a - 4) + 8$

Applying the Concepts

71. Area The length of a rectangle is 5 units more than twice the width. Write an expression for the area of the rectangle.

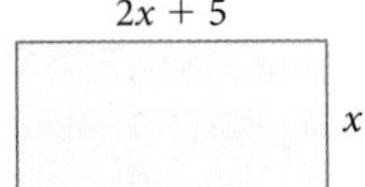

72. Area The length of a rectangle is 2 more than three times the width. Write an expression for the area of the rectangle.

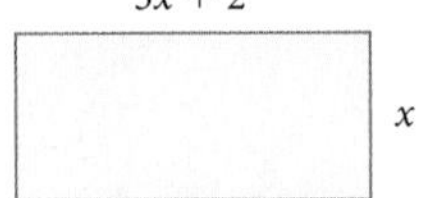

73. Area The width and length of a rectangle are given by two consecutive integers. Write an expression for the area of the rectangle.

74. Area The width and length of a rectangle are given by two consecutive even integers. Write an expression for the area of the rectangle.

Getting Ready for the Next Section

Simplify.

75. $13 \cdot 13$

76. $3x \cdot 3x$

77. $2(x)(-5)$

78. $2(2x)(-3)$

79. $6x + (-6x)$

80. $3x + (-3x)$

81. $(2x)(-3) + (2x)(3)$

82. $(2x)(-5y) + (2x)(5y)$

Multiply.

83. $-4(3x - 4)$

84. $-2x(2x + 7)$

85. $(x - 1)(x + 2)$

86. $(x + 5)(x - 6)$

87. $(x + 3)(x + 3)$

88. $(3x - 2)(3x - 2)$

Binomial Squares and Other Special Products

5.6

In this section we will combine the results of the last section with our definition of exponents to find some special products.

EXAMPLE 1 Find the square of $(3x - 2)$.

SOLUTION To square $(3x - 2)$, we multiply it by itself:

$$(3x - 2)^2 = (3x - 2)(3x - 2) \quad \text{Definition of exponents}$$
$$= 9x^2 - 6x - 6x + 4 \quad \text{FOIL method}$$
$$= 9x^2 - 12x + 4 \quad \text{Combine similar terms}$$

Notice that the first and last terms in the answer are the square of the first and last terms in the original problem and that the middle term is twice the product of the two terms in the original binomial.

EXAMPLE 2 Expand and multiply.

a. $(a + b)^2$ **b.** $(a - b)^2$

SOLUTION

a. $(a + b)^2 = (a + b)(a + b)$
$= a^2 + 2ab + b^2$

b. $(a - b)^2 = (a - b)(a - b)$
$= a^2 - 2ab + b^2$

Note: A very common mistake when squaring binomials is to write

$$(a + b)^2 = a^2 + b^2$$

which just isn't true. The mistake becomes obvious when we substitute 2 for a and 3 for b:

$$(2 + 3)^2 \neq 2^2 + 3^2$$
$$25 \neq 13$$

Exponents do not distribute over addition or subtraction.

Binomial squares having the form of Example 2 occur very frequently in algebra. It will be to your advantage to memorize the following rule for squaring a binomial.

RULE

The square of a binomial is the sum of the square of the first term, the square of the last term, and twice the product of the two original terms. In symbols this rule is written as follows:

$$(x + y)^2 = \underset{\text{Square of first term}}{x^2} + \underset{\text{Twice product of the two terms}}{2xy} + \underset{\text{Square of last term}}{y^2}$$

EXAMPLES Multiply using the preceding rule:

		First term squared		*Twice their product*		*Last term squared*		*Answer*
3. $(x - 5)^2$	$=$	x^2	$+$	$2(x)(-5)$	$+$	25	$=$	$x^2 - 10x + 25$
4. $(x + 2)^2$	$=$	x^2	$+$	$2(x)(2)$	$+$	4	$=$	$x^2 + 4x + 4$
5. $(2x - 3)^2$	$=$	$4x^2$	$+$	$2(2x)(-3)$	$+$	9	$=$	$4x^2 - 12x + 9$
6. $(5x - 4)^2$	$=$	$25x^2$	$+$	$2(5x)(-4)$	$+$	16	$=$	$25x^2 - 40x + 16$

Another special product that occurs frequently is $(a + b)(a - b)$. The only difference in the two binomials is the sign between the two terms. The interesting thing about this type of product is that the middle term is always zero. Here are some examples.

EXAMPLES Multiply using the FOIL method:

7. $(2x - 3)(2x + 3) = 4x^2 + 6x - 6x - 9$ Foil method

$= 4x^2 - 9$

8. $(x - 5)(x + 5) = x^2 + 5x - 5x - 25$ Foil method

$= x^2 - 25$

9. $(3x - 1)(3x + 1) = 9x^2 + 3x - 3x - 1$ Foil method

$= 9x^2 - 1$

Notice that in each case the middle term is zero and therefore doesn't appear in the answer. The answers all turn out to be the difference of two squares. Here is a rule to help you memorize the result.

RULE

When multiplying two binomials that differ only in the sign between their terms, subtract the square of the last term from the square of the first term.

$$(a - b)(a + b) = a^2 - b^2$$

Here are some problems that result in the difference of two squares.

EXAMPLES Multiply using the preceding rule:

10. $(x + 3)(x - 3) = x^2 - 9$

11. $(a + 2)(a - 2) = a^2 - 4$

12. $(9a + 1)(9a - 1) = 81a^2 - 1$

13. $(2x - 5y)(2x + 5y) = 4x^2 - 25y^2$

14. $(3a - 7b)(3a + 7b) = 9a^2 - 49b^2$

Although all the problems in this section can be worked correctly using the methods in the previous section, they can be done much faster if the two rules are *memorized*. Here is a summary of the two rules:

$$(a + b)^2 = (a + b)(a + b) = a^2 + 2ab + b^2$$

$$(a - b)^2 = (a - b)(a - b) = a^2 - 2ab + b^2$$

$$(a - b)(a + b) = a^2 - b^2$$

EXAMPLE 15 Write an expression in symbols for the sum of the squares of three consecutive even integers. Then, simplify that expression.

SOLUTION If we let $x =$ the first of the even integers, then $x + 2$ is the next consecutive even integer, and $x + 4$ is the one after that. An expression for the sum of their squares is

$$x^2 + (x + 2)^2 + (x + 4)^2 \quad \text{Sum of squares}$$
$$= x^2 + (x^2 + 4x + 4) + (x^2 + 8x + 16) \quad \text{Expand squares}$$
$$= 3x^2 + 12x + 20 \quad \text{Add similar terms}$$

GETTING READY FOR CLASS

After reading through the preceding section, respond in your own words and in complete sentences.

A. Explain why $(x + 3)^2$ cannot be $x^2 + 9$.
B. What kind of products result in the difference of two squares?
C. When multiplied out, how will $(x + 3)^2$ and $(x - 3)^2$ differ?

Problem Set 5.6

1. $x^2 - 4x + 4$
2. $x^2 + 4x + 4$
3. $a^2 + 6a + 9$
4. $a^2 - 6a + 9$
5. $x^2 - 10x + 25$
6. $x^2 - 8x + 16$
7. $a^2 - a + \frac{1}{4}$
8. $a^2 + a + \frac{1}{4}$
9. $x^2 + 20x + 100$
10. $x^2 - 20x + 100$
11. $a^2 + 1.6a + 0.64$
12. $a^2 - 0.8a + 0.16$
13. $4x^2 - 4x + 1$
14. $9x^2 + 12x + 4$
15. $16a^2 + 40a + 25$
16. $16a^2 - 40a + 25$
17. $9x^2 - 12x + 4$
18. $4x^2 - 12x + 9$
19. $9a^2 + 30ab + 25b^2$
20. $25a^2 - 30ab + 9b^2$
21. $16x^2 - 40xy + 25y^2$
22. $25x^2 + 40xy + 16y^2$
23. $49m^2 + 28mn + 4n^2$
24. $4m^2 - 28mn + 49n^2$
25. $36x^2 - 120xy + 100y^2$
26. $100x^2 + 120xy + 36y^2$
27. $x^4 + 10x^2 + 25$
28. $x^4 + 6x^2 + 9$
29. $a^4 + 2a^2 + 1$
30. $a^4 - 4a^2 + 4$

Perform the indicated operations.

1. $(x - 2)^2$ **2.** $(x + 2)^2$ **3.** $(a + 3)^2$ **4.** $(a - 3)^2$

5. $(x - 5)^2$ **6.** $(x - 4)^2$ **7.** $\left(a - \frac{1}{2}\right)^2$ **8.** $\left(a + \frac{1}{2}\right)^2$

9. $(x + 10)^2$ **10.** $(x - 10)^2$ **11.** $(a + 0.8)^2$ **12.** $(a - 0.4)^2$

13. $(2x - 1)^2$ **14.** $(3x + 2)^2$ **15.** $(4a + 5)^2$ **16.** $(4a - 5)^2$

17. $(3x - 2)^2$ **18.** $(2x - 3)^2$ **19.** $(3a + 5b)^2$ **20.** $(5a - 3b)^2$

21. $(4x - 5y)^2$ **22.** $(5x + 4y)^2$ **23.** $(7m + 2n)^2$ **24.** $(2m - 7n)^2$

25. $(6x - 10y)^2$ **26.** $(10x + 6y)^2$ **27.** $(x^2 + 5)^2$ **28.** $(x^2 + 3)^2$

29. $(a^2 + 1)^2$ **30.** $(a^2 - 2)^2$

Comparing Expressions Fill in each table.

31.

x	$(x + 3)^2$	$x^2 + 9$	$x^2 + 6x + 9$
1	16	10	16
2	25	13	25
3	36	18	36
4	49	25	49

32.

x	$(x - 5)^2$	$x^2 + 25$	$x^2 - 10x + 25$
1	16	26	16
2	9	29	9
3	4	34	4
4	1	41	1

33.

a	1	3	3	4
b	1	5	4	5
$(a + b)^2$	4	64	49	81
$a^2 + b^2$	2	34	25	41
$a^2 + ab + b^2$	3	49	37	61
$a^2 + 2ab + b^2$	4	64	49	81

34.

a	2	5	2	4
b	1	2	5	3
$(a - b)^2$	1	9	9	1
$a^2 - b^2$	3	21	−21	7
$a^2 - 2ab + b^2$	1	9	9	1

35. $a^2 - 25$
36. $a^2 - 36$
37. $y^2 - 1$
38. $y^2 - 4$
39. $81 - x^2$
40. $100 - x^2$
41. $4x^2 - 25$
42. $9x^2 - 25$
43. $16x^2 - \frac{1}{9}$
44. $36x^2 - \frac{1}{16}$
45. $4a^2 - 49$
46. $9a^2 - 100$
47. $36 - 49x^2$
48. $49 - 36x^2$
49. $x^4 - 9$
50. $x^4 - 4$
51. $a^4 - 16$
52. $a^4 - 81$
53. $25y^8 - 64$
54. $49y^{10} - 36$
55. $2x^2 - 34$
56. $2x^2 - 65$
57. $-12x^2 + 20x + 8$
58. $5x^2 - 42x + 16$
59. $a^2 + 4a + 6$
60. $a^2 - 4$
61. $8x^3 + 36x^2 + 54x + 27$
62. $27x^3 - 54x^2 + 36x - 8$
63. $(50 - 1)(50 + 1)$
$= 2500 - 1 = 2499$
64. $(100 - 1)(100 + 1)$
$= 10{,}000 - 1 = 9{,}999$
65. Both equal 25.
66. Both equal 11.
67. $x^2 + (x + 1)^2 = 2x^2 + 2x + 1$
68. $x^2 + (x + 2)^2 = 2x^2 + 4x + 4$
69. $x^2 + (x + 1)^2 + (x + 2)^2$
$= 3x^2 + 6x + 5$
70. $x^2 + (x + 2)^2 + (x + 4)^2$
$= 3x^2 + 12x + 20$

Multiply.

35. $(a + 5)(a - 5)$
36. $(a - 6)(a + 6)$
37. $(y - 1)(y + 1)$
38. $(y - 2)(y + 2)$
39. $(9 + x)(9 - x)$
40. $(10 - x)(10 + x)$
41. $(2x + 5)(2x - 5)$
42. $(3x + 5)(3x - 5)$
43. $\left(4x + \frac{1}{3}\right)\left(4x - \frac{1}{3}\right)$
44. $\left(6x + \frac{1}{4}\right)\left(6x - \frac{1}{4}\right)$
45. $(2a + 7)(2a - 7)$
46. $(3a + 10)(3a - 10)$
47. $(6 - 7x)(6 + 7x)$
48. $(7 - 6x)(7 + 6x)$
49. $(x^2 + 3)(x^2 - 3)$
50. $(x^2 + 2)(x^2 - 2)$
51. $(a^2 + 4)(a^2 - 4)$
52. $(a^2 + 9)(a^2 - 9)$
53. $(5y^4 - 8)(5y^4 + 8)$
54. $(7y^5 + 6)(7y^5 - 6)$

Multiply and simplify.

55. $(x + 3)(x - 3) + (x - 5)(x + 5)$
56. $(x - 7)(x + 7) + (x - 4)(x + 4)$
57. $(2x + 3)^2 - (4x - 1)^2$
58. $(3x - 5)^2 - (2x + 3)^2$
59. $(a + 1)^2 - (a + 2)^2 + (a + 3)^2$
60. $(a - 1)^2 + (a - 2)^2 - (a - 3)^2$
61. $(2x + 3)^3$
62. $(3x - 2)^3$

Applying the Concepts

63. Shortcut The formula for the difference of two squares can be used as a shortcut to multiplying certain whole numbers if they have the correct form. Use the difference of two squares formula to multiply 49(51) by first writing 49 as $(50 - 1)$ and 51 as $(50 + 1)$.

64. Shortcut Use the difference of two squares formula to multiply 101(99) by first writing 101 as $(100 + 1)$ and 99 as $(100 - 1)$.

65. Comparing Expressions Evaluate the expression $(x + 3)^2$ and the expression $x^2 + 6x + 9$ for $x = 2$.

66. Comparing Expressions Evaluate the expression $x^2 - 25$ and the expression $(x - 5)(x + 5)$ for $x = 6$.

67. Number Problem Write an expression for the sum of the squares of two consecutive integers. Then, simplify that expression.

68. Number Problem Write an expression for the sum of the squares of two consecutive odd integers. Then, simplify that expression.

69. Number Problem Write an expression for the sum of the squares of three consecutive integers. Then, simplify that expression.

70. Number Problem Write an expression for the sum of the squares of three consecutive odd integers. Then, simplify that expression.

71. $a^2 + ab + ba + b^2$
$= a^2 + 2ab + b^2$
72. $x^2 + 5x + 5x + 25$
$= x^2 + 10x + 25$
73. $2x^2$
74. $3x$
75. x^2
76. $2x^2$
77. $3x$
78. x^2
79. $3xy$
80. $\frac{y^2}{2}$

71. Area We can use the concept of area to further justify our rule for squaring a binomial. The length of each side of the square shown in the figure is $a + b$. (The longer line segment has length a and the shorter line segment has length b.) The area of the whole square is $(a + b)^2$. However, the whole area is the sum of the areas of the two smaller squares and the two smaller rectangles that make it up. Write the area of the two smaller squares and the two smaller rectangles and then add them together to verify the formula $(a + b)^2 = a^2 + 2ab + b^2$.

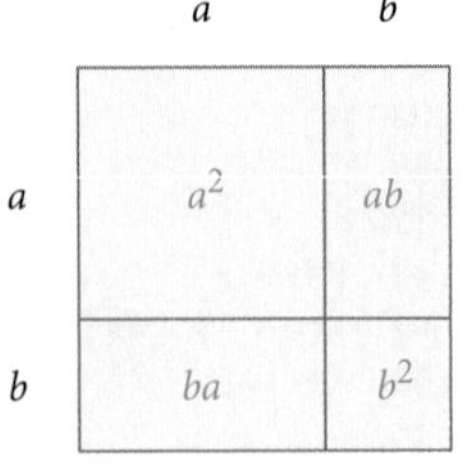

72. Area The length of each side of the large square shown in the figure is $x + 5$. Therefore, its area is $(x + 5)^2$. Find the area of the two smaller squares and the two smaller rectangles that make up the large square, then add them together to verify the formula $(x + 5)^2 = x^2 + 10x + 25$.

For each problem, fill in the area of each small rectangle and square, and then add the results together to find the indicated product.

Getting Ready for the Next Section

Simplify.

73. $\frac{10x^3}{5x}$ **74.** $\frac{15x^2}{5x}$ **75.** $\frac{3x^2}{3}$ **76.** $\frac{4x^2}{2}$

77. $\frac{9x^2}{3x}$ **78.** $\frac{3x^4}{3x^2}$ **79.** $\frac{24x^3y^2}{8x^2y}$ **80.** $\frac{4x^2y^3}{8x^2y}$

Dividing a Polynomial by a Monomial

5.7

To divide a polynomial by a monomial, we will use the definition of division and apply the distributive property. Follow the steps in this example closely.

EXAMPLE 1 Divide $10x^3 - 15x^2$ by $5x$.

SOLUTION

$$\frac{10x^3 - 15x^2}{5x} = (10x^3 - 15x^2)\frac{1}{5x}$$ Division by $5x$ is the same as multiplication by $\frac{1}{5x}$

$$= 10x^3\left(\frac{1}{5x}\right) - 15x^2\left(\frac{1}{5x}\right)$$ Distribute $\frac{1}{5x}$ to both terms

$$= \frac{10x^3}{5x} - \frac{15x^2}{5x}$$ Multiplication by $\frac{1}{5x}$ is the same as division by $5x$

$$= 2x^2 - 3x$$ Division of monomials as done in Section 5.3

If we were to leave out the first steps, the problem would look like this:

$$\frac{10x^3 - 15x^2}{5x} = \frac{10x^3}{5x} - \frac{15x^2}{5x}$$
$$= 2x^2 - 3x$$

The problem is much shorter and clearer this way. You may leave out the first two steps from Example 1 when working problems in this section. They are part of Example 1 only to help show you why the following rule is true.

RULE

To divide a polynomial by a monomial, simply divide each term in the polynomial by the monomial.

Here are some further examples using our rule for division of a polynomial by a monomial.

EXAMPLE 2 Divide $\frac{3x^2 - 6}{3}$.

SOLUTION We begin by writing the 3 in the denominator under each term in the numerator. Then we simplify the result:

$$\frac{3x^2 - 6}{3} = \frac{3x^2}{3} - \frac{6}{3}$$ Divide each term in the numerator by 3

$$= x^2 - 2$$ Simplify

EXAMPLE 3 Divide $\frac{4x^2 - 2}{2}$.

SOLUTION Dividing each term in the numerator by 2, we have

$$\frac{4x^2 - 2}{2} = \frac{4x^2}{2} - \frac{2}{2}$$ Divide each term in the numerator by 2

$$= 2x^2 - 1$$ Simplify

EXAMPLE 4 Find the quotient of $27x^3 - 9x^2$ and $3x$.

SOLUTION We again are asked to divide the first polynomial by the second one:

$$\frac{27x^3 - 9x^2}{3x} = \frac{27x^3}{3x} - \frac{9x^2}{3x} \quad \text{Divide each term by 3x}$$

$$= 9x^2 - 3x \quad \text{Simplify}$$

EXAMPLE 5 Divide $(15x^2y - 21xy^2) \div (-3xy)$.

SOLUTION This is the same type of problem we have shown in the first four examples; it is just worded a little differently. Note that when we divide each term in the first polynomial by $-3xy$, the negative sign must be taken into account:

$$\frac{15x^2y - 21xy^2}{-3xy} = \frac{15x^2y}{-3xy} - \frac{21xy^2}{-3xy} \quad \text{Divide each term by } -3xy$$

$$= -5x - (-7y) \quad \text{Simplify}$$

$$= -5x + 7y \quad \text{Simplify}$$

EXAMPLE 6 Divide $\dfrac{24x^3y^2 + 16x^2y^2 - 4x^2y^3}{8x^2y}$

SOLUTION Writing $8x^2y$ under each term in the numerator and then simplifying, we have

$$\frac{24x^3y^2 + 16x^2y^2 - 4x^2y^3}{8x^2y} = \frac{24x^3y^2}{8x^2y} + \frac{16x^2y^2}{8x^2y} - \frac{4x^2y^3}{8x^2y}$$

$$= 3xy + 2y - \frac{y^2}{2}.$$

From the examples in this section, it is clear that to divide a polynomial by a monomial, we must divide each term in the polynomial by the monomial. Often, students taking algebra for the first time will make the following mistake:

$$\frac{x+2}{2} = x + 1 \quad \text{Mistake}$$

The mistake here is in not dividing both terms in the numerator by 2. The correct way to divide $x + 2$ by 2 looks like this:

$$\frac{x+2}{2} = \frac{x}{2} + \frac{2}{2} = \frac{x}{2} + 1 \quad \text{Correct}$$

GETTING READY FOR CLASS

After reading through the preceding section, respond in your own words and in complete sentences.

A. What property of real numbers is the key to dividing a polynomial by a monomial?
B. Describe how you would divide a polynomial by 5x.
C. Why is our answer to Example 6 not a polynomial?

Problem Set 5.7

1. $x - 2$
2. $2x^2 - 3$
3. $3 - 2x^2$
4. $10x^2 - 4x$
5. $5xy - 2y$
6. $3y^2 + 4xy$
7. $7x^4 - 6x^3 + 5x^2$
8. $8x^3 - 6x^2 + 4x$
9. $10x^4 - 5x^2 + 1$
10. $15x^5 + 10x^2 - 5$
11. $-4a + 2$
12. $-\frac{a^2}{2} + 3a$
13. $-8a^4 - 12a^3$
14. $-15a^5 - 10a^2$
15. $-4b - 5a$
16. $-3ab + 5b^2$
17. $-6a^2b + 3ab^2 - 7b^3$
18. $-2b^3 + 8ab^2 + 11a^2b$
19. $-\frac{a}{2} - b - \frac{b^2}{2a}$
20. $-\frac{ab}{2} + b^2 - \frac{b^3}{2a}$
21. $3x + 4y$
22. $3x - y$
23. $-y + 3$
24. $7y - 6$
25. $5y - 4$
26. $-2y^2 + 3$
27. $xy - x^2y^2$
28. $y - xy^2$
29. $-1 + xy$
30. $1 + ab$
31. $-a + 1$
32. $ab - bc$
33. $x^2 - 3xy + y^2$
34. $x - 3y^2 + y^3$
35. $2 - 3b + 5b^2$
36. $-ab + 3$
37. $-2xy + 1$
38. $xy - \frac{1}{2}$
39. $xy - \frac{1}{2}$
40. $\frac{a}{6} + \frac{b}{3} - \frac{c}{6}$
41. $\frac{1}{4x} - \frac{1}{2a} + \frac{3}{4}$
42. $\frac{2a}{x} - \frac{3b}{2x} + \frac{3c}{x}$
43. $\frac{4x^2}{3} + \frac{2}{3x} + \frac{1}{x^2}$
44. $-\frac{3}{2x} + \frac{1}{2} + x$
45. $3a^{3m} - 9a^m$
46. $2 - 3a^{2m}$
47. $2x^{4m} - 5x^{2m} + 7$
48. $3 + 4x^{2m} - 5x^{4m}$

Divide the following polynomials by $5x$.

1. $5x^2 - 10x$
2. $10x^3 - 15x$
3. $15x - 10x^3$
4. $50x^3 - 20x^2$
5. $25x^2y - 10xy$
6. $15xy^2 + 20x^2y$
7. $35x^5 - 30x^4 + 25x^3$
8. $40x^4 - 30x^3 + 20x^2$
9. $50x^5 - 25x^3 + 5x$
10. $75x^6 + 50x^3 - 25x$

Divide the following by $-2a$.

11. $8a^2 - 4a$
12. $a^3 - 6a^2$
13. $16a^5 + 24a^4$
14. $30a^6 + 20a^3$
15. $8ab + 10a^2$
16. $6a^2b - 10ab^2$
17. $12a^3b - 6a^2b^2 + 14ab^3$
18. $4ab^3 - 16a^2b^2 - 22a^3b$
19. $a^2 + 2ab + b^2$
20. $a^2b - 2ab^2 + b^3$

Perform the following divisions (find the following quotients).

21. $\frac{6x + 8y}{2}$
22. $\frac{9x - 3y}{3}$
23. $\frac{7y - 21}{-7}$
24. $\frac{14y - 12}{2}$
25. $\frac{10xy - 8x}{2x}$
26. $\frac{12xy^2 - 18x}{-6x}$
27. $\frac{x^2y - x^3y^2}{x}$
28. $\frac{x^2y - x^3y^2}{x^2}$
29. $\frac{x^2y - x^3y^2}{-x^2y}$
30. $\frac{ab + a^2b^2}{ab}$
31. $\frac{a^2b^2 - ab^2}{-ab^2}$
32. $\frac{a^2b^2c - ab^2c^2}{abc}$
33. $\frac{x^3 - 3x^2y + xy^2}{x}$
34. $\frac{x^2 - 3xy^2 + xy^3}{x}$
35. $\frac{10a^2 - 15a^2b + 25a^2b^2}{5a^2}$
36. $\frac{11a^2b^2 - 33ab}{-11ab}$
37. $\frac{26x^2y^2 - 13xy}{-13xy}$
38. $\frac{6x^2y^2 - 3xy}{6xy}$
39. $\frac{4x^2y^2 - 2xy}{4xy}$
40. $\frac{6x^2a + 12x^2b - 6x^2c}{36x^2}$
41. $\frac{5a^2x - 10ax^2 + 15a^2x^2}{20a^2x^2}$
42. $\frac{12ax - 9bx + 18cx}{6x^2}$
43. $\frac{16x^5 + 8x^2 + 12x}{12x^3}$
44. $\frac{27x^2 - 9x^3 - 18x^4}{-18x^3}$

Divide. Assume all variables represent positive numbers.

45. $\frac{9a^{5m} - 27a^{3m}}{3a^{2m}}$
46. $\frac{26a^{3m} - 39a^{5m}}{13a^{3m}}$
47. $\frac{10x^{5m} - 25x^{3m} + 35x^m}{5x^m}$
48. $\frac{18x^{2m} + 24x^{4m} - 30x^{6m}}{6x^{2m}}$

49. $3x^2 - x + 6$
50. $6x^3 + 8x^2 - 5x$
51. 4
52. -4
53. $x + 5$
54. $x - 4$
55. Both equal 7.
56. Both equal 17.
57. $\frac{3(10) + 8}{2} = 19$; $3(10) + 4 = 34$
58. $\frac{5 + 10}{5} = 3$
59. $146\frac{20}{27}$
60. $1{,}444\frac{8}{13}$
61. $2x + 5$
62. $7x^2 + 9 + 3x^4$
63. $x^2 - 3x$
64. $-2x + 6$
65. $2x^3 - 10x^2$
66. $10x^2 - 50x$
67. $-2x$
68. $10x^2$
69. 2
70. -4

Simplify each numerator, and then divide.

49. $\dfrac{2x^3(3x + 2) - 3x^2(2x - 4)}{2x^2}$

50. $\dfrac{5x^2(6x - 3) + 6x^3(3x - 1)}{3x}$

51. $\dfrac{(x + 2)^2 - (x - 2)^2}{2x}$

52. $\dfrac{(x - 3)^2 - (x + 3)^2}{3x}$

53. $\dfrac{(x + 5)^2 + (x + 5)(x - 5)}{2x}$

54. $\dfrac{(x - 4)^2 + (x + 4)(x - 4)}{2x}$

55. Comparing Expressions Evaluate the expression $\frac{10x + 15}{5}$ and the expression $2x + 3$ when $x = 2$.

56. Comparing Expressions Evaluate the expression $\frac{6x^2 + 4x}{2x}$ and the expression $3x + 2$ when $x = 5$.

57. Comparing Expressions Show that the expression $\frac{3x + 8}{2}$ is not the same as the expression $3x + 4$ by replacing x with 10 in both expressions and simplifying the results.

58. Comparing Expressions Show that the expression $\frac{x + 10}{x}$ is not equal to 10 by replacing x with 5 and simplifying.

Getting Ready for the Next Section

Divide.

59. $27\overline{)3{,}962}$

60. $13\overline{)18{,}780}$

61. $\dfrac{2x^2 + 5x}{x}$

62. $\dfrac{7x^5 + 9x^3 + 3x^7}{x^3}$

Multiply.

63. $(x - 3)x$

64. $(x - 3)(-2)$

65. $2x^2(x - 5)$

66. $10x(x - 5)$

Subtract.

67. $(x^2 - 5x) - (x^2 - 3x)$

68. $(2x^3 + 0x^2) - (2x^3 - 10x^2)$

69. $(-2x + 8) - (-2x + 6)$

70. $(4x - 14) - (4x - 10)$

Dividing a Polynomial by a Polynomial

5.8

Since long division for polynomials is very similar to long division with whole numbers, we will begin by reviewing a division problem with whole numbers. You may realize when looking at Example 1 that you don't have a very good idea why you proceed as you do with long division. What you do know is that the process always works. We are going to approach the explanations in this section in much the same manner; that is, we won't always be sure why the steps we will use are important, only that they always produce the correct result.

EXAMPLE 1 Divide $27\overline{)3{,}962}$.

SOLUTION

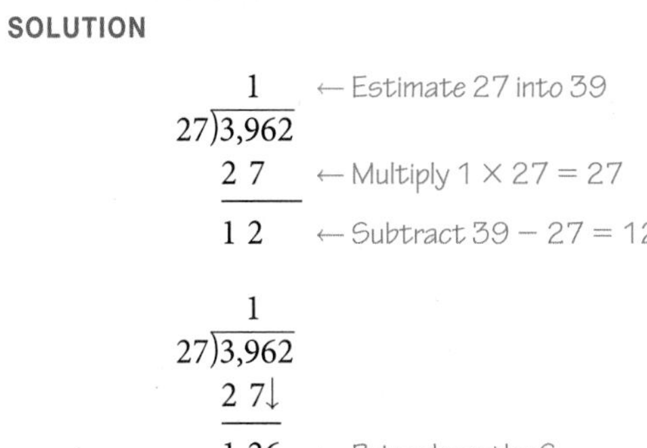

These are the four basic steps in long division. Estimate, multiply, subtract, and bring down the next term. To finish the problem, we simply perform the same four steps again:

14 ← 4 is the estimate
27)3,962
2 7↓
1 26
1 08 ← Multiply to get 108
182 ← Subtract to get 18, then bring down the 2

One more time.

146 ← 6 is the estimate
27)3,962
2 7
1 26
1 08
182
162 ← Multiply to get 162
20 ← Subtract to get 20

Since there is nothing left to bring down, we have our answer.

$$\frac{3{,}962}{27} = 146 + \frac{20}{27} \qquad \text{or} \qquad 146\frac{20}{27}$$

Here is how it works with polynomials.

EXAMPLE 2 Divide $\dfrac{x^2 - 5x + 8}{x - 3}$

SOLUTION

$$\begin{array}{r l}
x \phantom{{}-5x+8} & \leftarrow \text{Estimate } x^2 \div x = x \\
x - 3 \overline{) \; x^2 - 5x + 8} & \\
\cancel{+} x^2 \cancel{-} 3x \phantom{{}+8} & \leftarrow \text{Multiply } x(x-3) = x^2 - 3x \\
\hline
- 2x \phantom{{}+8} & \leftarrow \text{Subtract } (x^2 - 5x) - (x^2 - 3x) = -2x
\end{array}$$

$$\begin{array}{r l}
x \phantom{{}-5x+8} & \\
x - 3 \overline{) \; x^2 - 5x + 8} & \\
\cancel{+} x^2 \cancel{-} 3x \phantom{{}+8} & \\
\hline
- 2x + 8 & \leftarrow \text{Bring down the 8}
\end{array}$$

Notice that to subtract one polynomial from another, we add its opposite. That is why we change the signs on $x^2 - 3x$ and add what we get to $x^2 - 5x$. (To subtract the second polynomial, simply change the signs and add.)

We perform the same four steps again:

$$\begin{array}{r l}
x - 2 & \leftarrow -2 \text{ is the estimate } (-2x \div x = -2) \\
x - 3 \overline{) \; x^2 - 5x + 8} & \\
\cancel{+} x^2 \cancel{-} 3x \;\downarrow & \\
\hline
- 2x + 8 & \\
\cancel{-} 2x \cancel{+} 6 & \leftarrow \text{Multiply } -2(x-3) = -2x + 6. \\
\hline
2 & \leftarrow \text{Subtract } (-2x + 8) - (-2x + 6) = 2
\end{array}$$

Since there is nothing left to bring down, we have our answer:

$$\frac{x^2 - 5x + 8}{x - 3} = x - 2 + \frac{2}{x - 3}$$

To check our answer, we multiply $(x - 3)(x - 2)$ to get $x^2 - 5x + 6$. Then, adding on the remainder, 2, we have $x^2 - 5x + 8$.

EXAMPLE 3 Divide $\dfrac{6x^2 - 11x - 14}{2x - 5}$.

SOLUTION

$$\begin{array}{r}
3x + 2 \\
2x - 5 \overline{) \; 6x^2 - 11x - 14} \\
\cancel{+} 6x^2 \cancel{-} 15x \quad \downarrow \\
\hline
+ \;\; 4x - 14 \\
\cancel{+} \;\; 4x \cancel{-} 10 \\
\hline
- \;\; 4
\end{array}$$

$$\frac{6x^2 - 11x - 14}{2x - 5} = 3x + 2 + \frac{-4}{2x - 5}$$

One last step is sometimes necessary. The two polynomials in a division problem must both be in descending powers of the variable and cannot skip any powers from the highest power down to the constant term.

EXAMPLE 4 Divide $\dfrac{2x^3 - 3x + 2}{x - 5}$.

SOLUTION The problem will be much less confusing if we write $2x^3 - 3x + 2$ as $2x^3 + 0x^2 - 3x + 2$. Adding $0x^2$ does not change our original problem.

$$\begin{array}{r|l}
 & \quad 2x^2 \qquad\qquad \leftarrow \text{Estimate } 2x^3 \div x = 2x^2 \\
x - 5 & \; 2x^3 + 0x^2 - 3x + 2 \\
 & \underline{\not- 2x^3 \not+ 10x^2} \quad \downarrow \qquad \leftarrow \text{Multiply } 2x^2(x - 5) = 2x^3 - 10x^2 \\
 & \qquad + 10x^2 - 3x \qquad \leftarrow \text{Subtract: } (2x^3 + 0x^2) - (2x^3 - 10x^2) = 10x^2
\end{array}$$

↑ Bring down the next term

Adding the term $0x^2$ gives us a column in which to write $10x^2$. (Remember, you can add and subtract only similar terms.)

Here is the completed problem:

$$\begin{array}{r|l}
 & \quad 2x^2 + 10x + 47 \\
x - 5 & \; 2x^3 + 0x^2 - 3x + 2 \\
 & \underline{\not- 2x^3 \not+ 10x^2} \quad \downarrow \\
 & \qquad + 10x^2 - 3x \\
 & \qquad \underline{\not- 10x^2 \not+ 50x} \quad \downarrow \\
 & \qquad\qquad + 47x + 2 \\
 & \qquad\qquad \underline{\not- 47x \not+ 235} \\
 & \qquad\qquad\qquad 237
\end{array}$$

Our answer is $\dfrac{2x^3 - 3x + 2}{x - 5} = 2x^2 + 10x + 47 + \dfrac{237}{x - 5}$ ■

As you can see, long division with polynomials is a mechanical process. Once you have done it correctly a couple of times, it becomes very easy to produce the correct answer.

GETTING READY FOR CLASS

After reading through the preceding section, respond in your own words and in complete sentences.

A. What are the four steps used in long division with whole numbers?

B. How is division of two polynomials similar to long division with whole numbers?

C. What are the four steps used in long division with polynomials?

D. How do we use 0 when dividing the polynomial $2x^3 - 3x + 2$ by $x - 5$?

Problem Set 5.8

1. $x - 2$
2. $x - 3$
3. $a + 4$
4. $a + 5$
5. $x - 3$
6. $x + 5$
7. $x + 3$
8. $2x + 3$
9. $a - 5$
10. $2a - 5$
11. $x + 2 + \frac{2}{x + 3}$
12. $x + 2 + \frac{-2}{x + 3}$
13. $a - 2 + \frac{12}{a + 5}$
14. $a - 1 + \frac{8}{a + 5}$
15. $x + 4 + \frac{9}{x - 2}$
16. $x + 9 + \frac{36}{x - 3}$
17. $x + 4 + \frac{-10}{x + 1}$
18. $x - 2 + \frac{-4}{x + 1}$
19. $a + 1 + \frac{-1}{a + 2}$
20. $a - 2 + \frac{5}{a + 1}$
21. $x - 3 + \frac{17}{2x + 4}$
22. $5x - 7 + \frac{52}{3x + 8}$
23. $3a - 2 + \frac{7}{2a + 3}$
24. $2a + 1 + \frac{2}{2a + 1}$
25. $2a^2 - a - 3$
26. $a^2 - a + 2$
27. $x^2 - x + 5$
28. $x^2 + 2x - 4$
29. $x^2 + x + 1$
30. $x^2 - x + 1$
31. $x^2 + 2x + 4$
32. $x^2 - 3x + 9$

Divide.

1. $\dfrac{x^2 - 5x + 6}{x - 3}$
2. $\dfrac{x^2 - 5x + 6}{x - 2}$
3. $\dfrac{a^2 + 9a + 20}{a + 5}$
4. $\dfrac{a^2 + 9a + 20}{a + 4}$
5. $\dfrac{x^2 - 6x + 9}{x - 3}$
6. $\dfrac{x^2 + 10x + 25}{x + 5}$
7. $\dfrac{2x^2 + 5x - 3}{2x - 1}$
8. $\dfrac{4x^2 + 4x - 3}{2x - 1}$
9. $\dfrac{2a^2 - 9a - 5}{2a + 1}$
10. $\dfrac{4a^2 - 8a - 5}{2a + 1}$
11. $\dfrac{x^2 + 5x + 8}{x + 3}$
12. $\dfrac{x^2 + 5x + 4}{x + 3}$
13. $\dfrac{a^2 + 3a + 2}{a + 5}$
14. $\dfrac{a^2 + 4a + 3}{a + 5}$
15. $\dfrac{x^2 + 2x + 1}{x - 2}$
16. $\dfrac{x^2 + 6x + 9}{x - 3}$
17. $\dfrac{x^2 + 5x - 6}{x + 1}$
18. $\dfrac{x^2 - x - 6}{x + 1}$
19. $\dfrac{a^2 + 3a + 1}{a + 2}$
20. $\dfrac{a^2 - a + 3}{a + 1}$
21. $\dfrac{2x^2 - 2x + 5}{2x + 4}$
22. $\dfrac{15x^2 + 19x - 4}{3x + 8}$
23. $\dfrac{6a^2 + 5a + 1}{2a + 3}$
24. $\dfrac{4a^2 + 4a + 3}{2a + 1}$
25. $\dfrac{6a^3 - 13a^2 - 4a + 15}{3a - 5}$
26. $\dfrac{2a^3 - a^2 + 3a + 2}{2a + 1}$

Fill in the missing terms in the numerator, and then use long division to find the quotients (see Example 4).

27. $\dfrac{x^3 + 4x + 5}{x + 1}$
28. $\dfrac{x^3 + 4x^2 - 8}{x + 2}$
29. $\dfrac{x^3 - 1}{x - 1}$
30. $\dfrac{x^3 + 1}{x + 1}$
31. $\dfrac{x^3 - 8}{x - 2}$
32. $\dfrac{x^3 + 27}{x + 3}$

33. $491.17
34. $370.00
35. $331.42
36. $285.83
37. 47
38. 17
39. 14
40. 36
41. 70
42. 53
43. 35
44. 8
45. 5
46. 11,110
47. 6,540
48. 5,280
49. 1,760
50. 9
51. 20
52. 39
53. 63
54. 16
55. 53
56. 44

Long Division Use the information in the table to find the monthly payment for auto insurance for the cities below. Round to the nearest cent.

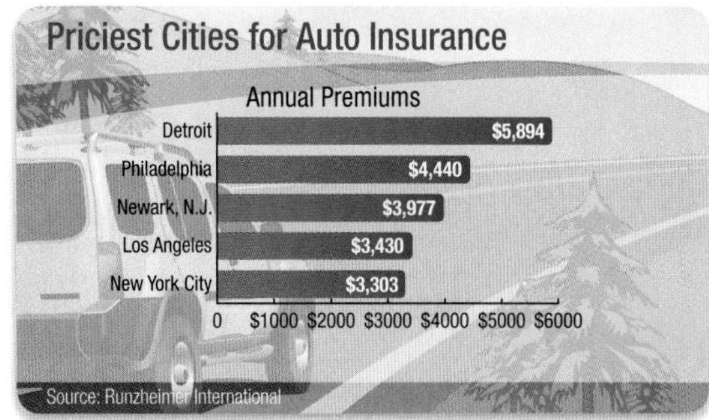

33. Detroit **34.** Philadelphia **35.** Newark, N.J. **36.** Los Angeles

Maintaining Your Skills

Simplify.

37. $6(3 + 4) + 5$

38. $[(1 + 2)(2 + 3)] + (4 \div 2)$

39. $1^2 + 2^2 + 3^2$

40. $(1 + 2 + 3)^2$

41. $5(6 + 3 \cdot 2) + 4 + 3 \cdot 2$

42. $(1 + 2)^3 + [(2 \cdot 3) + (4 \cdot 5)]$

43. $(1^3 + 2^3) + [(2 \cdot 3) + (4 \cdot 5)]$

44. $[2(3 + 4 + 5)] \div 3$

45. $(2 \cdot 3 + 4 + 5) \div 3$

46. $10^4 + 10^3 + 10^2 + 10^1$

47. $6 \cdot 10^3 + 5 \cdot 10^2 + 4 \cdot 10^1$

48. $5 \cdot 10^3 + 2 \cdot 10^2 + 8 \cdot 10^1$

49. $1 \cdot 10^3 + 7 \cdot 10^2 + 6 \cdot 10^1$

50. $4(2 - 1) + 5(3 - 2)$

51. $4 \cdot 2 - 1 + 5 \cdot 3 - 2$

52. $2^3 + 3^2 \cdot 4 - 5$

53. $(2^3 + 3^2) \cdot 4 - 5$

54. $4^2 - 2^4 + (2 \cdot 2)^2$

55. $2(2^2 + 3^2) + 3(3^2)$

56. $2 \cdot 2^2 + 3^2 + 3 \cdot 3^2$

SPOTLIGHT ON SUCCESS *Complaining*

iStockphoto © Nicholas Sutcliffe

Don't complain about anything, ever.

Do you complain to your classmates about your teacher? If you do, it could be getting in the way of your success in the class.

I have students that tell me that they like the way I teach and that they are enjoying my class. I have other students, in the same class, that complain to each other about me. They say I don't explain things well enough. Are the complaining students giving themselves a reason for not doing well in the class? I think so. They are shifting the responsibility for their success from themselves to me. It's not their fault they are not doing well, it's mine. When these students are alone, trying to do homework, they start thinking about how unfair everything is and they lose their motivation to study. Without intending to, they have set themselves up to fail by making their complaints more important than their progress in the class.

What happens when you stop complaining? You put yourself back in charge of your success. When there is no one to blame if things don't go well, you are more likely to do well. I have had students tell me that, once they stopped complaining about a class, the teacher became a better teacher and they started to actually enjoy going to class.

If you find yourself complaining to your friends about a class or a teacher, make a decision to stop. When other people start complaining to each other about the class or the teacher, walk away; don't participate in the complaining session. Try it for a day, or a week, or for the rest of the term. It may be difficult to do at first, but I'm sure you will like the results, and if you don't, you can always go back to complaining.

Chapter 5 Summary

Exponents: Definition and Properties [5.1, 5.2]

EXAMPLES

1. a. $2^3 = 2 \cdot 2 \cdot 2 = 8$

b. $x^5 \cdot x^3 = x^{5+3} = x^8$

c. $\frac{x^5}{x^3} = x^{5-3} = x^2$

d. $(3x)^2 = 3^2 \cdot x^2 = 9x^2$

e. $\left(\frac{2}{3}\right)^3 = \frac{2^3}{3^3} = \frac{8}{27}$

f. $(x^5)^3 = x^{5 \cdot 3} = x^{15}$

g. $3^{-2} = \frac{1}{3^2} = \frac{1}{9}$

Integer exponents indicate repeated multiplications.

$a^r \cdot a^s = a^{r+s}$ — To multiply with the same base, you add exponents

$\frac{a^r}{a^s} = a^{r-s}$ — To divide with the same base, you subtract exponents

$(ab)^r = a^r \cdot b^r$ — Exponents distribute over multiplication

$\left(\frac{a}{b}\right)^r = \frac{a^r}{b^r}$ — Exponents distribute over division

$(a^r)^s = a^{r \cdot s}$ — A power of a power is the product of the powers

$a^{-r} = \frac{1}{a^r}$ — Negative exponents imply reciprocals

Multiplication of Monomials [5.3]

2. $(5x^2)(3x^4) = 15x^6$

To multiply two monomials, multiply coefficients and add exponents.

Division of Monomials [5.3]

3. $\frac{12x^9}{4x^5} = 3x^4$

To divide two monomials, divide coefficients and subtract exponents.

Scientific Notation [5.1, 5.2, 5.3]

4. $768{,}000 = 7.68 \times 10^5$
$0.00039 = 3.9 \times 10^{-4}$

A number is in scientific notation when it is written as the product of a number between 1 and 10 and an integer power of 10.

Addition of Polynomials [5.4]

5. $(3x^2 - 2x + 1) + (2x^2 + 7x - 3)$
$= 5x^2 + 5x - 2$

To add two polynomials, add coefficients of similar terms.

Subtraction of Polynomials [5.4]

6. $(3x + 5) - (4x - 3)$
$= 3x + 5 - 4x + 3$
$= -x + 8$

To subtract one polynomial from another, add the opposite of the second to the first.

Multiplication of Polynomials [5.5]

To multiply a polynomial by a monomial, we apply the distributive property. To multiply two binomials we use the FOIL method. In other situations we use the COLUMN method. Each method achieves the same result: To multiply any two polynomials, we multiply each term in the first polynomial by each term in the second polynomial.

7. a. $2a^2(5a^2 + 3a - 2)$
   $= 10a^4 + 6a^3 - 4a^2$

   b. $(x + 2)(3x - 1)$
   $= 3x^2 - x + 6x - 2$
   $= 3x^2 + 5x - 2$

   c.
$$\begin{array}{r} 2x^2 - 3x + 4 \\ 3x - 2 \\ \hline 6x^3 - 9x^2 + 12x \phantom{{}-8} \\ - 4x^2 + 6x - 8 \\ \hline 6x^3 - 13x^2 + 18x - 8 \end{array}$$

Special Products [5.6]

$(a + b)^2 = a^2 + 2ab + b^2$
$(a - b)^2 = a^2 - 2ab + b^2$ } Binomial squares

$(a + b)(a - b) = a^2 - b^2$ Difference of two squares

8. $(x + 3)^2 = x^2 + 6x + 9$
   $(x - 3)^2 = x^2 - 6x + 9$
   $(x + 3)(x - 3) = x^2 - 9$

Dividing a Polynomial by a Monomial [5.7]

To divide a polynomial by a monomial, divide each term in the polynomial by the monomial.

9. $\dfrac{12x^3 - 18x^2}{6x} = 2x^2 - 3x$

Long Division with Polynomials [5.8]

Division with polynomials is similar to long division with whole numbers. The steps in the process are estimate, multiply, subtract, and bring down the next term. The divisors in all the long-division problems in this chapter were binomials.

10.
$$\begin{array}{r} x - 2 \\ x - 3\,\overline{)\,x^2 - 5x + 8} \\ \underline{\not{+}\,x^2 \not{-}\,3x \phantom{{}+8}} \\ -2x + 8 \\ \underline{\not{-}\,2x \not{+}\,6} \\ 2 \end{array}$$

Chapter 5 Test

Simplify each of the following expressions. [5.1]

1. $(-2)^5$

2. $\left(\frac{2}{3}\right)^3$

3. $(4x^2)^2(2x^3)^3$

Simplify each expression. Write all answers with positive exponents only. [5.2]

4. 4^{-2}

5. $(4a^5b^3)^0$

6. $\frac{x^{-4}}{x^{-7}}$

7. $\frac{(x^{-3})^2(x^{-5})^{-3}}{(x^{-3})^{-4}}$

8. Write 0.04307 in scientific notation. [5.2]

9. Write 7.63×10^6 in expanded form. [5.1]

Simplify. Write all answers with positive exponents only. [5.3]

10. $\frac{17x^2y^5z^3}{51x^4y^2z}$

11. $\frac{(3a^3b)(4a^2b^5)}{24a^2b^4}$

12. $\frac{28x^4}{4x} + \frac{30x^7}{6x^4}$

13. $\frac{(1.1 \times 10^5)(3 \times 10^{-2})}{4.4 \times 10^{-5}}$

Add or subtract as indicated. [5.4]

14. $9x^2 - 2x + 7x + 4$

15. $(4x^2 + 5x - 6) - (2x^2 - x - 4)$

16. Subtract $2x + 7$ from $7x + 3$. [5.4]

17. Find the value of $3a^2 + 4a + 6$ when a is -3. [5.4]

Multiply. [5.5]

18. $3x^2(5x^2 - 2x + 4)$

19. $\left(x + \frac{1}{4}\right)\left(x - \frac{1}{3}\right)$

20. $(2x - 3)(5x + 6)$

21. $(x + 4)(x^2 - 4x + 16)$

Multiply. [5.6]

22. $(x - 6)^2$

23. $(2a + 4b)^2$

24. $(3x - 6)(3x + 6)$

25. $(x^2 - 4)(x^2 + 4)$

26. Divide $18x^3 - 36x^2 + 6x$ by $6x$. [5.7]

Divide. [5.8]

27. $\frac{9x^2 - 6x - 4}{3x - 1}$

28. $\frac{4x^2 - 5x + 6}{x - 4}$

29. Volume Find the volume of a cube if the length of a side is 3.2 inches. [5.1]

30. Volume Find the volume of a rectangular solid if the length is three times the width, and the height is one third the width. [5.3]

Factoring

6

Chapter Outline

6.1 The Greatest Common Factor and Factoring by Grouping

6.2 Factoring Trinomials

6.3 More Trinomials to Factor

6.4 The Difference of Two Squares

6.5 The Sum and Difference of Two Cubes

6.6 Factoring: A General Review

6.7 Solving Equations by Factoring

6.8 Applications

If you watch professional football on television, you will hear the announcers refer to "hang time" when the punter punts the ball. Hang time is the amount of time the ball is in the air, and it depends on only one thing—the initial vertical velocity imparted to the ball by the kicker's foot. We can find the hang time of a football by solving equations. Table 1 shows the equations to solve for hang time, given various initial vertical velocities. Figure 1 is a visual representation of some equations associated with the ones in Table 1. In Figure 1, you can find hang time on the horizontal axis.

TABLE 1

Hang Time for a Football

Intial Vertical Velocity	Equation in factored form	Hang Time
16	$16t(1 - t) = 0$	1
32	$16t(2 - t) = 0$	2
48	$16t(3 - t) = 0$	3
64	$16t(4 - t) = 0$	4
80	$16t(5 - t) = 0$	5

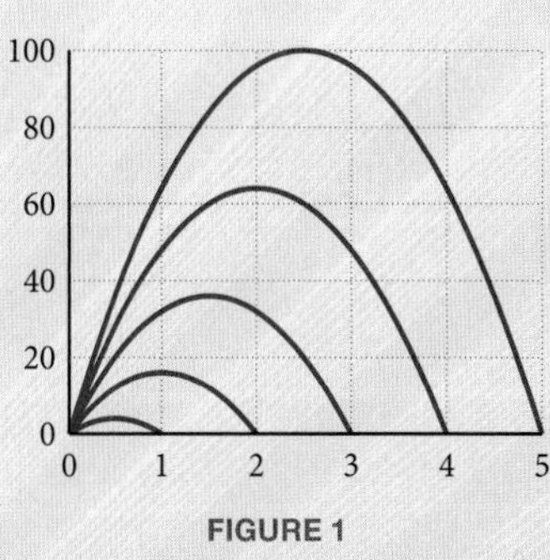

FIGURE 1

The equations in the second column of the table are in what is called "factored form." Once the equation is in factored form, hang time can be read from the second factor. In this chapter we develop techniques that allow us to factor a variety of polynomials. Factoring is the key to solving equations like the ones in Table 1.

Study Skills

This is the last chapter in which we will mention study skills. You know by now what works best for you and what you have to do to achieve your goals for this course. From now on, it is simply a matter of sticking with the things that work for you and avoiding the things that do not. It seems simple, but as with anything that takes effort, it is up to you to see that you maintain the skills that get you where you want to be in the course.

If you intend to take more classes in mathematics and want to ensure your success in those classes, then you can work toward this goal: ***Become the type of student who can learn mathematics on his or her own.*** Most people who have degrees in mathematics were students who could learn mathematics on their own. This doesn't mean that you have to learn it all on your own; it simply means that if you have to, you can learn it on your own. Attaining this goal gives you independence and puts you in control of your success in any math class you take.

The Greatest Common Factor and Factoring by Grouping

6.1

In Chapter 1 we used the following diagram to illustrate the relationship between multiplication and factoring.

$$\text{Factors} \longrightarrow 3 \cdot 5 = 15 \longleftarrow \text{Product}$$

Multiplication (left to right); Factoring (right to left)

A similar relationship holds for multiplication of polynomials. Reading the following diagram from left to right, we say the product of the binomials $x + 2$ and $x + 3$ is the trinomial $x^2 + 5x + 6$. However, if we read in the other direction, we can say that $x^2 + 5x + 6$ factors into the product of $x + 2$ and $x + 3$.

$$\text{Factors} \longrightarrow (x + 2)(x + 3) = x^2 + 5x + 6 \longleftarrow \text{Product}$$

Multiplication (left to right); Factoring (right to left)

In this chapter we develop a systematic method of factoring polynomials.

In this section we will apply the distributive property to polynomials to factor from them what is called the greatest common factor.

(def DEFINITION *greatest common factor*

The ***greatest common factor*** for a polynomial is the largest monomial that divides (is a factor of) each term of the polynomial.

We use the term *largest monomial* to mean the monomial with the greatest coefficient and highest power of the variable.

EXAMPLE 1 Find the greatest common factor for the polynomial:

$$3x^5 + 12x^2$$

SOLUTION The terms of the polynomial are $3x^5$ and $12x^2$. The largest number that divides the coefficients is 3, and the highest power of x that is a factor of x^5 and x^2 is x^2. Therefore, the greatest common factor for $3x^5 + 12x^2$ is $3x^2$; that is, $3x^2$ is the largest monomial that divides each term of $3x^5 + 12x^2$.

EXAMPLE 2 Find the greatest common factor for:

$$8a^3b^2 + 16a^2b^3 + 20a^3b^3$$

SOLUTION The largest number that divides each of the coefficients is 4. The highest power of the variable that is a factor of a^3b^2, a^2b^3, and a^3b^3 is a^2b^2. The greatest common factor for $8a^3b^2 + 16a^2b^3 + 20a^3b^3$ is $4a^2b^2$. It is the largest monomial that is a factor of each term.

Once we have recognized the greatest common factor of a polynomial, we can apply the distributive property and factor it out of each term. We rewrite the polynomial as the product of its greatest common factor with the polynomial that remains after the greatest common factor has been factored from each term in the original polynomial.

EXAMPLE 3 Factor the greatest common factor from $3x - 15$.

SOLUTION The greatest common factor for the terms $3x$ and 15 is 3. We can rewrite both $3x$ and 15 so that the greatest common factor 3 is showing in each term. It is important to realize that $3x$ means $3 \cdot x$. The 3 and the x are not "stuck" together:

$$3x - 15 = 3 \cdot x - 3 \cdot 5$$

Now, applying the distributive property, we have:

$$3 \cdot x - 3 \cdot 5 = 3(x - 5)$$

To check a factoring problem like this, we can multiply 3 and $x - 5$ to get $3x - 15$, which is what we started with. Factoring is simply a procedure by which we change sums and differences into products. In this case we changed the difference $3x - 15$ into the product $3(x - 5)$. Note, however, that we have not changed the meaning or value of the expression. The expression we end up with is equivalent to the expression we started with.

EXAMPLE 4 Factor the greatest common factor from:

$$5x^3 - 15x^2$$

SOLUTION The greatest common factor is $5x^2$. We rewrite the polynomial as:

$$5x^3 - 15x^2 = 5x^2 \cdot x - 5x^2 \cdot 3$$

Then we apply the distributive property to get:

$$5x^2 \cdot x - 5x^2 \cdot 3 = 5x^2(x - 3)$$

To check our work, we simply multiply $5x^2$ and $(x - 3)$ to get $5x^3 - 15x^2$, which is our original polynomial.

EXAMPLE 5 Factor the greatest common factor from:

$$16x^5 - 20x^4 + 8x^3$$

SOLUTION The greatest common factor is $4x^3$. We rewrite the polynomial so we can see the greatest common factor $4x^3$ in each term; then we apply the distributive property to factor it out.

$$\begin{aligned} 16x^5 - 20x^4 + 8x^3 &= 4x^3 \cdot 4x^2 - 4x^3 \cdot 5x + 4x^3 \cdot 2 \\ &= 4x^3(4x^2 - 5x + 2) \end{aligned}$$

EXAMPLE 6 Factor the greatest common factor from:

$$6x^3y - 18x^2y^2 + 12xy^3$$

SOLUTION The greatest common factor is $6xy$. We rewrite the polynomial in terms of $6xy$ and then apply the distributive property as follows:

$$\begin{aligned} 6x^3y - 18x^2y^2 + 12xy^3 &= 6xy \cdot x^2 - 6xy \cdot 3xy + 6xy \cdot 2y^2 \\ &= 6xy(x^2 - 3xy + 2y^2) \end{aligned}$$

EXAMPLE 7 Factor the greatest common factor from:

$$3a^2b - 6a^3b^2 + 9a^3b^3$$

SOLUTION The greatest common factor is $3a^2b$:

$$\begin{aligned} 3a^2b - 6a^3b^2 + 9a^3b^3 &= 3a^2b(1) - 3a^2b(2ab) + 3a^2b(3ab^2) \\ &= 3a^2b(1 - 2ab + 3ab^2) \end{aligned}$$

Factoring by Grouping

To develop our next method of factoring, called *factoring by grouping*, we start by examining the polynomial $xc + yc$. The greatest common factor for the two terms is c. Factoring c from each term we have:

$$xc + yc = c(x + y)$$

But suppose that c itself was a more complicated expression, such as $a + b$, so that the expression we were trying to factor was $x(a + b) + y(a + b)$, instead of $xc + yc$. The greatest common factor for $x(a + b) + y(a + b)$ is $(a + b)$. Factoring this common factor from each term looks like this:

$$x(a + b) + y(a + b) = (a + b)(x + y)$$

To see how all of this applies to factoring polynomials, consider the polynomial

$$xy + 3x + 2y + 6$$

There is no greatest common factor other than the number 1. However, if we group the terms together two at a time, we can factor an x from the first two terms and a 2 from the last two terms:

$$xy + 3x + 2y + 6 = x(y + 3) + 2(y + 3)$$

The expression on the right can be thought of as having two terms: $x(y + 3)$ and $2(y + 3)$. Each of these expressions contains the common factor $y + 3$, which can be factored out using the distributive property:

$$x(y + 3) + 2(y + 3) = (y + 3)(x + 2)$$

This last expression is in factored form. The process we used to obtain it is called factoring by grouping. Here are some additional examples.

EXAMPLE 8 Factor $ax + bx + ay + by$.

SOLUTION We begin by factoring x from the first two terms and y from the last two terms:

$$\begin{aligned} ax + bx + ay + by &= x(a + b) + y(a + b) \\ &= (a + b)(x + y) \end{aligned}$$

To convince yourself that this is factored correctly, multiply the two factors $(a + b)$ and $(x + y)$.

EXAMPLE 9 Factor by grouping: $3ax - 2a + 15x - 10$.

SOLUTION First, we factor a from the first two terms and 5 from the last two terms. Then, we factor $3x - 2$ from the remaining two expressions:

$$\begin{aligned} 3ax - 2a + 15x - 10 &= a(3x - 2) + 5(3x - 2) \\ &= (3x - 2)(a + 5) \end{aligned}$$

Again, multiplying $(3x - 2)$ and $(a + 5)$ will convince you that these are the correct factors.

EXAMPLE 10 Factor $2x^2 + 5ax - 2xy - 5ay$.

SOLUTION From the first two terms we factor x. From the second two terms we must factor $-y$ so that the binomial that remains after we do so matches the binomial produced by the first two terms:

$$\begin{aligned} 2x^2 + 5ax - 2xy - 5ay &= x(2x + 5a) - y(2x + 5a) \\ &= (2x + 5a)(x - y) \end{aligned}$$

Another way to accomplish the same result is to use the commutative property to interchange the middle two terms, and then factor by grouping:

$$\begin{aligned} 2x^2 + 5ax - 2xy - 5ay &= 2x^2 - 2xy + 5ax - 5ay && \text{Commutative property} \\ &= 2x(x - y) + 5a(x - y) \\ &= (x - y)(2x + 5a) \end{aligned}$$

This is the same result we obtained previously.

GETTING READY FOR CLASS

After reading through the preceding section, respond in your own words and in complete sentences.

A. What is the greatest common factor for a polynomial?
B. After factoring a polynomial, how can you check your result?
C. When would you try to factor by grouping?
D. What is the relationship between multiplication and factoring?

Problem Set 6.1

Factor the following by taking out the greatest common factor.

1. $15x + 25$

2. $14x + 21$

3. $6a + 9$

4. $8a + 10$

5. $4x - 8y$

6. $9x - 12y$

7. $3x^2 - 6x - 9$

8. $2x^2 + 6x + 4$

9. $3a^2 - 3a - 60$

10. $2a^2 - 18a + 28$

11. $24y^2 - 52y + 24$

12. $18y^2 + 48y + 32$

13. $9x^2 - 8x^3$

14. $7x^3 - 4x^2$

15. $13a^2 - 26a^3$

16. $5a^2 - 10a^3$

17. $21x^2y - 28xy^2$

18. $30xy^2 - 25x^2y$

19. $22a^2b^2 - 11ab^2$

20. $15x^3 - 25x^2 + 30x$

21. $7x^3 + 21x^2 - 28x$

22. $16x^4 - 20x^2 - 16x$

23. $121y^4 - 11x^4$

24. $25a^4 - 5b^4$

25. $100x^4 - 50x^3 + 25x^2$

26. $36x^5 + 72x^3 - 81x^2$

27. $8a^2 + 16b^2 + 32c^2$

28. $9a^2 - 18b^2 - 27c^2$

29. $4a^2b - 16ab^2 + 32a^2b^2$

30. $5ab^2 + 10a^2b^2 + 15a^2b$

31. $121a^3b^2 - 22a^2b^3 + 33a^3b^3$

32. $20a^4b^3 - 18a^3b^4 + 22a^4b^4$

33. $12x^2y^3 - 72x^5y^3 - 36x^4y^4$

34. $49xy - 21x^2y^2 + 35x^3y^3$

Factor by grouping.

35. $xy + 5x + 3y + 15$

36. $xy + 2x + 4y + 8$

37. $xy + 6x + 2y + 12$

38. $xy + 2y + 6x + 12$

39. $ab + 7a - 3b - 21$

40. $ab + 3b - 7a - 21$

41. $ax - bx + ay - by$

42. $ax - ay + bx - by$

43. $2ax + 6x - 5a - 15$

44. $3ax + 21x - a - 7$

45. $3xb - 4b - 6x + 8$

46. $3xb - 4b - 15x + 20$

47. $x^2 + ax + 2x + 2a$

48. $x^2 + ax + 3x + 3a$

49. $x^2 - ax - bx + ab$

50. $x^2 + ax - bx - ab$

Factor by grouping. You can group the terms together two at a time or three at a time. Either way will produce the same result.

51. $ax + ay + bx + by + cx + cy$

52. $ax + bx + cx + ay + by + cy$

Factor the following polynomials by grouping the terms together two at a time.

53. $6x^2 + 9x + 4x + 6$

54. $6x^2 - 9x - 4x + 6$

55. $20x^2 - 2x + 50x - 5$

56. $20x^2 + 25x + 4x + 5$

57. $20x^2 + 4x + 25x + 5$

58. $20x^2 + 4x - 25x - 5$

59. $x^3 + 2x^2 + 3x + 6$

60. $x^3 - 5x^2 - 4x + 20$

61. $6x^3 - 4x^2 + 15x - 10$

62. $8x^3 - 12x^2 + 14x - 21$

63. The greatest common factor of the binomial $3x + 6$ is 3. The greatest common factor of the binomial $2x + 4$ is 2. What is the greatest common factor of their product $(3x + 6)(2x + 4)$ when it has been multiplied out?

64. The greatest common factors of the binomials $4x + 2$ and $5x + 10$ are 2 and 5, respectively. What is the greatest common factor of their product $(4x + 2)(5x + 10)$ when it has been multiplied out?

65. The following factorization is incorrect. Find the mistake, and correct the right-hand side:

$$12x^2 + 6x + 3 = 3(4x^2 + 2x)$$

66. Find the mistake in the following factorization, and then rewrite the right-hand side correctly:

$$10x^2 + 2x + 6 = 2(5x^2 + 3)$$

Applying the Concepts

67. Investing If you invest \$1,000 in an account with an annual interest rate of r compounded annually, the amount of money you have in the account after one year is:

$$A = 1{,}000 + 1{,}000r$$

Write this formula again with the right side in factored form. Then, find the amount of money in this account at the end of one year if the interest rate is 12%.

68. Investing If you invest P dollars in an account with an annual interest rate of 8% compounded annually, then the amount of money in that account after one year is given by the formula:

$$A = P + 0.08P$$

Rewrite this formula with the right side in factored form, and then find the amount of money in the account at the end of one year if \$500 was the initial investment.

69. Biological Growth If 1,000,000 bacteria are placed in a petri dish and the bacteria have a growth rate of r (a percent expressed as a decimal) per hour, then 1 hour later the amount of bacteria will be $A = 1{,}000{,}000 + 1{,}000{,}000r$ bacteria.

a. Factor the right side of the equation.

b. If $r = 30\%$, find the number of bacteria present after one hour.

70. Biological Growth If there are B E. coli bacteria present initially in a petri dish and their growth rate is r (a percent expressed as a decimal) per hour, then after one hour there will be $A = B + Br$ bacteria present.

a. Factor the right side of this equation.

b. The following bar graph shows the number of E. coli bacteria present initially and the number of bacteria present hours later. Use the bar chart to find B and A in the preceding equation.

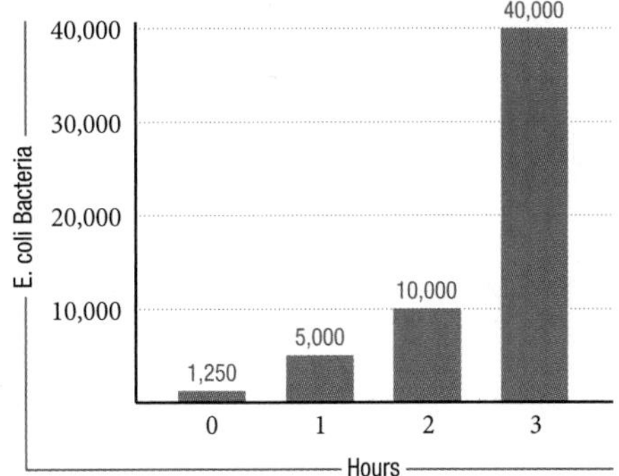

Getting Ready for the Next Section

Multiply.

71. $(x - 7)(x + 2)$

72. $(x - 7)(x - 2)$

73. $(x - 3)(x + 2)$

74. $(x + 3)(x - 2)$

75. $(x + 3)(x^2 - 3x + 9)$

76. $(x - 2)(x^2 + 2x + 4)$

77. $(2x + 1)(x^2 + 4x - 3)$

78. $(3x + 2)(x^2 - 2x - 4)$

79. $3x^4(6x^3 - 4x^2 + 2x)$

80. $2x^4(5x^3 + 4x^2 - 3x)$

81. $\left(x + \frac{1}{3}\right)\left(x + \frac{2}{3}\right)$

82. $\left(x + \frac{1}{4}\right)\left(x + \frac{3}{4}\right)$

83. $(6x + 4y)(2x - 3y)$

84. $(8a - 3b)(4a - 5b)$

85. $(9a + 1)(9a - 1)$

86. $(7b + 1)(7b + 1)$

87. $(x - 9)(x - 9)$

88. $(x - 8)(x - 8)$

89. $(x + 2)(x^2 - 2x + 4)$

90. $(x - 3)(x^2 + 3x + 9)$

Factoring Trinomials

6.2

Note: As you will see as we progress through the book, factoring is a tool that is used in solving a number of problems. Before seeing how it is used, however, we first must learn how to do it. So, in this section and the two sections that follow, we will be developing our factoring skills.

In this section we will factor trinomials in which the coefficient of the squared term is 1. The more familiar we are with multiplication of binomials the easier factoring trinomials will be.

Recall multiplication of binomials from Chapter 4:

$$(x + 3)(x + 4) = x^2 + 7x + 12$$

$$(x - 5)(x + 2) = x^2 - 3x - 10$$

The first term in the answer is the product of the first terms in each binomial. The last term in the answer is the product of the last terms in each binomial. The middle term in the answer comes from adding the product of the outside terms to the product of the inside terms.

Let's have a and b represent real numbers and look at the product of $(x + a)$ and $(x + b)$:

$$\begin{aligned}(x + a)(x + b) &= x^2 + ax + bx + ab \\ &= x^2 + (a + b)x + ab\end{aligned}$$

The coefficient of the middle term is the sum of a and b. The last term is the product of a and b. Writing this as a factoring problem, we have:

$$x^2 + \underbrace{(a + b)}_{\text{Sum}}x + \underbrace{ab}_{\text{Product}} = (x + a)(x + b)$$

To factor a trinomial in which the coefficient of x^2 is 1, we need only find the numbers a and b whose sum is the coefficient of the middle term and whose product is the constant term (last term).

EXAMPLE 1 Factor $x^2 + 8x + 12$.

SOLUTION The coefficient of x^2 is 1. We need two numbers whose sum is 8 and whose product is 12. The numbers are 6 and 2:

$$x^2 + 8x + 12 = (x + 6)(x + 2)$$

We can easily check our work by multiplying $(x + 6)$ and $(x + 2)$

$$\begin{aligned}\text{Check:}\quad (x + 6)(x + 2) &= x^2 + 6x + 2x + 12 \\ &= x^2 + 8x + 12\end{aligned}$$

EXAMPLE 2 Factor $x^2 - 2x - 15$.

SOLUTION The coefficient of x^2 is again 1. We need to find a pair of numbers whose sum is -2 and whose product is -15. Here are all the possibilities for products that are -15.

Products	Sums
$-1(15) = -15$	$-1 + 15 = 14$
$1(-15) = -15$	$1 + (-15) = -14$
$-5(3) = -15$	$-5 + 3 = -2$
$5(-3) = -15$	$5 + (-3) = 2$

Note: Again, we can check our results by multiplying our factors to see if their product is the original polynomial.

The third line gives us what we want. The factors of $x^2 - 2x - 15$ are $(x - 5)$ and $(x + 3)$:

$$x^2 - 2x - 15 = (x - 5)(x + 3)$$

EXAMPLE 3 Factor $2x^2 + 10x - 28$.

SOLUTION The coefficient of x^2 is 2. We begin by factoring out the greatest common factor, which is 2:

$$2x^2 + 10x - 28 = 2(x^2 + 5x - 14)$$

Now, we factor the remaining trinomial by finding a pair of numbers whose sum is 5 and whose product is -14. Here are the possibilities:

Products	Sums
$-1(14) = -14$	$-1 + 14 = 13$
$1(-14) = -14$	$1 + (-14) = -13$
$-7(2) = -14$	$-7 + 2 = -5$
$7(-2) = -14$	$7 + (-2) = 5$

Note: In Example 3 we began by factoring out the greatest common factor. The first step in factoring any trinomial is to look for the greatest common factor. If the trinomial in question has a greatest common factor other than 1, we factor it out first and then try to factor the trinomial that remains.

From the last line we see that the factors of $x^2 + 5x - 14$ are $(x + 7)$ and $(x - 2)$. Here is the complete problem:

$$\begin{aligned} 2x^2 + 10x - 28 &= 2(x^2 + 5x - 14) \\ &= 2(x + 7)(x - 2) \end{aligned}$$

EXAMPLE 4 Factor $3x^3 - 3x^2 - 18x$.

SOLUTION We begin by factoring out the greatest common factor, which is $3x$. Then we factor the remaining trinomial. Without showing the table of products and sums as we did in Examples 2 and 3, here is the complete problem:

$$\begin{aligned} 3x^3 - 3x^2 - 18x &= 3x(x^2 - x - 6) \\ &= 3x(x - 3)(x + 2) \end{aligned}$$

EXAMPLE 5 Factor $x^2 + 8xy + 12y^2$.

SOLUTION This time we need two expressions whose product is $12y^2$ and whose sum is $8y$. The two expressions are $6y$ and $2y$ (see Example 1 in this section):

$$x^2 + 8xy + 12y^2 = (x + 6y)(x + 2y)$$

You should convince yourself that these factors are correct by finding their product.

Note: Trinomials in which the coefficient of the second-degree term is 1 are the easiest to factor. Success in factoring any type of polynomial is directly related to the amount of time spent working the problems. The more we practice, the more accomplished we become at factoring.

GETTING READY FOR CLASS

After reading through the preceding section, respond in your own words and in complete sentences.

A. When the leading coefficient of a trinomial is 1, what is the relationship between the other two coefficients and the factors of the trinomial?

B. When factoring polynomials, what should you look for first?

C. How can you check to see that you have factored a trinomial correctly?

D. Describe how you would find the factors of $x^2 + 8x + 12$.

Factor the following trinomials.

1. $x^2 + 7x + 12$
2. $x^2 + 7x + 10$
3. $x^2 + 3x + 2$
4. $x^2 + 7x + 6$
5. $a^2 + 10a + 21$
6. $a^2 - 7a + 12$
7. $x^2 - 7x + 10$
8. $x^2 - 3x + 2$
9. $y^2 - 10y + 21$
10. $y^2 - 7y + 6$
11. $x^2 - x - 12$
12. $x^2 - 4x - 5$
13. $y^2 + y - 12$
14. $y^2 + 3y - 18$
15. $x^2 + 5x - 14$
16. $x^2 - 5x - 24$
17. $r^2 - 8r - 9$
18. $r^2 - r - 2$
19. $x^2 - x - 30$
20. $x^2 + 8x + 12$
21. $a^2 + 15a + 56$
22. $a^2 - 9a + 20$
23. $y^2 - y - 42$
24. $y^2 + y - 42$
25. $x^2 + 13x + 42$
26. $x^2 - 13x + 42$

Factor the following problems completely. First, factor out the greatest common factor, and then factor the remaining trinomial.

27. $2x^2 + 6x + 4$
28. $3x^2 - 6x - 9$
29. $3a^2 - 3a - 60$
30. $2a^2 - 18a + 28$
31. $100x^2 - 500x + 600$
32. $100x^2 - 900x + 2{,}000$
33. $100p^2 - 1{,}300p + 4{,}000$
34. $100p^2 - 1{,}200p + 3{,}200$
35. $x^4 - x^3 - 12x^2$
36. $x^4 - 11x^3 + 24x^2$
37. $2r^3 + 4r^2 - 30r$
38. $5r^3 + 45r^2 + 100r$
39. $2y^4 - 6y^3 - 8y^2$
40. $3r^3 - 3r^2 - 6r$
41. $x^5 + 4x^4 + 4x^3$
42. $x^5 + 13x^4 + 42x^3$
43. $3y^4 - 12y^3 - 15y^2$
44. $5y^4 - 10y^3 + 5y^2$
45. $4x^4 - 52x^3 + 144x^2$
46. $3x^3 - 3x^2 - 18x$

Factor the following trinomials.

47. $x^2 + 5xy + 6y^2$
48. $x^2 - 5xy + 6y^2$
49. $x^2 - 9xy + 20y^2$
50. $x^2 + 9xy + 20y^2$
51. $a^2 + 2ab - 8b^2$
52. $a^2 - 2ab - 8b^2$
53. $a^2 - 10ab + 25b^2$
54. $a^2 + 6ab + 9b^2$
55. $a^2 + 10ab + 25b^2$
56. $a^2 - 6ab + 9b^2$
57. $x^2 + 2xa - 48a^2$
58. $x^2 - 3xa - 10a^2$
59. $x^2 - 5xb - 36b^2$
60. $x^2 - 13xb + 36b^2$

Factor completely.

61. $x^4 - 5x^2 + 6$

62. $x^6 - 2x^3 - 15$

63. $x^2 - 80x - 2{,}000$

64. $x^2 - 190x - 2{,}000$

65. $x^2 - x + \frac{1}{4}$

66. $x^2 - \frac{2}{3}x + \frac{1}{9}$

67. $x^2 + 0.6x + 0.08$

68. $x^2 + 0.8x + 0.15$

69. If one of the factors of $x^2 + 24x + 128$ is $x + 8$, what is the other factor?

70. If one factor of $x^2 + 260x + 2{,}500$ is $x + 10$, what is the other factor?

71. What polynomial, when factored, gives $(4x + 3)(x - 1)$?

72. What polynomial factors to $(4x - 3)(x + 1)$?

Getting Ready for the Next Section

Multiply using the FOIL method.

73. $(6a + 1)(a + 2)$

74. $(6a - 1)(a - 2)$

75. $(3a + 2)(2a + 1)$

76. $(3a - 2)(2a - 1)$

77. $(6a + 2)(a + 1)$

78. $(3a + 1)(2a + 2)$

More Trinomials to Factor 6.3

We will now consider trinomials whose greatest common factor is 1 and whose leading coefficient (the coefficient of the squared term) is a number other than 1.

Suppose we want to factor the trinomial $2x^2 - 5x - 3$. We know the factors (if they exist) will be a pair of binomials. The product of their first terms is $2x^2$ and the product of their last term is -3. Let us list all the possible factors along with the trinomial that would result if we were to multiply them together. Remember, the middle term comes from the product of the inside terms plus the product of the outside terms.

Binomial Factors	First Term	Middle Term	Last Term
$(2x - 3)(x + 1)$	$2x^2$	$-x$	-3
$(2x + 3)(x - 1)$	$2x^2$	$+x$	-3
$(2x - 1)(x + 3)$	$2x^2$	$+5x$	-3
$(2x + 1)(x - 3)$	$2x^2$	$-5x$	-3

We can see from the last line that the factors of $2x^2 - 5x - 3$ are $(2x + 1)$ and $(x - 3)$. There is no straightforward way, as there was in the previous section, to find the factors, other than by trial and error or by simply listing all the possibilities. We look for possible factors that, when multiplied, will give the correct first and last terms, and then we see if we can adjust them to give the correct middle term.

EXAMPLE 1 Factor $6a^2 + 7a + 2$.

SOLUTION We list all the possible pairs of factors that, when multiplied together, give a trinomial whose first term is $6a^2$ and whose last term is $+2$.

Binomial Factors	First Term	Middle Term	Last Term
$(6a + 1)(a + 2)$	$6a^2$	$+13a$	$+2$
$(6a - 1)(a - 2)$	$6a^2$	$-13a$	$+2$
$(3a + 2)(2a + 1)$	$6a^2$	$+7a$	$+2$
$(3a - 2)(2a - 1)$	$6a^2$	$-7a$	$+2$

Note: Remember, we can always check our results by multiplying the factors we have and comparing that product with our original polynomial.

The factors of $6a^2 + 7a + 2$ are $(3a + 2)$ and $(2a + 1)$.

Check: $(3a + 2)(2a + 1) = 6a^2 + 7a + 2$

Notice that in the preceding list we did not include the factors $(6a + 2)$ and $(a + 1)$. We do not need to try these since the first factor has a 2 common to each term and so could be factored again, giving $2(3a + 1)(a + 1)$. Since our original trinomial, $6a^2 + 7a + 2$, did *not* have a greatest common factor of 2, neither of its factors will.

EXAMPLE 2 Factor $4x^2 - x - 3$.

SOLUTION We list all the possible factors that, when multiplied, give a trinomial whose first term is $4x^2$ and whose last term is -3.

Binomial Factors	First Term	Middle Term	Last Term
$(4x + 1)(x - 3)$	$4x^2$	$-11x$	-3
$(4x - 1)(x + 3)$	$4x^2$	$+11x$	-3
$(4x + 3)(x - 1)$	$4x^2$	$-x$	-3
$(4x - 3)(x + 1)$	$4x^2$	$+x$	-3
$(2x + 1)(2x - 3)$	$4x^2$	$-4x$	-3
$(2x - 1)(2x + 3)$	$4x^2$	$+4x$	-3

The third line shows that the factors are $(4x + 3)$ and $(x - 1)$.

$$\text{Check:}\quad (4x + 3)(x - 1) = 4x^2 - x - 3$$

You will find that the more practice you have at factoring this type of trinomial, the faster you will get the correct factors. You will pick up some shortcuts along the way, or you may come across a system of eliminating some factors as possibilities. Whatever works best for you is the method you should use. Factoring is a very important tool, and you must be good at it.

EXAMPLE 3 Factor $12y^3 + 10y^2 - 12y$.

SOLUTION We begin by factoring out the greatest common factor, $2y$:

$$12y^3 + 10y^2 - 12y = 2y(6y^2 + 5y - 6)$$

Note: Once again, the first step in any factoring problem is to factor out the greatest common factor if it is other than 1.

We now list all possible factors of a trinomial with the first term $6y^2$ and last term -6, along with the associated middle terms.

Possible Factors	Middle Term When Multiplied
$(3y + 2)(2y - 3)$	$-5y$
$(3y - 2)(2y + 3)$	$+5y$
$(6y + 1)(y - 6)$	$-35y$
$(6y - 1)(y + 6)$	$+35y$

The second line gives the correct factors. The complete problem is:

$$\begin{aligned} 12y^3 + 10y^2 - 12y &= 2y(6y^2 + 5y - 6) \\ &= 2y(3y - 2)(2y + 3) \end{aligned}$$

EXAMPLE 4 Factor $30x^2y - 5xy^2 - 10y^3$.

SOLUTION The greatest common factor is $5y$:

$$\begin{aligned} 30x^2y - 5xy^2 - 10y^3 &= 5y(6x^2 - xy - 2y^2) \\ &= 5y(2x + y)(3x - 2y) \end{aligned}$$

EXAMPLE 5 A ball is tossed into the air with an upward velocity of 16 feet per second from the top of a building 32 feet high. The equation that gives the height of the ball above the ground at any time t is

$$h = 32 + 16t - 16t^2$$

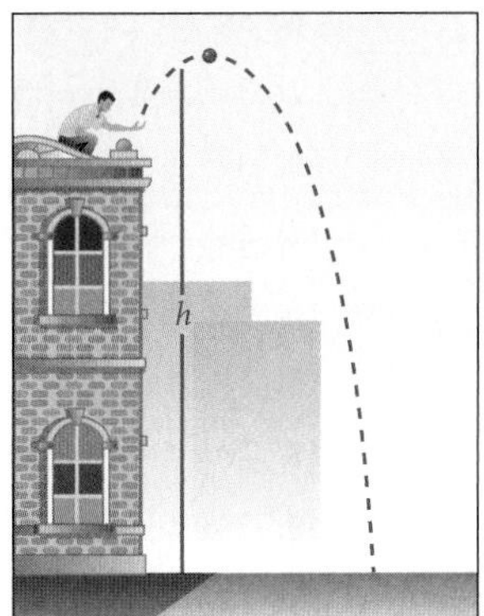

Factor the right side of this equation and then find h when t is 2.

SOLUTION We begin by factoring out the greatest common factor, 16. Then, we factor the trinomial that remains:

$$h = 32 + 16t - 16t^2$$

$$h = 16(2 + t - t^2)$$

$$h = 16(2 - t)(1 + t)$$

Letting $t = 2$ in the equation, we have

$$h = 16(0)(3) = 0$$

When t is 2, h is 0.

GETTING READY FOR CLASS

After reading through the preceding section, respond in your own words and in complete sentences.

A. What is the first step in factoring a trinomial?

B. Describe the criteria you would use to set up a table of possible factors of a trinomial.

C. What does it mean if you factor a trinomial and one of your factors has a greatest common factor of 3?

D. Describe how you would look for possible factors of $6a^2 + 7a + 2$.

Problem Set 6.3

Factor the following trinomials.

1. $2x^2 + 7x + 3$
2. $2x^2 + 5x + 3$
3. $2a^2 - a - 3$
4. $2a^2 + a - 3$
5. $3x^2 + 2x - 5$
6. $3x^2 - 2x - 5$
7. $3y^2 - 14y - 5$
8. $3y^2 + 14y - 5$
9. $6x^2 + 13x + 6$
10. $6x^2 - 13x + 6$
11. $4x^2 - 12xy + 9y^2$
12. $4x^2 + 12xy + 9y^2$
13. $4y^2 - 11y - 3$
14. $4y^2 + y - 3$
15. $20x^2 - 41x + 20$
16. $20x^2 + 9x - 20$
17. $20a^2 + 48ab - 5b^2$
18. $20a^2 + 29ab + 5b^2$
19. $20x^2 - 21x - 5$
20. $20x^2 - 48x - 5$
21. $12m^2 + 16m - 3$
22. $12m^2 + 20m + 3$
23. $20x^2 + 37x + 15$
24. $20x^2 + 13x - 15$
25. $12a^2 - 25ab + 12b^2$
26. $12a^2 + 7ab - 12b^2$
27. $3x^2 - xy - 14y^2$
28. $3x^2 + 19xy - 14y^2$
29. $14x^2 + 29x - 15$
30. $14x^2 + 11x - 15$
31. $6x^2 - 43x + 55$
32. $6x^2 - 7x - 55$
33. $15t^2 - 67t + 38$
34. $15t^2 - 79t - 34$

Factor each of the following completely. Look first for the greatest common factor.

35. $4x^2 + 2x - 6$
36. $6x^2 - 51x + 63$
37. $24a^2 - 50a + 24$
38. $18a^2 + 48a + 32$
39. $10x^3 - 23x^2 + 12x$
40. $10x^4 + 7x^3 - 12x^2$
41. $6x^4 - 11x^3 - 10x^2$
42. $6x^3 + 19x^2 + 10x$
43. $10a^3 - 6a^2 - 4a$
44. $6a^3 + 15a^2 + 9a$
45. $15x^3 - 102x^2 - 21x$
46. $2x^4 - 24x^3 + 64x^2$
47. $35y^3 - 60y^2 - 20y$
48. $14y^4 - 32y^3 + 8y^2$
49. $15a^4 - 2a^3 - a^2$
50. $10a^5 - 17a^4 + 3a^3$
51. $24x^2y - 6xy - 45y$
52. $8x^2y^2 + 26xy^2 + 15y^2$
53. $12x^2y - 34xy^2 + 14y^3$
54. $12x^2y - 46xy^2 + 14y^3$
55. Evaluate the expression $2x^2 + 7x + 3$ and the expression $(2x + 1)(x + 3)$ for $x = 2$.
56. Evaluate the expression $2a^2 - a - 3$ and the expression $(2a - 3)(a + 1)$ for $a = 5$.
57. What polynomial factors to $(2x + 3)(2x - 3)$?
58. What polynomial factors to $(5x + 4)(5x - 4)$?
59. What polynomial factors to $(x + 3)(x - 3)(x^2 + 9)$?
60. What polynomial factors to $(x + 2)(x - 2)(x^2 + 4)$?

Applying the Concepts

61. Archery Margaret shoots an arrow into the air. The equation for the height (in feet) of the tip of the arrow is:

$$h = 8 + 62t - 16t^2$$

Factor the right side of this equation. Then fill in the table for various heights of the arrow, using the factored form of the equation.

Time t (seconds)	0	1	2	3	4
Height h (feet)					

62. Coin Toss At the beginning of every football game, the referee flips a coin to see who will kick off. The equation that gives the height (in feet) of the coin tossed in the air is:

$$h = 6 + 29t - 16t^2$$

a. Factor this equation.

b. Use the factored form of the equation to find the height of the quarter after 0 seconds, 1 second, and 2 seconds.

63. Constructing a Box Yesterday I was experimenting with how to cut and fold a certain piece of cardboard to make a box with different volumes. Unfortunately, today I have lost both the cardboard and most of my notes. I remember that I made the box by cutting equal squares from the corners then folding up the side flaps:

I don't remember how big the cardboard was, and I can only find the last page of notes, which says that if x is the length of a side of a small square (in inches), then the volume is $V = 99x - 40x^2 + 4x^3$.

a. Factor the right side of this expression completely.

b. What were the dimensions of the original piece of cardboard?

64. Constructing a Box Repeat Problem 63 if the remaining formula is $V = 15x - 16x^2 + 4x^3$.

Getting Ready for the Next Section

Multiply each of the following.

65. $(x + 3)(x - 3)$

66. $(x - 4)(x + 4)$

67. $(x + 5)(x - 5)$

68. $(x - 6)(x + 6)$

69. $(x + 7)(x - 7)$

70. $(x - 8)(x + 8)$

71. $(x + 9)(x - 9)$

72. $(x - 10)(x + 10)$

73. $(2x - 3y)(2x + 3y)$

74. $(5x - 6y)(5x + 6y)$

75. $(x^2 + 4)(x + 2)(x - 2)$

76. $(x^2 + 9)(x + 3)(x - 3)$

77. $(x + 3)^2$

78. $(x - 4)^2$

79. $(x + 5)^2$

80. $(x - 6)^2$

81. $(x + 7)^2$

82. $(x - 8)^2$

83. $(x + 9)^2$

84. $(x - 10)^2$

85. $(2x + 3)^2$

86. $(3x - y)^2$

87. $(4x - 2y)^2$

88. $(5x - 6y)^2$

The Difference of Two Squares 6.4

In Chapter 4 we listed the following three special products:

$$(a + b)^2 = (a + b)(a + b) = a^2 + 2ab + b^2$$
$$(a - b)^2 = (a - b)(a - b) = a^2 - 2ab + b^2$$
$$(a + b)(a - b) = a^2 - b^2$$

Since factoring is the reverse of multiplication, we can also consider the three special products as three special factorings:

$$a^2 + 2ab + b^2 = (a + b)^2$$
$$a^2 - 2ab + b^2 = (a - b)^2$$
$$a^2 - b^2 = (a + b)(a - b)$$

Any trinomial of the form $a^2 + 2ab + b^2$ or $a^2 - 2ab + b^2$ can be factored by the methods of Section 4.6. The last line is the factoring to obtain the difference of two squares. The difference of two squares always factors in this way. Again, these are patterns you must be able to recognize on sight.

EXAMPLE 1 Factor $16x^2 - 25$.

SOLUTION We can see that the first term is a perfect square, and the last term is also. This fact becomes even more obvious if we rewrite the problem as:

$$16x^2 - 25 = (4x)^2 - (5)^2$$

The first term is the square of the quantity $4x$, and the last term is the square of 5. The completed problem looks like this:

$$\begin{aligned} 16x^2 - 25 &= (4x)^2 - (5)^2 \\ &= (4x + 5)(4x - 5) \end{aligned}$$

To check our results, we multiply:

$$\begin{aligned} (4x + 5)(4x - 5) &= 16x^2 + 20x - 20x - 25 \\ &= 16x^2 - 25 \end{aligned}$$

EXAMPLE 2 Factor $36a^2 - 1$.

SOLUTION We rewrite the two terms to show they are perfect squares and then factor. Remember, 1 is its own square, $1^2 = 1$.

$$\begin{aligned} 36a^2 - 1 &= (6a)^2 - (1)^2 \\ &= (6a + 1)(6a - 1) \end{aligned}$$

To check our results, we multiply:

$$\begin{aligned} (6a + 1)(6a - 1) &= 36a^2 + 6a - 6a - 1 \\ &= 36a^2 - 1 \end{aligned}$$

EXAMPLE 3 Factor $x^4 - y^4$.

SOLUTION x^4 is the perfect square $(x^2)^2$, and y^4 is $(y^2)^2$:

$$\begin{aligned} x^4 - y^4 &= (x^2)^2 - (y^2)^2 \\ &= (x^2 - y^2)(x^2 + y^2) \end{aligned}$$

The factor $(x^2 - y^2)$ is itself the difference of two squares and therefore can be factored again. The factor $(x^2 + y^2)$ is the *sum* of two squares and cannot be factored again. The complete problem is this:

$$\begin{aligned} x^4 - y^4 &= (x^2)^2 - (y^2)^2 \\ &= (x^2 - y^2)(x^2 + y^2) \\ &= (x + y)(x - y)(x^2 + y^2) \end{aligned}$$

Note: If you think the sum of two squares $x^2 + y^2$ factors, you should try it. Write down the factors you think it has, and then multiply them using the foil method. You won't get $x^2 + y^2$.

EXAMPLE 4 Factor $25x^2 - 60x + 36$.

SOLUTION Although this trinomial can be factored by the method we used in Section 6.3, we notice that the first and last terms are the perfect squares $(5x)^2$ and $(6)^2$. Before going through the method for factoring trinomials by listing all possible factors, we can check to see if $25x^2 - 60x + 36$ factors to $(5x - 6)^2$. We need only multiply to check:

$$\begin{aligned} (5x - 6)^2 &= (5x - 6)(5x - 6) \\ &= 25x^2 - 30x - 30x + 36 \\ &= 25x^2 - 60x + 36 \end{aligned}$$

The trinomial $25x^2 - 60x + 36$ factors to $(5x - 6)(5x - 6) = (5x - 6)^2$.

EXAMPLE 5 Factor $5x^2 + 30x + 45$.

SOLUTION We begin by factoring out the greatest common factor, which is 5. Then we notice that the trinomial that remains is a perfect square trinomial:

$$\begin{aligned} 5x^2 + 30x + 45 &= 5(x^2 + 6x + 9) \\ &= 5(x + 3)^2 \end{aligned}$$

Note: As we have indicated before, perfect square trinomials like the ones in Examples 4 and 5 can be factored by the methods developed in previous sections. Recognizing that they factor to binomial squares simply saves time in factoring.

EXAMPLE 6 Factor $(x - 3)^2 - 25$.

SOLUTION This example has the form $a^2 - b^2$, where a is $x - 3$ and b is 5. We factor it according to the formula for the difference of two squares:

$$\begin{aligned} (x - 3)^2 - 25 &= (x - 3)^2 - 5^2 && \text{Write 25 as } 5^2 \\ &= [(x - 3) - 5][(x - 3) + 5] && \text{Factor} \\ &= (x - 8)(x + 2) && \text{Simplify} \end{aligned}$$

Notice in this example we could have expanded $(x - 3)^2$, subtracted 25, and then factored to obtain the same result:

$$\begin{aligned} (x - 3)^2 - 25 &= x^2 - 6x + 9 - 25 && \text{Expand } (x \cdot 3)^2 \\ &= x^2 - 6x - 16 && \text{Simplify} \\ &= (x - 8)(x + 2) && \text{Factor} \end{aligned}$$

GETTING READY FOR CLASS

After reading through the preceding section, respond in your own words and in complete sentences.

A. Describe how you factor the difference of two squares.
B. What is a perfect square trinomial?
C. How do you know when you've factored completely?
D. Describe how you would factor $25x^2 - 60x + 36$.

SPOTLIGHT ON SUCCESS *Student Instructor Stefanie*

Never confuse a single defeat with a final defeat.
—F. Scott Fitzgerald

The idea that has worked best for my success in college, and more specifically in my math courses, is to stay positive and be resilient. I have learned that a 'bad' grade doesn't make me a failure; if anything it makes me strive to do better. That is why I never let a bad grade on a test or even in a class get in the way of my overall success.

By sticking with this positive attitude, I have been able to achieve my goals. My grades have never represented how well I know the material. This is because I have struggled with test anxiety and it has consistently lowered my test scores in a number of courses. However, I have not let it defeat me. When I applied to graduate school, I did not meet the grade requirements for my top two schools, but that did not stop me from applying.

One school asked that I convince them that my knowledge of mathematics was more than my grades indicated. If I had let my grades stand in the way of my goals, I wouldn't have been accepted to both of my top two schools, and will be attending one of them in the Fall, on my way to becoming a mathematics teacher.

Problem Set 6.4

Factor the following.

1. $x^2 - 9$ **2.** $x^2 - 25$ **3.** $a^2 - 36$ **4.** $a^2 - 64$

5. $x^2 - 49$ **6.** $x^2 - 121$ **7.** $4a^2 - 16$ **8.** $4a^2 + 16$

9. $9x^2 + 25$ **10.** $16x^2 - 36$ **11.** $25x^2 - 169$ **12.** $x^2 - y^2$

13. $9a^2 - 16b^2$ **14.** $49a^2 - 25b^2$ **15.** $9 - m^2$ **16.** $16 - m^2$

17. $25 - 4x^2$ **18.** $36 - 49y^2$ **19.** $2x^2 - 18$ **20.** $3x^2 - 27$

21. $32a^2 - 128$ **22.** $3a^3 - 48a$ **23.** $8x^2y - 18y$ **24.** $50a^2b - 72b$

25. $a^4 - b^4$ **26.** $a^4 - 16$ **27.** $16m^4 - 81$ **28.** $81 - m^4$

29. $3x^3y - 75xy^3$ **30.** $2xy^3 - 8x^3y$

Factor the following.

31. $x^2 - 2x + 1$ **32.** $x^2 - 6x + 9$

33. $x^2 + 2x + 1$ **34.** $x^2 + 6x + 9$

35. $a^2 - 10a + 25$ **36.** $a^2 + 10a + 25$

37. $y^2 + 4y + 4$ **38.** $y^2 - 8y + 16$

39. $x^2 - 4x + 4$ **40.** $x^2 + 8x + 16$

41. $m^2 - 12m + 36$ **42.** $m^2 + 12m + 36$

43. $4a^2 + 12a + 9$ **44.** $9a^2 - 12a + 4$

45. $49x^2 - 14x + 1$ **46.** $64x^2 - 16x + 1$

47. $9y^2 - 30y + 25$ **48.** $25y^2 + 30y + 9$

49. $x^2 + 10xy + 25y^2$ **50.** $25x^2 + 10xy + y^2$

51. $9a^2 + 6ab + b^2$ **52.** $9a^2 - 6ab + b^2$

Factor the following by first factoring out the greatest common factor.

53. $3a^2 + 18a + 27$ **54.** $4a^2 - 16a + 16$

55. $2x^2 + 20xy + 50y^2$ **56.** $3x^2 + 30xy + 75y^2$

57. $5x^3 + 30x^2y + 45xy^2$ **58.** $12x^2y - 36xy^2 + 27y^3$

Factor by grouping the first three terms together.

59. $x^2 + 6x + 9 - y^2$ **60.** $x^2 + 10x + 25 - y^2$

61. $x^2 + 2xy + y^2 - 9$ **62.** $a^2 + 2ab + b^2 - 25$

63. Find a value for b so that the polynomial $x^2 + bx + 49$ factors to $(x + 7)^2$.

64. Find a value of b so that the polynomial $x^2 + bx + 81$ factors to $(x + 9)^2$.

65. Find the value of c for which the polynomial $x^2 + 10x + c$ factors to $(x + 5)^2$.

66. Find the value of a for which the polynomial $ax^2 + 12x + 9$ factors to $(2x + 3)^2$.

Applying the Concepts

67. Area

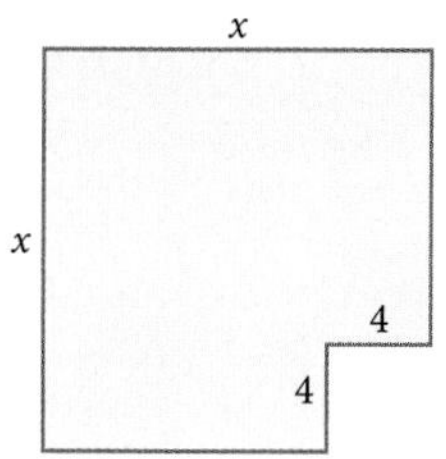

a. What is the area of the following figure?

b. Factor the answer from part **a.**

c. Find a way to cut the figure into two pieces and put them back together to show that the factorization in part **b.** is correct.

68. Area

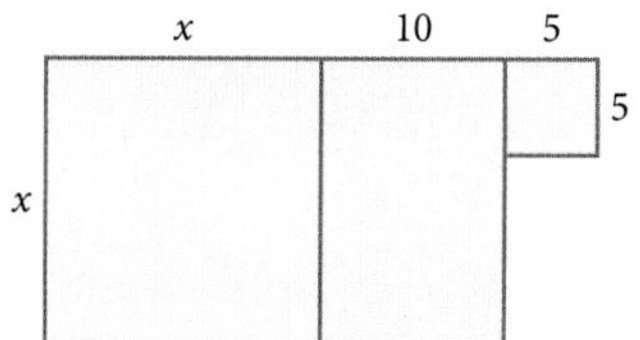

a. What is the area of the following figure?

b. Factor the expression from part **a.**

c. Cut and rearrange the figure to show that the factorization is correct.

Find the area for the shaded regions; then write your result in factored form.

69.

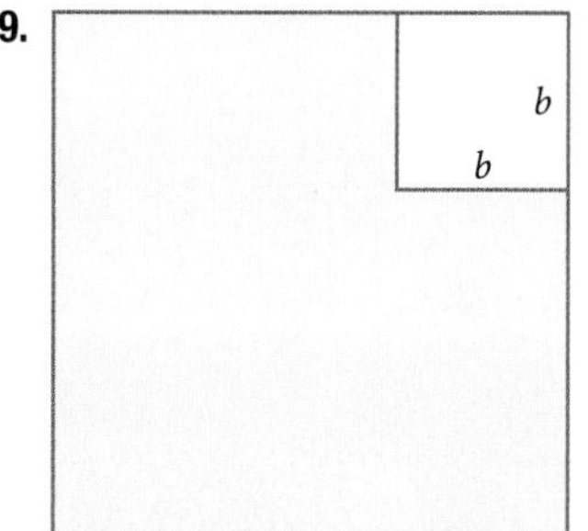

70.

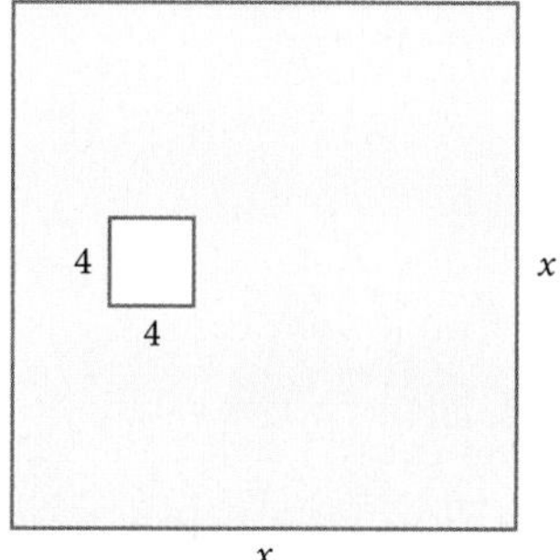

Getting Ready for the Next Section

Multiply each of the following:

71. **a.** 1^3 **b.** 2^3 **c.** 3^3 **d.** 4^3 **e.** 5^3

72. **a.** $(-1)^3$ **b.** $(-2)^3$ **c.** $(-3)^3$ **d.** $(-4)^3$ **e.** $(-5)^3$

73. **a.** $x(x^2 - x + 1)$ **b.** $1(x^2 - x + 1)$ **c.** $(x + 1)(x^2 - x + 1)$

74. **a.** $x(x^2 + x + 1)$ **b.** $-1(x^2 + x + 1)$ **c.** $(x - 1)(x^2 + x + 1)$

75. **a.** $x(x^2 - 2x + 4)$ **b.** $2(x^2 - 2x + 4)$ **c.** $(x + 2)(x^2 - 2x + 4)$

76. **a.** $x(x^2 + 2x + 4)$ **b.** $-2(x^2 + 2x + 4)$ **c.** $(x - 2)(x^2 + 2x + 4)$

77. **a.** $x(x^2 - 3x + 9)$ **b.** $3(x^2 - 3x + 9)$ **c.** $(x + 3)(x^2 - 3x + 9)$

78. **a.** $x(x^2 + 3x + 9)$ **b.** $-3(x^2 + 3x + 9)$ **c.** $(x - 3)(x^2 + 3x + 9)$

The Sum and Difference of Two Cubes 6.5

Previously, we factored a variety of polynomials. Among the polynomials we factored were polynomials that were the difference of two squares. The formula we used to factor the difference of two squares looks like this:

$$a^2 - b^2 = (a + b)(a - b)$$

If we ran across a binomial that had the form of the difference of two squares, we factored it by applying this formula. For example, to factor $x^2 - 25$, we simply notice that it can be written in the form $x^2 - 5^2$, which looks like the difference of two squares. According to the formula above, this binomial factors into $(x + 5)(x - 5)$.

In this section we want to use two new formulas that will allow us to factor the sum and difference of two cubes. For example, we want to factor the binomial $x^3 - 8$, which is the difference of two cubes. (To see that it is the differrence of two cubes, notice that it can be written $x^3 - 2^3$.) We also want to factor $y^3 + 27$, which is the sum of two cubes. (To see this, notice that $y^3 + 27$ can be written as $y^3 + 3^3$.)

The formulas that allow us to factor the sum of two cubes and the difference of two cubes are not as simple as the formula for factoring the difference of two squares. Here is what they look like:

$$a^3 + b^3 = (a + b)(a^2 - ab + b^2)$$

$$a^3 - b^3 = (a - b)(a^2 + ab + b^2)$$

Let's begin our work with these two formulas by showing that they are true. To do so, we multiply out the right side of each formula.

EXAMPLE 1 Verify the two formulas.

SOLUTION We verify the formulas by multiplying the right sides and comparing the results with the left sides:

$$\begin{array}{rrrr} a^2 & - ab & + b^2 & \\ & a & + b & \\ \hline a^3 & - a^2b & + ab^2 & \\ & a^2b & - ab^2 & + b^3 \\ \hline a^3 & & & + b^3 \end{array} \qquad \begin{array}{rrrr} a^2 & + ab & + b^2 & \\ & a & - b & \\ \hline a^3 & + a^2b & + ab^2 & \\ & - a^2b & - ab^2 & - b^3 \\ \hline a^3 & & & - b^3 \end{array}$$

The first formula is correct. The second formula is correct.

Here are some examples that use the formulas for factoring the sum and difference of two cubes.

EXAMPLE 2 Factor $x^3 - 8$.

SOLUTION Since the two terms are prefect cubes, we write them as such and apply the formula:

$$\begin{aligned} x^3 - 8 &= x^3 - 2^3 \\ &= (x - 2)(x^2 + 2x + 2^2) \\ &= (x - 2)(x^2 + 2x + 4) \end{aligned}$$

EXAMPLE 3 Factor $y^3 + 27$.

SOLUTION Proceeding as we did in Example 2, we first write 27 as 3^3. Then, we apply the formula for factoring the sum of two cubes, which is $a^3 + b^3 = (a + b)(a^2 - ab + b^2)$:

$$\begin{aligned} y^3 + 27 &= y^3 + 3^3 \\ &= (y + 3)(y^2 - 3y + 3^2) \\ &= (y + 3)(y^2 - 3y + 9) \end{aligned}$$

Here are some examples using the formulas for factoring the sum and difference of two cubes.

EXAMPLE 4 Factor $64 + t^3$.

SOLUTION The first term is the cube of 4 and the second term is the cube of t. Therefore,

$$\begin{aligned} 64 + t^3 &= 4^3 + t^3 \\ &= (4 + t)(16 - 4t + t^2) \end{aligned}$$

EXAMPLE 5 Factor $27x^3 + 125y^3$.

SOLUTION Writing both terms as perfect cubes, we have

$$\begin{aligned} 27x^3 + 125y^3 &= (3x)^3 + (5y)^3 \\ &= (3x + 5y)(9x^2 - 15xy + 25y^2) \end{aligned}$$

EXAMPLE 6 Factor $a^3 - \frac{1}{8}$.

SOLUTION The first term is the cube of a, whereas the second term is the cube of $\frac{1}{2}$:

$$\begin{aligned} a^3 - \frac{1}{8} &= a^3 - \left(\frac{1}{2}\right)^3 \\ &= \left(a - \frac{1}{2}\right)\left(a^2 + \frac{1}{2}a + \frac{1}{4}\right) \end{aligned}$$

EXAMPLE 7 Factor $x^6 - y^6$.

SOLUTION We have a choice of how we want to write the two terms to begin. We can write the expression as the difference of two squares, $(x^3)^2 - (y^3)^2$, or as the difference of two cubes, $(x^2)^3 - (y^2)^3$. It is better to use the difference of two squares if we have a choice:

$$\begin{aligned} x^6 - y^6 &= (x^3)^2 - (y^3)^2 \\ &= (x^3 - y^3)(x^3 + y^3) \\ &= (x - y)(x^2 + xy + y^2)(x + y)(x^2 - xy + y^2) \end{aligned}$$

GETTING READY FOR CLASS

After reading through the preceding section, respond in your own words and in complete sentences.

A. How can you check your work when factoring?
B. Why are the numbers 8, 27, 64, and 125 used so frequently in this section?
C. List the cubes of the numbers 1 through 10.
D. How are you going to remember that the sum of two cubes factors, while the sum of two squares is prime?

Problem Set 6.5

Factor each of the following.

1. $x^3 - y^3$
2. $x^3 + y^3$
3. $a^3 + 8$
4. $a^3 - 8$
5. $27 + x^3$
6. $27 - x^3$
7. $y^3 - 1$
8. $y^3 + 1$
9. $y^3 - 64$
10. $y^3 + 64$
11. $125h^3 - t^3$
12. $t^3 + 125h^3$
13. $x^3 - 216$
14. $216 + x^3$
15. $2y^3 - 54$
16. $81 + 3y^3$
17. $2a^3 - 128b^3$
18. $128a^3 + 2b^3$
19. $2x^3 + 432y^3$
20. $432x^3 - 2y^3$
21. $10a^3 - 640b^3$
22. $640a^3 + 10b^3$
23. $10r^3 - 1{,}250$
24. $10r^3 + 1{,}250$
25. $64 + 27a^3$
26. $27 - 64a^3$
27. $8x^3 - 27y^3$
28. $27x^3 - 8y^3$
29. $t^3 + \frac{1}{27}$
30. $t^3 - \frac{1}{27}$
31. $27x^3 - \frac{1}{27}$
32. $8x^3 + \frac{1}{8}$
33. $64a^3 + 125b^3$
34. $125a^3 - 27b^3$
35. $\frac{1}{8}x^3 - \frac{1}{27}y^3$
36. $\frac{1}{27}x^3 + \frac{1}{8}y^3$
37. $a^6 - b^6$
38. $x^6 - 64y^6$
39. $64x^6 - y^6$
40. $x^6 - (3y)^6$
41. $x^6 - (5y)^6$
42. $(4x)^6 - (7y)^6$

Getting Ready for the Next Section

Multiply each of the following.

43. $2x^3(x + 2)(x - 2)$
44. $3x^2(x + 3)(x - 3)$
45. $3x^2(x - 3)^2$
46. $2x^2(x + 5)^2$
47. $y(y^2 + 25)$
48. $y^3(y^2 + 36)$
49. $(5a - 2)(3a + 1)$
50. $(3a - 4)(2a - 1)$
51. $4x^2(x - 5)(x + 2)$
52. $6x(x - 4)(x + 2)$
53. $2ab^3(b^2 - 4b + 1)$
54. $2a^3b(a^2 + 3a + 1)$

Factoring: A General Review 6.6

In this section we will review the different methods of factoring that we presented in the previous sections of the chapter. This section is important because it will give you an opportunity to factor a variety of polynomials. Prior to this section, the polynomials you worked with were grouped together according to the method used to factor them; that is, in Section 6.4 all the polynomials you factored were either the difference of two squares or perfect square trinomials. What usually happens in a situation like this is that you become proficient at factoring the kind of polynomial you are working with at the time but have trouble when given a variety of polynomials to factor.

We begin this section with a checklist that can be used in factoring polynomials of any type. When you have finished this section and the problem set that follows, you want to be proficient enough at factoring that the checklist is second nature to you.

HOW TO *Factor a polynomial*

Step 1: If the polynomial has a greatest common factor other than 1, then factor out the greatest common factor.

Step 2: If the polynomial has two terms (it is a binomial), then see if it is the difference of two squares. Remember, if it is the sum of two squares, it will not factor.

Step 3: If the polynomial has three terms (a trinomial), then either it is a perfect square trinomial, which will factor into the square of a binomial, or it is not a perfect square trinomial, in which case you use the trial and error method developed in Section 5.3.

Step 4: If the polynomial has more than three terms, try to factor it by grouping.

Step 5: As a final check, see if any of the factors you have written can be factored further. If you have overlooked a common factor, you can catch it here.

Here are some examples illustrating how we use the checklist.

EXAMPLE 1 Factor $2x^5 - 8x^3$.

SOLUTION First, we check to see if the greatest common factor is other than 1. Since the greatest common factor is $2x^3$, we begin by factoring it out. Once we have done so, we notice that the binomial that remains is the difference of two squares:

$$2x^5 - 8x^3 = 2x^3(x^2 - 4)$$ Factor out the greatest common factor, $2x^3$

$$= 2x^3(x + 2)(x - 2)$$ Factor the difference of two squares

Note that the greatest common factor $2x^3$ that we factored from each term in the first step of Example 1 remains as part of the answer to the problem; that is because it is one of the factors of the original binomial. Remember, the expression we end up with when factoring must be equal to the expression we start with. We can't just drop a factor and expect the resulting expression to equal the original expression.

EXAMPLE 2 Factor $3x^4 - 18x^3 + 27x^2$.

SOLUTION Step 1 is to factor out the greatest common factor, $3x^2$. After we have done so, we notice that the trinomial that remains is a perfect square trinomial, which will factor as the square of a binomial:

$$3x^4 - 18x^3 + 27x^2 = 3x^2(x^2 - 6x + 9) \quad \text{Factor out } 3x^2$$
$$= 3x^2(x - 3)^2 \quad x^2 - 6x + 9 \text{ is the square of } x - 3$$

EXAMPLE 3 Factor $y^3 + 25y$.

SOLUTION We begin by factoring out the y that is common to both terms. The binomial that remains after we have done so is the sum of two squares, which does not factor, so after the first step we are finished:

$$y^3 + 25y = y(y^2 + 25)$$

Factor out the greatest common factor, y; then notice that $y^2 + 25$ cannot factored further

EXAMPLE 4 Factor $6a^2 - 11a + 4$.

SOLUTION Here we have a trinomial that does not have a greatest common factor other than 1. Since it is not a perfect square trinomial, we factor it by trial and error; that is, we look for binomial factors of the product whose first terms is $6a^2$ and of the product whose last terms is 4. Then we look for the combination of these types of binomials whose product gives us a middle term of $-11a$. Without showing all the different possibilities, here is the answer:

$$6a^2 - 11a + 4 = (3a - 4)(2a - 1)$$

EXAMPLE 5 Factor $6x^3 - 12x^2 - 48x$.

SOLUTION This trinomial has a greatest common factor of $6x$. The trinomial that remains after the $6x$ has been factored from each term must be factored by trial and error:

$$6x^3 - 12x^2 - 48x = 6x(x^2 - 2x - 8)$$
$$= 6x(x - 4)(x + 2)$$

EXAMPLE 6 Factor $2ab^5 + 8ab^4 + 2ab^3$.

SOLUTION The greatest common factor is $2ab^3$. We begin by factoring it from each term. After that we find the trinomial that remains cannot be factored further:

$$2ab^5 + 8ab^4 + 2ab^3 = 2ab^3(b^2 + 4b + 1)$$

EXAMPLE 7 Factor $xy + 8x + 3y + 24$.

SOLUTION Since our polynomial has four terms, we try factoring by grouping:

$$xy + 8x + 3y + 24 = x(y + 8) + 3(y + 8)$$
$$= (y + 8)(x + 3)$$

GETTING READY FOR CLASS

Respond in your own words and in complete sentences.

A. What is the first step in factoring any polynomial?

B. If a polynomial has four terms, what method of factoring should you try?

C. If a polynomial has two terms, what method of factoring should you try?

D. What is the last step in factoring any polynomial?

Problem Set 6.6

Factor each of the following polynomials completely; that is, once you are finished factoring, none of the factors you obtain should be factorable. Also, note that the even-numbered problems are not necessarily similar to the odd-numbered problems that precede them in this problem set.

1. $x^2 - 81$ **2.** $x^2 - 18x + 81$ **3.** $x^2 + 2x - 15$

4. $15x^2 + 11x - 6$ **5.** $x^2 + 6x + 9$ **6.** $12x^2 - 11x + 2$

7. $y^2 - 10y + 25$ **8.** $21y^2 - 25y - 4$ **9.** $2a^3b + 6a^2b + 2ab$

10. $6a^2 - ab - 15b^2$ **11.** $x^2 + x + 1$ **12.** $2x^2 - 4x + 2$

13. $12a^2 - 75$ **14.** $18a^2 - 50$ **15.** $9x^2 - 12xy + 4y^2$

16. $x^3 - x^2$ **17.** $4x^3 + 16xy^2$ **18.** $16x^2 + 49y^2$

19. $2y^3 + 20y^2 + 50y$ **20.** $3y^2 - 9y - 30$

21. $a^6 + 4a^4b^2$ **22.** $5a^2 - 45b^2$

23. $xy + 3x + 4y + 12$ **24.** $xy + 7x + 6y + 42$

25. $x^4 - 16$ **26.** $x^4 - 81$

27. $xy - 5x + 2y - 10$ **28.** $xy - 7x + 3y - 21$

29. $5a^2 + 10ab + 5b^2$ **30.** $3a^3b^2 + 15a^2b^2 + 3ab^2$

31. $x^2 + 49$ **32.** $16 - x^4$

33. $3x^2 + 15xy + 18y^2$ **34.** $3x^2 + 27xy + 54y^2$

35. $2x^2 + 15x - 38$ **36.** $2x^2 + 7x - 85$

37. $100x^2 - 300x + 200$ **38.** $100x^2 - 400x + 300$

39. $x^2 - 64$ **40.** $9x^2 - 4$

41. $x^2 + 3x + ax + 3a$ **42.** $x^2 + 4x + bx + 4b$

43. $49a^7 - 9a^5$ **44.** $a^4 - 1$

45. $49x^2 + 9y^2$ **46.** $12x^4 - 62x^3 + 70x^2$

47. $25a^3 + 20a^2 + 3a$ **48.** $36a^4 - 100a^2$

49. $xa - xb + ay - by$ **50.** $xy - bx + ay - ab$

51. $48a^4b - 3a^2b$ **52.** $18a^4b^2 - 12a^3b^3 + 8a^2b^4$

53. $20x^4 - 45x^2$ **54.** $16x^3 + 16x^2 + 3x$

55. $3x^2 + 35xy - 82y^2$ **56.** $3x^2 + 37xy - 86y^2$

57. $16x^5 - 44x^4 + 30x^3$ **58.** $16x^2 + 16x - 1$

59. $2x^2 + 2ax + 3x + 3a$ **60.** $2x^2 + 2ax + 5x + 5a$

61. $y^4 - 1$ **62.** $25y^7 - 16y^5$

63. $12x^4y^2 + 36x^3y^3 + 27x^2y^4$ **64.** $16x^3y^2 - 4xy^2$

Getting Ready for the Next Section

Solve each equation.

65. $3x - 6 = 9$ **66.** $5x - 1 = 14$ **67.** $2x + 3 = 0$

68. $4x - 5 = 0$ **69.** $4x + 3 = 0$ **70.** $3x - 1 = 0$

Solving Equations by Factoring 6.7

In this section we will use the methods of factoring developed in previous sections, along with a special property of 0, to solve quadratic equations.

DEFINITION *quadratic equation*

Any equation that can be put in the form $ax^2 + bx + c = 0$, where a, b, and c are real numbers ($a \neq 0$), is called a ***quadratic equation.*** The equation $ax^2 + bx + c = 0$ is called ***standard form*** for a quadratic equation:

$$\underset{\text{an } x^2 \text{ term}}{a(\text{variable})^2} + \underset{\text{an } x \text{ term}}{b(\text{variable})} + \underset{\text{and a constant term}}{c(\text{absence of the variable})} = 0$$

The number 0 has a special property. If we multiply two numbers and the product is 0, then one or both of the original two numbers must be 0. In symbols, this property looks like this.

PROPERTY *Zero-Factor Property*

Let a and b represent real numbers. If $a \cdot b = 0$, then $a = 0$ or $b = 0$.

Suppose we want to solve the quadratic equation $x^2 + 5x + 6 = 0$. We can factor the left side into $(x + 2)(x + 3)$. Then we have:

$$x^2 + 5x + 6 = 0$$
$$(x + 2)(x + 3) = 0$$

Now, $(x + 2)$ and $(x + 3)$ both represent real numbers. Their product is 0; therefore, either $(x + 3)$ is 0 or $(x + 2)$ is 0. Either way we have a solution to our equation. We use the property of 0 stated to finish the problem:

$$\begin{aligned} x^2 + 5x + 6 &= 0 \\ (x + 2)(x + 3) &= 0 \\ x + 2 = 0 \quad &\text{or} \quad x + 3 = 0 \\ x = -2 \quad &\text{or} \quad x = -3 \end{aligned}$$

Our solution set is $\{-2, -3\}$. Our equation has two solutions. To check our solutions we have to check each one separately to see that they both produce a true statement when used in place of the variable:

When	$x = -3$
the equation	$x^2 + 5x + 6 = 0$
becomes	$(-3)^2 + 5(-3) + 6 \stackrel{?}{=} 0$
	$9 + (-15) + 6 = 0$
	$0 = 0$

$$\begin{array}{ll} \text{When} & x = -2 \\ \text{the equation} & x^2 + 5x + 6 = 0 \\ \text{becomes} & (-2)^2 + 5(-2) + 6 \stackrel{?}{=} 0 \\ & 4 + (-10) + 6 = 0 \\ & 0 = 0 \end{array}$$

We have solved a quadratic equation by replacing it with two linear equations in one variable.

HOW TO ***Strategy for solving a quadratic equation by factoring***

Step 1: Put the equation in standard form; that is, 0 on one side and decreasing powers of the variable on the other.
Step 2: Factor completely.
Step 3: Use the zero-factor property to set each variable factor from step 2 to 0.
Step 4: Solve each equation produced in step 3.
Step 5: Check each solution, if necessary.

EXAMPLE 1 Solve the equation $2x^2 - 5x = 12$.

SOLUTION

Step 1: Begin by adding -12 to both sides, so the equation is in standard form:

$$2x^2 - 5x = 12$$
$$2x^2 - 5x - 12 = 0$$

Step 2: Factor the left side completely:

$$(2x + 3)(x - 4) = 0$$

Step 3: Set each factor to 0:

$$2x + 3 = 0 \quad \text{or} \quad x - 4 = 0$$

Step 4: Solve each of the equations from step 3:

$$\begin{array}{ll} 2x + 3 = 0 & x - 4 = 0 \\ 2x = -3 & x = 4 \\ x = -\frac{3}{2} & \end{array}$$

Step 5: Substitute each solution into $2x^2 - 5x = 12$ to check:

$$\begin{array}{ll} \text{Check: } -\frac{3}{2} & \text{Check: } 4 \\ 2\left(-\frac{3}{2}\right)^2 - 5\left(-\frac{3}{2}\right) \stackrel{?}{=} 12 & 2(4)^2 - 5(4) \stackrel{?}{=} 12 \\ 2\left(\frac{9}{4}\right) + 5\left(\frac{3}{2}\right) = 12 & 2(16) - 20 = 12 \\ \frac{9}{2} + \frac{15}{2} = 12 & 32 - 20 = 12 \\ \frac{24}{2} = 12 & 12 = 12 \\ 12 = 12 & \end{array}$$

EXAMPLE 2 Solve for a: $16a^2 - 25 = 0$.

SOLUTION The equation is already in standard form:

$$16a^2 - 25 = 0$$

$$(4a - 5)(4a + 5) = 0 \quad \text{Factor the left side}$$

$$4a - 5 = 0 \quad \text{or} \quad 4a + 5 = 0 \quad \text{Set each factor to 0}$$

$$4a = 5 \qquad 4a = -5 \quad \text{Solve the resulting equations}$$

$$a = \frac{5}{4} \qquad a = -\frac{5}{4}$$

The solutions are $\frac{5}{4}$ and $-\frac{5}{4}$.

EXAMPLE 3 Solve $4x^2 = 8x$.

SOLUTION We begin by adding $-8x$ to each side of the equation to put it in standard form. Then we factor the left side of the equation by factoring out the greatest common factor.

$$4x^2 = 8x$$

$$4x^2 - 8x = 0 \quad \text{Add } -8x \text{ to each side}$$

$$4x(x - 2) = 0 \quad \text{Factor the left side}$$

$$4x = 0 \quad \text{or} \quad x - 2 = 0 \quad \text{Set each factor to 0}$$

$$x = 0 \quad \text{or} \quad x = 2 \quad \text{Solve the resulting equations}$$

The solutions are 0 and 2.

EXAMPLE 4 Solve $x(2x + 3) = 44$.

SOLUTION We must multiply out the left side first and then put the equation in standard form:

$$x(2x + 3) = 44$$

$$2x^2 + 3x = 44 \quad \text{Multiply out the left side}$$

$$2x^2 + 3x - 44 = 0 \quad \text{Add } -44 \text{ to each side}$$

$$(2x + 11)(x - 4) = 0 \quad \text{Factor the left side}$$

$$2x + 11 = 0 \quad \text{or} \quad x - 4 = 0 \quad \text{Set each factor to 0}$$

$$2x = -11 \quad \text{or} \quad x = 4 \quad \text{Solve the resulting equations}$$

$$x = -\frac{11}{2}$$

The two solutions are $-\frac{11}{2}$ and 4.

EXAMPLE 5 Solve for x: $5^2 = x^2 + (x + 1)^2$.

SOLUTION Before we can put this equation in standard form we must square the binomial. Remember, to square a binomial, we use the formula $(a + b)^2 = a^2 + 2ab + b^2$:

$5^2 = x^2 + (x + 1)^2$	
$25 = x^2 + x^2 + 2x + 1$	Expand 5^2 and $(x + 1)^2$
$25 = 2x^2 + 2x + 1$	Simplify the right side
$0 = 2x^2 + 2x - 24$	Add -25 to each side
$0 = 2(x^2 + x - 12)$	Begin factoring
$0 = 2(x + 4)(x - 3)$	Factor completely
$x + 4 = 0$ or $x - 3 = 0$	Set each variable factor to 0
$x = -4$ or $x = 3$	

Note, in the second to the last line, that we do not set 2 equal to 0. That is because 2 can never be 0. It is always 2. We only use the zero-factor property to set variable factors to 0 because they are the only factors that can possibly be 0.

Also notice that it makes no difference which side of the equation is 0 when we write the equation in standard form.

Although the equation in the next example is not a quadratic equation, it can be solved by the method shown in the first five examples.

EXAMPLE 6 Solve $24x^3 = -10x^2 + 6x$ for x.

SOLUTION First, we write the equation in standard form:

$24x^3 + 10x^2 - 6x = 0$	Standard form
$2x(12x^2 + 5x - 3) = 0$	Factor out $2x$
$2x(3x - 1)(4x + 3) = 0$	Factor remaining trinomial
$2x = 0$ or $3x - 1 = 0$ or $4x + 3 = 0$	Set factors to 0
$x = 0$ or $x = \frac{1}{3}$ or $x = -\frac{3}{4}$	Solutions

GETTING READY FOR CLASS

After reading through the preceding section, respond in your own words and in complete sentences.

A. When is an equation in standard form?
B. What is the first step in solving an equation by factoring?
C. Describe the zero-factor property in your own words.
D. Describe how you would solve the equation $2x^2 - 5x = 12$.

The following equations are already in factored form. Use the special zero factor property to set the factors to 0 and solve.

1. $(x + 2)(x - 1) = 0$
2. $(x + 3)(x + 2) = 0$
3. $(a - 4)(a - 5) = 0$
4. $(a + 6)(a - 1) = 0$
5. $x(x + 1)(x - 3) = 0$
6. $x(2x + 1)(x - 5) = 0$
7. $(3x + 2)(2x + 3) = 0$
8. $(4x - 5)(x - 6) = 0$
9. $m(3m + 4)(3m - 4) = 0$
10. $m(2m - 5)(3m - 1) = 0$
11. $2y(3y + 1)(5y + 3) = 0$
12. $3y(2y - 3)(3y - 4) = 0$

Solve the following equations.

13. $x^2 + 3x + 2 = 0$
14. $x^2 - x - 6 = 0$
15. $x^2 - 9x + 20 = 0$
16. $x^2 + 2x - 3 = 0$
17. $a^2 - 2a - 24 = 0$
18. $a^2 - 11a + 30 = 0$
19. $100x^2 - 500x + 600 = 0$
20. $100x^2 - 300x + 200 = 0$
21. $x^2 = -6x - 9$
22. $x^2 = 10x - 25$
23. $a^2 - 16 = 0$
24. $a^2 - 36 = 0$
25. $2x^2 + 5x - 12 = 0$
26. $3x^2 + 14x - 5 = 0$
27. $9x^2 + 12x + 4 = 0$
28. $12x^2 - 24x + 9 = 0$
29. $a^2 + 25 = 10a$
30. $a^2 + 16 = 8a$
31. $2x^2 = 3x + 20$
32. $6x^2 = x + 2$
33. $3m^2 = 20 - 7m$
34. $2m^2 = -18 + 15m$
35. $4x^2 - 49 = 0$
36. $16x^2 - 25 = 0$
37. $x^2 + 6x = 0$
38. $x^2 - 8x = 0$
39. $x^2 - 3x = 0$
40. $x^2 + 5x = 0$
41. $2x^2 = 8x$
42. $2x^2 = 10x$
43. $3x^2 = 15x$
44. $5x^2 = 15x$
45. $1{,}400 = 400 + 700x - 100x^2$
46. $2{,}700 = 700 + 900x - 100x^2$
47. $6x^2 = -5x + 4$
48. $9x^2 = 12x - 4$
49. $x(2x - 3) = 20$
50. $x(3x - 5) = 12$
51. $t(t + 2) = 80$
52. $t(t + 2) = 99$
53. $4{,}000 = (1{,}300 - 100p)p$
54. $3{,}200 = (1{,}200 - 100p)p$
55. $x(14 - x) = 48$
56. $x(12 - x) = 32$
57. $(x + 5)^2 = 2x + 9$
58. $(x + 7)^2 = 2x + 13$
59. $(y - 6)^2 = y - 4$
60. $(y + 4)^2 = y + 6$
61. $10^2 = (x + 2)^2 + x^2$
62. $15^2 = (x + 3)^2 + x^2$
63. $2x^3 + 11x^2 + 12x = 0$
64. $3x^3 + 17x^2 + 10x = 0$
65. $4y^3 - 2y^2 - 30y = 0$
66. $9y^3 + 6y^2 - 24y = 0$
67. $8x^3 + 16x^2 = 10x$
68. $24x^3 - 22x^2 = -4x$

69. $20a^3 = -18a^2 + 18a$

70. $12a^3 = -2a^2 + 10a$

71. $16t^2 - 32t + 12 = 0$

72. $16t^2 - 64t + 48 = 0$

Simplify each side as much as possible, then solve the equation.

73. $(a - 5)(a + 4) = -2a$

74. $(a + 2)(a - 3) = -2a$

75. $3x(x + 1) - 2x(x - 5) = -42$

76. $4x(x - 2) - 3x(x - 4) = -3$

77. $2x(x + 3) = x(x + 2) - 3$

78. $3x(x - 3) = 2x(x - 4) + 6$

79. $a(a - 3) + 6 = 2a$

80. $a(a - 4) + 8 = 2a$

81. $15(x + 20) + 15x = 2x(x + 20)$

82. $15(x + 8) + 15x = 2x(x + 8)$

83. $15 = a(a + 2)$

84. $6 = a(a - 5)$

Use factoring by grouping to solve the following equations.

85. $x^3 + 3x^2 - 4x - 12 = 0$

86. $x^3 + 5x^2 - 9x - 45 = 0$

87. $x^3 + x^2 - 16x - 16 = 0$

88. $4x^3 + 12x^2 - 9x - 27 = 0$

Getting Ready for the Next Section

Write each sentence as an algebraic equation.

89. The product of two consecutive integers is 72.

90. The product of two consecutive even integers is 80.

91. The product of two consecutive odd integers is 99.

92. The product of two consecutive odd integers is 63.

93. The product of two consecutive even integers is 10 less than 5 times their sum.

94. The product of two consecutive odd integers is 1 less than 4 times their sum.

The following word problems are taken from the book *Academic Algebra*, written by William J. Milne and published by the American Book Company in 1901. Solve each problem.

95. Cost of a Bicycle and a Suit A bicycle and a suit cost \$90. How much did each cost, if the bicycle cost 5 times as much as the suit?

96. Cost of a Cow and a Calf A man bought a cow and a calf for \$36, paying 8 times as much for the cow as for the calf. What was the cost of each?

97. Cost of a House and a Lot A house and a lot cost \$3,000. If the house cost 4 times as much as the lot, what was the cost of each?

98. Daily Wages A plumber and two helpers together earned \$7.50 per day. How much did each earn per day, if the plumber earned 4 times as much as each helper?

Applications 6.8

In this section we will look at some application problems, the solutions to which require solving a quadratic equation. We will also introduce the Pythagorean theorem, one of the oldest theorems in the history of mathematics. The person whose name we associate with the theorem, Pythagoras (of Samos), was a Greek philosopher and mathematician who lived from about 560 B.C. to 480 B.C. According to the British philosopher Bertrand Russell, Pythagoras was "intellectually one of the most important men that ever lived."

Also in this section, the solutions to the examples show only the essential steps from our Blueprint for Problem Solving. Recall that step 1 is done mentally; we read the problem and mentally list the items that are known and the items that are unknown. This is an essential part of problem solving. However, now that you have had experience with application problems, you are doing step 1 automatically.

Number Problems

EXAMPLE 1 The product of two consecutive odd integers is 63. Find the integers.

SOLUTION Let $x =$ the first odd integer; then $x + 2 =$ the second odd integer. An equation that describes the situation is:

$$x(x + 2) = 63 \qquad \text{Their product is 63}$$

We solve the equation:

$$x(x + 2) = 63$$
$$x^2 + 2x = 63$$
$$x^2 + 2x - 63 = 0$$
$$(x - 7)(x + 9) = 0$$
$$x - 7 = 0 \quad \text{or} \quad x + 9 = 0$$
$$x = 7 \quad \text{or} \quad x = -9$$

If the first odd integer is 7, the next odd integer is $7 + 2 = 9$. If the first odd integer is -9, the next consecutive odd integer is $-9 + 2 = -7$. We have two pairs of consecutive odd integers that are solutions. They are 7, 9 and -9, -7.

We check to see that their products are 63:

$$7(9) = 63$$
$$-7(-9) = 63$$

Suppose we know that the sum of two numbers is 50. We want to find a way to represent each number using only one variable. If we let x represent one of the two numbers, how can we represent the other? Let's suppose for a moment that x turns out to be 30. Then the other number will be 20, because their sum is 50; that is, if two numbers add up to 50 and one of them is 30, then the other must be $50 - 30 = 20$. Generalizing this to any number x, we see that if two numbers have a sum of 50 and one of the numbers is x, then the other must be $50 - x$. The table that follows shows some additional examples.

If two numbers have a sum of	and one of them is	then the other must be
50	x	$50 - x$
100	x	$100 - x$
10	y	$10 - y$
12	n	$12 - n$

Now, let's look at an example that uses this idea.

EXAMPLE 2 The sum of two numbers is 13. Their product is 40. Find the numbers.

SOLUTION If we let x represent one of the numbers, then $13 - x$ must be the other number because their sum is 13. Since their product is 40, we can write:

$$x(13 - x) = 40 \quad \text{The product of the two numbers is 40}$$

$$13x - x^2 = 40 \quad \text{Multiply the left side}$$

$$x^2 - 13x = -40 \quad \text{Multiply both sides by } -1 \text{ and reverse the order of the terms on the left side}$$

$$x^2 - 13x + 40 = 0 \quad \text{Add 40 to each side}$$

$$(x - 8)(x - 5) = 0 \quad \text{Factor the left side}$$

$$x - 8 = 0 \quad \text{or} \quad x - 5 = 0$$

$$x = 8 \quad \text{or} \quad x = 5$$

The two solutions are 8 and 5. If x is 8, then the other number is $13 - x = 13 - 8 = 5$. Likewise, if x is 5, the other number is $13 - x = 13 - 5 = 8$. Therefore, the two numbers we are looking for are 8 and 5. Their sum is 13 and their product is 40.

Geometry Problems

Many word problems dealing with area can best be described algebraically by quadratic equations.

EXAMPLE 3 The length of a rectangle is 3 more than twice the width. The area is 44 square inches. Find the dimensions (find the length and width).

SOLUTION As shown in Figure 1, let $x =$ the width of the rectangle. Then $2x + 3 =$ the length of the rectangle because the length is three more than twice the width.

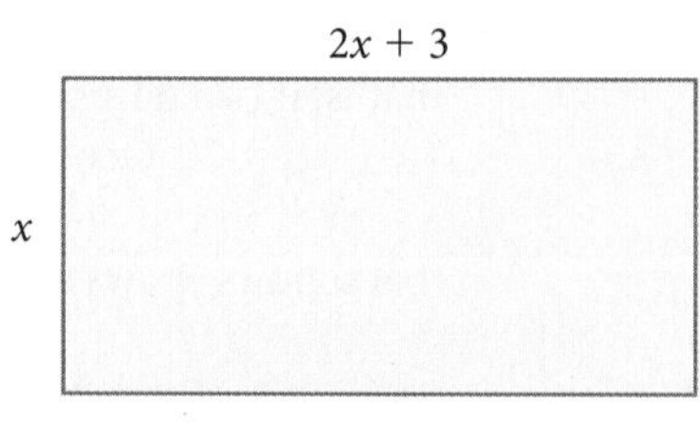

FIGURE 1

Since the area is 44 square inches, an equation that describes the situation is

$$x(2x + 3) = 44 \qquad \text{Length} \cdot \text{width} = \text{area}$$

We now solve the equation:

$$x(2x + 3) = 44$$
$$2x^2 + 3x = 44$$
$$2x^2 + 3x - 44 = 0$$
$$(2x + 11)(x - 4) = 0$$
$$2x + 11 = 0 \quad \text{or} \quad x - 4 = 0$$
$$x = -\frac{11}{2} \quad \text{or} \quad x = 4$$

The solution $x = -\frac{11}{2}$ cannot be used since length and width are always given in positive units. The width is 4. The length is 3 more than twice the width or $2(4) + 3 = 11$.

$$\text{Width} = 4 \text{ inches}$$
$$\text{Length} = 11 \text{ inches}$$

The solutions check in the original problem since $4(11) = 44$.

EXAMPLE 4 The numerical value of the area of a square is twice its perimeter. What is the length of its side?

SOLUTION As shown in Figure 2, let x = the length of its side. Then x^2 = the area of the square and $4x$ = the perimeter of the square:

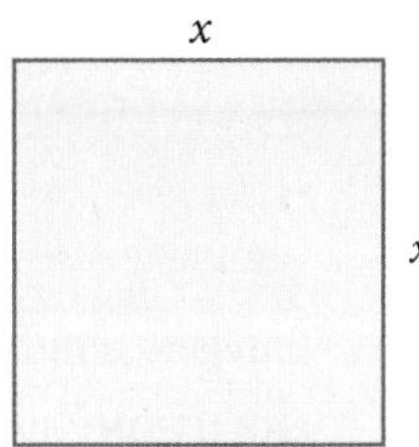

FIGURE 2

An equation that describes the situation is

$$x^2 = 2(4x) \qquad \text{The area is 2 times the perimeter}$$
$$x^2 = 8x$$
$$x^2 - 8x = 0$$
$$x(x - 8) = 0$$
$$x = 0 \quad \text{or} \quad x = 8$$

Since $x = 0$ does not make sense in our original problem, we use $x = 8$. If the side has length 8, then the perimeter is $4(8) = 32$ and the area is $8^2 = 64$. Since 64 is twice 32, our solution is correct.

FACTS FROM GEOMETRY *The Pythagorean Theorem*

Next, we will work some problems involving the Pythagorean theorem, which we mentioned in the introduction to this section. It may interest you to know that Pythagoras formed a secret society around the year 540 B.C. Known as the Pythagoreans, members kept no written record of their work; everything was handed down by spoken word. They influenced not only mathematics, but religion, science, medicine, and music as well. Among other things, they discovered the correlation between musical notes and the reciprocals of counting numbers, $\frac{1}{2}, \frac{1}{3}, \frac{1}{4}$, and so on. In their daily lives, they followed strict dietary and moral rules to achieve a higher rank in future lives.

PROPERTY *Pythagorean Theorem*

In any right triangle (Figure 3), the square of the longer side (called the hypotenuse) is equal to the sum of the squares of the other two sides (called legs).

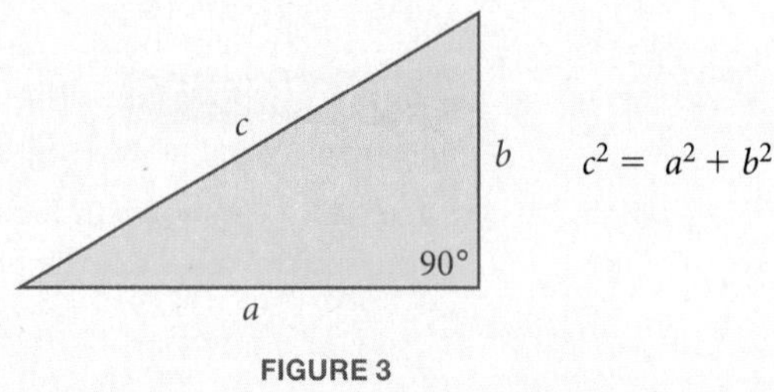

FIGURE 3

EXAMPLE 5 The three sides of a right triangle are three consecutive integers. Find the lengths of the three sides.

SOLUTION Let x = the first integer (shortest side)

then $x + 1$ = the next consecutive integer

and $x + 2$ = the last consecutive integer (longest side)

A diagram of the triangle is shown in Figure 4.

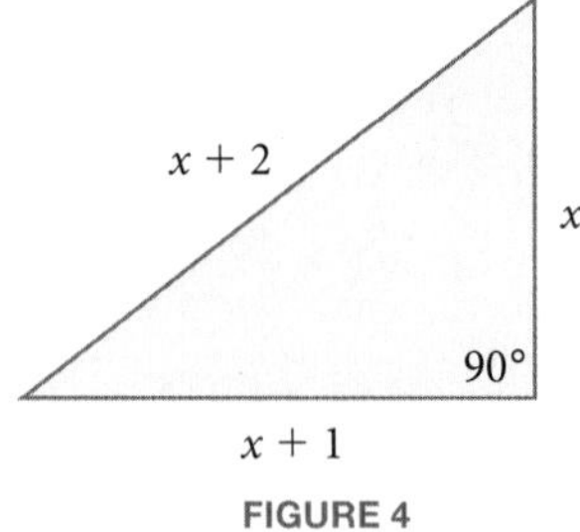

FIGURE 4

The Pythagorean theorem tells us that the square of the longest side $(x + 2)^2$ is equal to the sum of the squares of the two shorter sides, $(x + 1)^2 + x^2$. Here is the equation:

$$(x + 2)^2 = (x + 1)^2 + x^2$$

$$x^2 + 4x + 4 = x^2 + 2x + 1 + x^2 \quad \text{Expand squares}$$

$$x^2 - 2x - 3 = 0 \quad \text{Standard form}$$

$$(x - 3)(x + 1) = 0 \quad \text{Factor}$$

$$x - 3 = 0 \quad \text{or} \quad x + 1 = 0 \quad \text{Set factors to 0}$$

$$x = 3 \quad \text{or} \quad x = -1$$

Since a triangle cannot have a side with a negative number for its length, we must not use -1 for a solution to our original problem; therefore, the shortest side is 3. The other two sides are the next two consecutive integers, 4 and 5.

EXAMPLE 6 The hypotenuse of a right triangle is 5 inches, and the lengths of the two legs (the other two sides) are given by two consecutive integers. Find the lengths of the two legs.

SOLUTION If we let $x =$ the length of the shortest side, then the other side must be $x + 1$. A diagram of the triangle is shown in Figure 5.

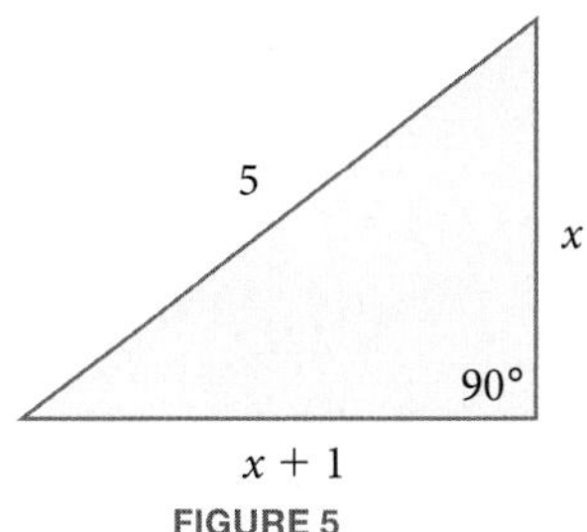

FIGURE 5

The Pythagorean theorem tells us that the square of the longest side, 5^2, is equal to the sum of the squares of the two shorter sides, $x^2 + (x + 1)^2$. Here is the equation:

$$5^2 = x^2 + (x + 1)^2 \quad \text{Pythagorean theorem}$$

$$25 = x^2 + x^2 + 2x + 1 \quad \text{Expand } 5^2 \text{ and } (x + 1)^2$$

$$25 = 2x^2 + 2x + 1 \quad \text{Simplify the right side}$$

$$0 = 2x^2 + 2x - 24 \quad \text{Add } -25 \text{ to each side}$$

$$0 = 2(x^2 + x - 12) \quad \text{Begin factoring}$$

$$0 = 2(x + 4)(x - 3) \quad \text{Factor completely}$$

$$x + 4 = 0 \quad \text{or} \quad x - 3 = 0 \quad \text{Set variable factors to 0}$$

$$x = -4 \quad \text{or} \quad x = 3$$

Since a triangle cannot have a side with a negative number for its length, we cannot use -4; therefore, the shortest side must be 3 inches. The next side is $x + 1 = 3 + 1 = 4$ inches. Since the hypotenuse is 5, we can check our solutions with the Pythagorean theorem as shown in Figure 6.

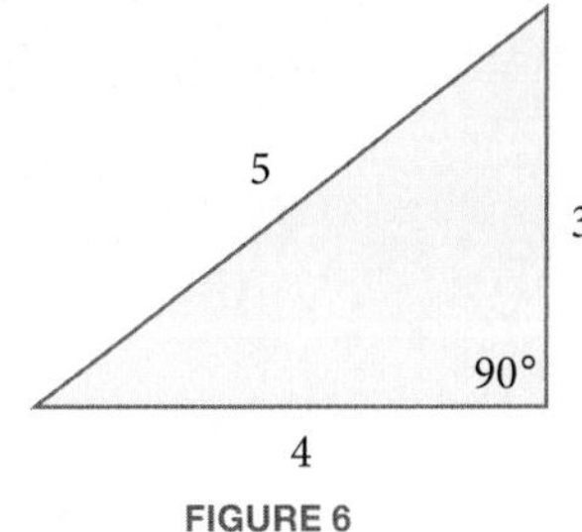

FIGURE 6

EXAMPLE 7 A company can manufacture x hundred items for a total cost of $C = 300 + 500x - 100x^2$. How many items were manufactured if the total cost is \$900?

SOLUTION We are looking for x when C is 900. We begin by substituting 900 for C in the cost equation. Then we solve for x:

$$\begin{array}{ll} \text{When} & C = 900 \\ \text{the equation} & C = 300 + 500x - 100x^2 \\ \text{becomes} & 900 = 300 + 500x - 100x^2 \end{array}$$

We can write this equation in standard form by adding -300, $-500x$, and $100x^2$ to each side. The result looks like this:

$$100x^2 - 500x + 600 = 0$$

$$100(x^2 - 5x + 6) = 0 \quad \text{Begin factoring}$$

$$100(x - 2)(x - 3) = 0 \quad \text{Factor completely}$$

$$x - 2 = 0 \quad \text{or} \quad x - 3 = 0 \quad \text{Set variable factors to 0}$$

$$x = 2 \quad \text{or} \quad x = 3$$

Our solutions are 2 and 3, which means that the company can manufacture 2 hundred items or 3 hundred items for a total cost of \$900.

EXAMPLE 8 A manufacturer of small portable radios knows that the number of radios she can sell each week is related to the price of the radios by the equation $x = 1{,}300 - 100p$ (x is the number of radios and p is the price per radio). What price should she charge for the radios to have a weekly revenue of \$4,000?

SOLUTION First, we must find the revenue equation. The equation for total revenue is $R = xp$, where x is the number of units sold and p is the price per unit. Since we want R in terms of p, we substitute $1{,}300 - 100p$ for x in the equation $R = xp$:

$$\begin{array}{ll} \text{If} & R = xp \\ \text{and} & x = 1{,}300 - 100p \\ \text{then} & R = (1{,}300 - 100p)p \end{array}$$

We want to find p when R is 4,000. Substituting 4,000 for R in the equation gives us:

$$4{,}000 = (1{,}300 - 100p)p$$

If we multiply out the right side, we have:

$$4{,}000 = 1{,}300p - 100p^2$$

To write this equation in standard form, we add $100p^2$ and $-1{,}300p$ to each side:

$$100p^2 - 1{,}300p + 4{,}000 = 0 \quad \text{Add } 100p^2 \text{ and } -1{,}300p$$

$$100(p^2 - 13p + 40) = 0 \quad \text{Begin factoring}$$

$$100(p - 5)(p - 8) = 0 \quad \text{Factor completely}$$

$$p - 5 = 0 \quad \text{or} \quad p - 8 = 0 \quad \text{Set variable factors to 0}$$

$$p = 5 \quad \text{or} \quad p = 8$$

If she sells the radios for \$5 each or for \$8 each, she will have a weekly revenue of \$4,000.

GETTING READY FOR CLASS

After reading through the preceding section, respond in your own words and in complete sentences.

A. What are consecutive integers?
B. Explain the Pythagorean theorem in words.
C. Write an application problem for which the solution depends on solving the equation $x(x + 1) = 12$.
D. Write an application problem for which the solution depends on solving the equation $x(2x - 3) = 40$.

SPOTLIGHT ON SUCCESS *Student Instructor Aaron*

Sometimes you have to take a step back in order to get a running start forward.
—Anonymous

As a high school senior I was encouraged to go to college immediately after graduating. I earned good grades in high school and I knew that I would have a pretty good group of schools to pick from. Even though I felt like "more school" was not quite what I wanted, the counselors had so much faith and had done this process so many times that it was almost too easy to get the applications out. I sent out applications to schools I knew I could get into and a "dream school."

One night in my email inbox there was a letter of acceptance from my dream school. There was just one problem with getting into this school. It was going to be difficult and I still had senioritis. Going into my first quarter of college was as exciting and difficult as I knew it would be. But after my first quarter I could see that this was not the time for me to be here. I was interested in the subject matter but I could not find my motivating purpose like I had in high school. Instead of dropping out completely, I decided a community college would be a good way for me to stay on track. Without necessarily knowing my direction, I could take the general education classes and get those out of the way while figuring out exactly what and where I felt a good place for me to be.

Now I know what I want to go to school for and the next time I walk onto a four year campus it will be on my terms with my reasons for being there driving me to succeed. I encourage everyone to continue school after high school, even if you have no clue as to what you want to study. There are always stepping stones, like community colleges, that can help you get a clearer picture of what you want to strive for.

Problem Set 6.8

Solve the following word problems. Be sure to show the equation used.

Number Problems

1. The product of two consecutive even integers is 80. Find the two integers.
2. The product of two consecutive integers is 72. Find the two integers.
3. The product of two consecutive odd integers is 99. Find the two integers.
4. The product of two consecutive integers is 132. Find the two integers.
5. The product of two consecutive even integers is 10 less than 5 times their sum. Find the two integers.
6. The product of two consecutive odd integers is 1 less than 4 times their sum. Find the two integers.
7. The sum of two numbers is 14. Their product is 48. Find the numbers.
8. The sum of two numbers is 12. Their product is 32. Find the numbers.
9. One number is 2 more than 5 times another. Their product is 24. Find the numbers.
10. One number is 1 more than twice another. Their product is 55. Find the numbers.
11. One number is 4 times another. Their product is 4 times their sum. Find the numbers.
12. One number is 2 more than twice another. Their product is 2 more than twice their sum. Find the numbers.

Geometry Problems

13. The length of a rectangle is 1 more than the width. The area is 12 square inches. Find the dimensions.
14. The length of a rectangle is 3 more than twice the width. The area is 44 square inches. Find the dimensions.
15. The height of a triangle is twice the base. The area is 9 square inches. Find the base.
16. The height of a triangle is 2 more than twice the base. The area is 20 square feet. Find the base.
17. The hypotenuse of a right triangle is 10 inches. The lengths of the two legs are given by two consecutive even integers. Find the lengths of the two legs.
18. The hypotenuse of a right triangle is 15 inches. One of the legs is 3 inches more than the other. Find the lengths of the two legs.
19. The shorter leg of a right triangle is 5 meters. The hypotenuse is 1 meter longer than the longer leg. Find the length of the longer leg.
20. The shorter leg of a right triangle is 12 yards. If the hypotenuse is 20 yards, how long is the other leg?

Business Problems

21. A company can manufacture x hundred items for a total cost of $C = 400 + 700x - 100x^2$. Find x if the total cost is $1,400.

22. If the total cost C of manufacturing x hundred items is given by the equation $C = 700 + 900x - 100x^2$, find x when C is $2,700.

23. The relationship between the number of calculators a company sells per week, x, and the price p of each calculator is given by the equation $x = 1{,}700 - 100p$. At what price should the calculators be sold if the weekly revenue is to be $7,000?

24. The relationship between the number of pencil sharpeners a company can sell each week, x, and the price p of each sharpener is given by the equation $x = 1{,}800 - 100p$. At what price should the sharpeners be sold if the weekly revenue is to be $7,200?

25. Pythagorean Theorem A 13-foot ladder is placed so that it reaches to a point on the wall that is 2 feet higher than twice the distance from the base of the wall to the base of the ladder.

a. How far from the wall is the base of the ladder?

b. How high does the ladder reach?

26. Constructing a Box I have a piece of cardboard that is twice as long as it is wide. If I cut a 2-inch by 2-inch square from each corner and fold up the resulting flaps, I get a box with a volume of 32 cubic inches. What are the dimensions of the cardboard.

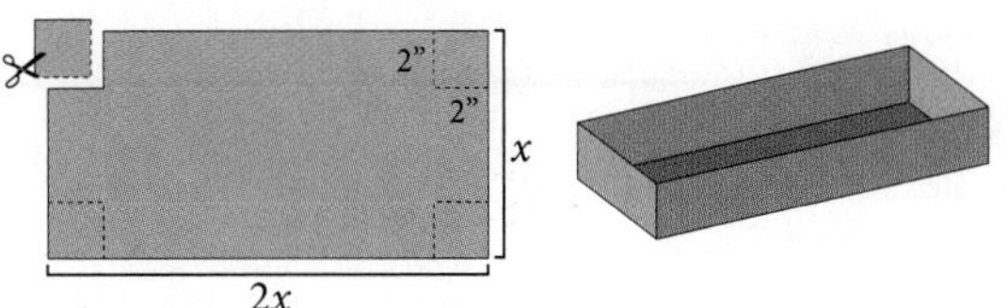

27. Projectile Motion A gun fires a bullet almost straight up from the edge of a 100-foot cliff. If the bullet leaves the gun with a speed of 396 feet per second, its height at time t is given by $h(t) = -16t^2 + 396t + 100$, measured from the ground below the cliff.

a. When will the bullet land on the ground below the cliff? (*Hint:* What is its height when it lands? Remember that we are measuring from the ground below, not from the cliff.)

b. Make a table showing the bullet's height every five seconds, from the time it is fired ($t = 0$) to the time it lands. (*Note:* It is faster to substitute into the factored form.)

28. Height of a Projectile If a rocket is fired vertically into the air with a speed of 240 feet per second, its height at time t seconds is given by $h(t) = -16t^2 + 240t$.

At what time(s) will the rocket be the following number of feet above the ground?

a. 704 feet

b. 896 feet

c. Why do parts **a.** and **b.** each have two answers?

d. How long will the rocket be in the air? (*Hint:* How high is it when it hits the ground?)

e. When the equation for part **d.** is solved, one of the answers is $t = 0$ seconds. What does this represent?

Maintaining Your Skills

Simplify each expression. (Write all answers with positive exponents only.)

29. $(5x^3)^2(2x^6)^3$

30. 2^{-3}

31. $\dfrac{x^4}{x^{-3}}$

32. $\dfrac{(20x^2y^3)(5x^4y)}{(2xy^5)(10x^2y^3)}$

33. $(2 \times 10^{-4})(4 \times 10^5)$

34. $\dfrac{9 \times 10^{-3}}{3 \times 10^{-2}}$

35. $20ab^2 - 16ab^2 + 6ab^2$

36. Subtract $6x^2 - 5x - 7$ from $9x^2 + 3x - 2$.

Multiply.

37. $2x^2(3x^2 + 3x - 1)$

38. $(2x + 3)(5x - 2)$

39. $(3y - 5)^2$

40. $(a - 4)(a^2 + 4a + 16)$

41. $(2a^2 + 7)(2a^2 - 7)$

42. Divide $15x^{10} - 10x^8 + 25x^6$ by $5x^6$.

Chapter 6 Summary

Greatest Common Factor [6.1]

EXAMPLES

1. $8x^4 - 10x^3 + 6x^2$
$= 2x^2 \cdot 4x^2 - 2x^2 \cdot 5x + 2x^2 \cdot 3$
$= 2x^2(4x^2 - 5x + 3)$

The largest monomial that divides each term of a polynomial is called the greatest common factor for that polynomial. We begin all factoring by factoring out the greatest common factor.

Factoring Trinomials [6.2, 6.3]

2. $x^2 + 5x + 6 = (x + 2)(x + 3)$
$x^2 - 5x + 6 = (x - 2)(x - 3)$
$6x^2 - x - 2 = (2x + 1)(3x - 2)$
$6x^2 + 7x + 2 = (2x + 1)(3x + 2)$

One method of factoring a trinomial is to list all pairs of binomials whose product of the first terms gives the first term of the trinomial and whose product of the last terms gives the last term of the trinomial. We then choose the pair that gives the correct middle term for the original trinomial.

Special Factoring [6.4]

3. $x^2 + 10x + 25 = (x + 5)^2$
$x^2 - 10x + 25 = (x - 5)^2$
$x^2 - 25 = (x + 5)(x - 5)$

$$a^2 + 2ab + b^2 = (a + b)^2$$
$$a^2 - 2ab + b^2 = (a - b)^2$$
$$a^2 - b^2 = (a + b)(a - b)$$

Sum and Difference of Two Cubes [6.5]

4. $x^3 - 27 = (x - 3)(x^2 + 3x + 9)$
$x^3 + 27 = (x + 3)(x^2 - 3x + 9)$

$a^3 - b^3 = (a - b)(a^2 + ab + b^2)$ Difference of two cubes

$a^3 + b^3 = (a + b)(a^2 - ab + b^2)$ Sum of two cubes

Strategy for Factoring a Polynomial [6.6]

5. a. $2x^5 - 8x^3 = 2x^3(x^2 - 4)$
$= 2x^3(x + 2)(x - 2)$

b. $3x^4 - 18x^3 + 27x^2$
$= 3x^2(x^2 - 6x + 9)$
$= 3x^2(x - 3)^2$

c. $6x^3 - 12x^2 - 48x$
$= 6x(x^2 - 2x - 8)$
$= 6x(x - 4)(x + 2)$

d. $x^2 + ax + bx + ab$
$= x(x + a) + b(x + a)$
$= (x + a)(x + b)$

Step 1: If the polynomial has a greatest common factor other than 1, then factor out the greatest common factor.

Step 2: If the polynomial has two terms (it is a binomial), then see if it is the difference of two squares or the sum or difference of two cubes, and then factor accordingly. Remember, if it is the sum of two squares, it will not factor.

Step 3: If the polynomial has three terms (a trinomial), then it is either a perfect square trinomial that will factor into the square of a binomial, or it is not a perfect square trinomial, in which case you use the trial and error method developed in Section 5.3.

Step 4: If the polynomial has more than three terms, then try to factor it by grouping.

Step 5: As a final check, see if any of the factors you have written can be factored further. If you have overlooked a common factor, you can catch it here.

Strategy for Solving a Quadratic Equation [6.7]

Step 1: Write the equation in standard form $ax^2 + bx + c = 0$

Step 2: Factor completely.

Step 3: Set each variable factor equal to 0.

Step 4: Solve the equations found in step 3.

Step 5: Check solutions, if necessary.

6. Solve $x^2 - 6x = -8$.

$$x^2 - 6x + 8 = 0$$
$$(x - 4)(x - 2) = 0$$
$$x - 4 = 0 \quad \text{or} \quad x - 2 = 0$$
$$x = 4 \quad \text{or} \quad x = 2$$

Both solutions check.

The Pythagorean Theorem [6.8]

In any right triangle, the square of the longest side (called the hypotenuse) is equal to the sum of the squares of the other two sides (called legs).

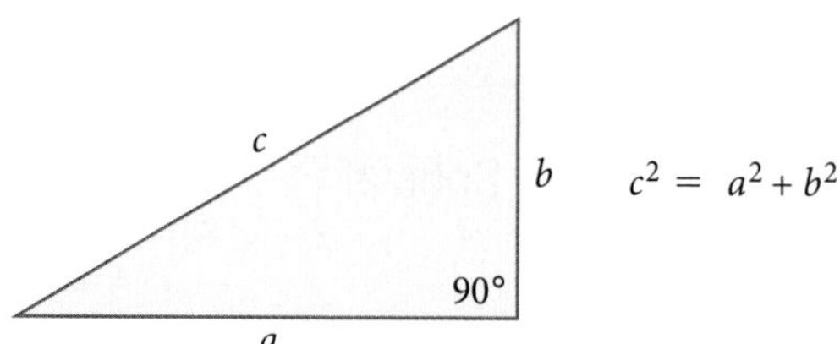

7. The hypotenuse of a right triangle is 5 inches, and the lengths of the two legs (the other two sides) are given by two consecutive integers. Find the lengths of the two legs.

If we let $x =$ the length of the shortest side, then the other side must be $x + 1$. The Pythagorean theorem tells us that

$$5^2 = x^2 + (x + 1)^2$$
$$25 = x^2 + x^2 + 2x + 1$$
$$25 = 2x^2 + 2x + 1$$
$$0 = 2x^2 + 2x - 24$$
$$0 = 2(x^2 + x - 12)$$
$$0 = 2(x + 4)(x - 3)$$
$$x + 4 = 0 \quad \text{or} \quad x - 3 = 0$$
$$x = -4 \quad \text{or} \quad x = 3$$

Since a triangle cannot have a side with a negative number for its length, we cannot use -4. One leg is $x = 3$ and the other leg is $x + 1 = 3 + 1 = 4$.

COMMON MISTAKE

It is a mistake to apply the zero-factor property to numbers other than zero. For example, consider the equation $(x - 3)(x + 4) = 18$. A fairly common mistake is to attempt to solve it with the following steps:

$$(x - 3)(x + 4) = 18$$
$$x - 3 = 18 \quad \text{or} \quad x + 4 = 18 \leftarrow \text{Mistake}$$
$$x = 21 \quad \text{or} \quad x = 14$$

These are obviously not solutions, as a quick check will verify:

Check: $x = 21$	Check: $x = 14$
$(21 - 3)(21 + 4) \stackrel{?}{=} 18$	$(14 - 3)(14 + 4) \stackrel{?}{=} 18$
$18 \cdot 25 = 18$	$11 \cdot 18 = 18$
$450 = 18$ ← false statements →	$198 = 18$

The mistake is in setting each factor equal to 18. It is not necessarily true that when the product of two numbers is 18, either one of them is itself 18. The correct solution looks like this:

$$(x - 3)(x + 4) = 18$$
$$x^2 + x - 12 = 18$$
$$x^2 + x - 30 = 0$$
$$(x + 6)(x - 5) = 0$$
$$x + 6 = 0 \quad \text{or} \quad x - 5 = 0$$
$$x = -6 \quad \text{or} \quad x = 5$$

To avoid this mistake, remember that before you factor a quadratic equation, you must write it in standard form. It is in standard form only when 0 is on one side and decreasing powers of the variable are on the other.

Factor out the greatest common factor. [6.1]

1. $6x + 18$

2. $12a^2b - 24ab + 8ab^2$

Factor by grouping. [6.1]

3. $x^2 + 3ax - 2bx - 6ab$

4. $15y - 5xy - 12 + 4x$

Factor the following completely. [6.2–6.6]

5. $x^2 + x - 12$

6. $x^2 - 4x - 21$

7. $x^2 - 25$

8. $x^4 - 16$

9. $x^2 + 36$

10. $18x^2 - 32y^2$

11. $x^3 + 4x^2 - 3x - 12$

12. $x^2 + bx - 3x - 3b$

13. $4x^2 - 6x - 10$

14. $4n^2 + 13n - 12$

15. $12c^2 + c - 6$

16. $12x^3 + 12x^2 - 9x$

17. $x^3 + 125y^3$

18. $54b^3 - 128$

Solve the following equations. [6.7]

19. $x^2 - 2x - 15 = 0$

20. $x^2 - 7x + 12 = 0$

21. $x^2 - 25 = 0$

22. $x^2 = 5x + 14$

23. $x^2 + x = 30$

24. $y^3 = 9y$

25. $2x^2 = -5x + 12$

26. $15x^3 - 65x^2 - 150x = 0$

Solve the following word problems. Be sure to show the system of equations used. [6.8]

27. Number Problem Two numbers have a sum of 18. Their product is 72. Find the numbers.

28. Consecutive Integers The product of two consecutive even integers is 14 more than their sum. Find the integers.

29. Geometry The length of a rectangle is 1 foot more than 3 times the width. The area is 52 square feet. Find the dimensions.

30. Geometry One leg of a right triangle is 2 feet more than the other. The hypotenuse is 10 feet. Find the lengths of the two legs.

31. Production Cost A company can manufacture x hundred items for a total cost C, given the equation $C = 100 + 500x - 100x^2$. How many items can be manufactured if the total cost is to be \$700?

32. Price and Revenue A manufacturer knows that the number of items he can sell each week, x, is related to the price p of each item by the equation $x = 800 - 100p$. What price should he charge for each item to have a weekly revenue of \$1,500? (*Remember:* $R = xp$.)

Quadratic Equations

7

Chapter Outline

7.1 Completing the Square

7.2 The Quadratic Formula

7.3 Additional Items Involving Solutions to Equations

7.4 More Equations

7.5 Graphing Parabolas

7.6 Quadratic Inequalities

Fir0002/Flagstaffotos
http://commons.wikimedia.org/wiki/Commons:GNU_Free_Documentation_License,_version_1.2

If you have been to the circus or the county fair recently, you may have witnessed one of the more spectacular acts, the human cannonball. The human cannonball shown in the photograph will reach a height of 70 feet, and travel a distance of 160 feet, before landing in a safety net. In this chapter, we use this information to derive the equation

$$f(x) = -\frac{7}{640}(x - 80)^2 + 70 \quad \text{for } 0 \le x \le 160$$

which describes the path flown by this particular cannonball. The table and graph below were constructed from this equation.

Path of a Human Cannonball

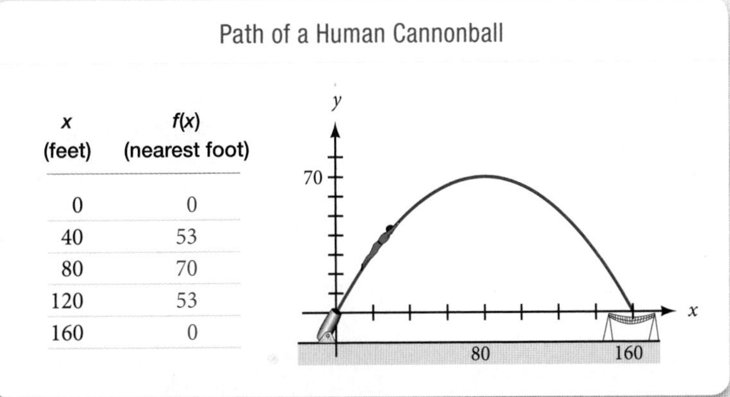

x (feet)	$f(x)$ (nearest foot)
0	0
40	53
80	70
120	53
160	0

All objects that are projected into the air, whether they are basketballs, bullets, arrows, or coins, follow parabolic paths like the one shown in the graph. Studying the material in this chapter will give you a more mathematical hold on the world around you.

Success Skills

If you have made it this far, then you have the study skills necessary to be successful in this course. Success skills are more general in nature and will help you with all your classes and ensure your success in college as well.

Let's start with a question:

Question: What quality is most important for success in any college course?

Answer: Independence. You want to become an independent learner.

We all know people like this. They are generally happy. They don't worry about getting the right instructor, or whether or not things work out every time. They have a confidence that comes from knowing that they are responsible for their success or failure in the goals they set for themselves.

Here are some of the qualities of an independent learner:

- Intends to succeed.
- Doesn't let setbacks deter them.
- Knows their resources.
  - Instructor's office hours
  - Math lab
  - Student Solutions Manual
  - Group study
  - Internet
- Doesn't mistake activity for achievement.
- Has a positive attitude.

There are other traits as well. The first step in becoming an independent learner is doing a little self-evaluation and then making of list of traits that you would like to acquire. What skills do you have that align with those of an independent learner? What attributes do you have that keep you from being an independent learner? What qualities would you like to obtain that you don't have now?

Completing the Square

7.1

Table 1 is taken from the trail map given to skiers at the Northstar at Tahoe Ski Resort in Lake Tahoe, California. The table gives the length of each chair lift at Northstar, along with the change in elevation from the beginning of the lift to the end of the lift.

Right triangles are good mathematical models for chair lifts. In this section, we will use our knowledge of right triangles, along with the new material developed in the section, to solve problems involving chair lifts and a variety of other examples.

TABLE 1 From the Trail Map for the Northstar at Tahoe Ski Resort

Lift Information		
Lift	Vertical Rise (feet)	Length (feet)
Big Springs Gondola	480	4,100
Bear Paw Double	120	790
Echo Triple	710	4,890
Aspen Express Quad	900	5,100
Forest Double	1,170	5,750
Lookout Double	960	4,330
Comstock Express Quad	1,250	5,900
Rendezvous Triple	650	2,900
Schaffer Camp Triple	1,860	6,150
Chipmunk Tow Lift	28	280
Bear Cub Tow Lift	120	750

In this section, we will develop the first of our new methods of solving quadratic equations. The new method is called *completing the square*. Completing the square on a quadratic equation allows us to obtain solutions, regardless of whether the equation can be factored. Before we solve equations by completing the square, we need to learn how to solve equations by taking square roots of both sides.

Consider the equation

$$x^2 = 16$$

We could solve it by writing it in standard form, factoring the left side, and proceeding as we did in Chapter 2. We can shorten our work considerably, however, if we simply notice that x must be either the positive square root of 16 or the negative square root of 16. That is,

If $\quad x^2 = 16$

Then $\quad x = \sqrt{16} \quad$ or $\quad x = -\sqrt{16}$

$\quad x = 4 \quad$ or $\quad x = -4$

We can generalize this result as follows.

PROPERTY ***Square Root Property for Equations***

If $a^2 = b$, where b is a real number, then $a = \sqrt{b}$ or $a = -\sqrt{b}$.

Notation The expression $a = \sqrt{b}$ or $a = -\sqrt{b}$ can be written in shorthand form as $a = \pm\sqrt{b}$. The symbol $\pm$ is read "plus or minus."

We can apply the Square Root Property for Equations to some fairly complicated quadratic equations.

EXAMPLE 1 Solve $(2x - 3)^2 = 25$.

SOLUTION

$$(2x - 3)^2 = 25$$

$2x - 3 = \pm\sqrt{25}$	Square Root Property for Equations
$2x - 3 = \pm 5$	$\sqrt{25} = 5$
$2x = 3 \pm 5$	Add 3 to both sides
$x = \dfrac{3 \pm 5}{2}$	Divide both sides by 2

The last equation can be written as two separate statements:

$$x = \frac{3+5}{2} \quad \text{or} \quad x = \frac{3-5}{2}$$

$$= \frac{8}{2} \qquad\qquad = \frac{-2}{2}$$

$$= 4 \quad \text{or} \quad = -1$$

The solution set is 4, -1.

Notice that we could have solved the equation in Example 1 by expanding the left side, writing the resulting equation in standard form, and then factoring. The problem would look like this:

$(2x - 3)^2 = 25$	Original equation
$4x^2 - 12x + 9 = 25$	Expand the left side
$4x^2 - 12x - 16 = 0$	Add -25 to each side
$4(x^2 - 3x - 4) = 0$	Begin factoring
$4(x - 4)(x + 1) = 0$	Factor completely
$x - 4 = 0$ or $x + 1 = 0$	Set variable factors equal to 0
$x = 4$ or $x = -1$	

As you can see, solving the equation by factoring leads to the same two solutions.

EXAMPLE 2 Solve for x: $(3x - 1)^2 = -12$

SOLUTION

$$\begin{aligned} (3x - 1)^2 &= -12 \\ 3x - 1 &= \pm\sqrt{-12} && \text{Square Root Property for Equations} \\ 3x - 1 &= \pm 2i\sqrt{3} && \sqrt{-12} = 2i\sqrt{3} \\ 3x &= 1 \pm 2i\sqrt{3} && \text{Add 1 to both sides} \\ x &= \frac{1 \pm 2i\sqrt{3}}{3} && \text{Divide both sides by 3} \end{aligned}$$

The solution set is $\left\{ \frac{1 + 2i\sqrt{3}}{3}, \frac{1 - 2i\sqrt{3}}{3} \right\}$.

Both solutions are complex. Here is a check of the first solution:

When $x = \frac{1 + 2i\sqrt{3}}{3}$

the equation $(3x - 1)^2 = -12$

becomes $\left(3 \cdot \frac{1 + 2i\sqrt{3}}{3} - 1\right)^2 \stackrel{?}{=} -12$

or

$$\begin{aligned} (1 + 2i\sqrt{3} - 1)^2 &\stackrel{?}{=} -12 \\ (2i\sqrt{3})^2 &\stackrel{?}{=} -12 \\ 4 \cdot i^2 \cdot 3 &\stackrel{?}{=} -12 \\ 12(-1) &\stackrel{?}{=} -12 \\ -12 &= -12 \end{aligned}$$

Note We cannot solve the equation in Example 2 by factoring. If we expand the left side and write the resulting equation in standard form, we are left with a quadratic equation that does not factor:

$(3x - 1)^2 = -12$

Equation from Example 2

$9x^2 - 6x + 1 = -12$

Expand the left side.

$9x^2 - 6x + 13 = 0$

Standard form, but not factorable

EXAMPLE 3 Solve $x^2 + 6x + 9 = 12$.

SOLUTION We can solve this equation as we have the equations in Examples 1 and 2 if we first write the left side as $(x + 3)^2$.

$$\begin{aligned} x^2 + 6x + 9 &= 12 && \text{Original equation} \\ (x + 3)^2 &= 12 && \text{Write } x^2 + 6x + 9 \text{ as } (x + 3)^2 \\ x + 3 &= \pm 2\sqrt{3} && \text{Square Root Property for Equations} \\ x &= -3 \pm 2\sqrt{3} && \text{Add } -3 \text{ to each side} \end{aligned}$$

We have two irrational solutions: $-3 + 2\sqrt{3}$ and $-3 - 2\sqrt{3}$. What is important about this problem, however, is the fact that the equation was easy to solve because the left side was a perfect square trinomial.

Method of Completing the Square

The method of completing the square is simply a way of transforming any quadratic equation into an equation of the form found in the preceding three examples. The key to understanding the method of completing the square lies in recognizing the relationship between the last two terms of any perfect square trinomial whose leading coefficient is 1.

Consider the following list of perfect square trinomials and their corresponding binomial squares:

$$x^2 - 6x + 9 = (x - 3)^2$$
$$x^2 + 8x + 16 = (x + 4)^2$$
$$x^2 - 10x + 25 = (x - 5)^2$$
$$x^2 + 12x + 36 = (x + 6)^2$$

In each case, the leading coefficient is 1. A more important observation comes from noticing the relationship between the linear and constant terms (middle and last terms) in each trinomial. Observe that the constant term in each case is the square of half the coefficient of x in the middle term. For example, in the last expression, the constant term 36 is the square of half of 12, where 12 is the coefficient of x in the middle term. (Notice also that the second terms in all the binomials on the right side are half the coefficients of the middle terms of the trinomials on the left side.) We can use these observations to build our own perfect square trinomials and, in doing so, solve some quadratic equations.

EXAMPLE 4 Solve $x^2 - 6x + 5 = 0$ by completing the square.

SOLUTION We begin by adding -5 to both sides of the equation. We want just $x^2 - 6x$ on the left side so that we can add on our own final term to get a perfect square trinomial:

$$x^2 - 6x + 5 = 0$$
$$x^2 - 6x \quad = -5 \qquad \text{Add } -5 \text{ to both sides}$$

Now we can add 9 to both sides and the left side will be a perfect square:

$$x^2 - 6x + 9 = -5 + 9$$
$$(x - 3)^2 = 4$$

The final line is in the form of the equations we solved previously:

$$x - 3 = \pm 2$$
$$x = 3 \pm 2 \qquad \text{Add 3 to both sides}$$
$$x = 3 + 2 \quad \text{or} \quad x = 3 - 2$$
$$x = 5 \quad \text{or} \quad x = 1$$

The two solutions are 5 and 1.

Note The equation in Example 4 can be solved quickly by factoring:

$$x^2 - 6x + 5 = 0$$
$$(x - 5)(x - 1) = 0$$
$$x - 5 = 0 \quad \text{or} \quad x - 1 = 0$$
$$x = 5 \quad \text{or} \quad x = 1$$

The reason we didn't solve it by factoring is we want to practice completing the square on some simple equations.

EXAMPLE 5 Solve by completing the square: $x^2 + 5x - 2 = 0$

SOLUTION We must begin by adding 2 to both sides. (The left side of the equation, as it is, is not a perfect square, because it does not have the correct constant term. We will simply "move" that term to the other side and use our own constant term.)

$$x^2 + 5x = 2 \qquad \text{Add 2 to each side}$$

We complete the square by adding the square of half the coefficient of the linear term to both sides:

$$x^2 + 5x + \frac{25}{4} = 2 + \frac{25}{4} \qquad \text{Half of 5 is } \frac{5}{2}\text{, the square of which is } \frac{25}{4}$$

$$\left(x + \frac{5}{2}\right)^2 = \frac{33}{4} \qquad 2 + \frac{25}{4} = \frac{8}{4} + \frac{25}{4} = \frac{33}{4}$$

$$x + \frac{5}{2} = \pm\sqrt{\frac{33}{4}} \qquad \text{Square Root Property for Equations}$$

$$x + \frac{5}{2} = \pm\frac{\sqrt{33}}{2} \qquad \text{Simplify the radical}$$

$$x = -\frac{5}{2} \pm \frac{\sqrt{33}}{2} \qquad \text{Add } -\frac{5}{2} \text{ to both sides}$$

$$x = \frac{-5 \pm \sqrt{33}}{2}$$

The solution set is $\left\{\frac{-5 + \sqrt{33}}{2}, \frac{-5 - \sqrt{33}}{2}\right\}$.

We can use a calculator to get decimal approximations to these solutions. If $\sqrt{33} \approx 5.74$, then

$$\frac{-5 + 5.74}{2} = 0.37$$

$$\frac{-5 - 5.74}{2} = -5.37$$

EXAMPLE 6 Solve for x: $3x^2 - 8x + 7 = 0$

SOLUTION

$$3x^2 - 8x + 7 = 0$$

$$3x^2 - 8x = -7 \qquad \text{Add } -7 \text{ to both sides}$$

We cannot complete the square on the left side because the leading coefficient is not 1. We take an extra step and divide both sides by 3:

$$\frac{3x^2}{3} - \frac{8x}{3} = -\frac{7}{3}$$

$$x^2 - \frac{8}{3}x = -\frac{7}{3}$$

Half of $\frac{8}{3}$ is $\frac{4}{3}$, the square of which is $\frac{16}{9}$:

$$x^2 - \frac{8}{3}x + \frac{16}{9} = -\frac{7}{3} + \frac{16}{9} \qquad \text{Add } \frac{16}{9} \text{ to both sides}$$

$$\left(x - \frac{4}{3}\right)^2 = -\frac{5}{9} \qquad \text{Simplify right side}$$

$$x - \frac{4}{3} = \pm\sqrt{-\frac{5}{9}} \qquad \text{Square Root Property for Equations}$$

$$x - \frac{4}{3} = \pm\frac{i\sqrt{5}}{3} \qquad \sqrt{-\frac{5}{9}} = \frac{\sqrt{-5}}{3} = \frac{i\sqrt{5}}{3}$$

$$x = \frac{4}{3} \pm \frac{i\sqrt{5}}{3} \qquad \text{Add } \frac{4}{3} \text{ to both sides}$$

$$x = \frac{4 \pm i\sqrt{5}}{3}$$

The solution set is $\left\{\frac{4 + i\sqrt{5}}{3}, \frac{4 - i\sqrt{5}}{3}\right\}$.

HOW TO *Solve a Quadratic Equation by Completing the Square*

To summarize the method used in the preceding two examples, we list the following steps:

Step 1: Write the equation in the form $ax^2 + bx = c$.

Step 2: If the leading coefficient is not 1, divide both sides by the coefficient so that the resulting equation has a leading coefficient of 1. That is, if $a \neq 1$, then divide both sides by a.

Step 3: Add the square of half the coefficient of the linear term to both sides of the equation.

Step 4: Write the left side of the equation as the square of a binomial, and simplify the right side if possible.

Step 5: Apply the Square Root Property for Equations, and solve as usual.

FACTS FROM GEOMETRY *More Special Triangles*

The triangles shown in Figures 1 and 2 occur frequently in mathematics.

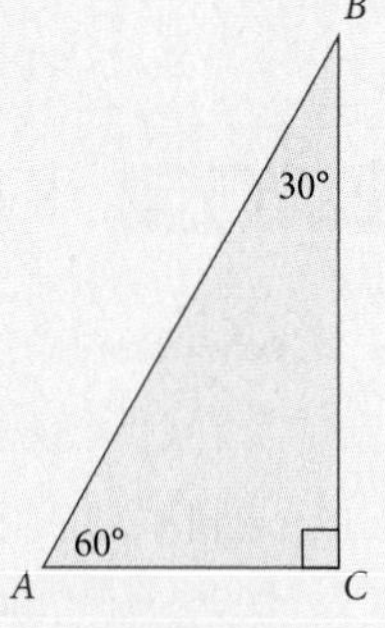

FIGURE 1

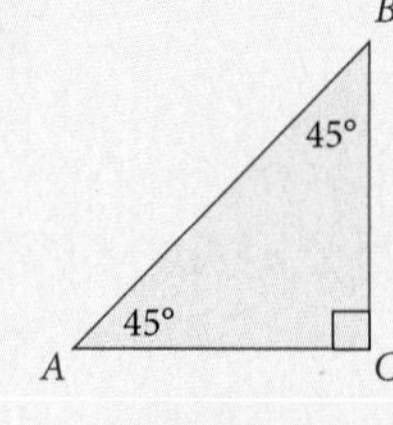

FIGURE 2

Note that both of the triangles are right triangles. We refer to the triangle in Figure 1 as a $30°–60°–90°$ triangle, and the triangle in Figure 2 as a $45°–45°–90°$ triangle.

EXAMPLE 7 If the shortest side in a 30°–60°–90° triangle is 1 inch, find the lengths of the other two sides.

SOLUTION In Figure 3, triangle ABC is a $30° - 60° - 90°$ triangle in which the shortest side AC is 1 inch long. Triangle DBC is also a $30° - 60° - 90°$ triangle in which the shortest side DC is 1 inch long.

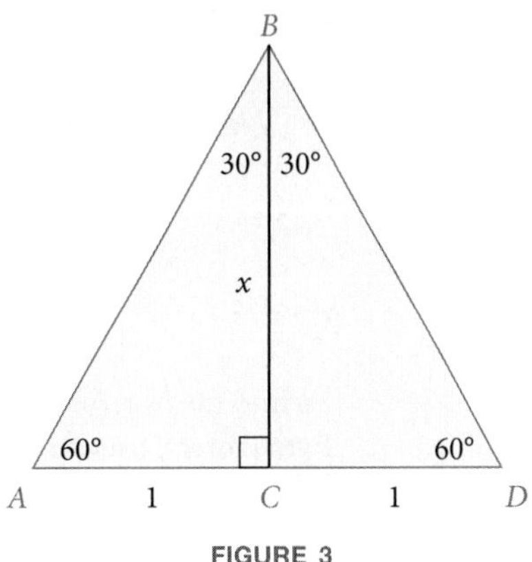

FIGURE 3

Notice that the large triangle ABD is an equilateral triangle because each of its interior angles is 60°. Each side of triangle ABD is 2 inches long. Side AB in triangle ABC is therefore 2 inches. To find the length of side BC, we use the Pythagorean theorem.

$$BC^2 + AC^2 = AB^2$$
$$x^2 + 1^2 = 2^2$$
$$x^2 + 1 = 4$$
$$x^2 = 3$$
$$x = \sqrt{3} \text{ inches}$$

Note that we write only the positive square root because x is the length of a side in a triangle and is therefore a positive number.

EXAMPLE 8 Table 1 in the introduction to this section gives the vertical rise of the Forest Double chair lift as 1,170 feet and the length of the chair lift as 5,750 feet. To the nearest foot, find the horizontal distance covered by a person riding this lift.

SOLUTION Figure 4 is a model of the Forest Double chair lift. A rider gets on the lift at point A and exits at point B. The length of the lift is AB.

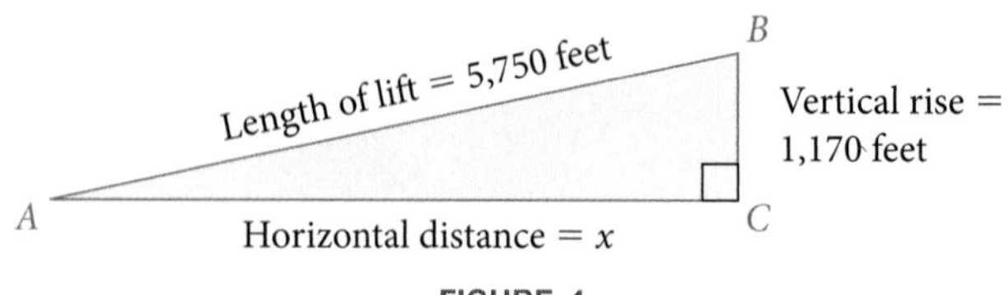

FIGURE 4

To find the horizontal distance covered by a person riding the chair lift, we use the Pythagorean theorem.

$$5{,}750^2 = x^2 + 1{,}170^2 \quad \text{Pythagorean theorem}$$

$$33{,}062{,}500 = x^2 + 1{,}368{,}900 \quad \text{Simplify squares}$$

$$x^2 = 33{,}062{,}500 - 1{,}368{,}900 \quad \text{Solve for } x^2$$

$$x^2 = 31{,}693{,}600 \quad \text{Simplify the right side}$$

$$x = \sqrt{31{,}693{,}600} \quad \text{Square Root Property for Equations}$$

$$= 5{,}630 \text{ feet} \quad \text{to the nearest foot}$$

A rider getting on the lift at point A and riding to point B will cover a horizontal distance of approximately 5,630 feet.

GETTING READY FOR CLASS

After reading through the preceding section, respond in your own words and in complete sentences.

A. What kind of equation do we solve using the method of completing the square?

B. Explain in words how you would complete the square on $x^2 - 16x = 4$.

C. What is the relationship between the shortest side and the longest side in a 30°–60°–90° triangle?

D. What two expressions together are equivalent to $x = \pm 4$?

Solve the following equations.

1. $x^2 = 25$ **2.** $x^2 = 16$ **3.** $a^2 = -9$ **4.** $a^2 = -49$

5. $y^2 = \frac{3}{4}$ **6.** $y^2 = \frac{5}{9}$ **7.** $x^2 + 12 = 0$ **8.** $x^2 + 8 = 0$

9. $4a^2 - 45 = 0$ **10.** $9a^2 - 20 = 0$ **11.** $(2y - 1)^2 = 25$ **12.** $(3y + 7)^2 = 1$

13. $(2a + 3)^2 = -9$ **14.** $(3a - 5)^2 = -49$

15. $(5x + 2)^2 = -8$ **16.** $(6x - 7)^2 = -75$

17. $x^2 + 8x + 16 = -27$ **18.** $x^2 - 12x + 36 = -8$

19. $4a^2 - 12a + 9 = -4$ **20.** $9a^2 - 12a + 4 = -9$

Copy each of the following, and fill in the blanks so the left side of each is a perfect square trinomial. That is, complete the square.

21. $x^2 + 12x + \_\_\_ = (x + \_\_\_)^2$ **22.** $x^2 + 6x + \_\_\_ = (x + \_\_\_)^2$

23. $x^2 - 4x + \_\_\_ = (x - \_\_\_)^2$ **24.** $x^2 - 2x + \_\_\_ = (x - \_\_\_)^2$

25. $a^2 - 10a + \_\_\_ = (a - \_\_\_)^2$ **26.** $a^2 - 8a + \_\_\_ = (a - \_\_\_)^2$

27. $x^2 + 5x + \_\_\_ = (x + \_\_\_)^2$ **28.** $x^2 + 3x + \_\_\_ = (x + \_\_\_)^2$

29. $y^2 - 7y + \_\_\_ = (y - \_\_\_)^2$ **30.** $y^2 - y + \_\_\_ = (y - \_\_\_)^2$

31. $x^2 + \frac{1}{2}x + \_\_\_ = (x + \_\_\_)^2$ **32.** $x^2 - \frac{3}{4}x + \_\_\_ = (x - \_\_\_)^2$

33. $x^2 + \frac{2}{3}x + \_\_\_ = (x + \_\_\_)^2$ **34.** $x^2 - \frac{4}{5}x + \_\_\_ = (x - \_\_\_)^2$

Solve each of the following quadratic equations by completing the square.

35. $x^2 + 4x = 12$ **36.** $x^2 - 2x = 8$ **37.** $x^2 + 12x = -27$

38. $x^2 - 6x = 16$ **39.** $a^2 - 2a + 5 = 0$ **40.** $a^2 + 10a + 22 = 0$

41. $y^2 - 8y + 1 = 0$ **42.** $y^2 + 6y - 1 = 0$ **43.** $x^2 - 5x - 3 = 0$

44. $x^2 - 5x - 2 = 0$ **45.** $2x^2 - 4x - 8 = 0$ **46.** $3x^2 - 9x - 12 = 0$

47. $3t^2 - 8t + 1 = 0$ **48.** $5t^2 + 12t - 1 = 0$ **49.** $4x^2 - 3x + 5 = 0$

50. $7x^2 - 5x + 2 = 0$ **51.** $3x^2 + 4x - 1 = 0$ **52.** $2x^2 + 6x - 1 = 0$

53. $2x^2 - 10x = 11$ **54.** $25x^2 - 20x = 1$ **55.** $4x^2 - 10x + 11 = 0$

56. $4x^2 - 6x + 1 = 0$

57. For the equation $x^2 = -9$

a. Can it be solved by factoring? **b.** Solve it.

58. For the equation $x^2 - 10x + 18 = 0$

a. Can it be solved by factoring? **b.** Solve it.

59. Solve the equation $x^2 - 6x = 0$

a. by factoring **b.** by completing the square

60. Solve the equation $x^2 + ax = 0$

a. by factoring **b.** by completing the square

61. Solve the equation $x^2 + 2x = 35$

a. by factoring **b.** by completing the square

62. Solve the equation $8x^2 - 10x - 25 = 0$

a. by factoring **b.** by completing the square

63. Is $x = -3 + \sqrt{2}$ a solution to $x^2 - 6x = 7$?

64. Is $x = 2 - \sqrt{5}$ a solution to $x^2 - 4x = 1$?

65. Solve each equation.

a. $5x - 7 = 0$ **b.** $5x - 7 = 8$ **c.** $(5x - 7)^2 = 8$

d. $\sqrt{5x - 7} = 8$ **e.** $\frac{5}{2} - \frac{7}{2x} = \frac{4}{x}$

66. Solve each equation.

a. $5x + 11 = 0$ **b.** $5x + 11 = 9$ **c.** $(5x + 11)^2 = 9$

d. $\sqrt{5x + 11} = 9$ **e.** $\frac{5}{3} - \frac{11}{3x} = \frac{3}{x}$

Applying the Concepts

67. Geometry If the shortest side in a 30°–60°–90° triangle is $\frac{1}{2}$ inch long, find the lengths of the other two sides.

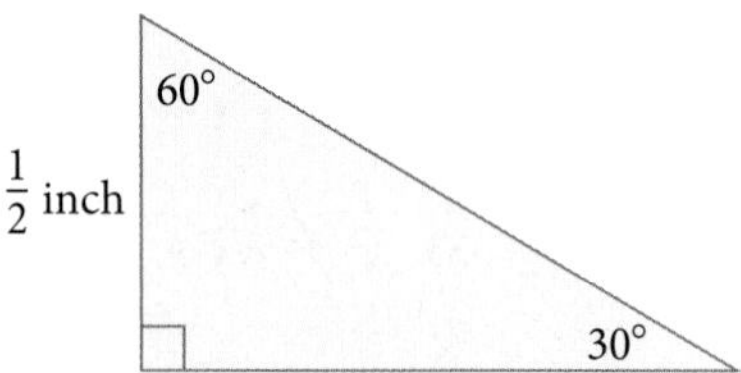

68. Geometry If the length of the longest side of a 30°–60°–90° triangle is x, find the lengths of the other two sides in terms of x.

69. Geometry If the length of the shorter sides of a 45°–45°–90° triangle is 1 inch, find the length of the hypotenuse.

70. Geometry If the length of the shorter sides of a 45°–45°–90° triangle is x, find the length of the hypotenuse, in terms of x.

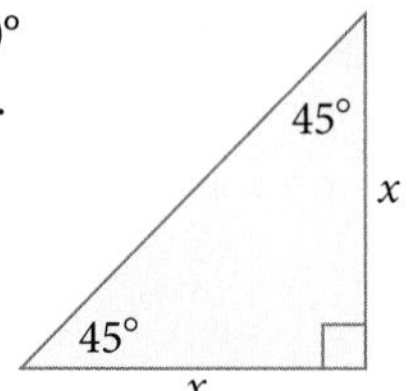

71. Chair Lift Use Table 1 from the introduction to this section to find the horizontal distance covered by a person riding the Bear Paw Double chair lift. Round your answer to the nearest foot.

72. Fermat's Last Theorem As mentioned in a previous chapter, the postage stamp shows Fermat's last theorem, which states that if n is an integer greater than 2, then there are no positive integers x, y, and z that will make the formula $x^n + y^n = z^n$ true. Use the formula $x^n + y^n = z^n$ to

a. find z if $n = 2$, $x = 6$, and $y = 8$. **b.** find y if $n = 2$, $x = 5$, and $z = 13$.

73. Interest Rate Suppose a deposit of $3,000 in a savings account that paid an annual interest rate r (compounded yearly) is worth $3,456 after 2 years. Using the formula $A = P(1 + r)^t$, we have

$$3{,}456 = 3{,}000(1 + r)^2$$

Solve for r to find the annual interest rate.

74. Special Triangles In Figure 5, triangle ABC has angles 45° and 30°, and height x. Find the lengths of sides AB, BC, and AC, in terms of x.

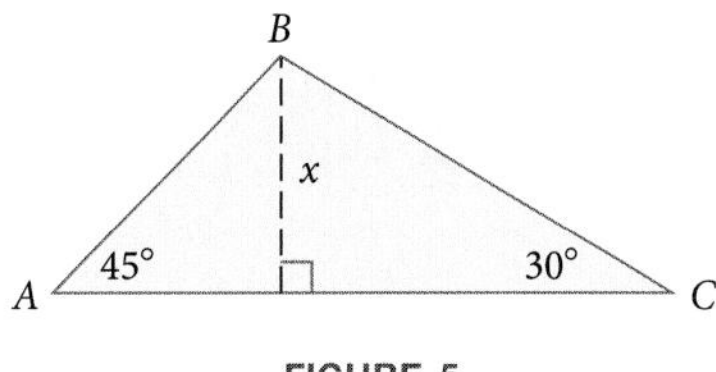

FIGURE 5

75. Length of an Escalator An escalator in a department store is made to carry people a vertical distance of 20 feet between floors. How long is the escalator if it makes an angle of 45° with the ground? (See Figure 6.)

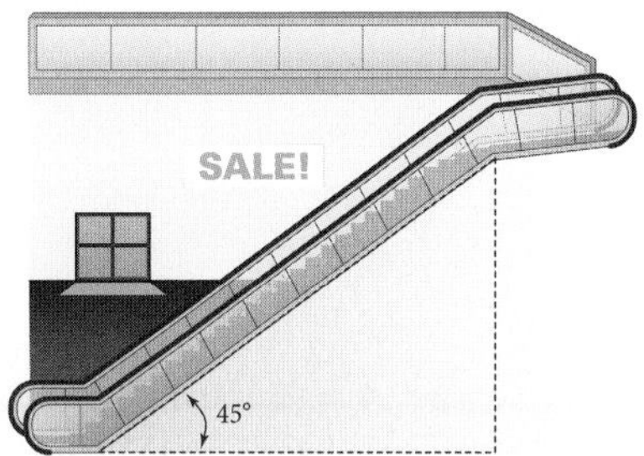

FIGURE 6

76. Dimensions of a Tent A two-person tent is to be made so the height at the center is 4 feet. If the sides of the tent are to meet the ground at an angle of 60° and the tent is to be 6 feet in length, how many square feet of material will be needed to make the tent? (Figure 7; assume that the tent has a floor and is closed at both ends.) Give your answer to the nearest tenth of a square foot.

FIGURE 7

Getting Ready for the Next Section

Simplify.

77. $49 - 4(6)(-5)$

78. $49 - 4(6)(2)$

79. $(-27)^2 - 4(0.1)(1{,}700)$

80. $25 - 4(4)(-10)$

81. $-7 + \dfrac{169}{12}$

82. $-7 - \dfrac{169}{12}$

Factor.

83. $27t^3 - 8$

84. $125t^3 + 1$

The Quadratic Formula

7.2

If you go on to take a business course or an economics course, you will find yourself spending lots of time with the three expressions that form the mathematical foundation of business: profit, revenue, and cost. Many times these expressions are given as polynomials, the topic of this section. The relationship between the three equations is known as the profit equation:

$$\text{Profit} = \text{Revenue} - \text{Cost}$$

$$P(x) = R(x) - C(x)$$

The table and graphs below were produced on a graphing calculator. They give numerical and graphical descriptions of revenue, profit, and cost for a company that manufactures and sells prerecorded videotapes according to the equations

$$R(x) = 11.5x - 0.05x^2 \qquad \text{and} \qquad C(x) = 200 + 2x$$

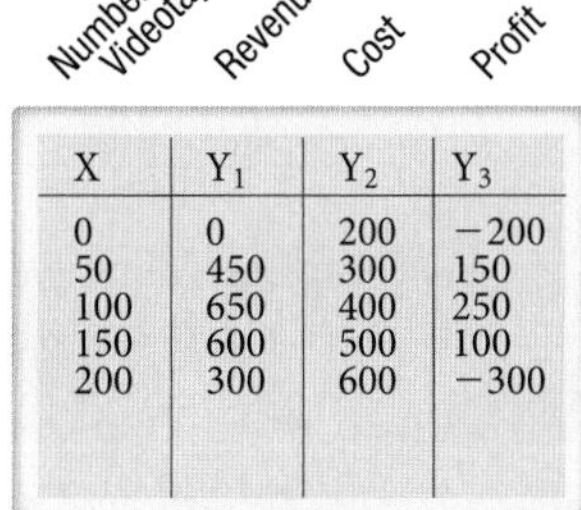

Number of Videotapes	Revenue	Cost	Profit
X	Y_1	Y_2	Y_3
0	0	200	−200
50	450	300	150
100	650	400	250
150	600	500	100
200	300	600	−300

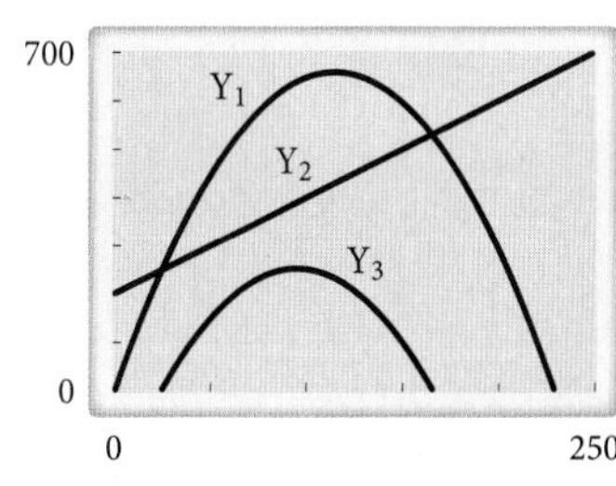

FIGURE 1

By studying the material in this section, you will get a more thorough look at the equations and relationships that are emphasized in business and economics.
In this section, we will use the method of completing the square from the preceding section to derive the quadratic formula. The *quadratic formula* is a very useful tool in mathematics. It allows us to solve all types of quadratic equations.

THEOREM *The Quadratic Theorem*

For any quadratic equation in the form $ax^2 + bx + c = 0$, $a \neq 0$, the two solutions are

$$x = \frac{-b + \sqrt{b^2 - 4ac}}{2a} \qquad \text{and} \qquad x = \frac{-b - \sqrt{b^2 - 4ac}}{2a}$$

Proof We will prove the quadratic theorem by completing the square on $ax^2 + bx + c = 0$:

$$ax^2 + bx + c = 0$$

$$ax^2 + bx = -c \qquad \text{Add } -c \text{ to both sides}$$

$$x^2 + \frac{b}{a}x = -\frac{c}{a} \qquad \text{Divide both sides by } a$$

To complete the square on the left side, we add the square of $\frac{1}{2}$ of $\frac{b}{a}$ to both sides $\left(\frac{1}{2} \text{ of } \frac{b}{a} \text{ is } \frac{b}{2a}\right)$.

$$x^2 + \frac{b}{a}x + \left(\frac{b}{2a}\right)^2 = -\frac{c}{a} + \left(\frac{b}{2a}\right)^2$$

We now simplify the right side as a separate step. We combine the two terms by writing each with the least common denominator $4a^2$:

$$-\frac{c}{a} + \left(\frac{b}{2a}\right)^2 = -\frac{c}{a} + \frac{b^2}{4a^2} = \frac{4a}{4a}\left(\frac{-c}{a}\right) + \frac{b^2}{4a^2} = \frac{-4ac + b^2}{4a^2}$$

It is convenient to write this last expression as

$$\frac{b^2 - 4ac}{4a^2}$$

Continuing with the proof, we have

$$x^2 + \frac{b}{a}x + \left(\frac{b}{2a}\right)^2 = \frac{b^2 - 4ac}{4a^2}$$

$$\left(x + \frac{b}{2a}\right)^2 = \frac{b^2 - 4ac}{4a^2} \qquad \text{Write left side as a binomial square}$$

$$x + \frac{b}{2a} = \pm\frac{\sqrt{b^2 - 4ac}}{2a} \qquad \text{Square Root Property for Equations}$$

$$x = -\frac{b}{2a} \pm \frac{\sqrt{b^2 - 4ac}}{2a} \qquad \text{Add } -\frac{b}{2a} \text{ to both sides}$$

$$= \frac{-b \pm \sqrt{b^2 - 4ac}}{2a}$$

Our proof is now complete. What we have is this: If our equation is in the form $ax^2 + bx + c = 0$ (standard form), where $a \neq 0$, the two solutions are always given by the formula

$$x = \frac{-b \pm \sqrt{b^2 - 4ac}}{2a}$$

This formula is known as the *quadratic formula*. If we substitute the coefficients a, b, and c of any quadratic equation in standard form into the formula, we need only perform some basic arithmetic to arrive at the solution set.

EXAMPLE 1 Solve $x^2 - 5x - 6 = 0$ by using the quadratic formula.

SOLUTION To use the quadratic formula, we must make sure the equation is in standard form; identify a, b, and c; substitute them into the formula; and work out the arithmetic.

For the equation $x^2 - 5x - 6 = 0$, $a = 1$, $b = -5$, and $c = -6$:

$$x = \frac{-b \pm \sqrt{b^2 - 4ac}}{2a}$$

$$= \frac{-(-5) \pm \sqrt{(-5)^2 - 4(1)(-6)}}{2(1)}$$

$$= \frac{5 \pm \sqrt{49}}{2}$$

$$= \frac{5 \pm 7}{2}$$

$$x = \frac{5+7}{2} \quad \text{or} \quad x = \frac{5-7}{2}$$

$$x = \frac{12}{2} \qquad x = -\frac{2}{2}$$

$$x = 6 \qquad x = -1$$

The two solutions are 6 and -1.

Note: Whenever the solutions to our quadratic equations turn out to be rational numbers, as in Example 1, it means the original equation could have been solved by factoring. (We didn't solve the equation in Example 1 by factoring because we were trying to get some practice with the quadratic formula.)

EXAMPLE 2 Solve for x: $2x^2 = -4x + 3$.

SOLUTION Before we can identify a, b, and c, we must write the equation in standard form. To do so, we add $4x$ and -3 to each side of the equation:

$$2x^2 = -4x + 3$$

$$2x^2 + 4x - 3 = 0 \qquad \text{Add } 4x \text{ and } -3 \text{ to each side}$$

Now that the equation is in standard form, we see that $a = 2$, $b = 4$, and $c = -3$. Using the quadratic formula we have:

$$\begin{aligned} x &= \frac{-b \pm \sqrt{b^2 - 4ac}}{2a} \\ &= \frac{-4 \pm \sqrt{4^2 - 4(2)(-3)}}{2(2)} \\ &= \frac{-4 \pm \sqrt{40}}{4} \\ &= \frac{-4 \pm 2\sqrt{10}}{4} \end{aligned}$$

We can reduce the final expression in the preceding equation to lowest terms by factoring 2 from the numerator and denominator and then dividing it out:

$$\begin{aligned} x &= \frac{2(-2 \pm \sqrt{10})}{2 \cdot 2} \\ &= \frac{-2 \pm \sqrt{10}}{2} \end{aligned}$$

Our two solutions are $\dfrac{-2 + \sqrt{10}}{2}$ and $\dfrac{-2 - \sqrt{10}}{2}$

EXAMPLE 3 Solve $x^2 - 6x = -7$.

SOLUTION We begin by writing the equation in standard form:

$$x^2 - 6x = -7$$

$$x^2 - 6x + 7 = 0 \qquad \text{Add 7 to each side}$$

Using $a = 1$, $b = -6$, and $c = 7$ in the quadratic formula

$$x = \frac{-b \pm \sqrt{b^2 - 4ac}}{2a}$$

we have:

$$\begin{aligned} x &= \frac{-(-6) \pm \sqrt{(-6)^2 - 4(1)(7)}}{2(1)} \\ &= \frac{6 \pm \sqrt{36 - 28}}{2} \\ &= \frac{6 \pm \sqrt{8}}{2} \\ &= \frac{6 \pm 2\sqrt{2}}{2} \end{aligned}$$

The two terms in the numerator have a 2 in common. We reduce to lowest terms by factoring the 2 from the numerator and then dividing numerator and denominator by 2:

$$\begin{aligned} &= \frac{2(3 \pm \sqrt{2})}{2} \\ &= 3 \pm \sqrt{2} \end{aligned}$$

The two solutions are $3 + \sqrt{2}$ and $3 - \sqrt{2}$. This time, let's check our solutions in the original equation $x^2 - 6x = -7$.

Checking $x = 3 + \sqrt{2}$, we have:

$$(3 + \sqrt{2})^2 - 6(3 + \sqrt{2}) \stackrel{?}{=} -7$$

$$9 + 6\sqrt{2} + 2 - 18 - 6\sqrt{2} = -7 \quad \text{Multiply}$$

$$11 - 18 + 6\sqrt{2} - 6\sqrt{2} = -7 \quad \text{Add 9 and 2}$$

$$-7 + 0 = -7 \quad \text{Subtraction}$$

$$-7 = -7 \quad \text{A true statement}$$

Checking $x = 3 - \sqrt{2}$, we have:

$$(3 - \sqrt{2})^2 - 6(3 - \sqrt{2}) \stackrel{?}{=} -7$$

$$9 - 6\sqrt{2} + 2 - 18 + 6\sqrt{2} = -7 \quad \text{Multiply}$$

$$11 - 18 - 6\sqrt{2} + 6\sqrt{2} = -7 \quad \text{Add 9 and 2}$$

$$-7 + 0 = -7 \quad \text{Subtraction}$$

$$-7 = -7 \quad \text{A true statement}$$

As you can see, both solutions yield true statements when used in place of the variable in the original equation.

EXAMPLE 4 Solve for x: $\frac{1}{10}x^2 - \frac{1}{5}x = -\frac{1}{2}$.

SOLUTION It will be easier to apply the quadratic formula if we clear the equation of fractions. Multiplying both sides of the equation by the LCD 10 gives us:

$$x^2 - 2x = -5$$

Next, we add 5 to both sides to put the equation into standard form:

$$x^2 - 2x + 5 = 0 \quad \text{Add 5 to both sides}$$

Applying the quadratic formula with $a = 1$, $b = -2$, and $c = 5$, we have:

$$x = \frac{-(-2) \pm \sqrt{(-2)^2 - 4(1)(5)}}{2(1)} = \frac{2 \pm \sqrt{-16}}{2} = \frac{2 \pm 4i}{2}$$

Dividing the numerator and denominator by 2, we have the two solutions:

$$x = 1 \pm 2i$$

The two solutions are $1 + 2i$ and $1 - 2i$.

EXAMPLE 5 Solve $(2x - 3)(2x - 1) = -4$.

SOLUTION We multiply the binomials on the left side and then add 4 to each side to write the equation in standard form. From there we identify a, b, and c and apply the quadratic formula:

$$(2x - 3)(2x - 1) = -4$$

$$4x^2 - 8x + 3 = -4 \quad \text{Multiply binomials on left side}$$

$$4x^2 - 8x + 7 = 0 \quad \text{Add 4 to each side}$$

Placing $a = 4$, $b = -8$, and $c = 7$ in the quadratic formula we have:

$$x = \frac{-(-8) \pm \sqrt{(-8)^2 - 4(4)(7)}}{2(4)}$$

$$= \frac{8 \pm \sqrt{64 - 112}}{8}$$

$$= \frac{8 \pm \sqrt{-48}}{8}$$

$$= \frac{8 \pm 4i\sqrt{3}}{8} \qquad \sqrt{-48} = i\sqrt{48} = i\sqrt{16}\sqrt{3} = 4i\sqrt{3}$$

To reduce this final expression to lowest terms, we factor a 4 from the numerator and then divide the numerator and denominator by 4:

$$= \frac{4(2 \pm i\sqrt{3})}{4 \cdot 2}$$

$$= \frac{2 \pm i\sqrt{3}}{2}$$

Note: It would be a mistake to try to reduce this final expression further. Sometimes first-year algebra students will try to divide the 2 in the denominator into the 2 in the numerator, which is a mistake. Remember, when we reduce to lowest terms, we do so by dividing the numerator and denominator by any factors they have in common. In this case 2 is not a factor of the numerator. This expression is in lowest terms.

Although the equation in our next example is not a quadratic equation, we solve it by using both factoring and the quadratic formula.

EXAMPLE 6 Solve $27t^3 - 8 = 0$.

SOLUTION It would be a mistake to add 8 to each side of this equation and then take the cube root of each side because we would lose two of our solutions. Instead, we factor the left side, and then set the factors equal to 0:

$$27t^3 - 8 = 0 \qquad \text{Equation in standard form}$$

$$(3t - 2)(9t^2 + 6t + 4) = 0 \qquad \text{Factor as the difference of two cubes.}$$

$$3t - 2 = 0 \quad \text{or} \quad 9t^2 + 6t + 4 = 0 \qquad \text{Set each factor equal to 0}$$

The first equation leads to a solution of $t = \frac{2}{3}$. The second equation does not factor, so we use the quadratic formula with $a = 9$, $b = 6$, and $c = 4$:

$$t = \frac{-6 \pm \sqrt{36 - 4(9)(4)}}{2(9)}$$

$$= \frac{-6 \pm \sqrt{36 - 144}}{18}$$

$$= \frac{-6 \pm \sqrt{-108}}{18}$$

$$= \frac{-6 \pm 6i\sqrt{3}}{18} \qquad \sqrt{-108} = i\sqrt{36 \cdot 3} = 6i\sqrt{3}$$

$$= \frac{6(-1 \pm i\sqrt{3})}{6 \cdot 3} \qquad \text{Factor 6 from the numerator and denominator}$$

$$= \frac{-1 \pm i\sqrt{3}}{3} \qquad \text{Divide out common factor 6}$$

The three solutions to our original equation are

$$\frac{2}{3}, \quad \frac{-1 + i\sqrt{3}}{3}, \quad \text{and} \quad \frac{-1 - i\sqrt{3}}{3}$$

EXAMPLE 7 If an object is thrown downward with an initial velocity of 20 feet per second, the distance $s(t)$, in feet, it travels in t seconds is given by the function $s(t) = 20t + 16t^2$. How long does it take the object to fall 40 feet?

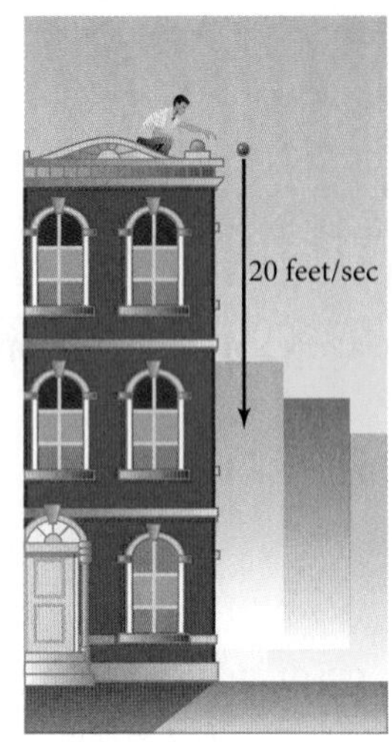

SOLUTION We let $s(t) = 40$, and solve for t:

When $\quad s(t) = 40$

the function $\quad s(t) = 20t + 16t^2$

becomes $\quad 40 = 20t + 16t^2$

or $\quad 16t^2 + 20t - 40 = 0$

$$4t^2 + 5t - 10 = 0 \qquad \text{Divide by 4}$$

Using the quadratic formula, we have

$$t = \frac{-5 \pm \sqrt{25 - 4(4)(-10)}}{2(4)}$$

$$= \frac{-5 \pm \sqrt{185}}{8}$$

$$= \frac{-5 + \sqrt{185}}{8} \quad \text{or} \quad t = \frac{-5 - \sqrt{185}}{8}$$

The second solution is impossible because it is a negative number and time t must be positive. It takes

$$t = \frac{-5 + \sqrt{185}}{8} \qquad \text{or approximately} \qquad \frac{-5 + 13.60}{8} \approx 1.08 \text{ seconds}$$

for the object to fall 40 feet.

The relationship between profit, revenue, and cost is given by the formula

$$P(x) = R(x) - C(x)$$

where $P(x)$ is the profit, $R(x)$ is the total revenue, and $C(x)$ is the total cost of producing and selling x items.

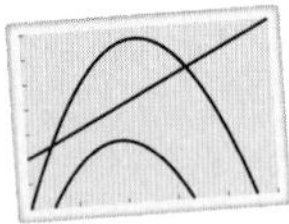

EXAMPLE 8 A company produces and sells copies of an accounting program for home computers. The total weekly cost (in dollars) to produce x copies of the program is $C(x) = 8x + 500$, and the weekly revenue for selling all x copies of the program is $R(x) = 35x - 0.1x^2$. How many programs must be sold each week for the weekly profit to be \$1,200?

SOLUTION Substituting the given expressions for $R(x)$ and $C(x)$ in the equation $P(x) = R(x) - C(x)$, we have a polynomial in x that represents the weekly profit $P(x)$:

$$P(x) = R(x) - C(x)$$

$$= 35x - 0.1x^2 - (8x + 500)$$

$$= 35x - 0.1x^2 - 8x - 500$$

$$= -500 + 27x - 0.1x^2$$

Setting this expression equal to 1,200, we have a quadratic equation to solve that gives us the number of programs x that need to be sold each week to bring in a profit of \$1,200:

$$1{,}200 = -500 + 27x - 0.1x^2$$

We can write this equation in standard form by adding the opposite of each term on the right side of the equation to both sides of the equation. Doing so produces the following equation:

$$0.1x^2 - 27x + 1{,}700 = 0$$

Applying the quadratic formula to this equation with $a = 0.1$, $b = -27$, and $c = 1{,}700$, we have

$$\begin{aligned} x &= \frac{27 \pm \sqrt{(-27)^2 - 4(0.1)(1{,}700)}}{2(0.1)} \\ &= \frac{27 \pm \sqrt{729 - 680}}{0.2} \\ &= \frac{27 \pm \sqrt{49}}{0.2} \\ &= \frac{27 \pm 7}{0.2} \end{aligned}$$

Writing this last expression as two separate expressions, we have our two solutions:

$$\begin{aligned} x &= \frac{27 + 7}{0.2} \quad &\text{or} \quad x &= \frac{27 - 7}{0.2} \\ &= \frac{34}{0.2} & &= \frac{20}{0.2} \\ &= 170 & &= 100 \end{aligned}$$

The weekly profit will be \$1,200 if the company produces and sells 100 programs or 170 programs. ■

What is interesting about this last example is that it has rational solutions, meaning it could have been solved by factoring. But looking back at the equation, factoring does not seem like a reasonable method of solution because the coefficients are either very large or very small. So, there are times when using the quadratic formula is a faster method of solution, even though the equation you are solving is factorable.

USING TECHNOLOGY *Graphing Calculators*

More About Example 7

We can solve the problem discussed in Example 7 by graphing the function $Y_1 = 20X + 16X^2$ in a window with X from 0 to 2 (because X is taking the place of t and we know t is a positive quantity) and Y from 0 to 50 (because we are looking for X when Y_1 is 40). Graphing Y_1 gives a graph similar to the graph in Figure 2. Using the Zoom and Trace features at $Y_1 = 40$ gives us $X = 1.08$ to the nearest hundredth, matching the results we obtained by solving the original equation algebraically.

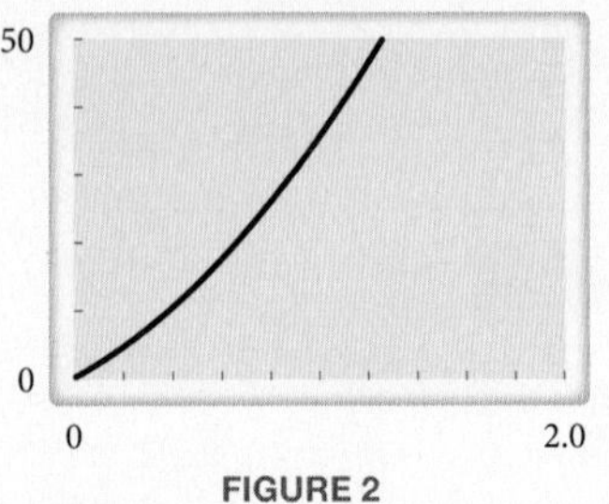

FIGURE 2

More About Example 8

To visualize the functions in Example 8, we set up our calculator this way:

$Y_1 = 35X - .1X^2$ Revenue function

$Y_2 = 8X + 500$ Cost function

$Y_3 = Y_1 - Y_2$ Profit function

Window: X from 0 to 350, Y from 0 to 3,500

Graphing these functions produces graphs similar to the ones shown in Figure 3. The lowest graph is the graph of the profit function. Using the Zoom and Trace features on the lowest graph at $Y_3 = 1{,}200$ produces two corresponding values of X, 170 and 100, which match the results in Example 8.

We will continue this discussion of the relationship between graphs of functions and solutions to equations in the Using Technology material in the next section.

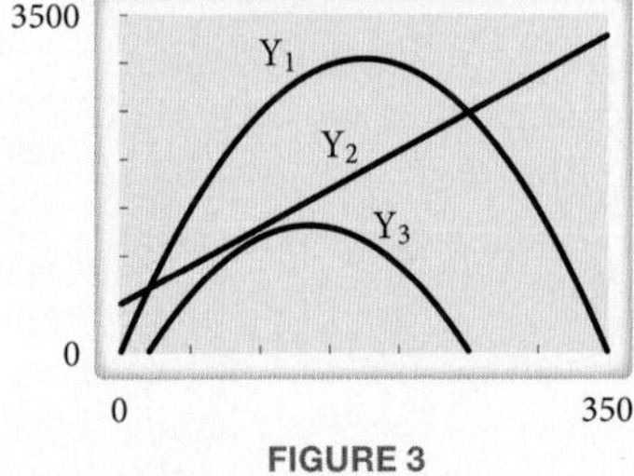

FIGURE 3

GETTING READY FOR CLASS

After reading through the preceding section, respond in your own words and in complete sentences.

A. What is the quadratic formula?
B. Under what circumstances should the quadratic formula be applied?
C. When would the quadratic formula result in complex solutions?
D. When will the quadratic formula result in only one solution?

Solve each equation. Use factoring or the quadratic formula, whichever is appropriate. (Try factoring first. If you have any difficulty factoring, then go right to the quadratic formula.)

1. $x^2 + 5x + 6 = 0$ **2.** $x^2 + 5x - 6 = 0$ **3.** $a^2 - 4a + 1 = 0$

4. $a^2 + 4a + 1 = 0$ **5.** $\frac{1}{6}x^2 - \frac{1}{2}x + \frac{1}{3} = 0$ **6.** $\frac{1}{6}x^2 + \frac{1}{2}x + \frac{1}{3} = 0$

7. $\frac{x^2}{2} + 1 = \frac{2x}{3}$ **8.** $\frac{x^2}{2} + \frac{2}{3} = -\frac{2x}{3}$ **9.** $y^2 - 5y = 0$

10. $2y^2 + 10y = 0$ **11.** $30x^2 + 40x = 0$ **12.** $50x^2 - 20x = 0$

13. $\frac{2t^2}{3} - t = -\frac{1}{6}$ **14.** $\frac{t^2}{3} - \frac{t}{2} = -\frac{3}{2}$

15. $0.01x^2 + 0.06x - 0.08 = 0$ **16.** $0.02x^2 - 0.03x + 0.05 = 0$

17. $2x + 3 = -2x^2$ **18.** $2x - 3 = 3x^2$

19. $100x^2 - 200x + 100 = 0$ **20.** $100x^2 - 600x + 900 = 0$

21. $\frac{1}{2}r^2 = \frac{1}{6}r - \frac{2}{3}$ **22.** $\frac{1}{4}r^2 = \frac{2}{5}r + \frac{1}{10}$

23. $(x - 3)(x - 5) = 1$ **24.** $(x - 3)(x + 1) = -6$

25. $(x + 3)^2 + (x - 8)(x - 1) = 16$ **26.** $(x - 4)^2 + (x + 2)(x + 1) = 9$

27. $\frac{x^2}{3} - \frac{5x}{6} = \frac{1}{2}$ **28.** $\frac{x^2}{6} + \frac{5}{6} = -\frac{x}{3}$

Multiply both sides of each equation by its LCD. Then solve the resulting equation.

29. $\frac{1}{x+1} - \frac{1}{x} = \frac{1}{2}$ **30.** $\frac{1}{x+1} + \frac{1}{x} = \frac{1}{3}$ **31.** $\frac{1}{y-1} + \frac{1}{y+1} = 1$

32. $\frac{2}{y+2} + \frac{3}{y-2} = 1$ **33.** $\frac{1}{x+2} + \frac{1}{x+3} = 1$ **34.** $\frac{1}{x+3} + \frac{1}{x+4} = 1$

35. $\frac{6}{r^2-1} - \frac{1}{2} = \frac{1}{r+1}$ **36.** $2 + \frac{5}{r-1} = \frac{12}{(r-1)^2}$

Solve each equation. In each case you will have three solutions.

37. $x^3 - 8 = 0$ **38.** $x^3 - 27 = 0$ **39.** $8a^3 + 27 = 0$

40. $27a^3 + 8 = 0$ **41.** $125t^3 - 1 = 0$ **42.** $64t^3 + 1 = 0$

Each of the following equations has three solutions. Look for the greatest common factor; then use the quadratic formula to find all solutions.

43. $2x^3 + 2x^2 + 3x = 0$ **44.** $6x^3 - 4x^2 + 6x = 0$ **45.** $3y^4 = 6y^3 - 6y^2$

46. $4y^4 = 16y^3 - 20y^2$ **47.** $6t^5 + 4t^4 = -2t^3$ **48.** $8t^5 + 2t^4 = -10t^3$

49. Which two of the expressions below are equivalent?

a. $\frac{6 + 2\sqrt{3}}{4}$ **b.** $\frac{3 + \sqrt{3}}{2}$ **c.** $6 + \frac{\sqrt{3}}{2}$

50. Which two of the expressions below are equivalent?

a. $\frac{8 - 4\sqrt{2}}{4}$ **b.** $2 - 4\sqrt{3}$ **c.** $2 - \sqrt{2}$

51. Solve $3x^2 - 5x = 0$

a. by factoring **b.** by the quadratic formula

52. Solve $3x^2 + 23x - 70 = 0$

a. by factoring **b.** by the quadratic formula

53. Can the equation $x^2 - 4x + 7 = 0$ be solved by factoring? Solve it.

54. Can the equation $x^2 = 5$ be solved by factoring? Solve it.

55. Is $x = -1 + i$ a solution to $x^2 + 2x = -2$.

56. Is $x = 2 + 2i$ a solution to $(x - 2)^2 = -4$.

Applying the Concepts

57. Falling Object An object is thrown downward with an initial velocity of 5 feet per second. The relationship between the distance s it travels and time t is given by $s = 5t + 16t^2$. How long does it take the object to fall 74 feet?

58. Coin Toss A coin is tossed upward with an initial velocity of 32 feet per second from a height of 16 feet above the ground. The equation giving the object's height h at any time t is $h = 16 + 32t - 16t^2$. Does the object ever reach a height of 32 feet?

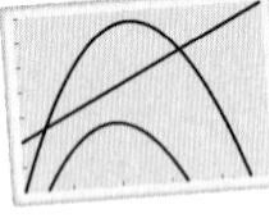

59. Profit The total cost (in dollars) for a company to manufacture and sell x items per week is $C = 60x + 300$, whereas the revenue brought in by selling all x items is $R = 100x - 0.5x^2$. How many items must be sold to obtain a weekly profit of \$300?

60. Profit Suppose a company manufactures and sells x picture frames each month with a total cost of $C = 1{,}200 + 3.5x$ dollars. If the revenue obtained by selling x frames is $R = 9x - 0.002x^2$, find the number of frames it must sell each month if its monthly profit is to be \$2,300.

Getting Ready for the Next Section

Find the value of $b^2 - 4ac$ when

61. $a = 1, b = -3, c = -40$

62. $a = 2, b = 3, c = 4$

63. $a = 4, b = 12, c = 9$

64. $a = -3, b = 8, c = -1$

Solve.

65. $k^2 - 144 = 0$

66. $36 - 20k = 0$

Multiply.

67. $(x - 3)(x + 2)$

68. $(t - 5)(t + 5)$

69. $(x - 3)(x - 3)(x + 2)$

70. $(t - 5)(t + 5)(t - 3)$

Additional Items Involving Solutions to Equations

7.3

In this section, we will do two things. First, we will define the discriminant and use it to find the kind of solutions a quadratic equation has without solving the equation. Second, we will use the zero-factor property to build equations from their solutions.

The Discriminant

The quadratic formula

$$x = \frac{-b \pm \sqrt{b^2 - 4ac}}{2a}$$

gives the solutions to any quadratic equation in standard form. There are times, when working with quadratic equations, that it is important only to know what kind of solutions the equation has.

DEFINITION *discriminant*

The expression under the radical in the quadratic formula is called the ***discriminant:***

$$\text{Discriminant} = D = b^2 - 4ac$$

The discriminant indicates the number and type of solutions to a quadratic equation, when the original equation has integer coefficients. For example, if we were to use the quadratic formula to solve the equation $2x^2 + 2x + 3 = 0$, we would find the discriminant to be

$$b^2 - 4ac = 2^2 - 4(2)(3) = -20$$

Because the discriminant appears under a square root symbol, we have the square root of a negative number in the quadratic formula. Our solutions would therefore be complex numbers. Similarly, if the discriminant were 0, the quadratic formula would yield

$$x = \frac{-b \pm \sqrt{0}}{2a} = \frac{-b \pm 0}{2a} = \frac{-b}{2a}$$

and the equation would have one rational solution, the number $\frac{-b}{2a}$.

The following table gives the relationship between the discriminant and the type of solutions to the equation.

For the equation $ax^2 + bx + c = 0$ where a, b, and c are integers and $a \neq 0$:

If the Discriminant $b^2 - 4ac$ Is	Then the Equation Will Have
Negative	Two complex solutions containing i
Zero	One rational solution
A positive number that is also a perfect square	Two rational solutions
A positive number that is not a perfect square	Two irrational solutions

In the second and third cases, when the discriminant is 0 or a positive perfect square, the solutions are rational numbers. The quadratic equations in these two cases are the ones that can be factored.

EXAMPLES For each equation, give the number and kind of solutions.

1. $x^2 - 3x - 40 = 0$

SOLUTION Using $a = 1, b = -3$, and $c = -40$ in $b^2 - 4ac$, we have

$$(-3)^2 - 4(1)(-40) = 9 + 160 = 169.$$

The discriminant is a perfect square. The equation therefore has two rational solutions.

2. $2x^2 - 3x + 4 = 0$

SOLUTION Using $a = 2, b = -3$, and $c = 4$, we have

$$b^2 - 4ac = (-3)^2 - 4(2)(4) = 9 - 32 = -23$$

The discriminant is negative, implying the equation has two complex solutions that contain i.

3. $4x^2 - 12x + 9 = 0$

SOLUTION Using $a = 4, b = -12$, and $c = 9$, the discriminant is

$$b^2 - 4ac = (-12)^2 - 4(4)(9) = 144 - 144 = 0$$

Because the discriminant is 0, the equation will have one rational solution.

4. $x^2 + 6x = 8$

SOLUTION We must first put the equation in standard form by adding -8 to each side. If we do so, the resulting equation is

$$x^2 + 6x - 8 = 0$$

Now we identify a, b, and c as 1, 6, and -8, respectively:

$$b^2 - 4ac = 6^2 - 4(1)(-8) = 36 + 32 = 68$$

The discriminant is a positive number, but not a perfect square. The equation will therefore have two irrational solutions.

EXAMPLE 5 Find an appropriate k so that the equation $4x^2 - kx = -9$ has exactly one rational solution.

SOLUTION We begin by writing the equation in standard form:

$$4x^2 - kx + 9 = 0$$

Using $a = 4, b = -k$, and $c = 9$, we have

$$\begin{aligned} b^2 - 4ac &= (-k)^2 - 4(4)(9) \\ &= k^2 - 144 \end{aligned}$$

An equation has exactly one rational solution when the discriminant is 0. We set the discriminant equal to 0 and solve:

$$\begin{aligned} k^2 - 144 &= 0 \\ k^2 &= 144 \\ k &= \pm 12 \end{aligned}$$

Choosing k to be 12 or -12 will result in an equation with one rational solution.

Building Equations From Their Solutions

Suppose we know that the solutions to an equation are $x = 3$ and $x = -2$. We can find equations with these solutions by using the zero-factor property. First, let's write our solutions as equations with 0 on the right side:

If	$x = 3$	First solution
then	$x - 3 = 0$	Add -3 to each side
and if	$x = -2$	Second solution
then	$x + 2 = 0$	Add 2 to each side

Now, because both $x - 3$ and $x + 2$ are 0, their product must be 0 also. We can therefore write

$$(x - 3)(x + 2) = 0 \qquad \text{Zero-factor property}$$

$$x^2 - x - 6 = 0 \qquad \text{Multiply out the left side}$$

Many other equations have 3 and -2 as solutions. For example, any constant multiple of $x^2 - x - 6 = 0$, such as $5x^2 - 5x - 30 = 0$, also has 3 and -2 as solutions. Similarly, any equation built from positive integer powers of the factors $x - 3$ and $x + 2$ will also have 3 and -2 as solutions. One such equation is

$$(x - 3)^2(x + 2) = 0$$

$$(x^2 - 6x + 9)(x + 2) = 0$$

$$x^3 - 4x^2 - 3x + 18 = 0$$

In mathematics, we distinguish between the solutions to this last equation and those to the equation $x^2 - x - 6 = 0$ by saying $x = 3$ is a solution of *multiplicity* 2 in the equation $x^3 - 4x^2 - 3x + 18 = 0$, and a solution of *multiplicity* 1 in the equation $x^2 - x - 6 = 0$.

EXAMPLE 6 Find an equation that has solutions $t = 5$, $t = -5$, and $t = 3$.

SOLUTION First, we use the given solutions to write equations that have 0 on their right sides:

If	$t = 5$	$t = -5$	$t = 3$
then	$t - 5 = 0$	$t + 5 = 0$	$t - 3 = 0$

Since $t - 5$, $t + 5$, and $t - 3$ are all 0, their product is also 0 by the zero-factor property. An equation with solutions of 5, -5, and 3 is

$$(t - 5)(t + 5)(t - 3) = 0 \qquad \text{Zero-factor property}$$

$$(t^2 - 25)(t - 3) = 0 \qquad \text{Multiply first two binomials}$$

$$t^3 - 3t^2 - 25t + 75 = 0 \qquad \text{Complete the multiplication}$$

The last line $t^3 - 3t^2 - 25t + 75 = 0$ gives us an equation with solutions of 5, -5, and 3. Remember, many other equations have these same solutions.

EXAMPLE 7 Find an equation with solutions $x = -\frac{2}{3}$ and $x = \frac{4}{5}$.

SOLUTION The solution $x = -\frac{2}{3}$ can be rewritten as $3x + 2 = 0$ as follows:

$$x = -\frac{2}{3} \quad \text{The first solution}$$

$$3x = -2 \quad \text{Multiply each side by 3}$$

$$3x + 2 = 0 \quad \text{Add 2 to each side}$$

Similarly, the solution $x = \frac{4}{5}$ can be rewritten as $5x - 4 = 0$:

$$x = \frac{4}{5} \quad \text{The second solution}$$

$$5x = 4 \quad \text{Multiply each side by 5}$$

$$5x - 4 = 0 \quad \text{Add } -4 \text{ to each side}$$

Because both $3x + 2$ and $5x - 4$ are 0, their product is 0 also, giving us the equation we are looking for:

$$(3x + 2)(5x - 4) = 0 \quad \text{Zero-factor property}$$

$$15x^2 - 2x - 8 = 0 \quad \text{Multiplication}$$

USING TECHNOLOGY *Graphing Calculators*

Solving Equations

Now that we have explored the relationship between equations and their solutions, we can look at how a graphing calculator can be used in the solution process. To begin, let's solve the equation $x^2 = x + 2$ using techniques from algebra: writing it in standard form, factoring, and then setting each factor equal to 0.

$$x^2 - x - 2 = 0 \quad \text{Standard form}$$

$$(x - 2)(x + 1) = 0 \quad \text{Factor}$$

$$x - 2 = 0 \quad \text{or} \quad x + 1 = 0 \quad \text{Set each factor equal to 0}$$

$$x = 2 \quad \text{or} \quad x = -1 \quad \text{Solve}$$

Our original equation, $x^2 = x + 2$, has two solutions: $x = 2$ and $x = -1$. To solve the equation using a graphing calculator, we need to associate it with an equation (or equations) in two variables. One way to do this is to associate the left side with the equation $y = x^2$ and the right side of the equation with $y = x + 2$. To do so, we set up the functions list in our calculator this way:

$$Y_1 = X^2$$

$$Y_2 = X + 2$$

Window: X from -5 to 5, Y from -5 to 5

Graphing these functions in this window will produce a graph similar to the one shown in Figure 1.

If we use the Trace feature to find the coordinates of the points of intesection, we find that the two curves intersect at $(-1, 1)$ and $(2, 4)$. We note that the x-coordinates of these two points match the solutions to the equation $x^2 = x + 2$, which we found using algebraic techniques. This makes sense because if two graphs intersect at a point (x, y), then the coordinates of that point satisfy both equations. If a point (x, y) satisfies both $y = x^2$ and $y = x + 2$, then for that particular point, $x^2 = x + 2$. From this, we conclude that the x-coordinates of the points of intersection are solutions to our original equation. Here is a summary of what we have discovered:

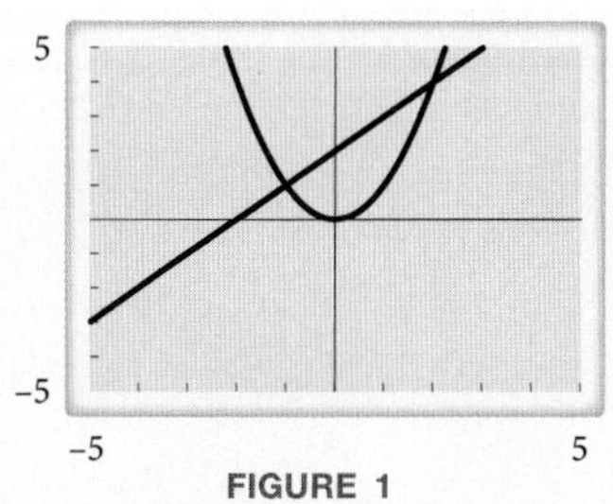

FIGURE 1

Conclusion 1 If the graphs of two functions $y = f(x)$ and $y = g(x)$ intersect in the coordinate plane, then the x-coordinates of the points of intersection are solutions to the equation $f(x) = g(x)$.

A second method of solving our original equation $x^2 = x + 2$ graphically requires the use of one function instead of two. To begin, we write the equation in standard form as $x^2 - x - 2 = 0$. Next, we graph the function $y = x^2 - x - 2$. The x-intercepts of the graph are the points with y-coordinates of 0. They therefore satisfy the equation $0 = x^2 - x - 2$, which is equivalent to our original equation. The graph in Figure 2 shows $\text{Y}_1 = \text{X}^2 - \text{X} - 2$ in a window with X from -5 to 5 and Y from -5 to 5.

Using the Trace feature, we find that the x-intercepts of the graph are $x = -1$ and $x = 2$, which match the solutions to our original equation $x^2 = x + 2$. We can summarize the relationship between solutions to an equation and the intercepts of its associated graph this way:

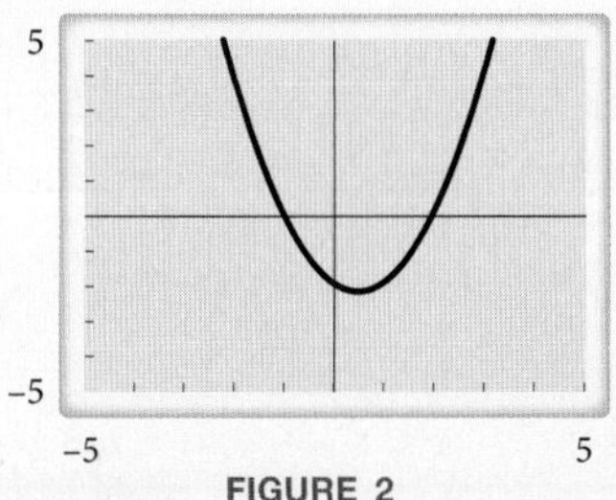

FIGURE 2

Conclusion 2 If $y = f(x)$ is a function, then any x-intercept on the graph of $y = f(x)$ is a solution to the equation $f(x) = 0$.

GETTING READY FOR CLASS

After reading through the preceding section, respond in your own words and in complete sentences.

A. What is the discriminant?
B. What kind of solutions do we get to a quadratic equation when the discriminant is negative?
C. What does it mean for a solution to have multiplicity 3?
D. When will a quadratic equation have two rational solutions?

Problem Set 7.3

Use the discriminant to find the number and kind of solutions for each of the following equations.

1. $x^2 - 6x + 5 = 0$
2. $x^2 - x - 12 = 0$
3. $4x^2 - 4x = -1$
4. $9x^2 + 12x = -4$
5. $x^2 + x - 1 = 0$
6. $x^2 - 2x + 3 = 0$
7. $2y^2 = 3y + 1$
8. $3y^2 = 4y - 2$
9. $x^2 - 9 = 0$
10. $4x^2 - 81 = 0$
11. $5a^2 - 4a = 5$
12. $3a = 4a^2 - 5$

Determine k so that each of the following has exactly one rational solution.

13. $x^2 - kx + 25 = 0$
14. $x^2 + kx + 25 = 0$
15. $x^2 = kx - 36$
16. $x^2 = kx - 49$
17. $4x^2 - 12x + k = 0$
18. $9x^2 + 30x + k = 0$
19. $kx^2 - 40x = 25$
20. $kx^2 - 2x = -1$
21. $3x^2 - kx + 2 = 0$
22. $5x^2 + kx + 1 = 0$

For each of the following problems, find an equation that has the given solutions.

23. $x = 5, x = 2$
24. $x = -5, x = -2$
25. $t = -3, t = 6$
26. $t = -4, t = 2$
27. $y = 2, y = -2, y = 4$
28. $y = 1, y = -1, y = 3$
29. $x = \frac{1}{2}, x = 3$
30. $x = \frac{1}{3}, x = 5$
31. $t = -\frac{3}{4}, t = 3$
32. $t = -\frac{4}{5}, t = 2$
33. $x = 3, x = -3, x = \frac{5}{6}$
34. $x = 5, x = -5, x = \frac{2}{3}$
35. $a = -\frac{1}{2}, a = \frac{3}{5}$
36. $a = -\frac{1}{3}, a = \frac{4}{7}$
37. $x = -\frac{2}{3}, x = \frac{2}{3}, x = 1$
38. $x = -\frac{4}{5}, x = \frac{4}{5}, x = -1$
39. $x = 2, x = -2, x = 3, x = -3$
40. $x = 1, x = -1, x = 5, x = -5$
41. $x = \sqrt{7}, x = -\sqrt{7}$
42. $x = -\sqrt{3}, x = \sqrt{3}$
43. $x = 5i, x = -5i$
44. $x = -2i, x = 2i$
45. $x = 1 + i, x = 1 - i$
46. $x = 2 + 3i, x = 2 - 3i$
47. $x = -2 - 3i, x = -2 + 3i$
48. $x = -1 + i, x = -1 - i$

49. Find an equation that has a solution of $x = 3$ of multiplicity 1 and a solution $x = -5$ of multiplicity 2.

50. Find an equation that has a solution of $x = 5$ of multiplicity 1 and a solution $x = -3$ of multiplicity 2.

51. Find an equation that has solutions $x = 3$ and $x = -3$ both of multiplicity 2.

52. Find an equation that has solutions $x = 4$ and $x = -4$, both of multiplicity 2.

53. Find all solutions to $x^3 + 6x^2 + 11x + 6 = 0$, if $x = -3$ is one of its solutions.

54. Find all solutions to $x^3 + 10x^2 + 29x + 20 = 0$, if $x = -4$ is one of its solutions.

55. One solution to $y^3 + 5y^2 - 2y - 24 = 0$ is $y = -3$. Find all solutions.

56. One solution to $y^3 + 3y^2 - 10y - 24 = 0$ is $y = -2$. Find all solutions.

57. If $x = 3$ is one solution to $x^3 - 5x^2 + 8x = 6$, find the other solutions.

58. If $x = 2$ is one solution to $x^3 - 6x^2 + 13x = 10$, find the other solutions.

59. Find all solutions to $t^3 = 13t^2 - 65t + 125$, if $t = 5$ is one of the solutions.

60. Find all solutions to $t^3 = 8t^2 - 25t + 26$, if $t = 2$ is one of the solutions.

Getting Ready for the Next Section

Simplify.

61. $(x + 3)^2 - 2(x + 3) - 8$

62. $(x - 2)^2 - 3(x - 2) - 10$

63. $(2a - 3)^2 - 9(2a - 3) + 20$

64. $(3a - 2)^2 + 2(3a - 2) - 3$

65. $2(4a + 2)^2 - 3(4a + 2) - 20$

66. $6(2a + 4)^2 - (2a + 4) - 2$

Solve.

67. $x^2 = \frac{1}{4}$

68. $x^2 = -2$

69. $\sqrt{x} = -3$

70. $\sqrt{x} = 2$

71. $x + 3 = 4$

72. $x + 3 = -2$

73. $y^2 - 2y - 8 = 0$

74. $y^2 + y - 6 = 0$

75. $4y^2 + 7y - 2 = 0$

76. $6x^2 - 13x - 5 = 0$

More Equations 7.4

We are now in a position to put our knowledge of quadratic equations to work to solve a variety of equations.

EXAMPLE 1 Solve $(x + 3)^2 - 2(x + 3) - 8 = 0$.

SOLUTION We can see that this equation is quadratic in form by replacing $x + 3$ with another variable, say, y. Replacing $x + 3$ with y we have

$$y^2 - 2y - 8 = 0$$

We can solve this equation by factoring the left side and then setting each factor equal to 0.

$$y^2 - 2y - 8 = 0$$
$$(y - 4)(y + 2) = 0 \quad \text{Factor}$$
$$y - 4 = 0 \quad \text{or} \quad y + 2 = 0 \quad \text{Set factors to 0}$$
$$y = 4 \quad \text{or} \quad y = -2$$

Because our original equation was written in terms of the variable x, we want our solutions in terms of x also. Replacing y with $x + 3$ and then solving for x, we have

$$x + 3 = 4 \quad \text{or} \quad x + 3 = -2$$
$$x = 1 \quad \text{or} \quad x = -5$$

The solutions to our original equation are 1 and -5.

The method we have just shown lends itself well to other types of equations that are quadratic in form, as we will see. In this example, however, there is another method that works just as well. Let's solve our original equation again, but this time, let's begin by expanding $(x + 3)^2$ and $2(x + 3)$.

$$(x + 3)^2 - 2(x + 3) - 8 = 0$$
$$x^2 + 6x + 9 - 2x - 6 - 8 = 0 \quad \text{Multiply}$$
$$x^2 + 4x - 5 = 0 \quad \text{Combine similar terms}$$
$$(x - 1)(x + 5) = 0 \quad \text{Factor}$$
$$x - 1 = 0 \quad \text{or} \quad x + 5 = 0 \quad \text{Set factors to 0}$$
$$x = 1 \quad \text{or} \quad x = -5$$

As you can see, either method produces the same result.

EXAMPLE 2 Solve $4x^4 + 7x^2 = 2$.

SOLUTION This equation is quadratic in x^2. We can make it easier to look at by using the substitution $y = x^2$. (The choice of the letter y is arbitrary. We could just as easily use the substitution $m = x^2$.) Making the substitution $y = x^2$ and then solving the resulting equation we have

$$4y^2 + 7y = 2$$
$$4y^2 + 7y - 2 = 0 \quad \text{Standard form}$$
$$(4y - 1)(y + 2) = 0 \quad \text{Factor}$$
$$4y - 1 = 0 \quad \text{or} \quad y + 2 = 0 \quad \text{Set factors to 0}$$
$$y = \frac{1}{4} \quad \text{or} \quad y = -2$$

Now we replace y with x^2 to solve for x:

$$x^2 = \frac{1}{4} \quad \text{or} \quad x^2 = -2$$

$$x = \pm\sqrt{\frac{1}{4}} \quad \text{or} \quad x = \pm\sqrt{-2} \qquad \text{Square Root Property for Equations}$$

$$x = \pm\frac{1}{2} \quad \text{or} \quad = \pm i\sqrt{2}$$

The solution set is $\left\{\frac{1}{2}, -\frac{1}{2}, i\sqrt{2}, -i\sqrt{2}\right\}$.

EXAMPLE 3 Solve for x: $x + \sqrt{x} - 6 = 0$

SOLUTION To see that this equation is quadratic in form, we have to notice that $(\sqrt{x})^2 = x$. That is, the equation can be rewritten as

$$(\sqrt{x})^2 + \sqrt{x} - 6 = 0$$

Replacing $\sqrt{x}$ with y and solving as usual, we have

$$y^2 + y - 6 = 0$$

$$(y + 3)(y - 2) = 0$$

$$y + 3 = 0 \quad \text{or} \quad y - 2 = 0$$

$$y = -3 \quad \text{or} \quad y = 2$$

Again, to find x, we replace y with x and solve:

$$\sqrt{x} = -3 \quad \text{or} \quad \sqrt{x} = 2$$

$$x = 9 \quad \text{or} \quad x = 4 \qquad \text{Square both sides of each equation}$$

Because we squared both sides of each equation, we have the possibility of obtaining extraneous solutions. We have to check both solutions in our original equation.

When	$x = 9$	When	$x = 4$
the equation	$x + \sqrt{x} - 6 = 0$	the equation	$x + \sqrt{x} - 6 = 0$
becomes	$9 + \sqrt{9} - 6 \stackrel{?}{=} 0$	becomes	$4 + \sqrt{4} - 6 \stackrel{?}{=} 0$
	$9 + 3 - 6 \stackrel{?}{=} 0$		$4 + 2 - 6 \stackrel{?}{=} 0$
	$6 \neq 0$		$0 = 0$
	This means 9 is extraneous		This means 4 is a solution

The only solution to the equation $x + \sqrt{x} - 6 = 0$ is $x = 4$.

We should note here that the two possible solutions, 9 and 4, to the equation in Example 3 can be obtained by another method. Instead of substituting for x, we can isolate it on one side of the equation and then square both sides to clear the equation of radicals.

$$
\begin{aligned}
x + \sqrt{x} - 6 &= 0 \\
\sqrt{x} &= -x + 6 && \text{Isolate } \sqrt{x} \\
x &= x^2 - 12x + 36 && \text{Square both sides} \\
0 &= x^2 - 13x + 36 && \text{Add } -x \text{ to both sides} \\
0 &= (x - 4)(x - 9) && \text{Factor}
\end{aligned}
$$

$$x - 4 = 0 \quad \text{or} \quad x - 9 = 0$$

$$x = 4 \qquad\qquad x = 9$$

We obtain the same two possible solutions. Because we squared both sides of the equation to find them, we would have to check each one in the original equation. As was the case in Example 3, only $x = 4$ is a solution; $x = 9$ is extraneous.

EXAMPLE 4 If an object is tossed into the air with an upward velocity of 12 feet per second from the top of a building h feet high, the time it takes for the object to hit the ground below is given by the formula

$$16t^2 - 12t - h = 0$$

Solve this formula for t.

SOLUTION The formula is in standard form and is quadratic in t. The coefficients a, b, and c that we need to apply to the quadratic formula are $a = 16$, $b = -12$, and $c = -h$. Substituting these quantities into the quadratic formula, we have

$$
\begin{aligned}
t &= \frac{12 \pm \sqrt{144 - 4(16)(-h)}}{2(16)} \\
&= \frac{12 \pm \sqrt{144 + 64h}}{32}
\end{aligned}
$$

We can factor the perfect square 16 from the two terms under the radical and simplify our radical somewhat:

$$
\begin{aligned}
t &= \frac{12 \pm \sqrt{16(9 + 4h)}}{32} \\
&= \frac{12 \pm 4\sqrt{9 + 4h}}{32}
\end{aligned}
$$

Now we can reduce to lowest terms by factoring a 4 from the numerator and denominator:

$$
\begin{aligned}
t &= \frac{4(3 \pm \sqrt{9 + 4h})}{4 \cdot 8} \\
&= \frac{3 \pm \sqrt{9 + 4h}}{8}
\end{aligned}
$$

If we were given a value of h, we would find that one of the solutions to this last formula would be a negative number. Because time is always measured in positive units, we wouldn't use that solution. ■

USING TECHNOLOGY *Graphing Calculators*

More About Example 1

As we mentioned before, algebraic expressions entered into a graphing calculator do not have to be simplified to be evaluated. This fact also applies to equations. We can graph the equation $y = (x + 3)^2 - 2(x + 3) - 8$ to assist us in solving the equation in Example 1. The graph is shown in Figure 1. Using the Zoom and Trace features at the x-intercepts gives us $x = 1$ and $x = -5$ as the solutions to the equation $0 = (x + 3)^2 - 2(x + 3) - 8$.

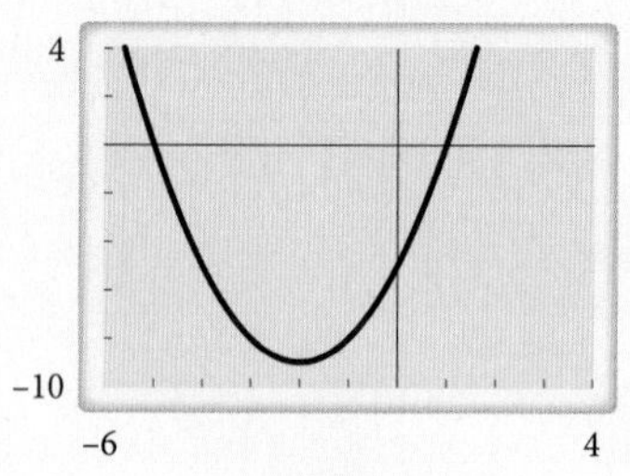

FIGURE 1

More About Example 2

Figure 2 shows the graph of $y = 4x^4 + 7x^2 - 2$. As we expect, the x-intercepts give the real number solutions to the equation $0 = 4x^4 + 7x^2 - 2$. The complex solutions do not appear on the graph.

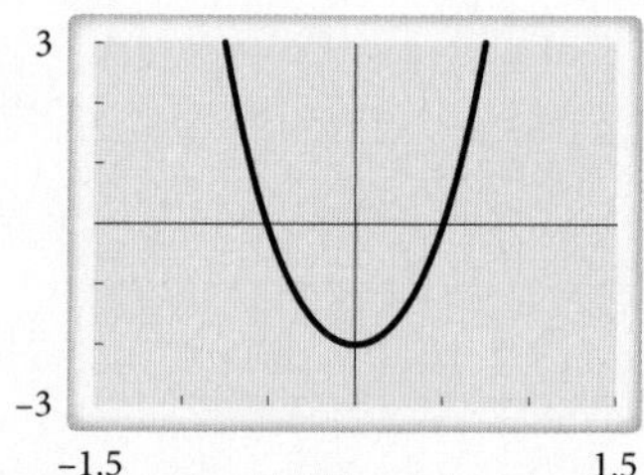

FIGURE 2

More About Example 3

In solving the equation in Example 3, we found that one of the possible solutions was an extraneous solution. If we solve the equation $x + \sqrt{x} - 6 = 0$ by graphing the function $y = x + \sqrt{x} - 6$, we find that the extraneous solution, 9, is not an x-intercept. Figure 3 shows that the only solution to the equation occurs at the x-intercept 4.

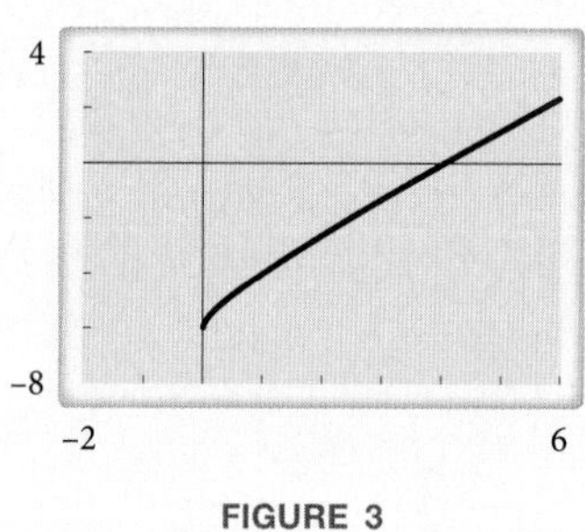

FIGURE 3

GETTING READY FOR CLASS

After reading through the preceding section, respond in your own words and in complete sentences.

A. What does it mean for an equation to be quadratic in form?

B. What are all the circumstances in solving equations (that we have studied) in which it is necessary to check for extraneous solutions?

C. How would you start to solve the equation $x + \sqrt{x} - 6 = 0$?

D. Is 9 a solution to $x + \sqrt{x} - 6 = 0$?

Solve each equation.

1. $(x - 3)^2 + 3(x - 3) + 2 = 0$
2. $(x + 4)^2 - (x + 4) - 6 = 0$
3. $2(x + 4)^2 + 5(x + 4) - 12 = 0$
4. $3(x - 5)^2 + 14(x - 5) - 5 = 0$
5. $x^4 - 6x^2 - 27 = 0$
6. $x^4 + 2x^2 - 8 = 0$
7. $x^4 + 9x^2 = -20$
8. $x^4 - 11x^2 = -30$
9. $(2a - 3)^2 - 9(2a - 3) = -20$
10. $(3a - 2)^2 + 2(3a - 2) = 3$
11. $2(4a + 2)^2 = 3(4a + 2) + 20$
12. $6(2a + 4)^2 = (2a + 4) + 2$
13. $6t^4 = -t^2 + 5$
14. $3t^4 = -2t^2 + 8$
15. $9x^4 - 49 = 0$
16. $25x^4 - 9 = 0$

Solve each of the following equations. Remember, if you square both sides of an equation in the process of solving it, you have to check all solutions in the original equation.

17. $x - 7\sqrt{x} + 10 = 0$
18. $x - 6\sqrt{x} + 8 = 0$
19. $t - 2\sqrt{t} - 15 = 0$
20. $t - 3\sqrt{t} - 10 = 0$
21. $6x + 11\sqrt{x} = 35$
22. $2x + \sqrt{x} = 15$
23. $(a - 2) - 11\sqrt{a - 2} + 30 = 0$
24. $(a - 3) - 9\sqrt{a - 3} + 20 = 0$
25. $(2x + 1) - 8\sqrt{2x + 1} + 15 = 0$
26. $(2x - 3) - 7\sqrt{2x - 3} + 12 = 0$
27. Solve the formula $16t^2 - vt - h = 0$ for t.
28. Solve the formula $16t^2 + vt + h = 0$ for t.
29. Solve the formula $kx^2 + 8x + 4 = 0$ for x.
30. Solve the formula $k^2x^2 + kx + 4 = 0$ for x.
31. Solve $x^2 + 2xy + y^2 = 0$ for x by using the quadratic formula with $a = 1$, $b = 2y$, and $c = y^2$.
32. Solve $x^2 - 2xy + y^2 = 0$ for x by using the quadratic formula, with $a = 1$, $b = -2y$, $c = y^2$.

Applying the Concepts

For Problems 33 and 34, t is in seconds.

33. **Falling Object** An object is tossed into the air with an upward velocity of 8 feet per second from the top of a building h feet high. The time it takes for the object to hit the ground below is given by the formula $16t^2 - 8t - h = 0$. Solve this formula for t.
34. **Falling Object** An object is tossed into the air with an upward velocity of 6 feet per second from the top of a building h feet high. The time it takes for the object to hit the ground below is given by the formula $16t^2 - 6t - h = 0$. Solve this formula for t.

iStockPhoto.com/©Jeremy Edwards

35. **Saint Louis Arch** The shape of the famous "Gateway to the West" arch in Saint Louis can be modeled by a parabola. The equation for one such parabola is:

$$y = -\frac{1}{150}x^2 + \frac{21}{5}x$$

a. Sketch the graph of the arch's equation on a coordinate axis.

b. Approximately how far do you have to walk to get from one side of the arch to the other?

36. **Area** In the following diagram, *ABCD* is a rectangle with diagonal *AC*. Find its area.

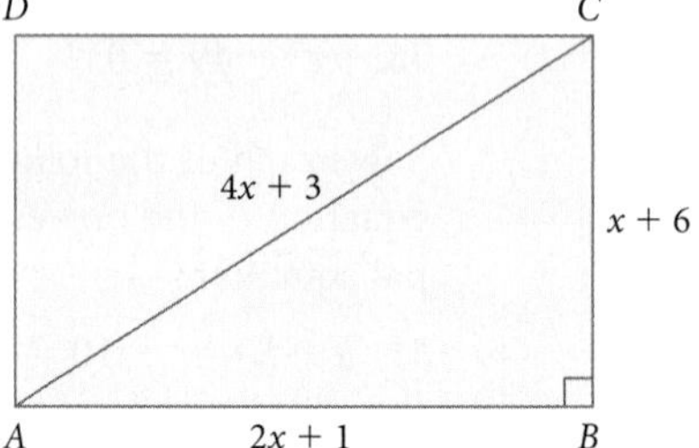

37. **Area and Perimeter** A total of 160 yards of fencing is to be used to enclose part of a lot that borders on a river. This situation is shown in the following diagram.

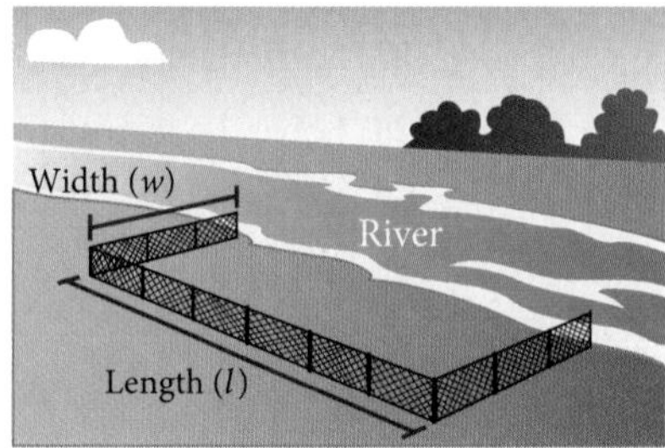

a. Write an equation that gives the relationship between the length and width and the 160 yards of fencing.

b. The formula for the area that is enclosed by the fencing and the river is $A = lw$. Solve the equation in part **a** for l, and then use the result to write the area in terms of w only.

c. Make a table that gives at least five possible values of w and associated area A.

d. From the pattern in your table shown in part **c**, what is the largest area that can be enclosed by the 160 yards of fencing? (Try some other table values if necessary.)

38. Area and Perimeter Rework all four parts of the preceding problem if it is desired to have an opening 2 yards wide in one of the shorter sides, as shown in the diagram.

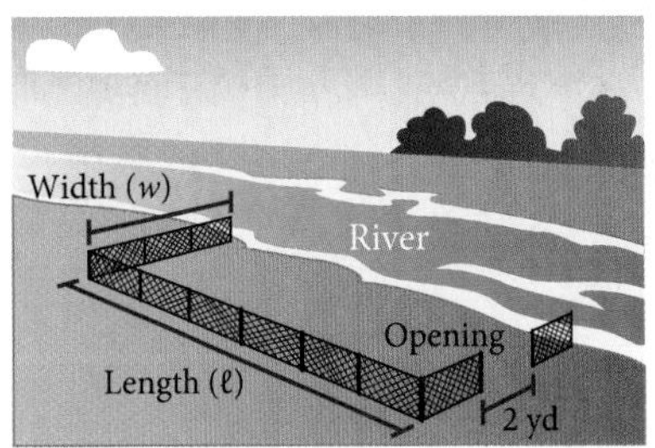

Getting Ready for the Next Section

39. Evaluate $y = 3x^2 - 6x + 1$ for $x = 1$.

40. Evaluate $y = -2x^2 + 6x - 5$ for $x = \frac{3}{2}$.

41. Let $P(x) = -0.1x^2 + 27x - 500$ and find $P(135)$.

42. Let $P(x) = -0.1x^2 + 12x - 400$ and find $P(600)$.

Solve.

43. $0 = a(80)^2 + 70$

44. $0 = a(80)^2 + 90$

45. $x^2 - 6x + 5 = 0$

46. $x^2 - 3x - 4 = 0$

47. $-x^2 - 2x + 3 = 0$

48. $-x^2 + 4x + 12 = 0$

49. $2x^2 - 6x + 5 = 0$

50. $x^2 - 4x + 5 = 0$

Fill in the blanks to complete the square.

51. $x^2 - 6x + \square = (x - \square)^2$

52. $x^2 - 10x + \square = (x - \square)^2$

53. $y^2 + 2y + \square = (y + \square)^2$

54. $y^2 - 12y + \square = (x - \square)^2$

Graphing Parabolas 7.5

The solution set to the equation

$$y = x^2 - 3$$

consists of ordered pairs. One method of graphing the solution set is to find a number of ordered pairs that satisfy the equation and to graph them. We can obtain some ordered pairs that are solutions to $y = x^2 - 3$ by use of a table as follows:

x	$y = x^2 - 3$	y	Solutions
-3	$y = (-3)^2 - 3 = 9 - 3 = 6$	6	$(-3, 6)$
-2	$y = (-2)^2 - 3 = 4 - 3 = 1$	1	$(-2, 1)$
-1	$y = (-1)^2 - 3 = 1 - 3 = -2$	-2	$(-1, -2)$
0	$y = 0^2 - 3 = 0 - 3 = -3$	-3	$(0, -3)$
1	$y = 1^2 - 3 = 1 - 3 = -2$	-2	$(1, -2)$
2	$y = 2^2 - 3 = 4 - 3 = 1$	1	$(2, 1)$
3	$y = 3^2 - 3 = 9 - 3 = 6$	6	$(3, 6)$

Graphing these solutions and then connecting them with a smooth curve, we have the graph of $y = x^2 - 3$. (See Figure 1.)

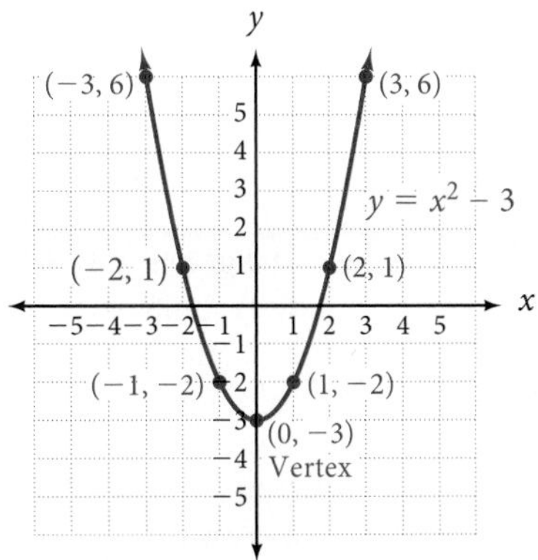

FIGURE 1

This graph is an example of a *parabola*. All equations of the form $y = ax^2 + bx + c$, $a \neq 0$, have parabolas for graphs.

Although it is always possible to graph parabolas by making a table of values of x and y that satisfy the equation, there are other methods that are faster and, in some cases, more accurate.

The important points associated with the graph of a parabola are the highest (or lowest) point on the graph and the x-intercepts. The y-intercepts can also be useful.

Intercepts for Parabolas

The graph of the equation $y = ax^2 + bx + c$ crosses the y-axis at $y = c$, because substituting $x = 0$ into $y = ax^2 + bx + c$ yields $y = c$.

Because the graph crosses the x-axis when $y = 0$, the x-intercepts are those values of x that are solutions to the quadratic equation $0 = ax^2 + bx + c$.

The Vertex of a Parabola

The highest or lowest point on a parabola is called the *vertex*. The vertex for the graph of $y = ax^2 + bx + c$ will always occur when

$$x = \frac{-b}{2a}$$

To see this, we must transform the right side of $y = ax^2 + bx + c$ into an expression that contains x in just one of its terms. This is accomplished by completing the square on the first two terms. Here is what it looks like:

$$\begin{aligned} y &= ax^2 + bx + c \\ &= a\left(x^2 + \frac{b}{a}x\right) + c \\ &= a\left[x^2 + \frac{b}{a}x + \left(\frac{b}{2a}\right)^2\right] + c - a\left(\frac{b}{2a}\right)^2 \\ &= a\left(x + \frac{b}{2a}\right)^2 + \frac{4ac - b^2}{4a} \end{aligned}$$

It may not look like it, but this last line indicates that the vertex of the graph of $y = ax^2 + bx + c$ has an x-coordinate of $\frac{-b}{2a}$. Because a, b, and c are constants, the only quantity that is varying in the last expression is the x in $\left(x + \frac{b}{2a}\right)^2$. Because the quantity $\left(x + \frac{b}{2a}\right)^2$ is the square of $x + \frac{b}{2a}$, the smallest it will ever be is 0, and that will happen when $x = \frac{-b}{2a}$.

We can use the vertex point along with the x- and y-intercepts to sketch the graph of any equation of the form $y = ax^2 + bx + c$. Here is a summary of the preceding information.

Graphing Parabolas I

The graph of $y = ax^2 + bx + c$, $a \neq 0$, will be a parabola with

1. A y-intercept at $y = c$
2. x-intercepts (if they exist) at

$$x = \frac{-b \pm \sqrt{b^2 - 4ac}}{2a}$$

3. A vertex when $x = \dfrac{-b}{2a}$

EXAMPLE 1 Sketch the graph of $y = x^2 - 6x + 5$.

SOLUTION To find the x-intercepts, we let $y = 0$ and solve for x:

$$\begin{aligned} 0 &= x^2 - 6x + 5 \\ 0 &= (x - 5)(x - 1) \\ x &= 5 \quad \text{or} \quad x = 1 \end{aligned}$$

To find the coordinates of the vertex, we first find

$$x = \frac{-b}{2a} = \frac{-(-6)}{2(1)} = 3$$

The x-coordinate of the vertex is 3. To find the y-coordinate, we substitute 3 for x in our original equation:

$$y = 3^2 - 6(3) + 5 = 9 - 18 + 5 = -4$$

The graph crosses the x-axis at 1 and 5 and has its vertex at $(3, -4)$. Plotting these points and connecting them with a smooth curve, we have the graph shown in Figure 2.

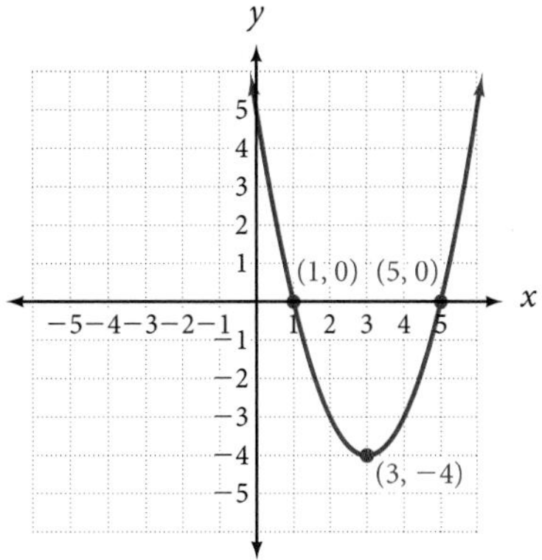

FIGURE 2

The graph is a parabola that opens up, so we say the graph is *concave up*. The vertex is the lowest point on the graph. (Note that the graph crosses the y-axis at 5, which is the value of y we obtain when we let $x = 0$.)

Finding the Vertex by Completing the Square

Another way to locate the vertex of the parabola in Example 1 is by completing the square on the first two terms on the right side of the equation $y = x^2 - 6x + 5$. In this case, we would do so by adding 9 to and subtracting 9 from the right side of the equation. This amounts to adding 0 to the equation, so we know we haven't changed its solutions. This is what it looks like:

$$\begin{aligned} y &= (x^2 - 6x \qquad) + 5 \\ &= (x^2 - 6x + 9) + 5 - 9 \\ &= (x - 3)^2 - 4 \end{aligned}$$

You may have to look at this last equation awhile to see this, but when $x = 3$, then $y = (x - 3)^2 - 4 = 0^2 - 4 = -4$ is the smallest y will ever be. That is why the vertex is at $(3, -4)$. As a matter of fact, this is the same kind of reasoning we used when we derived the formula $x = -\frac{b}{2a}$ for the x-coordinate of the vertex.

EXAMPLE 2 Graph $y = -x^2 - 2x + 3$.

SOLUTION To find the x-intercepts, we let $y = 0$:

$$0 = -x^2 - 2x + 3$$

$$0 = x^2 + 2x - 3 \qquad \text{Multiply each side by } -1$$

$$0 = (x + 3)(x - 1)$$

$$x = -3 \quad \text{or} \quad x = 1$$

The x-coordinate of the vertex is given by

$$x = \frac{-b}{2a} = \frac{-(-2)}{2(-1)} = \frac{2}{-2} = -1$$

To find the y-coordinate of the vertex, we substitute -1 for x in our original equation to get

$$y = -(-1)^2 - 2(-1) + 3 = -1 + 2 + 3 = 4$$

Our parabola has x-intercepts at -3 and 1, and a vertex at $(-1, 4)$. Figure 3 shows the graph.

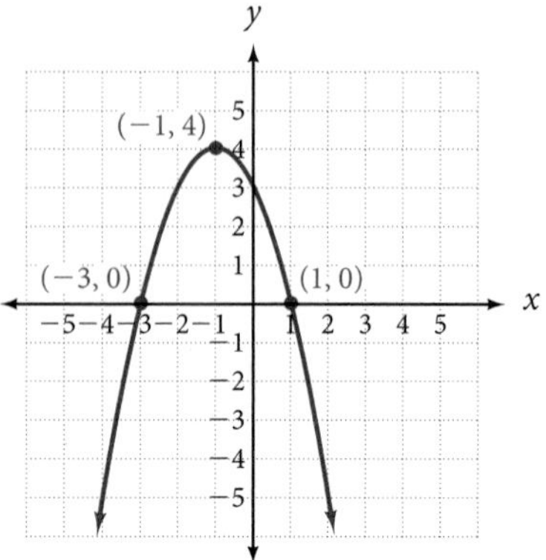

FIGURE 3

We say the graph is *concave down* because it opens downward. Again, we could have obtained the coordinates of the vertex by completing the square on the first two terms on the right side of our equation. To do so, we must first factor -1 from the first two terms. (Remember, the leading coefficient must be 1 to complete the square.) When we complete the square, we add 1 inside the parentheses, which actually decreases the right side of the equation by -1 because everything in the parentheses is multiplied by -1. To make up for it, we add 1 outside the parentheses.

$$\begin{aligned} y &= -1(x^2 + 2x \qquad) + 3 \\ &= -1(x^2 + 2x + 1) + 3 + 1 \\ &= -1(x + 1)^2 + 4 \end{aligned}$$

The last line tells us that the *largest* value of y will be 4, and that will occur when $x = -1$.

EXAMPLE 3 Graph $y = 3x^2 - 6x + 1$.

SOLUTION To find the x-intercepts, we let $y = 0$ and solve for x:

$$0 = 3x^2 - 6x + 1$$

Because the right side of this equation does not factor, we can look at the discrim-inant to see what kind of solutions are possible. The discriminant for this equation is

$$b^2 - 4ac = 36 - 4(3)(1) = 24$$

Because the discriminant is a positive number but not a perfect square, the equation will have irrational solutions. This means that the x-intercepts are irrational numbers and will have to be approximated with decimals using the quadratic formula. Rather than use the quadratic formula, we will find some other points on the graph, but first let's find the vertex.

Here are both methods of finding the vertex:

Using the formula that gives us the x-coordinate of the vertex, we have:

$$x = \frac{-b}{2a} = \frac{-(-6)}{2(3)} = 1$$

Substituting 1 for x in the equation gives us the y-coordinate of the vertex:

$$y = 3 \cdot 1^2 - 6 \cdot 1 + 1 = -2$$

To complete the square on the right side of the equation, we factor 3 from the first two terms, add 1 inside the parentheses, and add -3 outside the parentheses (this amounts to adding 0 to the right side):

$$\begin{aligned} y &= 3(x^2 - 2x \qquad) + 1 \\ &= 3(x^2 - 2x + 1) + 1 - 3 \\ &= 3(x - 1)^2 - 2 \end{aligned}$$

In either case, the vertex is $(1, -2)$.

If we can find two points, one on each side of the vertex, we can sketch the graph. Let's let $x = 0$ and $x = 2$, because each of these numbers is the same distance from $x = 1$, and $x = 0$ will give us the y-intercept.

When $x = 0$

$$\begin{aligned} y &= 3(0)^2 - 6(0) + 1 \\ &= 0 - 0 + 1 \\ &= 1 \end{aligned}$$

When $x = 2$

$$\begin{aligned} y &= 3(2)^2 - 6(2) + 1 \\ &= 12 - 12 + 1 \\ &= 1 \end{aligned}$$

The two points just found are $(0, 1)$ and $(2, 1)$. Plotting these two points along with the vertex $(1, -2)$, we have the graph shown in Figure 4.

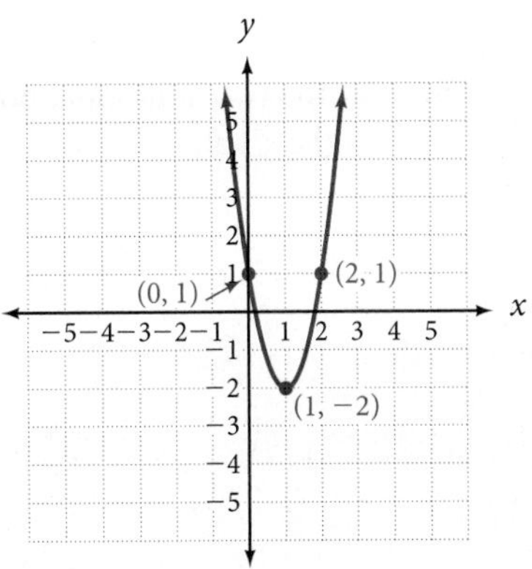

FIGURE 4

EXAMPLE 4 Graph $y = -2x^2 + 6x - 5$.

SOLUTION Letting $y = 0$, we have

$$0 = -2x^2 + 6x - 5$$

Again, the right side of this equation does not factor. The discriminant is $b^2 - 4ac = 36 - 4(-2)(-5) = -4$, which indicates that the solutions are complex numbers. This means that our original equation does not have x-intercepts. The graph does not cross the x-axis.

Let's find the vertex.

Using our formula for the x-coordinate of the vertex, we have

$$x = \frac{-b}{2a} = \frac{-6}{2(-2)} = \frac{6}{4} = \frac{3}{2}$$

To find the y-coordinate, we let $x = \frac{3}{2}$:

$$\begin{aligned} y &= -2\left(\frac{3}{2}\right)^2 + 6\left(\frac{3}{2}\right) - 5 \\ &= \frac{-18}{4} + \frac{18}{2} - 5 \\ &= \frac{-18 + 36 - 20}{4} \\ &= -\frac{1}{2} \end{aligned}$$

Finding the vertex by completing the square is a more complicated matter. To make the coefficient of x^2 a 1, we must factor -2 from the first two terms. To complete the square inside the parentheses, we add $\frac{9}{4}$. Since each term inside the parentheses is multiplied by -2, we add $\frac{9}{2}$ outside the parentheses so that the net result is the same as adding 0 to the right side:

$$\begin{aligned} y &= -2(x^2 - 3x \qquad) - 5 \\ &= -2\left(x^2 - 3x + \frac{9}{4}\right) - 5 + \frac{9}{2} \\ &= -2\left(x - \frac{3}{2}\right)^2 - \frac{1}{2} \end{aligned}$$

The vertex is $\left(\frac{3}{2}, -\frac{1}{2}\right)$. Because this is the only point we have so far, we must find two others. Let's let $x = 3$ and $x = 0$, because each point is the same distance from $x = \frac{3}{2}$ and on either side:

When $x = 3$

$$\begin{aligned} y &= -2(3)^2 + 6(3) - 5 \\ &= -18 + 18 - 5 \\ &= -5 \end{aligned}$$

When $x = 0$

$$\begin{aligned} y &= -2(0)^2 + 6(0) - 5 \\ &= 0 + 0 - 5 \\ &= -5 \end{aligned}$$

The two additional points on the graph are $(3, -5)$ and $(0, -5)$. Figure 5 shows the graph.

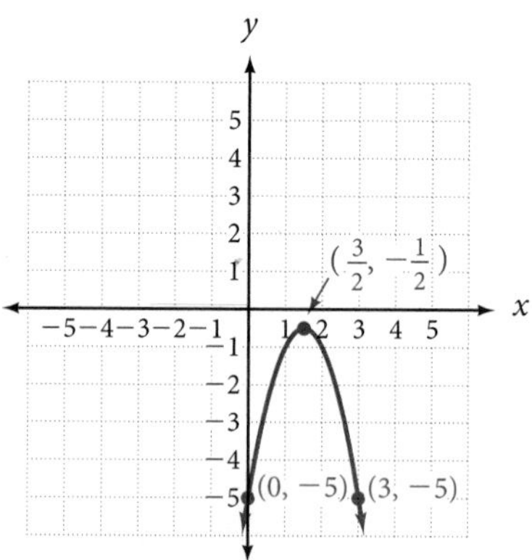

FIGURE 5

The graph is concave down. The vertex is the highest point on the graph.

By looking at the equations and graphs in Examples 1 through 4, we can conclude that the graph of $y = ax^2 + bx + c$ will be concave up when a is positive, and concave down when a is negative. Taking this even further, if $a > 0$, then the vertex is the lowest point on the graph, and if $a < 0$, the vertex is the highest point on the graph. Finally, if we complete the square on x in the equation $y = ax^2 + bx + c$, $a \neq 0$, we can rewrite the equation of our parabola as $y = a(x - h)^2 + k$. When the equation is in this form, the vertex is at the point (h, k). Here is a summary:

[Δ≠Σ] *Graphing Parabolas II*

The graph of

$$y = a(x - h)^2 + k, a \neq 0$$

will be a parabola with a vertex at (h, k). The vertex will be the highest point on the graph when $a < 0$, and the lowest point on the graph when $a > 0$.

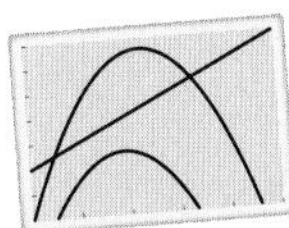

EXAMPLE 5 A company selling copies of an accounting program for home computers finds that it will make a weekly profit of P dollars from selling x copies of the program, according to the equation

$$P(x) = -0.1x^2 + 27x - 500$$

How many copies of the program should it sell to make the largest possible profit, and what is the largest possible profit?

SOLUTION Because the coefficient of x^2 is negative, we know the graph of this parabola will be concave down, meaning that the vertex is the highest point of the curve. We find the vertex by first finding its x-coordinate:

$$x = \frac{-b}{2a} = \frac{-27}{2(-0.1)} = \frac{27}{0.2} = 135$$

This represents the number of programs the company needs to sell each week to make a maximum profit. To find the maximum profit, we substitute 135 for x in the original equation. (A calculator is helpful for these kinds of calculations.)

$$\begin{aligned} P(135) &= -0.1(135)^2 + 27(135) - 500 \\ &= -0.1(18{,}225) + 3{,}645 - 500 \\ &= -1{,}822.5 + 3{,}645 - 500 \\ &= 1{,}322.5 \end{aligned}$$

The maximum weekly profit is \$1,322.50 and is obtained by selling 135 programs a week.

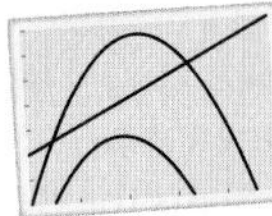

EXAMPLE 6 An art supply store finds that they can sell x sketch pads each week at p dollars each, according to the equation $x = 900 - 300p$. Graph the revenue equation $R = xp$. Then use the graph to find the price p that will bring in the maximum revenue. Finally, find the maximum revenue.

SOLUTION As it stands, the revenue equation contains three variables. Because we are asked to find the value of p that gives us the maximum value of R, we rewrite the equation using just the variables R and p. Because $x = 900 - 300p$, we have

$$R = xp = (900 - 300p)p$$

The graph of this equation is shown in Figure 6. The graph appears in the first quadrant only, because R and p are both positive quantities.

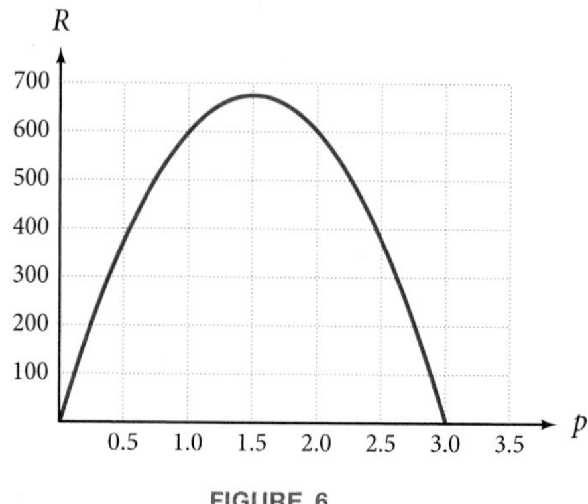

FIGURE 6

From the graph, we see that the maximum value of R occurs when $p = \$1.50$. We can calculate the maximum value of R from the equation:

When	$p = 1.5$
the equation	$R = (900 - 300p)p$
becomes	$R = (900 - 300 \cdot 1.5)1.5$
	$= (900 - 450)1.5$
	$= 450 \cdot 1.5$
	$= 675$

The maximum revenue is \$675. It is obtained by setting the price of each sketch pad at $p = \$1.50$.

USING TECHNOLOGY *Graphing Calculators*

If you have been using a graphing calculator for some of the material in this course, you are well aware that your calculator can draw all the graphs in this section very easily. It is important, however, that you be able to recognize and sketch the graph of any parabola by hand. It is a skill that all successful intermediate algebra students should possess, even if they are proficient in the use of a graphing calculator. My suggestion is that you work the problems in this section and problem set without your calculator. Then use your calculator to check your results.

Finding the Equation from the Graph

EXAMPLE 7 At the 1997 Washington County Fair in Oregon, David Smith, Jr., The Bullet, was shot from a cannon. As a human cannonball, he reached a height of 70 feet before landing in a net 160 feet from the cannon. Sketch the graph of his path, and then find the equation of the graph.

SOLUTION We assume that the path taken by the human cannonball is a parabola. If the origin of the coordinate system is at the opening of the cannon, then the net that catches him will be at 160 on the x-axis. Figure 7 shows the graph.

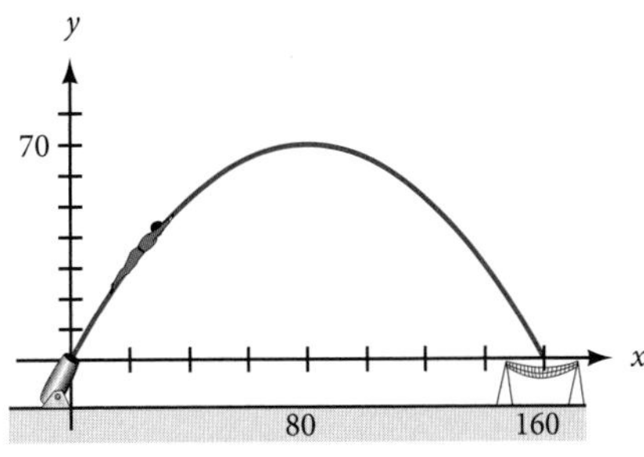

FIGURE 7

Because the curve is a parabola, we know the equation will have the form

$$y = a(x - h)^2 + k$$

Because the vertex of the parabola is at (80, 70), we can fill in two of the three constants in our equation, giving us

$$y = a(x - 80)^2 + 70$$

To find a, we note that the landing point will be (160, 0). Substituting the coordinates of this point into the equation, we solve for a:

$$0 = a(160 - 80)^2 + 70$$

$$0 = a(80)^2 + 70$$

$$0 = 6{,}400a + 70$$

$$a = -\frac{70}{6{,}400} = -\frac{7}{640}$$

The equation that describes the path of the human cannonball is

$$y = -\frac{7}{640}(x - 80)^2 + 70 \quad \text{for} \quad 0 \le x \le 160$$

USING TECHNOLOGY *Graphing Calculators*

Graph the equation found in Example 7 on a graphing calculator using the window shown here. (We will use this graph later in the book to find the angle between the cannon and the horizontal.)

Window: X from 0 to 180, increment 20
Y from 0 to 80, increment 10

On the TI-83, an increment of 20 for X means Xscl = 20.

GETTING READY FOR CLASS

After reading through the preceding section, respond in your own words and in complete sentences.

A. What is a parabola?

B. What part of the equation of a parabola determines whether the graph is concave up or concave down?

C. Suppose $f(x) = ax^2 + bx + c$ is the equation of a parabola. Explain how $f(4) = 1$ relates to the graph of the parabola.

D. A line can be graphed with two points. How many points are necessary to get a reasonable sketch of a parabola? Explain.

For each of the following equations, give the x-intercepts and the coordinates of the vertex, and sketch the graph.

1. $y = x^2 + 2x - 3$
2. $y = x^2 - 2x - 3$
3. $y = -x^2 - 4x + 5$
4. $y = x^2 + 4x - 5$
5. $y = x^2 - 1$
6. $y = x^2 - 4$
7. $y = -x^2 + 9$
8. $y = -x^2 + 1$
9. $y = 2x^2 - 4x - 6$
10. $y = 2x^2 + 4x - 6$
11. $y = x^2 - 2x - 4$
12. $y = x^2 - 2x - 2$

Graph each parabola. Label the vertex and any intercepts that exist.

13. $y = 2(x - 1)^2 + 3$
14. $y = 2(x + 1)^2 - 3$
15. $f(x) = -(x + 2)^2 + 4$
16. $f(x) = -(x - 3)^2 + 1$
17. $g(x) = \frac{1}{2}(x - 2)^2 - 4$
18. $g(x) = \frac{1}{3}(x - 3)^2 - 3$
19. $f(x) = -2(x - 4)^2 - 1$
20. $f(x) = -4(x - 1)^2 + 4$

Find the vertex and any two convenient points to sketch the graphs of the following equations.

21. $y = x^2 - 4x - 4$
22. $y = x^2 - 2x + 3$
23. $y = -x^2 + 2x - 5$
24. $y = -x^2 + 4x - 2$
25. $f(x) = x^2 + 1$
26. $f(x) = x^2 + 4$
27. $y = -x^2 - 3$
28. $y = -x^2 - 2$
29. $g(x) = 3x^2 + 4x + 1$
30. $g(x) = 2x^2 + 4x + 3$

For each of the following equations, find the coordinates of the vertex, and indicate whether the vertex is the highest point on the graph or the lowest point on the graph. (Do not graph.)

31. $y = x^2 - 6x + 5$
32. $y = -x^2 + 6x - 5$
33. $y = -x^2 + 2x + 8$
34. $y = x^2 - 2x - 8$
35. $y = 12 + 4x - x^2$
36. $y = -12 - 4x + x^2$
37. $y = -x^2 - 8x$
38. $y = x^2 + 8x$

Applying the Concepts

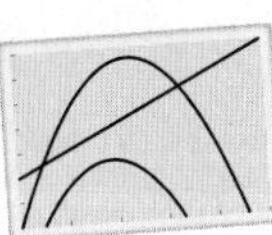

39. Maximum Profit A company finds that it can make a profit of P dollars each month by selling x patterns, according to the formula $P(x) = -0.002x^2 + 3.5x - 800$. How many patterns must it sell each month to have a maximum profit? What is the maximum profit?

40. Maximum Profit A company selling picture frames finds that it can make a profit of P dollars each month by selling x frames, according to the formula $P(x) = -0.002x^2 + 5.5x - 1{,}200$. How many frames must it sell each month to have a maximum profit? What is the maximum profit?

41. Maximum Height Chaudra is tossing a softball into the air with an underhand motion. The distance of the ball above her hand at any time is given by the function

$$h(t) = 32t - 16t^2 \quad \text{for} \quad 0 \le t \le 2$$

where $h(t)$ is the height of the ball (in feet) and t is the time (in seconds). Find the times at which the ball is in her hand, and the maximum height of the ball.

42. Maximum Area Justin wants to fence three sides of a rectangular exercise yard for his dog. The fourth side of the exercise yard will be a side of the house. He has 80 feet of fencing available. Find the dimensions of the exercise yard that will enclose the maximum area.

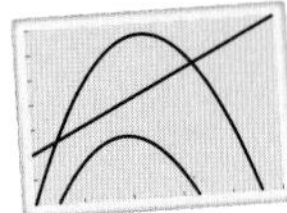

43. Maximum Revenue A company that manufactures typewriter ribbons knows that the number of ribbons x it can sell each week is related to the price p of each ribbon by the equation $x = 1{,}200 - 100p$. Graph the revenue equation $R = xp$. Then use the graph to find the price p that will bring in the maximum revenue. Finally, find the maximum revenue.

44. Maximum Revenue A company that manufactures diskettes for home computers finds that it can sell x diskettes each day at p dollars per diskette, according to the equation $x = 800 - 100p$. Graph the revenue equation $R = xp$. Then use the graph to find the price p that will bring in the maximum revenue. Finally, find the maximum revenue.

45. Maximum Revenue The relationship between the number of calculators x a company sells each day and the price p of each calculator is given by the equation $x = 1{,}700 - 100p$. Graph the revenue equation $R = xp$, and use the graph to find the price p that will bring in the maximum revenue. Then find the maximum revenue.

46. Maximum Revenue The relationship between the number x of pencil sharpeners a company sells each week and the price p of each sharpener is given by the equation $x = 1{,}800 - 100p$. Graph the revenue equation $R = xp$, and use the graph to find the price p that will bring in the maximum revenue. Then find the maximum revenue.

47. Human Cannonball A human cannonball is shot from a cannon at the county fair. He reaches a height of 60 feet before landing in a net 180 feet from the cannon. Sketch the graph of his path, and then find the equation of the graph.

48. Interpreting Graphs The graph below shows the different paths taken by the human cannonball when his velocity out of the cannon is 50 miles/hour, and his cannon is inclined at varying angles.

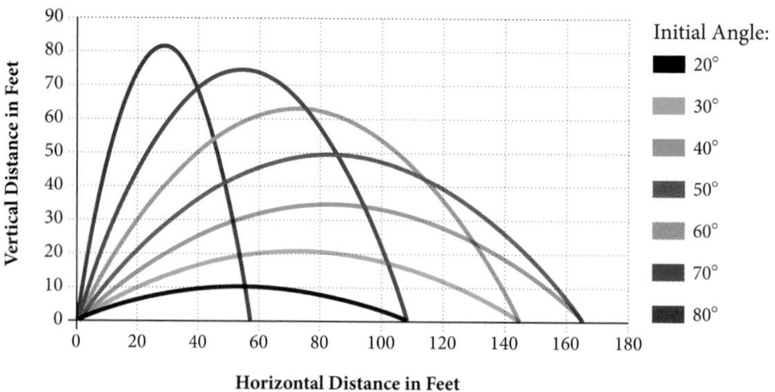

Initial Velocity: 50 miles per hour
Angle: 20°, 30°, 40°, 50°, 60°, 70°, 80°

a. If his landing net is placed 104 feet from the cannon, at what angle should the cannon be inclined so that he lands in the net?

b. Approximately where do you think he would land if the cannon was inclined at 45°?

c. If the cannon was inclined at 45°, approximately what height do you think he would attain?

d. Do you think there is another angle for which he would travel the same distance he travels at 80°? Give an estimate of that angle.

e. The fact that every landing point can come from two different paths makes us think that the equations that give us the landing points must be what type of equations?

Getting Ready for the Next Section

Solve.

49. $x^2 - 2x - 8 = 0$

50. $x^2 - x - 12 = 0$

51. $6x^2 - x = 2$

52. $3x^2 - 5x = 2$

53. $x^2 - 6x + 9 = 0$

54. $x^2 + 8x + 16 = 0$

Quadratic Inequalities 7.6

Quadratic inequalities in one variable are inequalities of the form

$$ax^2 + bx + c < 0 \qquad ax^2 + bx + c > 0$$

$$ax^2 + bx + c \leq 0 \qquad ax^2 + bx + c \geq 0$$

where a, b, and c are constants, with $a \neq 0$. The technique we will use to solve inequalities of this type involves graphing. Suppose, for example, we want to find the solution set for the inequality $x^2 - x - 6 > 0$. We begin by factoring the left side to obtain

$$(x - 3)(x + 2) > 0$$

We have two real numbers $x - 3$ and $x + 2$ whose product $(x - 3)(x + 2)$ is greater than zero. That is, their product is positive. The only way the product can be positive is either if both factors, $(x - 3)$ and $(x + 2)$, are positive or if they are both negative. To help visualize where $x - 3$ is positive and where it is negative, we draw a real number line and label it accordingly:

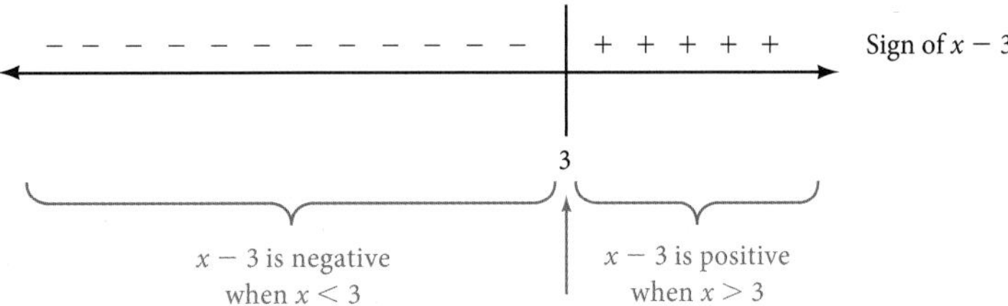

Here is a similar diagram showing where the factor $x + 2$ is positive and where it is negative:

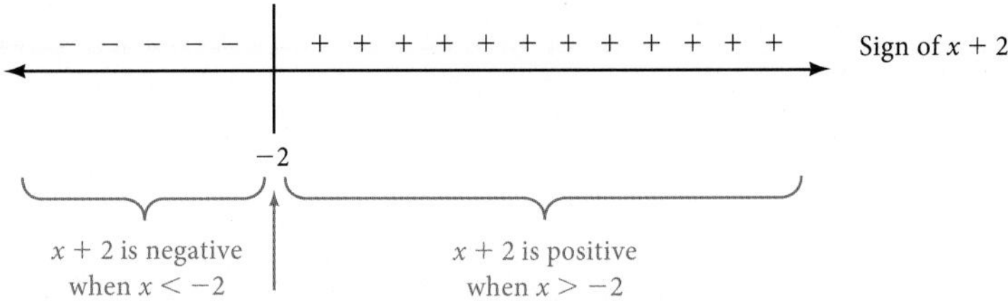

Drawing the two number lines together and eliminating the unnecessary numbers, we have

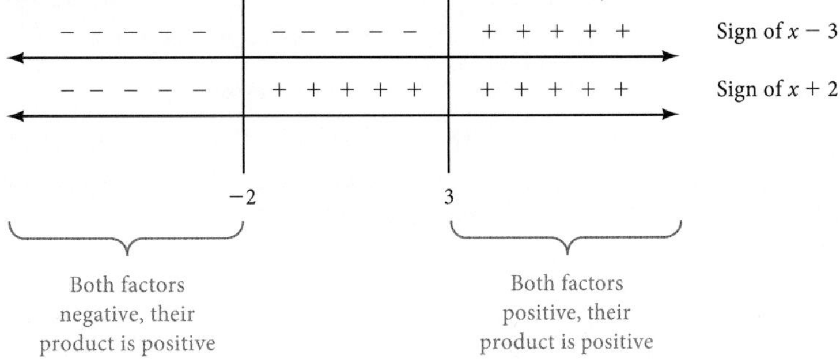

We can see from the preceding diagram that the graph of the solution to $x^2 - x - 6 > 0$ is

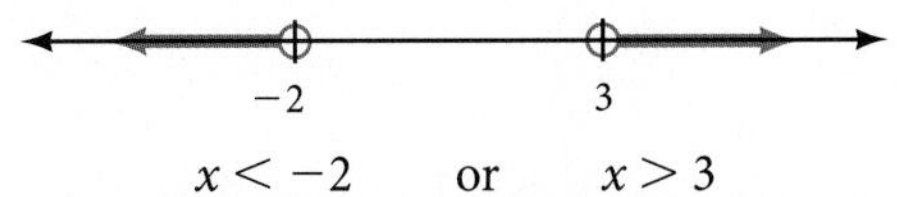

$$x < -2 \quad \text{or} \quad x > 3$$

USING TECHNOLOGY — *Graphical Solutions to Quadratic Inequalities*

We can solve the preceding problem by using a graphing calculator to visualize where the product $(x - 3)(x + 2)$ is positive. First, we graph the function $y = (x - 3)(x + 2)$ as shown in Figure 1.

Next, we observe where the graph is above the x-axis. As you can see, the graph is above the x-axis to the right of 3 and to the left of -2, as shown in Figure 2.

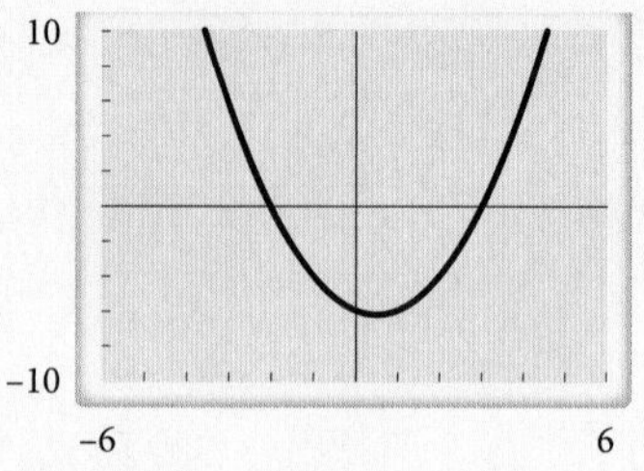

FIGURE 1

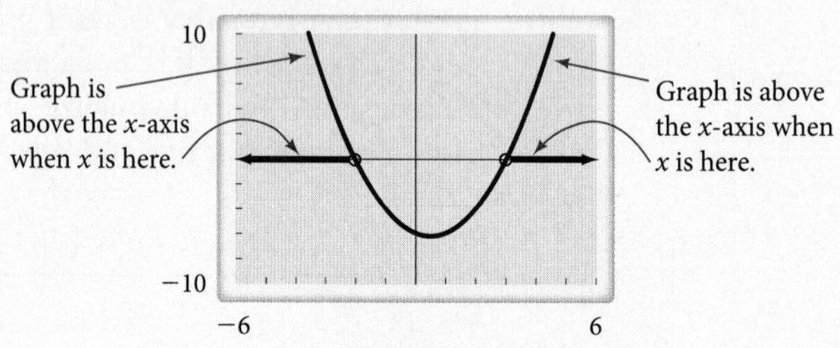

FIGURE 2

When the graph is above the x-axis, we have points whose y-coordinates are positive. Because these y-coordinates are the same as the expression $(x - 3)(x + 2)$, the values of x for which the graph of $y = (x - 3)(x + 2)$ is above the x-axis are the values of x for which the inequality $(x - 3)(x + 2) > 0$ is true. Our solution set is therefore

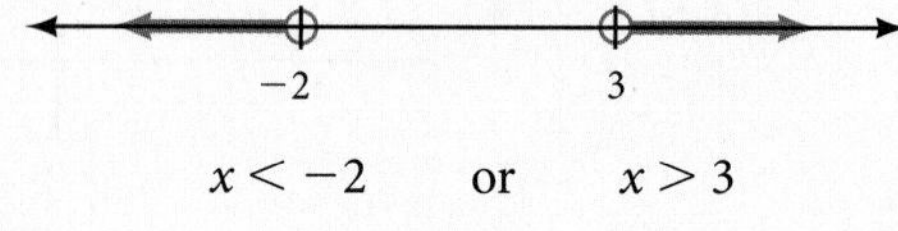

$$x < -2 \quad \text{or} \quad x > 3$$

EXAMPLE 1 Solve for x: $x^2 - 2x - 8 \le 0$.

ALGEBRAIC SOLUTION We begin by factoring:

$$x^2 - 2x - 8 \le 0$$

$$(x - 4)(x + 2) \le 0$$

The product $(x - 4)(x + 2)$ is negative or zero. The factors must have opposite signs. We draw a diagram showing where each factor is positive and where each factor is negative:

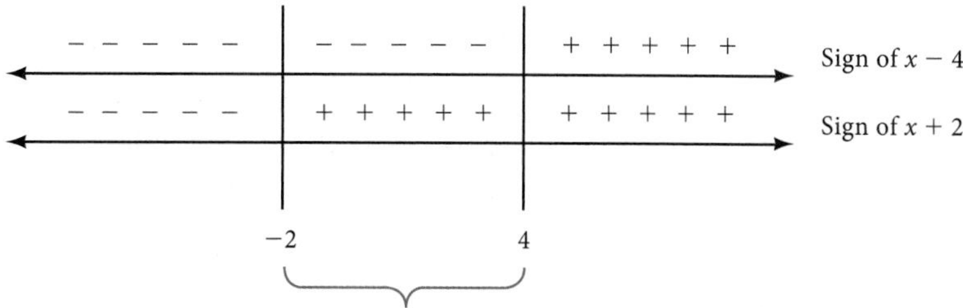

From the diagram, we have the graph of the solution set:

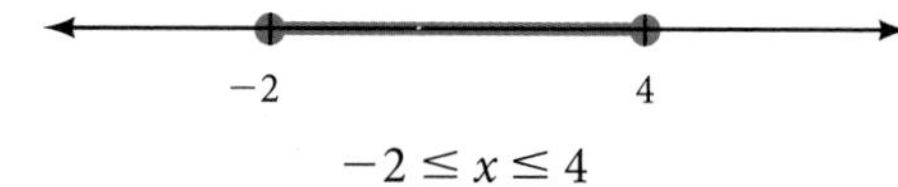

$$-2 \le x \le 4$$

GRAPHICAL SOLUTION To solve this inequality with a graphing calculator, we graph the function $y = (x - 4)(x + 2)$ and observe where the graph is below the x-axis. These points have negative y-coordinates, which means that the product $(x - 4)(x + 2)$ is negative for these points. Figure 3 shows the graph of $y = (x - 4)(x + 2)$, along with the region on the x-axis where the graph contains points with negative y-coordinates.

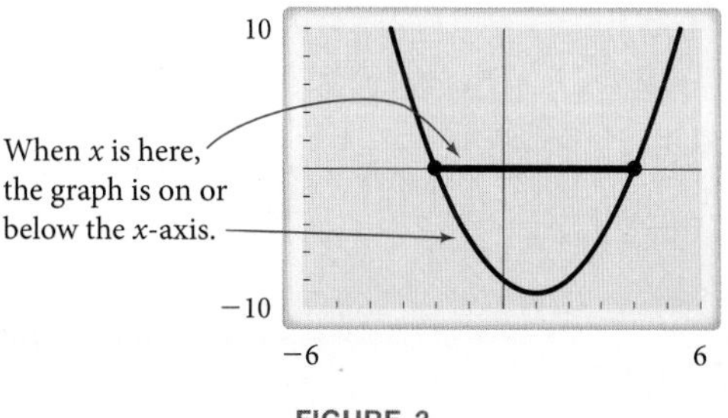

FIGURE 3

As you can see, the graph is below the x-axis when x is between -2 and 4. Because our original inequality includes the possibility that $(x - 4)(x + 2)$ is 0, we include the endpoints, -2 and 4, with our solution set.

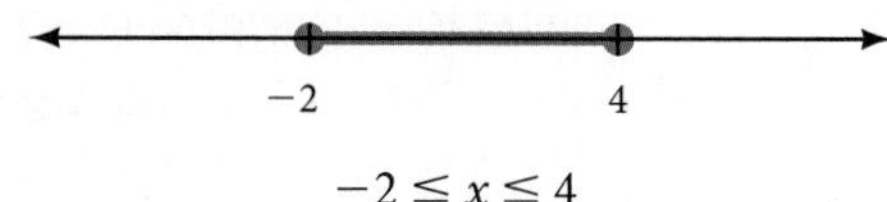

$$-2 \le x \le 4$$

EXAMPLE 2 Solve for x: $6x^2 - x \ge 2$.

ALGEBRAIC SOLUTION

$$6x^2 - x \ge 2$$

$$6x^2 - x - 2 \ge 0 \leftarrow \text{Standard form}$$

$$(3x - 2)(2x + 1) \ge 0$$

The product is positive or zero, so the factors must agree in sign. Here is the diagram showing where that occurs:

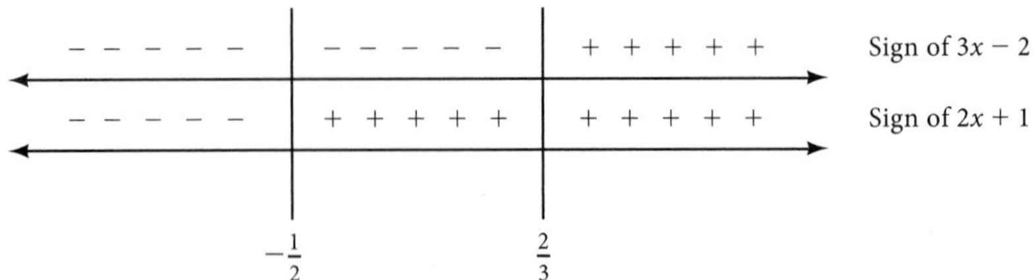

Because the factors agree in sign below $-\frac{1}{2}$ and above $\frac{2}{3}$, the graph of the solution set is

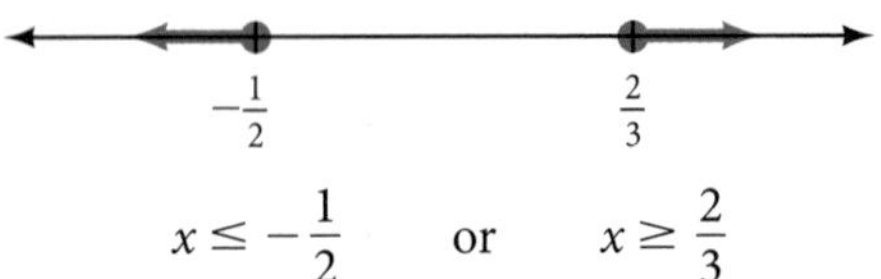

$$x \leq -\frac{1}{2} \quad \text{or} \quad x \geq \frac{2}{3}$$

GRAPHICAL SOLUTION To solve this inequality with a graphing calculator, we graph the function $y = (3x - 2)(2x + 1)$ and observe where the graph is above the x-axis. These are the points that have positive y-coordinates, which means that the product $(3x - 2)(2x + 1)$ is positive for these points. Figure 4 shows the graph of $y = (3x - 2)(2x + 1)$, along with the regions on the x-axis where the graph is on or above the x-axis.

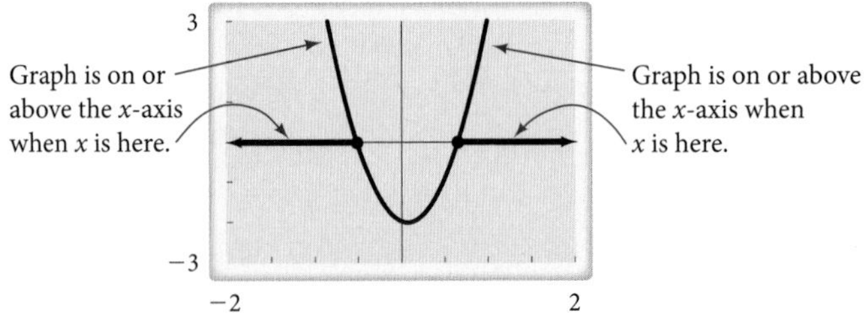

FIGURE 4

To find the points where the graph crosses the x-axis, we need to use either the Trace and Zoom features to zoom in on each point, or the calculator function that finds the intercepts automatically (on the TI-82/83 this is the root/zero function under the CALC key). Whichever method we use, we will obtain the following result:

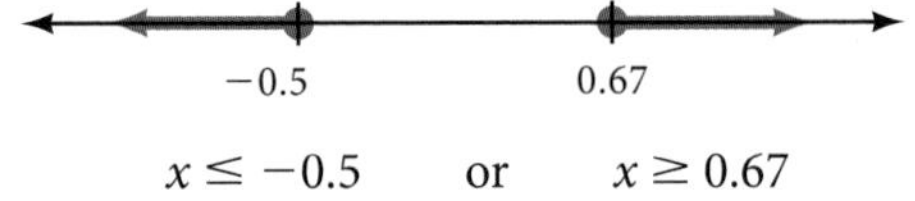

$$x \leq -0.5 \quad \text{or} \quad x \geq 0.67$$

EXAMPLE 3 Solve $x^2 - 6x + 9 \geq 0$.

ALGEBRAIC SOLUTION

$$x^2 - 6x + 9 \geq 0$$
$$(x - 3)^2 \geq 0$$

This is a special case in which both factors are the same. Because $(x - 3)^2$ is always positive or zero, the solution set is all real numbers. That is, any real number that is used in place of x in the original inequality will produce a true statement.

GRAPHICAL SOLUTION The graph of $y = (x - 3)^2$ is shown in Figure 5.

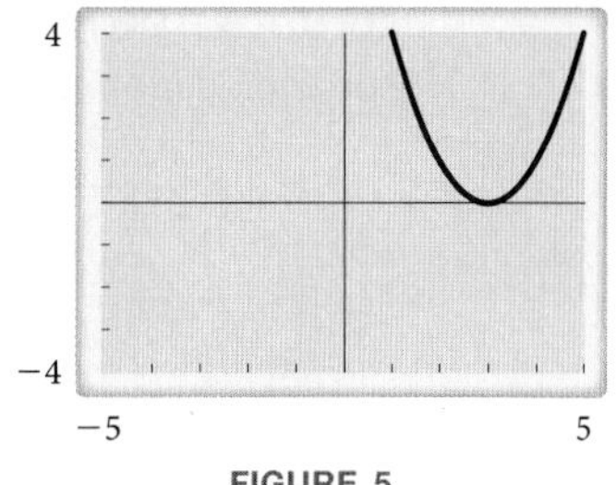

FIGURE 5

Notice that it touches the x-axis at 3 and is above the x-axis everywhere else. This means that every point on the graph has a y-coordinate greater than or equal to 0, no matter what the value of x. The conclusion that we draw from the graph is that the inequality $(x - 3)^2 \geq 0$ is true for all values of x.

Our next two examples involve inequalities that contain rational expressions.

EXAMPLE 4 Solve: $\dfrac{x - 4}{x + 1} \leq 0$.

SOLUTION The inequality indicates that the quotient of $(x - 4)$ and $(x + 1)$ is negative or 0 (less than or equal to 0). We can use the same reasoning we used to solve the first three examples, because quotients are positive or negative under the same conditions that products are positive or negative. Here is the diagram that shows where each factor is positive and where each factor is negative:

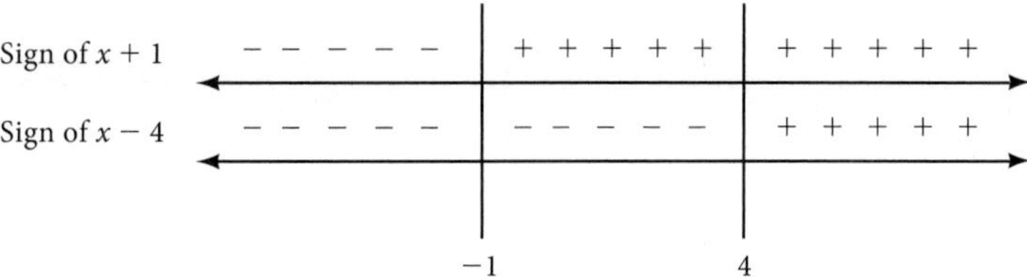

Between −1 and 4 the factors have opposite signs, making the quotient negative. Thus, the region between −1 and 4 is where the solutions lie, because the original inequality indicates the quotient $\frac{x - 4}{x + 1}$ is negative. The solution set and its graph are shown here:

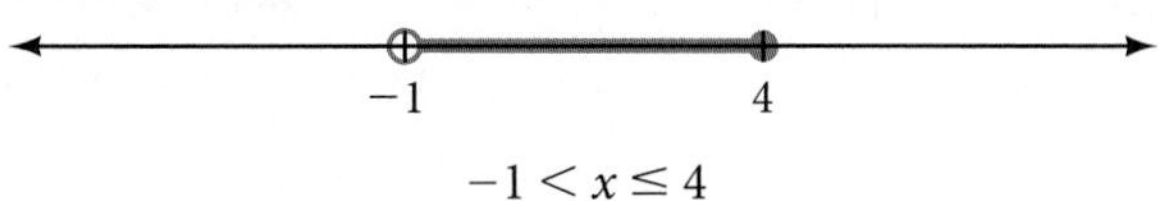

$$-1 < x \leq 4$$

Notice that the left endpoint is open—that is, it is not included in the solution set—because $x = -1$ would make the denominator in the original inequality 0. It is important to check all endpoints of solution sets to inequalities that involve rational expressions.

EXAMPLE 5 Solve: $\frac{3}{x-2} - \frac{2}{x-3} > 0$.

SOLUTION We begin by adding the two rational expressions on the left side. The common denominator is $(x-2)(x-3)$:

$$\frac{3}{x-2} \cdot \frac{(x-3)}{(x-3)} - \frac{2}{x-3} \cdot \frac{(x-2)}{(x-2)} > 0$$

$$\frac{3x-9-2x+4}{(x-2)(x-3)} > 0$$

$$\frac{x-5}{(x-2)(x-3)} > 0$$

This time the quotient involves three factors. Here is the diagram that shows the signs of the three factors:

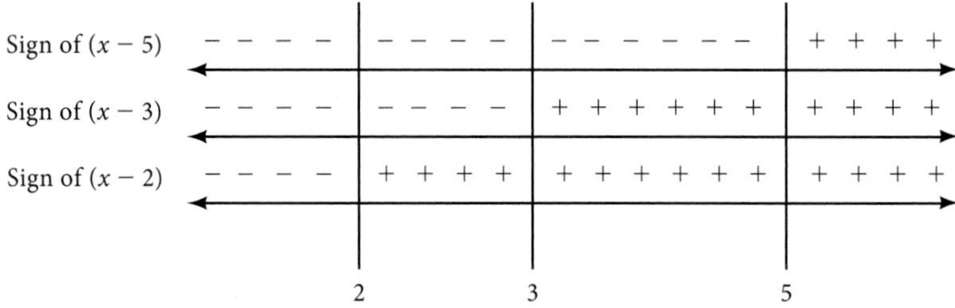

The original inequality indicates that the quotient is positive. For this to happen, either all three factors must be positive, or exactly two factors must be negative. Looking back at the diagram, we see the regions that satisfy these conditions are between 2 and 3 or above 5. Here is our solution set:

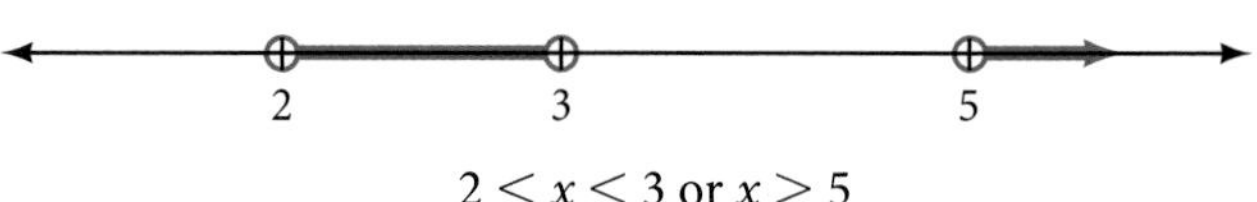

$$2 < x < 3 \text{ or } x > 5$$

GETTING READY FOR CLASS

After reading through the preceding section, respond in your own words and in complete sentences.

A. What is the first step in solving a quadratic inequality?

B. How do you show that the endpoint of a line segment is not part of the graph of a quadratic inequality?

C. How would you use the graph of $y = ax^2 + bx + c$ to help you find the graph of $ax^2 + bx + c < 0$?

D. Can a quadratic inequality have exactly one solution? Give an example.

Solve each of the following inequalities, and graph the solution set.

1. $x^2 + x - 6 > 0$
2. $x^2 + x - 6 < 0$
3. $x^2 - x - 12 \le 0$
4. $x^2 - x - 12 \ge 0$
5. $x^2 + 5x \ge -6$
6. $x^2 - 5x > 6$
7. $6x^2 < 5x - 1$
8. $4x^2 \ge -5x + 6$
9. $x^2 - 9 < 0$
10. $x^2 - 16 \ge 0$
11. $4x^2 - 9 \ge 0$
12. $9x^2 - 4 < 0$
13. $2x^2 - x - 3 < 0$
14. $3x^2 + x - 10 \ge 0$
15. $x^2 - 4x + 4 \ge 0$
16. $x^2 - 4x + 4 < 0$
17. $x^2 - 10x + 25 < 0$
18. $x^2 - 10x + 25 > 0$
19. $(x - 2)(x - 3)(x - 4) > 0$
20. $(x - 2)(x - 3)(x - 4) < 0$
21. $(x + 1)(x + 2)(x + 3) \le 0$
22. $(x + 1)(x + 2)(x + 3) \ge 0$
23. $\dfrac{x - 1}{x + 4} \le 0$
24. $\dfrac{x + 4}{x - 1} \le 0$
25. $\dfrac{3x}{x + 6} - \dfrac{8}{x + 6} < 0$
26. $\dfrac{5x}{x + 1} - \dfrac{3}{x + 1} < 0$
27. $\dfrac{4}{x - 6} + 1 > 0$
28. $\dfrac{2}{x - 3} + 1 \ge 0$
29. $\dfrac{x - 2}{(x + 3)(x - 4)} < 0$
30. $\dfrac{x - 1}{(x + 2)(x - 5)} < 0$
31. $\dfrac{2}{x - 4} - \dfrac{1}{x - 3} > 0$
32. $\dfrac{4}{x + 3} - \dfrac{3}{x + 2} > 0$
33. $\dfrac{x + 7}{2x + 12} + \dfrac{6}{x^2 - 36} \le 0$
34. $\dfrac{x + 1}{2x - 2} - \dfrac{2}{x^2 - 1} \le 0$
35. The graph of $y = x^2 - 4$ is shown in Figure 6. Use the graph to write the solution set for each of the following:

    a. $x^2 - 4 < 0$ b. $x^2 - 4 > 0$ c. $x^2 - 4 = 0$

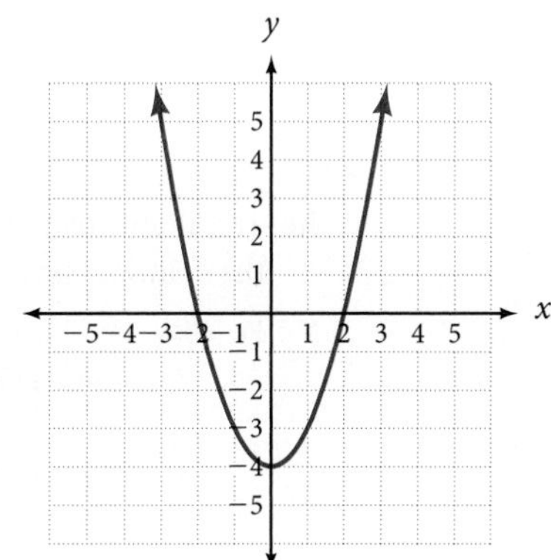

FIGURE 6

36. The graph of $y = 4 - x^2$ is shown in Figure 7. Use the graph to write the solution set for each of the following:

    a. $4 - x^2 < 0$ b. $4 - x^2 > 0$ c. $4 - x^2 = 0$

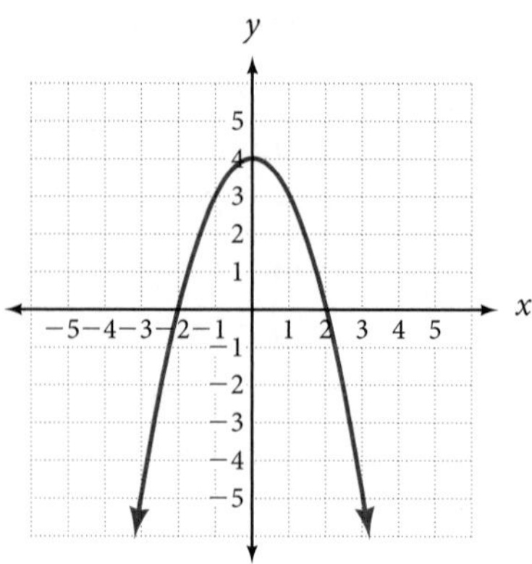

FIGURE 7

37. The graph of $y = x^2 - 3x - 10$ is shown in Figure 8. Use the graph to write the solution set for each of the following:

a. $x^2 - 3x - 10 < 0$ **b.** $x^2 - 3x - 10 > 0$ **c.** $x^2 - 3x - 10 = 0$

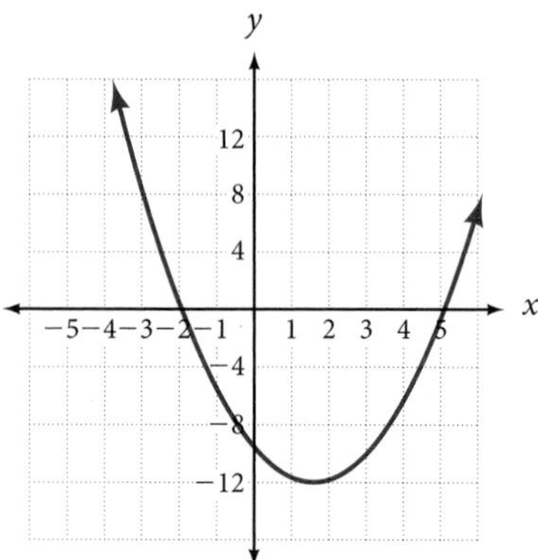

FIGURE 8

38. The graph of $y = x^2 + x - 12$ is shown in Figure 9. Use the graph to write the solution set for each of the following:

a. $x^2 + x - 12 < 0$ **b.** $x^2 + x - 12 > 0$ **c.** $x^2 + x - 12 = 0$

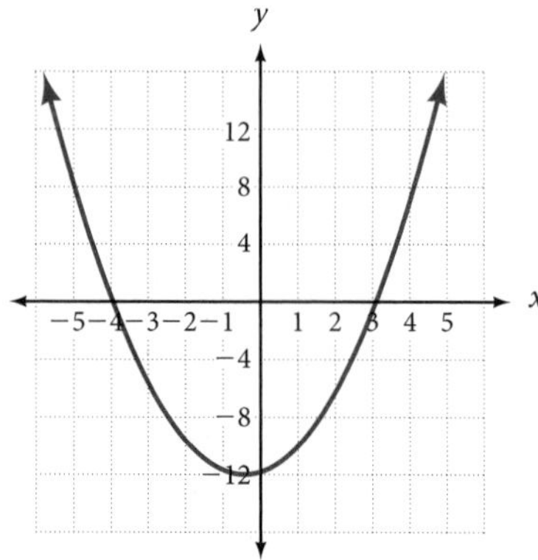

FIGURE 9

39. The graph of $y = x^3 - 3x^2 - x + 3$ is shown in Figure 10. Use the graph to write the solution set for each of the following:

a. $x^3 - 3x^2 - x + 3 < 0$ **b.** $x^3 - 3x^2 - x + 3 > 0$

c. $x^3 - 3x^2 - x + 3 = 0$

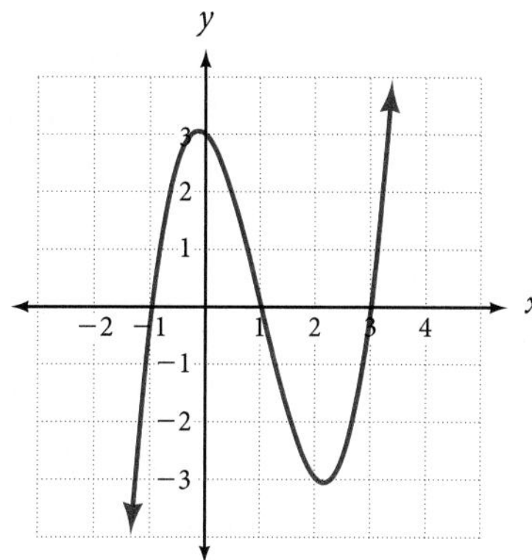

FIGURE 10

40. The graph of $y = x^3 + 4x^2 - 4x - 16$ is shown in Figure 11. Use the graph to write the solution set for each of the following:

a. $x^3 + 4x^2 - 4x - 16 < 0$ **b.** $x^3 + 4x^2 - 4x - 16 > 0$

c. $x^3 + 4x^2 - 4x - 16 = 0$

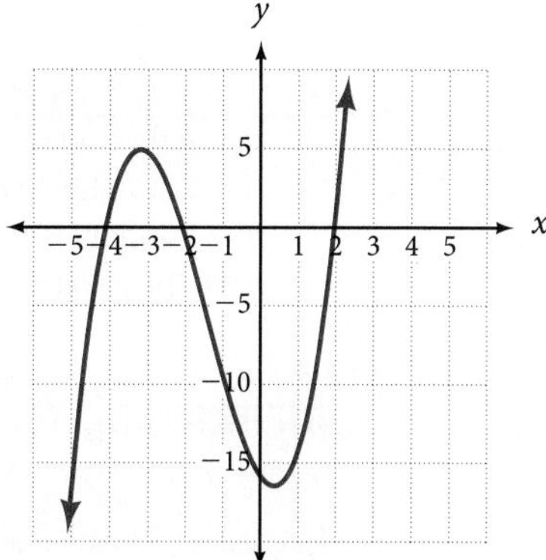

FIGURE 11

Applying the Concepts

41. Dimensions of a Rectangle The length of a rectangle is 3 inches more than twice the width. If the area is to be at least 44 square inches, what are the possibilities for the width?

42. Dimensions of a Rectangle The length of a rectangle is 5 inches less than three times the width. If the area is to be less than 12 square inches, what are the possibilities for the width?

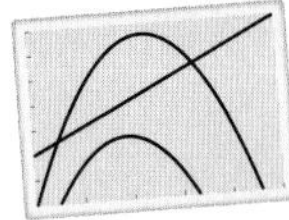

43. Revenue A manufacturer of portable radios knows that the weekly revenue produced by selling x radios is given by the equation $R = 1{,}300p - 100p^2$, where p is the price of each radio (in dollars). What price should be charged for each radio if the weekly revenue is to be at least \$4,000?

44. Revenue A manufacturer of small calculators knows that the weekly revenue produced by selling x calculators is given by the equation $R = 1{,}700p - 100p^2$, where p is the price of each calculator (in dollars). What price should be charged for each calculator if the revenue is to be at least \$7,000 each week?

45. Union Dues A labor union has 10,000 members. For every \$10 increase in union dues, membership is decreased by 200 people. If the current dues are \$100, what should be the new dues (to the nearest multiple of \$10) so income from dues is greatest, and what is that income? *Hint:* Because Income = (membership)(dues), we can let x = the number of \$10 increases in dues, and then this will give us income of $y = (10{,}000 - 200x)(100 + 10x)$.

46. Bookstore Receipts The owner of a used book store charges \$2 for quality paperbacks and usually sells 40 per day. For every 10-cent increase in the price of these paperbacks, he thinks that he will sell two fewer per day. What is the price he should charge (to the nearest 10 cents) for these books to maximize his income, and what would be that income? *Hint:* Let x = the number of 10-cent increases in price.

47. Jiffy-Lube The owner of a quick oil-change business charges \$20 per oil change and has 40 customers per day. If each increase of \$2 results in 2 fewer daily customers, what price should the owner charge (to the nearest \$2) for an oil change if the income from this business is to be as great as possible?

48. Computer Sales A computer manufacturer charges \$2,200 for its basic model and sells 1,500 computers per month at this price. For every \$200 increase in price, it is believed that 75 fewer computers will be sold. What price should the company place on its basic model of computer (to the nearest \$100) to have the greatest income?

Maintaining Your Skills

Use a calculator to evaluate, give answers to 4 decimal places

49. $\dfrac{50{,}000}{32{,}000}$ **50.** $\dfrac{2.4362}{1.9758} - 1$ **51.** $\dfrac{1}{2}\left(\dfrac{4.5926}{1.3876} - 2\right)$ **52.** $1 + \dfrac{0.06}{12}$

Solve each equation

53. $2\sqrt{3t - 1} = 2$ **54.** $\sqrt{4t + 5} + 7 = 3$

55. $\sqrt{x + 3} = x - 3$ **56.** $\sqrt{x + 3} = \sqrt{x} - 3$

Graph each equation

57. $y = \sqrt[3]{x - 1}$ **58.** $y = \sqrt[3]{x} - 1$

Chapter 7 Summary

The Square Root Property for Equations [7.1]

EXAMPLES

1. If $(x - 3)^2 = 25$
then $x - 3 = \pm 5$
$x = 3 \pm 5$
$x = 8$ or $x = -2$

If $a^2 = b$, where b is a real number, then

$$a = \sqrt{b} \quad \text{or} \quad a = -\sqrt{b}$$

which can be written as $a = \pm\sqrt{b}$.

To Solve a Quadratic Equation by Completing the Square [7.1]

2. Solve $x^2 - 6x - 6 = 0$
$x^2 - 6x = 6$
$x^2 - 6x + 9 = 6 + 9$
$(x - 3)^2 = 15$
$x - 3 = \pm\sqrt{15}$
$x = 3 \pm \sqrt{15}$

Step 1: Write the equation in the form $ax^2 + bx = c$.
Step 2: If $a \neq 1$, divide through by the constant a so the coefficient of x^2 is 1.
Step 3: Complete the square on the left side by adding the square of $\frac{1}{2}$ the coefficient of x to both sides.
Step 4: Write the left side of the equation as the square of a binomial. Simplify the right side if possible.
Step 5: Apply the square root property for equations, and solve as usual.

The Quadratic Theorem [7.2]

3. If $2x^2 + 3x - 4 = 0$, then

$$x = \frac{-3 \pm \sqrt{9 - 4(2)(-4)}}{2(2)}$$
$$= \frac{-3 \pm \sqrt{41}}{4}$$

For any quadratic equation in the form $ax^2 + bx + c = 0$, $a \neq 0$, the two solutions are

$$x = \frac{-b \pm \sqrt{b^2 - 4ac}}{2a}$$

This last equation is known as the *quadratic formula*.

The Discriminant [7.3]

4. The discriminant for
$x^2 + 6x + 9 = 0$
is $D = 36 - 4(1)(9) = 0$, which means the equation has one rational solution.

The expression $b^2 - 4ac$ that appears under the radical sign in the quadratic formula is known as the *discriminant*.

We can classify the solutions to $ax^2 + bx + c = 0$:

The solutions are	When the discriminant is
Two complex numbers containing i	Negative
One rational number	Zero
Two rational numbers	A positive perfect square
Two irrational numbers	A positive number, but not a perfect square

Equations Quadratic in Form [7.4]

There are a variety of equations whose form is quadratic. We solve most of them by making a substitution so the equation becomes quadratic, and then solving the equation by factoring or the quadratic formula. For example,

The equation	*is quadratic in*
$(2x-3)^2 + 5(2x-3) - 6 = 0$	$2x - 3$
$4x^4 - 7x^2 - 2 = 0$	x^2
$2x - 7\sqrt{x} + 3 = 0$	$\sqrt{x}$

5. The equation $x^4 - x^2 - 12 = 0$ is quadratic in x^2. Letting $y = x^2$ we have

$$y^2 - y - 12 = 0$$
$$(y-4)(y+3) = 0$$
$$y = 4 \quad \text{or} \quad y = -3$$

Resubstituting x^2 for y, we have

$$x^2 = 4 \quad \text{or} \quad x^2 = -3$$
$$x = \pm 2 \quad \text{or} \quad x = \pm i\sqrt{3}$$

Graphing Parabolas [7.5]

The graph of any equation of the form

$$y = ax^2 + bx + c \qquad a \neq 0$$

is a *parabola*. The graph is *concave up* if $a > 0$ and *concave down* if $a < 0$. The highest or lowest point on the graph is called the *vertex* and always has an x-coordinate of

$$x = \frac{-b}{2a}.$$

6. The graph of $y = x^2 - 4$ will be a parabola. It will cross the x-axis at 2 and -2, and the vertex will be $(0, -4)$.

Quadratic Inequalities [7.6]

We solve quadratic inequalities by manipulating the inequality to get 0 on the right side and then factoring the left side. We then make a diagram that indicates where the factors are positive and where they are negative. From this sign diagram and the original inequality we graph the appropriate solution set.

7. Solve $x^2 - 2x - 8 > 0$. We factor and draw the sign diagram:

$$(x-4)(x+2) > 0$$

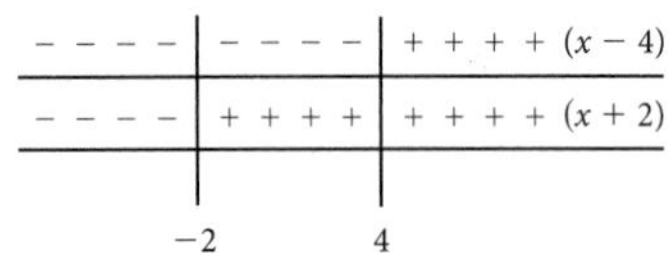

The solution is $x < -2$ or $x > 4$.

Solve each equation. [7.1, 7.2]

1. $(2x + 4)^2 = 25$

2. $(2x - 6)^2 = -8$

3. $y^2 - 10y + 25 = -4$

4. $(y + 1)(y - 3) = -6$

5. $8t^3 - 125 = 0$

6. $\frac{1}{a+2} - \frac{1}{3} = \frac{1}{a}$

7. Solve the formula $64(1 + r)^2 = A$ for r. [7.1]

8. Solve $x^2 - 4x = -2$ by completing the square. [7.1]

9. **Projectile Motion** An object projected upward with an initial velocity of 32 feet per second will rise and fall according to the equation $s(t) = 32t - 16t^2$, where s is its distance above the ground at time t. At what times will the object be 12 feet above the ground? [7.2]

10. **Revenue** The total weekly cost for a company to make x ceramic coffee cups is given by the formula $C(x) = 2x + 100$. If the weekly revenue from selling all x cups is $R(x) = 25x - 0.2x^2$, how many cups must it sell a week to make a profit of $200 a week? [7.2]

11. Find k so that $kx^2 = 12x - 4$ has one rational solution. [7.3]

12. Use the discriminant to identify the number and kind of solutions to $2x^2 - 5x = 7$. [7.3]

Find equations that have the given solutions. [7.3]

13. $x = 5, x = -\frac{2}{3}$

14. $x = 2, x = -2, x = 7$

Solve each equation. [7.4]

15. $4x^4 - 7x^2 - 2 = 0$

16. $(2t + 1)^2 - 5(2t + 1) + 6 = 0$

17. $2t - 7\sqrt{t} + 3 = 0$

18. **Projectile Motion** An object is tossed into the air with an upward velocity of 14 feet per second from the top of a building h feet high. The time it takes for the object to hit the ground below is given by the formula $16t^2 - 14t - h = 0$. Solve this formula for t. [7.4]

Sketch the graph of each of the following equations. Give the coordinates of the vertex in each case. [7.5]

19. $y = x^2 - 2x - 3$

20. $y = -x^2 + 2x + 8$

Graph each of the following inequalities. [7.6]

21. $x^2 - x - 6 \leq 0$

22. $2x^2 + 5x > 3$

23. **Profit** Find the maximum weekly profit for a company with weekly costs of $C = 5x + 100$ and weekly revenue of $R = 25x - 0.1x^2$. [7.5]

Rational Expressions and Rational Functions

8

Chapter Outline

8.1 Basic Properties and Reducing to Lowest Terms
8.2 Multiplication and Division of Rational Expressions
8.3 Addition and Subtraction of Rational Expressions
8.4 Complex Fractions
8.5 Equations With Rational Expressions
8.6 Applications
8.7 Division of Polynomials

iStockphoto.com © furabolo

If you have ever put yourself on a weight loss diet, you know that you lose more weight at the beginning of the diet than you do later. If we let $W(x)$ represent a person's weight after x weeks on the diet, then the rational function

$$W(x) = \frac{80(2x + 15)}{x + 6}$$

is a mathematical model of the person's weekly progress on a diet intended to take them from 200 pounds to about 160 pounds. Rational functions are good models for quantities that fall off rapidly to begin with, and then level off over time. The table shows some values for this function, along with the graph of this function.

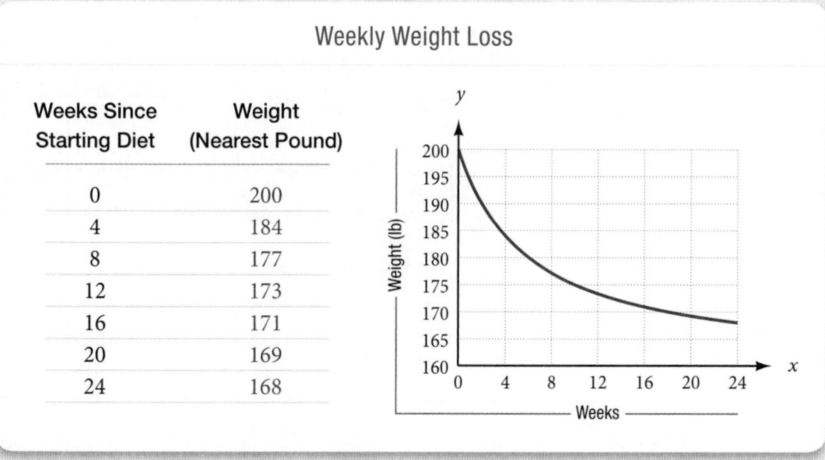

Weekly Weight Loss

Weeks Since Starting Diet	Weight (Nearest Pound)
0	200
4	184
8	177
12	173
16	171
20	169
24	168

As you progress through this chapter, you will acquire an intuitive feel for these types of functions, and as a result, you will see why they are good models for situations such as dieting.

Success Skills

Never mistake activity for achievement.

— John Wooden, legendary UCLA basketball coach

You may think that the John Wooden quote above has to do with being productive and efficient, or using your time wisely, but it is really about being honest with yourself. I have had students come to me after failing a test saying, "I can't understand why I got such a low grade after I put so much time in studying." One student even had help from a tutor and felt she understood everything that we covered. After asking her a few questions, it became clear that she spent all her time studying with a tutor and the tutor was doing most of the work. The tutor can work all the homework problems, but the student cannot. She has mistaken activity for achievement.

Can you think of situations in your life when you are mistaking activity for achievement?

How would you describe someone who is mistaking activity for achievement in the way they study for their math class?

Which of the following best describes the idea behind the John Wooden quote?

- Always be efficient.
- Don't kid yourself.
- Take responsibility for your own success.
- Study with purpose.

Basic Properties and Reducing to Lowest Terms

8.1

We will begin this section with the definition of a rational expression. We will then state the two basic properties associated with rational expressions and go on to apply one of the properties to reduce rational expressions to lowest terms.

Recall from Chapter 1 that a *rational number* is any number that can be expressed as the ratio of two integers:

$$\text{Rational numbers} = \left\{ \frac{a}{b} \middle| a \text{ and } b \text{ are integers, } b \neq 0 \right\}$$

A *rational expression* is defined similarly as any expression that can be written as the ratio of two polynomials:

$$\text{Rational expressions} = \left\{ \frac{P}{Q} \middle| P \text{ and } Q \text{ are polynominals, } Q \neq 0 \right\}$$

Some examples of rational expressions are

$$\frac{2x-3}{x+5} \qquad \frac{x^2-5x-6}{x^2-1} \qquad \frac{a-b}{b-a}$$

Basic Properties

For rational expressions, multiplying the numerator and denominator by the same nonzero expression may change the form of the rational expression, but it will always produce an expression equivalent to the original one. The same is true when dividing the numerator and denominator by the same nonzero quantity.

[Δ≠Σ PROPERTY *Properties of Rational Expressions*

If P, Q, and K are polynomials with $Q \neq 0$ and $K \neq 0$, then

$$\frac{P}{Q} = \frac{PK}{QK} \quad \text{and} \quad \frac{P}{Q} = \frac{\frac{P}{K}}{\frac{Q}{K}}$$

Reducing to Lowest Terms

The fraction $\frac{6}{8}$ can be written in lowest terms as $\frac{3}{4}$. The process is shown here:

$$\frac{6}{8} = \frac{3 \cdot 2}{4 \cdot 2} = \frac{3}{4}$$

Reducing $\frac{6}{8}$ to $\frac{3}{4}$ involves dividing the numerator and denominator by 2, the factor they have in common. Before dividing out the common factor 2, we must notice that the common factor *is* 2. (This may not be obvious because we are very familiar with the numbers 6 and 8 and therefore do not have to put much thought into finding what number divides both of them.)

We reduce rational expressions to lowest terms by first factoring the numerator and denominator and then dividing both numerator and denominator by any factors they have in common.

EXAMPLE 1 Reduce $\frac{x^2 - 9}{x - 3}$ to lowest terms.

SOLUTION Factoring, we have

$$\frac{x^2 - 9}{x - 3} = \frac{(x + 3)(x - 3)}{x - 3}$$

The numerator and denominator have the factor $x - 3$ in common. Dividing the numerator and denominator by $x - 3$, we have

$$\frac{(x + 3)\cancel{(x - 3)}}{\cancel{x - 3}} = \frac{x + 3}{1} = x + 3$$

Note The lines drawn through the $(x - 3)$ in the numerator and denominator indicate that we have divided through by $(x - 3)$. As the problems become more involved, these lines will help keep track of which factors have been divided out and which have not.

For the problem in Example 1, there is an implied restriction on the variable x: It cannot be 3. If x were 3, the expression $\frac{(x^2 - 9)}{(x - 3)}$ would become $\frac{0}{0}$, an expression that we cannot associate with a real number. For all problems involving rational expressions, we restrict the variable to only those values that result in a nonzero denominator. When we state the relationship

$$\frac{x^2 - 9}{x - 3} = x + 3$$

we are assuming that it is true for all values of x except $x = 3$.

Here are some other examples of reducing rational expressions to lowest terms.

EXAMPLE 2 Reduce $\frac{y^2 - 5y - 6}{y^2 - 1}$ to lowest terms.

SOLUTION

$$\frac{y^2 - 5y - 6}{y^2 - 1} = \frac{(y - 6)\cancel{(y + 1)}}{(y - 1)\cancel{(y + 1)}}$$

$$= \frac{y - 6}{y - 1}$$

EXAMPLE 3 Reduce $\frac{2a^3 - 16}{4a^2 - 12a + 8}$ to lowest terms.

SOLUTION

$$\frac{2a^3 - 16}{4a^2 - 12a + 8} = \frac{2(a^3 - 8)}{4(a^2 - 3a + 2)}$$

$$= \frac{2\cancel{(a - 2)}(a^2 + 2a + 4)}{4\cancel{(a - 2)}(a - 1)}$$

$$= \frac{a^2 + 2a + 4}{2(a - 1)}$$

EXAMPLE 4 Reduce $\frac{x^2 - 3x + ax - 3a}{x^2 - ax - 3x + 3a}$ to lowest terms.

SOLUTION

$$\frac{x^2 - 3x + ax - 3a}{x^2 - ax - 3x + 3a} = \frac{x(x - 3) + a(x - 3)}{x(x - a) - 3(x - a)}$$

$$= \frac{\cancel{(x - 3)}(x + a)}{(x - a)\cancel{(x - 3)}}$$

$$= \frac{x + a}{x - a}$$

The answer to Example 4 cannot be reduced further. It is a fairly common mistake to attempt to divide out an x or an a in this last expression. Remember, we can divide out only the factors common to the numerator and denominator of a rational expression.

The next example involves what we call a trick. The trick is to reverse the order of the terms in a difference by factoring -1 from each term in either the numerator or the denominator. The next examples illustrate how this is done.

EXAMPLE 5 Reduce to lowest terms: $\dfrac{a-b}{b-a}$

SOLUTION The relationship between $a - b$ and $b - a$ is that they are opposites. We can show this fact by factoring -1 from each term in the numerator:

$$\frac{a-b}{b-a} = \frac{-1(-a+b)}{b-a} \quad \text{Factor } -1 \text{ from each term in the numerator}$$

$$= \frac{-1(b-a)}{b-a} \quad \text{Reverse the order of the terms in the numerator}$$

$$= -1 \quad \text{Divide out common factor } b-a$$

EXAMPLE 6 Reduce to lowest terms: $\dfrac{x^2-25}{5-x}$

SOLUTION We begin by factoring the numerator:

$$\frac{x^2-25}{5-x} = \frac{(x-5)(x+5)}{5-x}$$

The factors $x - 5$ and $5 - x$ are similar but are not exactly the same. We can reverse the order of either by factoring -1 from it.

That is: $5 - x = -1(-5 + x) = -1(x - 5)$.

$$\frac{(x-5)(x+5)}{5-x} = \frac{(x-5)(x+5)}{-1(x-5)}$$

$$= \frac{x+5}{-1}$$

$$= -(x+5)$$

Rational Functions

We can extend our knowledge of rational expressions to rational functions with the following definition:

DEFINITION *rational function*

A ***rational function*** is any function that can be written in the form

$$f(x) = \frac{P(x)}{Q(x)}$$

where $P(x)$ and $Q(x)$ are polynomials and $Q(x) \neq 0$.

EXAMPLE 7 For the rational function $f(x) = \frac{x-4}{x-2}$, find $f(0)$, $f(-4)$, $f(4)$, $f(-2)$, and $f(2)$.

SOLUTION To find these function values, we substitute the given value of x into the rational expression, and then simplify if possible.

$$f(0) = \frac{0-4}{0-2} = \frac{-4}{-2} = 2$$

$$f(-4) = \frac{-4-4}{-4-2} = \frac{-8}{-6} = \frac{4}{3}$$

$$f(4) = \frac{4-4}{4-2} = \frac{0}{2} = 0$$

$$f(-2) = \frac{-2-4}{-2-2} = \frac{-6}{-4} = \frac{3}{2}$$

$$f(2) = \frac{2-4}{2-2} = \frac{-2}{0} \quad \text{Undefined}$$

Because the rational function in Example 7 is not defined when x is 2, the domain of that function does not include 2. We have more to say about the domain of a rational function next.

The Domain of a Rational Function

If the domain of a rational function is not specified, it is assumed to be all real numbers for which the function is defined. That is, the domain of the rational function

$$f(x) = \frac{P(x)}{Q(x)}$$

is all x for which $Q(x)$ is nonzero.

EXAMPLE 8 Find the domain for each function.

a. $f(x) = \frac{x-4}{x-2}$ **b.** $g(x) = \frac{x^2+5}{x+1}$ **c.** $h(x) = \frac{x}{x^2-9}$

SOLUTION

a. The domain for $f(x) = \frac{x-4}{x-2}$ is $\{x \mid x \neq 2\}$.

b. The domain for $g(x) = \frac{x^2+5}{x+1}$ is $\{x \mid x \neq -1\}$.

c. The domain for $h(x) = \frac{x}{x^2-9}$ is $\{x \mid x \neq -3, x \neq 3\}$.

Notice that, for these functions, $f(2)$, $g(-1)$, $h(-3)$, and $h(3)$ are all undefined, and that is why the domains are written as shown.

Difference Quotients

The diagram in Figure 1 is an important diagram from calculus. Although it may look complicated, the point of it is simple: The slope of the line passing through the points P and Q is given by the formula

$$\text{Slope of line through } PQ = m = \frac{f(x) - f(a)}{x - a}$$

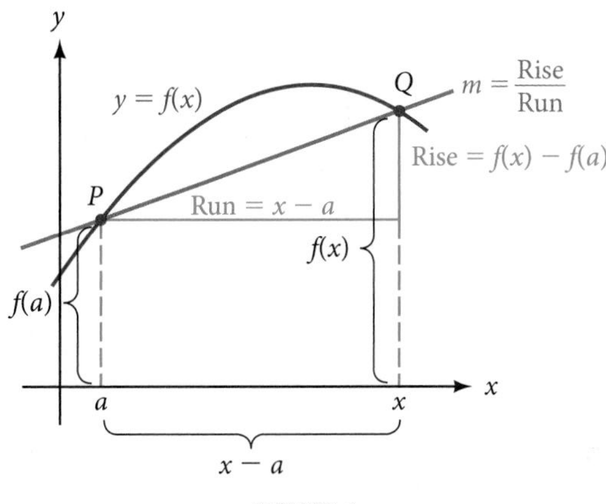

FIGURE 1

The expression $\frac{f(x) - f(a)}{x - a}$ is called a *difference quotient*. When $f(x)$ is a polynomial, it will be a rational expression.

EXAMPLE 9 If $f(x) = 3x - 5$, find $\frac{f(x) - f(a)}{x - a}$.

SOLUTION

$$\begin{aligned}\frac{f(x) - f(a)}{x - a} &= \frac{(3x - 5) - (3a - 5)}{x - a} \\ &= \frac{3x - 3a}{x - a} \\ &= \frac{3(x - a)}{x - a} \\ &= 3\end{aligned}$$

EXAMPLE 10 If $f(x) = x^2 - 4$, find $\frac{f(x) - f(a)}{x - a}$ and simplify.

SOLUTION Because $f(x) = x^2 - 4$ and $f(a) = a^2 - 4$, we have

$$\begin{aligned}\frac{f(x) - f(a)}{x - a} &= \frac{(x^2 - 4) - (a^2 - 4)}{x - a} \\ &= \frac{x^2 - 4 - a^2 + 4}{x - a} \\ &= \frac{x^2 - a^2}{x - a} \\ &= \frac{(x + a)\cancel{(x - a)}}{\cancel{x - a}} && \text{Factor and divide out common factor} \\ &= x + a\end{aligned}$$

The diagram in Figure 2 is similar to the one in Figure 1. The main difference is in how we label the points. From Figure 2, we can see another difference quotient that gives us the slope of the line through the points P and Q.

$$\text{Slope of line through } PQ = m = \frac{f(x+h) - f(x)}{h}$$

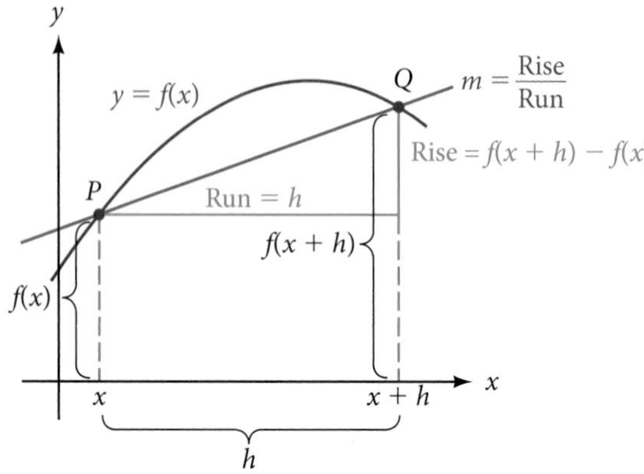

FIGURE 2

Examples 11 and 12 use the same functions used in Examples 9 and 10, but this time the new difference quotient is used.

EXAMPLE 11 If $f(x) = 3x - 5$, find $\dfrac{f(x+h) - f(x)}{h}$.

SOLUTION The expression $f(x + h)$ is given by

$$\begin{aligned} f(x+h) &= 3(x+h) - 5 \\ &= 3x + 3h - 5 \end{aligned}$$

Using this result gives us

$$\begin{aligned} \frac{f(x+h) - f(x)}{h} &= \frac{(3x + 3h - 5) - (3x - 5)}{h} \\ &= \frac{3h}{h} \\ &= 3 \end{aligned}$$

EXAMPLE 12 If $f(x) = x^2 - 4$, find $\dfrac{f(x+h) - f(x)}{h}$.

SOLUTION The expression $f(x + h)$ is given by

$$\begin{aligned} f(x+h) &= (x+h)^2 - 4 \\ &= x^2 + 2xh + h^2 - 4 \end{aligned}$$

Using this result gives us

$$\begin{aligned} \frac{f(x+h) - f(x)}{h} &= \frac{(x^2 + 2xh + h^2 - 4) - (x^2 - 4)}{h} \\ &= \frac{2xh + h^2}{h} \\ &= \frac{h(2x + h)}{h} \\ &= 2x + h \end{aligned}$$

GETTING READY FOR CLASS

After reading through the preceding section, respond in your own words and in complete sentences.

A. What is a rational expression?

B. Explain how to determine if a rational expression is in "lowest terms."

C. When is a rational expression undefined?

D. Explain the process we use to reduce a rational expression or a fraction to lowest terms.

Problem Set 8.1

1. If $g(x) = \frac{x+3}{x-1}$, find $g(0)$, $g(-3)$, $g(3)$, $g(-1)$, and $g(1)$, if possible.
2. If $g(x) = \frac{x-2}{x-1}$, find $g(0)$, $g(-2)$, $g(2)$, $g(-1)$, and $g(1)$, if possible.
3. If $h(t) = \frac{t-3}{t+1}$, find $h(0)$, $h(-3)$, $h(3)$, $h(-1)$, and $h(1)$, if possible.
4. If $h(t) = \frac{t-2}{t+1}$, find $h(0)$, $h(-2)$, $h(2)$, $h(-1)$, and $h(1)$, if possible.

State the domain for each rational function.

5. $f(x) = \dfrac{x-3}{x-1}$
6. $f(x) = \dfrac{x+4}{x-2}$
7. $g(x) = \dfrac{x^2-4}{x-2}$
8. $g(x) = \dfrac{x^2-9}{x-3}$
9. $h(t) = \dfrac{t-4}{t^2-16}$
10. $h(t) = \dfrac{t-5}{t^2-25}$

Reduce each rational expression to lowest terms.

11. $\dfrac{x^2-16}{6x+24}$
12. $\dfrac{12x-9y}{3x^2+3xy}$
13. $\dfrac{a^4-81}{a-3}$
14. $\dfrac{a^2-4a-12}{a^2+8a+12}$
15. $\dfrac{20y^2-45}{10y^2-5y-15}$
16. $\dfrac{20x^2-93x+34}{4x^2-9x-34}$
17. $\dfrac{12y-2xy-2x^2y}{6y-4xy-2x^2y}$
18. $\dfrac{250a+100ax+10ax^2}{50a-2ax^2}$
19. $\dfrac{(x-3)^2(x+2)}{(x+2)^2(x-3)}$
20. $\dfrac{(x-4)^3(x+3)}{(x+3)^2(x-4)}$
21. $\dfrac{x^3+1}{x^2-1}$
22. $\dfrac{x^3-1}{x^2-1}$
23. $\dfrac{4am-4an}{3n-3m}$
24. $\dfrac{ad-ad^2}{d-1}$
25. $\dfrac{ab-a+b-1}{ab+a+b+1}$
26. $\dfrac{6cd-4c-9d+6}{6d^2-13d+6}$
27. $\dfrac{21x^2-23x+6}{21x^2+x-10}$
28. $\dfrac{36x^2-11x-12}{20x^2-39x+18}$
29. $\dfrac{8x^2-6x-9}{8x^2-18x+9}$
30. $\dfrac{42x^2+23x-10}{14x^2+45x-14}$
31. $\dfrac{4x^2+29x+45}{8x^2-10x-63}$
32. $\dfrac{30x^2-61x+30}{60x^2+22x-60}$
33. $\dfrac{a^3+b^3}{a^2-b^2}$
34. $\dfrac{a^2-b^2}{a^3-b^3}$
35. $\dfrac{8x^4-8x}{4x^4+4x^3+4x^2}$
36. $\dfrac{6x^5-48x^3}{12x^3+24x^2+48x}$
37. $\dfrac{ax+2x+3a+6}{ay+2y-4a-8}$
38. $\dfrac{x^2-3ax-2x+6a}{x^2-3ax+2x-6a}$
39. $\dfrac{x^3+3x^2-4x-12}{x^2+x-6}$
40. $\dfrac{x^3+5x^2-4x-20}{x^2+7x+10}$

41. $\dfrac{x^3 - 8}{x^2 - 4}$

42. $\dfrac{y^2 - 9}{y^3 + 27}$

43. $\dfrac{8x^3 - 27}{4x^2 - 9}$

44. $\dfrac{25y^2 - 4}{125y^3 + 8}$

Refer to Examples 5 and 6 in this section, and reduce the following to lowest terms.

45. $\dfrac{x - 4}{4 - x}$

46. $\dfrac{6 - x}{x - 6}$

47. $\dfrac{y^2 - 36}{6 - y}$

48. $\dfrac{1 - y}{y^2 - 1}$

49. $\dfrac{1 - 9a^2}{9a^2 - 6a + 1}$

50. $\dfrac{1 - a^2}{a^2 - 2a + 1}$

Simplify each expression.

51. $\dfrac{(3x - 5) - (3a - 5)}{x - a}$

52. $\dfrac{(2x + 3) - (2a + 3)}{x - a}$

53. $\dfrac{(x^2 - 4) - (a^2 - 4)}{x - a}$

54. $\dfrac{(x^2 - 1) - (a^2 - 1)}{x - a}$

For the functions below, evaluate

a. $\dfrac{f(x) - f(a)}{x - a}$ **b.** $\dfrac{f(x + h) - f(x)}{h}$

55. $f(x) = 4x$

56. $f(x) = -3x$

57. $f(x) = 5x + 3$

58. $f(x) = 6x - 5$

59. $f(x) = x^2$

60. $f(x) = 3x^2$

61. $f(x) = x^2 + 1$

62. $f(x) = x^2 - 3$

63. $f(x) = x^2 - 3x + 4$

64. $f(x) = x^2 + 4x - 7$

The graphs of two rational functions are given in Figures 3 and 4. Use the graphs to find the following.

65. **a.** $f(2)$ **b.** $f(-1)$ **c.** $f(0)$ **d.** $g(3)$

66. **a.** $g(6)$ **b.** $g(-1)$ **c.** $f(g(6))$ **d.** $g(f(-2))$

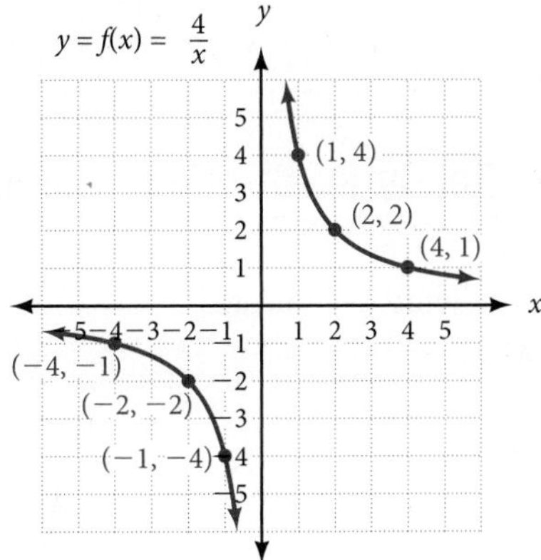

FIGURE 3

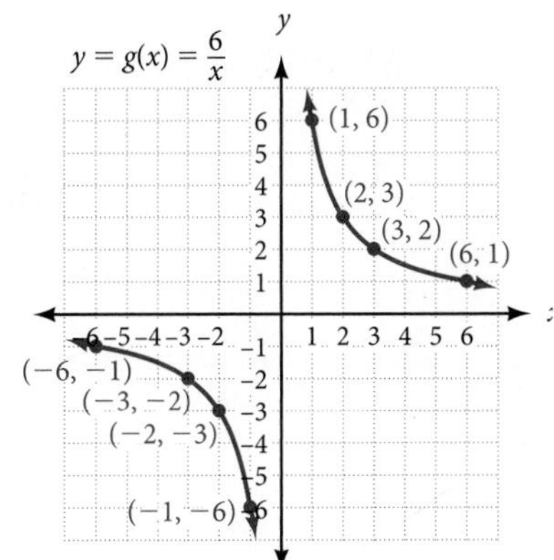

FIGURE 4

Applying the Concepts

67. Diet The following rational function is the one we mentioned in the introduction to this chapter. The quantity $W(x)$ is the weight (in pounds) of the person after x weeks of dieting. Use the function to fill in the table. Then compare your results with the graph in the chapter introduction.

$$W(x) = \frac{80(2x + 15)}{x + 6}$$

Weeks	Weight (lb)
x	$W(x)$
0	
1	
4	
12	
24	

68. Drag Racing The following rational function gives the speed $V(x)$, in miles per hour, of a dragster at each second x during a quarter-mile race.

Use the function to fill in the table.

$$V(x) = \frac{340x}{x + 3}$$

Time (sec)	Speed (mi/hr)
x	$V(x)$
0	
1	
2	
3	
4	
5	
6	

Getting Ready for the Next Section

Multiply or divide, as indicated.

69. $\frac{6}{7} \cdot \frac{14}{18}$

70. $\frac{6}{8} \div \frac{3}{5}$

71. $5y^2 \cdot 4x^2$

72. $4y^3 \cdot 3x^2$

73. $9x^4 \cdot 8y^5$

74. $6x^4 \cdot 12y^5$

Factor.

75. $x^2 - 4$

76. $x^2 - 6x + 9$

77. $x^3 - x^2y$

78. $a^2 - 5a + 6$

79. $2y^2 - 2$

80. $xa + xb + ya + yb$

Multiplication and Division of Rational Expressions

8.2

If you have ever taken a home videotape to be transferred to DVD, you know the amount you pay for the transfer depends on the number of copies you have made: The more copies you have made, the lower the charge per copy. The following demand function gives the price (in dollars) per tape $p(x)$ a company charges for making x DVDs. As you can see, it is a rational function.

$$p(x) = \frac{2(x + 60)}{x + 5}$$

The graph in Figure 1 shows this function from $x = 0$ to $x = 100$. As you can see, the more copies that are made, the lower the price per copy.

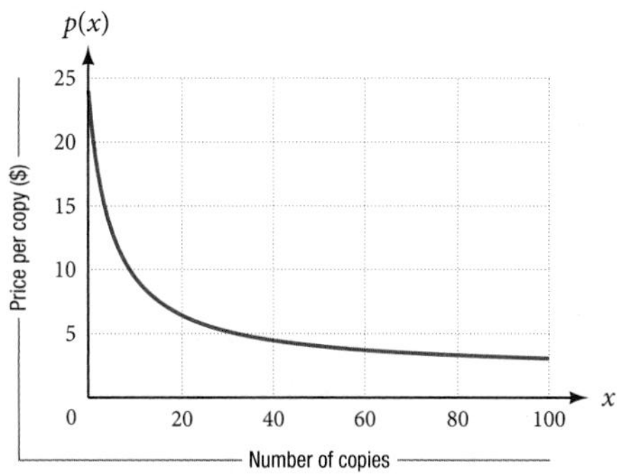

FIGURE 1

If we were interested in finding the revenue function for this situation, we would multiply the number of copies made x by the price per copy $p(x)$. This involves multiplication with a rational expression, which is one of the topics we cover in this section.

In Section 5.1, we found the process of reducing rational expressions to lowest terms to be the same process used in reducing fractions to lowest terms. The similarity also holds for the process of multiplication or division of rational expressions.

Multiplication with fractions is the simplest of the four basic operations. To multiply two fractions, we simply multiply numerators and multiply denominators. That is, if a, b, c, and d are real numbers, with $b \neq 0$ and $d \neq 0$, then

$$\frac{a}{b} \cdot \frac{c}{d} = \frac{ac}{bd}$$

EXAMPLE 1 Multiply $\frac{6}{7} \cdot \frac{14}{18}$.

SOLUTION

$$\frac{6}{7} \cdot \frac{14}{18} = \frac{6(14)}{7(18)}$$ Multiply numerators and denominators

$$= \frac{2 \cdot 3(2 \cdot 7)}{7(2 \cdot 3 \cdot 3)}$$ Factor

$$= \frac{2}{3}$$ Divide out common factors

Our next example is similar to some of the problems we worked in Chapter 1. We multiply fractions whose numerators and denominators are monomials by multiplying numerators and multiplying denominators and then reducing to lowest terms. Here is how it looks.

EXAMPLE 2 Multiply $\frac{8x^3}{27y^8} \cdot \frac{9y^3}{12x^2}$.

SOLUTION We multiply numerators and denominators without actually carrying out the multiplication:

$$\frac{8x^3}{27y^8} \cdot \frac{9y^3}{12x^2} = \frac{8 \cdot 9x^3y^3}{27 \cdot 12x^2y^8} \quad \text{Multiply Numerators, Multiply Denominators}$$

$$= \frac{4 \cdot 2 \cdot 9x^3y^3}{9 \cdot 3 \cdot 4 \cdot 3x^2y^8} \quad \text{Factor coefficients}$$

$$= \frac{2x}{9y^5} \quad \text{Divide out common factors}$$

The product of two rational expressions is the product of their numerators over the product of their denominators.

Once again, we should mention that the little slashes we have drawn through the factors are simply used to denote the factors we have divided out of the numerator and denominator.

EXAMPLE 3 Multiply $\frac{x-3}{x^2-4} \cdot \frac{x+2}{x^2-6x+9}$.

SOLUTION We begin by multiplying numerators and denominators. We then factor all polynomials and divide out factors common to the numerator and denominator:

$$\frac{x-3}{x^2-4} \cdot \frac{x+2}{x^2-6x+9} = \frac{(x-3)(x+2)}{(x^2-4)(x^2-6x+9)} \quad \text{Multiply}$$

$$= \frac{(x-3)(x+2)}{(x+2)(x-2)(x-3)(x-3)} \quad \text{Factor}$$

$$= \frac{1}{(x-2)(x-3)}$$

The first two steps can be combined to save time. We can perform the multiplication and factoring steps together.

EXAMPLE 4 Multiply $\frac{2y^2-4y}{2y^2-2} \cdot \frac{y^2-2y-3}{y^2-5y+6}$.

SOLUTION

$$\frac{2y^2-4y}{2y^2-2} \cdot \frac{y^2-2y-3}{y^2-5y+6} = \frac{2y(y-2)(y-3)(y+1)}{2(y+1)(y-1)(y-3)(y-2)}$$

$$= \frac{y}{y-1}$$

Notice in both of the preceding examples that we did not actually multiply the polynomials as we did in Chapter 1. It would be senseless to do that because we would then have to factor each of the resulting products to reduce them to lowest terms.

The quotient of two rational expressions is the product of the first and the reciprocal of the second. That is, we find the quotient of two rational expressions the same way we find the quotient of two fractions. Here is an example that reviews division with fractions.

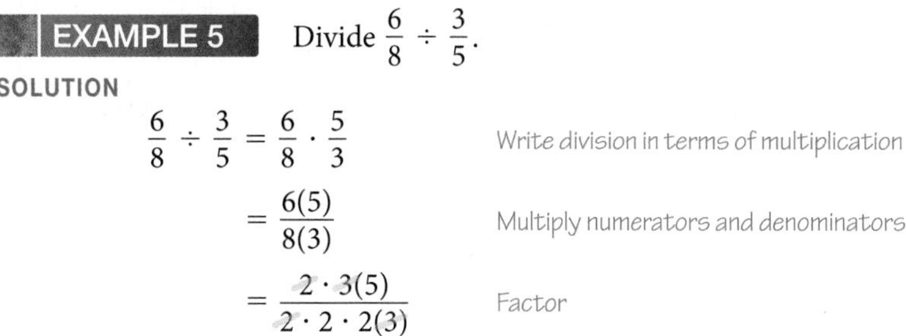

EXAMPLE 5 Divide $\frac{6}{8} \div \frac{3}{5}$.

SOLUTION

$$\frac{6}{8} \div \frac{3}{5} = \frac{6}{8} \cdot \frac{5}{3}$$ Write division in terms of multiplication

$$= \frac{6(5)}{8(3)}$$ Multiply numerators and denominators

$$= \frac{2 \cdot 3(5)}{2 \cdot 2 \cdot 2(3)}$$ Factor

$$= \frac{5}{4}$$ Divide out common factors

To divide one rational expression by another, we use the definition of division to multiply by the reciprocal of the expression that follows the division symbol.

EXAMPLE 6 Divide $\frac{8x^3}{5y^2} \div \frac{4x^2}{10y^6}$.

SOLUTION First, we rewrite the problem in terms of multiplication. Then we multiply.

$$\frac{8x^3}{5y^2} \div \frac{4x^2}{10y^6} = \frac{8x^3}{5y^2} \cdot \frac{10y^6}{4x^2}$$

$$= \frac{\overset{2}{8} \cdot \overset{2}{10}x^3y^6}{4 \cdot 5x^2y^2}$$

$$= 4xy^4$$

EXAMPLE 7 Divide $\frac{x^2 - y^2}{x^2 - 2xy + y^2} \div \frac{x^3 + y^3}{x^3 - x^2y}$.

SOLUTION We begin by writing the problem as the product of the first and the reciprocal of the second and then proceed as in the previous two examples:

$$\frac{x^2 - y^2}{x^2 - 2xy + y^2} \div \frac{x^3 + y^3}{x^3 - x^2y}$$ Multiply by the reciprocal of the divisor

$$= \frac{x^2 - y^2}{x^2 - 2xy + y^2} \cdot \frac{x^3 - x^2y}{x^3 + y^3}$$

$$= \frac{(x - y)(x + y)(x^2)(x - y)}{(x - y)(x - y)(x + y)(x^2 - xy + y^2)}$$ Factor and multiply

$$= \frac{x^2}{x^2 - xy + y^2}$$ Divide out common factors

Here are some more examples of multiplication and division with rational expressions.

EXAMPLE 8 Perform the indicated operations.

$$\frac{a^2-8a+15}{a+4}\cdot\frac{a+2}{a^2-5a+6}\div\frac{a^2-3a-10}{a^2+2a-8}$$

SOLUTION First, we rewrite the division as multiplication by the reciprocal. Then we proceed as usual.

$$\frac{a^2-8a+15}{a+4}\cdot\frac{a+2}{a^2-5a+6}\div\frac{a^2-3a-10}{a^2+2a-8}$$ Change division to multiplication by the reciprocal

$$=\frac{(a^2-8a+15)(a+2)(a^2+2a-8)}{(a+4)(a^2-5a+6)(a^2-3a-10)}$$ Factor

$$=\frac{(a-5)(a-3)(a+2)(a+4)(a-2)}{(a+4)(a-3)(a-2)(a-5)(a+2)}$$ Divide out common factors

$$=1$$

Our next example involves factoring by grouping. As you may have noticed, working the problems in this chapter gives you a very detailed review of factoring.

EXAMPLE 9 Multiply $\dfrac{xa+xb+ya+yb}{xa-xb-ya+yb}\cdot\dfrac{xa+xb-ya-yb}{xa-xb+ya-yb}$.

SOLUTION We will factor each polynomial by grouping, which takes two steps.

$$\frac{xa+xb+ya+yb}{xa-xb-ya+yb}\cdot\frac{xa+xb-ya-yb}{xa-xb+ya-yb}$$

$$=\frac{x(a+b)+y(a+b)}{x(a-b)-y(a-b)}\cdot\frac{x(a+b)-y(a+b)}{x(a-b)+y(a-b)}$$

$$=\frac{(a+b)(x+y)(a+b)(x-y)}{(a-b)(x-y)(a-b)(x+y)}$$

Factor by grouping

$$=\frac{(a+b)^2}{(a-b)^2}$$

EXAMPLE 10 Multiply $(4x^2-36)\cdot\dfrac{12}{4x+12}$.

SOLUTION We can think of $4x^2-36$ as having a denominator of 1. Thinking of it in this way allows us to proceed as we did in the previous examples.

$$(4x^2-36)\cdot\frac{12}{4x+12}$$

$$=\frac{4x^2-36}{1}\cdot\frac{12}{4x+12}$$ Write $4x^2-36$ with denominator 1

$$=\frac{4(x-3)(x+3)12}{4(x+3)}$$ Factor

$$=12(x-3)$$ Divide out common factors

EXAMPLE 11 Multiply $3(x-2)(x-1) \cdot \dfrac{5}{x^2-3x+2}$.

SOLUTION This problem is very similar to the problem in Example 10. Writing the first rational expression with a denominator of 1, we have

$$\frac{3(x-2)(x-1)}{1} \cdot \frac{5}{x^2-3x+2} = \frac{3(x-2)(x-1)5}{(x-2)(x-1)}$$

$$= 3 \cdot 5$$

$$= 15$$

GETTING READY FOR CLASS

After reading through the preceding section, respond in your own words and in complete sentences.

A. Summarize the steps used to multiply fractions.

B. What is the first step in multiplying two rational expressions?

C. Why is factoring important when multiplying and dividing rational expressions?

D. How is division with rational expressions different than multiplication of rational expressions?

Problem Set 8.2

Perform the indicated operations.

1. $\frac{2}{9} \cdot \frac{3}{4}$
2. $\frac{5}{6} \cdot \frac{7}{8}$
3. $\frac{3}{4} \div \frac{1}{3}$
4. $\frac{3}{8} \div \frac{5}{4}$
5. $\frac{3}{7} \cdot \frac{14}{24} \div \frac{1}{2}$
6. $\frac{6}{5} \cdot \frac{10}{36} \div \frac{3}{4}$
7. $\frac{10x^2}{5y^2} \cdot \frac{15y^3}{2x^4}$
8. $\frac{8x^3}{7y^4} \cdot \frac{14y^6}{16x^2}$
9. $\frac{11a^2b}{5ab^2} \div \frac{22a^3b^2}{10ab^4}$
10. $\frac{8ab^3}{9a^2b} \div \frac{16a^2b^2}{18ab^3}$
11. $\frac{6x^2}{5y^3} \cdot \frac{11z^2}{2x^2} \div \frac{33z^5}{10y^8}$
12. $\frac{4x^3}{7y^2} \cdot \frac{6z^5}{5x^6} \div \frac{24z^2}{35x^6}$

Perform the indicated operations. Be sure to write all answers in lowest terms.

13. $\frac{x^2 - 9}{x^2 - 4} \cdot \frac{x - 2}{x - 3}$
14. $\frac{x^2 - 16}{x^2 - 25} \cdot \frac{x - 5}{x - 4}$
15. $\frac{y^2 - 1}{y + 2} \cdot \frac{y^2 + 5y + 6}{y^2 + 2y - 3}$
16. $\frac{y - 1}{y^2 - y - 6} \cdot \frac{y^2 + 5y + 6}{y^2 - 1}$
17. $\frac{3x - 12}{x^2 - 4} \cdot \frac{x^2 + 6x + 8}{x - 4}$
18. $\frac{x^2 + 5x + 1}{4x - 4} \cdot \frac{x - 1}{x^2 + 5x + 1}$
19. $\frac{xy}{xy + 1} \div \frac{x}{y}$
20. $\frac{y}{x} \div \frac{xy}{xy - 1}$
21. $\frac{1}{x^2 - 9} \div \frac{1}{x^2 + 9}$
22. $\frac{1}{x^2 - 9} \div \frac{1}{(x - 3)^2}$
23. $\frac{y - 3}{y^2 - 6y + 9} \cdot \frac{y - 3}{4}$
24. $\frac{y - 3}{y^2 - 6y + 9} \div \frac{y - 3}{4}$
25. $\frac{5x + 2y}{25x^2 - 5xy - 6y^2} \cdot \frac{20x^2 - 7xy - 3y^2}{4x + y}$
26. $\frac{7x + 3y}{42x^2 - 17xy - 15y^2} \cdot \frac{12x^2 - 4xy - 5y^2}{2x + y}$
27. $\frac{a^2 - 5a + 6}{a^2 - 2a - 3} \div \frac{a - 5}{a^2 + 3a + 2}$
28. $\frac{a^2 + 7a + 12}{a - 5} \div \frac{a^2 + 9a + 18}{a^2 - 7a + 10}$
29. $\frac{4t^2 - 1}{6t^2 + t - 2} \div \frac{8t^3 + 1}{27t^3 + 8}$
30. $\frac{9t^2 - 1}{6t^2 + 7t - 3} \div \frac{27t^3 + 1}{8t^3 + 27}$
31. $\frac{2x^2 - 5x - 12}{4x^2 + 8x + 3} \div \frac{x^2 - 16}{2x^2 + 7x + 3}$
32. $\frac{x^2 - 2x + 1}{3x^2 + 7x - 20} \div \frac{x^2 + 3x - 4}{3x^2 - 2x - 5}$
33. $\frac{2a^2 - 21ab - 36b^2}{a^2 - 11ab - 12b^2} \div \frac{10a + 15b}{a^2 - b^2}$
34. $\frac{3a^2 + 7ab - 20b^2}{a^2 + 5ab + 4b^2} \div \frac{3a^2 - 17ab + 20b^2}{3a - 12b}$
35. $\frac{6c^2 - c - 15}{9c^2 - 25} \cdot \frac{15c^2 + 22c - 5}{6c^2 + 5c - 6}$
36. $\frac{m^2 + 4m - 21}{m^2 - 12m + 27} \cdot \frac{m^2 - 7m + 12}{m^2 + 3m - 28}$
37. $\frac{6a^2b + 2ab^2 - 20b^3}{4a^2b - 16b^3} \cdot \frac{10a^2 - 22ab + 4b^2}{27a^3 - 125b^3}$
38. $\frac{12a^2b - 3ab^2 - 42b^3}{9a^2 - 36b^2} \cdot \frac{6a^2 - 15ab + 6b^2}{8a^3b - b^4}$
39. $\frac{360x^3 - 490x}{36x^2 + 84x + 49} \cdot \frac{30x^2 + 83x + 56}{150x^3 + 65x^2 - 280x}$

40. $\dfrac{490x^2 - 640}{49x^2 - 112x + 64} \cdot \dfrac{28x^2 - 95x + 72}{56x^3 - 62x^2 - 144x}$

41. $\dfrac{x^5 - x^2}{5x^2 - 5x} \cdot \dfrac{10x^4 - 10x^2}{2x^4 + 2x^3 + 2x^2}$

42. $\dfrac{2x^4 - 16x}{3x^6 - 48x^2} \cdot \dfrac{6x^5 + 24x^3}{4x^4 + 8x^3 + 16x^2}$

43. $\dfrac{a^2 - 16b^2}{a^2 - 8ab + 16b^2} \cdot \dfrac{a^2 - 9ab + 20b^2}{a^2 - 7ab + 12b^2} \div \dfrac{a^2 - 25b^2}{a^2 - 6ab + 9b^2}$

44. $\dfrac{a^2 - 6ab + 9b^2}{a^2 - 4b^2} \cdot \dfrac{a^2 - 5ab + 6b^2}{(a - 3b)^2} \div \dfrac{a^2 - 9b^2}{a^2 - ab - 6b^2}$

45. $\dfrac{2y^2 - 7y - 15}{42y^2 - 29y - 5} \cdot \dfrac{12y^2 - 16y + 5}{7y^2 - 36y + 5} \div \dfrac{4y^2 - 9}{49y^2 - 1}$

46. $\dfrac{8y^2 + 18y - 5}{21y^2 - 16y + 3} \cdot \dfrac{35y^2 - 22y + 3}{6y^2 + 17y + 5} \div \dfrac{16y^2 - 1}{9y^2 - 1}$

47. $\dfrac{xy - 2x + 3y - 6}{xy + 2x - 4y - 8} \cdot \dfrac{xy + x - 4y - 4}{xy - x + 3y - 3}$

48. $\dfrac{ax + bx + 2a + 2b}{ax - 3a + bx - 3b} \cdot \dfrac{ax - bx - 3a + 3b}{ax - bx - 2a + 2b}$

49. $\dfrac{xy^2 - y^2 + 4xy - 4y}{xy - 3y + 4x - 12} \div \dfrac{xy^3 + 2xy^2 + y^3 + 2y^2}{xy^2 - 3y^2 + 2xy - 6y}$

50. $\dfrac{4xb - 8b + 12x - 24}{xb^2 + 3b^2 + 3xb + 9b} \div \dfrac{4xb - 8b - 8x + 16}{xb^2 + 3b^2 - 2xb - 6b}$

51. $\dfrac{2x^3 + 10x^2 - 8x - 40}{x^3 + 4x^2 - 9x - 36} \cdot \dfrac{x^2 + x - 12}{2x^2 + 14x + 20}$

52. $\dfrac{x^3 + 2x^2 - 9x - 18}{x^4 + 3x^3 - 4x^2 - 12x} \cdot \dfrac{x^3 + 5x^2 + 6x}{x^2 - x - 6}$

53. $\dfrac{w^3 - w^2x}{wy - w} \div \left(\dfrac{w - x}{y - 1}\right)^2$

54. $\dfrac{a^3 - a^2b}{ac - a} \div \left(\dfrac{a - b}{c - 1}\right)^2$

55. $\dfrac{mx + my + 2x + 2y}{6x^2 - 5xy - 4y^2} \div \dfrac{2mx - 4x + my - 2y}{3mx - 6x - 4my + 8y}$

56. $\dfrac{ax - 2a + 2xy - 4y}{ax + 2a - 2xy - 4y} \div \dfrac{ax + 2a + 2xy + 4y}{ax - 2a - 2xy + 4y}$

57. $(3x - 6) \cdot \dfrac{x}{x - 2}$

58. $(4x + 8) \cdot \dfrac{x}{x + 2}$

59. $(x^2 - 25) \cdot \dfrac{2}{x - 5}$

60. $(x^2 - 49) \cdot \dfrac{5}{x + 7}$

61. $(x^2 - 3x + 2) \cdot \dfrac{3}{3x - 3}$

62. $(x^2 - 3x + 2) \cdot \dfrac{-1}{x - 2}$

63. $(y - 3)(y - 4)(y + 3) \cdot \dfrac{-1}{y^2 - 9}$

64. $(y + 1)(y + 4)(y - 1) \cdot \dfrac{3}{y^2 - 1}$

65. $a(a + 5)(a - 5) \cdot \dfrac{a + 1}{a^2 + 5a}$

66. $a(a + 3)(a - 3) \cdot \dfrac{a - 1}{a^2 - 3a}$

The next two problems are intended to give you practice reading, and paying attention to, the instructions that accompany the problems you are working. Working these problems is an excellent way to get ready for a test or a quiz.

67. Work each problem according to the instructions given.

a. Simplify: $\dfrac{16 - 1}{64 - 1}$ **b.** Reduce: $\dfrac{25x^2 - 9}{125x^3 - 27}$

c. Multiply: $\dfrac{25x^2 - 9}{125x^3 - 27} \cdot \dfrac{5x - 3}{5x + 3}$

d. Divide: $\dfrac{25x^2 - 9}{125x^3 - 27} \div \dfrac{5x - 3}{25x^2 + 15x + 9}$

68. Work each problem according to the instructions given.

a. Simplify: $\dfrac{64 - 49}{64 + 112 + 49}$ **b.** Reduce: $\dfrac{9x^2 - 49}{9x^2 + 42x + 49}$

c. Multiply: $\dfrac{9x^2 - 49}{9x^2 + 42x + 49} \cdot \dfrac{3x + 7}{3x - 7}$

d. Divide: $\dfrac{9x^2 - 49}{9x^2 + 42x + 49} \div \dfrac{3x + 7}{3x - 7}$

Getting Ready for the Next Section

Combine.

69. $\dfrac{4}{9} + \dfrac{2}{9}$ **70.** $\dfrac{3}{8} + \dfrac{1}{8}$ **71.** $\dfrac{3}{14} + \dfrac{7}{30}$ **72.** $\dfrac{3}{10} + \dfrac{11}{42}$

Multiply.

73. $-1(7 - x)$ **74.** $-1(3 - x)$

Factor.

75. $x^2 - 1$ **76.** $x^2 - 2x - 3$ **77.** $2x + 10$

78. $x^2 + 4x + 3$ **79.** $a^3 - b^3$ **80.** $8y^3 - 27$

Addition and Subtraction of Rational Expressions

8.3

This section is concerned with addition and subtraction of rational expressions. In the first part of this section, we will look at addition of expressions that have the same denominator. In the second part of this section, we will look at addition of expressions that have different denominators.

Addition and Subtraction with the Same Denominator

To add two expressions that have the same denominator, we simply add numerators and put the sum over the common denominator. Because the process we use to add and subtract rational expressions is the same process used to add and subtract fractions, we will begin with an example involving fractions.

EXAMPLE 1 Add $\frac{4}{9} + \frac{2}{9}$.

SOLUTION We add fractions with the same denominator by using the distributive property. Here is a detailed look at the steps involved.

$$\frac{4}{9} + \frac{2}{9} = 4\left(\frac{1}{9}\right) + 2\left(\frac{1}{9}\right)$$

$$= (4 + 2)\left(\frac{1}{9}\right) \quad \text{Distributive property}$$

$$= 6\left(\frac{1}{9}\right)$$

$$= \frac{6}{9}$$

$$= \frac{2}{3} \quad \text{Divide numerator and denominator by common factor 3}$$

Note that the important thing about the fractions in this example is that they each have a denominator of 9. If they did not have the same denominator, we could not have written them as two terms with a factor of $\frac{1}{9}$ in common. Without the $\frac{1}{9}$ common to each term, we couldn't apply the distributive property. Without the distributive property, we would not have been able to add the two fractions in this form.

In the following examples, we will not show all the steps we showed in Example 1. The steps are shown in Example 1 so you will see why both fractions must have the same denominator before we can add them. In practice, we simply add numerators and place the result over the common denominator.

We add and subtract rational expressions with the same denominator by combining numerators and writing the result over the common denominator. Then we reduce the result to lowest terms, if possible. Example 2 shows this process in detail. If you see the similarities between operations on rational numbers and operations on rational expressions, this chapter will look like an extension of rational numbers rather than a completely new set of topics.

EXAMPLE 2 Add $\frac{x}{x^2 - 1} + \frac{1}{x^2 - 1}$.

SOLUTION Because the denominators are the same, we simply add numerators:

$$\frac{x}{x^2 - 1} + \frac{1}{x^2 - 1} = \frac{x + 1}{x^2 - 1}$$ Add numerators

$$= \frac{x + 1}{(x - 1)(x + 1)}$$ Factor denominator

$$= \frac{1}{x - 1}$$ Divide out common factor $x + 1$

Our next example involves subtraction of rational expressions. Pay careful attention to what happens to the signs of the terms in the numerator of the second expression when we subtract it from the first expression.

EXAMPLE 3 Subtract $\frac{2x - 5}{x - 2} - \frac{x - 3}{x - 2}$.

SOLUTION Because each expression has the same denominator, we simply subtract the numerator in the second expression from the numerator in the first expression and write the difference over the common denominator $x - 2$. We must be careful, however, that we subtract both terms in the second numerator. To ensure that we do, we will enclose that numerator in parentheses.

$$\frac{2x - 5}{x - 2} - \frac{x - 3}{x - 2} = \frac{2x - 5 - (x - 3)}{x - 2}$$ Subtract numerators

$$= \frac{2x - 5 - x + 3}{x - 2}$$ Remove parentheses

$$= \frac{x - 2}{x - 2}$$ Combine similar terms in the numerator

$$= 1$$ Reduce (or divide)

Note the $+3$ in the numerator of the second step. It is a common mistake to write this as -3, by forgetting to subtract both terms in the numerator of the second expression. Whenever the expression we are subtracting has two or more terms in its numerator, we have to watch for this mistake.

Next we consider addition and subtraction of fractions and rational expressions that have different denominators.

Addition and Subtraction With Different Denominators

Before we look at an example of addition of fractions with different denominators, we need to review the definition for the least common denominator (LCD).

DEFINITION *least common denominator*

The ***least common denominator*** for a set of denominators is the smallest expression that is divisible by each of the denominators.

The first step in combining two fractions is to find the LCD. Once we have the common denominator, we rewrite each fraction as an equivalent fraction with the common denominator. After that, we simply add or subtract as we did in our first three examples.

Example 4 is a review of the step-by-step procedure used to add two fractions with different denominators.

EXAMPLE 4 Add $\frac{3}{14} + \frac{7}{30}$.

SOLUTION

Step 1: *Find the LCD.*

To do this, we first factor both denominators into prime factors.

Factor 14: $14 = 2 \cdot 7$

Factor 30: $30 = 2 \cdot 3 \cdot 5$

Because the LCD must be divisible by 14, it must have factors of 2 and 7. It must also be divisible by 30 and, therefore, have factors of 2, 3, and 5. We do not need to repeat the 2 that appears in both the factors of 14 and those of 30. Therefore,

$$\text{LCD} = 2 \cdot 3 \cdot 5 \cdot 7 = 210$$

Step 2: *Change to equivalent fractions.*

Because we want each fraction to have a denominator of 210 and at the same time keep its original value, we multiply each by 1 in the appropriate form.

Change $\frac{3}{14}$ to a fraction with denominator 210:

$$\frac{3}{14} \cdot \frac{15}{15} = \frac{45}{210}$$

Change $\frac{7}{30}$ to a fraction with denominator 210:

$$\frac{7}{30} \cdot \frac{7}{7} = \frac{49}{210}$$

Step 3: *Add numerators of equivalent fractions found in step 2:*

$$\frac{45}{210} + \frac{49}{210} = \frac{94}{210}$$

Step 4: *Reduce to lowest terms, if necessary:*

$$\frac{94}{210} = \frac{47}{105}$$

The main idea in adding fractions is to write each fraction again with the LCD for a denominator. In doing so, we must be sure not to change the value of either of the original fractions.

EXAMPLE 5 Add $\frac{-2}{x^2 - 2x - 3} + \frac{3}{x^2 - 9}$.

SOLUTION

Step 1: ***Factor each denominator and build the LCD from the factors:***

$$\begin{aligned} x^2 - 2x - 3 &= (x - 3)(x + 1) \\ x^2 - 9 &= (x - 3)(x + 3) \end{aligned} \qquad LCD = (x - 3)(x + 3)(x + 1)$$

Step 2: ***Change each rational expression to an equivalent expression that has the LCD for a denominator:***

$$\frac{-2}{x^2 - 2x - 3} = \frac{-2}{(x - 3)(x + 1)} \cdot \frac{(x + 3)}{(x + 3)} = \frac{-2x - 6}{(x - 3)(x + 3)(x + 1)}$$

$$\frac{3}{x^2 - 9} = \frac{3}{(x - 3)(x + 3)} \cdot \frac{(x + 1)}{(x + 1)} = \frac{3x + 3}{(x - 3)(x + 3)(x + 1)}$$

Step 3: ***Add numerators of the rational expressions found in step 2:***

$$\frac{-2x - 6}{(x - 3)(x + 3)(x + 1)} + \frac{3x + 3}{(x - 3)(x + 3)(x + 1)} = \frac{x - 3}{(x - 3)(x + 3)(x + 1)}$$

Step 4: ***Reduce to lowest terms by dividing out the common factor*** $x - 3$***:***

$$\frac{\cancel{x - 3}}{(\cancel{x - 3})(x + 3)(x + 1)} = \frac{1}{(x + 3)(x + 1)}$$

EXAMPLE 6 Subtract $\frac{x + 4}{2x + 10} - \frac{5}{x^2 - 25}$.

SOLUTION We begin by factoring each denominator:

$$\frac{x + 4}{2x + 10} - \frac{5}{x^2 - 25} = \frac{x + 4}{2(x + 5)} - \frac{5}{(x + 5)(x - 5)}$$

The LCD is $2(x + 5)(x - 5)$. Completing the problem, we have

$$= \frac{x + 4}{2(x + 5)} \cdot \frac{(x - 5)}{(x - 5)} - \frac{5}{(x + 5)(x - 5)} \cdot \frac{2}{2}$$

$$= \frac{x^2 - x - 20}{2(x + 5)(x - 5)} - \frac{10}{2(x + 5)(x - 5)}$$

$$= \frac{x^2 - x - 30}{2(x + 5)(x - 5)}$$

To see if this expression will reduce, we factor the numerator into $(x - 6)(x + 5)$.

$$= \frac{(x - 6)(\cancel{x + 5})}{2(\cancel{x + 5})(x - 5)}$$

$$= \frac{x - 6}{2(x - 5)}$$

EXAMPLE 7 Subtract $\frac{2x-2}{x^2+4x+3} - \frac{x-1}{x^2+5x+6}$.

SOLUTION We factor each denominator and build the LCD from those factors:

$$\frac{2x-2}{x^2+4x+3} - \frac{x-1}{x^2+5x+6}$$

$$= \frac{2x-2}{(x+3)(x+1)} - \frac{x-1}{(x+3)(x+2)}$$

$$= \frac{2x-2}{(x+3)(x+1)} \cdot \frac{(x+2)}{(x+2)} - \frac{x-1}{(x+3)(x+2)} \cdot \frac{(x+1)}{(x+1)}$$ The LCD is $(x+1)(x+2)(x+3)$

$$= \frac{2x^2+2x-4}{(x+1)(x+2)(x+3)} - \frac{x^2-1}{(x+1)(x+2)(x+3)}$$ Multiply out each numerator

$$= \frac{(2x^2+2x-4)-(x^2-1)}{(x+1)(x+2)(x+3)}$$ Subtract numerators

$$= \frac{x^2+2x-3}{(x+1)(x+2)(x+3)}$$ Factor numerator to see if we can rdeuce

$$= \frac{(x+3)(x-1)}{(x+1)(x+2)(x+3)}$$ Reduce

$$= \frac{x-1}{(x+1)(x+2)}$$

EXAMPLE 8 Add $\frac{x^2}{x-7} + \frac{6x+7}{7-x}$.

SOLUTION In Section 5.1, we were able to reverse the terms in a factor such as $7 - x$ by factoring -1 from each term. In a problem like this, the same result can be obtained by multiplying the numerator and denominator by -1:

$$\frac{x^2}{x-7} + \frac{6x+7}{7-x} \cdot \frac{-1}{-1} = \frac{x^2}{x-7} + \frac{-6x-7}{x-7}$$

$$= \frac{x^2-6x-7}{x-7}$$ Add numerators

$$= \frac{(x-7)(x+1)}{(x-7)}$$ Factor numerator

$$= x + 1$$ Divide out $x - 7$

For our next example, we will look at a problem in which we combine a whole number and a rational expression.

EXAMPLE 9 Subtract $2 - \frac{9}{3x + 1}$.

SOLUTION To subtract these two expressions, we think of 2 as a rational expression with a denominator of 1.

$$2 - \frac{9}{3x + 1} = \frac{2}{1} - \frac{9}{3x + 1}$$

The LCD is $3x + 1$. Multiplying the numerator and denominator of the first expression by $3x + 1$ gives us a rational expression equivalent to 2, but with a denominator of $3x + 1$.

$$\frac{2}{1} \cdot \frac{(3x + 1)}{(3x + 1)} - \frac{9}{3x + 1} = \frac{6x + 2 - 9}{3x + 1}$$

$$= \frac{6x - 7}{3x + 1}$$

The numerator and denominator of this last expression do not have any factors in common other than 1, so the expression is in lowest terms.

EXAMPLE 10 Write an expression for the sum of a number and twice its reciprocal. Then, simplify that expression.

SOLUTION If x is the number, then its reciprocal is $\frac{1}{x}$. Twice its reciprocal is $\frac{2}{x}$. The sum of the number and twice its reciprocal is

$$x + \frac{2}{x}$$

To combine these two expressions, we think of the first term x as a rational expression with a denominator of 1. The LCD is x:

$$x + \frac{2}{x} = \frac{x}{1} + \frac{2}{x}$$

$$= \frac{x}{1} \cdot \frac{x}{x} + \frac{2}{x}$$

$$= \frac{x^2 + 2}{x}$$

GETTING READY FOR CLASS

After reading through the preceding section, respond in your own words and in complete sentences.

A. Briefly describe how you would add two rational expressions that have the same denominator.

B. Why is factoring important in finding a least common denominator?

C. What is the last step in adding or subtracting two rational expressions?

D. Explain how you would change the fraction $\frac{5}{x - 3}$ to an equivalent fraction with denominator $x^2 - 9$.

Combine the following fractions.

1. $\frac{3}{4}+\frac{1}{2}$ **2.** $\frac{5}{6}+\frac{1}{3}$ **3.** $\frac{2}{5}-\frac{1}{15}$ **4.** $\frac{5}{8}-\frac{1}{4}$

5. $\frac{5}{6}+\frac{7}{8}$ **6.** $\frac{3}{4}+\frac{2}{3}$ **7.** $\frac{9}{48}-\frac{3}{54}$ **8.** $\frac{6}{28}-\frac{5}{42}$

9. $\frac{3}{4}-\frac{1}{8}+\frac{2}{3}$ **10.** $\frac{1}{3}-\frac{5}{6}+\frac{5}{12}$

Combine the following rational expressions. Reduce all answers to lowest terms.

11. $\frac{x}{x+3}+\frac{3}{x+3}$ **12.** $\frac{5x}{5x+2}+\frac{2}{5x+2}$ **13.** $\frac{4}{y-4}-\frac{y}{y-4}$

14. $\frac{8}{y+8}+\frac{y}{y+8}$ **15.** $\frac{x}{x^2-y^2}-\frac{y}{x^2-y^2}$ **16.** $\frac{x}{x^2-y^2}+\frac{y}{x^2-y^2}$

17. $\frac{2x-3}{x-2}-\frac{x-1}{x-2}$ **18.** $\frac{2x-4}{x+2}-\frac{x-6}{x+2}$ **19.** $\frac{1}{a}+\frac{2}{a^2}-\frac{3}{a^3}$

20. $\frac{3}{a}+\frac{2}{a^2}-\frac{1}{a^3}$ **21.** $\frac{7x-2}{2x+1}-\frac{5x-3}{2x+1}$ **22.** $\frac{7x-1}{3x+2}-\frac{4x-3}{3x+2}$

23. Work each problem according to the instructions given.

a. Multiply: $\frac{3}{8}\cdot\frac{1}{6}$ **b.** Divide: $\frac{3}{8}\div\frac{1}{6}$

c. Add: $\frac{3}{8}+\frac{1}{6}$ **d.** Multiply: $\frac{x+3}{x-3}\cdot\frac{5x+15}{x^2-9}$

e. Divide: $\frac{x+3}{x-3}\div\frac{5x+15}{x^2-9}$ **f.** Subtract: $\frac{x+3}{x-3}-\frac{5x+15}{x^2-9}$

24. Work each problem according to the instructions given.

a. Multiply: $\frac{16}{49}\cdot\frac{1}{28}$ **b.** Divide: $\frac{16}{49}\div\frac{1}{28}$

c. Subtract: $\frac{16}{49}-\frac{1}{28}$ **d.** Multiply: $\frac{3x-2}{3x+2}\cdot\frac{15x+6}{9x^2-4}$

e. Divide: $\frac{3x-2}{3x+2}\div\frac{15x+6}{9x^2-4}$ **f.** Subtract: $\frac{3x+2}{3x-2}-\frac{15x+6}{9x^2-4}$

Combine the following rational expressions. Reduce all answers to lowest terms.

25. $\frac{3x+1}{2x-6}-\frac{x+2}{x-3}$ **26.** $\frac{x+1}{x-2}-\frac{4x+7}{5x-10}$

27. $\frac{6x+5}{5x-25}-\frac{x+2}{x-5}$ **28.** $\frac{4x+2}{3x+12}-\frac{x-2}{x+4}$

29. $\frac{x+1}{2x-2}-\frac{2}{x^2-1}$ **30.** $\frac{x+7}{2x+12}+\frac{6}{x^2-36}$

31. $\frac{1}{a-b}-\frac{3ab}{a^3-b^3}$ **32.** $\frac{1}{a+b}+\frac{3ab}{a^3+b^3}$

33. $\frac{1}{2y-3}-\frac{18y}{8y^3-27}$ **34.** $\frac{1}{3y-2}-\frac{18y}{27y^3-8}$

35. $\frac{x}{x^2-5x+6}-\frac{3}{3-x}$ **36.** $\frac{x}{x^2+4x+4}-\frac{2}{2+x}$

37. $\frac{2}{4t-5}+\frac{9}{8t^2-38t+35}$ **38.** $\frac{3}{2t-5}+\frac{21}{8t^2-14t-15}$

39. $\dfrac{1}{a^2 - 5a + 6} + \dfrac{3}{a^2 - a - 2}$

40. $\dfrac{-3}{a^2 + a - 2} + \dfrac{5}{a^2 - a - 6}$

41. $\dfrac{1}{8x^3 - 1} - \dfrac{1}{4x^2 - 1}$

42. $\dfrac{1}{27x^3 - 1} - \dfrac{1}{9x^2 - 1}$

43. $\dfrac{4}{4x^2 - 9} - \dfrac{6}{8x^2 - 6x - 9}$

44. $\dfrac{9}{9x^2 + 6x - 8} - \dfrac{6}{9x^2 - 4}$

45. $\dfrac{4a}{a^2 + 6a + 5} - \dfrac{3a}{a^2 + 5a + 4}$

46. $\dfrac{3a}{a^2 + 7a + 10} - \dfrac{2a}{a^2 + 6a + 8}$

47. $\dfrac{2x - 1}{x^2 + x - 6} - \dfrac{x + 2}{x^2 + 5x + 6}$

48. $\dfrac{4x + 1}{x^2 + 5x + 4} - \dfrac{x + 3}{x^2 + 4x + 3}$

49. $\dfrac{2x - 8}{3x^2 + 8x + 4} + \dfrac{x + 3}{3x^2 + 5x + 2}$

50. $\dfrac{5x + 3}{2x^2 + 5x + 3} - \dfrac{3x + 9}{2x^2 + 7x + 6}$

51. $\dfrac{2}{x^2 + 5x + 6} - \dfrac{4}{x^2 + 4x + 3} + \dfrac{3}{x^2 + 3x + 2}$

52. $\dfrac{-5}{x^2 + 3x - 4} + \dfrac{5}{x^2 + 2x - 3} + \dfrac{1}{x^2 + 7x + 12}$

53. $\dfrac{2x + 8}{x^2 + 5x + 6} - \dfrac{x + 5}{x^2 + 4x + 3} - \dfrac{x - 1}{x^2 + 3x + 2}$

54. $\dfrac{2x + 11}{x^2 + 9x + 20} - \dfrac{x + 1}{x^2 + 7x + 12} - \dfrac{x + 6}{x^2 + 8x + 15}$

55. $2 + \dfrac{3}{2x + 1}$

56. $3 - \dfrac{2}{2x + 3}$

57. $5 + \dfrac{2}{4 - t}$

58. $7 + \dfrac{3}{5 - t}$

59. $x - \dfrac{4}{2x + 3}$

60. $x - \dfrac{5}{3x + 4} + 1$

61. $\dfrac{x}{x + 2} + \dfrac{1}{2x + 4} - \dfrac{3}{x^2 + 2x}$

62. $\dfrac{x}{x + 3} + \dfrac{7}{3x + 9} - \dfrac{2}{x^2 + 3x}$

63. $\dfrac{1}{x} + \dfrac{x}{2x + 4} - \dfrac{2}{x^2 + 2x}$

64. $\dfrac{1}{x} + \dfrac{x}{3x + 9} - \dfrac{3}{x^2 + 3x}$

65. Let $f(x) = \dfrac{2}{x + 4}$ and $g(x) = \dfrac{x - 1}{x^2 + 3x - 4}$; find $f(x) + g(x)$

66. Let $f(t) = \dfrac{5}{3t - 2}$ and $g(t) = \dfrac{t - 3}{3t^2 + 7t - 6}$; find $f(t) - g(t)$

67. Let $f(x) = \dfrac{2x}{x^2 - x - 2}$ and $g(x) = \dfrac{5}{x^2 + x - 6}$; find $f(x) + g(x)$

68. Let $f(x) = \dfrac{7}{x^2 - x - 12}$ and $g(x) = \dfrac{5}{x^2 + x - 6}$; find $f(x) - g(x)$

Applying the Concepts

69. Optometry The formula

$$P = \frac{1}{a} + \frac{1}{b}$$

is used by optometrists to help determine how strong to make the lenses for a pair of eyeglasses. If a is 10 and b is 0.2, find the corresponding value of P.

70. Quadratic Formula Later in the book we will work with the quadratic formula. The derivation of the formula requires that you can add the fractions below. Add the fractions.

$$\frac{-c}{a} + \frac{b^2}{(2a)^2}$$

71. Number Problem Write an expression for the sum of a number and 4 times its reciprocal. Then, simplify that expression.

72. Number Problem Write an expression for the sum of a number and 3 times its reciprocal. Then, simplify that expression.

73. Number Problem Write an expression for the sum of the reciprocals of two consecutive integers. Then, simplify that expression.

74. Number Problem Write an expression for the sum of the reciprocals of two consecutive even integers. Then, simplify that expression.

Getting Ready for the Next Section

Divide.

75. $\frac{3}{4} \div \frac{5}{8}$

76. $\frac{2}{3} \div \frac{5}{6}$

Multiply.

77. $x\left(1 + \frac{2}{x}\right)$

78. $3\left(x + \frac{1}{3}\right)$

79. $3x\left(\frac{1}{x} - \frac{1}{3}\right)$

80. $3x\left(\frac{1}{x} + \frac{1}{3}\right)$

Factor.

81. $x^2 - 4$

82. $x^2 - x - 6$

Complex Fractions 8.4

The quotient of two fractions or two rational expressions is called a *complex fraction*. This section is concerned with the simplification of complex fractions.

EXAMPLE 1 Simplify $\dfrac{\frac{3}{4}}{\frac{5}{8}}$.

SOLUTION There are generally two methods that can be used to simplify complex fractions.

Method 1 We can multiply the numerator and denominator of the complex fractions by the LCD for both of the fractions, which in this case is 8.

$$\frac{\frac{3}{4}}{\frac{5}{8}} = \frac{\frac{3}{4} \cdot 8}{\frac{5}{8} \cdot 8} = \frac{6}{5}$$

Method 2 Instead of dividing by $\frac{5}{8}$ we can multiply by $\frac{8}{5}$.

$$\frac{\frac{3}{4}}{\frac{5}{8}} = \frac{3}{4} \cdot \frac{8}{5} = \frac{24}{20} = \frac{6}{5}$$

Here are some examples of complex fractions involving rational expressions. Most can be solved using either of the two methods shown in Example 1.

EXAMPLE 2 Simplify $\dfrac{\frac{1}{x} + \frac{1}{y}}{\frac{1}{x} - \frac{1}{y}}$.

SOLUTION This problem is most easily solved using Method 1. We begin by multiplying both the numerator and denominator by the quantity xy, which is the LCD for all the fractions:

$$\frac{\frac{1}{x} + \frac{1}{y}}{\frac{1}{x} - \frac{1}{y}} = \frac{\left(\frac{1}{x} + \frac{1}{y}\right) \cdot xy}{\left(\frac{1}{x} - \frac{1}{y}\right) \cdot xy}$$

$$= \frac{\frac{1}{x}(xy) + \frac{1}{y}(xy)}{\frac{1}{x}(xy) - \frac{1}{y}(xy)}$$

Apply the distributive property to distribute xy over both term in the numerator and denominator.

$$= \frac{y + x}{y - x}$$

EXAMPLE 3 Simplify $\dfrac{\dfrac{x-2}{x^2-9}}{\dfrac{x^2-4}{x+3}}$.

SOLUTION Applying Method 2, we have

$$\frac{\dfrac{x-2}{x^2-9}}{\dfrac{x^2-4}{x+3}} = \frac{x-2}{x^2-9} \cdot \frac{x+3}{x^2-4}$$

$$= \frac{(x-2)(x+3)}{(x+3)(x-3)(x+2)(x-2)}$$

$$= \frac{1}{(x-3)(x+2)}$$

EXAMPLE 4 Simplify $\dfrac{1-\dfrac{4}{x^2}}{1-\dfrac{1}{x}-\dfrac{6}{x^2}}$.

SOLUTION The simplest way to simplify this complex fraction is to multiply the numerator and denominator by the LCD, x^2:

$$\frac{1-\dfrac{4}{x^2}}{1-\dfrac{1}{x}-\dfrac{6}{x^2}} = \frac{x^2\left(1-\dfrac{4}{x^2}\right)}{x^2\left(1-\dfrac{1}{x}-\dfrac{6}{x^2}\right)}$$ Multiply numerator and denominator by x^2

$$= \frac{x^2 \cdot 1 - x^2 \cdot \dfrac{4}{x^2}}{x^2 \cdot 1 - x^2 \cdot \dfrac{1}{x} - x^2 \cdot \dfrac{6}{x^2}}$$ Distributive property

$$= \frac{x^2-4}{x^2-x-6}$$ Simplify

$$= \frac{(x-2)(x+2)}{(x-3)(x+2)}$$ Factor

$$= \frac{x-2}{x-3}$$ Reduce

EXAMPLE 5 Simplify $2 - \dfrac{3}{x+\frac{1}{3}}$.

SOLUTION First, we simplify the expression that follows the subtraction sign.

$$2 - \frac{3}{x+\frac{1}{3}} = 2 - \frac{3 \cdot 3}{3\left(x+\frac{1}{3}\right)} = 2 - \frac{9}{3x+1}$$

Now we subtract by rewriting the first term, 2, with the LCD, $3x + 1$.

$$2 - \frac{9}{3x+1} = \frac{2}{1} \cdot \frac{3x+1}{3x+1} - \frac{9}{3x+1}$$

$$= \frac{6x+2-9}{3x+1} = \frac{6x-7}{3x+1}$$

GETTING READY FOR CLASS

Respond in your own words and in complete sentences.

A. What is a complex fraction?

B. Explain how a least common denominator can be used to simplify a complex fraction.

C. Explain how some complex fractions can be converted to division problems. When is it more efficient to convert a complex fraction to a division problem of rational expressions?

D. Which method of simplifying complex fractions do you prefer? Why?

Problem Set 8.4

Simplify each of the following as much as possible.

1. $\dfrac{\frac{3}{4}}{\frac{2}{3}}$

2. $\dfrac{\frac{5}{9}}{\frac{7}{12}}$

3. $\dfrac{\frac{1}{3} - \frac{1}{4}}{\frac{1}{2} + \frac{1}{8}}$

4. $\dfrac{\frac{1}{6} - \frac{1}{3}}{\frac{1}{4} - \frac{1}{8}}$

5. $\dfrac{3 + \frac{2}{5}}{1 - \frac{3}{7}}$

6. $\dfrac{2 + \frac{5}{6}}{1 - \frac{7}{8}}$

7. $\dfrac{\frac{1}{x}}{1 + \frac{1}{x}}$

8. $\dfrac{1 - \frac{1}{x}}{\frac{1}{x}}$

9. $\dfrac{1 + \frac{1}{a}}{1 - \frac{1}{a}}$

10. $\dfrac{1 - \frac{2}{a}}{1 - \frac{3}{a}}$

11. $\dfrac{\frac{1}{x} - \frac{1}{y}}{\frac{1}{x} + \frac{1}{y}}$

12. $\dfrac{\frac{1}{x} + \frac{2}{y}}{\frac{2}{x} + \frac{1}{y}}$

13. $\dfrac{\frac{x - 5}{x^2 - 4}}{\frac{x^2 - 25}{x + 2}}$

14. $\dfrac{\frac{3x + 1}{x^2 - 49}}{\frac{9x^2 - 1}{x - 7}}$

15. $\dfrac{\frac{4a}{2a^3 + 2}}{\frac{8a}{4a + 4}}$

16. $\dfrac{\frac{2a}{3a^3 - 3}}{\frac{4a}{6a - 6}}$

17. $\dfrac{1 - \frac{9}{x^2}}{1 - \frac{1}{x} - \frac{6}{x^2}}$

18. $\dfrac{4 - \frac{1}{x^2}}{4 + \frac{4}{x} + \frac{1}{x^2}}$

19. $\dfrac{2 + \frac{5}{a} - \frac{3}{a^2}}{2 - \frac{5}{a} + \frac{2}{a^2}}$

20. $\dfrac{3 + \frac{5}{a} - \frac{2}{a^2}}{3 - \frac{10}{a} + \frac{3}{a^2}}$

21. $\dfrac{2 + \frac{3}{x} - \frac{18}{x^2} - \frac{27}{x^3}}{2 + \frac{9}{x} + \frac{9}{x^2}}$

22. $\dfrac{3 + \frac{5}{x} - \frac{12}{x^2} - \frac{20}{x^3}}{3 + \frac{11}{x} + \frac{10}{x^2}}$

23. $\dfrac{1 + \frac{1}{x + 3}}{1 - \frac{1}{x + 3}}$

24. $\dfrac{1 + \frac{1}{x - 2}}{1 - \frac{1}{x - 2}}$

25. $\dfrac{1 + \frac{1}{x + 3}}{1 + \frac{7}{x - 3}}$

26. $\dfrac{1 + \frac{1}{x - 2}}{1 - \frac{3}{x + 2}}$

27. $\dfrac{1 - \frac{1}{a + 1}}{1 + \frac{1}{a - 1}}$

28. $\dfrac{\frac{1}{a - 1} + 1}{\frac{1}{a + 1} - 1}$

29. $\dfrac{\frac{1}{x + 3} + \frac{1}{x - 3}}{\frac{1}{x + 3} - \frac{1}{x - 3}}$

30. $\dfrac{\frac{1}{x + a} + \frac{1}{x - a}}{\frac{1}{x + a} - \frac{1}{x - a}}$

31. $\dfrac{\dfrac{y+1}{y-1}+\dfrac{y-1}{y+1}}{\dfrac{y+1}{y-1}-\dfrac{y-1}{y+1}}$

32. $\dfrac{\dfrac{y-1}{y+1}-\dfrac{y+1}{y-1}}{\dfrac{y-1}{y+1}+\dfrac{y+1}{y-1}}$

33. $1-\dfrac{x}{1-\dfrac{1}{x}}$

34. $x-\dfrac{1}{x-\dfrac{1}{2}}$

35. $1+\dfrac{1}{1+\dfrac{1}{1+1}}$

36. $1-\dfrac{1}{1-\dfrac{1}{1-\dfrac{1}{2}}}$

37. $\dfrac{1-\dfrac{1}{x+\dfrac{1}{2}}}{1+\dfrac{1}{x+\dfrac{1}{2}}}$

38. $\dfrac{2+\dfrac{1}{x-\dfrac{1}{3}}}{2-\dfrac{1}{x-\dfrac{1}{3}}}$

39. $\dfrac{\dfrac{1}{x+h}-\dfrac{1}{x}}{h}$

40. $\dfrac{\dfrac{1}{(x+h)^2}-\dfrac{1}{x^2}}{h}$

41. $\dfrac{\dfrac{3}{ab}+\dfrac{4}{bc}-\dfrac{2}{ac}}{\dfrac{5}{abc}}$

42. $\dfrac{\dfrac{x}{yz}-\dfrac{y}{xz}+\dfrac{z}{xy}}{\dfrac{1}{x^2y^2}-\dfrac{1}{x^2z^2}+\dfrac{1}{y^2z^2}}$

43. $\dfrac{\dfrac{t^2-2t-8}{t^2+7t+6}}{\dfrac{t^2-t-6}{t^2+2t+1}}$

44. $\dfrac{\dfrac{y^2-5y-14}{y^2+3y-10}}{\dfrac{y^2-8y+7}{y^2+6y+5}}$

45. $\dfrac{5+\dfrac{4}{b-1}}{\dfrac{7}{b+5}-\dfrac{3}{b-1}}$

46. $\dfrac{\dfrac{6}{x+5}-7}{\dfrac{8}{x+5}-\dfrac{9}{x+3}}$

47. $\dfrac{\dfrac{3}{x^2-x-6}}{\dfrac{2}{x+2}-\dfrac{4}{x-3}}$

48. $\dfrac{\dfrac{9}{a-7}+\dfrac{8}{2a+3}}{\dfrac{10}{2a^2-11a-21}}$

49. $\dfrac{\dfrac{1}{m-4}+\dfrac{1}{m-5}}{\dfrac{1}{m^2-9m+20}}$

50. $\dfrac{\dfrac{1}{k^2-7k+12}}{\dfrac{1}{k-3}+\dfrac{1}{k-4}}$

Applying the Concepts

51. **Difference Quotient** For each rational function below, find the difference quotient

$$\frac{f(x)-f(a)}{x-a}$$

a. $f(x)=\dfrac{4}{x}$ b. $f(x)=\dfrac{1}{x+1}$ c. $f(x)=\dfrac{1}{x^2}$

52. **Difference Quotient** For each rational function below, find the difference quotient

$$\frac{f(x+h)-f(x)}{h}$$

a. $f(x)=\dfrac{4}{x}$ b. $f(x)=\dfrac{1}{x+1}$ c. $f(x)=\dfrac{1}{x^2}$

53. **Doppler Effect** The change in the pitch of a sound (such as a train whistle) as an object passes is called the Doppler effect, named after C. J. Doppler (1803 – 1853). A person will *hear* a sound with a frequency, h, according to the formula

$$h = \frac{f}{1 + \frac{v}{s}}$$

where f is the actual frequency of the sound being produced, s is the speed of sound (about 740 miles per hour), and v is the velocity of the moving object.

a. Examine this fraction, and then explain why h and f approach the same value as v becomes smaller and smaller.

b. Solve this formula for v.

54. **Work Problem** A water storage tank has two drains. It can be shown that the time it takes to empty the tank if both drains are open is given by the formula

$$\frac{1}{\frac{1}{a} + \frac{1}{b}}$$

where a = time it takes for the first drain to empty the tank, and b = time for the second drain to empty the tank.

a. Simplify this complex fraction.

b. Find the amount of time needed to empty the tank using both drains if, used alone, the first drain empties the tank in 4 hours and the second drain can empty the tank in 3 hours.

Getting Ready for the Next Section

Multiply.

55. $x(y - 2)$

56. $x(y - 1)$

57. $6\left(\frac{x}{2} - 3\right)$

58. $6\left(\frac{x}{3} + 1\right)$

59. $xab \cdot \frac{1}{x}$

60. $xab\left(\frac{1}{b} + \frac{1}{a}\right)$

Factor.

61. $y^2 - 25$

62. $x^2 - 3x + 2$

63. $xa + xb$

64. $xy - y$

Solve.

65. $5x - 4 = 6$

66. $y^2 + y - 20 = 2y$

Equations With Rational Expressions 8.5

The first step in solving an equation that contains one or more rational expressions is to find the LCD for all denominators in the equation. We then multiply both sides of the equation by the LCD to clear the equation of all fractions. That is, after we have multiplied through by the LCD, each term in the resulting equation will have a denominator of 1.

EXAMPLE 1 Solve $\frac{x}{2} - 3 = \frac{2}{3}$.

SOLUTION The LCD for 2 and 3 is 6. Multiplying both sides by 6, we have

$$6\left(\frac{x}{2} - 3\right) = 6\left(\frac{2}{3}\right)$$

$$6\left(\frac{x}{2}\right) - 6(3) = 6\left(\frac{2}{3}\right)$$

$$3x - 18 = 4$$

$$3x = 22$$

$$x = \frac{22}{3}$$

Multiplying both sides of an equation by the LCD clears the equation of fractions because the LCD has the property that all the denominators divide it evenly.

EXAMPLE 2 Solve $\frac{6}{a-4} = \frac{3}{8}$.

SOLUTION The LCD for $a - 4$ and 8 is $8(a - 4)$. Multiplying both sides by this quantity yields

$$8(a-4) \cdot \frac{6}{a-4} = 8(a-4) \cdot \frac{3}{8}$$

$$48 = (a - 4) \cdot 3$$

$$48 = 3a - 12$$

$$60 = 3a$$

$$20 = a$$

The solution set is 20, which checks in the original equation.

When we multiply both sides of an equation by an expression containing the variable, we must be sure to check our solutions. The multiplication property of equality does not allow multiplication by 0. If the expression we multiply by contains the variable, then it has the possibility of being 0. In the last example, we multiplied both sides by $8(a - 4)$. This gives a restriction $a \neq 4$ for any solution we come up with.

EXAMPLE 3 Solve $\frac{x}{x-2} + \frac{2}{3} = \frac{2}{x-2}$.

SOLUTION The LCD is $3(x-2)$. We are assuming $x \neq 2$ when we multiply both sides of the equation by $3(x-2)$:

$$3(x-2) \cdot \left(\frac{x}{x-2} + \frac{2}{3}\right) = 3(x-2) \cdot \frac{2}{x-2}$$

$$3x + (x-2) \cdot 2 = 3 \cdot 2$$

$$3x + 2x - 4 = 6$$

$$5x - 4 = 6$$

$$5x = 10$$

$$x = 2$$

Note In the process of solving the equation, we multiplied both sides by $3(x-2)$, solved for x, and got $x = 2$ for our solution. But when x is 2, the quantity $3(x-2) = 3(2-2) = 3(0) = 0$, which means we multiplied both sides of our equation by 0, which is not allowed under the multiplication property of equality.

The only possible solution is $x = 2$. Checking this value back in the original equation gives

$$\frac{2}{2-2} + \frac{2}{3} \stackrel{?}{=} \frac{2}{2-2}$$

$$\frac{2}{0} + \frac{2}{3} \stackrel{?}{=} \frac{2}{0}$$

The first and last terms are undefined. The proposed solution, $x = 2$, does not check in the original equation. The solution set is the empty set. There is no solution to the original equation.

When the proposed solution to an equation is not actually a solution, it is called an *extraneous* solution. In the last example, $x = 2$ is an extraneous solution.

EXAMPLE 4 Solve $\frac{5}{x^2 - 3x + 2} - \frac{1}{x-2} = \frac{1}{3x-3}$.

SOLUTION Writing the equation again with the denominators in factored form, we have

$$\frac{5}{(x-2)(x-1)} - \frac{1}{x-2} = \frac{1}{3(x-1)}$$

Note We can check the proposed solution in any of the equations obtained before multiplying through by the LCD. We cannot check the proposed solution in an equation obtained after multiplying through by the LCD because, if we have multiplied by 0, the resulting equations will not be equivalent to the original one.

The LCD is $3(x-2)(x-1)$. Multiplying through by the LCD, we have

$$3(x-2)(x-1)\frac{5}{(x-2)(x-1)} - 3(x-2)(x-1) \cdot \frac{1}{(x-2)}$$

$$= 3(x-2)(x-1) \cdot \frac{1}{3(x-1)}$$

$$3 \cdot 5 - 3(x-1) \cdot 1 = (x-2) \cdot 1$$

$$15 - 3x + 3 = x - 2$$

$$-3x + 18 = x - 2$$

$$-4x + 18 = -2$$

$$-4x = -20$$

$$x = 5$$

Checking the proposed solution $x = 5$ in the original equation yields a true statement. Try it and see.

EXAMPLE 5 Solve $3 + \frac{1}{x} = \frac{10}{x^2}$.

SOLUTION To clear the equation of denominators, we multiply both sides by x^2:

$$x^2\left(3 + \frac{1}{x}\right) = x^2\left(\frac{10}{x^2}\right)$$

$$3(x^2) + \left(\frac{1}{x}\right)(x^2) = \left(\frac{10}{x^2}\right)(x^2)$$

$$3x^2 + x = 10$$

Rewrite in standard form, and solve:

$$3x^2 + x - 10 = 0$$

$$(3x - 5)(x + 2) = 0$$

$$3x - 5 = 0 \quad \text{or} \quad x + 2 = 0$$

$$x = \frac{5}{3} \quad \text{or} \quad x = -2$$

The solution set is $\left\{-2, \frac{5}{3}\right\}$. Both solutions check in the original equation. Remember: We have to check all solutions any time we multiply both sides of the equation by an expression that contains the variable, just to be sure we haven't multiplied by 0.

EXAMPLE 6 Solve $\frac{y - 4}{y^2 - 5y} = \frac{2}{y^2 - 25}$.

SOLUTION Factoring each denominator, we find the LCD is $y(y - 5)(y + 5)$. Multiplying each side of the equation by the LCD clears the equation of denominators and leads us to our possible solutions:

$$y(y-5)(y+5) \cdot \frac{y - 4}{y(y - 5)} = \frac{2}{(y - 5)(y + 5)} \cdot y(y-5)(y+5)$$

$$(y + 5)(y - 4) = 2y$$

$$y^2 + y - 20 = 2y \qquad \text{Multiply out the left side}$$

$$y^2 - y - 20 = 0 \qquad \text{Add } -2y \text{ to each side}$$

$$(y - 5)(y + 4) = 0$$

$$y - 5 = 0 \quad \text{or} \quad y + 4 = 0$$

$$y = 5 \quad \text{or} \quad y = -4$$

The two possible solutions are 5 and -4. If we substitute -4 for y in the original equation, we find that it leads to a true statement. It is therefore a solution. On the other hand, if we substitute 5 for y in the original equation, we find that both sides of the equation are undefined. The only solution to our original equation is $y = -4$. The other possible solution $y = 5$ is extraneous.

EXAMPLE 7 Solve for y: $x = \frac{y - 4}{y - 2}$

SOLUTION To solve for y, we first multiply each side by $y - 2$ to obtain

$$x(y - 2) = y - 4$$

$$xy - 2x = y - 4 \quad \text{Distributive property}$$

$$xy - y = 2x - 4 \quad \text{Collect all terms containing } y \text{ on the left side}$$

$$y(x - 1) = 2x - 4 \quad \text{Factor } y \text{ from each term on the left side}$$

$$y = \frac{2x - 4}{x - 1} \quad \text{Divide each side by } x - 1$$

EXAMPLE 8 Solve the formula $\frac{1}{x} = \frac{1}{b} + \frac{1}{a}$ for x.

SOLUTION We begin by multiplying both sides by the least common denominator xab. As you can see from our previous examples, multiplying both sides of an equation by the LCD is equivalent to multiplying each term of both sides by the LCD:

$$xab \cdot \frac{1}{x} = \frac{1}{b} \cdot xab + \frac{1}{a} \cdot xab$$

$$ab = xa + xb$$

$$ab = (a + b)x \quad \text{Factor } x \text{ from the right side}$$

$$\frac{ab}{a + b} = x$$

We know we are finished because the variable we were solving for is alone on one side of the equation and does not appear on the other side.

Graphing Rational Functions

In our next example, we investigate the graph of a rational function.

EXAMPLE 9 Graph the rational function $f(x) = \frac{6}{x - 2}$.

SOLUTION To find the y-intercept, we let x equal 0.

$$\text{When } x = 0: \quad y = \frac{6}{0 - 2} = \frac{6}{-2} = -3 \quad y\text{-intercept}$$

The graph will not cross the x-axis. If it did, we would have a solution to the equation

$$0 = \frac{6}{x - 2}$$

which has no solution because there is no number to divide 6 by to obtain 0.

The graph of our equation is shown in Figure 1 along with a table giving values of x and y that satisfy the equation. Notice that y is undefined when x is 2. This means that the graph will not cross the vertical line $x = 2$. (If it did, there would be a value of y for $x = 2$.) The line $x = 2$ is called a *vertical asymptote* of the graph. The graph will get very close to the vertical asymptote, but will never touch or cross it.

x	y
-4	-1
-1	-2
0	-3
1	-6
2	Undefined
3	6
4	3
5	2

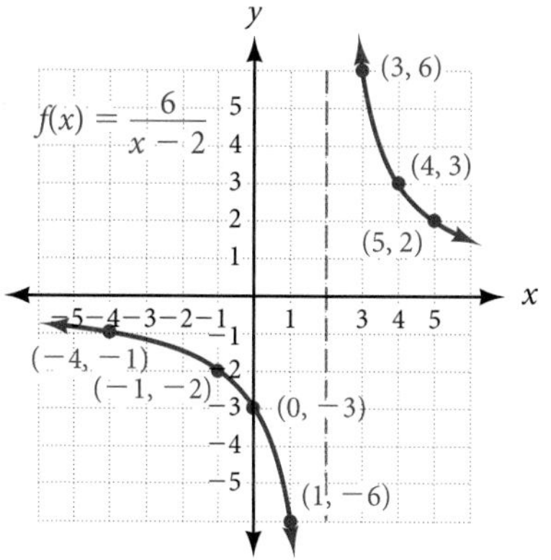

FIGURE 1 *The graph of* $f(x) = \dfrac{6}{x-2}$

If you were to graph $y = \frac{6}{x}$ on the coordinate system in Figure 1, you would see that the graph of $y = \frac{6}{x-2}$ is the graph of $y = \frac{6}{x}$ with all points shifted 2 units to the right.

USING TECHNOLOGY *More About Example 9*

We know the graph of $f(x) = \frac{6}{x-2}$ will not cross the vertical asymptote $x = 2$ because replacing x with 2 in the equation gives us an undefined expression, meaning there is no value of y to associate with $x = 2$. We can use a graphing calculator to explore the behavior of this function when x gets closer and closer to 2 by using the table function on the calculator. We want to put our own values for X into the table, so we set the independent variable to Ask. (On a TI-82/83, use the TBLSET key to set up the table.) To see how the function behaves as x gets close to 2, we let X take on values of 1.9, 1.99, and 1.999. Then we move to the other side of 2 and let X become 2.1, 2.01, and 2.001.

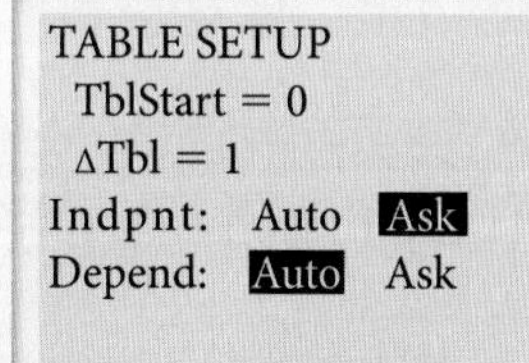

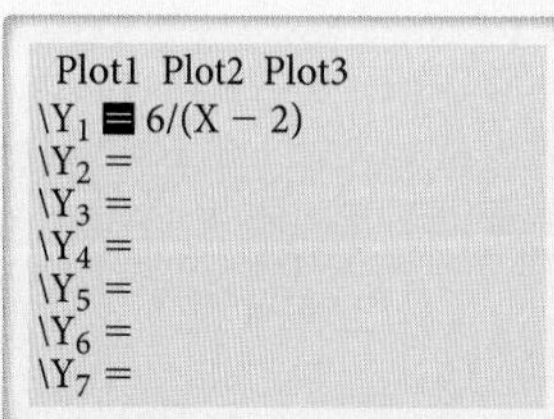

The table will look like this:

X	Y_1	
1.9	−60	
1.99	−600	
1.999	−6000	
2.1	60	
2.01	600	
2.001	6000	

Again, the calculator asks us for a table increment. Because we are inputting the x values ourselves, the increment value does not matter.

As you can see, the values in the table support the shape of the curve in Figure 1 around the vertical asymptote $x = 2$.

EXAMPLE 10 Graph: $g(x) = \dfrac{6}{x + 2}$

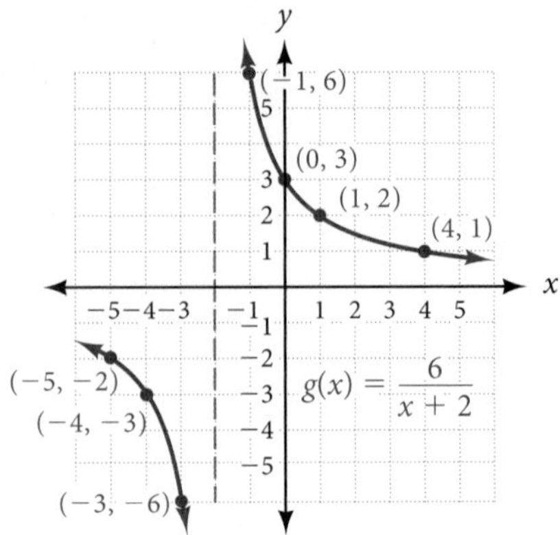

FIGURE 2 *The graph of* $g(x) = \dfrac{6}{x + 2}$

SOLUTION The only difference between this equation and the equation in Example 9 is in the denominator. This graph will have the same shape as the graph in Example 9, but the vertical asymptote will be $x = -2$ instead of $x = 2$. Figure 2 shows the graph.

Notice that the graphs shown in Figures 1 and 2 are both graphs of functions because no vertical line will cross either graph in more than one place. Notice the similarities and differences in our two functions,

$$f(x) = \frac{6}{x - 2} \quad \text{and} \quad g(x) = \frac{6}{x + 2}$$

and their graphs. The vertical asymptotes shown in Figures 1 and 2 correspond to the fact that both $f(2)$ and $g(-2)$ are undefined. The domain for the function f is all real numbers except $x = 2$, while the domain for g is all real numbers except $x = -2$.

GETTING READY FOR CLASS

After reading through the preceding section, respond in your own words and in complete sentences.

A. Explain how a least common denominator can be used to simplify an equation.
B. What is an extraneous solution?
C. How does the location of the vertical asymptote in the graph of a rational function relate to the equation of the function?
D. What is the last step in solving an equation that contains rational expressions?

Solve each of the following equations.

1. $\frac{x}{5} + 4 = \frac{5}{3}$

2. $\frac{x}{5} = \frac{x}{2} - 9$

3. $\frac{a}{3} + 2 = \frac{4}{5}$

4. $\frac{a}{4} + \frac{1}{2} = \frac{2}{3}$

5. $\frac{y}{2} + \frac{y}{4} + \frac{y}{6} = 3$

6. $\frac{y}{3} - \frac{y}{6} + \frac{y}{2} = 1$

7. $\frac{5}{2x} = \frac{1}{x} + \frac{3}{4}$

8. $\frac{1}{2a} = \frac{2}{a} - \frac{3}{8}$

9. $\frac{1}{x} = \frac{1}{3} - \frac{2}{3x}$

10. $\frac{5}{2x} = \frac{2}{x} - \frac{1}{12}$

11. $\frac{2x}{x-3} + 2 = \frac{2}{x-3}$

12. $\frac{2}{x+5} = \frac{2}{5} - \frac{x}{x+5}$

13. $1 - \frac{1}{x} = \frac{12}{x^2}$

14. $2 + \frac{5}{x} = \frac{3}{x^2}$

15. $y - \frac{4}{3y} = -\frac{1}{3}$

16. $\frac{y}{2} - \frac{4}{y} = -\frac{7}{2}$

17. $\frac{x+2}{x+1} = \frac{1}{x+1} + 2$

18. $\frac{x+6}{x+3} = \frac{3}{x+3} + 2$

19. $\frac{3}{a-2} = \frac{2}{a-3}$

20. $\frac{5}{a+1} = \frac{4}{a+2}$

21. $6 - \frac{5}{x^2} = \frac{7}{x}$

22. $10 - \frac{3}{x^2} = -\frac{1}{x}$

23. $\frac{1}{x-1} - \frac{1}{x+1} = \frac{3x}{x^2-1}$

24. $\frac{5}{x-1} + \frac{2}{x-1} = \frac{4}{x+1}$

25. $\frac{2}{x-3} + \frac{x}{x^2-9} = \frac{4}{x+3}$

26. $\frac{2}{x+5} + \frac{3}{x+4} = \frac{2x}{x^2+9x+20}$

27. $\frac{3}{2} - \frac{1}{x-4} = \frac{-2}{2x-8}$

28. $\frac{2}{x} - \frac{1}{x+1} = \frac{-2}{5x+5}$

29. $\frac{t-4}{t^2-3t} = \frac{-2}{t^2-9}$

30. $\frac{t+3}{t^2-2t} = \frac{10}{t^2-4}$

31. $\frac{3}{y-4} - \frac{2}{y+1} = \frac{5}{y^2-3y-4}$

32. $\frac{1}{y+2} - \frac{2}{y-3} = \frac{-2y}{y^2-y-6}$

33. $\frac{2}{1+a} = \frac{3}{1-a} + \frac{5}{a}$

34. $\frac{1}{a+3} - \frac{a}{a^2-9} = \frac{2}{3-a}$

35. $\frac{3}{2x-6} - \frac{x+1}{4x-12} = 4$

36. $\frac{2x-3}{5x+10} + \frac{3x-2}{4x+8} = 1$

37. $\frac{y+2}{y^2-y} - \frac{6}{y^2-1} = 0$

38. $\frac{y+3}{y^2-y} - \frac{8}{y^2-1} = 0$

39. $\frac{4}{2x-6} - \frac{12}{4x+12} = \frac{12}{x^2-9}$

40. $\frac{1}{x+2} + \frac{1}{x-2} = \frac{4}{x^2-4}$

41. $\frac{2}{y^2-7y+12} - \frac{1}{y^2-9} = \frac{4}{y^2-y-12}$

42. $\frac{1}{y^2+5y+4} + \frac{3}{y^2-1} = \frac{-1}{y^2+3y-4}$

43. Let $f(x) = \dfrac{1}{x-3}$ and $g(x) = \dfrac{1}{x+3}$ and find x if

a. $f(x) + g(x) = \dfrac{5}{8}$ **b.** $\dfrac{f(x)}{g(x)} = 5$ **c.** $f(x) = g(x)$

44. Let $f(x) = \dfrac{4}{x+2}$ and $g(x) = \dfrac{4}{x-2}$ and find x if

a. $f(x) - g(x) = -\dfrac{4}{3}$ **b.** $\dfrac{g(x)}{f(x)} = -7$ **c.** $f(x) = -g(x)$

45. Solve each equation.

a. $6x - 2 = 0$ **b.** $\dfrac{6}{x} - 2 = 0$ **c.** $\dfrac{x}{6} - 2 = -\dfrac{1}{2}$

d. $\dfrac{6}{x} - 2 = -\dfrac{1}{2}$ **e.** $\dfrac{6}{x^2} + 6 = \dfrac{20}{x}$

46. Solve each equation.

a. $5x - 2 = 0$ **b.** $5 - \dfrac{2}{x} = 0$ **c.** $\dfrac{x}{2} - 5 = -\dfrac{3}{4}$

d. $\dfrac{2}{x} - 5 = -\dfrac{3}{4}$ **e.** $-\dfrac{3}{x} + \dfrac{2}{x^2} = 5$

47. Work each problem according to the instructions given.

a. Divide: $\dfrac{6}{x^2 - 2x - 8} \div \dfrac{x+3}{x+2}$

b. Add: $\dfrac{6}{x^2 - 2x - 8} + \dfrac{x+3}{x+2}$

c. Solve: $\dfrac{6}{x^2 - 2x - 8} + \dfrac{x+3}{x+2} = 2$

48. Work each problem according to the instructions given.

a. Divide: $\dfrac{-10}{x^2 - 25} \div \dfrac{x-4}{x-5}$

b. Add: $\dfrac{-10}{x^2 - 25} + \dfrac{x-4}{x-5}$

c. Solve: $\dfrac{-10}{x^2 - 25} + \dfrac{x-4}{x-5} = \dfrac{4}{5}$

49. Solve $\dfrac{1}{x} = \dfrac{1}{b} - \dfrac{1}{a}$ for x.

50. Solve $\dfrac{1}{x} = \dfrac{1}{a} - \dfrac{1}{b}$ for x.

Solve for y.

51. $x = \dfrac{y-3}{y-1}$ **52.** $x = \dfrac{y-2}{y-3}$ **53.** $x = \dfrac{2y+1}{3y+1}$ **54.** $x = \dfrac{3y+2}{5y+1}$

Graph each function. Show the vertical asymptote.

55. $f(x) = \dfrac{1}{x-3}$ **56.** $f(x) = \dfrac{1}{x+3}$ **57.** $f(x) = \dfrac{4}{x+2}$ **58.** $f(x) = \dfrac{4}{x-2}$

59. $g(x) = \dfrac{2}{x-4}$ **60.** $g(x) = \dfrac{2}{x+4}$ **61.** $g(x) = \dfrac{6}{x+1}$ **62.** $g(x) = \dfrac{6}{x-1}$

Applying the Concepts

63. Geometry From plane geometry and the principle of similar triangles, the relationship between y_1, y_2, and h shown in Figure 3 can be expressed as

$$\frac{1}{h} = \frac{1}{y_1} + \frac{1}{y_2}$$

Two poles are 12 feet high and 8 feet high. If a cable is attached to the top of each one and stretched to the bottom of the other, what is the height above the ground at which the two wires will meet?

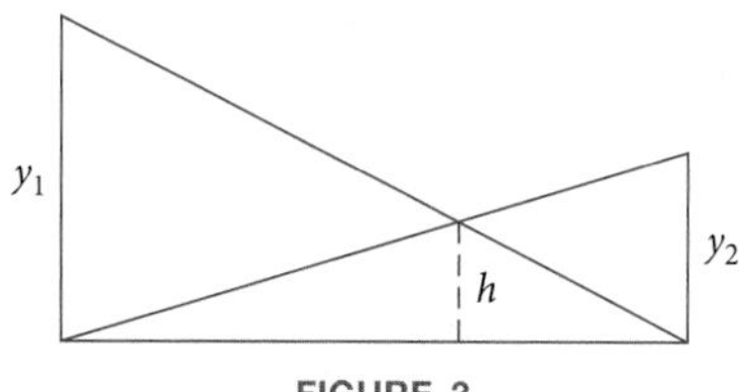

FIGURE 3

64. Kayak Race In a kayak race, the participants must paddle a kayak 450 meters down a river and then return 450 meters up the river to the starting point (Figure 4). Susan has correctly deduced that the total time t (in seconds) depends on the speed c (in meters per second) of the water according to the following expression:

$$t = \frac{450}{v + c} + \frac{450}{v - c}$$

where v is the speed of the kayak relative to the water (the speed of the kayak in still water).

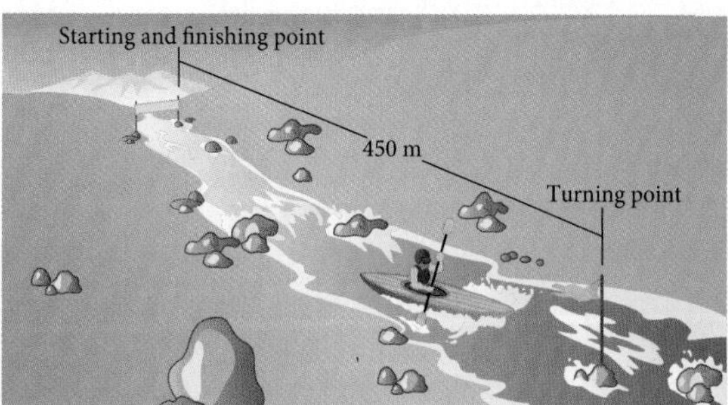

FIGURE 4

a. Fill in the following table.

Time	Speed of Kayak Relative to the Water	Current of the River
t(sec)	*v*(m/sec)	*c*(m/sec)
240		1
300		2
	4	3
	3	1
540	3	
	3	3

b. If the kayak race were conducted in the still waters of a lake, do you think that the total time of a given participant would be greater than, equal to, or smaller than the time in the river? Justify your answer.

c. Suppose Peter can drive his kayak at 4.1 meters per second and that the speed of the current is 4.1 meters per second. What will happen when Peter makes the turn and tries to come back up the river? How does this situation show up in the equation for total time?

Getting Ready for the Next Section

Multiply.

65. $39.3 \cdot 60$

66. $1{,}100 \cdot 60 \cdot 60$

Divide. Round to the nearest tenth, if necessary.

67. $65{,}000 \div 5{,}280$

68. $3{,}960{,}000 \div 5{,}280$

Multiply.

69. $2x\left(\frac{1}{x} + \frac{1}{2x}\right)$

70. $3x\left(\frac{1}{x} + \frac{1}{3x}\right)$

Solve.

71. $12(x + 3) + 12(x - 3) = 3(x^2 - 9)$

72. $40 + 2x = 60 - 3x$

73. $\frac{1}{10} - \frac{1}{12} = \frac{1}{x}$

74. $\frac{1}{x} + \frac{1}{2x} = 2$

Applications 8.6

We begin this section with some application problems, the solutions to which involve equations that contain rational expressions. As you will see, the solutions to the examples show only the essential steps from our Blueprint for Problem Solving. Recall that step 1 was done mentally; we read the problem and mentally list the items that are known and the items that are unknown. This is an essential part of problem solving. Now that you have had experience with application problems, however, you are doing step 1 automatically.

EXAMPLE 1 One number is twice another. The sum of their reciprocals is 2. Find the numbers.

SOLUTION Let $x =$ the smaller number. The larger number is $2x$. Their reciprocals are $\frac{1}{x}$ and $\frac{1}{2x}$. The equation that describes the situation is

$$\frac{1}{x} + \frac{1}{2x} = 2$$

Multiplying both sides by the LCD $2x$, we have

$$2x \cdot \frac{1}{x} + 2x \cdot \frac{1}{2x} = 2x(2)$$

$$2 + 1 = 4x$$

$$3 = 4x$$

$$x = \frac{3}{4}$$

The smaller number is $\frac{3}{4}$. The larger is $2\left(\frac{3}{4}\right) = \frac{6}{4} = \frac{3}{2}$. Adding their reciprocals, we have

$$\frac{4}{3} + \frac{2}{3} = \frac{6}{3} = 2$$

The sum of the reciprocals of $\frac{3}{4}$ and $\frac{3}{2}$ is 2.

EXAMPLE 2 Two families from the same neighborhood plan a ski trip together. The first family is driving a newer vehicle and makes the 455-mile trip at a speed 5 miles per hour faster than the second family who is traveling in an older vehicle. The second family takes a half-hour longer to make the trip. What are the speeds of the two families?

SOLUTION The following table will be helpful in finding the equation necessary to solve this problem.

	d(distance)	*r*(rate)	*t*(time)
First Family			
Second Family			

If we let x be the speed of the second family, then the speed of the first family will be $x + 5$. Both families travel the same distance of 455 miles. Putting this information into the table we have

	d	*r*	*t*
First Family	455	$x + 5$	
Second Family	455	x	

To fill in the last two spaces in the table, we use the relationship $d = r \cdot t$. Since the last column of the table is the time, we solve the equation $d = r \cdot t$ for t and get

$$t = \frac{d}{r}$$

Taking the distance and dividing by the rate (speed) for each family, we complete the table.

	d	*r*	*t*
First Family	455	$x + 5$	$\frac{455}{x+5}$
Second Family	455	x	$\frac{455}{x}$

Reading the problem again, we find that the time for the second family is longer than the time for the first family by one-half hour. In other words, the time for the second family can be found by adding one-half hour to the time for the first family, or

$$\frac{455}{x+5} + \frac{1}{2} = \frac{455}{x}$$

Multiplying both sides by the LCD of $2x(x + 5)$ gives

$$2x \cdot (455) + x(x + 5) \cdot 1 = 455 \cdot 2(x + 5)$$

$$910x + x^2 + 5x = 910x + 4550$$

$$x^2 + 5x - 4550 = 0$$

$$(x + 70)\,(x - 65) = 0$$

$$x = -70 \quad \text{or} \quad x = 65$$

Since we cannot have a negative speed, the only solution is $x = 65$. Then

$$x + 5 = 65 + 5 = 70$$

The speed of the first family is 70 miles per hour, and the speed of the second family is 65 miles per hour. ■

EXAMPLE 3 The current of a river is 3 miles per hour. It takes a motorboat a total of 3 hours to travel 12 miles upstream and return 12 miles downstream. What is the speed of the boat in still water?

SOLUTION This time we let $x =$ the speed of the boat in still water. Then, we fill in as much of the table as possible using the information given in the problem. For instance, because we let $x =$ the speed of the boat in still water, the rate upstream (against the current) must be $x - 3$. The rate downstream (with the current) is $x + 3$.

	d	r	t
Upstream	12	$x - 3$	
Downstream	12	$x + 3$	

The last two boxes can be filled in using the relationship

$$t = \frac{d}{r}$$

	d	r	t
Upstream	12	$x - 3$	$\frac{12}{x - 3}$
Downstream	12	$x + 3$	$\frac{12}{x + 3}$

The total time for the trip up and back is 3 hours:

$$\text{Time upstream} + \text{Time downstream} = \text{Total time}$$

$$\frac{12}{x-3} \quad + \quad \frac{12}{x+3} \quad = \quad 3$$

Multiplying both sides by $(x - 3)(x + 3)$, we have

$$12(x + 3) + 12(x - 3) = 3(x^2 - 9)$$

$$12x + 36 + 12x - 36 = 3x^2 - 27$$

$$3x^2 - 24x - 27 = 0$$

$$x^2 - 8x - 9 = 0 \qquad \text{Divide both sides by 3}$$

$$(x - 9)(x + 1) = 0$$

$$x = 9 \quad \text{or} \quad x = -1$$

The speed of the motorboat in still water is 9 miles per hour. (We don't use $x = -1$ because the speed of the motorboat cannot be a negative number.)

EXAMPLE 4 An inlet pipe can fill a pool in 10 hours, while the drain can empty it in 12 hours. If the pool is empty and both the inlet pipe and drain are open, how long will it take to fill the pool?

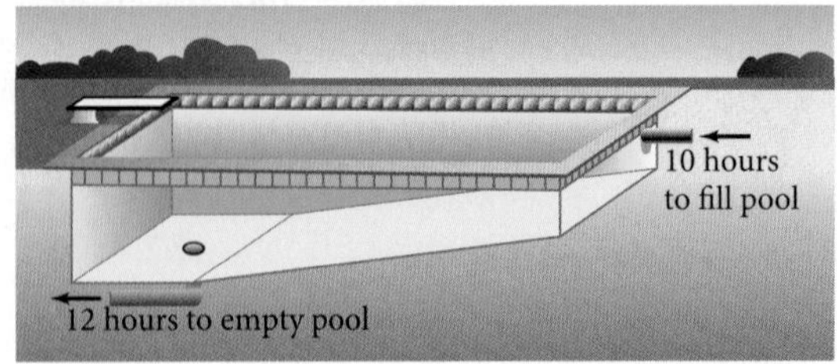

SOLUTION It is helpful to think in terms of how much work is done by each pipe in 1 hour.

Let $x =$ the time it takes to fill the pool with both pipes open.

If the inlet pipe can fill the pool in 10 hours, then in 1 hour it is $\frac{1}{10}$ full. If the outlet pipe empties the pool in 12 hours, then in 1 hour it is $\frac{1}{12}$ empty. If the pool can be filled in x hours with both the inlet pipe and the drain open, then in 1 hour it is $\frac{1}{x}$ full when both pipes are open.

Here is the equation:

In 1 hour

$$\begin{bmatrix}\text{Amount filled}\\ \text{by inlet pipe}\end{bmatrix} - \begin{bmatrix}\text{Amount emptied}\\ \text{by the drain}\end{bmatrix} = \begin{bmatrix}\text{Fraction of pool}\\ \text{filled with both pipes}\end{bmatrix}$$

$$\frac{1}{10} - \frac{1}{12} = \frac{1}{x}$$

Multiplying through by $60x$, we have

$$60x \cdot \frac{1}{10} - 60x \cdot \frac{1}{12} = 60x \cdot \frac{1}{x}$$

$$6x - 5x = 60$$

$$x = 60$$

It takes 60 hours to fill the pool if both the inlet pipe and the drain are open.

More About Graphing Rational Functions

We continue our investigation of the graphs of rational functions by considering the graph of a rational function with binomials in the numerator and denominator.

EXAMPLE 5 Graph the rational function $y = \dfrac{x-4}{x-2}$.

SOLUTION In addition to making a table to find some points on the graph, we can analyze the graph as follows:

1. The graph will have a y-intercept of 2, because when $x = 0$, $y = \dfrac{-4}{-2} = 2$.
2. To find the x-intercept, we let $y = 0$ to get

   $$0 = \frac{x-4}{x-2}$$

   The only way this expression can be 0 is if the numerator is 0, which happens when $x = 4$. (If you want to solve this equation, multiply both sides by $x - 2$. You will get the same solution, $x = 4$.)
3. The graph will have a vertical asymptote at $x = 2$, because $x = 2$ will make the denominator of the function 0, meaning y is undefined when x is 2.
4. The graph will have a *horizontal asymptote* at $y = 1$ because for very large values of x, $\frac{x-4}{x-2}$ is very close to 1. The larger x is, the closer $\frac{x-4}{x-2}$ is to 1. The same is true for very small values of x, such as $-1{,}000$ and $-10{,}000$.

Putting this information together with the ordered pairs in the table next to the figure, we have the graph shown in Figure 1.

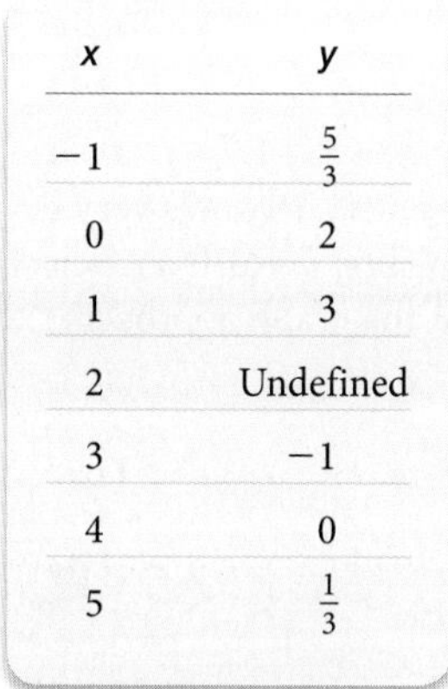

x	y
-1	$\frac{5}{3}$
0	2
1	3
2	Undefined
3	-1
4	0
5	$\frac{1}{3}$

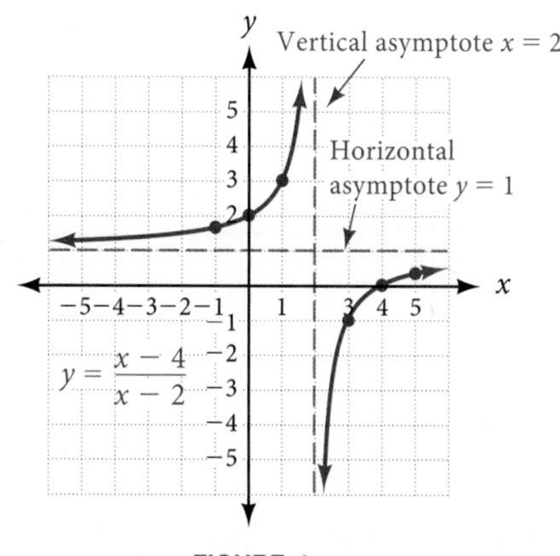

FIGURE 1

USING TECHNOLOGY *More About Example 5*

In the previous section, we used technology to explore the graph of a rational function around a vertical asymptote. This time, we are going to explore the graph near the horizontal asymptote. In Figure 1, the horizontal asymptote is at $y = 1$. To show that the graph approaches this line as x becomes very large, we use the table function on our graphing calculator, with X taking values of 100, 1,000, and 10,000. To show that the graph approaches the line $y = 1$ on the left side of the coordinate system, we let X become -100, $-1{,}000$ and $-10{,}000$.

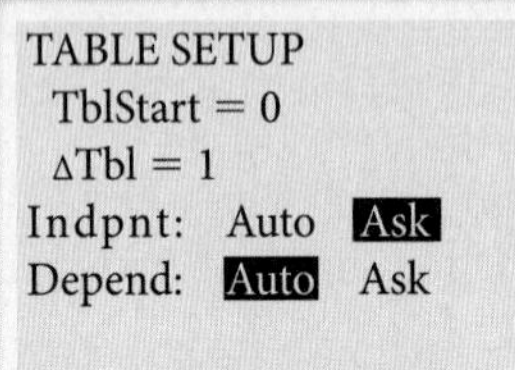

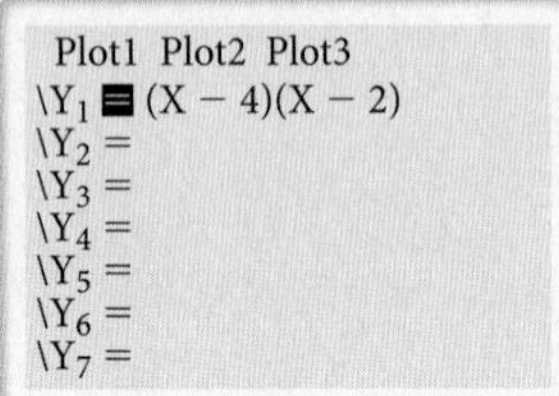

The table will look like this:

X	Y_1	
100	.97959	
1000	.998	
10000	.9998	
−100	1.0196	
−1000	1.002	
−100000	1.0002	

As you can see, as x becomes very large in the positive direction, the graph approaches the line $y = 1$ from below. As x becomes very small in the negative direction, the graph approaches the line $y = 1$ from above.

GETTING READY FOR CLASS

After reading through the preceding section, respond in your own words and in complete sentences.

A. Briefly list the steps in the Blueprint for Problem Solving that you have used previously to solve application problems.

B. Write an application problem for which the solution depends on solving the equation $\frac{1}{2} + \frac{1}{3} = \frac{1}{x}$.

C. One number is twice another, write an expression for the sum of their reciprocals.

D. Write a formula for the relationship between distance, rate, and time.

Solve the following word problems. Be sure to show the equation in each case.

Number Problems

1. One number is 3 times another. The sum of their reciprocals is $\frac{20}{3}$. Find the numbers.

2. One number is 3 times another. The sum of their reciprocals is $\frac{4}{9}$. Find the numbers.

3. The sum of a number and its reciprocal is $\frac{10}{3}$. Find the number.

4. The sum of a number and twice its reciprocal is $\frac{27}{5}$. Find the number.

5. The sum of the reciprocals of two consecutive integers is $\frac{7}{12}$. Find the two integers.

6. Find two consecutive even integers, the sum of whose reciprocals is $\frac{3}{4}$.

7. If a certain number is added to the numerator and denominator of $\frac{7}{9}$, the result is $\frac{5}{6}$. Find the number.

8. Find the number you would add to both the numerator and denominator of $\frac{8}{11}$ so that the result would be $\frac{6}{7}$.

9. The speed of a boat in still water is 5 miles per hour. If the boat travels 3 miles downstream in the same amount of time it takes to travel 1.5 miles upstream, what is the speed of the current?

a. Let x be the speed of the current. Complete the distance and rate columns in the table.

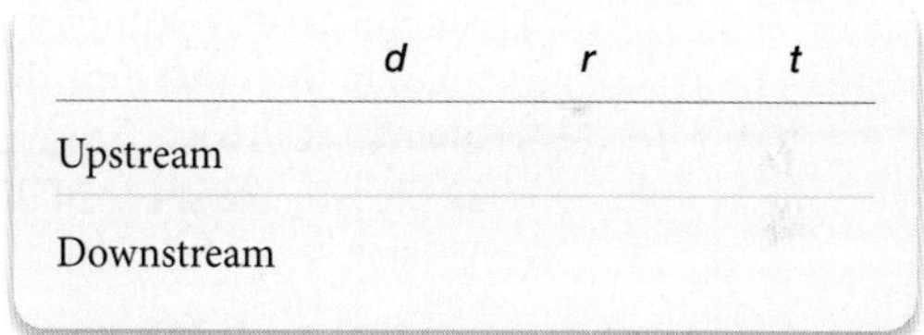

	d	*r*	*t*
Upstream			
Downstream			

b. Now use the distance and rate information to complete the time column.

c. What does the problem tell us about the two times? Use this fact to write an equation involving the two expressions for time.

d. Solve the equation. Write your answer as a complete sentence.

10. A boat, which moves at 18 miles per hour in still water, travels 14 miles downstream in the same amount of time it takes to travel 10 miles upstream. Find the speed of the current.

a. Let x be the speed of the current. Complete the distance and rate columns in the table.

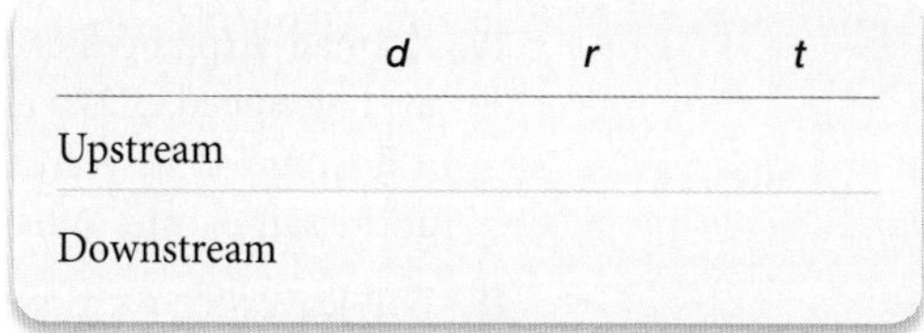

	d	*r*	*t*
Upstream			
Downstream			

b. Now use the distance and rate information to complete the time column.

c. What does the problem tell us about the two times? Use this fact to write an equation involving the two expressions for time.

d. Solve the equation. Write your answer as a complete sentence.

Rate Problems

11. The current of a river is 2 miles per hour. A boat travels to a point 8 miles upstream and back again in 3 hours. What is the speed of the boat in still water?

12. A motorboat travels at 4 miles per hour in still water. It goes 12 miles upstream and 12 miles back again in a total of 8 hours. Find the speed of the current of the river.

13. Train A has a speed 15 miles per hour greater than that of train B. If train A travels 150 miles in the same time train B travels 120 miles, what are the speeds of the two trains?

a. Let x be the speed of the train B. Complete the distance and rate columns in the table.

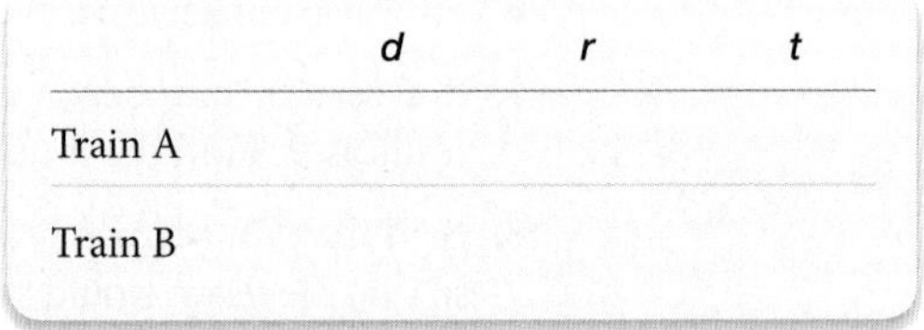

	d	r	t
Train A			
Train B			

b. Now use the distance and rate information to complete the time column.

c. What does the problem tell us about the two times? Use this fact to write an equation involving the two expressions for time.

d. Solve the equation. Write your answer as a complete sentence.

14. A train travels 30 miles per hour faster than a car. If the train covers 120 miles in the same time the car covers 80 miles, what are the speeds of each of them?

a. Let x be the speed of the car. Complete the distance and rate columns in the table.

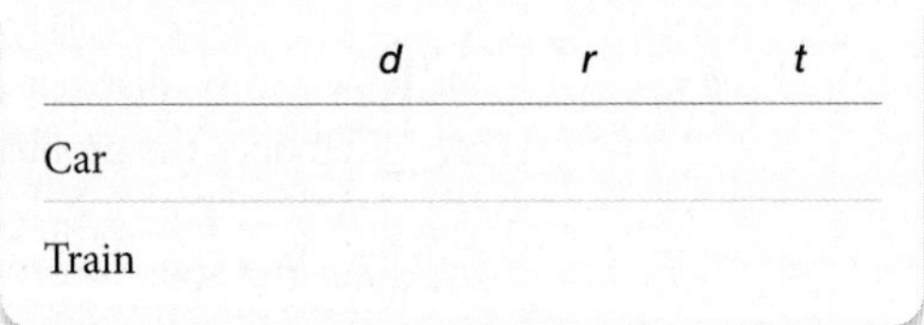

	d	r	t
Car			
Train			

b. Now use the distance and rate information to complete the time column.

c. What does the problem tell us about the two times? Use this fact to write an equation involving the two expressions for time.

d. Solve the equation. Write your answer as a complete sentence.

15. A small airplane flies 810 miles from Los Angeles to Portland, OR, with an average speed of 270 miles per hour. An hour and a half after the plane leaves, a Boeing 747 leaves Los Angeles for Portland. Both planes arrive in Portland at the same time. What was the average speed of the 747?

16. Lou leaves for a cross-country excursion on a bicycle traveling at 20 miles per hour. His friends are driving the trip and will meet him at several rest stops along the way. The first stop is scheduled 30 miles from the original starting point. If the people driving leave 15 minutes after Lou from the same place, how fast will they have to drive to reach the first rest stop at the same time as Lou?

17. A tour bus leaves Sacramento every Friday evening at 5:00 P.M. for a 270-mile trip to Las Vegas. This week, however, the bus leaves at 5:30 P.M. To arrive in Las Vegas on time, the driver drives 6 miles per hour faster than usual. What is the bus' usual speed?

18. A bakery delivery truck leaves the bakery at 5:00 A.M. each morning on its 140-mile route. One day the driver gets a late start and does not leave the bakery until 5:30 A.M. To finish her route on time the driver drives 5 miles per hour faster than usual. At what speed does she usually drive?

Work Problems

19. A water tank can be filled by an inlet pipe in 8 hours. It takes twice that long for the outlet pipe to empty the tank. How long will it take to fill the tank if both pipes are open?

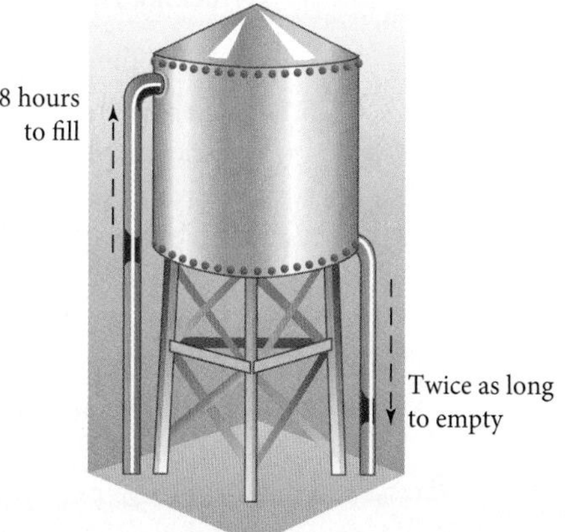

20. A sink can be filled from the faucet in 5 minutes. It takes only 3 minutes to empty the sink when the drain is open. If the sink is full and both the faucet and the drain are open, how long will it take to empty the sink?

21. It takes 10 hours to fill a pool with the inlet pipe. It can be emptied in 15 hours with the outlet pipe. If the pool is half full to begin with, how long will it take to fill it from there if both pipes are open?

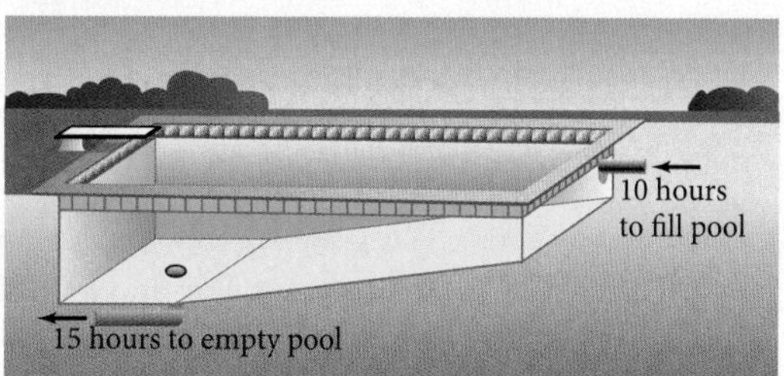

22. A sink is one-quarter full when both the faucet and the drain are opened. The faucet alone can fill the sink in 6 minutes, while it takes 8 minutes to empty it with the drain. How long will it take to fill the remaining three quarters of the sink?

23. A sink has two faucets: one for hot water and one for cold water. The sink can be filled by a cold water faucet in 3.5 minutes. If both faucets are open, the sink is filled in 2.1 minutes. How long does it take to fill the sink with just the hot water faucet open?

24. A water tank is being filled by two inlet pipes. Pipe A can fill the tank in $4\frac{1}{2}$ hours, but both pipes together can fill the tank in 2 hours. How long does it take to fill the tank using only pipe B?

Miscellaneous Problems

25. Rhind Papyrus Nearly 4,000 years ago, Egyptians worked mathematical exercises involving reciprocals. The *Rhind Papyrus* contains a wealth of such problems, and one of them is as follows:

> "A quantity and its two thirds are added together, one third of this is added, then one third of the sum is taken, and the result is 10."

Write an equation and solve this exercise.

26. Photography For clear photographs, a camera must be properly focused. Professional photographers use a mathematical relationship relating the distance from the camera lens to the object being photographed, *a*; the distance from the lens to the film, *b*; and the focal length of the lens, *f*. These quantities, *a*, *b*, and *f*, are related by the equation

$$\frac{1}{a} + \frac{1}{b} = \frac{1}{f}$$

A camera has a focal length of 3 inches. If the lens is 5 inches from the film, how far should the lens be placed from the object being photographed for the camera to be perfectly focused?

The Periodic Table If you take a chemistry class, you will work with the Periodic Table of Elements. Figure 3 shows three of the elements listed in the periodic table. As you can see, the bottom number in each figure is the molecular weight of the element. In chemistry, a mole is the amount of a substance that will give the weight in grams equal to the molecular weight. For example, 1 mole of lead is 207.2 grams.

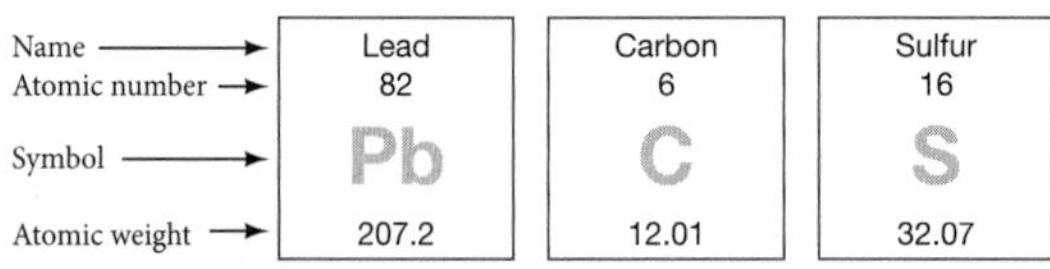

FIGURE 3

27. Chemistry For the element carbon, 1 mole = 12.01 grams.

a. To the nearest gram, how many grams of carbon are in 2.5 moles of carbon?

b. How many moles of carbon are in 39 grams of carbon? Round to the nearest hundredth.

28. Chemistry For the element sulfur, 1 mole = 32.07 grams.

a. How many grams of sulfur are in 3 moles of sulfur?

b. How many moles of sulfur are found in 80.2 grams of sulfur?

Graph each rational function. In each case, show the vertical asymptote, the horizontal asymptote, and any intercepts that exist.

29. $f(x) = \dfrac{x - 3}{x - 1}$

30. $f(x) = \dfrac{x + 4}{x - 2}$

31. $f(x) = \dfrac{x + 3}{x - 1}$

32. $f(x) = \dfrac{x - 2}{x - 1}$

33. $g(x) = \dfrac{x - 3}{x + 1}$

34. $g(x) = \dfrac{x - 2}{x + 1}$

Getting Ready for the Next Section

Divide.

35. $\dfrac{10x^2}{5x^2}$

36. $\dfrac{-15x^4}{5x^2}$

37. $\dfrac{4x^4y^3}{-2x^2y}$

38. $\dfrac{10a^4b^2}{4a^2b^2}$

39. $4{,}628 \div 25$

40. $7{,}546 \div 35$

Multiply.

41. $2x^2(2x - 4)$

42. $3x^2(x - 2)$

43. $(2x - 4)(2x^2 + 4x + 5)$

44. $(x - 2)(3x^2 + 6x + 15)$

Subtract.

45. $(2x^2 - 7x + 9) - (2x^2 - 4x)$

46. $(x^2 - 6xy - 7y^2) - (x^2 + xy)$

Factor.

47. $x^2 - a^2$

48. $x^2 - 1$

49. $x^2 - 6xy - 7y^2$

50. $2x^2 - 5xy + 3y^2$

Division of Polynomials 8.7

First Bank of San Luis Obispo charges \$2.00 per month and \$0.15 per check for a regular checking account. So, if you write x checks in one month, the total monthly cost of the checking account will be $C(x) = 2.00 + 0.15x$. From this formula, we see that the more checks we write in a month, the more we pay for the account. But it is also true that the more checks we write in a month, the lower the cost per check. To find the cost per check, we use the *average cost* function. To find the average cost function, we divide the total cost by the number of checks written.

$$\text{Average Cost} = \overline{C}(x) = \frac{C(x)}{x} = \frac{2.00 + 0.15x}{x}$$

This last expression gives us the average cost per check for each of the x checks written. To work with this last expression, we need to know something about division with polynomials, and that is what we will cover in this section.

We begin this section by considering division of a polynomial by a monomial. This is the simplest kind of polynomial division. The rest of the section is devoted to division of a polynomial by a polynomial. This kind of division is similar to long division with whole numbers.

Dividing a Polynomial by a Monomial

To divide a polynomial by a monomial, we use the definition of division and apply the distributive property. The following example illustrates the procedure.

EXAMPLE 1 Divide $\dfrac{10x^5 - 15x^4 + 20x^3}{5x^2}$.

SOLUTION

$$= (10x^5 - 15x^4 + 20x^3) \cdot \frac{1}{5x^2}$$ Dividing by $5x^2$ is the same as multiplying by $\frac{1}{5x^2}$

$$= 10x^5 \cdot \frac{1}{5x^2} - 15x^4 \cdot \frac{1}{5x^2} + 20x^3 \cdot \frac{1}{5x^2}$$ Distributive property

$$= \frac{10x^5}{5x^2} - \frac{15x^4}{5x^2} + \frac{20x^3}{5x^2}$$ Multiplying by $\frac{1}{5x^2}$ is the same as dividing by $5x^2$

$$= 2x^3 - 3x^2 + 4x$$

Notice that division of a polynomial by a monomial is accomplished by dividing each term of the polynomial by the monomial. The first two steps are usually not shown in a problem like this. They are part of Example 1 to justify distributing $5x^2$ under all three terms of the polynomial $10x^5 - 15x^4 + 20x^3$.

Here are some more examples of this kind of division.

EXAMPLE 2 Divide $\dfrac{8x^3y^5 - 16x^2y^2 + 4x^4y^3}{-2x^2y}$. Write the result with positive exponents.

SOLUTION

$$\frac{8x^3y^5 - 16x^2y^2 + 4x^4y^3}{-2x^2y} = \frac{8x^3y^5}{-2x^2y} + \frac{-16x^2y^2}{-2x^2y} + \frac{4x^4y^3}{-2x^2y}$$

$$= -4xy^4 + 8y - 2x^2y^2$$

EXAMPLE 3 Divide $\dfrac{10a^4b^2 + 8ab^3 - 12a^3b + 6ab}{4a^2b^2}$. Write the result with positive exponents.

SOLUTION

$$\frac{10a^4b^2 + 8ab^3 - 12a^3b + 6ab}{4a^2b^2} = \frac{10a^4b^2}{4a^2b^2} + \frac{8ab^3}{4a^2b^2} - \frac{12a^3b}{4a^2b^2} + \frac{6ab}{4a^2b^2}$$
$$= \frac{5a^2}{2} + \frac{2b}{a} - \frac{3a}{b} + \frac{3}{2ab}$$

Notice in Example 3 that the result is not a polynomial because of the last three terms. If we were to write each as a product, some of the variables would have negative exponents. For example, the second term would be

$$\frac{2b}{a} = 2a^{-1}b$$

The divisor in each of the preceding examples was a monomial. We now want to turn our attention to division of polynomials in which the divisor has two or more terms.

Dividing a Polynomial by a Polynomial

EXAMPLE 4 Divide: $\dfrac{x^2 - 6xy - 7y^2}{x + y}$

SOLUTION In this case, we can factor the numerator and perform division by simply dividing out common factors, just like we did in previous sections:

$$\frac{x^2 - 6xy - 7y^2}{x + y} = \frac{\cancel{(x + y)}(x - 7y)}{\cancel{x + y}}$$
$$= x - 7y$$

Long Division

For the type of division shown in Example 4, the denominator must be a factor of the numerator. When the denominator is not a factor of the numerator, or in the case where we can't factor the numerator, the method used in Example 4 won't work. We need to develop a new method for these cases. Because this new method is very similar to *long division* with whole numbers, we will review the method of long division here.

EXAMPLE 5 Divide $25\overline{)4,628}$.

SOLUTION

$$\begin{array}{r} 1 \\ 25\overline{)4,628} \\ \underline{25} \\ 21 \end{array}$$

Estimate 25 into 46.

Multiply $1 \times 25 = 25$

Subtract $46 - 25 = 21$

$$\begin{array}{r} 1 \\ 25\overline{)4,628} \\ \underline{25\downarrow} \\ 212 \end{array}$$

Bring down the 2.

These are the four basic steps in long division: estimate, multiply, subtract, and bring down the next term. To complete the problem, we simply perform the same four steps:

```
      18        8 is the estimate
25)4,628
   25
   212
   200↓         Multiply to get 200
    128         Subtract to get 12, then bring down the 8
```

One more time:

```
     185        5 is the estimate
25)4,628
   25
   212
   200↓
    128
    125         Multiply to get 125
      3         Subtract to get 3
```

Because 3 is less than 25 and we have no more terms to bring down, we have our answer:

$$\frac{4{,}628}{25} = 185 + \frac{3}{25}$$

To check our answer, we multiply 185 by 25 and then add 3 to the result:

$$25(185) + 3 = 4{,}625 + 3 = 4{,}628$$

Long division with polynomials is similar to long division with whole numbers. Both use the same four basic steps: estimate, multiply, subtract, and bring down the next term. We use long division with polynomials when the denominator has two or more terms and is not a factor of the numerator. Here is an example.

EXAMPLE 6 Divide $\dfrac{2x^2 - 7x + 9}{x - 2}$.

SOLUTION

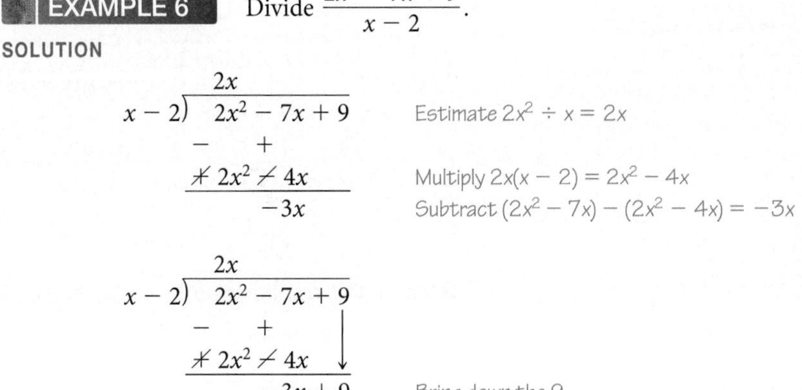
Estimate $2x^2 \div x = 2x$
Multiply $2x(x - 2) = 2x^2 - 4x$
Subtract $(2x^2 - 7x) - (2x^2 - 4x) = -3x$
Bring down the 9

Notice we change the signs on $2x^2 - 4x$ and add in the subtraction step. Subtracting a polynomial is equivalent to adding its opposite.

We repeat the four steps.

$$
\begin{array}{r}
2x - 3 \\
x - 2\,\overline{)\;2x^2 - 7x + 9} \\
- \quad + \\
\cancel{+}\,2x^2 \cancel{-}\,4x \\
\hline
-3x + 9 \\
+ \quad - \\
\cancel{-}3x \cancel{+}\,6 \\
\hline
3
\end{array}
$$

-3 is the estimate: $-3x \div x = -3$

Multiply $-3(x - 2) = -3x + 6$

Subtract $(-3x + 9) - (-3x + 6) = 3$

Because we have no other term to bring down, we have our answer:

$$\frac{2x^2 - 7x + 9}{x - 2} = 2x - 3 + \frac{3}{x - 2}$$

To check, we multiply $(2x - 3)(x - 2)$ to get $2x^2 - 7x + 6$; then, adding the remainder 3 to this result, we have $2x^2 - 7x + 9$. ■

In setting up a division problem involving two polynomials, we must remember two things: (1) Both polynomials should be in decreasing powers of the variable, and (2) neither should skip any powers from the highest power down to the constant term. If there are any missing terms, they can be filled in using a coefficient of 0.

EXAMPLE 7 Divide $2x - 4\overline{)4x^3 - 6x - 11}$.

SOLUTION Because the trinomial is missing a term in x^2, we can fill it in with $0x^2$:

$$4x^3 - 6x - 11 = 4x^3 + 0x^2 - 6x - 11$$

Adding $0x^2$ does not change our original problem.

$$
\begin{array}{r}
2x^2 + 4x + 5 \\
2x - 4\,\overline{)\;4x^3 + 0x^2 - 6x - 11} \\
- \quad + \qquad\qquad \\
\cancel{+}4x^3 \cancel{-}\,8x^2 \qquad\qquad \\
\hline
+ 8x^2 - 6x \qquad \\
- \quad + \qquad \\
\cancel{+}\,8x^2 \cancel{-}\,16x \qquad \\
\hline
+ 10x - 11 \\
- \quad + \\
\cancel{+}\,10x \cancel{-}\,20 \\
\hline
+ 9
\end{array}
$$

Note Adding the $0x^2$ term gives us a column in which to write $-8x^2$.

$$\frac{4x^3 - 6x - 11}{2x - 4} = 2x^2 + 4x + 5 + \frac{9}{2x - 4}$$

To check this result, we multiply $2x - 4$ and $2x^2 + 4x + 5$:

$$
\begin{array}{r}
2x^2 + 4x + 5 \\
\times \qquad 2x - 4 \\
\hline
4x^3 + 8x^2 + 10x \qquad \\
+ \quad - 8x^2 - 16x - 20 \\
\hline
4x^3 \qquad\quad - 6x - 20
\end{array}
$$

Adding 9 (the remainder) to this result gives us the polynomial $4x^3 - 6x - 11$. Our answer checks.

For our next example, let's do Example 4 again, but this time use long division.

EXAMPLE 8 Divide $\dfrac{x^2 - 6xy - 7y^2}{x + y}$.

SOLUTION

$$
\begin{array}{r|l}
 & \quad x - 7y \\
x + y & \overline{\; x^2 - 6xy - 7y^2} \\
 & \;\; - \quad\; - \\
 & \not+ x^2 \not+ \;xy \quad \downarrow \\
\hline
 & \qquad - 7xy - 7y^2 \\
 & \qquad\; + \qquad + \\
 & \qquad \underline{- 7xy - 7y^2} \\
 & \qquad\qquad\qquad 0
\end{array}
$$

In this case, the remainder is 0, and we have

$$\frac{x^2 - 6xy - 7y^2}{x + y} = x - 7y$$

which is easy to check because

$$(x + y)(x - 7y) = x^2 - 6xy - 7y^2$$

EXAMPLE 9 Factor $x^3 + 9x^2 + 26x + 24$ completely if $x + 2$ is one of its factors.

SOLUTION Because $x + 2$ is one of the factors of the polynomial we are trying to factor, it must divide that polynomial evenly—that is, without a remainder. Therefore, we begin by dividing the polynomial by $x + 2$:

$$
\begin{array}{r|l}
 & \quad x^2 + 7x + 12 \\
x + 2 & \overline{\; x^3 + 9x^2 + 26x + 24} \\
 & \;\; - \quad\; - \\
 & \not+ x^3 \not+ 2x^2 \quad \downarrow \quad\; \downarrow \\
\hline
 & \qquad + 7x^2 + 26x \\
 & \qquad\; - \qquad - \\
 & \qquad \not+ 7x^2 \not+ 14x \quad \downarrow \\
\hline
 & \qquad\qquad + 12x + 24 \\
 & \qquad\qquad\; - \qquad - \\
 & \qquad\qquad \underline{\not+ 12x \not+ 24} \\
 & \qquad\qquad\qquad\quad 0
\end{array}
$$

Now we know that the polynomial we are trying to factor is equal to the product of $x + 2$ *and* $x^2 + 7x + 12$. To factor completely, we simply factor $x^2 + 7x + 12$:

$$
\begin{aligned}
x^3 + 9x^2 + 26x + 24 &= (x + 2)(x^2 + 7x + 12) \\
&= (x + 2)(x + 3)(x + 4)
\end{aligned}
$$

GETTING READY FOR CLASS

After reading through the preceding section, respond in your own words and in complete sentences.

A. What are the four steps used in long division with polynomials?
B. What does it mean to have a remainder of 0?
C. When must long division be performed, and when can factoring be used to divide polynomials?
D. What property of real numbers is the key to dividing a polynomial by a monomial?

Find the following quotients.

1. $\dfrac{4x^3 - 8x^2 + 6x}{2x}$

2. $\dfrac{6x^3 + 12x^2 - 9x}{3x}$

3. $\dfrac{10x^4 + 15x^3 - 20x^2}{-5x^2}$

4. $\dfrac{12x^5 - 18x^4 - 6x^3}{6x^3}$

5. $\dfrac{8y^5 + 10y^3 - 6y}{4y^3}$

6. $\dfrac{6y^4 - 3y^3 + 18y^2}{9y^2}$

7. $\dfrac{5x^3 - 8x^2 - 6x}{-2x^2}$

8. $\dfrac{-9x^5 + 10x^3 - 12x}{-6x^4}$

9. $\dfrac{28a^3b^5 + 42a^4b^3}{7a^2b^2}$

10. $\dfrac{a^2b + ab^2}{ab}$

11. $\dfrac{10x^3y^2 - 20x^2y^3 - 30x^3y^3}{-10x^2y}$

12. $\dfrac{9x^4y^4 + 18x^3y^4 - 27x^2y^4}{-9xy^3}$

Divide by factoring numerators and then dividing out common factors.

13. $\dfrac{x^2 - x - 6}{x - 3}$

14. $\dfrac{x^2 - x - 6}{x + 2}$

15. $\dfrac{2a^2 - 3a - 9}{2a + 3}$

16. $\dfrac{2a^2 + 3a - 9}{2a - 3}$

17. $\dfrac{5x^2 - 14xy - 24y^2}{x - 4y}$

18. $\dfrac{5x^2 - 26xy - 24y^2}{5x + 4y}$

19. $\dfrac{x^3 - y^3}{x - y}$

20. $\dfrac{x^3 + 8}{x + 2}$

21. $\dfrac{y^4 - 16}{y - 2}$

22. $\dfrac{y^4 - 81}{y - 3}$

23. $\dfrac{x^3 + 2x^2 - 25x - 50}{x - 5}$

24. $\dfrac{x^3 + 2x^2 - 25x - 50}{x + 5}$

25. $\dfrac{4x^3 + 12x^2 - 9x - 27}{x + 3}$

26. $\dfrac{9x^3 + 18x^2 - 4x - 8}{x + 2}$

Divide using the long division method.

27. $\dfrac{x^2 - 5x - 7}{x + 2}$

28. $\dfrac{x^2 + 4x - 8}{x - 3}$

29. $\dfrac{6x^2 + 7x - 18}{3x - 4}$

30. $\dfrac{8x^2 - 26x - 9}{2x - 7}$

31. $\dfrac{2x^3 - 3x^2 - 4x + 5}{x + 1}$

32. $\dfrac{3x^3 - 5x^2 + 2x - 1}{x - 2}$

33. $\dfrac{2y^3 - 9y^2 - 17y + 39}{2y - 3}$

34. $\dfrac{3y^3 - 19y^2 + 17y + 4}{3y - 4}$

35. $\dfrac{2x^3 - 9x^2 + 11x - 6}{2x^2 - 3x + 2}$

36. $\dfrac{6x^3 + 7x^2 - x + 3}{3x^2 - x + 1}$

37. $\dfrac{6y^3 - 8y + 5}{2y - 4}$

38. $\dfrac{9y^3 - 6y^2 + 8}{3y - 3}$

39. $\dfrac{a^4 - 2a + 5}{a - 2}$

40. $\dfrac{a^4 + a^3 - 1}{a + 2}$

41. $\dfrac{y^4 - 16}{y - 2}$

42. $\dfrac{y^4 - 81}{y - 3}$

43. $\dfrac{x^4 + x^3 - 3x^2 - x + 2}{x^2 + 3x + 2}$

44. $\dfrac{2x^4 + x^3 + 4x - 3}{2x^2 - x + 3}$

45. Factor $x^3 + 6x^2 + 11x + 6$ completely if one of its factors is $x + 3$.

46. Factor $x^3 + 10x^2 + 29x + 20$ completely if one of its factors is $x + 4$.

47. Factor $x^3 + 5x^2 - 2x - 24$ completely if one of its factors is $x + 3$.

48. Factor $x^3 + 3x^2 - 10x - 24$ completely if one of its factors is $x + 2$.

49. Problems 21 and 41 are the same problem. Are the two answers you obtained equivalent?

50. Problems 22 and 42 are the same problem. Are the two answers you obtained equivalent?

51. Find $P(-2)$ if $P(x) = x^2 - 5x - 7$. Compare it with the remainder in Problem 27.

52. Find $P(3)$ if $P(x) = x^2 + 4x - 8$. Compare it with the remainder in Problem 28.

Applying the Concepts

53. The Factor Theorem The factor theorem of algebra states that if $x - a$ is a factor of a polynomial, $P(x)$, then $P(a) = 0$. Verify the following.

a. That $x - 2$ is a factor of $P(x) = x^3 - 3x^2 + 5x - 6$, and that $P(2) = 0$

b. That $x - 5$ is a factor of $P(x) = x^4 - 5x^3 - x^2 + 6x - 5$, and that $P(5) = 0$

54. The Remainder Theorem The remainder theorem of algebra states that if a polynomial, $P(x)$, is divided by $x - a$, then the remainder is $P(a)$. Verify the remainder theorem by showing that when $P(x) = x^2 - x + 3$ is divided by $x - 2$ the remainder is 5, and that $P(2) = 5$.

55. Checking Account First Bank of San Luis Obispo charges \$2.00 per month and \$0.15 per check for a regular checking account. As we mentioned in the introduction to this section, the total monthly cost of this account is $C(x) = 2.00 + 0.15x$. To find the average cost of each of the x checks, we divide the total cost by the number of checks written. That is,

$$\overline{C}(x) = \frac{C(x)}{x}$$

a. Use the total cost function to fill in the following table.

x	1	5	10	15	20
$C(x)$					

b. Find the formula for the average cost function, $\overline{C}(x)$.

c. Use the average cost function to fill in the following table.

x	1	5	10	15	20
$\overline{C}(x)$					

d. What happens to the average cost as more checks are written?

e. Give the domain and range of each of the functions.

56. Average Cost A company that manufactures computer diskettes uses the function $C(x) = 200 + 2x$ to represent the daily cost of producing x diskettes.

a. Find the average cost function, $\overline{C}(x)$.

b. Use the average cost function to fill in the following table:

x	1	5	10	15	20
$\overline{C}(x)$					

c. What happens to the average cost as more items are produced?

d. Graph the function $y = \overline{C}(x)$ for $x > 0$.

e. What is the domain of this function?

f. What is the range of this function?

57. Average Cost For long distance service, a particular phone company charges a monthly fee of \$4.95 plus \$0.07 per minute of calling time used. The relationship between the number of minutes of calling time used, m, and the amount of the monthly phone bill $T(m)$ is given by the function $T(m) = 4.95 + 0.07m$.

a. Find the total cost when 100, 400, and 500 minutes of calling time is used in 1 month.

b. Find a formula for the average cost per minute function $\overline{T}(m)$.

c. Find the average cost per minute of calling time used when 100, 400, and 500 minutes are used in 1 month.

58. Average Cost A company manufactures electric pencil sharpeners. Each month they have fixed costs of \$40,000 and variable costs of \$8.50 per sharpener. Therefore, the total monthly cost to manufacture x sharpeners is given by the function $C(x) = 40{,}000 + 8.5x$.

a. Find the total cost to manufacture 1,000, 5,000, and 10,000 sharpeners a month.

b. Write an expression for the average cost per sharpener function $\overline{C}(x)$.

c. Find the average cost per sharpener to manufacture 1,000, 5,000, and 10,000 sharpeners per month.

Maintaining Your Skills

Reviewing these problems will help clarify the different methods we have used in this chapter.

Perform the indicated operations.

59. $\dfrac{2a + 10}{a^3} \cdot \dfrac{a^2}{3a + 15}$

60. $\dfrac{4a + 8}{a^2 - a - 6} \div \dfrac{a^2 + 7a + 12}{a^2 - 9}$

61. $(x^2 - 9)\left(\dfrac{x + 2}{x + 3}\right)$

62. $\dfrac{1}{x + 4} + \dfrac{8}{x^2 - 16}$

63. $\dfrac{2x - 7}{x - 2} - \dfrac{x - 5}{x - 2}$

64. $2 + \dfrac{25}{5x - 1}$

Simplify each expression.

65. $\dfrac{\dfrac{1}{x} - \dfrac{1}{3}}{\dfrac{1}{x} + \dfrac{1}{3}}$

66. $\dfrac{1 - \dfrac{9}{x^2}}{1 - \dfrac{1}{x} - \dfrac{6}{x^2}}$

Solve each equation.

67. $\dfrac{x}{x - 3} + \dfrac{3}{2} = \dfrac{3}{x - 3}$

68. $1 - \dfrac{3}{x} = \dfrac{-2}{x^2}$

Chapter 8 Summary

EXAMPLES

1. $\frac{3}{4}$ is a rational number. $\frac{x-3}{x^2-9}$ is a rational expression.

Rational Numbers and Expressions [8.1]

A *rational number* is any number that can be expressed as the ratio of two integers:

$$\text{Rational numbers} = \left\{ \frac{a}{b} \,\middle|\, a \text{ and } b \text{ are integers, } b \neq 0 \right\}$$

A *rational expression* is any quantity that can be expressed as the ratio of two polynomials:

$$\text{Rational expressions} = \left\{ \frac{P}{Q} \,\middle|\, P \text{ and } Q \text{ are polynomials, } Q \neq 0 \right\}$$

Properties of Rational Expressions [8.1]

If P, Q, and K are polynomials with $Q \neq 0$ and $K \neq 0$, then

$$\frac{P}{Q} = \frac{PK}{QK} \quad \text{and} \quad \frac{P}{Q} = \frac{\frac{P}{K}}{\frac{Q}{K}}$$

which is to say that multiplying or dividing the numerator and denominator of a rational expression by the same nonzero quantity always produces an equivalent rational expression.

Reducing to Lowest Terms [8.1]

2. $\frac{x-3}{x^2-9} = \frac{\cancel{x-3}}{(\cancel{x-3})(x+3)}$

$= \frac{1}{x+3}$

To reduce a rational expression to lowest terms, we first factor the numerator and denominator and then divide the numerator and denominator by any factors they have in common.

Multiplication [8.2]

3. $\frac{x+1}{x^2-4} \cdot \frac{x+2}{3x+3}$

$= \frac{\cancel{(x+1)}\cancel{(x+2)}}{(x-2)\cancel{(x+2)}(3)\cancel{(x+1)}}$

$= \frac{1}{3(x-2)}$

To multiply two rational numbers or rational expressions, multiply numerators and multiply denominators. In symbols,

$$\frac{P}{Q} \cdot \frac{R}{S} = \frac{PR}{QS} \qquad (Q \neq 0 \text{ and } S \neq 0)$$

In practice, we don't really multiply, but rather, we factor and then divide out common factors.

Division [8.2]

4. $\frac{x^2-y^2}{x^3+y^3} \div \frac{x-y}{x^2-xy+y^2}$

$= \frac{x^2-y^2}{x^3+y^3} \cdot \frac{x^2-xy+y^2}{x-y}$

$= \frac{\cancel{(x+y)}\cancel{(x-y)}\cancel{(x^2-xy+y^2)}}{\cancel{(x+y)}\cancel{(x^2-xy+y^2)}\cancel{(x-y)}}$

$= 1$

To divide one rational expression by another, we use the definition of division to rewrite our division problem as an equivalent multiplication problem. To divide by a rational expression we multiply by its reciprocal. In symbols,

$$\frac{P}{Q} \div \frac{R}{S} = \frac{P}{Q} \cdot \frac{S}{R} = \frac{PS}{QR} \qquad (Q \neq 0, S \neq 0, R \neq 0)$$

Least Common Denominator [8.3]

5. The LCD for $\frac{2}{x-3}$ and $\frac{3}{5}$ is $5(x-3)$.

The *least common denominator,* LCD, for a set of denominators is the smallest quantity divisible by each of the denominators.

Addition and Subtraction [8.3]

6. $\frac{2}{x-3} + \frac{3}{5}$

$= \frac{2}{x-3} \cdot \frac{5}{5} + \frac{3}{5} \cdot \frac{x-3}{x-3}$

$= \frac{3x+1}{5(x-3)}$

If P, Q, and R represent polynomials, $R \neq 0$, then

$$\frac{P}{R} + \frac{Q}{R} = \frac{P+Q}{R} \quad \text{and} \quad \frac{P}{R} - \frac{Q}{R} = \frac{P-Q}{R}$$

When adding or subtracting rational expressions with different denominators, we must find the LCD for all denominators and change each rational expression to an equivalent expression that has the LCD.

Complex Fractions [8.4]

7. $\dfrac{\frac{1}{x} + \frac{1}{y}}{\frac{1}{x} - \frac{1}{y}} = \dfrac{xy\left(\frac{1}{x} + \frac{1}{y}\right)}{xy\left(\frac{1}{x} - \frac{1}{y}\right)}$

$= \frac{y+x}{y-x}$

A rational expression that contains, in its numerator or denominator, other rational expressions is called a *complex fraction*. One method of simplifying a complex fraction is to multiply the numerator and denominator by the LCD for all denominators.

Equations Involving Rational Expressions [8.5]

8. Solve $\frac{x}{2} + 3 = \frac{1}{3}$.

$6\left(\frac{x}{2}\right) + 6 \cdot 3 = 6 \cdot \frac{1}{3}$

$3x + 18 = 2$

$x = -\frac{16}{3}$

To solve an equation involving rational expressions, we first find the LCD for all denominators appearing on either side of the equation. We then multiply both sides by the LCD to clear the equation of all fractions and solve as usual.

Dividing a Polynomial by a Monomial [8.7]

9. $\frac{15x^3 - 20x^2 + 10x}{5x}$

$= 3x^2 - 4x + 2$

To divide a polynomial by a monomial, divide each term of the polynomial by the monomial.

Long Division with Polynomials [8.7]

10.

$$\begin{array}{r} x - 2 \\ x - 3 \overline{)\, x^2 - 5x + 8} \\ \underline{\not- x^2 \not+ 3x} \\ -2x + 8 \\ \underline{\not+ 2x \not- 6} \\ 2 \end{array}$$

If division with polynomials cannot be accomplished by dividing out factors common to the numerator and denominator, then we use a process similar to long division with whole numbers. The steps in the process are estimate, multiply, subtract, and bring down the next term.

COMMON MISTAKES

1. Attempting to divide the numerator and denominator of a rational expression by a quantity that is not a factor of both. Like this:

$$\frac{x^2 - \overset{3}{\cancel{9x}} + \overset{2}{\cancel{20}}}{x^2 - \underset{1}{\cancel{3x}} - \underset{1}{\cancel{10}}} \quad \text{Mistake}$$

This makes no sense at all. The numerator and denominator must be factored completely before any factors they have in common can be recognized:

$$\frac{x^2 - 9x + 20}{x^2 - 3x - 10} = \frac{\cancel{(x-5)}(x-4)}{\cancel{(x-5)}(x+2)}$$

$$= \frac{x-4}{x+2}$$

2. Forgetting to check solutions to equations involving rational expressions. When we multiply both sides of an equation by a quantity containing the variable, we must be sure to check for extraneous solutions.

Chapter 8 Test

Reduce to lowest terms. [8.1]

1. $\dfrac{x^2 - y^2}{x - y}$

2. $\dfrac{2x^2 - 5x + 3}{2x^2 - x - 3}$

Multiply and divide as indicated. [8.2]

3. $\dfrac{a^2 - 16}{5a - 15} \cdot \dfrac{10(a - 3)^2}{a^2 - 7a + 12}$

4. $\dfrac{a^4 - 81}{a^2 + 9} \div \dfrac{a^2 - 8a + 15}{4a - 20}$

5. $\dfrac{x^3 - 8}{2x^2 - 9x + 10} \div \dfrac{x^2 + 2x + 4}{2x^2 + x - 15}$

Add and subtract as indicated. [8.3]

6. $\dfrac{4}{21} + \dfrac{6}{35}$

7. $\dfrac{3}{4} - \dfrac{1}{2} + \dfrac{5}{8}$

8. $\dfrac{a}{a^2 - 9} + \dfrac{3}{a^2 - 9}$

9. $\dfrac{1}{x} + \dfrac{2}{x - 3}$

10. $\dfrac{4x}{x^2 + 6x + 5} - \dfrac{3x}{x^2 + 5x + 4}$

11. $\dfrac{2x + 8}{x^2 + 4x + 3} - \dfrac{x + 4}{x^2 + 5x + 6}$

Simplify each complex fraction. [8.4]

12. $\dfrac{3 - \dfrac{1}{a + 3}}{3 + \dfrac{1}{a + 3}}$

13. $\dfrac{1 - \dfrac{9}{x^2}}{1 + \dfrac{1}{x} - \dfrac{6}{x^2}}$

Solve each of the following equations. [8.5]

14. $\dfrac{1}{x} + 3 = \dfrac{4}{3}$

15. $\dfrac{x}{x - 3} + 3 = \dfrac{3}{x - 3}$

16. $\dfrac{y + 3}{2y} + \dfrac{5}{y - 1} = \dfrac{1}{2}$

17. $1 - \dfrac{1}{x} = \dfrac{6}{x^2}$

18. Graph $f(x) = \dfrac{x + 4}{x - 1}$.

Solve the following applications. Be sure to show the equation in each case. [8.6]

19. Number Problem What number must be subtracted from the denominator of $\frac{10}{23}$ to make the result $\frac{1}{3}$

20. Speed of a Boat The current of a river is 2 miles per hour. It takes a motorboat a total of 3 hours to travel 8 miles upstream and return 8 miles downstream. What is the speed of the boat in still water?

21. Filling a Pool An inlet pipe can fill a pool in 10 hours, and the drain can empty it in 15 hours. If the pool is half full and both the inlet pipe and the drain are left open, how long will it take to fill the pool the rest of the way?

22. Unit Analysis The top of Mount Whitney, the highest point in California, is 14,494 feet above sea level. Give this height in miles to the nearest tenth of a mile.

23. Unit Analysis A bullet fired from a gun travels a distance of 4,750 feet in 3.2 seconds. Find the average speed of the bullet in miles per hour. Round to the nearest whole number.

Divide. [8.7]

24. $\dfrac{24x^3y + 12x^2y^2 - 16xy^3}{4xy}$

25. $\dfrac{2x^3 - 9x^2 + 10}{2x - 1}$

9 Rational Exponents and Roots

Chapter Outline

9.1 Rational Exponents

9.2 Simplified Form for Radicals

9.3 Addition and Subtraction of Radical Expressions

9.4 Multiplication and Division of Radical Expressions

9.5 Equations Involving Radicals

9.6 Complex Numbers

iStockphoto.com © trait2lumiere

Ecology and conservation are topics that interest most college students. If our rivers and oceans are to be preserved for future generations, we need to work to eliminate pollution from our waters. If a river is flowing at 1 meter per second and a pollutant is entering the river at a constant rate, the shape of the pollution plume can often be modeled by the simple equation

$$y = \sqrt{x}$$

The following table and graph were produced from the equation.

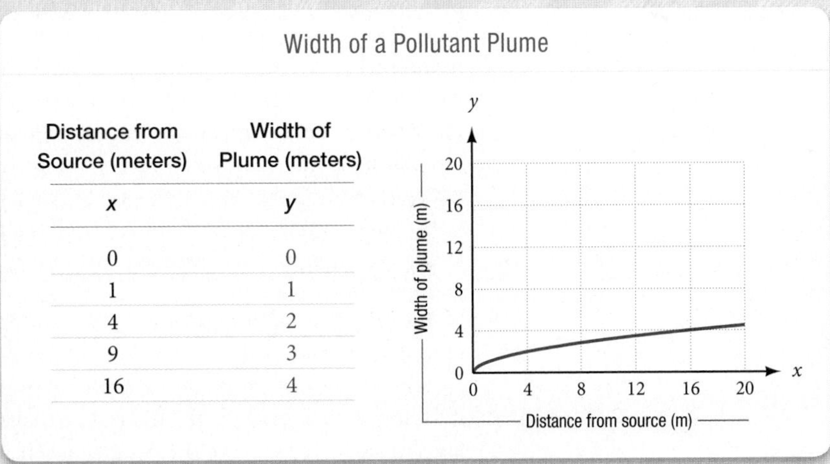

Width of a Pollutant Plume

Distance from Source (meters)	Width of Plume (meters)
x	y
0	0
1	1
4	2
9	3
16	4

To visualize how the graph models the pollutant plume, imagine that the river is flowing from left to right, parallel to the x-axis, with the x-axis as one of its banks. The pollutant is entering the river from the bank at (0, 0).

By modeling pollution with mathematics, we can use our knowledge of mathematics to help control and eliminate pollution.

Success Skills

iStockphoto © Silvia Boratti

Don't complain about anything, ever.

Do you complain to your classmates about your teacher? If you do, it could be getting in the way of your success in the class.

I have students that tell me that they like the way I teach and that they are enjoying my class. I have other students, in the same class, that complain to each other about me. They say I don't explain things well enough. Are the complaining students giving themselves a reason for not doing well in the class? I think so. They are shifting the responsibility for their success from themselves to me. It's not their fault they are not doing well, it's mine. When these students are alone, trying to do homework, they start thinking about how unfair everything is and they lose their motivation to study. Without intending to, they have set themselves up to fail by making their complaints more important than their progress in the class.

What happens when you stop complaining? You put yourself back in charge of your success. When there is no one to blame if things don't go well, you are more likely to do well. I have had students tell me that, once they stopped complaining about a class, the teacher became a better teacher and they started to actually enjoy going to class.

If you find yourself complaining to your friends about a class or a teacher, make a decision to stop. When other people start complaining to each other about the class or the teacher, walk away; don't participate in the complaining session. Try it for a day, or a week, or for the rest of the term. It may be difficult to do at first, but I'm sure you will like the results, and if you don't, you can always go back to complaining.

Rational Exponents 9.1

Figure 1 shows a square in which each of the four sides is 1 inch long. To find the square of the length of the diagonal c, we apply the Pythagorean theorem:

$$c^2 = 1^2 + 1^2$$
$$c^2 = 2$$

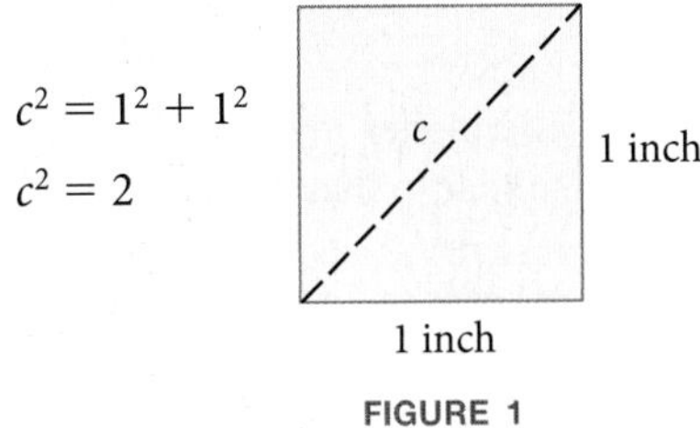

FIGURE 1

Because we know that c is positive and that its square is 2, we call c the *positive square root* of 2, and we write $c = \sqrt{2}$. Associating numbers, such as $\sqrt{2}$, with the diagonal of a square or rectangle allows us to analyze some interesting items from geometry. One particularly interesting geometric object that we will study in this section is shown in Figure 2. It is constructed from a right triangle, and the length of the diagonal is found from the Pythagorean theorem. We will come back to this figure at the end of this section.

The Golden Rectangle

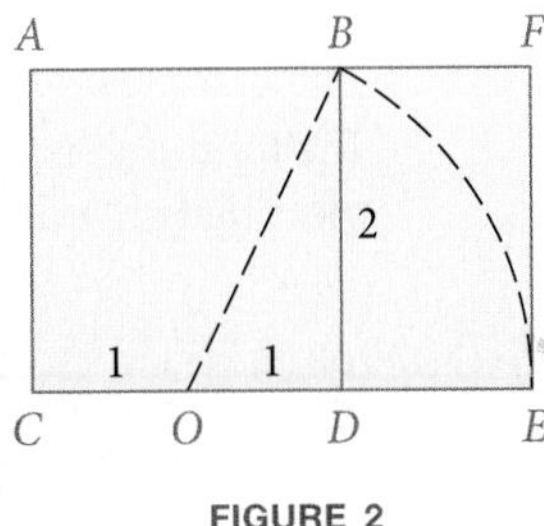

FIGURE 2

In Chapter 1, we developed notation (exponents) to give us the square, cube, or any other power of a number. For instance, if we wanted the square of 3, we wrote $3^2 = 9$. If we wanted the cube of 3, we wrote $3^3 = 27$. In this section, we will develop notation that will take us in the reverse direction, that is, from the square of a number, say 25, back to the original number, 5.

(def) DEFINITION *positive square root*

If x is a nonnegative real number, then the expression $\sqrt{x}$ is called the ***positive square root*** of x and is such that

$$(\sqrt{x})^2 = x$$

In words: $\sqrt{x}$ is the positive number we square to get x.

The negative square root of x, $-\sqrt{x}$, is defined in a similar manner.

EXAMPLE 1 The positive square root of 64 is 8 because 8 is the positive number with the property $8^2 = 64$. The negative square root of 64 is -8 because -8 is the negative number whose square is 64. We can summarize both these facts by saying

$$\sqrt{64} = 8 \quad \text{and} \quad -\sqrt{64} = -8$$

Note It is a common mistake to assume that an expression like $\sqrt{25}$ indicates both square roots, 5 and -5. The expression $\sqrt{25}$ indicates only the positive square root of 25, which is 5. If we want the negative square root, we must use a negative sign: $-\sqrt{25} = -5$.

The higher roots, cube roots, fourth roots, and so on, are defined by definitions similar to that of square roots.

Note We have restricted the even roots in this definition to nonnegative numbers. Even roots of negative numbers exist, but are not represented by real numbers. That is, $\sqrt{-4}$ is not a real number because there is no real number whose square is -4.

DEFINITION

If x is a real number and n is a positive integer, then

Positive square root of x, $\sqrt{x}$, is such that $(\sqrt{x})^2 = x$ $\quad x \geq 0$

Cube root of x, $\sqrt[3]{x}$, is such that $(\sqrt[3]{x})^3 = x$

Positive fourth root of x, $\sqrt[4]{x}$, is such that $(\sqrt[4]{x})^4 = x$ $\quad x \geq 0$

Fifth root of x, $\sqrt[5]{x}$, is such that $(\sqrt[5]{x})^5 = x$

⋮

The nth root of x, $\sqrt[n]{x}$, is such that $(\sqrt[n]{x})^n = x$ $\quad x \geq 0$ if n is even

The following is a table of the most common roots used in this book. Any of the roots that are unfamiliar should be memorized.

Square Roots		Cube Roots	Fourth Roots
$\sqrt{0} = 0$	$\sqrt{49} = 7$	$\sqrt[3]{0} = 0$	$\sqrt[4]{0} = 0$
$\sqrt{1} = 1$	$\sqrt{64} = 8$	$\sqrt[3]{1} = 1$	$\sqrt[4]{1} = 1$
$\sqrt{4} = 2$	$\sqrt{81} = 9$	$\sqrt[3]{8} = 2$	$\sqrt[4]{16} = 2$
$\sqrt{9} = 3$	$\sqrt{100} = 10$	$\sqrt[3]{27} = 3$	$\sqrt[4]{81} = 3$
$\sqrt{16} = 4$	$\sqrt{121} = 11$	$\sqrt[3]{64} = 4$	
$\sqrt{25} = 5$	$\sqrt{144} = 12$	$\sqrt[3]{125} = 5$	
$\sqrt{36} = 6$	$\sqrt{169} = 13$		

Notation An expression like $\sqrt[3]{8}$ that involves a root is called a *radical expression*. In the expression $\sqrt[3]{8}$, the ^{3} is called the *index*, the $\sqrt{}$ is the *radical sign*, and 8 is called the *radicand*. The index of a radical must be a positive integer greater than 1. If no index is written, it is assumed to be 2.

Roots and Negative Numbers

When dealing with negative numbers and radicals, the only restriction concerns negative numbers under even roots. We can have negative signs in front of radicals and negative numbers under odd roots and still obtain real numbers. Here are some examples to help clarify this. In the last section of this chapter, we will see how to deal with even roots of negative numbers.

EXAMPLES Simplify each expression, if possible.

2. $\sqrt[3]{-8} = -2$ because $(-2)^3 = -8$.

3. $\sqrt{-4}$ is not a real number because there is no real number whose square is -4.

4. $-\sqrt{25} = -5$, because -5 is the negative square root of 25.

5. $\sqrt[5]{-32} = -2$ because $(-2)^5 = -32$.

6. $\sqrt[4]{-81}$ is not a real number because there is no real number we can raise to the fourth power and obtain -81.

Variables Under a Radical

From the preceding examples, it is clear that we must be careful that we do not try to take an even root of a negative number. For this reason, we will assume that all variables appearing under a radical sign represent nonnegative numbers.

EXAMPLES Assume all variables represent nonnegative numbers, and simplify each expression as much as possible.

7. $\sqrt{25a^4b^6} = 5a^2b^3$ because $(5a^2b^3)^2 = 25a^4b^6$.

8. $\sqrt[3]{x^6y^{12}} = x^2y^4$ because $(x^2y^4)^3 = x^6y^{12}$.

9. $\sqrt[4]{81r^8s^{20}} = 3r^2s^5$ because $(3r^2s^5)^4 = 81r^8s^{20}$.

Rational Numbers as Exponents

We will now develop a second kind of notation involving exponents that will allow us to designate square roots, cube roots, and so on in another way.

Consider the equation $x = 8^{1/3}$. Although we have not encountered fractional exponents before, let's assume that all the properties of exponents hold in this case. Cubing both sides of the equation, we have

$$x^3 = (8^{1/3})^3$$
$$x^3 = 8^{(1/3)(3)}$$
$$x^3 = 8^1$$
$$x^3 = 8$$

The last line tells us that x is the number whose cube is 8. It must be true, then, that x is the cube root of 8, $x = \sqrt[3]{8}$. Because we started with $x = 8^{1/3}$, it follows that

$$8^{1/3} = \sqrt[3]{8}$$

It seems reasonable, then, to define fractional exponents as indicating roots. Here is the formal definition.

(def) DEFINITION

If x is a real number and n is a positive integer greater than 1, then

$$x^{1/n} = \sqrt[n]{x} \qquad (x \geq 0 \text{ when } n \text{ is even})$$

In words: The quantity $x^{1/n}$ is the nth root of x.

With this definition, we have a way of representing roots with exponents. Here are some examples.

EXAMPLES Write each expression as a root and then simplify, if possible.

10. $8^{1/3} = \sqrt[3]{8} = 2$

11. $36^{1/2} = \sqrt{36} = 6$

12. $-25^{1/2} = -\sqrt{25} = -5$

13. $(-25)^{1/2} = \sqrt{-25}$, which is not a real number

14. $\left(\frac{4}{9}\right)^{1/2} = \sqrt{\frac{4}{9}} = \frac{2}{3}$

The properties of exponents developed in Chapter 1 were applied to integer exponents only. We will now extend these properties to include rational exponents also. We do so without proof.

PROPERTY ***Properties of Exponents***

If a and b are real numbers and r and s are rational numbers, and a and b are nonnegative whenever r and s indicate even roots, then

1. $a^r \cdot a^s = a^{r+s}$ **4.** $a^{-r} = \frac{1}{a^r}$ $(a \neq 0)$

2. $(a^r)^s = a^{rs}$ **5.** $\left(\frac{a}{b}\right)^r = \frac{a^r}{b^r}$ $(b \neq 0)$

3. $(ab)^r = a^r b^r$ **6.** $\frac{a^r}{a^s} = a^{r-s}$ $(a \neq 0)$

Sometimes rational exponents can simplify our work with radicals. Here are Examples 8 and 9 again, but this time we will work them using rational exponents.

EXAMPLES Write each radical with a rational exponent, then simplify.

15.
$$\begin{aligned}\sqrt[3]{x^6y^{12}} &= (x^6y^{12})^{1/3} \\ &= (x^6)^{1/3}(y^{12})^{1/3} \\ &= x^2y^4\end{aligned}$$

16.
$$\begin{aligned}\sqrt[4]{81r^8s^{20}} &= (81r^8s^{20})^{1/4} \\ &= 81^{1/4}(r^8)^{1/4}(s^{20})^{1/4} \\ &= 3r^2s^5\end{aligned}$$

So far, the numerators of all the rational exponents we have encountered have been 1. The next theorem extends the work we can do with rational exponents to rational exponents with numerators other than 1.

We can extend our properties of exponents with the following theorem.

Theorem 6.1

If a is a nonnegative real number, m is an integer, and n is a positive integer, then

$$a^{m/n} = (a^{1/n})^m = (a^m)^{1/n}$$

Proof We can prove Theorem 6.1 using the properties of exponents. Because $m/n = m(1/n)$, we have

$$a^{m/n} = a^{m(1/n)} = (a^m)^{1/n} \qquad a^{m/n} = a^{(1/n)(m)} = (a^{1/n})^m$$

Here are some examples that illustrate how we use this theorem.

EXAMPLES Simplify as much as possible.

Note On a scientific calculator, Example 17 would look like this:

8 [y^x] [(] 2 [÷] 3 [)] [=]

17. $8^{2/3} = (8^{1/3})^2$ Theorem 6.1

$= 2^2$ Definition of fractional exponents

$= 4$ The square of 2 is 4.

18. $25^{(3/2)} = (25^{1/2})^3$ Theorem 6.1

$= 5^3$ Definition of fractional exponents

$= 125$ The cube of 5 is 125.

19. $9^{-3/2} = (9^{1/2})^{-3}$ Theorem 6.1

$= 3^{-3}$ Definition of fractional exponents

$= \dfrac{1}{3^3}$ Property 4 for exponents

$= \dfrac{1}{27}$ The cube of 3 is 27

20. $\left(\dfrac{27}{8}\right)^{-4/3} = \left[\left(\dfrac{27}{8}\right)^{1/3}\right]^{-4}$ Theorem 6.1

$= \left(\dfrac{3}{2}\right)^{-4}$ Definition of fractional exponents

$= \left(\dfrac{2}{3}\right)^{4}$ Property 4 for exponents

$= \dfrac{16}{81}$ The fourth power of $\frac{2}{3}$ is $\frac{16}{81}$

The following examples show the application of the properties of exponents to rational exponents.

EXAMPLES Assume all variables represent positive quantities, and simplify as much as possible.

21. $x^{1/3} \cdot x^{5/6} = x^{1/3 + 5/6}$ Property 1

$= x^{2/6 + 5/6}$ LCD is 6

$= x^{7/6}$ Add fractions

22. $(y^{2/3})^{3/4} = y^{(2/3)(3/4)}$ Property 2

$= y^{1/2}$ Multiply fractions: $\frac{2}{3} \cdot \frac{3}{4} = \frac{6}{12} = \frac{1}{2}$

23. $\frac{z^{1/3}}{z^{1/4}} = z^{1/3 - 1/4}$ Property 6

$= z^{4/12 - 3/12}$ LCD is 12

$= z^{1/12}$ Subtract fractions

24. $\left(\frac{a^{-1/3}}{b^{1/2}}\right)^6 = \frac{(a^{-1/3})^6}{(b^{1/2})^6}$ Property 5

$= \frac{a^{-2}}{b^3}$ Property 2

$= \frac{1}{a^2b^3}$ Property 4

25. $\frac{(x^{-3}y^{1/2})^4}{x^{10}y^{3/2}} = \frac{(x^{-3})^4(y^{1/2})^4}{x^{10}y^{3/2}}$ Property 3

$= \frac{x^{-12}y^2}{x^{10}y^{3/2}}$ Property 2

$= x^{-22}y^{1/2}$ Property 6

$= \frac{y^{1/2}}{x^{22}}$ Property 4

FACTS FROM GEOMETRY ***The Pythagorean Theorem (Again) and the Golden Rectangle***

Now that we have had some experience working with square roots, we can rewrite the Pythagorean theorem using a square root. If triangle ABC is a right triangle with $C = 90°$, then the length of the longest side is the ***positive square root*** of the sum of the squares of the other two sides (see Figure 3).

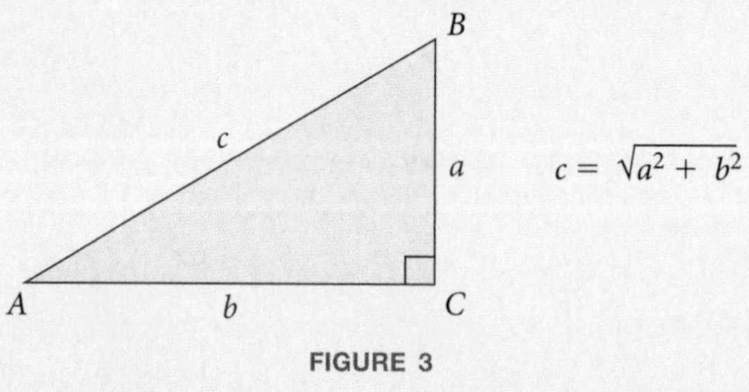

FIGURE 3

In the introduction to this chapter, we mentioned the golden rectangle. Its origins can be traced back over 2,000 years to the Greek civilization that produced Pythagoras, Socrates, Plato, Aristotle, and Euclid. The most important mathematical work to come from that Greek civilization was Euclid's *Elements,* an elegantly written summary of all that was known about geometry at that time in history. Euclid's *Elements,* according to Howard Eves, an authority on the history of mathematics, exercised a greater influence on scientific thinking than any other work. Here is how we construct a golden rectangle from a square of side 2, using the same method that Euclid used in his *Elements.*

Constructing a Golden Rectangle From a Square of Side 2

Step 1: Draw a square with a side of length 2. Connect the midpoint of side *CD* to corner *B*. (Note that we have labeled the midpoint of segment *CD* with the letter *O*.)

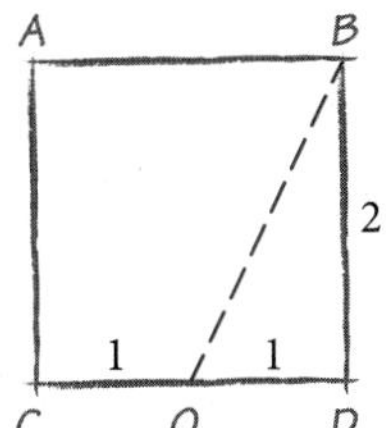

Step 2: Drop the diagonal from step 1 down so it aligns with side *CD*.

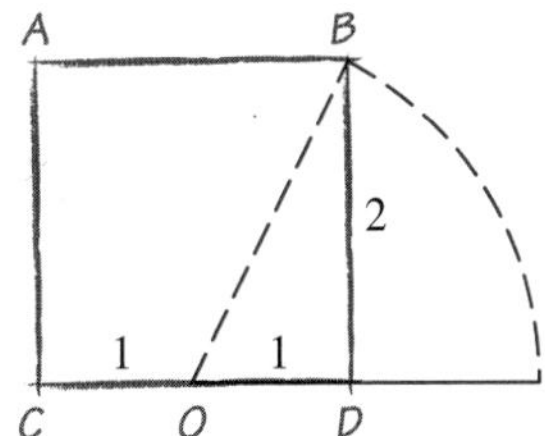

Step 3: Form rectangle *ACEF*. This is a golden rectangle.

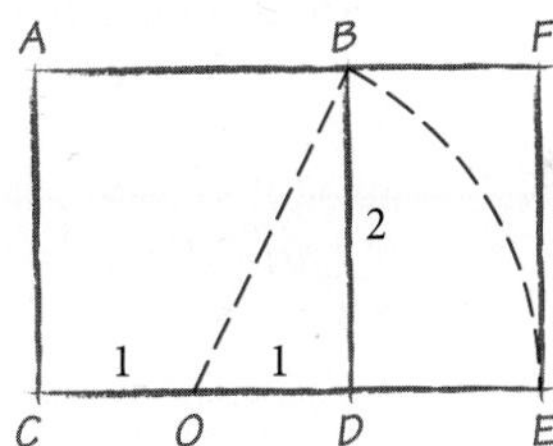

All golden rectangles are constructed from squares. Every golden rectangle, no matter how large or small it is, will have the same shape. To associate a number with the shape of the golden rectangle, we use the ratio of its length to its width. This ratio is called the *golden ratio*. To calculate the golden ratio, we must first find the length of the diagonal we used to construct the golden rectangle. Figure 4 shows the golden rectangle we constructed from a square of side 2. The length of the diagonal *OB* is found by applying the Pythagorean theorem to triangle *OBD*.

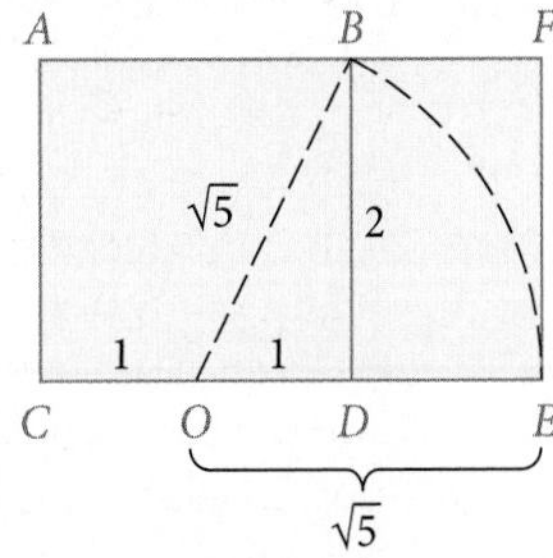

FIGURE 4

The length of segment OE is equal to the length of diagonal OB; both are $\sqrt{5}$. Because the distance from C to O is 1, the length CE of the golden rectangle is $1 + \sqrt{5}$. Now we can find the golden ratio:

$$\text{Golden ratio} = \frac{\text{length}}{\text{width}} = \frac{CE}{EF} = \frac{1 + \sqrt{5}}{2}$$

USING TECHNOLOGY *Graphing Calculators — A Word of Caution*

Some graphing calculators give surprising results when evaluating expressions such as $(-8)^{2/3}$. As you know from reading this section, the expression $(-8)^{2/3}$ simplifies to 4, either by taking the cube root first and then squaring the result, or by squaring the base first and then taking the cube root of the result. Here are three different ways to evaluate this expression on your calculator:

1. (−8)^(2/3) To evaluate $(-8)^{2/3}$
2. ((−8)^2)^(1/3) To evaluate $((-8)^2)^{1/3}$
3. ((−8)^(1/3))^2 To evaluate $((-8)^{1/3})^2$

Note any differences in the results.

Next, graph each of the following functions, one at a time.

1. $Y_1 = X^{2/3}$ 2. $Y_2 = (X^2)^{1/3}$ 3. $Y_3 = (X^{1/3})^2$

The correct graph is shown in Figure 5. Note which of your graphs match the correct graph.

Different calculators evaluate exponential expressions in different ways. You should use the method (or methods) that gave you the correct graph.

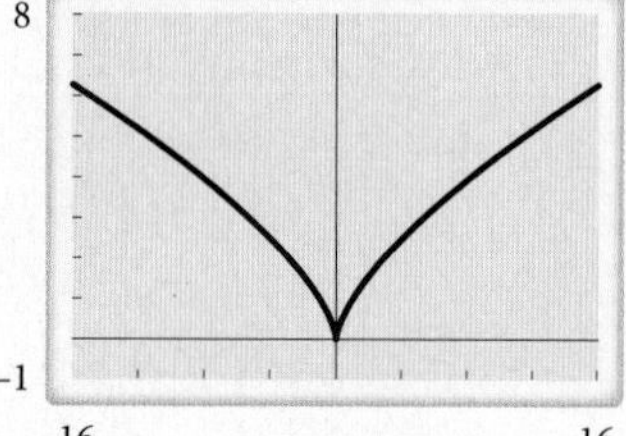

FIGURE 5

GETTING READY FOR CLASS

After reading through the preceding section, respond in your own words and in complete sentences.

A. Every real number has two square roots. Explain the notation we use to tell them apart. Use the square roots of 3 for examples.

B. Explain why a square root of −4 is not a real number.

C. We use the notation $\sqrt{2}$ to represent the positive square root of 2. Explain why there isn't a simpler way to express the positive square root of 2.

D. For the expression $a^{m/n}$, explain the significance of the numerator m and the significance of the denominator n in the exponent.

1. 12
2. -12
3. Not a real number
4. Not a real number
5. -7
6. 7
7. -3
8. -3
9. 2
10. -2
11. Not a real number
12. Not a real number
13. 0.2
14. 0.9
15. 0.2
16. 0.5
17. $6a^4$
18. $7a^5$
19. $3a^4$
20. $2a^5$
21. xy^2
22. x^2y
23. $2x^2y$
24. $2xy^2$
25. $2a^3b^5$
26. $3a^6b^2$
27. 6
28. 7
29. -3
30. -4
31. 2
32. -2
33. -2
34. -3
35. 2
36. 3
37. $\frac{9}{5}$
38. $\frac{3}{4}$
39. $\frac{4}{5}$
40. $\frac{2}{3}$
41. 9
42. 16
43. 125
44. 27
45. 8
46. 27
47. $\frac{1}{3}$
48. $\frac{1}{3}$
49. $\frac{1}{27}$
50. $\frac{1}{8}$
51. $\frac{6}{5}$
52. $\frac{7}{4}$
53. $\frac{8}{27}$
54. $\frac{4}{9}$

Find each of the following roots, if possible.

1. $\sqrt{144}$ **2.** $-\sqrt{144}$ **3.** $\sqrt{-144}$ **4.** $\sqrt{-49}$

5. $-\sqrt{49}$ **6.** $\sqrt{49}$ **7.** $\sqrt[3]{-27}$ **8.** $-\sqrt[3]{27}$

9. $\sqrt[4]{16}$ **10.** $-\sqrt[4]{16}$ **11.** $\sqrt[4]{-16}$ **12.** $-\sqrt[4]{-16}$

13. $\sqrt{0.04}$ **14.** $\sqrt{0.81}$ **15.** $\sqrt[3]{0.008}$ **16.** $\sqrt[3]{0.125}$

Simplify each expression. Assume all variables represent nonnegative numbers.

17. $\sqrt{36a^8}$ **18.** $\sqrt{49a^{10}}$ **19.** $\sqrt[3]{27a^{12}}$ **20.** $\sqrt[3]{8a^{15}}$

21. $\sqrt[3]{x^3y^6}$ **22.** $\sqrt[3]{x^6y^3}$ **23.** $\sqrt[5]{32x^{10}y^5}$ **24.** $\sqrt[5]{32x^5y^{10}}$

25. $\sqrt[4]{16a^{12}b^{20}}$ **26.** $\sqrt[4]{81a^{24}b^8}$

Use the definition of rational exponents to write each of the following with the appropriate root. Then simplify.

27. $36^{1/2}$ **28.** $49^{1/2}$ **29.** $-9^{1/2}$ **30.** $-16^{1/2}$

31. $8^{1/3}$ **32.** $-8^{1/3}$ **33.** $(-8)^{1/3}$ **34.** $-27^{1/3}$

35. $32^{1/5}$ **36.** $81^{1/4}$ **37.** $\left(\frac{81}{25}\right)^{1/2}$ **38.** $\left(\frac{9}{16}\right)^{1/2}$

39. $\left(\frac{64}{125}\right)^{1/3}$ **40.** $\left(\frac{8}{27}\right)^{1/3}$

Use Theorem 6.1 to simplify each of the following as much as possible.

41. $27^{2/3}$ **42.** $8^{4/3}$ **43.** $25^{3/2}$ **44.** $9^{3/2}$

45. $16^{3/4}$ **46.** $81^{3/4}$

Simplify each expression. Remember, negative exponents give reciprocals.

47. $27^{-1/3}$ **48.** $9^{-1/2}$ **49.** $81^{-3/4}$ **50.** $4^{-3/2}$

51. $\left(\frac{25}{36}\right)^{-1/2}$ **52.** $\left(\frac{16}{49}\right)^{-1/2}$ **53.** $\left(\frac{81}{16}\right)^{-3/4}$ **54.** $\left(\frac{27}{8}\right)^{-2/3}$

55. $16^{1/2} + 27^{1/3}$ **56.** $25^{1/2} + 100^{1/2}$ **57.** $8^{-2/3} + 4^{-1/2}$ **58.** $49^{-1/2} + 25^{-1/2}$

Use the properties of exponents to simplify each of the following as much as possible. Assume all bases are positive.

59. $x^{3/5} \cdot x^{1/5}$ **60.** $x^{3/4} \cdot x^{5/4}$ **61.** $(a^{3/4})^{4/3}$ **62.** $(a^{2/3})^{3/4}$

63. $\frac{x^{1/5}}{x^{3/5}}$ **64.** $\frac{x^{2/7}}{x^{5/7}}$ **65.** $\frac{x^{5/6}}{x^{2/3}}$ **66.** $\frac{x^{7/8}}{x^{8/7}}$

67. $(x^{3/5}y^{5/6}z^{1/3})^{3/5}$ **68.** $(x^{3/4}y^{1/8}z^{5/6})^{4/5}$ **69.** $\frac{a^{3/4}b^2}{a^{7/8}b^{1/4}}$ **70.** $\frac{a^{1/3}b^4}{a^{3/5}b^{1/3}}$

71. $\frac{(y^{2/3})^{3/4}}{(y^{1/3})^{3/5}}$ **72.** $\frac{(y^{5/4})^{2/5}}{(y^{1/4})^{4/3}}$ **73.** $\left(\frac{a^{-1/4}}{b^{1/2}}\right)^8$ **74.** $\left(\frac{a^{-1/5}}{b^{1/3}}\right)^{15}$

Simplify. (Assume all variables are nonnegative.)

75. a. $\sqrt{25}$ **b.** $\sqrt{0.25}$ **c.** $\sqrt{2500}$ **d.** $\sqrt{0.0025}$

76. a. $\sqrt[3]{8}$ **b.** $\sqrt[3]{0.008}$ **c.** $\sqrt[3]{8{,}000}$ **d.** $\sqrt[3]{8 \times 10^{-6}}$

77. a. $\sqrt{16a^4b^8}$ **b.** $\sqrt[3]{16a^4b^8}$ **c.** $\sqrt[4]{16a^4b^8}$

78. a. $\sqrt{64x^5y^{10}}$ **b.** $\sqrt[3]{64x^5y^{10}}$ **c.** $\sqrt[4]{64x^5y^{10}}$

55. 7
56. 15
57. $\frac{3}{4}$
58. $\frac{12}{35}$
59. $x^{4/5}$
60. x^2
61. a
62. $a^{1/2}$
63. $\frac{1}{x^{2/5}}$
64. $\frac{1}{x^{3/7}}$
65. $x^{1/6}$
66. $\frac{1}{x^{15/56}}$
67. $x^{9/25}y^{1/2}z^{1/5}$
68. $x^{3/5}x^{1/10}z^{2/3}$
69. $\frac{b^{7/4}}{a^{1/8}}$
70. $\frac{b^{11/3}}{a^{4/15}}$
71. $y^{3/10}$
72. $y^{1/6}$
73. $\frac{1}{a^2b^4}$
74. $\frac{1}{a^3b^5}$
75. a. 5
b. 0.5
c. 50
d. 0.05
76. a. 2
b. 0.2
c. 20
d. 0.02
77 a. $4a^2b^4$
b. $2ab^2\sqrt[3]{2ab^2}$
c. $2ab^2$
78. a. $8x^2y^5\sqrt{x}$
b. $4xy^3\sqrt[3]{x^2y}$
c. $2xy^2\sqrt[4]{4xy^2}$

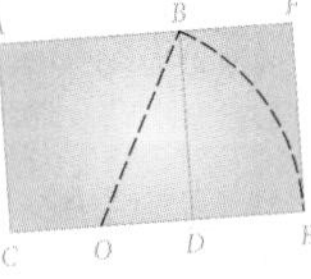

79. Show that the expression $(a^{1/2} + b^{1/2})^2$ is not equal to $a + b$ by replacing a with 9 and b with 4 in both expressions and then simplifying each.

80. Show that the statement $(a^2 + b^2)^{1/2} = a + b$ is not, in general, true by replacing a with 3 and b with 4 and then simplifying both sides.

81. You may have noticed, if you have been using a calculator to find roots, that you can find the fourth root of a number by pressing the square root button twice. Written in symbols, this fact looks like this:

$$\sqrt{\sqrt{a}} = \sqrt[4]{a} \qquad (a \geq 0)$$

Show that this statement is true by rewriting each side with exponents instead of radical notation and then simplifying the left side.

82. Show that the statement is true by rewriting each side with exponents instead of radical notation and then simplifying the left side.

$$\sqrt[3]{\sqrt{a}} = \sqrt[6]{a} \qquad (a \geq 0)$$

Applying the Concepts

83. **Maximum Speed** The maximum speed (v) that an automobile can travel around a curve of radius r without skidding is given by the equation

$$v = \left(\frac{5r}{2}\right)^{1/2}$$

where v is in miles per hour and r is measured in feet. What is the maximum speed a car can travel around a curve with a radius of 250 feet without skidding?

84. **Relativity** The equation

$$L = \left(1 - \frac{v^2}{c^2}\right)^{1/2}$$

gives the relativistic length of a 1-foot ruler traveling with velocity v. Find L if

$$\frac{v}{c} = \frac{3}{5}$$

85. **Golden Ratio** The golden ratio is the ratio of the length to the width in any golden rectangle. The exact value of this number is $\frac{1+\sqrt{5}}{2}$. Use a calculator to find a decimal approximation to this number and round it to the nearest thousandth.

86. **Golden Ratio** The reciprocal of the golden ratio is $\frac{2}{1+\sqrt{5}}$. Find a decimal approximation to this number that is accurate to the nearest thousandth.

87. **Sequences** Find the next term in the following sequence. Then explain how this sequence is related to the Fibonacci sequence.

$$\frac{3}{2}, \frac{5}{3}, \frac{8}{5}, \ldots$$

88. **Sequences** Write the first 10 terms in the sequence shown in Problem 87. Then find a decimal approximation to each of the 10 terms, rounding each to the nearest thousandth.

83. 25 mph
84. 0.8 feet
85. 1.618
86. 0.618
87. $\frac{13}{8}$. The denominator is the sum of the 2 previous denominators, and the numerator is the sum of the 2 previous numerators.

88. $\frac{3}{2}, \frac{5}{3}, \frac{8}{5}, \frac{13}{8}, \frac{21}{13}, \frac{34}{21}, \frac{55}{34}, \frac{89}{55}, \frac{144}{89}, \frac{233}{144}$,
1.5, 1.667, 1.6, 1.625, 1.615, 1.619, 1.618, 1.618, 1.618, 1.618
89. a. 420 picometers
b. 594.0 pico meters
c. 5.94×10^{-10} meters
90. a. $\sqrt{2}$
b. $\sqrt{3}$
91. 5
92. 2
93. 6
94. 3
95. $4x^2y$
96. $2x^3y^4$
97. $5y$
98. $8x^3$
99. 3
100. −2
101. 2
102. −5
103. $2ab$
104. $4a^2b$
105. 25
106. 4
107. $48x^4y^2$
108. $8a^3b^3$
109. $4x^6y^6$
110. $27a^6c^3$

89. Chemistry Figure 6 shows part of a model of a magnesium oxide (MgO) crystal. Each corner of the square is at the center of one oxygen ion ($O^{2}-$), and the center of the middle ion is at the center of the square. The radius for each oxygen ion is 60 picometers (pm), and the radius for each magnesium ion ($Mg^{2}+$) is 150 picometers.

a. Find the length of the side of the square. Write your answer in picometers.

b. Find the length of the diagonal of the square. Write your answer in picometers.

c. If 1 meter is 10^{12} picometers, give the length of the diagonal of the square in meters.

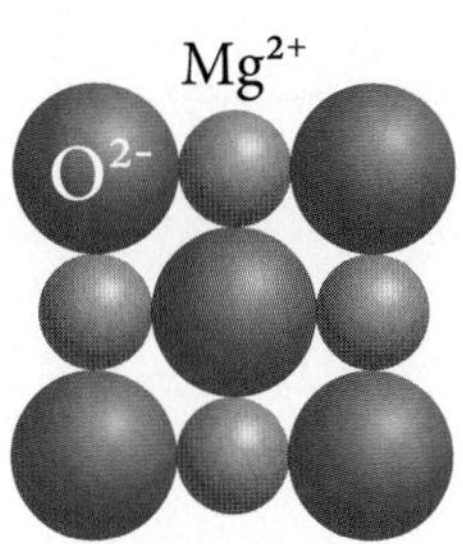

FIGURE 6

90. Geometry The length of each side of the cube shown in Figure 7 is 1 inch.

a. Find the length of the diagonal CH.

b. Find the length of the diagonal CF.

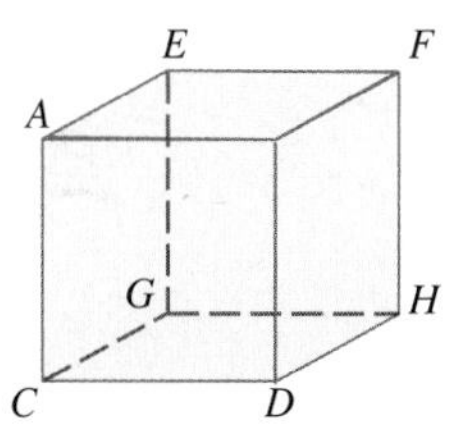

FIGURE 7

Getting Ready for the Next Section

Simplify. Assume all variable are positive real numbers.

91. $\sqrt{25}$ **92.** $\sqrt{4}$ **93.** $\sqrt{6^2}$ **94.** $\sqrt{3^2}$

95. $\sqrt{16x^4y^2}$ **96.** $\sqrt{4x^6y^8}$ **97.** $\sqrt{(5y)^2}$ **98.** $\sqrt{(8x^3)^2}$

99. $\sqrt[3]{27}$ **100.** $\sqrt[3]{-8}$ **101.** $\sqrt[3]{2^3}$ **102.** $\sqrt[3]{(-5)^3}$

103. $\sqrt[3]{8a^3b^3}$ **104.** $\sqrt[3]{64a^6b^3}$

Fill in the blank.

105. $50 = \_\_\_\_ \cdot 2$ **106.** $12 = \_\_\_\_ \cdot 3$

107. $48x^4y^3 = \_\_\_\_ \cdot y$ **108.** $40a^5b^4 = \_\_\_\_ \cdot 5a^2b$

109. $12x^7y^6 = \_\_\_\_ \cdot 3x$ **110.** $54a^6b^2c^4 = \_\_\_\_ \cdot 2b^2c$

Simplified Form for Radicals

9.2

The Spiral of Roots

Earlier in this chapter, we showed how the Pythagorean theorem can be used to construct a golden rectangle. In a similar manner, the Pythagorean theorem can be used to contruct the attractive spiral shown here.

This spiral is called the Spiral of Roots because each of the diagonals is the positive square root of one of the positive integers. At the end of this section, we will use the Pythagorean theorem and some of the material in this section to construct this spiral.

In this section, we will use radical notation instead of rational exponents. We will begin by stating two properties of radicals. Following this, we will give a definition for simplified form for radical expressions. The examples in this section show how we use the properties of radicals to write radical expresions in simplified form.

Here are the first two properties of radicals. For these two properties, we will assume a and b are nonnegative real numbers whenever n is an even number.

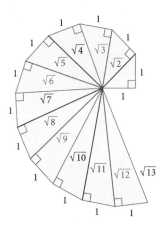

Note There is not a property for radicals that says the nth root of a sum is the sum of the nth roots. That is,

$$\sqrt[n]{a+b} \neq \sqrt[n]{a} + \sqrt[n]{b}$$

PROPERTY *Property 1 for Radicals*

$$\sqrt[n]{ab} = \sqrt[n]{a}\sqrt[n]{b}$$

In words: The nth root of a product is the product of the nth roots.

Proof of Property 1

$$\sqrt[n]{ab} = (ab)^{1/n}$$ Definition of fractional exponents

$$= a^{1/n}b^{1/n}$$ Exponents distribute over products

$$= \sqrt[n]{a}\sqrt[n]{b}$$ Definition of fractional exponents

PROPERTY *Property 2 for Radicals*

$$\sqrt[n]{\frac{a}{b}} = \frac{\sqrt[n]{a}}{\sqrt[n]{b}} = \quad (b \neq 0)$$

In words: The nth root of a quotient is the quotient of the nth roots.

The proof of Property 2 is similar to the proof of Property 1.

These two properties of radicals allow us to change the form of and simplify radical expressions without changing their value.

RULE *Simplified Form for Radical Expressions*

A radical expression is in ***simplified form*** if

1. None of the factors of the radicand (the quantity under the radical sign) can be written as powers greater than or equal to the index—that is, no perfect squares can be factors of the quantity under a square root sign, no perfect cubes can be factors of what is under a cube root sign, and so forth.
2. There are no fractions under the radical sign.
3. There are no radicals in the denominator.

Satisfying the first condition for simplified form actually amounts to taking as much out from under the radical sign as possible. The following examples illustrate the first condition for simplified form.

EXAMPLE 1 Write $\sqrt{50}$ in simplified form.

SOLUTION The largest perfect square that divides 50 is 25. We write 50 as $25 \cdot 2$ and apply Property 1 for radicals:

$$\begin{aligned} \sqrt{50} &= \sqrt{25 \cdot 2} && 50 = 25 \cdot 2 \\ &= \sqrt{25}\sqrt{2} && \text{Property 1} \\ &= 5\sqrt{2} && \sqrt{25} = 5 \end{aligned}$$

We have taken as much as possible out from under the radical sign—in this case, factoring 25 from 50 and then writing $\sqrt{25}$ as 5.

As we progress through this chapter you will see more and more expressions that involve the product of a number and a radical. Here are some examples:

$$3\sqrt{2} \qquad \frac{1}{2}\sqrt{5} \qquad 5\sqrt{7} \qquad 3x\sqrt{2x} \qquad 2ab\sqrt{5a}$$

All of these are products. The first expression $3\sqrt{2}$ is the product of 3 and $\sqrt{2}$. That is,

$$3\sqrt{2} = 3 \cdot \sqrt{2}$$

The 3 and the $\sqrt{2}$ are not stuck together is some mysterious way. The expression $3\sqrt{2}$ is simply the product of two numbers, one of which is rational, and the other is irrational.

EXAMPLE 2 Write in simplified form: $\sqrt{48x^4y^3}$, where $x, y \geq 0$

SOLUTION The largest perfect square that is a factor of the radicand is $16x^4y^2$. Applying Property 1 again, we have

$$\begin{aligned} \sqrt{48x^4y^3} &= \sqrt{16x^4y^2 \cdot 3y} \\ &= \sqrt{16x^4y^2}\sqrt{3y} \\ &= 4x^2y\sqrt{3y} \end{aligned}$$

EXAMPLE 3 Write $\sqrt[3]{40a^5b^4}$ in simplified form.

SOLUTION We now want to factor the largest perfect cube from the radicand. We write $40a^5b^4$ as $8a^3b^3 \cdot 5a^2b$ and proceed as we did in Examples 1 and 2.

$$\begin{aligned}\sqrt[3]{40a^5b^4} &= \sqrt[3]{8a^3b^3 \cdot 5a^2b}\\ &= \sqrt[3]{8a^3b^3}\sqrt[3]{5a^2b}\\ &= 2ab\sqrt[3]{5a^2b}\end{aligned}$$

Our next examples involve fractions and simplified form for radicals.

EXAMPLE 4 Simplify each expression.

a. $\dfrac{\sqrt{12}}{6}$ **b.** $\dfrac{5\sqrt{18}}{15}$ **c.** $\dfrac{6+\sqrt{8}}{2}$ **d.** $\dfrac{-1+\sqrt{45}}{2}$

SOLUTION In each case, we simplify the radical first, then we factor and reduce to lowest terms.

a. $\dfrac{\sqrt{12}}{6} = \dfrac{2\sqrt{3}}{6}$ Simplify the radical $\sqrt{12} = \sqrt{4 \cdot 3} = \sqrt{4}\sqrt{3} = 2\sqrt{3}$

$= \dfrac{2\sqrt{3}}{2 \cdot 3}$ Factor denominator

$= \dfrac{\sqrt{3}}{3}$ Divide out common factors

b. $\dfrac{5\sqrt{18}}{15} = \dfrac{5 \cdot 3\sqrt{2}}{15}$ $\sqrt{18} = \sqrt{9 \cdot 2} = \sqrt{9}\sqrt{2} = 3\sqrt{2}$

$= \dfrac{5 \cdot 3\sqrt{2}}{3 \cdot 5}$ Factor denominator

$= \sqrt{2}$ Divide out common factors

c. $\dfrac{6+\sqrt{8}}{2} = \dfrac{6+2\sqrt{2}}{2}$ $\sqrt{8} = \sqrt{4 \cdot 2} = \sqrt{4}\sqrt{2} = 2\sqrt{2}$

$= \dfrac{2(3+\sqrt{2})}{2}$ Factor numerator

$= 3 + \sqrt{2}$ Divide out common factors

d. $\dfrac{-1+\sqrt{45}}{2} = \dfrac{-1+3\sqrt{5}}{2}$ $\sqrt{45} = \sqrt{9 \cdot 5} = \sqrt{9}\sqrt{5} = 3\sqrt{5}$

This expression cannot be simplified further because $-1 + 3\sqrt{5}$ and 2 have no factors in common.

Rationalizing the Denominator

EXAMPLE 5 Simplify $\sqrt{\dfrac{3}{4}}$.

SOLUTION Applying Property 2 for radicals, we have

$\sqrt{\dfrac{3}{4}} = \dfrac{\sqrt{3}}{\sqrt{4}}$ Property 2

$= \dfrac{\sqrt{3}}{2}$ $\sqrt{4} = 2$

The last expression is in simplified form because it satisfies all three conditions for simplified form.

EXAMPLE 6 Write $\sqrt{\frac{5}{6}}$ in simplified form.

SOLUTION Proceeding as in Example 5, we have

$$\sqrt{\frac{5}{6}} = \frac{\sqrt{5}}{\sqrt{6}}$$

The resulting expression satisfies the second condition for simplified form because neither radical contains a fraction. It does, however, violate Condition 3 because it has a radical in the denominator. Getting rid of the radical in the denominator is called *rationalizing the denominator* and is accomplished, in this case, by multiplying the numerator and denominator by $\sqrt{6}$:

$$\begin{aligned} \frac{\sqrt{5}}{\sqrt{6}} &= \frac{\sqrt{5}}{\sqrt{6}} \cdot \frac{\sqrt{6}}{\sqrt{6}} \\ &= \frac{\sqrt{30}}{\sqrt{6^2}} \\ &= \frac{\sqrt{30}}{6} \end{aligned}$$

EXAMPLES Rationalize the denominator.

7. $$\begin{aligned} \frac{4}{\sqrt{3}} &= \frac{4}{\sqrt{3}} \cdot \frac{\sqrt{3}}{\sqrt{3}} \\ &= \frac{4\sqrt{3}}{\sqrt{3^2}} \\ &= \frac{4\sqrt{3}}{3} \end{aligned}$$

8. $$\begin{aligned} \frac{2\sqrt{3x}}{\sqrt{5y}} &= \frac{2\sqrt{3x}}{\sqrt{5y}} \cdot \frac{\sqrt{5y}}{\sqrt{5y}} \\ &= \frac{2\sqrt{15xy}}{\sqrt{(5y)^2}} \\ &= \frac{2\sqrt{15xy}}{5y} \end{aligned}$$

When the denominator involves a cube root, we must multiply by a radical that will produce a perfect cube under the cube root sign in the denominator, as Example 9 illustrates.

EXAMPLE 9 Rationalize the denominator in $\frac{7}{\sqrt[3]{4}}$.

SOLUTION Because $4 = 2^2$, we can multiply both numerator and denominator by $\sqrt[3]{2}$ and obtain $\sqrt[3]{2^3}$ in the denominator.

$$\begin{aligned} \frac{7}{\sqrt[3]{4}} &= \frac{7}{\sqrt[3]{2^2}} \\ &= \frac{7}{\sqrt[3]{2^2}} \cdot \frac{\sqrt[3]{2}}{\sqrt[3]{2}} \\ &= \frac{7\sqrt[3]{2}}{\sqrt[3]{2^3}} \\ &= \frac{7\sqrt[3]{2}}{2} \end{aligned}$$

EXAMPLE 10 Simplify $\sqrt{\dfrac{12x^5y^3}{5z}}$.

SOLUTION We use Property 2 to write the numerator and denominator as two separate radicals:

$$\sqrt{\frac{12x^5y^3}{5z}} = \frac{\sqrt{12x^5y^3}}{\sqrt{5z}}$$

Simplifying the numerator, we have

$$\frac{\sqrt{12x^5y^3}}{\sqrt{5z}} = \frac{\sqrt{4x^4y^2}\sqrt{3xy}}{\sqrt{5z}}$$

$$= \frac{2x^2y\sqrt{3xy}}{\sqrt{5z}}$$

To rationalize the denominator, we multiply the numerator and denominator by $\sqrt{5z}$:

$$\frac{2x^2y\sqrt{3xy}}{\sqrt{5z}} \cdot \frac{\sqrt{5z}}{\sqrt{5z}} = \frac{2x^2y\sqrt{15xyz}}{\sqrt{(5z)^2}}$$

$$= \frac{2x^2y\sqrt{15xyz}}{5z}$$

Square Root of a Perfect Square

So far in this chapter, we have assumed that all our variables are nonnegative when they appear under a square root symbol. There are times, however, when this is not the case.

Consider the following two statements:

$$\sqrt{3^2} = \sqrt{9} = 3 \qquad \text{and} \qquad \sqrt{(-3)^2} = \sqrt{9} = 3$$

Whether we operate on 3 or -3, the result is the same: Both expressions simplify to 3. The other operation we have worked with in the past that produces the same result is absolute value. That is,

$$|3| = 3 \qquad \text{and} \qquad |-3| = 3$$

This leads us to the next property of radicals.

PROPERTY *Property 3 for Radicals*

If a is a real number, then $\sqrt{a^2} = |a|$.

The result of this discussion and Property 3 is simply this:

If we know a is positive, then $\sqrt{a^2} = a$.

If we know a is negative, then $\sqrt{a^2} = |a|$.

If we don't know if a is positive or negative, then $\sqrt{a^2} = |a|$.

EXAMPLES Simplify each expression. Do *not* assume the variables represent positive numbers.

11. $\sqrt{9x^2} = 3|x|$

12. $\sqrt{x^3} = |x|\sqrt{x}$

13. $\sqrt{x^2 - 6x + 9} = \sqrt{(x - 3)^2} = |x - 3|$

14. $\sqrt{x^3 - 5x^2} = \sqrt{x^2(x - 5)} = |x|\sqrt{x - 5}$

As you can see, we must use absolute value symbols when we take a square root of a perfect square, unless we know the base of the perfect square is a positive number. The same idea holds for higher even roots, but not for odd roots. With odd roots, no absolute value symbols are necessary.

EXAMPLES Simplify each expression.

15 $\sqrt[3]{(-2)^3} = \sqrt[3]{-8} = -2$

16. $\sqrt[3]{(-5)^3} = \sqrt[3]{-125} = -5$

We can extend this discussion to all roots as follows:

PROPERTY *Extending Property 3 for Radicals*

If a is a real number, then

$$\sqrt[n]{a^n} = |a| \quad \text{if} \quad n \text{ is even}$$

$$\sqrt[n]{a^n} = a \quad \text{if} \quad n \text{ is odd}$$

The Spiral of Roots

To visualize the square roots of the positive integers, we can construct the spiral of roots that we mentioned in the introduction to this section. To begin, we draw two line segments, each of length 1, at right angles to each other. Then we use the Pythagorean theorem to find the length of the diagonal. Figure 1 illustrates this procedure.

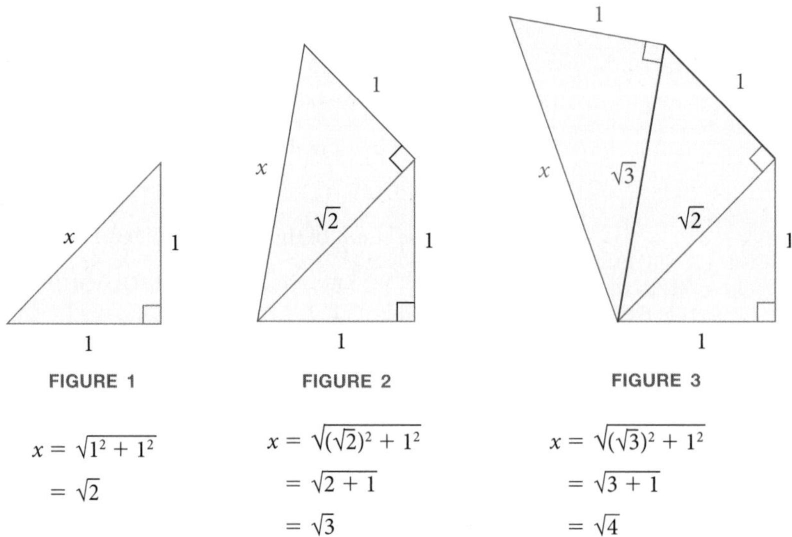

FIGURE 1

$$x = \sqrt{1^2 + 1^2} = \sqrt{2}$$

FIGURE 2

$$x = \sqrt{(\sqrt{2})^2 + 1^2} = \sqrt{2 + 1} = \sqrt{3}$$

FIGURE 3

$$x = \sqrt{(\sqrt{3})^2 + 1^2} = \sqrt{3 + 1} = \sqrt{4}$$

Next, we construct a second triangle by connecting a line segment of length 1 to the end of the first diagonal so that the angle formed is a right angle. We find the length of the second diagonal using the Pythagorean theorem. Figure 2 illustrates this procedure. Continuing to draw new triangles by connecting line segments of length 1 to the end of each new diagonal, so that the angle formed is a right angle, the spiral of roots begins to appear (Figure 3).

The Spiral of Roots and Function Notation

Looking over the diagrams and calculations in the preceding discussion, we see that each diagonal in the spiral of roots is found by using the length of the previous diagonal.

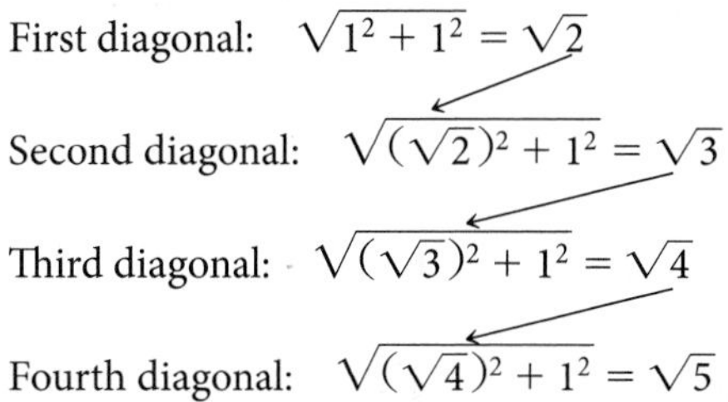

First diagonal: $\sqrt{1^2 + 1^2} = \sqrt{2}$
Second diagonal: $\sqrt{(\sqrt{2})^2 + 1^2} = \sqrt{3}$
Third diagonal: $\sqrt{(\sqrt{3})^2 + 1^2} = \sqrt{4}$
Fourth diagonal: $\sqrt{(\sqrt{4})^2 + 1^2} = \sqrt{5}$

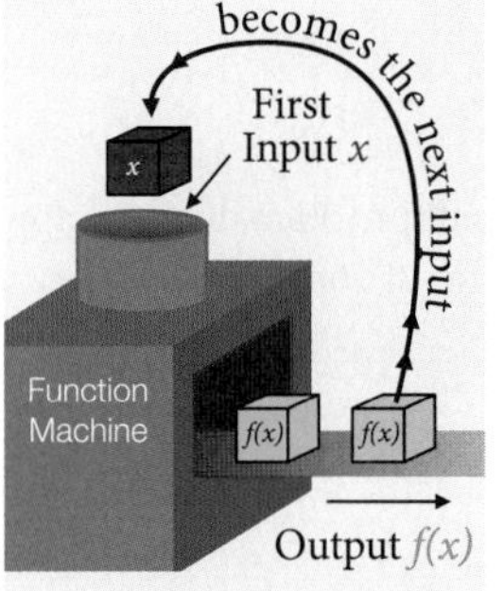

A process like this one, in which the answer to one calculation is used to find the answer to the next calculation, is called a *recursive* process. In this particular case, we can use function notation to model the process. If we let x represent the length of any diagonal, then the length of the next diagonal is given by

$$f(x) = \sqrt{x^2 + 1}$$

To begin the process of finding the diagonals, we let $x = 1$:

$$f(1) = \sqrt{1^2 + 1} = \sqrt{2}$$

To find the next diagonal, we substitue $\sqrt{2}$ for x to obtain

$$f[f(1)] = f(\sqrt{2}) = \sqrt{(\sqrt{2})^2 + 1} = \sqrt{3}$$

$$f(f[f(1)]) = f(\sqrt{3}) = \sqrt{(\sqrt{3})^2 + 1} = \sqrt{4}$$

We can describe this process of finding the diagonals of the spiral of roots very concisely this way:

$$f(1), f[f(1)], f(f[f(1)]), \ldots \qquad \text{where } f(x) = \sqrt{x^2 + 1}$$

This sequence of function values is a special case of a general category of similar sequences that are closely connected to *fractals* and *chaos*, two topics in mathematics that are currently receiving a good deal of attention.

USING TECHNOLOGY

As our preceding discussion indicates, the length of each diagonal in the spiral of roots is used to calculate the length of the next diagonal. The [ANS] key on a graphing calculator can be used effectively in a situation like this. To begin, we store the number 1 in the variable ANS. Next, we key in the fomula used to produce each diagonal using ANS for the variable. After that, it is simply a matter of pressing [ENTER], as many times as we like, to produce the lengths of as many diagonals as we like. Here is a summary of what we do:

Enter This	*Display Shows*
1 [ENTER]	1.000
$\sqrt{\ }$ (ANS^2 + 1) [ENTER]	1.414
[ENTER]	1.732
[ENTER]	2.000
[ENTER]	2.236

If you continue to press the [ENTER] key, you will produce decimal approximations for as many of the diagonals in the spiral of roots as you like.

GETTING READY FOR CLASS

After reading through the preceding section, respond in your own words and in complete sentences.

A. Explain why this statement is false: "The square root of a sum is the sum of the square roots."

B. What is simplified form for an expression that contains a square root?

C. Why is it not necessarily true that $\sqrt{a^2} = a$?

D. What does it mean to rationalize the denominator in an expression?

Use Property 1 for radicals to write each of the following expressions in simplified form. (Assume all variables are nonnegative through Problem 70.)

1. $\sqrt{8}$ **2.** $\sqrt{32}$ **3.** $\sqrt{98}$ **4.** $\sqrt{75}$

5. $\sqrt{288}$ **6.** $\sqrt{128}$ **7.** $\sqrt{80}$ **8.** $\sqrt{200}$

9. $\sqrt{48}$ **10.** $\sqrt{27}$ **11.** $\sqrt{675}$ **12.** $\sqrt{972}$

13. $\sqrt[3]{54}$ **14.** $\sqrt[3]{24}$ **15.** $\sqrt[3]{128}$ **16.** $\sqrt[3]{162}$

17. $\sqrt[3]{432}$ **18.** $\sqrt[3]{1{,}536}$ **19.** $\sqrt[5]{64}$ **20.** $\sqrt[4]{48}$

21. $\sqrt{18x^3}$ **22.** $\sqrt{27x^5}$ **23.** $\sqrt[4]{32y^7}$ **24.** $\sqrt[5]{32y^7}$

25. $\sqrt[3]{40x^4y^7}$ **26.** $\sqrt[3]{128x^6y^2}$ **27.** $\sqrt{48a^2b^3c^4}$ **28.** $\sqrt{72a^4b^3c^2}$

29. $\sqrt[3]{48a^2b^3c^4}$ **30.** $\sqrt[3]{72a^4b^3c^2}$ **31.** $\sqrt[5]{64x^8y^{12}}$ **32.** $\sqrt[4]{32x^9y^{10}}$

33. $\sqrt[5]{243x^7y^{10}z^5}$ **34.** $\sqrt[5]{64x^8y^4z^{11}}$

Substitute the given numbers into the expression $\sqrt{b^2 - 4ac}$, and then simplify.

35. $a = 2, b = -6, c = 3$ **36.** $a = 6, b = 7, c = -5$

37. $a = 1, b = 2, c = 6$ **38.** $a = 2, b = 5, c = 3$

39. $a = \frac{1}{2}, b = -\frac{1}{2}, c = -\frac{5}{4}$ **40.** $a = \frac{7}{4}, b = -\frac{3}{4}, c = -2$

41. Simplify each expression.

a. $\frac{\sqrt{20}}{4}$ **b.** $\frac{3\sqrt{20}}{15}$ **c.** $\frac{4 + \sqrt{12}}{2}$ **d.** $\frac{2 + \sqrt{9}}{5}$

42. Simplify each expression.

a. $\frac{\sqrt{12}}{4}$ **b.** $\frac{2\sqrt{32}}{8}$ **c.** $\frac{9 + \sqrt{27}}{3}$ **d.** $\frac{-6 - \sqrt{64}}{2}$

43. Simplify each expression.

a. $\frac{10 + \sqrt{75}}{5}$ **b.** $\frac{-6 + \sqrt{45}}{3}$ **c.** $\frac{-2 - \sqrt{27}}{6}$

44. Simplify each expression.

a. $\frac{12 - \sqrt{12}}{6}$ **b.** $\frac{-4 - \sqrt{8}}{2}$ **c.** $\frac{6 - \sqrt{48}}{8}$

Rationalize the denominator in each of the following expressions

45. $\frac{2}{\sqrt{3}}$ **46.** $\frac{3}{\sqrt{2}}$ **47.** $\frac{5}{\sqrt{6}}$ **48.** $\frac{7}{\sqrt{5}}$

49. $\sqrt{\frac{1}{2}}$ **50.** $\sqrt{\frac{1}{3}}$ **51.** $\sqrt{\frac{1}{5}}$ **52.** $\sqrt{\frac{1}{6}}$

53. $\frac{4}{\sqrt[3]{2}}$ **54.** $\frac{5}{\sqrt[3]{3}}$ **55.** $\frac{2}{\sqrt[3]{9}}$ **56.** $\frac{3}{\sqrt[3]{4}}$

57. $\sqrt[4]{\frac{3}{2x^2}}$ **58.** $\sqrt[4]{\frac{5}{3x^2}}$ **59.** $\sqrt[4]{\frac{8}{y}}$ **60.** $\sqrt[4]{\frac{27}{y}}$

61. $\sqrt[3]{\frac{4x}{3y}}$ **62.** $\sqrt[3]{\frac{7x}{6y}}$ **63.** $\sqrt[3]{\frac{2x}{9y}}$ **64.** $\sqrt[3]{\frac{5x}{4y}}$

1. $2\sqrt{2}$
2. $4\sqrt{2}$
3. $7\sqrt{2}$
4 $5\sqrt{3}$
5. $12\sqrt{2}$
6. $8\sqrt{2}$
7. $4\sqrt{5}$
8. $10\sqrt{2}$
9. $4\sqrt{3}$
10. $3\sqrt{3}$
11. $15\sqrt{3}$
12. $18\sqrt{3}$
13. $3\sqrt[3]{2}$
14. $2\sqrt[3]{3}$
15. $4\sqrt[3]{2}$
16. $3\sqrt[3]{6}$
17. $6\sqrt[3]{2}$
18. $8\sqrt[3]{3}$
19. $2\sqrt[5]{2}$
20. $2\sqrt[4]{3}$
21. $3x\sqrt{2x}$
22. $3x^2\sqrt{3x}$
23. $2y\sqrt[4]{2y^3}$
24. $2y\sqrt[5]{y^2}$
25. $2xy^2\sqrt[3]{5xy}$
26. $4x^2\sqrt[3]{2y^2}$
27. $4abc^2\sqrt{3b}$
28. $6a^2bc\sqrt{2b}$
29. $2bc\sqrt[3]{6a^2c}$
30. $2ab\sqrt[3]{9ac^2}$
31. $2xy^2\sqrt[5]{2x^3y^2}$
32. $2x^2y^2\sqrt[4]{2xy^2}$
33. $3xy^2z\sqrt[5]{x^2}$
34. $2xz^2\sqrt[5]{2x^3y^4z}$
35. $2\sqrt{3}$
36. 13
37. $\sqrt{-20}$; not real number
38. 1
39. $\frac{\sqrt{11}}{2}$
40. $\frac{\sqrt{233}}{4}$
41. a. $\frac{\sqrt{5}}{2}$
b. $\frac{2\sqrt{5}}{5}$
c. $2 + \sqrt{3}$
d. 1
42. a. $\frac{\sqrt{3}}{2}$
b. $\sqrt{2}$
c. $3 + \sqrt{3}$
d. -7
43. a. $2 + \sqrt{3}$
b. $-2 + \sqrt{5}$
c. $\frac{-2 - 3\sqrt{3}}{6}$
44. a. $\frac{6 - \sqrt{3}}{3}$
b. $-2 - \sqrt{2}$
c. $\frac{3 - 2\sqrt{3}}{4}$

45. $\frac{2\sqrt{3}}{3}$
46. $\frac{3\sqrt{2}}{2}$
47. $\frac{5\sqrt{6}}{6}$
48. $\frac{7\sqrt{5}}{5}$
49. $\frac{\sqrt{2}}{2}$
50. $\frac{\sqrt{3}}{3}$
51. $\frac{\sqrt{5}}{5}$
52. $\frac{\sqrt{6}}{6}$
53. $2\sqrt[3]{4}$
54. $\frac{5\sqrt[3]{9}}{3}$
55. $\frac{2\sqrt[3]{3}}{3}$
56. $\frac{3\sqrt[3]{2}}{2}$
57. $\frac{\sqrt[4]{24x^2}}{2x}$
58. $\frac{\sqrt[4]{135x^2}}{3x}$
59. $\frac{\sqrt[4]{8y^3}}{y}$
60. $\frac{\sqrt[4]{27y^3}}{y}$
61. $\frac{\sqrt[3]{36xy^2}}{3y}$
62. $\frac{\sqrt[3]{252xy^2}}{6y}$
63. $\frac{\sqrt[3]{6xy^2}}{3y}$
64. $\frac{\sqrt[3]{10xy^2}}{2y}$
65. $\frac{3x\sqrt{15xy}}{5y}$
66. $\frac{2x^2\sqrt{21xy}}{7y}$
67. $\frac{5xy\sqrt{6xz}}{2z}$
68. $\frac{5xy\sqrt{6yz}}{3z}$
69. a. $\frac{\sqrt{2}}{2}$ b. $\frac{\sqrt[3]{4}}{2}$ c. $\frac{\sqrt[4]{8}}{2}$
70. a. $\frac{\sqrt{3}}{3}$ b. $\frac{\sqrt[3]{3}}{3}$ c. $\frac{\sqrt[4]{3}}{3}$
71. $5|x|$
72. $7|x|$
73. $3|xy|\sqrt{3x}$
74. $2|xy|\sqrt{10x}$
75. $|x-5|$
76. $|x-8|$
77. $|2x+3|$
78. $|4x+5|$
79. $2|a(a+2)|$
80. $3|a(a+1)|$
81. $2|x|\sqrt{x-2}$
82. $3|x|\sqrt{2x-1}$

Write each of the following in simplified form.

65. $\sqrt{\frac{27x^3}{5y}}$ **66.** $\sqrt{\frac{12x^5}{7y}}$ **67.** $\sqrt{\frac{75x^3y^2}{2z}}$ **68.** $\sqrt{\frac{50x^2y^3}{3z}}$

Rationalize the denominator.

69. a. $\frac{1}{\sqrt{2}}$ **b.** $\frac{1}{\sqrt[3]{2}}$ **c.** $\frac{1}{\sqrt[4]{2}}$

70. a. $\frac{1}{\sqrt{3}}$ **b.** $\frac{1}{\sqrt[3]{9}}$ **c.** $\frac{1}{\sqrt[4]{27}}$

Simplify each expression. Do *not* assume the variables represent positive numbers.

71. $\sqrt{25x^2}$ **72.** $\sqrt{49x^2}$ **73.** $\sqrt{27x^3y^2}$ **74.** $\sqrt{40x^3y^2}$

75. $\sqrt{x^2-10x+25}$ **76.** $\sqrt{x^2-16x+64}$

77. $\sqrt{4x^2+12x+9}$ **78.** $\sqrt{16x^2+40x+25}$

79. $\sqrt{4a^4+16a^3+16a^2}$ **80.** $\sqrt{9a^4+18a^3+9a^2}$

81. $\sqrt{4x^3-8x^2}$ **82.** $\sqrt{18x^3-9x^2}$

83. Show that the statement $\sqrt{a+b}=\sqrt{a}+\sqrt{b}$ is not true by replacing a with 9 and b with 16 and simplifying both sides.

84. Find a pair of values for a and b that will make the statement $\sqrt{a+b}=\sqrt{a}+\sqrt{b}$ true.

Applying the Concepts

85. Diagonal Distance The distance d between opposite corners of a rectangular room with length l and width w is given by

$$d=\sqrt{l^2+w^2}$$

How far is it between opposite corners of a living room that measures 10 by 15 feet?

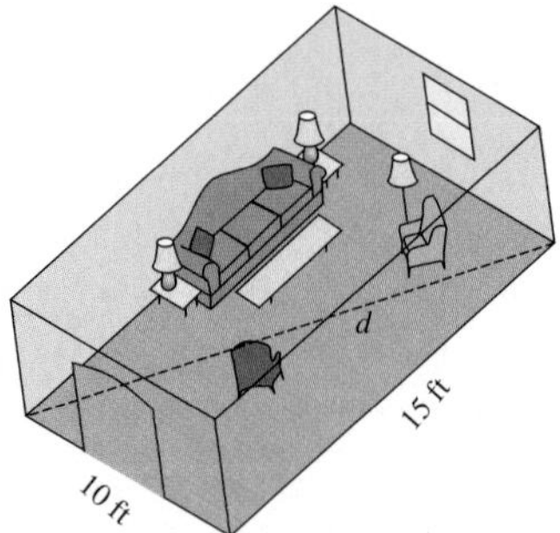

86. Radius of a Sphere The radius r of a sphere with volume V can be found by using the formula

$$r=\sqrt[3]{\frac{3V}{4\pi}}$$

Find the radius of a sphere with volume 9 cubic feet. Write your answer in simplified form. (Use $\frac{22}{7}$ for π.)

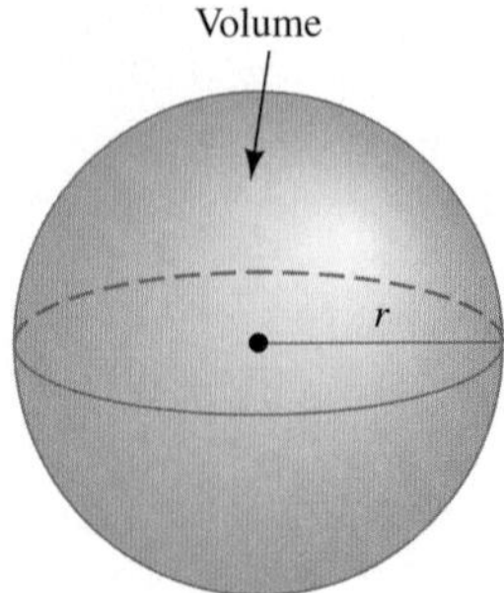

87. Distance to the Horizon If you are at a point k miles above the surface of the Earth, the distance you can see, in miles, is approximated by the equation $d=\sqrt{8000k+k^2}$.

83. $\sqrt{9+16} \stackrel{?}{=} \sqrt{9} + \sqrt{16}$
$\sqrt{25} \stackrel{?}{=} 3 + 4$
$5 \neq 7$
84. 0,0; 0,1; 1,0
85. $5\sqrt{13}$ feet
86. $\frac{3\sqrt[3]{847}}{22}$ feet
87. a. $\approx$ 89.4 miles
b. $\approx$ 126.5 miles
c. $\approx$ 154.9 miles
88. 7.2%
91. $\sqrt{2}, \sqrt{3}, 2, \sqrt{5}, \sqrt{6}, \sqrt{7}$;
$a_{10} = \sqrt{11}$; $a_{100} = \sqrt{101}$
92. $2\sqrt{2}, 2\sqrt{3}, 4, 2\sqrt{5}, 2\sqrt{6}, 2\sqrt{7}$;
$a_{10} = 2\sqrt{11}$; $a_{100} = 2\sqrt{101}$
93. $7x$
94. $13x$
95. $27xy^2$
96. $53a^2b$
97. $\frac{5}{6}x$
98. $\frac{31}{24}x$
99. $3\sqrt{2}$
100. $2\sqrt{2}$
101. $5y\sqrt{3xy}$
102. $2\sqrt{3xy}$
103. $2a\sqrt[3]{ab^2}$
104. $3\sqrt[3]{ab^2}$

a. How far can you see from a point that is 1 mile above the surface of the Earth?

b. How far can you see from a point that is 2 miles above the surface of the Earth?

c. How far can you see from a point that is 3 miles above the surface of the Earth?

88. Investing If you invest P dollars and you want the investment to grow to A dollars in t years, the interest rate that must be earned if interest is compounded annually is given by the formula

$$r = \sqrt[t]{\frac{A}{P}} - 1.$$

If you invest \$4,000 and want to have \$7,000 in 8 years, what interest rate must be earned?

89. Spiral of Roots Construct your own spiral of roots by using a ruler. Draw the first triangle by using two 1-inch lines. The first diagonal will have a length of $\sqrt{2}$ inches. Each new triangle will be formed by drawing a 1-inch line segment at the end of the previous diagonal so the angle formed is 90°.

90. Spiral of Roots Construct a spiral of roots by using line segments of length 2 inches. The length of the first diagonal will be $2\sqrt{2}$ inches. The length of the second diagonal will be $2\sqrt{3}$ inches.

91. Spiral of Roots If $f(x) = \sqrt{x^2 + 1}$, find the first six terms in the following sequence. Use your results to predict the value of the 10th term and the 100th term.

$$f(1), f[f(1)], f(f[f(1)]), \ldots$$

92. Spiral of Roots If $f(x) = \sqrt{x^2 + 4}$, find the first six terms in the following sequence. Use your results to predict the value of the 10th term and the 100th term. (The numbers in this sequence are the lengths of the diagonals of the spiral you drew in Problem 90.)

$$f(2), f[f(2)], f(f[f(2)]), \ldots$$

Getting Ready for the Next Section

Simplify the following.

93. $5x - 4x + 6x$

94. $12x + 8x - 7x$

95. $35xy^2 - 8xy^2$

96. $20a^2b + 33a^2b$

97. $\frac{1}{2}x + \frac{1}{3}x$

98. $\frac{2}{3}x + \frac{5}{8}x$

Write in simplified form for radicals.

99. $\sqrt{18}$

100. $\sqrt{8}$

101. $\sqrt{75xy^3}$

102. $\sqrt{12xy}$

103. $\sqrt[3]{8a^4b^2}$

104. $\sqrt[3]{27ab^2}$

Addition and Subtraction of Radical Expressions 9.3

In Chapter 1, we found we could add similar terms when combining polynomials. The same idea applies to addition and subtraction of radical expressions.

DEFINITION *similar radicals*

Two radicals are said to be ***similar radicals*** if they have the same index and the same radicand.

The expressions $5\sqrt[3]{7}$ and $-8\sqrt[3]{7}$ are similar since the index is ^{3} in both cases and the radicands are 7. The expressions $3\sqrt[4]{5}$ and $7\sqrt[3]{5}$ are not similar because they have different indices, and the expressions $2\sqrt[5]{8}$ and $3\sqrt[5]{9}$ are not similar because the radicands are not the same.

We add and subtract radical expressions in the same way we add and subtract polynomials—by combining similar terms under the distributive property.

EXAMPLE 1 Combine $5\sqrt{3} - 4\sqrt{3} + 6\sqrt{3}$.

SOLUTION All three radicals are similar. We apply the distributive property to get

$$\begin{aligned} 5\sqrt{3} - 4\sqrt{3} + 6\sqrt{3} &= (5 - 4 + 6)\sqrt{3} \\ &= 7\sqrt{3} \end{aligned}$$

EXAMPLE 2 Combine $3\sqrt{8} + 5\sqrt{18}$.

SOLUTION The two radicals do not seem to be similar. We must write each in simplified form before applying the distributive property.

$$\begin{aligned} 3\sqrt{8} + 5\sqrt{18} &= 3\sqrt{4 \cdot 2} + 5\sqrt{9 \cdot 2} \\ &= 3\sqrt{4}\,\sqrt{2} + 5\sqrt{9}\,\sqrt{2} \\ &= 3 \cdot 2\,\sqrt{2} + 5 \cdot 3\,\sqrt{2} \\ &= 6\,\sqrt{2} + 15\,\sqrt{2} \\ &= (6 + 15)\,\sqrt{2} \\ &= 21\,\sqrt{2} \end{aligned}$$

The result of Example 2 can be generalized to the following rule for sums and differences of radical expressions.

RULE

To add or subtract radical expressions, put each in simplified form and apply the distributive property, if possible. We can add only similar radicals. We must write each expression in simplified form for radicals before we can tell if the radicals are similar.

EXAMPLE 3 Combine $7\sqrt{75xy^3} - 4y\sqrt{12xy}$, where $x, y \geq 0$.

SOLUTION We write each expression in simplified form and combine similar radicals:

$$\begin{aligned}7\sqrt{75xy^3} - 4y\sqrt{12xy} &= 7\sqrt{25y^2}\,\sqrt{3xy} - 4y\sqrt{4}\,\sqrt{3xy}\\ &= 35y\sqrt{3xy} - 8y\sqrt{3xy}\\ &= (35y - 8y)\sqrt{3xy}\\ &= 27y\sqrt{3xy}\end{aligned}$$

EXAMPLE 4 Combine $10\sqrt[3]{8a^4b^2} + 11a\sqrt[3]{27ab^2}$.

SOLUTION Writing each radical in simplified form and combining similar terms, we have

$$\begin{aligned}10\sqrt[3]{8a^4b^2} + 11a\sqrt[3]{27ab^2} &= 10\sqrt[3]{8a^3}\,\sqrt[3]{ab^2} + 11a\sqrt[3]{27}\,\sqrt[3]{ab^2}\\ &= 20a\sqrt[3]{ab^2} + 33a\sqrt[3]{ab^2}\\ &= 53a\sqrt[3]{ab^2}\end{aligned}$$

EXAMPLE 5 Combine $\frac{\sqrt{3}}{2} + \frac{1}{\sqrt{3}}$.

SOLUTION We begin by writing the second term in simplified form.

$$\begin{aligned}\frac{\sqrt{3}}{2} + \frac{1}{\sqrt{3}} &= \frac{\sqrt{3}}{2} + \frac{1}{\sqrt{3}} \cdot \frac{\sqrt{3}}{\sqrt{3}}\\ &= \frac{\sqrt{3}}{2} + \frac{\sqrt{3}}{3}\\ &= \frac{1}{2}\sqrt{3} + \frac{1}{3}\sqrt{3}\\ &= \left(\frac{1}{2} + \frac{1}{3}\right)\sqrt{3}\\ &= \frac{5}{6}\sqrt{3} = \frac{5\sqrt{3}}{6}\end{aligned}$$

EXAMPLE 6 Construct a golden rectangle from a square of side 4. Then show that the ratio of the length to the width is the golden ratio $\frac{1+\sqrt{5}}{2}$.

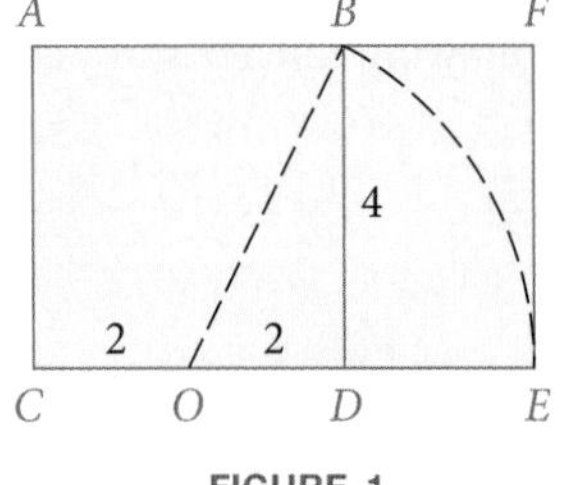

FIGURE 1

SOLUTION Figure 1 shows the golden rectangle constructed from a square of side 4. The length of the diagonal OB is found from the Pythagorean theorem.

$$OB = \sqrt{2^2 + 4^2} = \sqrt{4 + 16} = \sqrt{20} = 2\sqrt{5}$$

The ratio of the length to the width for the rectangle is the golden ratio.

$$\text{Golden ratio} = \frac{CE}{EF} = \frac{2 + 2\sqrt{5}}{4} = \frac{2(1 + \sqrt{5})}{2 \cdot 2} = \frac{1 + \sqrt{5}}{2}$$

As you can see, showing that the ratio of length to width in this rectangle is the golden ratio depends on our ability to write $\sqrt{20}$ as $2\sqrt{5}$ and our ability to reduce to lowest terms by factoring and then dividing out the common factor 2 from the numerator and denominator.

GETTING READY FOR CLASS

After reading through the preceding section, respond in your own words and in complete sentences.

A. What are similar radicals?
B. When can we add two radical expressions?
C. What is the first step when adding or subtracting expressions containing radicals?
D. What is the golden ratio, and where does it come from?

Problem Set 9.3

Combine the following expressions. (Assume any variables under an even root are nonnegative.)

1. $3\sqrt{5} + 4\sqrt{5}$
2. $6\sqrt{3} - 5\sqrt{3}$
3. $3x\sqrt{7} - 4x\sqrt{7}$
4. $6y\sqrt{a} + 7y\sqrt{a}$
5. $5\sqrt[3]{10} - 4\sqrt[3]{10}$
6. $6\sqrt[4]{2} + 9\sqrt[4]{2}$
7. $8\sqrt[5]{6} - 2\sqrt[5]{6} + 3\sqrt[5]{6}$
8. $7\sqrt[6]{7} - \sqrt[6]{7} + 4\sqrt[6]{7}$
9. $3x\sqrt{2} - 4x\sqrt{2} + x\sqrt{2}$
10. $5x\sqrt{6} - 3x\sqrt{6} - 2x\sqrt{6}$
11. $\sqrt{20} - \sqrt{80} + \sqrt{45}$
12. $\sqrt{8} - \sqrt{32} - \sqrt{18}$
13. $4\sqrt{8} - 2\sqrt{50} - 5\sqrt{72}$
14. $\sqrt{48} - 3\sqrt{27} + 2\sqrt{75}$
15. $5x\sqrt{8} + 3\sqrt{32x^2} - 5\sqrt{50x^2}$
16. $2\sqrt{50x^2} - 8x\sqrt{18} - 3\sqrt{72x^2}$
17. $5\sqrt[3]{16} - 4\sqrt[3]{54}$
18. $\sqrt[3]{81} + 3\sqrt[3]{24}$
19. $\sqrt[3]{x^4y^2} + 7x\sqrt[3]{xy^2}$
20. $2\sqrt[3]{x^8y^6} - 3y^2\sqrt[3]{8x^8}$
21. $5a^2\sqrt{27ab^3} - 6b\sqrt{12a^5b}$
22. $9a\sqrt{20a^3b^2} + 7b\sqrt{45a^5}$
23. $b\sqrt[3]{24a^5b} + 3a\sqrt[3]{81a^2b^4}$
24. $7\sqrt[3]{a^4b^3c^2} - 6ab\sqrt[3]{ac^2}$
25. $5x\sqrt[4]{3y^5} + y\sqrt[4]{243x^4y} + \sqrt[4]{48x^4y^5}$
26. $x\sqrt[4]{5xy^8} + y\sqrt[4]{405x^5y^4} + y^2\sqrt[4]{80x^5}$
27. $\dfrac{\sqrt{2}}{2} + \dfrac{1}{\sqrt{2}}$
28. $\dfrac{\sqrt{3}}{3} + \dfrac{1}{\sqrt{3}}$
29. $\dfrac{\sqrt{5}}{3} + \dfrac{1}{\sqrt{5}}$
30. $\dfrac{\sqrt{6}}{2} + \dfrac{1}{\sqrt{6}}$
31. $\sqrt{x} - \dfrac{1}{\sqrt{x}}$
32. $\sqrt{x} + \dfrac{1}{\sqrt{x}}$
33. $\dfrac{\sqrt{18}}{6} + \sqrt{\dfrac{1}{2}} + \dfrac{\sqrt{2}}{2}$
34. $\dfrac{\sqrt{12}}{6} + \sqrt{\dfrac{1}{3}} + \dfrac{\sqrt{3}}{3}$
35. $\sqrt{6} - \sqrt{\dfrac{2}{3}} + \sqrt{\dfrac{1}{6}}$
36. $\sqrt{15} - \sqrt{\dfrac{3}{5}} + \sqrt{\dfrac{5}{3}}$
37. $\sqrt[3]{25} + \dfrac{3}{\sqrt[3]{5}}$
38. $\sqrt[4]{8} + \dfrac{1}{\sqrt[4]{2}}$
39. Use a calculator to find a decimal approximation for $\sqrt{12}$ and for $2\sqrt{3}$.
40. Use a calculator to find decimal approximations for $\sqrt{50}$ and $5\sqrt{2}$.
41. Use a calculator to find a decimal approximation for $\sqrt{8} + \sqrt{18}$. Is it equal to the decimal approximation for $\sqrt{26}$ or $\sqrt{50}$?
42. Use a calculator to find a decimal approximation for $\sqrt{3} + \sqrt{12}$. Is it equal to the decimal approximation for $\sqrt{15}$ or $\sqrt{27}$?

Each of the following statements is false. Correct the right side of each one to make the statement true.

43. $3\sqrt{2x} + 5\sqrt{2x} = 8\sqrt{4x}$
44. $5\sqrt{3} - 7\sqrt{3} = -2\sqrt{9}$
45. $\sqrt{9 + 16} = 3 + 4$
46. $\sqrt{36 + 64} = 6 + 8$

1. $7\sqrt{5}$
2. $\sqrt{3}$
3. $-x\sqrt{7}$
4. $13y\sqrt{a}$
5. $\sqrt[3]{10}$
6. $15\sqrt[4]{2}$
7. $9\sqrt[5]{6}$
8. $10\sqrt[6]{7}$
9. 0
10. 0
11. $\sqrt{5}$
12. $-5\sqrt{2}$
13. $-32\sqrt{2}$
14. $5\sqrt{3}$
15. $-3x\sqrt{2}$
16. $-32x\sqrt{2}$
17. $-2\sqrt[3]{2}$
18. $9\sqrt[3]{3}$
19. $8x\sqrt[3]{xy^2}$
20. $-4x^2y^2\sqrt[3]{x^2}$
21. $3a^2b\sqrt{3ab}$
22. $39a^2b\sqrt{5a}$
23. $11ab\sqrt[3]{3a^2b}$
24. $ab\sqrt[3]{ac^2}$
25. $10xy\sqrt[4]{3y}$
26. $6xy^2\sqrt[4]{5x}$
27. $\sqrt{2}$
28. $\dfrac{2\sqrt{3}}{3}$
29. $\dfrac{8\sqrt{5}}{15}$
30. $\dfrac{2\sqrt{6}}{3}$
31. $\dfrac{(x-1)\sqrt{x}}{x}$
32. $\dfrac{(x+1)\sqrt{x}}{x}$
33. $\dfrac{3\sqrt{2}}{2}$
34. $\sqrt{3}$
35. $\dfrac{5\sqrt{6}}{6}$
36. $\dfrac{17\sqrt{15}}{15}$
37. $\dfrac{8\sqrt[3]{25}}{5}$
38. $\dfrac{3\sqrt[4]{8}}{2}$
39. $\sqrt{12} \approx 3.464$; $2\sqrt{3} \approx 2(1.732) = 3.464$
40. $\sqrt{50} \approx 7.071$; $5\sqrt{2} \approx 7.071$
41. $\sqrt{8} + \sqrt{18} \approx 2.828 + 4.243 = 7.071$; $\sqrt{50} \approx 7.071$; $\sqrt{26} \approx 5.099$
42. $\sqrt{3} + \sqrt{12} \approx 1.732 + 3.464 = 5.196$; $\sqrt{15} \approx 3.873$; $\sqrt{27} \approx 5.196$
43. $8\sqrt{2x}$
44. $-2\sqrt{3}$
45. 5
46. 10

53. $\sqrt{2}$
54. $\frac{\sqrt{3}}{2}$

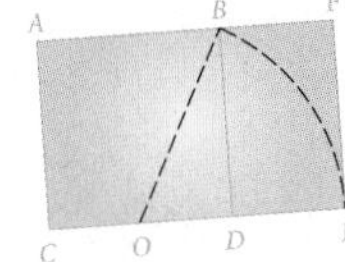

55. a. $\sqrt{2}:1 \approx 1.414:1$
b. $5:\sqrt{2}$
c. 5:4

Applying the Concepts

47. Golden Rectangle Construct a golden rectangle from a square of side 8. Then show that the ratio of the length to the width is the golden ratio $\frac{1+\sqrt{5}}{2}$.

48. Golden Rectangle Construct a golden rectangle from a square of side 10. Then show that the ratio of the length to the width is the golden ratio $\frac{1+\sqrt{5}}{2}$.

49. Golden Rectangle Use a ruler to construct a golden rectangle from a square of side 1 inch. Then show that the ratio of the length to the width is the golden ratio.

50. Golden Rectangle Use a ruler to construct a golden rectangle from a square of side $\frac{2}{3}$ inch. Then show that the ratio of the length to the width is the golden ratio.

51. Golden Rectangle To show that all golden rectangles have the same ratio of length to width, construct a golden rectangle from a square of side $2x$. Then show that the ratio of the length to the width is the golden ratio.

52. Golden Rectangle To show that all golden rectangles have the same ratio of length to width, construct a golden rectangle from a square of side x. Then show that the ratio of the length to the width is the golden ratio.

53. Isosceles Right Triangles A triangle is isosceles if it has two equal sides, and a triangle is a right triangle if it has a right angle in it. Sketch an isosceles right triangle, and find the ratio of the hypotenuse to a leg.

54. Equilateral Triangles A triangle is equilateral if it has three equal sides. The triangle in the figure is equilateral with each side of length $2x$. Find the ratio of the height to a side.

$2x$ $2x$

x x

55. Pyramids The following solid is called a regular square pyramid because its base is a square and all eight edges are the same length, 5. It is also true that the vertex, V, is directly above the center of the base.

a. Find the ratio of a diagonal of the base to the length of a side.

b. Find the ratio of the area of the base to the diagonal of the base.

c. Find the ratio of the area of the base to the perimeter of the base.

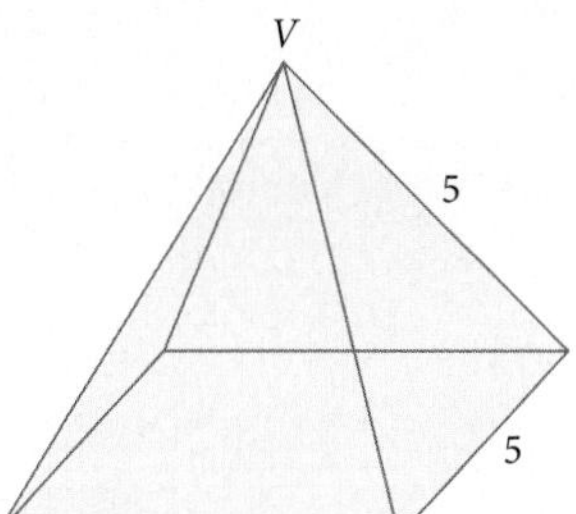

56. $\frac{\sqrt{6}}{3}$
57. 6
58. 35
59. $4x^2 + 3xy - y^2$
60. $2x^2 - xy - y^2$
61. $x^2 + 6x + 9$
62. $9x^2 - 12xy + 4y^2$
63. $x^2 - 4$
64. $4x^2 - 25$
65. $6\sqrt{2}$
66. 30
67. 6
68. 2
69. $9x$
70. $4y$
71. $\frac{\sqrt{6}}{2}$
72. $\frac{\sqrt{30}}{6}$

56. Pyramids Refer to this diagram of a square pyramid. Find the ratio of the height h of the pyramid to the altitude a.

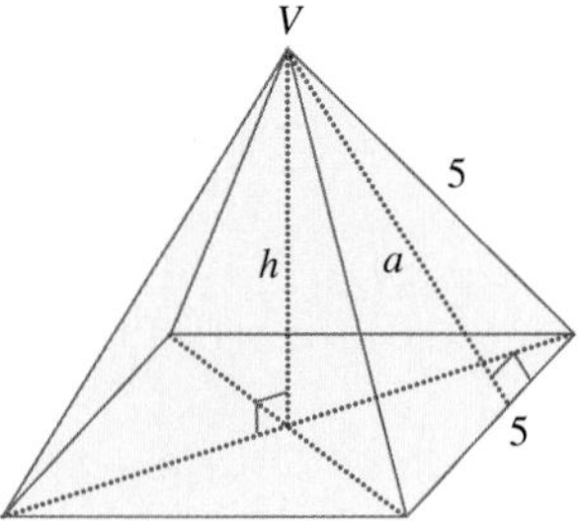

Getting Ready for the Next Section

Simplify the following.

57. $3 \cdot 2$

58. $5 \cdot 7$

59. $(x + y)(4x - y)$

60. $(2x + y)(x - y)$

61. $(x + 3)^2$

62. $(3x - 2y)^2$

63. $(x - 2)(x + 2)$

64. $(2x + 5)(2x - 5)$

Simplify the following expressions.

65. $2\sqrt{18}$

66. $5\sqrt{36}$

67. $(\sqrt{6})^2$

68. $(\sqrt{2})^2$

69. $(3\sqrt{x})^2$

70. $(2\sqrt{y})^2$

Rationalize the denominator.

71. $\frac{\sqrt{3}}{\sqrt{2}}$

72. $\frac{\sqrt{5}}{\sqrt{6}}$

Multiplication and Division of Radical Expressions

9.4

We have worked with the golden rectangle more than once in this chapter. The following is one such golden rectangle.

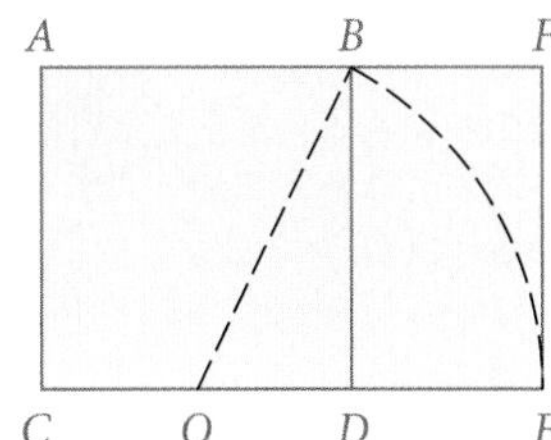

By now you know that, in any golden rectangle constructed from a square (of any size), the ratio of the length to the width will be

$$\frac{1 + \sqrt{5}}{2}$$

which we call the golden ratio. What is interesting is that the smaller rectangle on the right, *BFED*, is also a golden rectangle. We will use the mathematics developed in this section to confirm this fact.

In this section, we will look at multiplication and division of expressions that contain radicals. As you will see, multiplication of expressions that contain radicals is very similar to multiplication of polynomials. The division problems in this section are just an extension of the work we did previously when we rationalized denominators.

EXAMPLE 1 Multiply $(3\sqrt{5})(2\sqrt{7})$.

SOLUTION We can rearrange the order and grouping of the numbers in this product by applying the commutative and associative properties. Following this, we apply Property 1 for radicals and multiply:

$$(3\sqrt{5})(2\sqrt{7}) = (3 \cdot 2)(\sqrt{5}\sqrt{7})$$ Communicative and associative properties

$$= (3 \cdot 2)(\sqrt{5 \cdot 7})$$ Property 1 for radicals

$$= 6\sqrt{35}$$ Multiplication

In practice, it is not necessary to show the first two steps.

EXAMPLE 2 Multiply $\sqrt{3}(2\sqrt{6} - 5\sqrt{12})$.

SOLUTION Applying the distributive property, we have

$$\sqrt{3}(2\sqrt{6} - 5\sqrt{12}) = \sqrt{3} \cdot 2\sqrt{6} - \sqrt{3} \cdot 5\sqrt{12}$$

$$= 2\sqrt{18} - 5\sqrt{36}$$

Writing each radical in simplified form gives

$$2\sqrt{18} - 5\sqrt{36} = 2\sqrt{9}\sqrt{2} - 5\sqrt{36}$$

$$= 6\sqrt{2} - 30$$

EXAMPLE 3 Multiply $(\sqrt{3} + \sqrt{5})(4\sqrt{3} - \sqrt{5})$.

SOLUTION The same principle that applies when multiplying two binomials applies to this product. We must multiply each term in the first expression by each term in the second one. Any convenient method can be used. Let's use the FOIL method.

$$\begin{aligned}(\sqrt{3} + \sqrt{5})(4\sqrt{3} - \sqrt{5}) &= \overset{\text{F}}{\sqrt{3} \cdot 4\sqrt{3}} - \overset{\text{O}}{\sqrt{3}\cdot\sqrt{5}} + \overset{\text{I}}{\sqrt{5} \cdot 4\sqrt{3}} - \overset{\text{L}}{\sqrt{5}\cdot\sqrt{5}} \\ &= 4 \cdot 3 - \sqrt{15} + 4\sqrt{15} - 5 \\ &= 12 + 3\sqrt{15} - 5 \\ &= 7 + 3\sqrt{15}\end{aligned}$$

EXAMPLE 4 Expand and simplify $(\sqrt{x} + 3)^2$.

SOLUTION 1 We can write this problem as a multiplication problem and proceed as we did in Example 3:

$$\begin{aligned}(\sqrt{x} + 3)^2 &= (\sqrt{x} + 3)(\sqrt{x} + 3) \\ &= \overset{\text{F}}{\sqrt{x} \cdot \sqrt{x}} + \overset{\text{O}}{3\sqrt{x}} + \overset{\text{I}}{3\sqrt{x}} + \overset{\text{L}}{3 \cdot 3} \\ &= x + 3\sqrt{x} + 3\sqrt{x} + 9 \\ &= x + 6\sqrt{x} + 9\end{aligned}$$

SOLUTION 2 We can obtain the same result by applying the formula for the square of a sum: $(a + b)^2 = a^2 + 2ab + b^2$.

$$\begin{aligned}(\sqrt{x} + 3)^2 &= (\sqrt{x})^2 + 2(\sqrt{x})(3) + 3^2 \\ &= x + 6\sqrt{x} + 9\end{aligned}$$

EXAMPLE 5 Expand $(3\sqrt{x} - 2\sqrt{y})^2$ and simplify the result.

SOLUTION Let's apply the formula for the square of a difference, $(a - b)^2 = a^2 - 2ab + b^2$.

$$\begin{aligned}(3\sqrt{x} - 2\sqrt{y})^2 &= (3\sqrt{x})^2 - 2(3\sqrt{x})(2\sqrt{y}) + (2\sqrt{y})^2 \\ &= 9x - 12\sqrt{xy} + 4y\end{aligned}$$

EXAMPLE 6 Expand and simplify $(\sqrt{x + 2} - 1)^2$.

SOLUTION Applying the formula $(a - b)^2 = a^2 - 2ab + b^2$, we have

$$\begin{aligned}(\sqrt{x + 2} - 1)^2 &= (\sqrt{x + 2})^2 - 2\sqrt{x + 2}(1) + 1^2 \\ &= x + 2 - 2\sqrt{x + 2} + 1 \\ &= x + 3 - 2\sqrt{x + 2}\end{aligned}$$

EXAMPLE 7 Multiply $(\sqrt{6} + \sqrt{2})(\sqrt{6} - \sqrt{2})$.

SOLUTION We notice the product is of the form $(a + b)(a - b)$, which always gives the difference of two squares, $a^2 - b^2$:

$$\begin{aligned}(\sqrt{6} + \sqrt{2})(\sqrt{6} - \sqrt{2}) &= (\sqrt{6})^2 - (\sqrt{2})^2 \\ &= 6 - 2 \\ &= 4\end{aligned}$$

In Example 7, the two expressions $(\sqrt{6} + \sqrt{2})$ and $(\sqrt{6} - \sqrt{2})$ are called *conjugates*. In general, the conjugate of $\sqrt{a} + \sqrt{b}$ is $\sqrt{a} - \sqrt{b}$. If a and b are integers, multiplying conjugates of this form always produces a rational number. That is, if a and b are positive integers, then

$$\begin{aligned}(\sqrt{a} + \sqrt{b})(\sqrt{a} - \sqrt{b}) &= \sqrt{a}\sqrt{a} - \sqrt{a}\sqrt{b} + \sqrt{a}\sqrt{b} - \sqrt{b}\sqrt{b} \\ &= a - \sqrt{ab} + \sqrt{ab} - b \\ &= a - b\end{aligned}$$

which is rational if a and b are rational.

Division with radical expressions is the same as rationalizing the denominator. In Section 6.2, we were able to divide $\sqrt{3}$ by $\sqrt{2}$ by rationalizing the denominator:

$$\frac{\sqrt{3}}{\sqrt{2}} = \frac{\sqrt{3}}{\sqrt{2}} \cdot \frac{\sqrt{2}}{\sqrt{2}} = \frac{\sqrt{6}}{2}$$

We can accomplish the same result with expressions such as

$$\frac{6}{\sqrt{5} - \sqrt{3}}$$

by multiplying the numerator and denominator by the conjugate of the denominator.

EXAMPLE 8 Divide $\dfrac{6}{\sqrt{5} - \sqrt{3}}$. (Rationalize the denominator.)

SOLUTION Because the product of two conjugates is a rational number, we multiply the numerator and denominator by the conjugate of the denominator.

$$\begin{aligned}\frac{6}{\sqrt{5} - \sqrt{3}} &= \frac{6}{\sqrt{5} - \sqrt{3}} \cdot \frac{(\sqrt{5} + \sqrt{3})}{(\sqrt{5} + \sqrt{3})} \\ &= \frac{6\sqrt{5} + 6\sqrt{3}}{(\sqrt{5})^2 - (\sqrt{3})^2} \\ &= \frac{6\sqrt{5} + 6\sqrt{3}}{5 - 3} \\ &= \frac{6\sqrt{5} + 6\sqrt{3}}{2}\end{aligned}$$

The numerator and denominator of this last expression have a factor of 2 in common. We can reduce to lowest terms by factoring 2 from the numerator and then dividing both the numerator and denominator by 2:

$$\begin{aligned}&= \frac{2(3\sqrt{5} + 3\sqrt{3})}{2} \\ &= 3\sqrt{5} + 3\sqrt{3}\end{aligned}$$

EXAMPLE 9 Rationalize the denominator $\dfrac{\sqrt{5}-2}{\sqrt{5}+2}$.

SOLUTION To rationalize the denominator, we multiply the numerator and denominator by the conjugate of the denominator:

$$\begin{aligned}\frac{\sqrt{5}-2}{\sqrt{5}+2} &= \frac{\sqrt{5}-2}{\sqrt{5}+2}\cdot\frac{(\sqrt{5}-2)}{(\sqrt{5}-2)}\\ &= \frac{5-2\sqrt{5}-2\sqrt{5}+4}{(\sqrt{5})^2-2^2}\\ &= \frac{9-4\sqrt{5}}{5-4}\\ &= \frac{9-4\sqrt{5}}{1}\\ &= 9-4\sqrt{5}\end{aligned}$$

EXAMPLE 10 A golden rectangle constructed from a square of side 2 is shown in Figure 1. Show that the smaller rectangle *BDEF* is also a golden rectangle by finding the ratio of its length to its width.

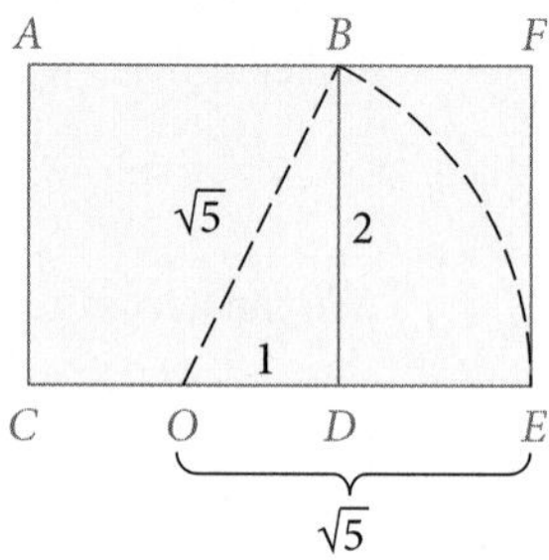

FIGURE 1

SOLUTION First, find expressions for the length and width of the smaller rectangle.

$$\text{Length} = EF = 2$$
$$\text{Width} = DE = \sqrt{5}-1$$

Next, we find the ratio of length to width.

$$\text{Ratio of length to width} = \frac{EF}{DE} = \frac{2}{\sqrt{5}-1}$$

To show that the small rectangle is a golden rectangle, we must show that the ratio of length to width is the golden ratio. We do so by rationalizing the denominator.

$$\begin{aligned}\frac{2}{\sqrt{5}-1} &= \frac{2}{\sqrt{5}-1}\cdot\frac{\sqrt{5}+1}{\sqrt{5}+1}\\ &= \frac{2(\sqrt{5}+1)}{5-1}\\ &= \frac{2(\sqrt{5}+1)}{4}\\ &= \frac{\sqrt{5}+1}{2} \qquad \text{Divide out common factor 2}\end{aligned}$$

Because addition is commutative, this last expression is the golden ratio. Therefore, the small rectangle in Figure 1 is a golden rectangle. ■

GETTING READY FOR CLASS

After reading through the preceding section, respond in your own words and in complete sentences.

A. Explain why $(\sqrt{5} + \sqrt{2})^2 \neq 5 + 2$.

B. Explain in words how you would rationalize the denominator in the expression $\frac{\sqrt{3}}{\sqrt{5} - \sqrt{2}}$.

C. What are conjugates?

D. What result is guaranteed when multiplying radical expressions that are conjugates?

Problem Set 9.4

Multiply. (Assume all expressions appearing under a square root symbol represent nonnegative numbers throughout this problem set.)

1. $\sqrt{6}\sqrt{3}$

2. $\sqrt{6}\sqrt{2}$

3. $(2\sqrt{3})(5\sqrt{7})$

4. $(3\sqrt{5})(2\sqrt{7})$

5. $(4\sqrt{6})(2\sqrt{15})(3\sqrt{10})$

6. $(4\sqrt{35})(2\sqrt{21})(5\sqrt{15})$

7. $(3\sqrt[3]{3})(6\sqrt[3]{9})$

8. $(2\sqrt[3]{2})(6\sqrt[3]{4})$

9. $\sqrt{3}(\sqrt{2} - 3\sqrt{3})$

10. $\sqrt{2}(5\sqrt{3} + 4\sqrt{2})$

11. $6\sqrt[3]{4}(2\sqrt[3]{2} + 1)$

12. $7\sqrt[3]{5}(3\sqrt[3]{25} - 2)$

13. $(\sqrt{3} + \sqrt{2})(3\sqrt{3} - \sqrt{2})$

14. $(\sqrt{5} - \sqrt{2})(3\sqrt{5} + 2\sqrt{2})$

15. $(\sqrt{x} + 5)(\sqrt{x} - 3)$

16. $(\sqrt{x} + 4)(\sqrt{x} + 2)$

17. $(3\sqrt{6} + 4\sqrt{2})(\sqrt{6} + 2\sqrt{2})$

18. $(\sqrt{7} - 3\sqrt{3})(2\sqrt{7} - 4\sqrt{3})$

19. $(\sqrt{3} + 4)^2$

20. $(\sqrt{5} - 2)^2$

21. $(\sqrt{x} - 3)^2$

22. $(\sqrt{x} + 4)^2$

23. $(2\sqrt{a} - 3\sqrt{b})^2$

24. $(5\sqrt{a} - 2\sqrt{b})^2$

25. $(\sqrt{x - 4} + 2)^2$

26. $(\sqrt{x - 3} + 2)^2$

27. $(\sqrt{x - 5} - 3)^2$

28. $(\sqrt{x - 3} - 4)^2$

29. $(\sqrt{3} - \sqrt{2})(\sqrt{3} + \sqrt{2})$

30. $(\sqrt{5} - \sqrt{2})(\sqrt{5} + \sqrt{2})$

31. $(\sqrt{a} + 7)(\sqrt{a} - 7)$

32. $(\sqrt{a} + 5)(\sqrt{a} - 5)$

33. $(5 - \sqrt{x})(5 + \sqrt{x})$

34. $(3 - \sqrt{x})(3 + \sqrt{x})$

35. $(\sqrt{x - 4} + 2)(\sqrt{x - 4} - 2)$

36. $(\sqrt{x + 3} + 5)(\sqrt{x + 3} - 5)$

37. $(\sqrt{3} + 1)^3$

38. $(\sqrt{5} - 2)^3$

Rationalize the denominator in each of the following.

39. $\dfrac{\sqrt{2}}{\sqrt{6} - \sqrt{2}}$

40. $\dfrac{\sqrt{5}}{\sqrt{5} + \sqrt{3}}$

41. $\dfrac{\sqrt{5}}{\sqrt{5} + 1}$

42. $\dfrac{\sqrt{7}}{\sqrt{7} - 1}$

43. $\dfrac{\sqrt{x}}{\sqrt{x} - 3}$

44. $\dfrac{\sqrt{x}}{\sqrt{x} + 2}$

45. $\dfrac{\sqrt{5}}{2\sqrt{5} - 3}$

46. $\dfrac{\sqrt{7}}{3\sqrt{7} - 2}$

47. $\dfrac{3}{\sqrt{x} - \sqrt{y}}$

48. $\dfrac{2}{\sqrt{x} + \sqrt{y}}$

49. $\dfrac{\sqrt{6} + \sqrt{2}}{\sqrt{6} - \sqrt{2}}$

50. $\dfrac{\sqrt{5} - \sqrt{3}}{\sqrt{5} + \sqrt{3}}$

51. $\dfrac{\sqrt{7} - 2}{\sqrt{7} + 2}$

52. $\dfrac{\sqrt{11} + 3}{\sqrt{11} - 3}$

53. Work each problem according to the instructions given.

a. Add: $(\sqrt{x} + 2) + (\sqrt{x} - 2)$

b. Multiply: $(\sqrt{x} + 2)(\sqrt{x} - 2)$

c. Square: $(\sqrt{x} + 2)^2$

d. Divide: $\dfrac{\sqrt{x} + 2}{\sqrt{x} - 2}$

54. Work each problem according to the instructions given.

a. Add: $(\sqrt{x} - 3) + (\sqrt{x} + 3)$

b. Multiply: $(\sqrt{x} - 3)(\sqrt{x} + 3)$

c. Square: $(\sqrt{x} + 3)^2$

d. Divide: $\dfrac{\sqrt{x} + 3}{\sqrt{x} - 3}$

55. Work each problem according to the instructions given.

a. Add: $(5 + \sqrt{2}) + (5 - \sqrt{2})$

b. Multiply: $(5 + \sqrt{2})(5 - \sqrt{2})$

1. $3\sqrt{2}$
2. $2\sqrt{3}$
3. $10\sqrt{21}$
4. $6\sqrt{35}$
5. 720
6. 4,200
7. 54
8. 24
9. $\sqrt{6} - 9$
10. $5\sqrt{6} + 8$
11. $24 + 6\sqrt[3]{4}$
12. $105 - 14\sqrt[3]{5}$
13. $7 + 2\sqrt{6}$
14. $11 - \sqrt{10}$
15. $x + 2\sqrt{x} - 15$
16. $x + 6\sqrt{x} + 8$
17. $34 + 20\sqrt{3}$
18. $50 - 10\sqrt{21}$
19. $19 + 8\sqrt{3}$
20. $9 - 4\sqrt{5}$
21. $x - 6\sqrt{x} + 9$
22. $x + 8\sqrt{x} + 16$
23. $4a - 12\sqrt{ab} + 9b$
24. $25a - 20\sqrt{ab} + 4b$
25. $x + 4\sqrt{x - 4}$
26. $x + 4\sqrt{x - 3} + 1$
27. $x - 6\sqrt{x - 5} + 4$
28. $x - 8\sqrt{x - 3} + 13$
29. 1
30. 3
31. $a - 49$
32. $a - 25$
33. $25 - x$
34. $9 - x$
35. $x - 8$
36. $x - 22$
37. $10 + 6\sqrt{3}$
38. $17\sqrt{5} - 38$
39. $\dfrac{\sqrt{3} + 1}{2}$
40. $\dfrac{5 - \sqrt{15}}{2}$
41. $\dfrac{5 - \sqrt{5}}{4}$
42. $\dfrac{7 + \sqrt{7}}{6}$
43. $\dfrac{x + 3\sqrt{x}}{x - 9}$
44. $\dfrac{x - 2\sqrt{x}}{x - 4}$
45. $\dfrac{10 + 3\sqrt{5}}{11}$
46. $\dfrac{21 + 2\sqrt{7}}{59}$
47. $\dfrac{3\sqrt{x} + 3\sqrt{y}}{x - y}$
48. $\dfrac{2\sqrt{x} - 2\sqrt{y}}{x - y}$
49. $2 + \sqrt{3}$
50. $4 - \sqrt{15}$
51. $\dfrac{11 - 4\sqrt{7}}{3}$
52. $10 + 3\sqrt{11}$

c. Square: $(5 + \sqrt{2})^2$

d. Divide: $\dfrac{5 + \sqrt{2}}{5 - \sqrt{2}}$

56. Work each problem according to the instructions given.

a. Add: $(2 + \sqrt{3}) + (2 - \sqrt{3})$

b. Multiply: $(2 + \sqrt{3})(2 - \sqrt{3})$

c. Square: $(2 + \sqrt{3})^2$

d. Divide: $\dfrac{2 + \sqrt{3}}{2 - \sqrt{3}}$

57. Work each problem according to the instructions given.

a. Add: $\sqrt{2} + (\sqrt{6} + \sqrt{2})$

b. Multiply: $\sqrt{2}(\sqrt{6} + \sqrt{2})$

c. Divide: $\dfrac{\sqrt{6} + \sqrt{2}}{\sqrt{2}}$

d. Divide: $\dfrac{\sqrt{2}}{\sqrt{6} + \sqrt{2}}$

58. Work each problem according to the instructions given.

a. Add: $\sqrt{5} + (\sqrt{5} + \sqrt{10})$

b. Multiply: $\sqrt{5}(\sqrt{5} + \sqrt{10})$

c. Divide: $\dfrac{\sqrt{5} + \sqrt{10}}{\sqrt{5}}$

d. Divide: $\dfrac{\sqrt{5}}{\sqrt{5} + \sqrt{10}}$

59. Work each problem according to the instructions given.

a. Add: $\left(\dfrac{1 + \sqrt{5}}{2}\right) + \left(\dfrac{1 - \sqrt{5}}{2}\right)$

b. Multiply: $\left(\dfrac{1 + \sqrt{5}}{2}\right)\left(\dfrac{1 - \sqrt{5}}{2}\right)$

60. Work each problem according to the instructions given.

a. Add: $\left(\dfrac{1 + \sqrt{3}}{2}\right) + \left(\dfrac{1 - \sqrt{3}}{2}\right)$

b. Multiply: $\left(\dfrac{1 + \sqrt{3}}{2}\right)\left(\dfrac{1 - \sqrt{3}}{2}\right)$

61. Show that the product below is 5:

$$(\sqrt[3]{2} + \sqrt[3]{3})(\sqrt[3]{4} - \sqrt[3]{6} + \sqrt[3]{9})$$

62. Show that the product below is $x + 8$:

$$(\sqrt[3]{x} + 2)(\sqrt[3]{x^2} - 2\sqrt[3]{x} + 4)$$

Each of the following statements below is false. Correct the right side of each one to make it true.

63. $5(2\sqrt{3}) = 10\sqrt{15}$ **64.** $3(2\sqrt{x}) = 6\sqrt{3x}$ **65.** $(\sqrt{x} + 3)^2 = x + 9$

66. $(\sqrt{x} - 7)^2 = x - 49$ **67.** $(5\sqrt{3})^2 = 15$ **68.** $(3\sqrt{5})^2 = 15$

Applying the Concepts

69. Gravity If an object is dropped from the top of a 100-foot building, the amount of time t (in seconds) that it takes for the object to be h feet from the ground is given by the formula

$$t = \frac{\sqrt{100 - h}}{4}$$

How long does it take before the object is 50 feet from the ground? How long does it take to reach the ground? (When it is on the ground, h is 0.)

70. Gravity Use the formula given in Problem 69 to determine h if t is 1.25 seconds.

53. a. $2\sqrt{x}$
b. $x - 4$
c. $x + 4\sqrt{x} + 4$
d. $\frac{x + 4\sqrt{x} + 4}{x - 4}$

54. a. $2\sqrt{x}$
b. $x - 9$
c. $x + 6\sqrt{x} + 9$
d. $\frac{x + 6\sqrt{x} + 9}{x - 9}$

55. a. 10
b. 23
c. $27 + 10\sqrt{2}$
d. $\frac{27 + 10\sqrt{2}}{23}$

56. a. 4
b. 1
c. $7 + 4\sqrt{3}$
d. $7 + 4\sqrt{3}$

57. a. $\sqrt{6} + 2\sqrt{2}$
b. $2 + 2\sqrt{3}$
c. $1 + \sqrt{3}$
d. $\frac{-1 + \sqrt{3}}{2}$

58. a. $2\sqrt{5} + \sqrt{10}$
b. $5 + 5\sqrt{2}$
c. $1 + \sqrt{2}$
d. $-1 + \sqrt{2}$

59. a. 1
b. -1

60. a. 1
b. $-\frac{1}{2}$

61. $(\sqrt[3]{2} + \sqrt[3]{3})(\sqrt[3]{4} - \sqrt[3]{6} + \sqrt[3]{9}) = \sqrt[3]{8} - \sqrt[3]{12} + \sqrt[3]{18} + \sqrt[3]{12} - \sqrt[3]{18} + \sqrt[3]{27} = 2 + 3 = 5$

62. $(\sqrt[3]{x} + 2)(\sqrt[3]{x^2} - 2\sqrt[3]{x} + 4) = \sqrt[3]{x^3} - 2\sqrt[3]{x^2} + 4\sqrt[3]{x} + 2\sqrt[3]{x^2} - 4\sqrt[3]{x} + 8 = x + 8$

63. $10\sqrt{3}$

64. $6\sqrt{x}$

65. $x + 6\sqrt{x} + 9$

66. $x - 14\sqrt{x} + 49$

67. 75

68. 45

69. $\frac{5\sqrt{2}}{4}$ second; $\frac{5}{2}$ second

70. 75 feet

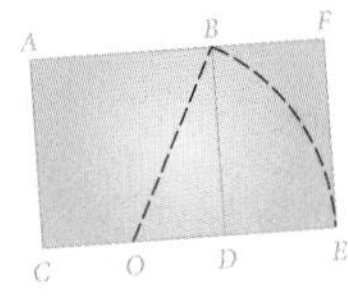

71. Answers will vary
72. Answers will vary
73. Answers will vary
74. Answers will vary
75. $t^2 + 10t + 25$
76. $x^2 - 8x + 16$
77. x
78. $3x$
79. 7
80. 4
81. $-4, -3$
82. $-2, 5$
83. $-6, -3$
84. 2, 3
85. $-5, -2$
86. 5, 8
87. Yes
88. No
89. No
90. Yes

71. Golden Rectangle Rectangle *ACEF* in Figure 2 is a golden rectangle. If side *AC* is 6 inches, show that the smaller rectangle *BDEF* is also a golden rectangle.

72. Golden Rectangle Rectangle *ACEF* in Figure 2 is a golden rectangle. If side *AC* is 1 inch, show that the smaller rectangle *BDEF* is also a golden rectangle.

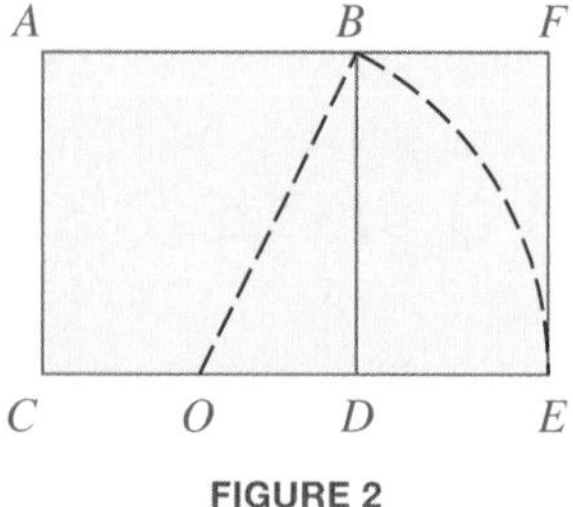

FIGURE 2

73. Golden Rectangle If side *AC* in Figure 2 is $2x$, show that rectangle *BDEF* is a golden rectangle.

74. Golden Rectangle If side *AC* in Figure 2 is x, show that rectangle *BDEF* is a golden rectangle.

Getting Ready for the Next Section

Simplify.

75. $(t + 5)^2$ **76.** $(x - 4)^2$ **77.** $\sqrt{x} \cdot \sqrt{x}$ **78.** $\sqrt{3x} \cdot \sqrt{3x}$

Solve.

79. $3x + 4 = 5^2$

80. $4x - 7 = 3^2$

81. $t^2 + 7t + 12 = 0$

82. $x^2 - 3x - 10 = 0$

83. $t^2 + 10t + 25 = t + 7$

84. $x^2 - 4x + 4 = x - 2$

85. $(x + 4)^2 = x + 6$

86. $(x - 6)^2 = x - 4$

87. Is $x = 7$ a solution to $\sqrt{3x + 4} = 5$?

88. Is $x = 4$ a solution to $\sqrt{4x - 7} = -3$?

89. Is $t = -6$ a solution to $t + 5 = \sqrt{t + 7}$?

90. Is $t = -3$ a solution to $t + 5 = \sqrt{t + 7}$?

Equations Involving Radicals 9.5

This section is concerned with solving equations that involve one or more radicals. The first step in solving an equation that contains a radical is to eliminate the radical from the equation. To do so, we need an additional property.

PROPERTY *Squaring Property of Equality*

If both sides of an equation are squared, the solutions to the original equation are solutions to the resulting equation.

We will never lose solutions to our equations by squaring both sides. We may, however, introduce *extraneous solutions*. Extraneous solutions satisfy the equation obtained by squaring both sides of the original equation, but do not satisfy the original equation.

We know that if two real numbers a and b are equal, then so are their squares:

$$\text{If} \quad a = b$$

$$\text{then} \quad a^2 = b^2$$

On the other hand, extraneous solutions are introduced when we square opposites. That is, even though opposites are not equal, their squares are. For example,

$$5 = -5 \quad \text{A false statement}$$

$$(5)^2 = (-5)^2 \quad \text{Square both sides}$$

$$25 = 25 \quad \text{A true statement}$$

We are free to square both sides of an equation any time it is convenient. We must be aware, however, that doing so may introduce extraneous solutions. We must, therefore, check all our solutions in the original equation if at any time we square both sides of the original equation.

EXAMPLE 1 Solve for x: $\sqrt{3x + 4} = 5$.

SOLUTION We square both sides and proceed as usual:

$$\sqrt{3x + 4} = 5$$

$$(\sqrt{3x + 4})^2 = 5^2$$

$$3x + 4 = 25$$

$$3x = 21$$

$$x = 7$$

Checking $x = 7$ in the original equation, we have

$$\sqrt{3(7) + 4} \stackrel{?}{=} 5$$

$$\sqrt{21 + 4} \stackrel{?}{=} 5$$

$$\sqrt{25} \stackrel{?}{=} 5$$

$$5 = 5$$

The solution $x = 7$ satisfies the original equation.

EXAMPLE 2 Solve $\sqrt{4x - 7} = -3$.

SOLUTION Squaring both sides, we have

$$\sqrt{4x - 7} = -3$$
$$(\sqrt{4x - 7})^2 = (-3)^2$$
$$4x - 7 = 9$$
$$4x = 16$$
$$x = 4$$

Checking $x = 4$ in the original equation gives

$$\sqrt{4(4) - 7} \stackrel{?}{=} -3$$
$$\sqrt{16 - 7} \stackrel{?}{=} -3$$
$$\sqrt{9} \stackrel{?}{=} -3$$
$$3 = -3 \quad \text{A false statement}$$

The solution $x = 4$ produces a false statement when checked in the original equation. Because $x = 4$ was the only possible solution, there is no solution to the original equation. The possible solution $x = 4$ is an extraneous solution. It satisfies the equation obtained by squaring both sides of the original equation, but does not satisfy the original equation.

Note The fact that there is no solution to the equation in Example 2 was obvious to begin with. Notice that the left side of the equation is the positive square root of $4x - 7$, which must be a positive number or 0. The right side of the equation is -3. Because we cannot have a number that is either positive or zero equal to a negative number, there is no solution to the equation.

EXAMPLE 3 Solve $\sqrt{5x - 1} + 3 = 7$.

SOLUTION We must isolate the radical on the left side of the equation. If we attempt to square both sides without doing so, the resulting equation will also contain a radical. Adding -3 to both sides, we have

$$\sqrt{5x - 1} + 3 = 7$$
$$\sqrt{5x - 1} = 4$$

We can now square both sides and proceed as usual:

$$(\sqrt{5x - 1})^2 = 4^2$$
$$5x - 1 = 16$$
$$5x = 17$$
$$x = \frac{17}{5}$$

Checking $x = \frac{17}{5}$, we have

$$\sqrt{5\left(\frac{17}{5}\right) - 1} + 3 \stackrel{?}{=} 7$$
$$\sqrt{17 - 1} + 3 \stackrel{?}{=} 7$$
$$\sqrt{16} + 3 \stackrel{?}{=} 7$$
$$4 + 3 \stackrel{?}{=} 7$$
$$7 = 7$$

EXAMPLE 4 Solve $t + 5 = \sqrt{t + 7}$.

SOLUTION This time, squaring both sides of the equation results in a quadratic equation:

$$(t + 5)^2 = (\sqrt{t + 7})^2 \qquad \text{Square both sides}$$
$$t^2 + 10t + 25 = t + 7$$
$$t^2 + 9t + 18 = 0 \qquad \text{Standard form}$$
$$(t + 3)(t + 6) = 0 \qquad \text{Factor the left side}$$
$$t + 3 = 0 \quad \text{or} \quad t + 6 = 0 \qquad \text{Set factors equal to 0}$$
$$t = -3 \quad \text{or} \quad t = -6$$

We must check each solution in the original equation:

Check $t = -3$

$$-3 + 5 \stackrel{?}{=} \sqrt{-3 + 7}$$
$$2 \stackrel{?}{=} \sqrt{4}$$
$$2 = 2 \qquad \text{A true statement}$$

Check $t = -6$

$$-6 + 5 \stackrel{?}{=} \sqrt{-6 + 7}$$
$$-1 \stackrel{?}{=} \sqrt{1}$$
$$-1 = 1 \qquad \text{A false statement}$$

Because $t = -6$ does not check, our only solution is $t = -3$.

EXAMPLE 5 Solve $\sqrt{x - 3} = \sqrt{x} - 3$.

SOLUTION We begin by squaring both sides. Note carefully what happens when we square the right side of the equation, and compare the square of the right side with the square of the left side. You must convince yourself that these results are correct. (The note in the margin will help if you are having trouble convincing yourself that what is written below is true.)

$$(\sqrt{x - 3})^2 = (\sqrt{x} - 3)^2$$
$$x - 3 = x - 6\sqrt{x} + 9$$

Now we still have a radical in our equation, so we will have to square both sides again. Before we do, though, let's isolate the remaining radical.

$$x - 3 = x - 6\sqrt{x} + 9$$
$$-3 = -6\sqrt{x} + 9 \qquad \text{Add } -x \text{ to each side}$$
$$-12 = -6\sqrt{x} \qquad \text{Add } -9 \text{ to each side}$$
$$2 = \sqrt{x} \qquad \text{Divide each side by } -6$$
$$4 = x \qquad \text{Square each side}$$

Note It is very important that you realize that the square of $(\sqrt{x} - 3)$ is not $x + 9$. Remember, when we square a difference with two terms, we use the formula

$$(a - b)^2 = a^2 - 2ab + b^2$$

Applying this formula to $(\sqrt{x} - 3)^2$ we have

$$(\sqrt{x} - 3)^2 = (\sqrt{x})^2 - 2(\sqrt{x})(3) + 3^2 = x - 6\sqrt{x} + 9$$

Our only possible solution is $x = 4$, which we check in our original equation as follows:

$$\sqrt{4 - 3} \stackrel{?}{=} \sqrt{4} - 3$$
$$\sqrt{1} \stackrel{?}{=} 2 - 3$$
$$1 = -1 \qquad \text{A false statement}$$

Substituting 4 for x in the original equation yields a false statement. Because 4 was our only possible solution, there is no solution to our equation.

Here is another example of an equation for which we must apply our squaring property twice before all radicals are eliminated.

EXAMPLE 6 Solve $\sqrt{x+1} = 1 - \sqrt{2x}$.

SOLUTION This equation has two separate terms involving radical signs. Squaring both sides gives

$$x + 1 = 1 - 2\sqrt{2x} + 2x$$

$$-x = -2\sqrt{2x} \quad \text{Add } -2x \text{ and } -1 \text{ to both sides}$$

$$x^2 = 4(2x) \quad \text{Square both sides}$$

$$x^2 - 8x = 0 \quad \text{Standard form}$$

Our equation is a quadratic equation in standard form. To solve for x, we factor the left side and set each factor equal to 0:

$$x(x - 8) = 0 \quad \text{Factor left side}$$

$$x = 0 \quad \text{or} \quad x - 8 = 0$$

$$x = 8 \quad \text{Set factors equal to 0}$$

Because we squared both sides of our equation, we have the possibility that one or both of the solutions are extraneous. We must check each one in the original equation:

Check $x = 8$		Check $x = 0$	
$\sqrt{8+1} \stackrel{?}{=} 1 - \sqrt{2 \cdot 8}$		$\sqrt{0+1} \stackrel{?}{=} 1 - \sqrt{2 \cdot 0}$	
$\sqrt{9} \stackrel{?}{=} 1 - \sqrt{16}$		$\sqrt{1} \stackrel{?}{=} 1 - \sqrt{0}$	
$3 \stackrel{?}{=} 1 - 4$		$1 \stackrel{?}{=} 1 - 0$	
$3 = -3$	A false statement	$1 = 1$	A true statement

Because $x = 8$ does not check, it is an extraneous solution. Our only solution is $x = 0$.

EXAMPLE 7 Solve $\sqrt{x+1} = \sqrt{x+2} - 1$.

SOLUTION Squaring both sides we have

$$(\sqrt{x+1})^2 = (\sqrt{x+2} - 1)^2$$

$$x + 1 = x + 2 - 2\sqrt{x+2} + 1$$

Once again, we are left with a radical in our equation. Before we square each side again, we must isolate the radical on the right side of the equation.

$$x + 1 = x + 3 - 2\sqrt{x+2} \quad \text{Simplify the right side}$$

$$1 = 3 - 2\sqrt{x+2} \quad \text{Add } -x \text{ to each side}$$

$$-2 = -2\sqrt{x+2} \quad \text{Add } -3 \text{ to each side}$$

$$1 = \sqrt{x+2} \quad \text{Divide each side by } -2$$

$$1 = x + 2 \quad \text{Square both sides}$$

$$-1 = x \quad \text{Add } -2 \text{ to each side}$$

Checking our only possible solution, $x = -1$, in our original equation, we have

$$\sqrt{-1+1} \stackrel{?}{=} \sqrt{-1+2} - 1$$
$$\sqrt{0} \stackrel{?}{=} \sqrt{1} - 1$$
$$0 \stackrel{?}{=} 1 - 1$$
$$0 = 0 \quad \text{A true statement}$$

Our solution checks.

It is also possible to raise both sides of an equation to powers greater than 2. We only need to check for extraneous solutions when we raise both sides of an equation to an even power. Raising both sides of an equation to an odd power will not produce extraneous solutions.

EXAMPLE 8 Solve $\sqrt[3]{4x+5} = 3$.

SOLUTION Cubing both sides, we have

$$(\sqrt[3]{4x+5})^3 = 3^3$$
$$4x + 5 = 27$$
$$4x = 22$$
$$x = \frac{22}{4}$$
$$x = \frac{11}{2}$$

We do not need to check $x = \frac{11}{2}$ because we raised both sides to an odd power.

We end this section by looking at graphs of some equations that contain radicals.

EXAMPLE 9 Graph $y = \sqrt{x}$ and $y = \sqrt[3]{x}$.

SOLUTION The graphs are shown in Figures 1 and 2. Notice that the graph of $y = \sqrt{x}$ appears in the first quadrant only, because in the equation $y = \sqrt{x}$, x and y cannot be negative.

The graph of $y = \sqrt[3]{x}$ appears in Quadrants 1 and 3 because the cube root of a positive number is also a positive number, and the cube root of a negative number is a negative number. That is, when x is positive, y will be positive, and when x is negative, y will be negative.

The graphs of both equations will contain the origin, because $y = 0$ when $x = 0$ in both equations.

x	y
−4	Undefined
−1	Undefined
0	0
1	1
4	2
9	3
16	4

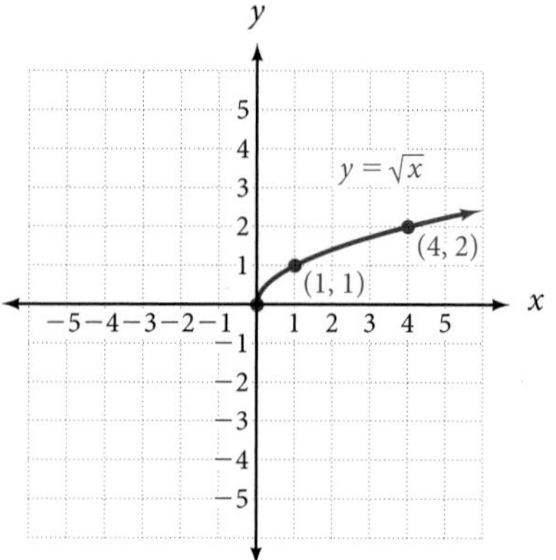

FIGURE 1

x	y
−27	−3
−8	−2
−1	−1
0	0
1	1
8	2
27	3

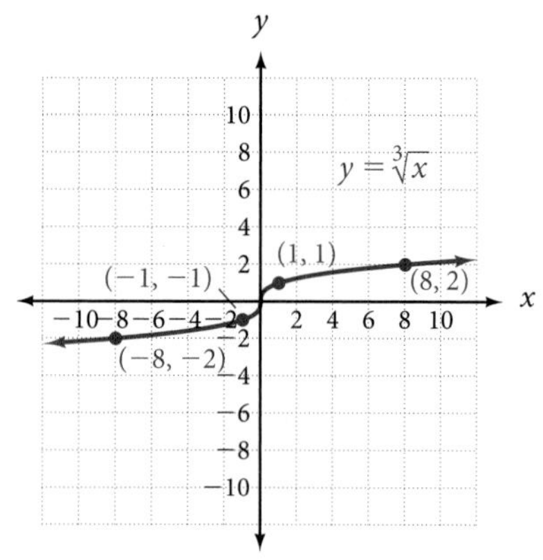

FIGURE 2

GETTING READY FOR CLASS

Respond in your own words and in complete sentences.

A. What is the squaring property of equality?

B. Under what conditions do we obtain extraneous solutions to equations that contain radical expressions?

C. If we have raised both sides of an equation to a power, when is it not necessary to check for extraneous solutions?

D. When will you need to apply the squaring property of equality twice in the process of solving an equation containing radicals?

1. 4
2. 5
3. $\varnothing$
4. $\varnothing$
5. 5
6. $\frac{5}{3}$
7. $\varnothing$
8. $\varnothing$
9. $\frac{39}{2}$
10. 8
11. $\varnothing$
12. $\varnothing$
13. 5
14. 20
15. 3
16. $\frac{1}{5}$
17. $-\frac{32}{3}$
18. $-\frac{15}{2}$
19. 3, 4
20. Possible solutions 1 and 6, only 6 checks
21. $-1, -2$
22. Possible solutions -1 and 6; only 6 checks
23. -1
24. $\varnothing$
25. $\varnothing$
26. 1
27. 7
28. -1
29. 0, 3
30. only 8 checks
31. -4
32. -6
33. 8
34. 1
35. 0
36. -1
37. 9
38. $\varnothing$
39. 0
40. 1
41. 8
42. 4
43. Possible solution 9, which does not check; $\varnothing$
44. Possible solution 7, which does not check;
45. **a.** 100 **b.** 40 **c.** $\varnothing$ **d.** Possible solutions 5, 8; only 8 checks
46. **a.** $\varnothing$ **b.** 17 **c.** Possible -3, 5; only 5 checks **d.** Possible -3, -7; only -3 checks

Problem Set 9.5

Solve each of the following equations.

1. $\sqrt{2x+1} = 3$ **2.** $\sqrt{3x+1} = 4$ **3.** $\sqrt{4x+1} = -5$

4. $\sqrt{6x+1} = -5$ **5.** $\sqrt{2y-1} = 3$ **6.** $\sqrt{3y-1} = 2$

7. $\sqrt{5x-7} = -1$ **8.** $\sqrt{8x+3} = -6$ **9.** $\sqrt{2x-3} - 2 = 4$

10. $\sqrt{3x+1} - 4 = 1$ **11.** $\sqrt{4a+1} + 3 = 2$ **12.** $\sqrt{5a-3} + 6 = 2$

13. $\sqrt[4]{3x+1} = 2$ **14.** $\sqrt[4]{4x+1} = 3$ **15.** $\sqrt[3]{2x-5} = 1$

16. $\sqrt[3]{5x+7} = 2$ **17.** $\sqrt[3]{3a+5} = -3$ **18.** $\sqrt[3]{2a+7} = -2$

19. $\sqrt{y-3} = y-3$ **20.** $\sqrt{y+3} = y-3$ **21.** $\sqrt{a+2} = a+2$

22. $\sqrt{a+10} = a-2$ **23.** $\sqrt{2x+4} = \sqrt{1-x}$

24. $\sqrt{3x+4} = -\sqrt{2x+3}$ **25.** $\sqrt{4a+7} = -\sqrt{a+2}$

26. $\sqrt{7a-1} = \sqrt{2a+4}$ **27.** $\sqrt[4]{5x-8} = \sqrt[4]{4x-1}$

28. $\sqrt[4]{6x+7} = \sqrt[4]{x+2}$ **29.** $x+1 = \sqrt{5x+1}$

30. $x-1 = \sqrt{6x+1}$ **31.** $t+5 = \sqrt{2t+9}$

32. $t+7 = \sqrt{2t+13}$ **33.** $\sqrt{y-8} = \sqrt{8-y}$

34. $\sqrt{2y+5} = \sqrt{5y+2}$ **35.** $\sqrt[3]{3x+5} = \sqrt[3]{5-2x}$

36. $\sqrt[3]{4x+9} = \sqrt[3]{3-2x}$

The following equations will require that you square both sides twice before all the radicals are eliminated. Solve each equation using the methods shown in Examples 5, 6, and 7.

37. $\sqrt{x-8} = \sqrt{x} - 2$ **38.** $\sqrt{x+3} = \sqrt{x} - 3$

39. $\sqrt{x+1} = \sqrt{x} + 1$ **40.** $\sqrt{x-1} = \sqrt{x} - 1$

41. $\sqrt{x+8} = \sqrt{x-4} + 2$ **42.** $\sqrt{x+5} = \sqrt{x-3} + 2$

43. $\sqrt{x-5} - 3 = \sqrt{x-8}$ **44.** $\sqrt{x-3} - 4 = \sqrt{x-3}$

45. Solve each equation.

a. $\sqrt{y} - 4 = 6$ **b.** $\sqrt{y-4} = 6$

c. $\sqrt{y-4} = -6$ **d.** $\sqrt{y-4} = y-6$

46. Solve each equation.

a. $\sqrt{2y} + 15 = 7$ **b.** $\sqrt{2y+15} = 7$

c. $\sqrt{2y+15} = y$ **d.** $\sqrt{2y+15} = y+6$

47. Solve each equation.

a. $x-3 = 0$ **b.** $\sqrt{x} - 3 = 0$

c. $\sqrt{x-3} = 0$ **d.** $\sqrt{x} + 3 = 0$

e. $\sqrt{x} + 3 = 5$ **f.** $\sqrt{x} + 3 = -5$

g. $x-3 = \sqrt{5-x}$

48. Solve each equation.

a. $x-2 = 0$ **b.** $\sqrt{x} - 2 = 0$ **c.** $\sqrt{x} + 2 = 0$

d. $\sqrt{x+2} = 0$ **e.** $\sqrt{x} + 2 = 7$ **f.** $x-2 = \sqrt{2x-1}$

47. a. 3
b. 9
c. 3
d. ∅
e. 4
f. ∅
g. Possible 1, 4; only 4 checks

48. a. 2
b. 4
c. ∅
d. −2
e. 25
f. Possible 1, 5; only 5 checks

49. $h = 100 - 16t^2$

50. $h = \dfrac{gt^2 + 40t}{2}$

51. $\frac{392}{121} \approx 3.24$ feet

52. **a.** 5 meters
b. 10 meters
c. 2,500 meters
d. 10,000 meters

53.-64. See answer section

Applying the Concepts

49. Solving a Formula Solve the following formula for h:

$$t = \frac{\sqrt{100 - h}}{4}$$

50. Solving a Formula Solve the following formula for h:

$$t = \sqrt{\frac{2h - 40t}{g}}$$

51. Pendulum Clock The length of time (T) in seconds it takes the pendulum of a clock to swing through one complete cycle is given by the formula

$$T = 2\pi\sqrt{\frac{L}{32}}$$

where L is the length, in feet, of the pendulum, and π is approximately $\frac{22}{7}$. How long must the pendulum be if one complete cycle takes 2 seconds?

52. Pollution A long straight river, 100 meters wide, is flowing at 1 meter per second. A pollutant is entering the river at a constant rate from one of its banks. As the pollutant disperses in the water, it forms a plume that is modeled by the equation $y = \sqrt{x}$. Use this information to answer the following questions.

a. How wide is the plume 25 meters down river from the source of the pollution?

b. How wide is the plume 100 meters down river from the source of the pollution?

c. How far down river from the source of the pollution does the plume reach halfway across the river?

d. How far down the river from the source of the pollution does the plume reach the other side of the river?

Graph each equation.

53. $y = 2\sqrt{x}$	**54.** $y = -2\sqrt{x}$	**55.** $y = \sqrt{x} - 2$	**56.** $y = \sqrt{x} + 2$
57. $y = \sqrt{x - 2}$	**58.** $y = \sqrt{x + 2}$	**59.** $y = 3\sqrt[3]{x}$	**60.** $y = -3\sqrt[3]{x}$
61. $y = \sqrt[3]{x} + 3$	**62.** $y = \sqrt[3]{x} - 3$	**63.** $y = \sqrt[3]{x + 3}$	**64.** $y = \sqrt[3]{x - 3}$

65. 5
66. 7
67. $2\sqrt{3}$
68. $5\sqrt{2}$
69. -1
70. 1
71. 1
72. -1
73. 4
74. $\frac{1}{2}$
75. 2
76. 4
77. $10 - 2x$
78. $1 + 2x$
79. $2 - 3x$
80. $-4 + 2x$
81. $6 + 7x - 20x^2$
82. $56 - 17x - 3x^2$
83. $8x - 12x^2$
84. $21x + 6x^2$
85. $4 + 12x + 9x^2$
86. $9 + 30x + 25x^2$
87. $4 - 9x^2$
88. $16 - 25x^2$

Getting Ready for the Next Section

Simplify.

65. $\sqrt{25}$ **66.** $\sqrt{49}$ **67.** $\sqrt{12}$ **68.** $\sqrt{50}$

69. $(-1)^{15}$ **70.** $(-1)^{20}$ **71.** $(-1)^{50}$ **72.** $(-1)^5$

Solve.

73. $3x = 12$ **74.** $4 = 8y$ **75.** $4x - 3 = 5$ **76.** $7 = 2y - 1$

Perform the indicated operation.

77. $(3 + 4x) + (7 - 6x)$

78. $(2 - 5x) + (-1 + 7x)$

79. $(7 + 3x) - (5 + 6x)$

80. $(5 - 2x) - (9 - 4x)$

81. $(3 - 4x)(2 + 5x)$

82. $(8 + x)(7 - 3x)$

83. $2x(4 - 6x)$

84. $3x(7 + 2x)$

85. $(2 + 3x)^2$

86. $(3 + 5x)^2$

87. $(2 - 3x)(2 + 3x)$

88. $(4 - 5x)(4 + 5x)$

Complex Numbers

9.6

The equation $x^2 = -9$ has no real number solutions because the square of a real number is always positive. We have been unable to work with square roots of negative numbers like $\sqrt{-25}$ and $\sqrt{-16}$ for the same reason. Complex numbers allow us to expand our work with radicals to include square roots of negative numbers and to solve equations like $x^2 = -9$ and $x^2 = -64$. Our work with complex numbers is based on the following definition.

DEFINITION *the number i*

The ***number i*** is such that $i = \sqrt{-1}$ (which is the same as saying $i^2 = -1$).

The number i, as we have defined it here, is not a real number. Because of the way we have defined i, we can use it to simplify square roots of negative numbers.

Square Roots of Negative Numbers

If a is a positive number, then $\sqrt{-a}$ can always be written as $i\sqrt{a}$. That is,

$$\sqrt{-a} = i\sqrt{a} \qquad \text{if } a \text{ is a positive number}$$

To justify our rule, we simply square the quantity $i\sqrt{a}$ to obtain $-a$. Here is what it looks like when we do so:

$$\begin{aligned}(i\sqrt{a})^2 &= i^2 \cdot (\sqrt{a})^2\\ &= -1 \cdot a\\ &= -a\end{aligned}$$

Here are some examples that illustrate the use of our new rule.

EXAMPLES Write each square root in terms of the number i.

1. $\sqrt{-25} = i\sqrt{25} = i \cdot 5 = 5i$

2. $\sqrt{-49} = i\sqrt{49} = i \cdot 7 = 7i$

3. $\sqrt{-12} = i\sqrt{12} = i \cdot 2\sqrt{3} = 2i\sqrt{3}$

4. $\sqrt{-17} = i\sqrt{17}$

Note In Examples 3 and 4, we wrote i before the radical simply to avoid confusion. If we were to write the answer to 3 as $2\sqrt{3}i$, some people would think the i was under the radical sign, but it is not.

If we assume all the properties of exponents hold when the base is i, we can write any power of i as i, -1, $-i$, or 1. Using the fact that $i^2 = -1$, we have

$$\begin{aligned}i^1 &= i\\ i^2 &= -1\\ i^3 &= i^2 \cdot i = -1(i) = -i\\ i^4 &= i^2 \cdot i^2 = -1(-1) = 1\end{aligned}$$

Because $i^4 = 1$, i^5 will simplify to i, and we will begin repeating the sequence i, -1, $-i$, 1 as we simplify higher powers of i: Any power of i simplifies to i, -1, $-i$, or 1. The easiest way to simplify higher powers of i is to write them in terms of i^2. For instance, to simplify i^{21}, we would write it as

$$(i^2)^{10} \cdot i \qquad \text{because} \qquad 2 \cdot 10 + 1 = 21$$

Then, because $i^2 = -1$, we have

$$(-1)^{10} \cdot i = 1 \cdot i = i$$

EXAMPLES Simplify as much as possible.

5. $i^{30} = (i^2)^{15} = (-1)^{15} = -1$

6. $i^{11} = (i^2)^5 \cdot i = (-1)^5 \cdot i = (-1)i = -i$

7. $i^{40} = (i^2)^{20} = (-1)^{20} = 1$

DEFINITION *complex number*

A ***complex number*** is any number that can be put in the form

$$a + bi$$

where a and b are real numbers and $i = \sqrt{-1}$. The form $a + bi$ is called ***standard form*** for complex numbers. The number a is called the ***real part*** of the complex number. The number b is called the ***imaginary part*** of the complex number.

Every real number is a complex number. For example, 8 can be written as $8 + 0i$. Likewise, $-\frac{1}{2}$, π, $\sqrt{3}$, and 29 are complex numbers because they can all be written in the form $a + bi$:

$$-\frac{1}{2} = -\frac{1}{2} + 0i \qquad \pi = \pi + 0i \qquad \sqrt{3} = \sqrt{3} + 0i \qquad -9 = -9 + 0i$$

The rest of the complex numbers that are not real numbers, are divided into two additional categories; *compound numbers* and *pure imaginary numbers*. The diagram below shows all three subsets of the complex numbers, along with examples of the type of numbers that fall into those subsets.

Subsets of the Complex Numbers

All numbers of the form $a + bi$ fall into one of the following categories. Each category is a subset of the complex numbers.

Real Numbers	Compound Numbers	Pure Imaginary Numbers
When $a \neq 0$ and $b = 0$ Examples include: $-10, 0, 1, \sqrt{3}, \frac{5}{8}, \pi$	When neither a nor b is 0 Examples include: $5 + 4i, \frac{1}{3} + 4i, \sqrt{5} - i,$ $-6 + i\sqrt{5}$	When $a = 0$ and $b \neq 0$ Examples include: $-4i, i\sqrt{3}, -5i\sqrt{7}, \frac{3}{4}i$

Note See Section 1.1 for a review of the subsets of real numbers.

Note: The definition for compound numbers is from Jim Metz of Kapiolani Community College in Hawaii. Some textbooks use the phrase *imaginary numbers* to represent both the compound numbers and the pure imaginary numbers. In those books, the pure imaginary numbers are a subset of the imaginary numbers. We like the definition from Mr. Metz because it keeps the three subsets from overlapping.

$$\begin{aligned}
&\quad\quad\quad\quad\quad\quad\quad\quad\quad\quad\ \text{F}\quad\quad\ \text{O}\quad\quad\ \text{I}\quad\quad\quad \text{L}\\
(3 - 4i)(2 + 5i) &= 3 \cdot 2 + 3 \cdot 5i - 2 \cdot 4i - 4i(5i)\\
&= 6 + 15i - 8i - 20i^2
\end{aligned}$$

Combining similar terms and using the fact that $i^2 = -1$, we can simplify as follows:

$$\begin{aligned}
6 + 15i - 8i - 20i^2 &= 6 + 7i - 20(-1)\\
&= 6 + 7i + 20\\
&= 26 + 7i
\end{aligned}$$

The product of the complex numbers $3 - 4i$ and $2 + 5i$ is the complex number $26 + 7i$.

EXAMPLE 14 Multiply $2i(4 - 6i)$.

SOLUTION Applying the distributive property gives us

$$\begin{aligned}
2i(4 - 6i) &= 2i \cdot 4 - 2i \cdot 6i\\
&= 8i - 12i^2\\
&= 12 + 8i
\end{aligned}$$

EXAMPLE 15 Expand $(3 + 5i)^2$.

SOLUTION We treat this like the square of a binomial. Remember, $(a + b)^2 = a^2 + 2ab + b^2$:

$$\begin{aligned}
(3 + 5i)^2 &= 3^2 + 2(3)(5i) + (5i)^2\\
&= 9 + 30i + 25i^2\\
&= 9 + 30i - 25\\
&= -16 + 30i
\end{aligned}$$

EXAMPLE 16 Multiply $(2 - 3i)(2 + 3i)$.

SOLUTION This product has the form $(a - b)(a + b)$, which we know results in the difference of two squares, $a^2 - b^2$:

$$\begin{aligned}
(2 - 3i)(2 + 3i) &= 2^2 - (3i)^2\\
&= 4 - 9i^2\\
&= 4 + 9\\
&= 13
\end{aligned}$$

The product of the two complex numbers $2 - 3i$ and $2 + 3i$ is the real number 13. The two complex numbers $2 - 3i$ and $2 + 3i$ are called complex conjugates. The fact that their product is a real number is very useful.

Equality for Complex Numbers

Two complex numbers are equal if and only if their real parts are equal and their imaginary parts are equal. That is, for real numbers a, b, c, and d,

$$a + bi = c + di \quad \text{if and only if} \quad a = c \quad \text{and} \quad b = d$$

EXAMPLE 8 Find x and y if $3x + 4i = 12 - 8yi$.

SOLUTION Because the two complex numbers are equal, their real parts are equal and their imaginary parts are equal:

$$\begin{aligned} 3x &= 12 \quad \text{and} \quad & 4 &= -8y \\ x &= 4 & y &= -\frac{1}{2} \end{aligned}$$

EXAMPLE 9 Find x and y if $(4x - 3) + 7i = 5 + (2y - 1)i$.

SOLUTION The real parts are $4x - 3$ and 5. The imaginary parts are 7 and $2y - 1$:

$$\begin{aligned} 4x - 3 &= 5 \quad \text{and} \quad & 7 &= 2y - 1 \\ 4x &= 8 & 8 &= 2y \\ x &= 2 & y &= 4 \end{aligned}$$

Addition and Subtraction of Complex Numbers

To add two complex numbers, add their real parts and their imaginary parts. That is, if a, b, c, and d are real numbers, then

$$(a + bi) + (c + di) = (a + c) + (b + d)i$$

If we assume that the commutative, associative, and distributive properties hold for the number i, then the definition of addition is simply an extension of these properties.

We define subtraction in a similar manner. If a, b, c, and d are real numbers, then

$$(a + bi) - (c + di) = (a - c) + (b - d)i$$

EXAMPLES Add or subtract as indicated.

10. $(3 + 4i) + (7 - 6i) = (3 + 7) + (4 - 6)i = 10 - 2i$

11. $(7 + 3i) - (5 + 6i) = (7 - 5) + (3 - 6)i = 2 - 3i$

12. $(5 - 2i) - (9 - 4i) = (5 - 9) + (-2 + 4)i = -4 + 2i$

Multiplication of Complex Numbers

Because complex numbers have the same form as binomials, we find the product of two complex numbers the same way we find the product of two binomials.

EXAMPLE 13 Multiply $(3 - 4i)(2 + 5i)$.

SOLUTION Multiplying each term in the second complex number by each term in the first, we have (dēf DEFINITION *complex conjugates*

The complex numbers $a + bi$ and $a - bi$ are called ***complex conjugates***. One important property they have is that their product is the real number $a^2 + b^2$. Here's why:

$$\begin{aligned}(a + bi)(a - bi) &= a^2 - (bi)^2 \\ &= a^2 - b^2i^2 \\ &= a^2 - b^2(-1) \\ &= a^2 + b^2\end{aligned}$$

Division With Complex Numbers

The fact that the product of two complex conjugates is a real number is the key to division with complex numbers.

EXAMPLE 17 Divide $\frac{2 + i}{3 - 2i}$.

SOLUTION We want a complex number in standard form that is equivalent to the quotient $\frac{2 + i}{3 - 2i}$. We need to eliminate i from the denominator. Multiplying the numerator and denominator by $3 + 2i$ will give us what we want:

$$\begin{aligned}\frac{2 + i}{3 - 2i} &= \frac{2 + i}{3 - 2i} \cdot \frac{(3 + 2i)}{(3 + 2i)} \\ &= \frac{6 + 4i + 3i + 2i^2}{9 - 4i^2} \\ &= \frac{6 + 7i - 2}{9 + 4} \\ &= \frac{4 + 7i}{13} \\ &= \frac{4}{13} + \frac{7}{13}i\end{aligned}$$

Dividing the complex number $2 + i$ by $3 - 2i$ gives the complex number $\frac{4}{13} + \frac{7}{13}i$.

EXAMPLE 18 Divide $\frac{7 - 4i}{i}$.

SOLUTION The conjugate of the denominator is $-i$. Multiplying numerator and denominator by this number, we have

$$\begin{aligned}\frac{7 - 4i}{i} &= \frac{7 - 4i}{i} \cdot \frac{-i}{-i} \\ &= \frac{-7i + 4i^2}{-i^2} \\ &= \frac{-7i + 4(-1)}{-(-1)} \\ &= -4 - 7i\end{aligned}$$

GETTING READY FOR CLASS

After reading through the preceding section, respond in your own words and in complete sentences.

A. What is the number i?
B. What is a complex number?
C. What kind of number will always result when we multiply complex conjugates?
D. Explain how to divide complex numbers.

Problem Set 9.6

Write the following in terms of i, and simplify as much as possible.

1. $\sqrt{-36}$ **2.** $\sqrt{-49}$ **3.** $-\sqrt{-25}$ **4.** $-\sqrt{-81}$

5. $\sqrt{-72}$ **6.** $\sqrt{-48}$ **7.** $-\sqrt{-12}$ **8.** $-\sqrt{-75}$

Write each of the following as i, -1, $-i$, or 1.

9. i^{28} **10.** i^{31} **11.** i^{26} **12.** i^{37}

13. i^{75} **14.** i^{42}

Find x and y so each of the following equations is true.

15. $2x + 3yi = 6 - 3i$ **16.** $4x - 2yi = 4 + 8i$

17. $2 - 5i = -x + 10yi$ **18.** $4 + 7i = 6x - 14yi$

19. $2x + 10i = -16 - 2yi$ **20.** $4x - 5i = -2 + 3yi$

21. $(2x - 4) - 3i = 10 - 6yi$ **22.** $(4x - 3) - 2i = 8 + yi$

23. $(7x - 1) + 4i = 2 + (5y + 2)i$ **24.** $(5x + 2) - 7i = 4 + (2y + 1)i$

Combine the following complex numbers.

25. $(2 + 3i) + (3 + 6i)$ **26.** $(4 + i) + (3 + 2i)$

27. $(3 - 5i) + (2 + 4i)$ **28.** $(7 + 2i) + (3 - 4i)$

29. $(5 + 2i) - (3 + 6i)$ **30.** $(6 + 7i) - (4 + i)$

31. $(3 - 5i) - (2 + i)$ **32.** $(7 - 3i) - (4 + 10i)$

33. $[(3 + 2i) - (6 + i)] + (5 + i)$ **34.** $[(4 - 5i) - (2 + i)] + (2 + 5i)$

35. $[(7 - i) - (2 + 4i)] - (6 + 2i)$ **36.** $[(3 - i) - (4 + 7i)] - (3 - 4i)$

37. $(3 + 2i) - [(3 - 4i) - (6 + 2i)]$ **38.** $(7 - 4i) - [(-2 + i) - (3 + 7i)]$

39. $(4 - 9i) + [(2 - 7i) - (4 + 8i)]$ **40.** $(10 - 2i) - [(2 + i) - (3 - i)]$

Find the following products.

41. $3i(4 + 5i)$ **42.** $2i(3 + 4i)$ **43.** $6i(4 - 3i)$

44. $11i(2 - i)$ **45.** $(3 + 2i)(4 + i)$ **46.** $(2 - 4i)(3 + i)$

47. $(4 + 9i)(3 - i)$ **48.** $(5 - 2i)(1 + i)$ **49.** $(1 + i)^3$

50. $(1 - i)^3$ **51.** $(2 - i)^3$ **52.** $(2 + i)^3$

53. $(2 + 5i)^2$ **54.** $(3 + 2i)^2$ **55.** $(1 - i)^2$

56. $(1 + i)^2$ **57.** $(3 - 4i)^2$ **58.** $(6 - 5i)^2$

59. $(2 + i)(2 - i)$ **60.** $(3 + i)(3 - i)$ **61.** $(6 - 2i)(6 + 2i)$

62. $(5 + 4i)(5 - 4i)$ **63.** $(2 + 3i)(2 - 3i)$ **64.** $(2 - 7i)(2 + 7i)$

65. $(10 + 8i)(10 - 8i)$ **66.** $(11 - 7i)(11 + 7i)$

1. $6i$
2. $7i$
3. $-5i$
4. $-9i$
5. $6i\sqrt{2}$
6. $4i\sqrt{3}$
7. $-2i\sqrt{3}$
8. $-5i\sqrt{3}$
9. 1
10. $-i$
11. -1
12. i
13. $-i$
14. -1
15. $x = 3, y = -1$
16. $x = 1, y = -4$
17. $x = -2, y = -\frac{1}{2}$
18. $x = \frac{2}{3}, y = -\frac{1}{2}$
19. $x = -8, y = -5$
20. $x = -\frac{1}{2}, y = -\frac{5}{3}$
21. $x = 7, y = \frac{1}{2}$
22. $x = \frac{11}{4}, y = -2$
23. $x = \frac{3}{7}, y = \frac{2}{5}$
24. $x = \frac{2}{5}, y = -4$
25. $5 + 9i$
26. $7 + 3i$
27. $5 - i$
28. $10 - 2i$
29. $2 - 4i$
30. $2 + 6i$
31. $1 - 6i$
32. $3 - 13i$
33. $2 + 2i$
34. $4 - i$
35. $-1 - 7i$
36. $-4 - 4i$
37. $6 + 8i$
38. $12 + 2i$
39. $2 - 24i$
40. $11 - 4i$
41. $-15 + 12i$
42. $-8 + 6i$
43. $18 + 24i$
44. $11 + 22i$
45. $10 + 11i$
46. $10 - 10i$
47. $21 + 23i$
48. $7 + 3i$
49. $-2 + 2i$
50. $-2 - 2i$
51. $2 - 11i$
52. $2 + 11i$
53. $-21 + 20i$
54. $5 + 12i$
55. $-2i$
56. $2i$
57. $-7 - 24i$

58. $11 - 60i$
59. 5
60. 10
61. 40
62. 41
63. 13
64. 53
65. 164
66. 170
67. $-3 - 2i$
68. $4 - 3i$
69. $-2 + 5i$
70. $3 + 4i$
71. $\frac{8}{13} + \frac{12}{13}i$
72. $\frac{12}{41} + \frac{15}{41}i$
73. $-\frac{18}{13} - \frac{12}{13}i$
74. $\frac{2}{29} - \frac{5}{29}i$
75. $-\frac{5}{13} + \frac{12}{13}i$
76. $-\frac{33}{65} - \frac{56}{65}i$
77. $\frac{13}{15} - \frac{2}{5}i$
78. $\frac{4}{61} + \frac{17}{61}i$
79. $R = -11 - 7i$ ohms
80. $-\frac{25}{58} - \frac{39}{58}i$ amps
81. $-\frac{3}{2}$
82. Possible solution 2, which does not check; ∅
83. $-3, \frac{1}{2}$
84. $-3, 4$
85. $\frac{5}{4}$ or $\frac{4}{5}$
86. 24 hours

Find the following quotients. Write all answers in standard form for complex numbers.

67. $\frac{2 - 3i}{i}$
68. $\frac{3 + 4i}{i}$
69. $\frac{5 + 2i}{-i}$
70. $\frac{4 - 3i}{-i}$
71. $\frac{4}{2 - 3i}$
72. $\frac{3}{4 - 5i}$
73. $\frac{6}{-3 + 2i}$
74. $\frac{-1}{-2 - 5i}$
75. $\frac{2 + 3i}{2 - 3i}$
76. $\frac{4 - 7i}{4 + 7i}$
77. $\frac{5 + 4i}{3 + 6i}$
78. $\frac{2 + i}{5 - 6i}$

Applying the Concepts

79. **Electric Circuits** Complex numbers may be applied to electrical circuits. Electrical engineers use the fact that resistance R to electrical flow of the electrical current I and the voltage V are related by the formula $V = RI$. (Voltage is measured in volts, resistance in ohms, and current in amperes.) Find the resistance to electrical flow in a circuit that has a voltage $V = (80 + 20i)$ volts and current $I = (-6 + 2i)$ amps.

80. **Electric Circuits** Refer to the information about electrical circuits in Problem 79, and find the current in a circuit that has a resistance of $(4 + 10i)$ ohms and a voltage of $(5 - 7i)$ volts.

Maintaining Your Skills

The following problems review material we covered in Sections 5.5 and 5.6.

Solve each equation. [5.5]

81. $\frac{t}{3} - \frac{1}{2} = -1$
82. $\frac{x}{x - 2} + \frac{2}{3} = \frac{2}{x - 2}$
83. $2 + \frac{5}{y} = \frac{3}{y^2}$
84. $1 - \frac{1}{y} = \frac{12}{y^2}$

Solve each application problem. [5.6]

85. The sum of a number and its reciprocal is $\frac{41}{20}$. Find the number.

86. It takes an inlet pipe 8 hours to fill a tank. The drain can empty the tank in 6 hours. If the tank is full and both the inlet pipe and drain are open, how long will it take to drain the tank?

Chapter 9 Summary

The numbers in brackets refer to the section(s) in which the topic can be found.

Square Roots [9.1]

Every positive real number x has two square roots. The *positive square root* of x is written $\sqrt{x}$, and the *negative square root* of x is written $-\sqrt{x}$. Both the positive and the negative square roots of x are numbers we square to get x; that is,

$$\left.\begin{aligned}(\sqrt{x})^2 &= x\\ \text{and}\qquad (-\sqrt{x})^2 &= x\end{aligned}\right\} \text{ for } x \geq 0$$

EXAMPLES

1. The number 49 has two square roots, 7 and -7. They are written like this:

$$\sqrt{49} = 7 \qquad -\sqrt{49} = -7$$

Higher Roots [9.1]

In the expression $\sqrt[n]{a}$, n is the *index*, a is the *radicand*, and $\sqrt{}$ is the *radical sign*. The expression $\sqrt[n]{a}$ is such that

$(\sqrt[n]{a})^n = a \qquad a \geq 0$ when n is even

2. $\sqrt[3]{8} = 2$

$\sqrt[3]{-27} = -3$

Rational Exponents [9.1]

Rational exponents are used to indicate roots. The relationship between rational exponents and roots is as follows:

$$a^{1/n} = \sqrt[n]{a} \qquad \text{and} \qquad a^{m/n} = (a^{1/n})^m = (a^m)^{1/n}$$

$a \geq 0$ when n is even

3. $25^{1/2} = \sqrt{25} = 5$

$8^{2/3} = (\sqrt[3]{8})^2 = 2^2 = 4$

$9^{3/2} = (\sqrt{9})^3 = 3^3 = 27$

Properties of Radicals [9.2]

If a and b are nonnegative real numbers whenever n is even, then

1. $\sqrt[n]{ab} = \sqrt[n]{a}\sqrt[n]{b}$
2. $\sqrt[n]{\dfrac{a}{b}} = \dfrac{\sqrt[n]{a}}{\sqrt[n]{b}} \qquad (b \neq 0)$

4. $\sqrt{4 \cdot 5} = \sqrt{4}\,\sqrt{5} = 2\sqrt{5}$

$$\sqrt{\frac{7}{9}} = \frac{\sqrt{7}}{\sqrt{9}} = \frac{\sqrt{7}}{3}$$

Simplified Form for Radicals [9.2]

A radical expression is said to be in *simplified form*

1. If there is no factor of the radicand that can be written as a power greater than or equal to the index;
2. If there are no fractions under the radical sign; and
3. If there are no radicals in the denominator.

5. $$\sqrt{\frac{4}{5}} = \frac{\sqrt{4}}{\sqrt{5}} = \frac{2}{\sqrt{5}} \cdot \frac{\sqrt{5}}{\sqrt{5}} = \frac{2\sqrt{5}}{5}$$

Addition and Subtraction of Radical Expressions [9.3]

We add and subtract radical expressions by using the distributive property to combine similar radicals. Similar radicals are radicals with the same index and the same radicand.

6. $5\sqrt{3} - 7\sqrt{3} = (5-7)\sqrt{3}$
$= -2\sqrt{3}$

$\sqrt{20} + \sqrt{45} = 2\sqrt{5} + 3\sqrt{5}$
$= (2+3)\sqrt{5}$
$= 5\sqrt{5}$

Multiplication of Radical Expressions [9.4]

We multiply radical expressions in the same way that we multiply polynomials. We can use the distributive property and the FOIL method.

7. $(\sqrt{x}+2)(\sqrt{x}+3)$
$= \sqrt{x}\sqrt{x} + 3\sqrt{x} + 2\sqrt{x} + 2 \cdot 3$
$= x + 5\sqrt{x} + 6$

Rationalizing the Denominator [9.2, 9.4]

When a fraction contains a square root in the denominator, we rationalize the denominator by multiplying numerator and denominator by

1. The square root itself if there is only one term in the denominator, or
2. The conjugate of the denominator if there are two terms in the denominator.

Rationalizing the denominator is also called division of radical expressions.

8. $\frac{3}{\sqrt{2}} = \frac{3}{\sqrt{2}} \cdot \frac{\sqrt{2}}{\sqrt{2}} = \frac{3\sqrt{2}}{2}$

$\frac{3}{\sqrt{5}-\sqrt{3}} = \frac{3}{\sqrt{5}-\sqrt{3}} \cdot \frac{\sqrt{5}+\sqrt{3}}{\sqrt{5}+\sqrt{3}}$
$= \frac{3\sqrt{5}+3\sqrt{3}}{5-3}$
$= \frac{3\sqrt{5}+3\sqrt{3}}{2}$

Squaring Property of Equality [9.5]

We may square both sides of an equation any time it is convenient to do so, as long as we check all resulting solutions in the original equation.

9. $\sqrt{2x+1} = 3$
$(\sqrt{2x+1})^2 = 3^2$
$2x + 1 = 9$
$x = 4$

Complex Numbers [9.6]

A *complex number* is any number that can be put in the form

$$a + bi$$

where a and b are real numbers and $i = \sqrt{-1}$. The *real part* of the complex number is a, and b is the *imaginary part.*

If a, b, c, and d are real numbers, then we have the following definitions associated with complex numbers:

1. Equality

$$a + bi = c + di \quad \text{if and only if} \quad a = c \text{ and } b = d$$

2. Addition and subtraction

$$(a + bi) + (c + di) = (a + c) + (b + d)i$$
$$(a + bi) - (c + di) = (a - c) + (b - d)i$$

3. Multiplication

$$(a + bi)(c + di) = (ac - bd) + (ad + bc)i$$

4. Division is similar to rationalizing the denominator.

10. $3 + 4i$ is a complex number.

Addition
$(3 + 4i) + (2 - 5i) = 5 - i$

Multiplication
$(3 + 4i)(2 - 5i)$
$= 6 - 15i + 8i - 20i^2$
$= 6 - 7i + 20$
$= 26 - 7i$

Division
$\frac{2}{3+4i} = \frac{2}{3+4i} \cdot \frac{3-4i}{3-4i}$
$= \frac{6-8i}{9+16}$
$= \frac{6}{25} - \frac{8}{25}i$

Simplify each of the following. (Assume all variable bases are positive integers and all variable exponents are positive real numbers throughout this test.) [9.1]

1. $27^{-2/3}$ **2.** $\left(\frac{25}{49}\right)^{-1/2}$ **3.** $a^{3/4} \cdot a^{-1/3}$ **4.** $\frac{(x^{2/3}y^{-3})^{1/2}}{(x^{3/4}y^{1/2})^{-1}}$

5. $\sqrt{49x^8y^{10}}$ **6.** $\sqrt[5]{32x^{10}y^{20}}$ **7.** $\frac{(36a^8b^4)^{1/2}}{(27a^9b^6)^{1/3}}$ **8.** $\frac{(x^ny^{1/n})^n}{(x^{1/n}y^n)^{n^2}}$

Multiply. [9.1]

9. $2a^{1/2}(3a^{3/2} - 5a^{1/2})$ **10.** $(4a^{3/2} - 5)^2$

Factor. [9.1]

11. $3x^{2/3} + 5x^{1/3} - 2$ **12.** $9x^{2/3} - 49$

Combine. [9.3]

13. $\frac{4}{x^{1/2}} + x^{1/2}$ **14.** $\frac{x^2}{(x^2 - 3)^{1/2}} - (x^2 - 3)^{1/2}$

Write in simplified form. [9.2]

15. $\sqrt{125x^3y^5}$ **16.** $\sqrt[3]{40x^7y^8}$ **17.** $\sqrt{\frac{2}{3}}$ **18.** $\sqrt{\frac{12a^4b^3}{5c}}$

Combine. [9.3]

19. $3\sqrt{12} - 4\sqrt{27}$ **20.** $\sqrt[3]{24a^3b^3} - 5a\sqrt[3]{3b^3}$

Multiply. [9.4]

21. $(\sqrt{x} + 7)(\sqrt{x} - 4)$ **22.** $(3\sqrt{2} - \sqrt{3})^2$

Rationalize the denominator. [9.4]

23. $\frac{5}{\sqrt{3} - 1}$ **24.** $\frac{\sqrt{x} - \sqrt{2}}{\sqrt{x} + \sqrt{2}}$

Solve for x. [9.5]

25. $\sqrt{3x + 1} = x - 3$ **26.** $\sqrt[3]{2x + 7} = -1$

27. $\sqrt{x + 3} = \sqrt{x + 4} - 1$

Graph. [9.5]

28. $y = \sqrt{x - 2}$ **29.** $y = \sqrt[3]{x} + 3$

30. Solve for x and y so that the following equation is true [9.6]:

$$(2x + 5) - 4i = 6 - (y - 3)i$$

Perform the indicated operations. [9.6]

31. $(3 + 2i) - [(7 - i) - (4 + 3i)]$ **32.** $(2 - 3i)(4 + 3i)$

33. $(5 - 4i)^2$ **34.** $\frac{2 - 3i}{2 + 3i}$

35. Show that i^{38} can be written as -1. [9.6}

Exponential and Logarithmic Functions

10

Chapter Outline

10.1 Exponential Functions

10.2 The Inverse of a Function

10.3 Logarithms are Exponents

10.4 Properties of Logarithms

10.5 Common Logarithms and Natural Logarithms

10.6 Exponential Equations and Change of Base

If you have had any problems with or had testing done on your thyroid gland, then you may have come in contact with radioactive iodine-131. Like all radioactive elements, iodine-131 decays naturally. The half-life of iodine-131 is 8 days, which means that every 8 days a sample of iodine-131 will decrease to half of its original amount. The following table and graph show what happens to a 1,600-microgram sample of iodine-131 over time.

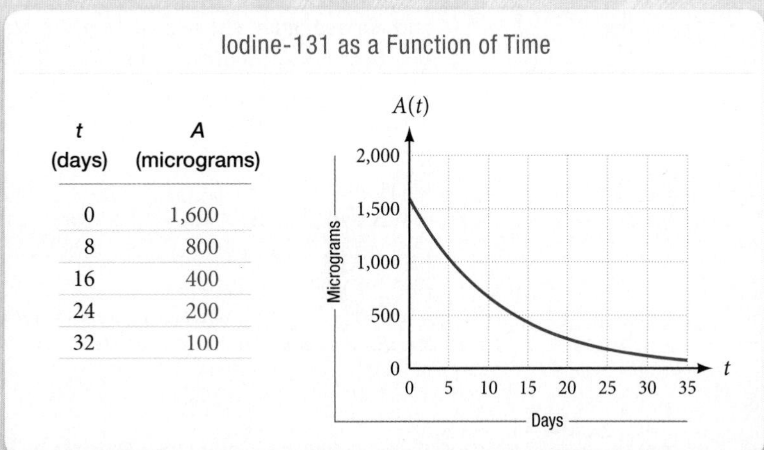

Iodine-131 as a Function of Time

t (days)	A (micrograms)
0	1,600
8	800
16	400
24	200
32	100

The function represented by the information in the table and graph is

$$A(t) = 1{,}600 \cdot 2^{-t/8}$$

It is one of the types of functions we will study in this chapter.

Success Skills

Dear Student,

Now that you are close to finishing this course, I want to pass on a couple of things that have helped me a great deal with my career. I'll introduce each one with a quote:

Do something for the person you will be 5 years from now.

I have always made sure that I arranged my life so that I was doing something for the person I would be 5 years later. For example, when I was 20 years old, I was in college. I imagined that the person I would be as a 25-year-old, would want to have a college degree, so I made sure I stayed in school. That's all there is to this. It is not a hard, rigid philosophy. It is a soft, behind the scenes, foundation. It does not include ideas such as "Five years from now I'm going to graduate at the top of my class from the best college in the country." Instead, you think, "five years from now I will have a college degree, or I will still be in school working towards it."

This philosophy led to a community college teaching job, writing textbooks, doing videos with the textbooks, then to MathTV and the book you are reading right now. Along the way there were many other options and directions that I didn't take, but all the choices I made were due to keeping the person I would be in 5 years in mind.

It's easier to ride a horse in the direction it is going.

I started my college career thinking that I would become a dentist. I enrolled in all the courses that were required for dental school. When I completed the courses, I applied to a number of dental schools, but wasn't accepted. I kept going to school, and applied again the next year, again, without success. My life was not going in the direction of dental school, even though I had worked hard to put it in that direction. So I did a little inventory of the classes I had taken and the grades I earned, and realized that I was doing well in mathematics. My life was actually going in that direction so I decided to see where that would take me. It was a good decision.

It is a good idea to work hard toward your goals, but it is also a good idea to take inventory every now and then to be sure you are headed in the direction that is best for you.

I wish you good luck with the rest of your college years, and with whatever you decide you want to do as a career.

Pat McKeague
Fall 2010

Exponential Functions

10.1

To obtain an intuitive idea of how exponential functions behave, we can consider the heights attained by a bouncing ball. When a ball used in the game of racquetball is dropped from any height, the first bounce will reach a height that is $\frac{2}{3}$ of the original height. The second bounce will reach $\frac{2}{3}$ of the height of the first bounce, and so on, as shown in Figure 1.

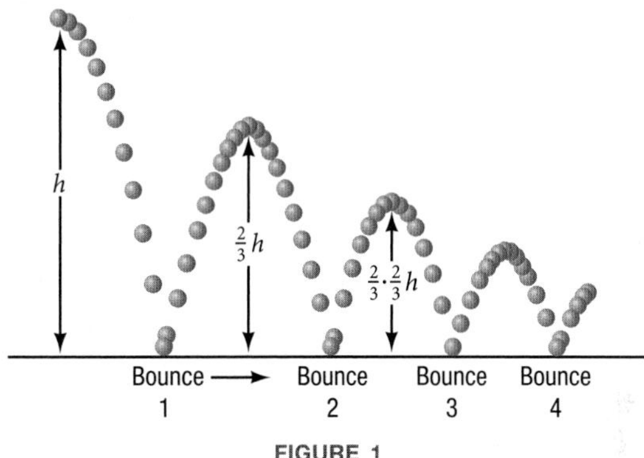

FIGURE 1

If the ball is initially dropped from a height of 1 meter, then during the first bounce it will reach a height of $\frac{2}{3}$ meter. The height of the second bounce will reach $\frac{2}{3}$ of the height reached on the first bounce. The maximum height of any bounce is $\frac{2}{3}$ of the height of the previous bounce.

Initial height: $h = 1$

Bounce 1: $h = \frac{2}{3}(1) = \frac{2}{3}$

Bounce 2: $h = \frac{2}{3}\left(\frac{2}{3}\right) = \left(\frac{2}{3}\right)^2$

Bounce 3: $h = \frac{2}{3}\left(\frac{2}{3}\right)^2 = \left(\frac{2}{3}\right)^3$

Bounce 4: $h = \frac{2}{3}\left(\frac{2}{3}\right)^3 = \left(\frac{2}{3}\right)^4$

$\vdots$

Bounce n: $h = \frac{2}{3}\left(\frac{2}{3}\right)^{n-1} = \left(\frac{2}{3}\right)^n$

This last equation is exponential in form. We classify all exponential functions together with the following definition.

(dēf **DEFINITION** *exponential function*

An ***exponential function*** is any function that can be written in the form

$$f(x) = b^x$$

where b is a positive real number other than 1.

Each of the following is an exponential function:

$$f(x) = 2^x \qquad y = 3^x \qquad f(x) = \left(\frac{1}{4}\right)^x$$

The first step in becoming familiar with exponential functions is to find some values for specific exponential functions.

EXAMPLE 1 If the exponential functions f and g are defined by

$$f(x) = 2^x \quad \text{and} \quad g(x) = 3^x$$

then

$$f(0) = 2^0 = 1 \qquad g(0) = 3^0 = 1$$

$$f(1) = 2^1 = 2 \qquad g(1) = 3^1 = 3$$

$$f(2) = 2^2 = 4 \qquad g(2) = 3^2 = 9$$

$$f(3) = 2^3 = 8 \qquad g(3) = 3^3 = 27$$

$$f(-2) = 2^{-2} = \frac{1}{2^2} = \frac{1}{4} \qquad g(-2) = 3^{-2} = \frac{1}{3^2} = \frac{1}{9}$$

$$f(-3) = 2^{-3} = \frac{1}{2^3} = \frac{1}{8} \qquad g(-3) = 3^{-3} = \frac{1}{3^3} = \frac{1}{27}$$

In the introduction to this chapter, we indicated that the half-life of iodine-131 is 8 days, which means that every 8 days a sample of iodine-131 will decrease to half of its original amount. If we start with A_0 micrograms of iodine-131, then after t days the sample will contain

$$A(t) = A_0 \cdot 2^{-t/8}$$

micrograms of iodine-131.

EXAMPLE 2 A patient is administered a 1,200-microgram dose of iodine-131. How much iodine-131 will be in the patient's system after 10 days, and after 16 days?

SOLUTION The initial amount of iodine-131 is $A_0 = 1{,}200$, so the function that gives the amount left in the patient's system after t days is

$$A(t) = 1{,}200 \cdot 2^{-t/8}$$

After 10 days, the amount left in the patient's system is

$$A(10) = 1{,}200 \cdot 2^{-10/8} = 1{,}200 \cdot 2^{-1.25} \approx 504.5 \text{ micrograms}$$

After 16 days, the amount left in the patient's system is

$$A(16) = 1{,}200 \cdot 2^{-16/8} = 1{,}200 \cdot 2^{-2} = 300 \text{ micrograms}$$

Note Recall that the symbol $\approx$ is read "is approximately equal to".

We will now turn our attention to the graphs of exponential functions. Because the notation y is easier to use when graphing, and $y = f(x)$, for convenience we will write the exponential functions as

$$y = b^x$$

EXAMPLE 3 Sketch the graph of the exponential function $y = 2^x$.

SOLUTION Using the results of Example 1, we produce the following table. Graphing the ordered pairs given in the table and connecting them with a smooth curve, we have the graph of $y = 2^x$ shown in Figure 2.

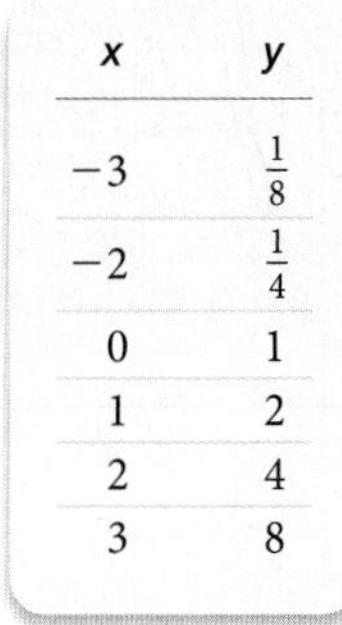

x	y
-3	$\frac{1}{8}$
-2	$\frac{1}{4}$
0	1
1	2
2	4
3	8

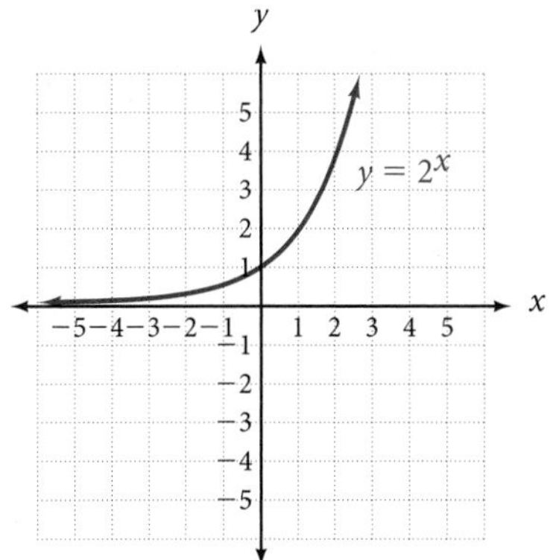

FIGURE 2

Notice that the graph does not cross the x-axis. It *approaches* the x-axis — in fact, we can get it as close to the x-axis as we want without it actually intersecting the x-axis. For the graph of $y = 2^x$ to intersect the x-axis, we would have to find a value of x that would make $2^x = 0$. Because no such value of x exists, the graph of $y = 2^x$ cannot intersect the x-axis.

EXAMPLE 4 Sketch the graph of $y = \left(\frac{1}{3}\right)^x$.

SOLUTION The table beside Figure 3 gives some ordered pairs that satisfy the equation. Using the ordered pairs from the table, we have the graph shown in Figure 3.

x	y
23	27
22	9
-1	3
0	1
1	$\frac{1}{3}$
2	$\frac{1}{9}$
3	$\frac{1}{27}$

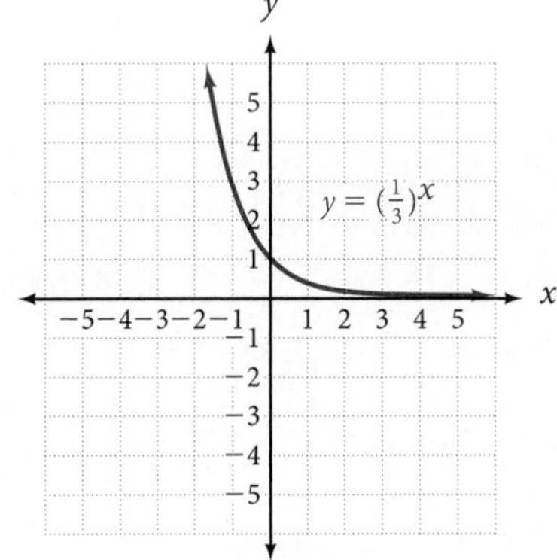

FIGURE 3

The graphs of all exponential functions have two things in common: (1) Each crosses the y-axis at $(0, 1)$ because $b^0 = 1$; and (2) none can cross the x-axis because $b^x = 0$ is impossible due to the restrictions on b.

Figures 4 and 5 show some families of exponential curves to help you become more familiar with them on an intuitive level.

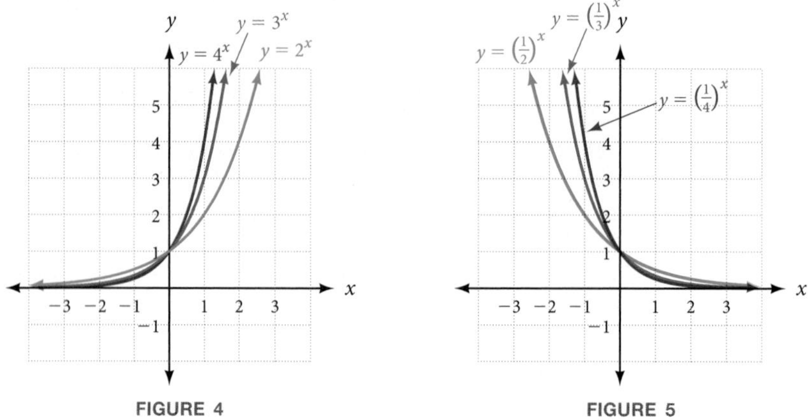

FIGURE 4

FIGURE 5

Among the many applications of exponential functions are the applications having to do with interest-bearing accounts. Here are the details.

Compound Interest If P dollars are deposited in an account with annual interest rate r, compounded n times per year, then the amount of money in the account after t years is given by the formula

$$A(t) = P\left(1 + \frac{r}{n}\right)^{nt}$$

EXAMPLE 5 Suppose you deposit \$500 in an account with an annual interest rate of 8% compounded quarterly. Find an equation that gives the amount of money in the account after t years. Then find

a. The amount of money in the account after 5 years.

b. The number of years it will take for the account to contain \$1,000.

SOLUTION First, we note that $P = 500$ and $r = 0.08$. Interest that is compounded quarterly is compounded four times a year, giving us $n = 4$. Substituting these numbers into the preceding formula, we have our function

$$A(t) = 500\left(1 + \frac{0.08}{4}\right)^{4t} = 500(1.02)^{4t}$$

a. To find the amount after 5 years, we let $t = 5$:

$$A(5) = 500(1.02)^{4 \cdot 5} = 500(1.02)^{20} \approx \$742.97$$

Our answer is found on a calculator, and then rounded to the nearest cent.

b. To see how long it will take for this account to total \$1,000, we graph the equation $Y_1 = 500(1.02)^{4X}$ on a graphing calculator, and then look to see where it intersects the line $Y_2 = 1{,}000$. The two graphs are shown in Figure 6.

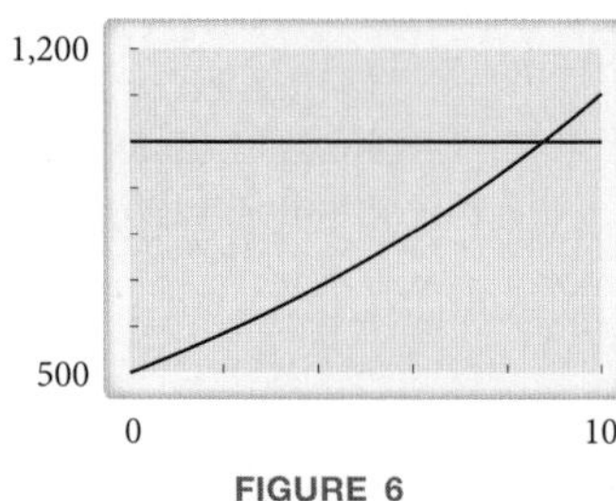

FIGURE 6

Using Zoom and Trace, or the Intersect function on the graphing calculator, we find that the two curves intersect at X ≈ 8.75 and Y = 1,000. This means that our account will contain \$1,000 after the money has been on deposit for 8.75 years. ■

The Natural Exponential Function

A commonly occurring exponential function is based on a special number we denote with the letter e. The number e is a number like π. It is irrational and occurs in many formulas that describe the world around us. Like π, it can be approximated with a decimal number. Whereas π is approximately 3.1416, e is approximately 2.7183. (If you have a calculator with a key labeled e^x, you can use it to find e^1 to find a more accurate approximation to e.) We cannot give a more precise definition of the *number e* without using some of the topics taught in calculus. For the work we are going to do with the number e, we only need to know that it is an irrational number that is approximately 2.7183.

Here are a table and graph (Figure 7) for the natural exponential function

$$y = f(x) = e^x$$

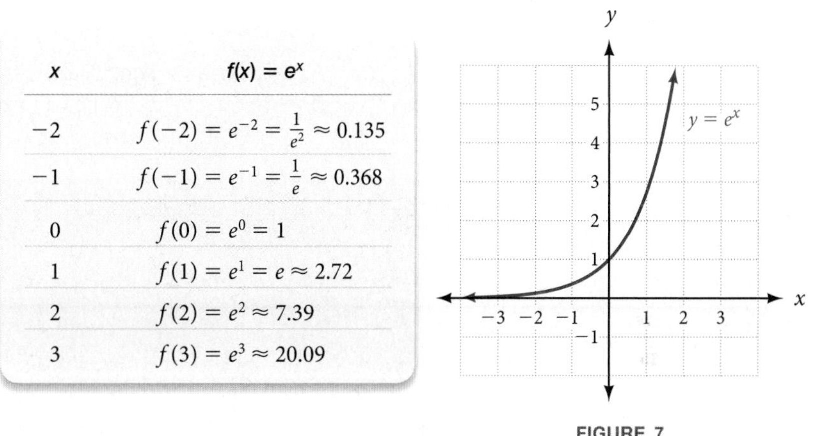

x	$f(x) = e^x$
-2	$f(-2) = e^{-2} = \frac{1}{e^2} \approx 0.135$
-1	$f(-1) = e^{-1} = \frac{1}{e} \approx 0.368$
0	$f(0) = e^0 = 1$
1	$f(1) = e^1 = e \approx 2.72$
2	$f(2) = e^2 \approx 7.39$
3	$f(3) = e^3 \approx 20.09$

FIGURE 7

One common application of natural exponential functions is with interest-bearing accounts. In Example 5, we worked with the formula

$$A(t) = P\left(1 + \frac{r}{n}\right)^{nt}$$

that gives the amount of money in an account if P dollars are deposited for t years at annual interest rate r, compounded n times per year. In Example 5, the number of compounding periods was four. What would happen if we let the number of compounding periods become larger and larger, so that we compounded the interest every day, then every hour, then every second, and so on? If we take this as far as it can go, we end up compounding the interest every moment. When this happens, we have an account with interest that is compounded continuously, and the amount of money in such an account depends on the number e. Here are the details.

Continuously Compounded Interest If P dollars are deposited in an account with annual interest rate r, compounded continuously, then the amount of money in the account after t years is given by the formula

$$A(t) = Pe^{rt}$$

EXAMPLE 6 Suppose you deposit \$500 in an account with an annual interest rate of 8% compounded continuously. Find an equation that gives the amount of money in the account after t years. Then find the amount of money in the account after 5 years.

SOLUTION Because the interest is compounded continuously, we use the formula $A(t) = Pe^{rt}$. Substituting $P = 500$ and $r = 0.08$ into this formula, we have

$$A(t) = 500e^{0.08t}$$

After 5 years, this account will contain

$$A(5) = 500e^{0.08 \cdot 5} = 500e^{0.4} \approx \$745.91$$

to the nearest cent. Compare this result with the answer to Example 5a.

GETTING READY FOR CLASS

After reading through the preceding section, respond in your own words and in complete sentences.

A. What is an exponential function?

B. In an exponential function, explain why the base b cannot equal 1. (What kind of function would you get if the base was equal to 1?)

C. Explain continuously compounded interest.

D. What characteristics do the graphs of $y = 2^x$ and $y = \left(\frac{1}{2}\right)^x$ have in common?

1. 1
2. 1
3. 2
4. 16
5. $\frac{1}{27}$
6. $\frac{1}{3}$
7. 13
8. 5
9. $\frac{7}{12}$
10. 25
11. $\frac{3}{16}$
12. 63.9

13.-28. See answer section.

29. $h = 6 \cdot \left(\frac{2}{3}\right)^n$; 5th bounce: $6\left(\frac{2}{3}\right)^5 \approx 0.79$ feet
30. $h = A\left(\frac{1}{2}\right)^n$; $h(8) = 0.039$ feet
31. 4.27 days
32. In 1990, \$1.89 per pound; in 1995, \$5.77 per pound; in 2010, \$163.93 per pound

Let $f(x) = 3^x$ and $g(x) = \left(\frac{1}{2}\right)^x$, and evaluate each of the following.

1. $g(0)$ **2.** $f(0)$ **3.** $g(-1)$ **4.** $g(-4)$

5. $f(-3)$ **6.** $f(-1)$ **7.** $f(2) + g(-2)$ **8.** $f(2) - g(-2)$

Let $f(x) = 4^x$ and $g(x) = \left(\frac{1}{3}\right)^x$. Evaluate each of the following.

9. $f(-1) + g(1)$ **10.** $f(2) + g(-2)$ **11.** $\frac{f(-2)}{g(1)}$ **12.** $f(3) - f(2)$

Graph each of the following functions.

13. $y = 4^x$ **14.** $y = 2^{-x}$ **15.** $y = 3^{-x}$ **16.** $y = \left(\frac{1}{3}\right)^{-x}$

17. $y = 2^{x+1}$ **18.** $y = 2^{x-3}$ **19.** $y = e^x$ **20.** $y = e^{-x}$

21. $y = \left(\frac{1}{3}\right)^x$ **22.** $y = \left(\frac{1}{2}\right)^{-x}$ **23.** $y = 3^{x+2}$ **24.** $y = 2 \cdot 3^{-x}$

Graph each of the following functions on the same coordinate system for positive values of x only.

25. $y = 2x, y = x^2, y = 2^x$ **26.** $y = 3x, y = x^3, y = 3^x$

27. On a graphing calculator, graph the family of curves $y = b^x$, $b = 2, 4, 6, 8$.

28. On a graphing calculator, graph the family of curves $y = b^x$, $b = \frac{1}{2}, \frac{1}{4}, \frac{1}{6}, \frac{1}{8}$.

Applying the Concepts

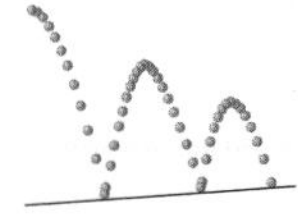

29. Bouncing Ball Suppose the ball mentioned in the introduction to this section is dropped from a height of 6 feet above the ground. Find an exponential equation that gives the height h the ball will attain during the nth bounce. How high will it bounce on the fifth bounce?

30. Bouncing Ball A golf ball is manufactured so that if it is dropped from A feet above the ground onto a hard surface, the maximum height of each bounce will be one half of the height of the previous bounce. Find an exponential equation that gives the height h the ball will attain during the nth bounce. If the ball is dropped from 10 feet above the ground onto a hard surface, how high will it bounce on the eighth bounce?

31. Exponential Decay Twinkies on the shelf of a convenience store lose their fresh tastiness over time. We say that the taste quality is 1 when the Twinkies are first put on the shelf at the store, and that the quality of tastiness declines according to the function $Q(t) = 0.85^t$ (t in days). Graph this function on a graphing calculator, and determine when the taste quality will be one half of its original value.

32. Exponential Growth Automobiles built before 1993 use Freon in their air conditioners. The federal government now prohibits the manufacture of Freon. Because the supply of Freon is decreasing, the price per pound is increasing exponentially. Current estimates put the formula for the price per pound of Freon at $p(t) = 1.89(1.25)^t$, where t is the number of years since 1990. Find the price of Freon in 1995 and 1990. How much will Freon cost in the year 2010?

33. a. $A(t) = 1{,}200\left(1 + \frac{.06}{4}\right)^{4t}$
b. \$1,932.39
c. About 11.64 years
d. \$1,939.29
34. a. $A(t) = 500\left(1 + \frac{.08}{12}\right)^{12t}$
b. $A(5) \approx \$744.92$
c. About 8.7 years
d. \$745.91
35. a. \$129,138.48
b. $\{t \mid 0 \le t \le 6\}$
c. See answer section.
d. $\{V(t) \mid 52{,}942.05 \le V(t) \le 450{,}000\}$
e. After approximately 4 years and 8 months
36. a. \$95,625.18
b. $\{t \mid 0 \le t \le 7\}$
d. $\{V(t) \mid 50{,}056.46 \le V(t) = 375{,}000\}$
e. About 6 years, 1 month
37. $f(1) = 200$, $f(2) = 800$, $f(3) = 3{,}200$
38. $f(1) = 200$; $f(2) = 400$; $f(3) = 800$; $f(4) = 1{,}600$ Will have 100,000 after 9.97 days.

33. Compound Interest Suppose you deposit \$1,200 in an account with an annual interest rate of 6% compounded quarterly.

a. Find an equation that gives the amount of money in the account after t years.

b. Find the amount of money in the account after 8 years.

c. How many years will it take for the account to contain \$2,400?

d. If the interest were compounded continuously, how much money would the account contain after 8 years?

34. Compound Interest Suppose you deposit \$500 in an account with an annual interest rate of 8% compounded monthly.

a. Find an equation that gives the amount of money in the account after t years.

b. Find the amount of money in the account after 5 years.

c. How many years will it take for the account to contain \$1,000?

d. If the interest were compounded continuously, how much money would the account contain after 5 years?

Declining-Balance Depreciation The declining-balance method of depreciation is an accounting method businesses use to deduct most of the cost of new equipment during the first few years of purchase. Unlike other methods, the declining-balance formula does not consider salvage value.

35. Value of a Crane The function

$$V(t) = 450{,}000\,(1 - 0.30)^t,$$

where V is value and t is time in years, can be used to find the value of a crane for the first 6 years of use.

a. What is the value of the crane after 3 years and 6 months?

b. State the domain of this function.

c. Sketch the graph of this function.

d. State the range of this function.

e. After how many years will the crane be worth only \$85,000?

36. Value of a Printing Press The function $V(t) = 375{,}000(1 - 0.25)^t$, where V is value and t is time in years, can be used to find the value of a printing press during the first 7 years of use.

a. What is the value of the printing press after 4 years and 9 months?

b. State the domain of this function.

c. Sketch the graph of this function.

d. State the range of this function.

e. After how many years will the printing press be worth only \$65,000?

37. Bacteria Growth Suppose it takes 12 hours for a certain strain of bacteria to reproduce by dividing in half. If 50 bacteria are present to begin with, then the total number present after x days will be $f(x) = 50 \cdot 4^x$. Find the total number present after 1 day, 2 days, and 3 days.

38. Bacteria Growth Suppose it takes 1 day for a certain strain of bacteria to reproduce by dividing in half. If 100 bacteria are present to begin with, then the total number present after x days will be $f(x) = 100 \cdot 2^x$. Find the total number present after 1 day, 2 days, 3 days, and 4 days. How many days must elapse before over 100,000 bacteria are present?

39.-40. See answer section.
41. a. \$0.42
b. \$1.00
c. \$1.78
d. \$17.84
42. a. Approximately 593 million passengers traveled by plane in 1997.
b. Approximately 1.101 billion passengers will travel by plane in 2015.
43. 1,258,525 bankruptcies, which is 58,474 less than the actual number.
44. a. The value of the car after 3 years in approximately \$12,148.53.
b. The depreciated amount (purchase price − value at 4 years = 25,600 − 9475.85) is approximately \$16,124.15.
45. a. 251,437 cells
b. 12,644 cells
c. 32 cells

39. Value of a Painting A painting is purchased as an investment for \$150. If the painting's value doubles every 3 years, then its value is given by the function

$$V(t) = 150 \cdot 2^{t/3} \quad \text{for } t \geq 0$$

where t is the number of years since it was purchased, and $V(t)$ is its value (in dollars) at that time. Graph this function.

40. Value of a Painting A painting is purchased as an investment for \$125. If the painting's value doubles every 5 years, then its value is given by the function

$$V(t) = 125 \cdot 2^{t/5} \quad \text{for } t \geq 0$$

where t is the number of years since it was purchased, and $V(t)$ is its value (in dollars) at that time. Graph this function.

41. Cost Increase The cost of a can of Coca Cola in 1960 was \$0.10. The exponential function that models the cost of a Coca Cola by year is given below, where t is the number of years since 1960.

$$C(t) = 0.10e^{0.0576t}$$

a. What was the expected cost of a can of Coca Cola in 1985?

b. What was the expected cost of a can of Coca Cola in 2000?

c. What is the expected cost of a can of Coca Cola in 2010?

d. What is the expected cost of a can of Coca Cola in 2050?

42. Airline Travel The number of airline passengers in 1990 was 466 million. The number of passengers traveling by airplane each year has increased exponentially according to the model, $P(t) = 466 \cdot 1.035^t$, where t is the number of years since 1990 (U.S. Census Bureau).

a. How many passengers traveled in 1997?

b. How many passengers will travel in 2015?

43. Bankruptcy Model In 1997, there were a total of 1,316,999 bankruptcies filed under the Bankruptcy Reform Act (Administrative Office of the U.S. Courts, Statistical Tables for the Federal Judiciary). The model for the number of bankruptcies filed is $B(t) = 0.798 \cdot 1.164^t$, where t is the number of years since 1994 and B is the number of bankruptcies filed in terms of millions. How close was the model in predicting the actual number of bankruptcies filed in 1997?

44. Value of a Car As a car ages, its value decreases. The value of a particular car with an original purchase price of \$25,600 is modeled by the following function, where c is the value at time t (Kelly Blue Book).

$$c(t) = 25{,}600(1 - 0.22)^t$$

a. What is the value of the car when it is 3 years old?

b. What is the total depreciation amount after 4 years?

45. Bacteria Decay You are conducting a biology experiment and begin with 5,000,000 cells, but some of those cells are dying each minute. The rate of death of the cells is modeled by the function $A(t) = A_0 \cdot e^{-0.598t}$, where A_0 is the original number of cells, t is time in minutes, and A is the number of cells remaining after t minutes.

a. How may cells remain after 5 minutes?

b. How many cells remain after 10 minutes?

c. How many cells remain after 20 minutes?

46. a. The function underestimated the expenditures by $69 billion.
b. $4,123 billion
$4,577 billion
$5,080 billion

47. $y = \frac{x+3}{2}$

48. $y = 5x - 7$

49. $y = \pm\sqrt{x+3}$

50. $y = \sqrt[3]{x} - 4$

51. $y = \frac{2x-4}{x-1}$

52. $y = \frac{3x+5}{x-1}$

53. $y = x^2 + 3$

54. $y = (x-5)^2$

46. Health Care In 1990, $699 billion were spent on health care expenditures. The amount of money, E, in billions spent on health care expenditures can be estimated using the function $E(t) = 78.16(1.11)^t$, where t is time in years since 1970 (U.S. Census Bureau).

a. How close was the estimate determined by the function in estimating the actual amount of money spent on health care expenditures in 1990?

b. What are the expected health care expenditures in 2008, 2009, and 2010?

Getting Ready for the Next Section

Solve each equation for *y*.

47. $x = 2y - 3$

48. $x = \frac{y+7}{5}$

49. $x = y^2 - 3$

50. $x = (y+4)^3$

51. $x = \frac{y-4}{y-2}$

52. $x = \frac{y+5}{y-3}$

53. $x = \sqrt{y-3}$

54. $x = \sqrt{y} + 5$

The Inverse of a Function 10.2

The following diagram (Figure 1) shows the route Justin takes to school. He leaves his home and drives 3 miles east, and then turns left and drives 2 miles north. When he leaves school to drive home, he drives the same two segments, but in the reverse order and the opposite direction; that is, he drives 2 miles south, turns right, and drives 3 miles west. When he arrives home from school, he is right where he started. His route home "undoes" his route to school, leaving him where he began.

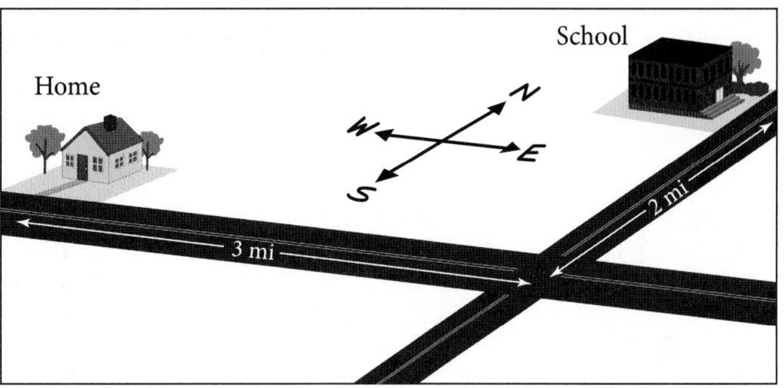

FIGURE 1

As you will see, the relationship between a function and its inverse function is similar to the relationship between Justin's route from home to school and his route from school to home.

Suppose the function f is given by

$$f = (1, 4), (2, 5), (3, 6), (4, 7)$$

The inverse of f is obtained by reversing the order of the coordinates in each ordered pair in f. The inverse of f is the relation given by

$$g = (4, 1), (5, 2), (6, 3), (7, 4)$$

It is obvious that the domain of f is now the range of g, and the range of f is now the domain of g. Every function (or relation) has an inverse that is obtained from the original function by interchanging the components of each ordered pair.

Suppose a function f is defined with an equation instead of a list of ordered pairs. We can obtain the equation of the inverse of f by interchanging the role of x and y in the equation for f.

EXAMPLE 1 If the function f is defined by $f(x) = 2x - 3$, find the equation that represents the inverse of f.

SOLUTION Because the inverse of f is obtained by interchanging the components of all the ordered pairs belonging to f, and each ordered pair in f satisfies the equation $y = 2x - 3$, we simply exchange x and y in the equation $y = 2x - 3$ to get the formula for the inverse of f:

$$x = 2y - 3$$

We now solve this equation for y in terms of x:

$$x + 3 = 2y$$

$$\frac{x + 3}{2} = y$$

$$y = \frac{x + 3}{2}$$

The last line gives the equation that defines the inverse of f. Let's compare the graphs of f and its inverse as given here. (See Figure 2.)

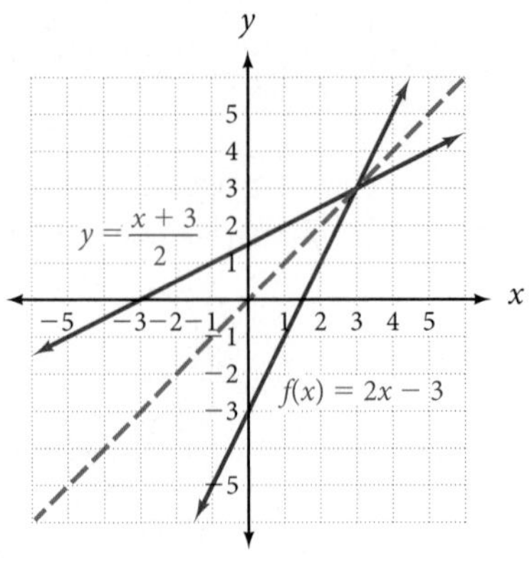

FIGURE 2

The graphs of f and its inverse have symmetry about the line $y = x$. This is a reasonable result since the one function was obtained from the other by interchanging x and y in the equation. The ordered pairs (a, b) and (b, a) always have symmetry about the line $y = x$.

EXAMPLE 2 Graph the function $y = x^2 - 2$ and its inverse. Give the equation for the inverse.

SOLUTION We can obtain the graph of the inverse of $y = x^2 - 2$ by graphing $y = x^2 - 2$ by the usual methods, and then reflecting the graph about the line $y = x$.

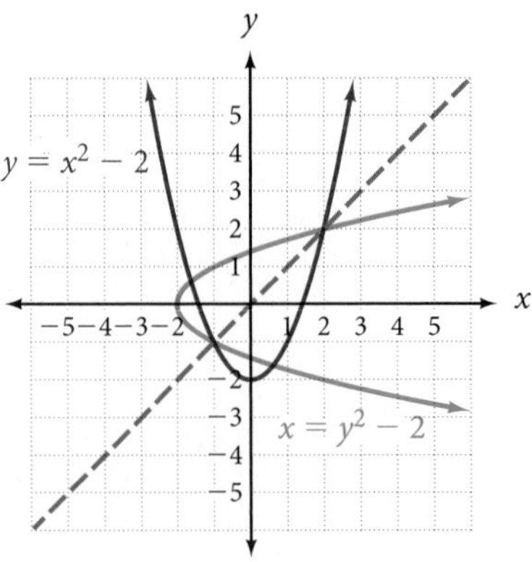

FIGURE 3

The equation that corresponds to the inverse of $y = x^2 - 2$ is obtained by interchanging x and y to get $x = y^2 - 2$.

We can solve the equation $x = y^2 - 2$ for y in terms of x as follows:

$$x = y^2 - 2$$
$$x + 2 = y^2$$
$$y = \pm\sqrt{x + 2}$$

Comparing the graphs from Examples 1 and 2, we observe that the inverse of a function is not always a function. In Example 1, both f and its inverse have graphs that are nonvertical straight lines and therefore both represent functions. In Example 2, the inverse of function f is not a function, since a vertical line crosses it in more than one place.

One-to-One Functions

We can distinguish between those functions with inverses that are also functions and those functions with inverses that are not functions with the following definition.

(dĕf DEFINITION *one-to-one functions*

A function is a ***one-to-one function*** if every element in the range comes from exactly one element in the domain.

This definition indicates that a one-to-one function will yield a set of ordered pairs in which no two different ordered pairs have the same second coordinates. For example, the function

$$f = \{(2, 3), (-1, 3), (5, 8)\}$$

is not one-to-one because the element 3 in the range comes from both 2 and -1 in the domain. On the other hand, the function

$$g = \{(5, 7), (3, -1), (4, 2)\}$$

is a one-to-one function because every element in the range comes from only one element in the domain.

Horizontal Line Test

If we have the graph of a function, we can determine if the function is one-to-one with the following test. If a horizontal line crosses the graph of a function in more than one place, then the function is not a one-to-one function because the points at which the horizontal line crosses the graph will be points with the same y-coordinates, but different x-coordinates. Therefore, the function will have an element in the range (the y-coordinate) that comes from more than one element in the domain (the x-coordinates).

Of the functions we have covered previously, all the linear functions and exponential functions are one-to-one functions because no horizontal lines can be found that will cross their graphs in more than one place.

Functions Whose Inverses Are Also Functions

Because one-to-one functions do not repeat second coordinates, when we reverse the order of the ordered pairs in a one-to-one function, we obtain a relation in which no two ordered pairs have the same first coordinate—by definition, this relation must be a function. In other words, every one-to-one function has an inverse that is itself a function. Because of this, we can use function notation to represent that inverse.

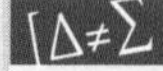

Inverse Function Notation

If $y = f(x)$ is a one-to-one function, then the inverse of f is also a function and can be denoted by $y = f^{-1}(x)$.

Note The notation f^{-1} does not represent the reciprocal of f. That is, the -1 in this notation is not an exponent. The notation f^{-1} is defined as representing the inverse function for a one-to-one function.

To illustrate, in Example 1 we found that the inverse of $f(x) = 2x - 3$ was the function $y = \frac{x+3}{2}$. We can write this inverse function with inverse function notation as

$$f^{-1}(x) = \frac{x+3}{2}$$

On the other hand, the inverse of the function in Example 2 is not itself a function, so we do not use the notation $f^{-1}(x)$ to represent it.

EXAMPLE 3 Find the inverse of $g(x) = \dfrac{x-4}{x-2}$.

SOLUTION To find the inverse for g, we begin by replacing $g(x)$ with y to obtain

$$y = \frac{x-4}{x-2} \qquad \text{The original function}$$

To find an equation for the inverse, we exchange x and y.

$$x = \frac{y-4}{y-2} \qquad \text{The inverse of the original function}$$

To solve for y, we first multiply each side by $y - 2$ to obtain

$$x(y-2) = y - 4$$

$$xy - 2x = y - 4 \qquad \text{Distributive property}$$

$$xy - y = 2x - 4 \qquad \text{Collect all terms containing } y \text{ on the left side}$$

$$y(x-1) = 2x - 4 \qquad \text{Factor } y \text{ from each term on the left side}$$

$$y = \frac{2x-4}{x-1} \qquad \text{Divide each side by } x - 1$$

Because our original function is one-to-one, as verified by the graph in Figure 4, its inverse is also a function. Therefore, we can use inverse function notation to write

$$g^{-1}(x) = \frac{2x-4}{x-1}$$

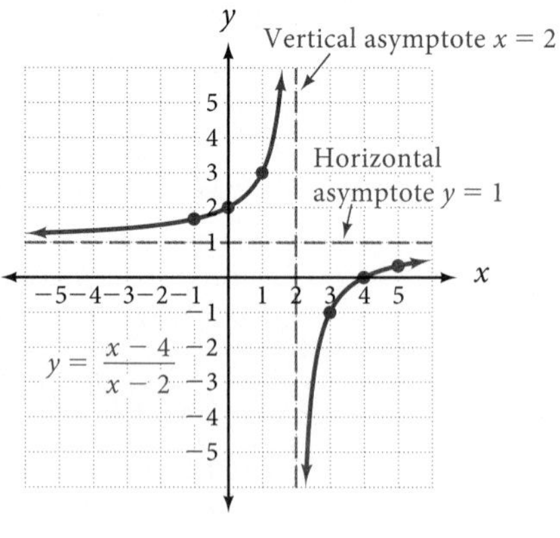

FIGURE 4

EXAMPLE 4 Graph the function $y = 2^x$ and its inverse $x = 2^y$.

SOLUTION We graphed $y = 2^x$ in the preceding section. We simply reflect its graph about the line $y = x$ to obtain the graph of its inverse $x = 2^y$.

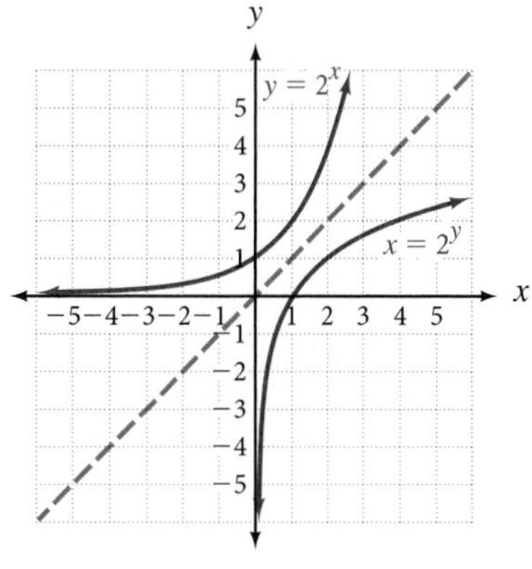

FIGURE 5

As you can see from the graph, $x = 2^y$ is a function. We do not have the mathematical tools to solve this equation for y, however. Therefore, we are unable to use the inverse function notation to represent this function. In the next section, we will give a definition that solves this problem. For now, we simply leave the equation as $x = 2^y$.

Functions, Relations, and Inverses—A Summary

Here is a summary of some of the things we know about functions, relations, and their inverses:

1. Every function is a relation, but not every relation is a function.
2. Every function has an inverse, but only one-to-one functions have inverses that are also functions.
3. The domain of a function is the range of its inverse, and the range of a function is the domain of its inverse.
4. If $y = f(x)$ is a one-to-one function, then we can use the notation $y = f^{-1}(x)$ to represent its inverse function.
5. The graph of a function and its inverse have symmetry about the line $y = x$.
6. If (a, b) belongs to the function f, then the point (b, a) belongs to its inverse.

GETTING READY FOR CLASS

After reading through the preceding section, respond in your own words and in complete sentences.

A. What is the inverse of a function?

B. What is the relationship between the graph of a function and the graph of its inverse?

C. Explain why only one-to-one functions have inverses that are also functions.

D. Describe the vertical line test, and explain the difference between the vertical line test and the horizontal line test.

Problem Set 10.2

For each of the following one-to-one functions, find the equation of the inverse. Write the inverse using the notation $f^{-1}(x)$.

1. $f(x) = 3x - 1$ **2.** $f(x) = 2x - 5$ **3.** $f(x) = x^3$

4. $f(x) = x^3 - 2$ **5.** $f(x) = \dfrac{x-3}{x-1}$ **6.** $f(x) = \dfrac{x-2}{x-3}$

7. $f(x) = \dfrac{x-3}{4}$ **8.** $f(x) = \dfrac{x+7}{2}$ **9.** $f(x) = \dfrac{1}{2}x - 3$

10. $f(x) = \dfrac{1}{3}x + 1$ **11.** $f(x) = \dfrac{2}{3}x - 3$ **12.** $f(x) = -\dfrac{1}{2}x + 4$

13. $f(x) = x^3 - 4$ **14.** $f(x) = -3x^3 + 2$ **15.** $f(x) = \dfrac{4x-3}{2x+1}$

16. $f(x) = \dfrac{3x-5}{4x+3}$ **17.** $f(x) = \dfrac{2x+1}{3x+1}$ **18.** $f(x) = \dfrac{3x+2}{5x+1}$

For each of the following relations, sketch the graph of the relation and its inverse, and write an equation for the inverse.

19. $y = 2x - 1$ **20.** $y = 3x + 1$ **21.** $y = x^2 - 3$ **22.** $y = x^2 + 1$

23. $y = x^2 - 2x - 3$ **24.** $y = x^2 + 2x - 3$

25. $y = 3^x$ **26.** $y = \left(\dfrac{1}{2}\right)^x$ **27.** $y = 4$ **28.** $y = -2$

29. $y = \dfrac{1}{2}x^3$ **30.** $y = x^3 - 2$ **31.** $y = \dfrac{1}{2}x + 2$ **32.** $y = \dfrac{1}{3}x - 1$

33. $y = \sqrt{x+2}$ **34.** $y = \sqrt{x} + 2$

35. Determine if the following functions are one-to-one.

a. **b.** **c.**

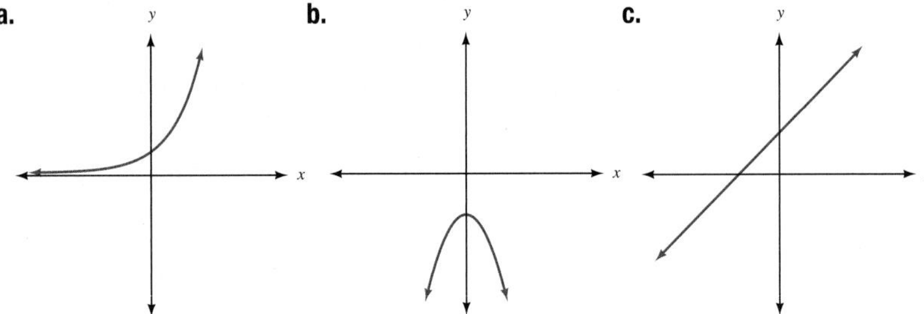

36. Could the following tables of values represent ordered pairs from one-to-one functions? Explain your answer.

a.

x	y
−2	5
−1	4
0	3
1	4
2	5

b.

x	y
1.5	0.1
2.0	0.2
2.5	0.3
3.0	0.4
3.5	0.5

37. If $f(x) = 3x - 2$, then $f^{-1}(x) = \dfrac{x+2}{3}$. Use these two functions to find

a. $f(2)$ **b.** $f^{-1}(2)$ **c.** $f[f^{-1}(2)]$ **d.** $f^{-1}[f(2)]$

---

1. $f^{-1}(x) = \frac{x+1}{3}$
2. $f^{-1}(x) = \frac{x+5}{2}$
3. $f^{-1}(x) = \sqrt[3]{x}$
4. $f^{-1}(x) = \sqrt[3]{x+2}$
5. $f^{-1}(x) = \frac{x-3}{x-1}$
6. $f^{-1}(x) = \frac{3x-2}{x-1}$
7. $f^{-1}(x) = 4x + 3$
8. $f^{-1}(x) = 2x - 7$
9. $f^{-1}(x) = 2(x+3) = 2x + 6$
10. $f^{-1}(x) = 3(x-1) = 3x - 3$
11. $f^{-1}(x) = \frac{3}{2}(x+3) = \frac{3}{2}x + \frac{9}{2}$
12. $f^{-1}(x) = -2(x-4) = -2x + 8$
13. $f^{-1}(x) = \sqrt[3]{x+4}$
14. $f^{-1}(x) = \sqrt[3]{\frac{2-x}{3}}$
15. $f^{-1}(x) = \frac{x+3}{4-2x}$
16. $f^{-1}(x) = \frac{3x+5}{3-4x}$
17. $f^{-1}(x) = \frac{1-x}{3x-2}$
18. $f^{-1}(x) = \frac{2-x}{5x-3}$
19. $f^{-1}(x) = \frac{x+1}{2}$
20. $f^{-1}(x) = \frac{x-1}{3}$
21. $f^{-1}(x) = \pm\sqrt{x+3}$
22. $f^{-1}(x) = \pm\sqrt{x-1}$
23. $f^{-1}(x) = 1 \pm \sqrt{x+4}$
24. $f^{-1}(x) = -1 \pm \sqrt{x+4}$
25. $x = 3^y$
26. $x = \left(\frac{1}{2}\right)^y$
27. $x = 4$
28. $x = -2$
29. $f^{-1}(x) = \sqrt[3]{2x}$
30. $f^{-1}(x) = \sqrt[3]{x+2}$
31. $f^{-1}(x) = 2(x-2)$
32. $f^{-1}(x) = 3(x+1)$
33. $f^{-1}(x) = x^2 - 2$
34. $f^{-1}(x) = x^2 - 4x + 4$
35. a. Yes
    b. No
    c. Yes
36. a. No
    b. Yes
37. a. 4
    b. $\frac{4}{3}$
    c. 2
    d. 2

38. a. 3
b. −18
c. −4
d. −4
39. $f^{-1}(x) = \frac{1}{x}$
40. $f^{-1}(x) = \frac{a}{x}$
41. $f^{-1}(x) = 7(x + 2) = 7x + 14$
42. a. $f^{-1}(x) = \frac{x-7}{2}$
b. $f^{-1}(x) = (x+9)^2$
c. $f^{-1}(x) = \sqrt[3]{x+4}$
d. $f^{-1}(x) = \sqrt[3]{x^2+4}$
43. a. −3
b. −6
c. 2
d. 3
e. −2
f. 3
g. inverses
44. a. Domain of $f(x)$: {−6, 2, 3, 6}; Range of $f(x)$: {−3, −2, 3, 4}
b. Domain of $g(x)$: {−3, −2, 3, 4}; Range of $g(x)$: {−6, 2, 3, 6}
c. Domain of $f(x)$ = Range of $g(x)$; Range of $f(x)$ = Domain $g(x)$
d. Yes
e. Yes

38. If $f(x) = \frac{1}{2}x + 5$, then $f^{-1}(x) = 2x - 10$. Use these two functions to find

a. $f(-4)$ **b.** $f^{-1}(-4)$ **c.** $f[f^{-1}(-4)]$ **d.** $f^{-1}[f(-4)]$

39. Let $f(x) = \frac{1}{x}$, and find $f^{-1}(x)$.

40. Let $f(x) = \frac{a}{x}$, and find $f^{-1}(x)$. (a is a real number constant.)

Applying the Concepts

41. Inverse Functions in Words Inverses may also be found by *inverse reasoning*. For example, to find the inverse of $f(x) = 3x + 2$, first list, in order, the operations done to variable x:

a. Multiply by 3. **b.** Add 2.

Then, to find the inverse, simply apply the inverse operations, in reverse order, to the variable x. That is:

c. Subtract 2. **d.** Divide by 3.

The inverse function then becomes $f^{-1}(x) = \frac{x-2}{3}$. Use this method of "inverse reasoning" to find the inverse of the function $f(x) = \frac{x}{7} - 2$.

42. Inverse Functions in Words Refer to the method of *inverse reasoning* explained in Problem 41. Use *inverse reasoning* to find the following inverses:

a. $f(x) = 2x + 7$ **b.** $f(x) = \sqrt{x} - 9$ **c.** $f(x) = x^3 - 4$ **d.** $f(x) = \sqrt{x^3 - 4}$

43. Reading Tables Evaluate each of the following functions using the functions defined by Tables 1 and 2.

a. $f[g(-3)]$ **b.** $g[f(-6)]$ **c.** $g[f(2)]$
d. $f[g(3)]$ **e.** $f[g(-2)]$ **f.** $g[f(3)]$
g. What can you conclude about the relationship between functions f and g?

TABLE 1

x	$f(x)$
−6	3
2	−3
3	−2
6	4

TABLE 2

x	$g(x)$
−3	2
−2	3
3	−6
4	6

44. Reading Tables Use the functions defined in Tables 1 and 2 in Problem 43 to answer the following questions.

a. What are the domain and range of f?
b. What are the domain and range of g?
c. How are the domain and range of f related to the domain and range of g?
d. Is f a one-to-one function?
e. Is g a one-to-one function?

45. a. 489.4
b. $s^{-1}(t) = \frac{t - 249.4}{16}$
c. 2006
46. a. 32.34%
b. $f^{-1}(x) = \frac{x - 24}{.417}$; 2002
47. a. 6629.33 ft/s
b. $f^{-1}(m) = \frac{15m}{22}$
c. 1.36 mph

45. Social Security A function that models the billions of dollars of Social Security payment (as shown in the chart) per year is $s(t) = 16t + 249.4$, where t is time in years since 1990 (U.S. Census Bureau).

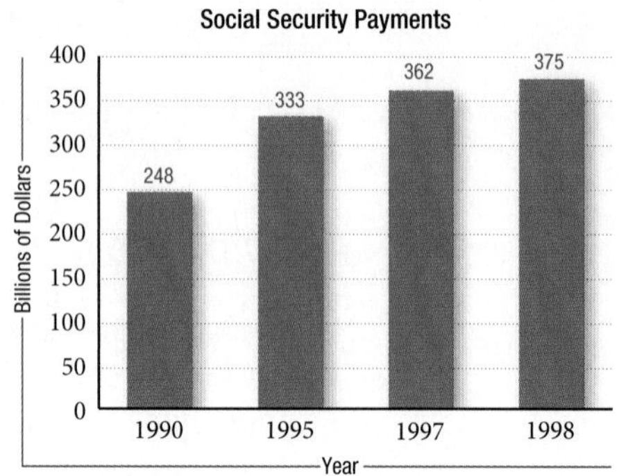

a. Use the model to estimate the amount of Social Security payments to be paid in 2005.

b. Write the inverse of the function.

c. Using the inverse function, estimate the year in which payments will reach \$507 billion.

46. Families The function for the percentage of one-parent families (as shown in the following chart) is $f(x) = 0.417x + 24$, when x is the time in years since 1990 (U.S. Census Bureau).

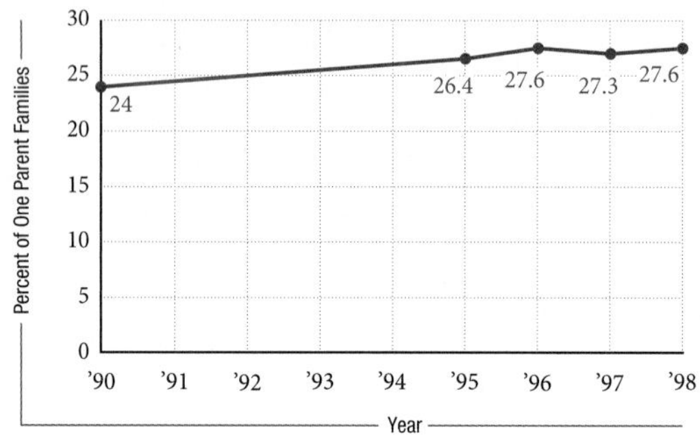

a. Use the function to predict the percentage of families with one parent in the year 2010.

b. Determine the inverse of the function, and estimate the year in which approximately 29% of the families are one-parent families.

47. Speed The fastest type of plane, a rocket plane, can travel at a speed of 4,520 miles per hour. The function $f(m) = \frac{22m}{15}$ converts miles per hour, m, to feet per second (World Book Encyclopedia).

a. Use the function to convert the speed of the rocket plane to feet per second.

b. Write the inverse of the function.

c. Using the inverse function, convert 2 feet per second to miles per hour.

48. a. 980 m/s
b. $s^{-1}(h) = 2.2379h$
c. 335.69 mph
49. $\frac{1}{9}$
50. 8
51. $\frac{2}{3}$
52. $\frac{3}{5}$
53. $\sqrt[3]{4}$
54. $\pm 2\sqrt{3}$
55. 3
56. 3
57. 4
58. 3
59. 4
60. 2
61. 1
62. 0

48. Speed A Lockheed SR-71A airplane set a world record (as reported by Air Force Armament Museum in 1996) with an absolute speed record of 2,193.167 miles per hour. The function $s(h) = 0.4468424h$ converts miles per hour, h, to meters per second, s (Air Force Armament Museum).

a. What is the absolute speed of the Lockheed SR-71A in meters per second?

b. What is the inverse of this function?

c. Using the inverse function, determine the speed of an airplane in miles per hour that flies 150 meters per second.

Getting Ready for the Next Section

Simplify.

49. 3^{-2}

50. 2^3

Solve.

51. $2 = 3x$

52. $3 = 5x$

53. $4 = x^3$

54. $12 = x^2$

Fill in the boxes to make each statement true.

55. $8 = 2^{\square}$

56. $27 = 3^{\square}$

57. $10{,}000 = 10^{\square}$

58. $1{,}000 = 10^{\square}$

59. $81 = 3^{\square}$

60. $81 = 9^{\square}$

61. $6 = 6^{\square}$

62. $1 = 5^{\square}$

Logarithms are Exponents

10.3

José Gomez/©Reuters

In January 1999, ABC News reported that an earthquake had occurred in Colombia, causing massive destruction. They reported the strength of the quake by indicating that it measured 6.0 on the Richter scale. For comparison, Table 1 gives the Richter magnitude of a number of other earthquakes.

Although the size of the numbers in the table do not seem to be very different, the intensity of the earthquakes they measure can be very different. For example, the 1989 San Francisco earthquake was more than 10 times stronger than the 1999 earthquake in Colombia. The reason behind this is that the Richter scale is a *logarithmic scale*.

TABLE 1 Earthquakes

Year	Earthquake	Richter Magnitude
1971	Los Angeles	6.6
1985	Mexico City	8.1
1989	San Francisco	7.1
1992	Kobe, Japan	7.2
1994	Northridge	6.6
1999	Armenia, Colombia	6.0

In this section, we start our work with logarithms, which will give you an understanding of the Richter scale. Let's begin.

As you know from your work in the previous sections, equations of the form

$$y = b^x \quad b > 0, b \neq 1$$

are called exponential functions. Because the equation of the inverse of a function can be obtained by exchanging x and y in the equation of the original function, the inverse of an exponential function must have the form

$$x = b^y \quad b > 0, b \neq 1$$

Now, this last equation is actually the equation of a logarithmic function, as the following definition indicates:

(def) DEFINITION

The expression $y = \log_b x$ is read "y is the logarithm to the base b of x" and is equivalent to the expression

$$x = b^y \qquad b > 0, b \neq 1$$

In words, we say "y is the number we raise b to in order to get x."

Notation When an expression is in the form $x = b^y$, it is said to be in exponential form. On the other hand, if an expression is in the form $y = \log_b x$, it is said to be in logarithmic form.

Here are some equivalent statements written in both forms.

Exponential Form		Logarithmic Form
$8 = 2^3$	$\Leftrightarrow$	$\log_2 8 = 3$
$25 = 5^2$	$\Leftrightarrow$	$\log_5 25 = 2$
$0.1 = 10^{-1}$	$\Leftrightarrow$	$\log_{10} 0.1 = -1$
$\frac{1}{8} = 2^{-3}$	$\Leftrightarrow$	$\log_2 \frac{1}{8} = -3$
$r = z^s$	$\Leftrightarrow$	$\log_z r = s$

EXAMPLE 1 Solve for x: $\log_3 x = -2$

SOLUTION In exponential form, the equation looks like this:

$$x = 3^{-2}$$

or

$$x = \frac{1}{9}$$

The solution is $\frac{1}{9}$.

EXAMPLE 2 Solve $\log_x 4 = 3$.

SOLUTION Again, we use the definition of logarithms to write the expression in exponential form:

$$4 = x^3$$

Taking the cube root of both sides, we have

$$\sqrt[3]{4} = \sqrt[3]{x^3}$$

$$x = \sqrt[3]{4}$$

The solution set is $\{\sqrt[3]{4}\}$.

EXAMPLE 3 Solve $\log_8 4 = x$.

SOLUTION We write the expression again in exponential form:

$$4 = 8^x$$

Because both 4 and 8 can be written as powers of 2, we write them in terms of powers of 2:

$$2^2 = (2^3)^x$$

$$2^2 = 2^{3x}$$

The only way the left and right sides of this last line can be equal is if the exponents are equal—that is, if

$$2 = 3x$$

or

$$x = \frac{2}{3}$$

The solution is $\frac{2}{3}$. We check as follows:

$$\log_8 4 = \frac{2}{3} \Leftrightarrow 4 = 8^{2/3}$$
$$4 = (\sqrt[3]{8})^2$$
$$4 = 2^2$$
$$4 = 4$$

The solution checks when used in the original equation.

Graphing Logarithmic Functions

Graphing logarithmic functions can be done using the graphs of exponential functions and the fact that the graphs of inverse functions have symmetry about the line $y = x$. Here's an example to illustrate.

EXAMPLE 4 Graph the equation $y = \log_2 x$.

SOLUTION The equation $y = \log_2 x$ is, by definition, equivalent to the exponential equation

$$x = 2^y$$

which is the equation of the inverse of the function

$$y = 2^x$$

The graph of $y = 2^x$ was given in Figure 2 of Section 8.1. We simply reflect the graph of $y = 2^x$ about the line $y = x$ to get the graph of $x = 2^y$, which is also the graph of $y = \log_2 x$. (See Figure 1.)

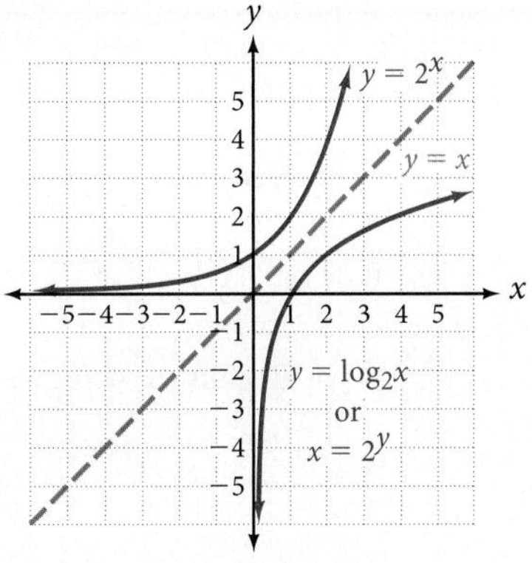

FIGURE 1

It is apparent from the graph that $y = \log_2 x$ is a function, because no vertical line will cross its graph in more than one place. The same is true for all logarithmic equations of the form $y = \log_b x$, where b is a positive number other than 1. Note also that the graph of $y = \log_b x$ will always appear to the right of the y-axis, meaning that x will always be positive in the expression $y = \log_b x$.

Two Special Identities

If b is a positive real number other than 1, then each of the following is a consequence of the definition of a logarithm:

$$(1)\ b^{\log_b x} = x \qquad \text{and} \qquad (2)\ \log_b b^x = x$$

The justifications for these identities are similar. Let's consider only the first one. Consider the expression

$$y = \log_b x$$

By definition, it is equivalent to

$$x = b^y$$

Substituting $\log_b x$ for y in the last line gives us

$$x = b^{\log_b x}$$

The next examples in this section show how these two special properties can be used to simplify expressions involving logarithms.

EXAMPLE 5 Simplify the following logarithmic expressions.

a. $\log_2 8$ **b.** $\log_{10} 10{,}000$ **c.** $\log_b b$

d. $\log_b 1$ **e.** $\log_4 (\log_5 5)$

SOLUTION

a. Substitute 2^3 for 8:

$$\begin{aligned}\log_2 8 &= \log_2 2^3 \\ &= 3\end{aligned}$$

b. 10,000 can be written as 10^4:

$$\begin{aligned}\log_{10} 10{,}000 &= \log_{10} 10^4 \\ &= 4\end{aligned}$$

c. Because $b^1 = b$, we have

$$\begin{aligned}\log_b b &= \log_b b^1 \\ &= 1\end{aligned}$$

d. Because $1 = b^0$, we have

$$\begin{aligned}\log_b 1 &= \log_b b^0 \\ &= 0\end{aligned}$$

e. Because $\log_5 5 = 1$,

$$\begin{aligned}\log_4(\log_5 5) &= \log_4 1 \\ &= 0\end{aligned}$$

Application

As we mentioned in the introduction to this section, one application of logarithms is in measuring the magnitude of an earthquake. If an earthquake has a shock wave T times greater than the smallest shock wave that can be measured on a seismograph, then the magnitude M of the earthquake, as measured on the Richter scale, is given by the formula

$$M = \log_{10} T$$

(When we talk about the size of a shock wave, we are talking about its amplitude. The amplitude of a wave is half the difference between its highest point and its lowest point.)

To illustrate the discussion, an earthquake that produces a shock wave that is 10,000 times greater than the smallest shock wave measurable on a seismograph will have a magnitude M on the Richter scale of

$$M = \log_{10} 10{,}000 = 4$$

EXAMPLE 6 If an earthquake has a magnitude of $M = 5$ on the Richter scale, what can you say about the size of its shock wave?

SOLUTION To answer this question, we put $M = 5$ into the formula $M = \log_{10} T$ to obtain

$$5 = \log_{10} T$$

Writing this expression in exponential form, we have

$$T = 10^5 = 100{,}000$$

We can say that an earthquake that measures 5 on the Richter scale has a shock wave 100,000 times greater than the smallest shock wave measurable on a seismograph.

From Example 6 and the discussion that preceded it, we find that an earthquake of magnitude 5 has a shock wave that is 10 times greater than an earthquake of magnitude 4, because 100,000 is 10 times 10,000.

GETTING READY FOR CLASS

After reading through the preceding section, respond in your own words and in complete sentences.

A. What is a logarithm?

B. What is the relationship between $y = 2^x$ and $y = \log_2 x$? How are their graphs related?

C. Will the graph of $y = \log_b x$ ever appear in the second or third quadrants? Explain why or why not.

D. Explain why $\log_2 0 = x$ has no solution for x.

Problem Set 10.3

Write each of the following expressions in logarithmic form.

1. $2^4 = 16$ **2.** $3^2 = 9$ **3.** $125 = 5^3$ **4.** $16 = 4^2$

5. $0.01 = 10^{-2}$ **6.** $0.001 = 10^{-3}$ **7.** $2^{-5} = \frac{1}{32}$ **8.** $4^{-2} = \frac{1}{16}$

9. $\left(\frac{1}{2}\right)^{-3} = 8$ **10.** $\left(\frac{1}{3}\right)^{-2} = 9$ **11.** $27 = 3^3$ **12.** $81 = 3^4$

Write each of the following expressions in exponential form.

13. $\log_{10} 100 = 2$ **14.** $\log_2 8 = 3$ **15.** $\log_2 64 = 6$

16. $\log_2 32 = 5$ **17.** $\log_8 1 = 0$ **18.** $\log_9 9 = 1$

19. $\log_{10} 0.001 = -3$ **20.** $\log_{10} 0.0001 = -4$ **21.** $\log_6 36 = 2$

22. $\log_7 49 = 2$ **23.** $\log_5 \frac{1}{25} = -2$ **24.** $\log_3 \frac{1}{81} = -4$

Solve each of the following equations for x.

25. $\log_3 x = 2$ **26.** $\log_4 x = 3$ **27.** $\log_5 x = -3$ **28.** $\log_2 x = -4$

29. $\log_2 16 = x$ **30.** $\log_3 27 = x$ **31.** $\log_8 2 = x$ **32.** $\log_{25} 5 = x$

33. $\log_x 4 = 2$ **34.** $\log_x 16 = 4$ **35.** $\log_x 5 = 3$ **36.** $\log_x 8 = 2$

37. $\log_5 25 = x$ **38.** $\log_5 x = -2$ **39.** $\log_x 36 = 2$ **40.** $\log_x \frac{1}{25} = 2$

41. $\log_8 4 = x$ **42.** $\log_{16} 8 = x$ **43.** $\log_9 \frac{1}{3} = x$ **44.** $\log_{27} 9 = x$

45. $\log_8 x = -2$ **46.** $\log_{36} \frac{1}{6} = x$

Sketch the graph of each of the following logarithmic equations.

47. $y = \log_3 x$ **48.** $y = \log_{1/2} x$ **49.** $y = \log_{1/3} x$ **50.** $y = \log_4 x$

51. $y = \log_5 x$ **52.** $y = \log_{1/5} x$ **53.** $y = \log_{10} x$ **54.** $y = \log_{1/4} x$

Each of the following graphs has an equation of the form $y = b^x$ or $y = \log_b x$. Find the equation for each graph.

55. **56.**

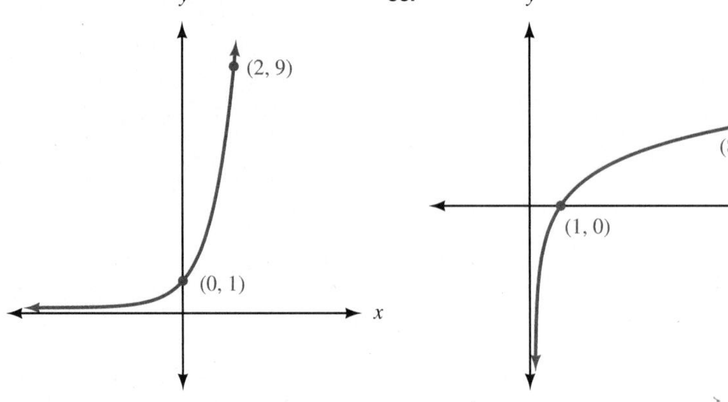

1. $\log_2 16 = 4$
2. $\log_3 9 = 2$
3. $\log_5 125 = 3$
4. $\log_4 16 = 2$
5. $\log_{10} 0.01 = -2$
6. $\log_{10} 0.001 = -3$
7. $\log_2 \frac{1}{32} = -5$
8. $\log_4 \frac{1}{16} = -2$
9. $\log_{1/2} 8 = -3$
10. $\log_{1/3} 9 = -2$
11. $\log_3 27 = 3$
12. $\log_3 81 = 4$
13. $10^2 = 100$
14. $2^3 = 8$
15. $2^6 = 64$
16. $2^5 = 32$
17. $8^0 = 1$
18. $9^1 = 9$
19. $10^{-3} = 0.001$
20. $10^{-4} = 0.0001$
21. $6^2 = 36$
22. $7^2 = 49$
23. $5^{-2} = \frac{1}{25}$
24. $3^{-4} = \frac{1}{81}$
25. 9
26. 64
27. $\frac{1}{125}$
28. $\frac{1}{16}$
29. 4
30. 3
31. $\frac{1}{3}$
32. $\frac{1}{2}$
33. 2
34. 2
35. $\sqrt[3]{5}$
36. $2^{3/2}$ or $\sqrt{8}$ or $2\sqrt{2}$
37. 2
38. $\frac{1}{25}$
39. 6
40. $\frac{1}{5}$
41. $\frac{2}{3}$
42. $\frac{3}{4}$
43. $-\frac{1}{2}$
44. $\frac{2}{3}$
45. $\frac{1}{64}$
46. $-\frac{1}{2}$

47.-54. See answer section.

55. $y = 3^x$
56. $y = \log_2 x$

57. $y = \log_{1/3}x$
58. $y = \left(\frac{1}{2}\right)^x$
59. 4
60. 2
61. $\frac{3}{2}$
62. $\frac{3}{2}$
63. 3
64. 4
65. 1
66. 1
67. 0
68. 0
69. 0
70. $\frac{3}{2}$
71. $\frac{1}{2}$
72. -4
73. $\frac{3}{2}$
74. $\frac{4}{5}$
75. 1
76. -1
77. -2
78. $-\frac{1}{2}$
79. 0
80. 0
81. $\frac{1}{2}$
82. 0
84. $y = 2^x$, $D: (-\infty, \infty)$, $R: (0, \infty)$; $y = \log_2 x$; $D: (0, \infty)$, $R: (-\infty, \infty)$; $D(y = 2^x) = R(y = \log_2 x)$; $R(y = 2^x) = D(y = \log_2 x)$
85. 2
86. 5

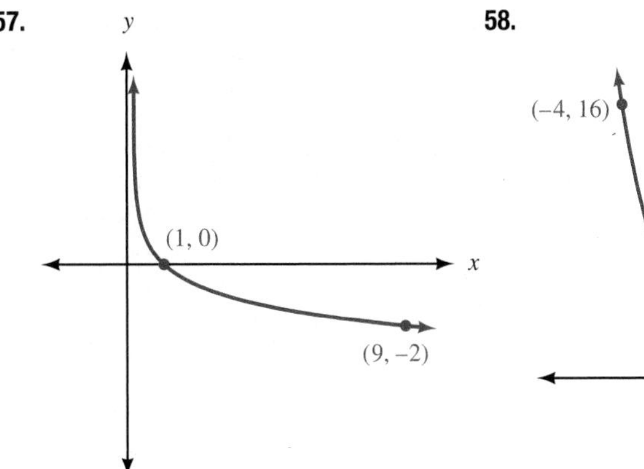

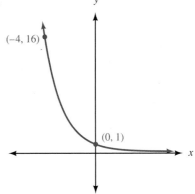

Simplify each of the following.

59. $\log_2 16$ **60.** $\log_3 9$ **61.** $\log_{25} 125$ **62.** $\log_9 27$

63. $\log_{10} 1{,}000$ **64.** $\log_{10} 10{,}000$ **65.** $\log_3 3$ **66.** $\log_4 4$

67. $\log_5 1$ **68.** $\log_{10} 1$ **69.** $\log_{17} 1$ **70.** $\log_4 8$

71. $\log_{16} 4$ **72.** $\log_{10} 0.0001$ **73.** $\log_{100} 1000$ **74.** $\log_{32} 16$

75. $\log_3 (\log_2 8)$ **76.** $\log_5 (\log_{32} 2)$ **77.** $\log_{1/2} (\log_3 81)$ **78.** $\log_9 (\log_8 2)$

79. $\log_3 (\log_6 6)$ **80.** $\log_5 (\log_3 3)$ **81.** $\log_4 [\log_2(\log_2 16)]$

82. $\log_4 [\log_3(\log_2 8)]$

Applying the Concepts

83. Metric System The metric system uses logical and systematic prefixes for multiplication. For instance, to multiply a unit by 100, the prefix "hecto" is applied, so a hectometer is equal to 100 meters. For each of the prefixes in the following table find the logarithm, base 10, of the multiplying factor.

Prefix	Multiplying Factor	$\log_{10}$ (Multiplying Factor)
Nano	0.000 000 001	-9
Micro	0.000 001	-6
Deci	0.1	-1
Giga	1,000,000,000	9
Peta	1,000,000,000,000,000	15

84. Domain and Range Use the graphs of $y = 2^x$ and $y = \log_2 x$ shown in Figure 1 of this section to find the domain and range for each function. Explain how the domain and range found for $y = 2^x$ relate to the domain and range found for $y = \log_2 x$.

85. Magnitude of an Earthquake Find the magnitude M of an earthquake with a shock wave that measures $T = 100$ on a seismograph.

86. Magnitude of an Earthquake Find the magnitude M of an earthquake with a shock wave that measures $T = 100{,}000$ on a seismograph.

87. 10^8 times as large
88. 10^6 times as large
89. 120
90. 139
91. 4
92. 9
93. $-4, 2$
94. $-4, 1$
95. $-\frac{11}{8}$
96. $\frac{101}{24}$
97. $2^3 = (x + 2)(x)$
98. $4^2 = x(x - 6)$
99. $3^4 = \frac{x-2}{x+1}$
100. $3^2 = \frac{x-1}{x-4}$

87. Shock Wave If an earthquake has a magnitude of 8 on the Richter scale, how many times greater is its shock wave than the smallest shock wave measurable on a seismograph?

88. Shock Wave If the 1999 Colombia earthquake had a magnitude of 6 on the Richter scale, how many times greater was its shock wave than the smallest shock wave measurable on a seismograph?

Earthquake The table below categorizes earthquake by the magnitude and identifies the average annual occurrence.

Earthquakes

Descriptor	Magnitude	Average Annual Occurrence
Great	≥8.0	1
Major	7–7.9	18
Strong	6–6.9	120
Moderate	5–5.9	800
Light	4–4.9	6,200
Minor	3–3.9	49,000
Very Minor	2–2.9	1,000 per day
Very Minor	1–1.9	8,000 per day

SOURCE: *USGS National Earthquake Information.*

89. What is the average number of earthquakes that occur per year when the number of times the associated shockwave is greater than the smallest measurable shockwave, T, is 1,000,000?

90. What is the average number of earthquakes that occur per year when $T = 1{,}000{,}000$ or greater?

Getting Ready for the Next Section

Simplify.

91. $8^{2/3}$

92. $27^{2/3}$

Solve.

93. $(x + 2)(x) = 2^3$

94. $(x + 3)(x) = 2^2$

95. $\frac{x - 2}{x + 1} = 9$

96. $\frac{x + 1}{x - 4} = 25$

Write in exponential form.

97. $\log_2 [(x + 2)(x)] = 3$

98. $\log_4 [x(x - 6)] = 2$

99. $\log_3\left(\frac{x - 2}{x + 1}\right) = 4$

100. $\log_3\left(\frac{x - 1}{x - 4}\right) = 2$

Properties of Logarithms 10.4

If we search for a definition of the word *decibel*, we find the following: A unit used to express relative difference in power or intensity, usually between two acoustic or electric signals, equal to ten times the common logarithm of the ratio of the two levels.

Decibels	Comparable to
10	A light whisper
20	Quiet conversation
30	Normal conversation
40	Light traffic
50	Typewriter, loud conversation
60	Noisy office
70	Normal traffic, quiet train
80	Rock music, subway
90	Heavy traffic, thunder
100	Jet plane at takeoff

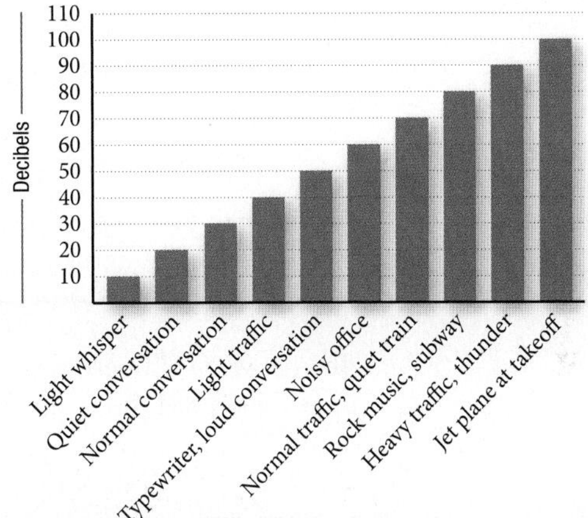

The precise definition for a *decibel* is

$$D = 10 \log_{10}\left(\frac{I}{I_0}\right)$$

where I is the intensity of the sound being measured, and I_0 is the intensity of the least audible sound. (Sound intensity is related to the amplitude of the sound wave that models the sound and is given in units of watts per meter^2.) In this section, we will see that the preceding formula can also be written as

$$D = 10(\log_{10} I - \log_{10} I_0)$$

The rules we use to rewrite expressions containing logarithms are called the *properties of logarithms*. There are three of them.

For the following three properties, x, y, and b are all positive real numbers, $b \neq 1$, and r is any real number.

PROPERTY *Property 1*

$$\log_b(xy) = \log_b x + \log_b y$$

In words: The logarithm of a ***product*** is the ***sum*** of the logarithms.

PROPERTY *Property 2*

$$\log_b\left(\frac{x}{y}\right) = \log_b x - \log_b y$$

In words: The logarithm of a ***quotient*** is the ***difference*** of the logarithms.

PROPERTY *Property 3*

$$\log_b x^r = r \log_b x$$

In words: The logarithm of a number raised to a *power* is the *product* of the power and the logarithm of the number.

Proof of Property 1 To prove Property 1, we simply apply the first identity for logarithms given in the preceding section:

$$b^{\log_b xy} = xy = (b^{\log_b x})(b^{\log_b y}) = b^{\log_b x + \log_b y}$$

Because the first and last expressions are equal and the bases are the same, the exponents $\log_b xy$ and $\log_b x + \log_b y$ must be equal. Therefore,

$$\log_b xy = \log_b x + \log_b y$$

The proofs of Properties 2 and 3 proceed in much the same manner, so we will omit them here. The examples that follow show how the three properties can be used.

EXAMPLE 1 Expand, using the properties of logarithms: $\log_5 \dfrac{3xy}{z}$

SOLUTION Applying Property 2, we can write the quotient of $3xy$ and z in terms of a difference:

$$\log_5 \frac{3xy}{z} = \log_5 3xy - \log_5 z$$

Applying Property 1 to the product $3xy$, we write it in terms of addition:

$$\log_5 \frac{3xy}{z} = \log_5 3 + \log_5 x + \log_5 y - \log_5 z$$

EXAMPLE 2 Expand, using the properties of logarithms:

$$\log_2 \frac{x^4}{\sqrt{y} \cdot z^3}$$

SOLUTION We write $\sqrt{y}$ as $y^{1/2}$ and apply the properties:

$$
\begin{aligned}
\log_2 \frac{x^4}{\sqrt{y} \cdot z^3} &= \log_2 \frac{x^4}{y^{1/2}z^3} && \sqrt{y} = y^{1/2} \\
&= \log_2 x^4 - \log_2(y^{1/2} \cdot z^3) && \text{Property 2} \\
&= \log_2 x^4 - (\log_2 y^{1/2} + \log_2 z^3) && \text{Property 1} \\
&= \log_2 x^4 - \log_2 y^{1/2} - \log_2 z^3 && \text{Remove parentheses and distribute } -1 \\
&= 4 \log_2 x - \frac{1}{2}\log_2 y - 3 \log_2 z && \text{Property 3}
\end{aligned}
$$

We can also use the three properties to write an expression in expanded form as just one logarithm.

EXAMPLE 3 Write as a single logarithm:

$$2 \log_{10} a + 3 \log_{10} b - \frac{1}{3}\log_{10} c$$

SOLUTION We begin by applying Property 3:

$$
\begin{aligned}
2 \log_{10} a + 3 \log_{10} b - \frac{1}{3}\log_{10} c &= \log_{10} a^2 + \log_{10} b^3 - \log_{10} c^{1/3} && \text{Property 3} \\
&= \log_{10} (a^2 \cdot b^3) - \log_{10} c^{1/3} && \text{Property 1} \\
&= \log_{10} \frac{a^2b^3}{c^{1/3}} && \text{Property 2} \\
&= \log_{10} \frac{a^2b^3}{\sqrt[3]{c}} && c^{1/3} = \sqrt[3]{c}
\end{aligned}
$$

The properties of logarithms along with the definition of logarithms are useful in solving equations that involve logarithms.

EXAMPLE 4 Solve for x: $\log_2(x + 2) + \log_2 x = 3$

SOLUTION Applying Property 1 to the left side of the equation allows us to write it as a single logarithm:

$$\log_2(x + 2) + \log_2 x = 3$$

$$\log_2[(x + 2)(x)] = 3$$

The last line can be written in exponential form using the definition of logarithms:

$$(x + 2)(x) = 2^3$$

Solve as usual:

$$x^2 + 2x = 8$$

$$x^2 + 2x - 8 = 0$$

$$(x + 4)(x - 2) = 0$$

$$x + 4 = 0 \quad \text{or} \quad x - 2 = 0$$

$$x = -4 \quad \text{or} \quad x = 2$$

In the previous section, we noted the fact that x in the expression $y = \log_b x$ cannot be a negative number. Because substitution of $x = -4$ into the original equation gives

$$\log_2(-2) + \log_2(-4) = 3$$

which contains logarithms of negative numbers, we cannot use -4 as a solution. The solution set is 2.

GETTING READY FOR CLASS

After reading through the preceding section, respond in your own words and in complete sentences.

A. Explain why the following statement is false: "The logarithm of a product is the product of the logarithms."
B. Explain why the following statement is false: "The logarithm of a quotient is the quotient of the logarithms."
C. Explain the difference between $\log_b m + \log_b n$ and $\log_b(m + n)$. Are they equivalent?
D. Explain the difference between $\log_b(mn)$ and $(\log_b m)(\log_b n)$ Are they equivalent?

Problem Set 10.4

Use the three properties of logarithms given in this section to expand each expression as much as possible.

1. $\log_3 4x$
2. $\log_2 5x$
3. $\log_6 \frac{5}{x}$
4. $\log_3 \frac{x}{5}$
5. $\log_2 y^5$
6. $\log_7 y^3$
7. $\log_9 \sqrt[3]{z}$
8. $\log_8 \sqrt{z}$
9. $\log_6 x^2y^4$
10. $\log_{10} x^2y^4$
11. $\log_5 \sqrt{x} \cdot y^4$
12. $\log_8 \sqrt[3]{xy^6}$
13. $\log_b \frac{xy}{z}$
14. $\log_b \frac{3x}{y}$
15. $\log_{10} \frac{4}{xy}$
16. $\log_{10} \frac{5}{4y}$
17. $\log_{10} \frac{x^2y}{\sqrt{z}}$
18. $\log_{10} \frac{\sqrt{x} \cdot y}{z^3}$
19. $\log_{10} \frac{x^3\sqrt{y}}{z^4}$
20. $\log_{10} \frac{x^4\sqrt[3]{y}}{\sqrt{z}}$
21. $\log_b \sqrt[3]{\frac{x^2y}{z^4}}$
22. $\log_b \sqrt[4]{\frac{x^4y^3}{z^5}}$
23. $\log_3 \sqrt[3]{\frac{x^2y}{z^6}}$
24. $\log_8 \sqrt[4]{\frac{x^5y^6}{z^3}}$
25. $\log_a \frac{4x^5}{9a^2}$
26. $\log_b \frac{16b^2}{25y^3}$

Write each expression as a single logarithm.

27. $\log_b x + \log_b z$
28. $\log_b x - \log_b z$
29. $2 \log_3 x - 3 \log_3 y$
30. $4 \log_2 x + 5 \log_2 y$
31. $\frac{1}{2} \log_{10} x + \frac{1}{3} \log_{10} y$
32. $\frac{1}{3} \log_{10} x - \frac{1}{4} \log_{10} y$
33. $3 \log_2 x + \frac{1}{2} \log_2 y - \log_2 z$
34. $2 \log_3 x + 3 \log_3 y - \log_3 z$
35. $\frac{1}{2} \log_2 x - 3 \log_2 y - 4 \log_2 z$
36. $3 \log_{10} x - \log_{10} y - \log_{10} z$
37. $\frac{3}{2} \log_{10} x - \frac{3}{4} \log_{10} y - \frac{4}{5} \log_{10} z$
38. $3 \log_{10} x - \frac{4}{3} \log_{10} y - 5 \log_{10} z$
39. $\frac{1}{2} \log_5 x + \frac{2}{3} \log_5 y - 4 \log_5 z$
40. $\frac{1}{4} \log_7 x + 5 \log_7 y - \frac{1}{3} \log_7 z$
41. $\log_3(x^2 - 16) - 2 \log_3(x + 4)$
42. $\log_4(x^2 - x - 6) - \log_4(x^2 - 9)$

Solve each of the following equations.

43. $\log_2 x + \log_2 3 = 1$
44. $\log_3 x + \log_3 3 = 1$
45. $\log_3 x - \log_3 2 = 2$
46. $\log_3 x + \log_3 2 = 2$
47. $\log_3 x + \log_3(x - 2) = 1$
48. $\log_6 x + \log_6(x - 1) = 1$
49. $\log_3(x + 3) - \log_3(x - 1) = 1$
50. $\log_4(x - 2) - \log_4(x + 1) = 1$
51. $\log_2 x + \log_2(x - 2) = 3$
52. $\log_4 x + \log_4(x + 6) = 2$
53. $\log_8 x + \log_8(x - 3) = \frac{2}{3}$
54. $\log_{27} x + \log_{27}(x + 8) = \frac{2}{3}$
55. $\log_3(x + 2) - \log_3 x = 1$
56. $\log_2(x + 3) - \log_2(x - 3) = 2$
57. $\log_2(x + 1) + \log_2(x + 2) = 1$
58. $\log_3 x + \log_3(x + 6) = 3$
59. $\log_9\sqrt{x} + \log_9\sqrt{2x + 3} = \frac{1}{2}$
60. $\log_8\sqrt{x} + \log_8\sqrt{5x + 2} = \frac{2}{3}$
61. $4 \log_3 x - \log_3 x^2 = 6$
62. $9 \log_4 x - \log_4 x^3 = 12$
63. $\log_5 \sqrt{x} + \log_5 \sqrt{6x + 5} = 1$
64. $\log_2 \sqrt{x} + \log_2 \sqrt{6x + 5} = 1$

1. $\log_3 4 + \log_3 x$
2. $\log_2 5 + \log_2 x$
3. $\log_6 5 - \log_6 x$
4. $\log_3 x - \log_3 5$
5. $5 \log_2 y$
6. $3 \log_7 y$
7. $\frac{1}{3} \log_9 z$
8. $\frac{1}{2} \log_8 z$
9. $2 \log_6 x + 4 \log_6 y$
10. $2 \log_{10} x + 4 \log_{10} y$
11. $\frac{1}{2} \log_5 x + 4 \log_5 y$
12. $\frac{1}{3} \log_8 x + 2 \log_8 y$
13. $\log_b x + \log_b y - \log_b z$
14. $\log_b 3 + \log_b x - \log_b y$
15. $\log_{10} 4 - \log_{10} x - \log_{10} y$
16. $\log_{10} 5 - \log_{10} 4 - \log_{10} y$
17. $2 \log_{10} x + \log_{10} y - \frac{1}{2} \log_{10} z$
18. $\frac{1}{2} \log_{10} x + \log_{10} y - 3 \log_{10} z$
19. $3 \log_{10} x + \frac{1}{2} \log_{10} y - 4 \log_{10} z$
20. $4 \log_{10} x + \frac{1}{3} \log_{10} y - \frac{1}{2} \log_{10} z$
21. $\frac{2}{3} \log_b x + \frac{1}{3} \log_b y - \frac{4}{3} \log_b z$
22. $\log_b x + \frac{3}{4} \log_b y - \frac{5}{4} \log_b z$
23. $\frac{2}{3}\log_3 x + \frac{1}{3}\log_3 y - 2\log_3 z$
24. $\frac{5}{4}\log_8 x + \frac{3}{2}\log_8 y - \frac{3}{4}\log_8 z$
25. $2\log_a 2 + 5\log_a x - 2\log_a 3 - 2$
26. $4\log_b 2 + 2 - 2\log_b 5 - 3\log_b y$
27. $\log_b xz$
28. $\log_b \frac{x}{z}$
29. $\log_3 \frac{x^2}{y^3}$
30. $\log_2 x^4y^5$
31. $\log_{10} \sqrt{x}\sqrt[3]{y}$
32. $\log_{10} \frac{\sqrt[3]{x}}{\sqrt[4]{y}}$
33. $\log_2 \frac{x^3\sqrt{y}}{z}$
34. $\log_3 \frac{x^2y^3}{z}$
35. $\log_2 \frac{\sqrt{x}}{y^3z^4}$
36. $\log_{10} \frac{x^3}{yz}$
37. $\log_{10} \frac{x^{3/2}}{y^{3/4}z^{4/5}}$
38. $\log_{10} \frac{x^3}{y^{4/3}z^5}$
39. $\log_5 \frac{\sqrt{x} \cdot \sqrt[3]{y^2}}{z^4}$
40. $\log_7 \frac{\sqrt[4]{x} \cdot y^5}{\sqrt[3]{z}}$
41. $\log_3 \frac{x - 4}{x + 4}$
42. $\log_4 \frac{x + 2}{x + 3}$
43. $\frac{2}{3}$
44. 1
45. 18

46. $\frac{9}{2}$
47. 3
48. 3
49. 3
50. $\varnothing$
51. 4
52. 2
53. 4
54. 1
55. 1
56. 5
57. 0
58. 3
59. $\frac{3}{2}$
60. $\frac{8}{5}$
61. 27
62. 16
63. $\frac{5}{3}$
64. $\frac{1}{2}$
67. a. 1.602
b. 2.505
c. 3.204
68. a. iii
b. iv
c. vi
d. v
e. i
f. ii
69. pH $= 6.1 + \log_{10}x - \log_{10}y$
70. 19.95
71. 2.52
72. 2

Applying the Concepts

65. Decibel Formula Use the properties of logarithms to rewrite the decibel formula $D = 10\log_{10}\left(\frac{I}{I_0}\right)$ as

$$D = 10(\log_{10} I - \log_{10} I_0).$$

66. Decibel Formula In the decibel formula $D = 10\log_{10}\left(\frac{I}{I_0}\right)$, the threshold of hearing, I_0, is

$$I_0 = 10^{-12} \text{ watts/meter}^2$$

Substitute 10^{-12} for I_0 in the decibel formula, then show that it simplifies to

$$D = 10(\log_{10} I + 12)$$

67. Finding Logarithms If $\log_{10} 8 = 0.903$ and $\log_{10} 5 = 0.699$, find the following without using a calculator.

a. $\log_{10} 40$ **b.** $\log_{10} 320$ **c.** $\log_{10} 1{,}600$

68. Matching Match each expression in the first column with an equivalent expression in the second column:

a. $\log_2(ab)$	**i.** b
b. $\log_2\left(\frac{a}{b}\right)$	**ii.** 2
c. $\log_5 a^b$	**iii.** $\log_2 a + \log_2 b$
d. $\log_a b^a$	**iv.** $\log_2 a - \log_2 b$
e. $\log_a a^b$	**v.** $a\log_a b$
f. $\log_3 9$	**vi.** $b\log_5 a$

69. Henderson–Hasselbalch Formula Doctors use the Henderson–Hasselbalch formula to calculate the pH of a person's blood. pH is a measure of the acidity and/or the alkalinity of a solution. This formula is represented as

$$\text{pH} = 6.1 + \log_{10}\left(\frac{x}{y}\right)$$

where x is the base concentration and y is the acidic concentration. Rewrite the Henderson–Hasselbalch formula so that the logarithm of a quotient is not involved.

70. Henderson–Hasselbalch Formula Refer to the information in the preceding problem about the Henderson–Hasselbalch formula. If most people have a blood pH of 7.4, use the Henderson–Hasselbalch formula to find the ratio of $\frac{x}{y}$ for an average person.

71. Food Processing The formula $M = 0.21(\log_{10} a - \log_{10} b)$ is used in the food processing industry to find the number of minutes M of heat processing a certain food should undergo at 250°F to reduce the probability of survival of *Clostridium botulinum* spores. The letter a represents the number of spores per can before heating, and b represents the number of spores per can after heating. Find M if $a = 1$ and $b = 10^{-12}$. Then find M using the same values for a and b in the formula $M = 0.21\log_{10}\frac{a}{b}$.

72. Acoustic Powers The formula $N = \log_{10}\frac{P_1}{P_2}$ is used in radio electronics to find the ratio of the acoustic powers of two electric circuits in terms of their electric powers. Find N if P_1 is 100 and P_2 is 1. Then use the same two values of P_1 and P_2 to find N in the formula $N = \log_{10} P_1 - \log_{10} P_2$.

73. 1
74. 4
75. 1
76. 1
77. 4
78. k
79. 2.5×10^{-6}
80. 7.9×10^{-5}
81. 51
82. 189

Getting Ready for the Next Section

Simplify.

73. 5^0 **74.** 4^1 **75.** $\log_3 3$ **76.** $\log_5 5$

77. $\log_b b^4$ **78.** $\log_a a^k$

Use a calculator to find each of the following. Write your answer in scientific notation with the first number in each answer rounded to the nearest tenth.

79. $10^{-5.6}$ **80.** $10^{-4.1}$

Divide and round to the nearest whole number

81. $\dfrac{2.00 \times 10^8}{3.96 \times 10^6}$ **82.** $\dfrac{3.25 \times 10^{12}}{1.72 \times 10^{10}}$

Common Logarithms and Natural Logarithms

10.5

Acid rain was first discovered in the 1960s by Gene Likens and his research team who studied the damage caused by acid rain to Hubbard Brook in New Hampshire. Acid rain is rain with a pH of 5.6 and below. As you will see as you work your way through this section, pH is defined in terms of common logarithms—one of the topics we present in this section. So, when you are finished with this section, you will have a more detailed knowledge of pH and acid rain.

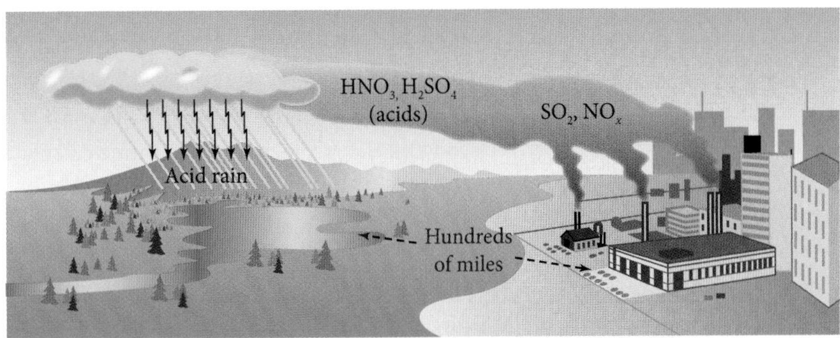

Two kinds of logarithms occur more frequently than other logarithms. Logarithms with a base of 10 are very common because our number system is a base-10 number system. For this reason, we call base-10 logarithms *common logarithms*.

(dĕf′ DEFINITION *common logarithms*

A ***common logarithm*** is a logarithm with a base of 10. Because common logarithms are used so frequently, it is customary, in order to save time, to omit notating the base. That is,

$$\log_{10} x = \log x$$

When the base is not shown, it is assumed to be 10.

Common Logarithms

Common logarithms of powers of 10 are simple to evaluate. We need only recognize that $\log 10 = \log_{10} 10 = 1$ and apply the third property of logarithms: $\log_b x^r = r \log_b x$.

$$
\begin{aligned}
\log 1{,}000 &= \log 10^3 = 3 \log 10 = 3(1) = 3 \\
\log 100 &= \log 10^2 = 2 \log 10 = 2(1) = 2 \\
\log 10 &= \log 10^1 = 1 \log 10 = 1(1) = 1 \\
\log 1 &= \log 10^0 = 0 \log 10 = 0(1) = 0 \\
\log 0.1 &= \log 10^{-1} = -1 \log 10 = -1(1) = -1 \\
\log 0.01 &= \log 10^{-2} = -2 \log 10 = -2(1) = -2 \\
\log 0.001 &= \log 10^{-3} = -3 \log 10 = -3(1) = -3
\end{aligned}
$$

To find common logarithms of numbers that are not powers of 10, we use a calculator with a $\boxed{\log}$ key.

Check the following logarithms to be sure you know how to use your calculator. (These answers have been rounded to the nearest ten-thousandth.)

$$\log 7.02 \approx 0.8463$$
$$\log 1.39 \approx 0.1430$$
$$\log 6.00 \approx 0.7782$$
$$\log 9.99 \approx 0.9996$$

EXAMPLE 1 Use a calculator to find log 2,760.

SOLUTION

$$\log 2{,}760 \approx 3.4409$$

To work this problem on a scientific calculator, we simply enter the number 2,760 and press the key labeled [log]. On a graphing calculator we press the [log] key first, then 2,760.

The 3 in the answer is called the *characteristic*, and the decimal part of the logarithm is called the *mantissa*.

EXAMPLE 2 Find log 0.0391.

SOLUTION $\log 0.0391 \approx -1.4078$

EXAMPLE 3 Find log 0.00523.

SOLUTION $\log 0.00523 \approx -2.2815$

EXAMPLE 4 Find x if $\log x = 3.8774$.

SOLUTION We are looking for the number whose logarithm is 3.8774. On a scientific calculator, we enter 3.8774 and press the key labeled [10^x]. On a graphing calculator we press [10^x] first, then 3.8774. The result is 7,540 to four significant digits. Here's why:

$$\begin{aligned} \text{If} \quad \log x &= 3.8774 \\ \text{then} \quad x &= 10^{3.8774} \\ &\approx 7{,}540 \end{aligned}$$

The number 7,540 is called the *antilogarithm* or just *antilog* of 3.8774. That is, 7,540 is the number whose logarithm is 3.8774.

EXAMPLE 5 Find x if $\log x = -2.4179$.

SOLUTION Using the [10^x] key, the result is 0.00382.

$$\begin{aligned} \text{If} \quad \log x &= -2.4179 \\ \text{then} \quad x &= 10^{-2.4179} \\ &\approx 0.00382 \end{aligned}$$

The antilog of -2.4179 is 0.00382. That is, the logarithm of 0.00382 is -2.4179.

In Section 8.3, we found that the magnitude M of an earthquake that produces a shock wave T times larger than the smallest shock wave that can be measured on a seismograph is given by the formula

$$\text{M} = \log_{10} T$$

We can rewrite this formula using our shorthand notation for common logarithms as

$$\text{M} = \log T$$

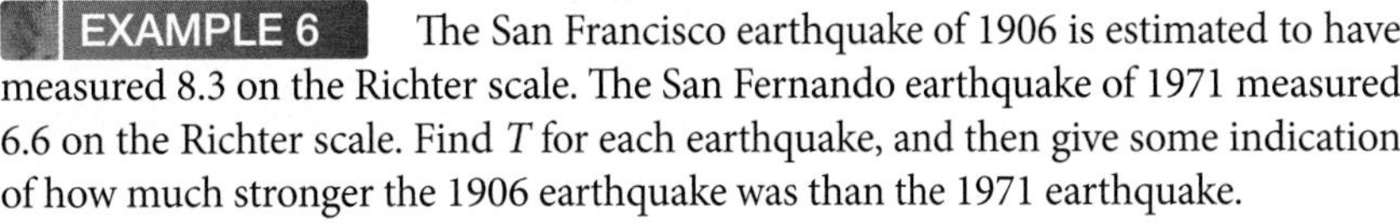

EXAMPLE 6 The San Francisco earthquake of 1906 is estimated to have measured 8.3 on the Richter scale. The San Fernando earthquake of 1971 measured 6.6 on the Richter scale. Find T for each earthquake, and then give some indication of how much stronger the 1906 earthquake was than the 1971 earthquake.

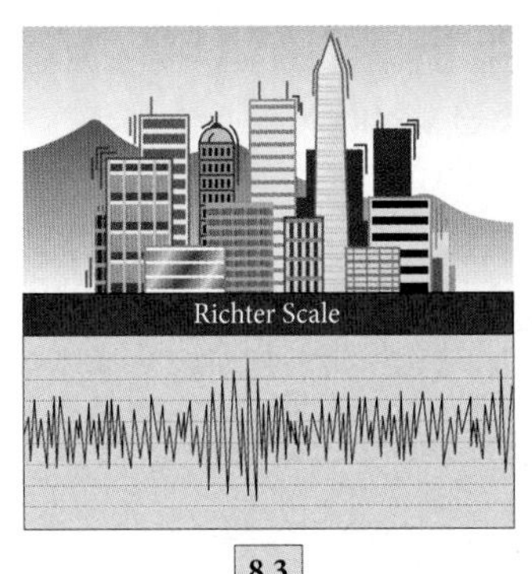

8.3

SOLUTION For the 1906 earthquake:

If $\log T = 8.3$, then $T = 2.00 \times 10^8$.

For the 1971 earthquake:

If $\log T = 6.6$, then $T = 3.98 \times 10^6$.

Dividing the two values of T and rounding our answer to the nearest whole number, we have

$$\frac{2.00 \times 10^8}{3.98 \times 10^6} \approx 50$$

The shock wave for the 1906 earthquake was approximately 50 times larger than the shock wave for the 1971 earthquake.

In chemistry, the pH of a solution is the measure of the acidity of the solution. The definition for pH involves common logarithms. Here it is:

$$\text{pH} = -\log[\text{H}^+]$$

where $[\text{H}^+]$ is the concentration of the hydrogen ion in moles per liter. The range for pH is from 0 to 14. Pure water, a neutral solution, has a pH of 7. An acidic solution, such as vinegar, will have a pH less than 7, and an alkaline solution, such as ammonia, has a pH above 7.

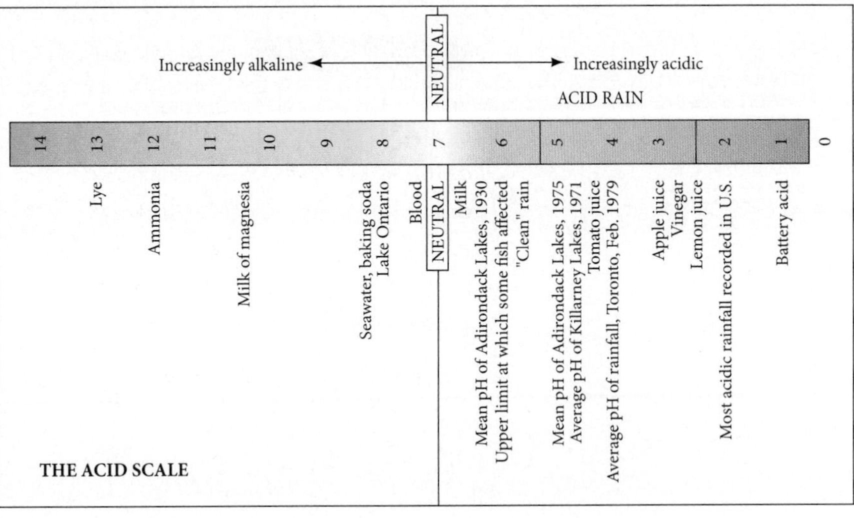

EXAMPLE 7 Normal rainwater has a pH of 5.6. What is the concentration of the hydrogen ion in normal rainwater?

SOLUTION Substituting 5.6 for pH in the formula $\text{pH} = -\log[\text{H}^+]$, we have

$5.6 = -\log[\text{H}^+]$	Substitution
$\log[\text{H}^+] = -5.6$	Isolate the logarithm
$[\text{H}^+] = 10^{-5.6}$	Write in exponential form
$\approx 2.5 \times 10^{-6}$ moles per liter	Answer in scientific notation

EXAMPLE 8 The concentration of the hydrogen ion in a sample of acid rain known to kill fish is 3.2×10^{-5} mole per liter. Find the pH of this acid rain to the nearest tenth.

SOLUTION Substituting 3.2×10^{-5} for $[\text{H}^+]$ in the formula $\text{pH} = -\log[\text{H}^+]$, we have

$\text{pH} = -\log[3.2 \times 10^{-5}]$	Substitution
$\approx -(-4.5)$	Evaluate the logarithm
≈ 4.5	Simplify

Natural Logarithms

DEFINITION *natural logarithms*

A ***natural logarithm*** is a logarithm with a base of e. The natural logarithm of x is denoted by $\ln x$. That is,

$$\ln x = \log_e x$$

We can assume that all our properties of exponents and logarithms hold for expressions with a base of e, because e is a real number. Here are some examples intended to make you more familiar with the number e and natural logarithms.

EXAMPLE 9 Simplify each of the following expressions.

a. $e^0 = 1$

b. $e^1 = e$

c. $\ln e = 1$ — In exponential form, $e^1 = e$

d. $\ln 1 = 0$ — In exponential form, $e^0 = 1$

e. $\ln e^3 = 3$

f. $\ln e^{-4} = -4$

g. $\ln e^t = t$

EXAMPLE 10 Use the properties of logarithms to expand the expression $\ln Ae^{5t}$.

SOLUTION Because the properties of logarithms hold for natural logarithms, we have

$$\begin{aligned} \ln Ae^{5t} &= \ln A + \ln e^{5t} \\ &= \ln A + 5t \ln e \\ &= \ln A + 5t \qquad \text{Because } \ln e = 1 \end{aligned}$$

EXAMPLE 11 If $\ln 2 = 0.6931$ and $\ln 3 = 1.0986$, find

a. $\ln 6$ **b.** $\ln 0.5$ **c.** $\ln 8$

SOLUTION

a. Because $6 = 2 \cdot 3$, we have

$$\begin{aligned} \ln 6 &= \ln 2 \cdot 3 \\ &= \ln 2 + \ln 3 \\ &= 0.6931 + 1.0986 \\ &= 1.7917 \end{aligned}$$

b. Writing 0.5 as $\frac{1}{2}$ and applying Property 2 for logarithms gives us

$$\begin{aligned} \ln 0.5 &= \ln \frac{1}{2} \\ &= \ln 1 - \ln 2 \\ &= 0 - 0.6931 \\ &= -0.6931 \end{aligned}$$

c. Writing 8 as 2^3 and applying Property 3 for logarithms, we have

$$\begin{aligned} \ln 8 &= \ln 2^3 \\ &= 3 \ln 2 \\ &= 3(0.6931) \\ &= 2.0793 \end{aligned}$$

GETTING READY FOR CLASS

After reading through the preceding section, respond in your own words and in complete sentences.

A. What is a common logarithm?
B. What is a natural logarithm?
C. Is e a rational number? Explain.
D. Find $\ln e$, and explain how you arrived at your answer.

Problem Set 10.5

Find the following logarithms.

1. log 378
2. log 426
3. log 37.8
4. log 42,600
5. log 3,780
6. log 0.4260
7. log 0.0378
8. log 0.0426
9. log 37,800
10. log 4,900
11. log 600
12. log 900
13. log 2,010
14. log 10,200
15. log 0.00971
16. log 0.0312
17. log 0.0314
18. log 0.00052
19. log 0.399
20. log 0.111

Find x in the following equations.

21. $\log x = 2.8802$
22. $\log x = 4.8802$
23. $\log x = -2.1198$
24. $\log x = -3.1198$
25. $\log x = 3.1553$
26. $\log x = 5.5911$
27. $\log x = -5.3497$
28. $\log x = -1.5670$
29. $\log x = -7.0372$
30. $\log x = -4.2000$
31. $\log x = 10$
32. $\log x = -1$
33. $\log x = -10$
34. $\log x = 1$
35. $\log x = 20$
36. $\log x = -20$
37. $\log x = -2$
38. $\log x = 4$
39. $\log x = \log_2 8$
40. $\log x = \log_3 9$
41. $\ln x = -1$
42. $\ln x = 4$
43. $\log x = 2 \log 5$
44. $\log x = -\log 4$
45. $\ln x = -3 \ln 2$
46. $\ln x = 5 \ln 3$

Simplify each of the following expressions.

47. $\ln e$
48. $\ln 1$
49. $\ln e^5$
50. $\ln e^{-3}$
51. $\ln e^x$
52. $\ln e^y$
53. $\log 10{,}000$
54. $\log 0.001$
55. $\ln \frac{1}{e^3}$
56. $\ln \sqrt{e}$
57. $\log \sqrt{1000}$
58. $\log \sqrt[3]{10{,}000}$

Use the properties of logarithms to expand each of the following expressions.

59. $\ln 10e^{3t}$
60. $\ln 10e^{4t}$
61. $\ln Ae^{-2t}$
62. $\ln Ae^{-3t}$
63. $\log [100(1.01)^{3t}]$
64. $\log \left[\frac{1}{10} (1.5)^{t+2} \right]$
65. $\ln (Pe^{rt})$
66. $\ln \left(\frac{1}{2} e^{-kt} \right)$
67. $-\log (4.2 \times 10^{-3})$
68. $-\log (5.7 \times 10^{-10})$

If $\ln 2 = 0.6931$, $\ln 3 = 1.0986$, and $\ln 5 = 1.6094$, find each of the following.

69. $\ln 15$
70. $\ln 10$
71. $\ln \frac{1}{3}$
72. $\ln \frac{1}{5}$
73. $\ln 9$
74. $\ln 25$
75. $\ln 16$
76. $\ln 81$

Applying the Concepts

77. **Atomic Bomb Tests** The formula for determining the magnitude, M, of an earthquake on the Richter Scale is $M = \log_{10} T$, where T is the number of times the shockwave is greater than the smallest measurable shockwave. The Bikini Atoll in the Pacific Ocean was used as a location for atomic bomb tests by the United States government in the 1950s. One such test resulted in an earthquake measurement of 5.0 on the Richter scale. Compare the 1906 San

1. 2.5775
2. 2.6294
3. 1.5775
4. 4.6294
5. 3.5775
6. −0.3706
7. −1.4225
8. −1.3706
9. 4.5775
10. 3.6902
11. 2.7782
12. 2.9542
13. 3.3032
14. 4.0086
15. −2.0128
16. −1.5058
17. −1.5031
18. −3.2840
19. −0.3990
20. −0.9547
21. 759
22. 75,893
23. 0.00759
24. 0.000759
25. 1,430
26. 390,032
27. 0.00000447
28. 0.0271
29. 0.0000000918
30. 0.0000631
31. 10^{10}
32. $\frac{1}{10}$
33. 10^{-10}
34. 10
35. 10^{20}
36. 10^{-20}
37. $\frac{1}{100}$
38. 10,000
39. 1,000
40. 100
41. $\frac{1}{e}$
42. e^4
43. 25
44. 0.25
45. $\frac{1}{8}$
46. 243
47. 1
48. 0
49. 5
50. −3
51. x
52. y
53. 4
54. −3
55. −3
56. $\frac{1}{2}$
57. $\frac{3}{2}$

58. $\frac{4}{3}$
59. ln 10 + 3t
60. ln 10 + 4t
61. ln A − 2t
62. ln A − 3t
63. 2 + 3$t$log1.01
64. −1 + (t + 2)log1.5
65. rt + lnP
66. $-kt + \ln\frac{1}{2}$
67. 3 − log4.2
68. 10 − log5.7
69. 2.7080
70. 2.3025
71. −1.0986
72. −1.6094
73. 2.1972
74. 3.2188
75. 2.7724
76. 4.3944
77. San Francisco was approx. 2,000 times greater.
78. 8.0
81. 2009
82. Over the 6-month period, the average score on the test decreased by approximatley 10 points.

Francisco earthquake of estimated magnitude 8.3 on the Richter scale to this atomic bomb test. Use the shock wave T for purposes of comparison.

78. Atomic Bomb Tests Today's nuclear weapons are 1,000 times more powerful than the atomic bombs tested in the Bikini Atoll mentioned in Problem 77. Use the shock wave T to determine the Richter scale measurement of a nuclear test today.

79. Getting Close to *e* Use a calculator to complete the following table.

x	$(1 + x)^{1/x}$
1	2
0.5	2.25
0.1	2.5937
0.01	2.7048
0.001	2.7169
0.0001	2.7181
0.00001	2.7183

What number does the expression $(1 + x)^{1/x}$ seem to approach as x gets closer and closer to zero?

80. Getting Close to *e* Use a calculator to complete the following table.

x	$\left(1 + \frac{1}{x}\right)^x$
1	2.0000
10	2.5937
50	2.6916
100	2.7048
500	2.7156
1,000	2.71692
10,000	2.71815
1,000,000	2.71828

What number does the expression $\left(1 + \frac{1}{x}\right)^x$ seem to approach as x gets larger and larger?

81. University Enrollment The percentage of students enrolled in a university who are between the ages of 25 and 34 can be modeled by the formula $s = 5 \ln x$, where s is the percentage of students and x is the number of years since 1989. Predict the year in which approximately 15% of students enrolled in a university are between the ages of 25 and 34.

82. Memory A class of students take a test on the mathematics concept of solving quadratic equations. That class agrees to take a similar form of the test each month for the next 6 months to test their memory of the topic since instruction. The function of the average score earned each month on the test is $m(x) = 75 - 5 \ln(x+1)$, where x

Time, x	Score, m
0	75
1	71.53
2	69.51
3	68.07
4	66.95
5	66.04
6	65.27

83. Approximately 3.19
84. 5.73
85. 1.78×10^{-5}
86. 1.78×10^{-6}
87. 3.16×10^{5}
88. 3.98×10^{6}
89. 2.00×10^{8}
90. 5.01×10^{8}

represents time in months. Complete the table to indicate the average score earned by the class at each month.

Use the following figure to solve Problems 83 – 86.

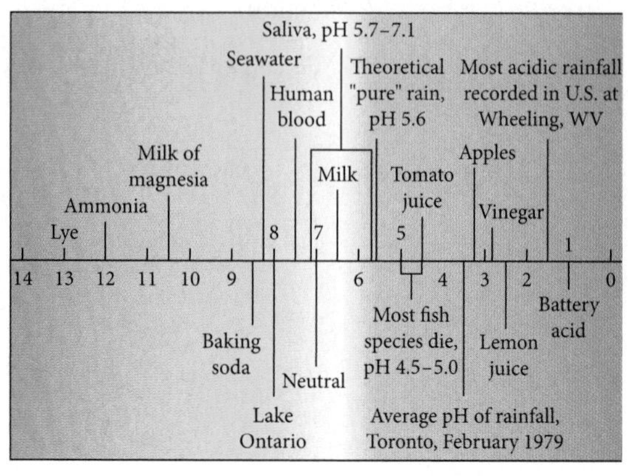

83. pH Find the pH of orange juice if the concentration of the hydrogen ion in the juice is $[H^+] = 6.50 \times 10^{-4}$.

84. pH Find the pH of milk if the concentration of the hydrogen ions in milk is $[H^+] = 1.88 \times 10^{-6}$.

85. pH Find the concentration of hydrogen ions in a glass of wine if the pH is 4.75.

86. pH Find the concentration of hydrogen ions in a bottle of vinegar if the pH is 5.75.

The Richter Scale Find the relative size T of the shock wave of earthquakes with the following magnitudes, as measured on the Richter scale.

87. 5.5 **88.** 6.6 **89.** 8.3 **90.** 8.7

91. Earthquake The chart below is a partial listing of earthquakes that were recorded in Canada during one year. Complete the chart by computing the magnitude on the Richter Scale, M, or the number of times the associated shockwave is larger than the smallest measurable shockwave, T.

Location	Date	Magnitude M	Shockwave T
Moresby Island	Jan. 23	4.0	1.00×10^{4}
Vancouver Island	Apr. 30	5.3	1.99×10^{5}
Quebec City	June 29	3.2	1.58×10^{3}
Mould Bay	Nov. 13	5.2	1.58×10^{5}
St. Lawrence	Dec. 14	3.7	5.01×10^{3}

SOURCE: *National Resources Canada, National Earthquake Hazards Program.*

92. 5.01×10^7
93. 12.9%
94. 24%
95. 5.3%
96. 8.7%
97. $\frac{7}{10}$
98. $\frac{29}{12}$
99. 3.1250
100. 3.2281
101. 1.2575
102. 1.0008
103. $t\log 1.05$
104. $t\log 1.033$
105. $.05t$
106. $-.000121t$

92. Earthquake On January 6, 2001, an earthquake with a magnitude of 7.7 on the Richter Scale hit southern India (*National Earthquake Information Center*). By what factor was this earthquake's shockwave greater than the smallest measurable shockwave?

Depreciation The annual rate of depreciation r on a car that is purchased for P dollars and is worth W dollars t years later can be found from the formula.

$$\log(1 - r) = \frac{1}{t} \log \frac{W}{P}$$

93. Find the annual rate of depreciation on a car that is purchased for \$9,000 and sold 5 years later for \$4,500.

94. Find the annual rate of depreciation on a car that is purchased for \$9,000 and sold 4 years later for \$3,000.

Two cars depreciate in value according to the following depreciation tables. In each case, find the annual rate of depreciation.

95.

Age in Years	Value in Dollars
New	7,550
5	5,750

96.

Age in Years	Value in Dollars
New	7,550
3	5,750

Getting Ready for the Next Section

Solve.

97. $5(2x + 1) = 12$

98. $4(3x - 2) = 21$

Use a calculator to evaluate, give answers to 4 decimal places.

99. $\dfrac{100{,}000}{32{,}000}$

100. $\dfrac{1.4982}{6.5681} + 3$

101. $\dfrac{1}{2}\left(\dfrac{-0.6931}{1.4289} + 3\right)$

102. $1 + \dfrac{0.04}{52}$

Use the power rule to rewrite the following logarithms.

103. $\log 1.05^t$

104. $\log 1.033^t$

Use identities to simplify.

105. $\ln e^{0.05t}$

106. $\ln e^{-0.000121t}$

Exponential Equations and Change of Base

10.6

For items involved in exponential growth, the time it takes for a quantity to double is called the *doubling time*. For example, if you invest \$5,000 in an account that pays 5% annual interest, compounded quarterly, you may want to know how long it will take for your money to double in value. You can find this doubling time if you can solve the equation

$$10{,}000 = 5{,}000\,(1.0125)^{4t}$$

As you will see as you progress through this section, logarithms are the key to solving equations of this type.

Logarithms are very important in solving equations in which the variable appears as an exponent. The equation

$$5^x = 12$$

is an example of one such equation. Equations of this form are called *exponential equations*. Because the quantities 5^x and 12 are equal, so are their common logarithms. We begin our solution by taking the logarithm of both sides:

$$\log 5^x = \log 12$$

We now apply Property 3 for logarithms, $\log x^r = r \log x$, to turn x from an exponent into a coefficient:

$$x \log 5 = \log 12$$

Dividing both sides by log 5 gives us

$$x = \frac{\log 12}{\log 5}$$

If we want a decimal approximation to the solution, we can find log 12 and log 5 on a calculator and divide:

$$\begin{aligned} x &\approx \frac{1.0792}{0.6990} \\ &\approx 1.5439 \end{aligned}$$

The complete problem looks like this:

$$\begin{aligned} 5^x &= 12 \\ \log 5^x &= \log 12 \\ x \log 5 &= \log 12 \\ x &= \frac{\log 12}{\log 5} \\ &\approx \frac{1.0792}{0.6990} \\ &\approx 1.5439 \end{aligned}$$

Here is another example of solving an exponential equation using logarithms.

EXAMPLE 1 Solve for x: $25^{2x+1} = 15$

SOLUTION Taking the logarithm of both sides and then writing the exponent $(2x + 1)$ as a coefficient, we proceed as follows:

$$25^{2x+1} = 15$$

$$\log 25^{2x+1} = \log 15 \quad \text{Take the log of both sides}$$

$$(2x + 1)\log 25 = \log 15 \quad \text{Property 3}$$

$$2x + 1 = \frac{\log 15}{\log 25} \quad \text{Divide by log 25}$$

$$2x = \frac{\log 15}{\log 25} - 1 \quad \text{Add } -1 \text{ to both sides}$$

$$x = \frac{1}{2}\left(\frac{\log 15}{\log 25} - 1\right) \quad \text{Multiply both sides by } \frac{1}{2}$$

Using a calculator, we can write a decimal approximation to the answer:

$$x \approx \frac{1}{2}\left(\frac{1.1761}{1.3979} - 1\right)$$

$$\approx \frac{1}{2}(0.8413 - 1)$$

$$\approx \frac{1}{2}(-0.1587)$$

$$\approx -0.079$$

If you invest P dollars in an account with an annual interest rate r that is compounded n times a year, then t years later the amount of money in that account will be

$$A = P\left(1 + \frac{r}{n}\right)^{nt}$$

EXAMPLE 2 How long does it take for \$5,000 to double if it is deposited in an account that yields 5% interest compounded once a year?

SOLUTION Substituting $P = 5{,}000$, $r = 0.05$, $n = 1$, and $A = 10{,}000$ into our formula, we have

$$10{,}000 = 5{,}000(1 + 0.05)^t$$

$$10{,}000 = 5{,}000(1.05)^t$$

$$2 = (1.05)^t \quad \text{Divide by 5,000}$$

This is an exponential equation. We solve by taking the logarithm of both sides:

$$\log 2 = \log(1.05)^t$$

$$= t \log 1.05$$

Dividing both sides by log 1.05, we have

$$t = \frac{\log 2}{\log 1.05}$$

$$\approx 14.2$$

It takes a little over 14 years for \$5,000 to double if it earns 5% interest per year, compounded once a year.

There is a fourth property of logarithms we have not yet considered. This last property allows us to change from one base to another and is therefore called the *change-of-base property*.

PROPERTY *Property 4 (Change of Base)*

If a and b are both positive numbers other than 1, and if $x > 0$, then

$$\log_a x = \frac{\log_b x}{\log_b a}$$

Base a Base b

The logarithm on the left side has a base of a, and both logarithms on the right side have a base of b. This allows us to change from base a to any other base b that is a positive number other than 1. Here is a proof of Property 4 for logarithms.

Proof We begin by writing the identity

$$a^{\log_a x} = x$$

Taking the logarithm base b of both sides and writing the exponent $\log_a x$ as a coefficient, we have

$$\log_b a^{\log_a x} = \log_b x$$

$$\log_a x \log_b a = \log_b x$$

Dividing both sides by $\log_b a$, we have the desired result:

$$\frac{\log_a x \log_b a}{\log_b a} = \frac{\log_b x}{\log_b a}$$

$$\log_a x = \frac{\log_b x}{\log_b a}$$

We can use this property to find logarithms we could not otherwise compute on our calculators—that is, logarithms with bases other than 10 or e. The next example illustrates the use of this property.

EXAMPLE 3 Find $\log_8 24$.

SOLUTION Because we do not have base-8 logarithms on our calculators, we can change this expression to an equivalent expression that contains only base-10 logarithms:

$$\log_8 24 = \frac{\log 24}{\log 8} \qquad \text{Property 4}$$

Don't be confused. We did not just drop the base, we changed to base 10. We could have written the last line like this:

$$\log_8 24 = \frac{\log_{10} 24}{\log_{10} 8}$$

From our calculators, we write

$$\log_8 24 \approx \frac{1.3802}{0.9031}$$

$$\approx 1.5283$$

Application

EXAMPLE 4 Suppose that the population in a small city is 32,000 in the beginning of 2010 and that the city council assumes that the population size t years later can be estimated by the equation

$$P = 32{,}000e^{0.05t}$$

Approximately when will the city have a population of 50,000?

SOLUTION We substitute 50,000 for P in the equation and solve for t:

$$50{,}000 = 32{,}000e^{0.05t}$$

$$1.5625 = e^{0.05t} \qquad \frac{50{,}000}{32{,}000} = 1.5625$$

To solve this equation for t, we can take the natural logarithm of each side:

$$\ln 1.5625 = \ln e^{0.05t}$$

$$= 0.05t \ln e \qquad \text{Property 3 for logarithms}$$

$$= 0.05t \qquad \text{Because } \ln e = 1$$

$$t = \frac{\ln 1.5625}{0.05} \qquad \text{Divide each side by 0.05}$$

$$\approx 8.93 \text{ years}$$

We can estimate that the population will reach 50,000 toward the end of 2018.

USING TECHNOLOGY *Graphing Calculators*

We can evaluate many logarithmic expressions on a graphing calculator by using the fact that logarithmic functions and exponential functions are inverses.

EXAMPLE 5 Evaluate the logarithmic expression $\log_3 7$ from the graph of an exponential function.

SOLUTION First, we let $\log_3 7 = x$. Next, we write this expression in exponential form as $3^x = 7$. We can solve this equation graphically by finding the intersection of the graphs $Y_1 = 3^x$ and $Y_2 = 7$, as shown in Figure 1.

Using the calculator, we find the two graphs intersect at (1.77, 7). Therefore, $\log_3 7 = 1.77$ to the nearest hundredth. We can check our work by evaluating the expression $3^{1.77}$ on our calculator with the key strokes

3 [^] 1.77 [ENTER]

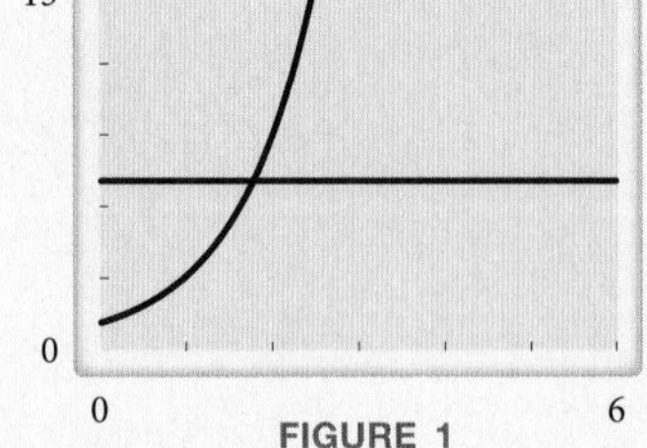

FIGURE 1

The result is 6.99 to the nearest hundredth, which seems reasonable since 1.77 is accurate to the nearest hundredth. To get a result closer to 7, we would need to find the intersection of the two graphs more accurately.

GETTING READY FOR CLASS

After reading through the preceding section, respond in your own words and in complete sentences.

A. What is an exponential equation?
B. How do logarithms help you solve exponential equations?
C. What is the change-of-base property?
D. Write an application modeled by the equation
$A = 10{,}000\left(1 + \frac{0.08}{2}\right)^{2 \cdot 5}$.

Problem Set 10.6

Solve each exponential equation. Use a calculator to write the answer in decimal form.

1. $3^x = 5$ **2.** $4^x = 3$ **3.** $5^x = 3$ **4.** $3^x = 4$

5. $5^{-x} = 12$ **6.** $7^{-x} = 8$ **7.** $12^{-x} = 5$ **8.** $8^{-x} = 7$

9. $8^{x+1} = 4$ **10.** $9^{x+1} = 3$ **11.** $4^{x-1} = 4$ **12.** $3^{x-1} = 9$

13. $3^{2x+1} = 2$ **14.** $2^{2x+1} = 3$ **15.** $3^{1-2x} = 2$ **16.** $2^{1-2x} = 3$

17. $15^{3x-4} = 10$ **18.** $10^{3x-4} = 15$ **19.** $6^{5-2x} = 4$ **20.** $9^{7-3x} = 5$

21. $3^{-4x} = 81$ **22.** $2^{5x} = \frac{1}{16}$ **23.** $5^{3x-2} = 15$ **24.** $7^{4x+3} = 200$

25. $100e^{3t} = 250$ **26.** $150e^{0.065t} = 400$

27. $1200\left(1 + \frac{0.072}{4}\right)^{4t} = 25000$ **28.** $2700\left(1 + \frac{0.086}{12}\right)^{12t} = 10000$

29. $50e^{-0.0742t} = 32$ **30.** $19e^{-0.000243t} = 12$

Use the change-of-base property and a calculator to find a decimal approximation to each of the following logarithms.

31. $\log_8 16$ **32.** $\log_9 27$ **33.** $\log_{16} 8$ **34.** $\log_{27} 9$

35. $\log_7 15$ **36.** $\log_3 12$ **37.** $\log_{15} 7$ **38.** $\log_{12} 3$

39. $\log_8 240$ **40.** $\log_6 180$ **41.** $\log_4 321$ **42.** $\log_5 462$

Find a decimal approximation to each of the following natural logarithms.

43. ln 345 **44.** ln 3,450 **45.** ln 0.345 **46.** ln 0.0345

47. ln 10 **48.** ln 100 **49.** ln 45,000 **50.** ln 450,000

Applying the Concepts

51. Compound Interest How long will it take for \$500 to double if it is invested at 6% annual interest compounded 2 times a year?

52. Compound Interest How long will it take for \$500 to double if it is invested at 6% annual interest compounded 12 times a year?

53. Compound Interest How long will it take for \$1,000 to triple if it is invested at 12% annual interest compounded 6 times a year?

54. Compound Interest How long will it take for \$1,000 to become \$4,000 if it is invested at 12% annual interest compounded 6 times a year?

55. Doubling Time How long does it take for an amount of money P to double itself if it is invested at 8% interest compounded 4 times a year?

56. Tripling Time How long does it take for an amount of money P to triple itself if it is invested at 8% interest compounded 4 times a year?

57. Tripling Time If a \$25 investment is worth \$75 today, how long ago must that \$25 have been invested at 6% interest compounded twice a year?

58. Doubling Time If a \$25 investment is worth \$50 today, how long ago must that \$25 have been invested at 6% interest compounded twice a year?

1. 1.4650
2. 0.7925
3. 0.6826
4. 1.2619
5. −1.5440
6. −1.0686
7. −0.6477
8. −0.9358
9. −0.3333
10. −0.5000
11. 2.000
12. 3.000
13. −0.1845
14. 0.2925
15. 0.1845
16. −0.2925
17. 1.6168
18. 1.7254
19. 2.1131
20. 2.0892
21. −1
22. $-\frac{4}{5}$
23. 1.2275
24. −.0693
25. .3054
26. 15.0897
27. 42.5528
28. 15.2793
29. 6.0147
30. 1891.08
31. 1.333
32. 1.500
33. 0.7500
34. 0.6667
35. 1.3917
36. 2.2619
37. 0.7186
38. 0.4421
39. 2.6356
40. 2.8982
41. 4.1632
42. 3.8122
43. 5.8435
44. 8.1461
45. −1.0642
46. −3.3668
47. 2.3026
48. 4.6052
49. 10.7144
50. 13.0170
51. 11.72 years
52. 11.58 years
53. 9.25 years
54. 11.67 years
55. 8.75 years
56. 13.87 years
57. 18.58 years
58. 11.72 years

59. 11.55 years
60. 11.55 years
61. 18.31 years
62. 9.16 years
63. 11.45 years
64. 20.12 years
65. October 2018
66. 100,000
67. 2009
68. 2012
69. 1992

Recall from Section 8.1 that if P dollars are invested in an account with annual interest rate r, compounded continuously, then the amount of money in the account after t years is given by the formula

$$A(t) = Pe^{rt}$$

59. **Continuously Compounded Interest** Repeat Problem 51 if the interest is compounded continuously.

60. **Continuously Compounded Interest** Repeat Problem 54 if the interest is compounded continuously.

61. **Continuously Compounded Interest** How long will it take \$500 to triple if it is invested at 6% annual interest, compounded continuously?

62. **Continuously Compounded Interest** How long will it take \$500 to triple if it is invested at 12% annual interest, compounded continuously?

63. **Continuously Compounded Interest** How long will it take for \$1,000 to be worth \$2,500 at 8% interest, compounded continuously?

64. **Continuously Compounded Interest** How long will it take for \$1,000 to be worth \$5,000 at 8% interest, compounded continuously?

65. **Exponential Growth** Suppose that the population in a small city is 32,000 at the beginning of 2005 and that the city council assumes that the population size t years later can be estimated by the equation

$$P(t) = 32{,}000e^{0.05t}$$

Approximately when will the city have a population of 64,000?

66. **Exponential Growth** Suppose the population of a city is given by the equation

$$P(t) = 100{,}000e^{0.05t}$$

where t is the number of years from the present time. How large is the population now? (*Now* corresponds to a certain value of t. Once you realize what that value of t is, the problem becomes very simple.)

67. **Airline Travel** The number of airline passengers in 1990 was 466 million. The number of passengers traveling by airplane each year has increased exponentially according to the model, $P(t) = 466 \cdot 1.035^t$, where t is the number of years since 1990 (U.S. Census Bureau). In what year is it predicted that 900 million passengers will travel by airline?

68. **Bankruptcy Model** In 1997, there were a total of 1,316,999 bankruptcies filed under the Bankruptcy Reform Act. The model for the number of bankruptcies filed is $B(t) = 0.798 \cdot 1.164^t$, where t is the number of years since 1994 and B is the number of bankruptcies filed in terms of millions (Administrative Office of the U.S. Courts, *Statistical Tables for the Federal Judiciary*). In what year is it predicted that 12 million bankruptcies will be filed?

69. **Health Care** In 1990, \$699 billion was spent on health care expenditures. The amount of money, E, in billions spent on health care expenditures can be estimated using the function $E(t) = 78.16(1.11)^t$, where t is time in years since 1970 (*U.S. Census Bureau*). In what year was it estimated that \$800 billion will be spent on health care expenditures?

70. 3.78 years old
71. 10.07 years
72. It has been approximately 4,127 years.
73. 2000
74. 2003
75. $(-2, -23)$, lowest
76. $\left(\frac{3}{2}, -\frac{67}{4}\right)$, lowest
77. $\left(\frac{3}{2}, 9\right)$, highest
78. $\left(\frac{3}{2}, \frac{27}{2}\right)$, highest
79. 2 seconds, 64 feet
80. 2 seconds, 104 feet

70. Value of a Car As a car ages, its value decreases. The value of a particular car with an original purchase price of \$25,600 is modeled by the function $c(t) = 25{,}600(1 - 0.22)^t$, where c is the value at time t (Kelly Blue Book). How old is the car when its value is \$10,000?

71. Compound Interest In 1986, the average cost of attending a public university through graduation was \$16,552 (U.S. Department of Education, National Center for Educational Statistics). If John's parents deposited that amount in an account in 1986 at an interest rate of 7% compounded semi-annually, how long will it take for the money to double?

72. Carbon Dating Scientists use Carbon-14 dating to find the age of fossils and other artifacts. The amount of Carbon-14 in an organism will yield information concerning its age. A formula used in Carbon-14 dating is $A(t) = A_0 \cdot 2^{-t/5600}$, where A_0 is the amount of carbon originally in the organism, t is time in years, and A is the amount of carbon remaining after t years. Determine the number of years since an organism died if it originally contained 1,000 gram of Carbon-14 and it currently contains 600 gram of Carbon-14.

73. Cost Increase The cost of a can of Coca Cola in 1960 was \$0.10. The function that models the cost of a Coca Cola by year is $C(t) = 0.10e^{0.0576t}$, where t is the number of years since 1960. In what year is it expected that a can of Coca Cola will cost \$1.00?

74. Online Banking Use The number of households using online banking services has increased from 754,000 in 1995 to 12,980,000 in 2000. The formula $H(t) = 0.76e^{0.55t}$ models the number of households, H, in millions when time is t years since 1995 according to the Home Banking Report. In what year is it estimated that 50,000,000 households will use online banking services?

Maintaining Your Skills

The following problems review material we covered in Section 7.5.

Find the vertex for each of the following parabolas, and then indicate if it is the highest or lowest point on the graph.

75. $y = 2x^2 + 8x - 15$

76. $y = 3x^2 - 9x - 10$

77. $y = 12x - 4x^2$

78. $y = 18x - 6x^2$

79. Maximum Height An object is projected into the air with an initial upward velocity of 64 feet per second. Its height h at any time t is given by the formula $h = 64t - 16t^2$. Find the time at which the object reaches its maximum height. Then, find the maximum height.

80. Maximum Height An object is projected into the air with an initial upward velocity of 64 feet per second from the top of a building 40 feet high. If the height h of the object t seconds after it is projected into the air is $h = 40 + 64t - 16t^2$, find the time at which the object reaches its maximum height. Then, find the maximum height it attains.

Chapter 10 Summary

EXAMPLES

Exponential Functions [10.1]

1. For the exponential function $f(x) = 2^x$,
$f(0) = 2^0 = 1$
$f(1) = 2^1 = 2$
$f(2) = 2^2 = 4$
$f(3) = 2^3 = 8$

Any function of the form

$$f(x) = b^x$$

where $b > 0$ and $b \neq 1$, is an *exponential function.*

One-to-One Functions [10.2]

2. The function $f(x) = x^2$ is not one-to-one because 9, which is in the range, comes from both 3 and -3 in the domain.

A function is a *one-to-one function* if every element in the range comes from exactly one element in the domain.

Inverse Functions [10.2]

3. The inverse of $f(x) = 2x - 3$ is
$$f^{-1}(x) = \frac{x + 3}{2}$$

The *inverse* of a function is obtained by reversing the order of the coordinates of the ordered pairs belonging to the function. Only one-to-one functions have inverses that are also functions.

Definition of Logarithms [10.3]

4. The definition allows us to write expressions like
$$y = \log_3 27$$
equivalently in exponential form as
$$3^y = 27$$
which makes it apparent that y is 3.

If b is a positive number not equal to 1, then the expression

$$y = \log_b x$$

is equivalent to $x = b^y$; that is, in the expression $y = \log_b x$, y is the number to which we raise b in order to get x. Expressions written in the form $y = \log_b x$ are said to be in *logarithmic form*. Expressions like $x = b^y$ are in *exponential form.*

Two Special Identities [10.3]

5. Examples of the two special identities are
$$5^{\log_5 12} = 12$$
and
$$\log_8 8^3 = 3$$

For $b > 0$, $b \neq 1$, the following two expressions hold for all positive real numbers x:

(1) $b^{\log_b x} = x$

(2) $\log_b b^x = x$

Properties of Logarithms [10.4]

6. We can rewrite the expression
$$\log_{10} \frac{45^6}{273}$$
using the properties of logarithms, as
$$6 \log_{10} 45 - \log_{10} 273$$

If x, y, and b are positive real numbers, $b \neq 1$, and r is any real number, then:

1. $\log_b(xy) = \log_b x + \log_b y$
2. $\log_b \left(\frac{x}{y}\right) = \log_b x - \log_b y$
3. $\log_b x^r = r \log_b x$

Common Logarithms [10.5]

7. $\log_{10} 10{,}000 = \log 10{,}000$
$= \log 10^4$
$= 4$

Common logarithms are logarithms with a base of 10. To save time in writing, we omit the base when working with common logarithms; that is,

$$\log x = \log_{10} x$$

Natural Logarithms [10.5]

8. $\ln e = 1$
$\ln 1 = 0$

Natural logarithms, written *ln x*, are logarithms with a base of e, where the number e is an irrational number (like the number π). A decimal approximation for e is 2.7183. All the properties of exponents and logarithms hold when the base is e.

Change of Base [10.6]

9. $\log_6 475 = \dfrac{\log 475}{\log 6}$
$\approx \dfrac{2.6767}{0.7782}$
≈ 3.44

If x, a, and b are positive real numbers, $a \neq 1$ and $b \neq 1$, then

$$\log_a x = \frac{\log_b x}{\log_b a}$$

COMMON MISTAKE

The most common mistakes that occur with logarithms come from trying to apply the three properties of logarithms to situations in which they don't apply. For example, a very common mistake looks like this:

$$\frac{\log 3}{\log 2} = \log 3 - \log 2 \quad \text{Mistake}$$

This is not a property of logarithms. To write the equation $\log 3 - \log 2$, we would have to start with

$$\log \frac{3}{2} \qquad \text{NOT} \qquad \frac{\log 3}{\log 2}$$

There is a difference.

Graph each exponential function. [10.1]

1. $f(x) = 2^x$

2. $g(x) = 3^{-x}$

Sketch the graph of each function and its inverse. Find $f^{-1}(x)$ for Problem 3. [10.2]

3. $f(x) = 2x - 3$

4. $f(x) = x^2 - 4$

Solve for x. [10.3]

5. $\log_4 x = 3$

6. $\log_x 5 = 2$

Graph each of the following [10.3]

7. $y = \log_2 x$

8. $y = \log_{1/2} x$

Evaluate each of the following. [10.3, 10.4, 10.5]

9. $\log_8 4$

10. $\log_7 21$

11. $\log 23{,}400$

12. $\log 0.0123$

13. $\ln 46.2$

14. $\ln 0.0462$

Use the properties of logarithms to expand each expression. [10.4]

15. $\log_2 \frac{8x^2}{y}$

16. $\log \frac{\sqrt{x}}{(y^4)\sqrt[5]{z}}$

Write each expression as a single logarithm. [10.4]

17. $2 \log_3 x - \frac{1}{2} \log_3 y$

18. $\frac{1}{3} \log x - \log y - 2 \log z$

Use a calculator to find x. [10.5]

19. $\log x = 4.8476$

20. $\log x = -2.6478$

Solve for x. [10.4, 10.6]

21. $5 = 3^x$

22. $4^{2x-1} = 8$

23. $\log_5 x - \log_5 3 = 1$

24. $\log_2 x + \log_2(x - 7) = 3$

25. pH Find the pH of a solution in which $[H^+] = 6.6 \times 10^{-7}$. [10.5]

26. Compound Interest If \$400 is deposited in an account that earns 10% annual interest compounded twice a year, how much money will be in the account after 5 years? [10.1]

27. Compound Interest How long will it take \$600 to become \$1,800 if the \$600 is deposited in an account that earns 8% annual interest compounded 4 times a year? [10.6]

28. Depreciation If a car depreciates in value 20% per year for the first 5 years after it is purchased for P_0 dollars, then its value after t years will be $V(t) = P_0(1 - r)^t$ for $0 \le t \le 5$. To the nearest dollar, find the value of a car 4 years after it is purchased for \$18,000. [10.1]

Answers to Odd-Numbered Problems

Chapter 1

PROBLEM SET 1.1

1. $-3x$ **3.** $-a$ **5.** $12x$ **7.** $6a$ **9.** $6x-3$ **11.** $7a+5$ **13.** $5x-5$ **15.** $4a+2$ **17.** $-9x-2$ **19.** $12a+3$ **21.** $10x-1$ **23.** $21y+6$ **25.** $-6x+8$ **27.** $-2a+3$ **29.** $-4x+26$ **31.** $4y-16$ **33.** $-6x-1$ **35.** $2x-12$ **37.** $10a+33$ **39.** $4x-9$ **41.** $7y-39$ **43.** $-19x-14$ **45.** 5 **47.** -9 **49.** 4 **51.** 4 **53.** -37 **55.** -41 **57.** 64 **59.** 64 **61.** 144 **63.** 144 **65.** 3 **67.** 0 **69.** 15 **71.** 6 **73. a.**

n	1	2	3	4
$3n$	3	6	9	12

b.

n	1	2	3	4
n^3	1	8	27	64

75. 1, 4, 7, 10, . . . an arithmetic sequence **77.** 0, 1, 4, 9, . . . a sequence of squares **79.** $-6y+4$ **81.** $0.17x$ **83.** $2x$ **85.** $5x-4$ **87.** $7x-5$ **89.** $-2x-9$ **91.** $7x+2$ **93.** $-7x+6$ **95.** $7x$ **97.** $-y$ **99.** $10y$ **101.** $0.17x+180$ **103.** $0.22x+60$ **105.** 49 **107.** 40 **109. a.** 42°F **b.** 28°F **c.** -14°F **111. a.** \$37.50 **b.** \$40.00 **c.** \$42.50 **113.** 12 **115.** -3 **117.** -9.7 **119.** $-\frac{5}{4}$ **121.** 53 **123.** $a-12$ **125.** 7

PROBLEM SET 1.2

1. 11 **3.** 4 **5.** $-\frac{3}{4}$ **7.** -5.8 **9.** -17 **11.** $-\frac{1}{8}$ **13.** -4 **15.** -3.6 **17.** 1 **19.** $-\frac{7}{45}$ **21.** 3 **23.** $\frac{11}{8}$ **25.** 21 **27.** 7 **29.** 3.5 **31.** 22 **33.** -2 **35.** -16 **37.** -3 **39.** 10 **41.** -12 **43.** 4 **45.** 2 **47.** -5 **49.** -1 **51.** -3 **53.** 8 **55.** -8 **57.** 2 **59.** 11 **61.** -5.8 **63. a.** 6% **b.** 5% **c.** 2% **d.** 75% **65.** y **67.** x **69.** 6 **71.** 6 **73.** -9 **75.** $-\frac{15}{8}$ **77.** 8 **79.** $-\frac{5}{4}$ **81.** $3x$

PROBLEM SET 1.3

1. 2 **3.** 4 **5.** $-\frac{1}{2}$ **7.** -2 **9.** 3 **11.** 4 **13.** 0 **15.** 0 **17.** 6 **19.** -50 **21.** $\frac{3}{2}$ **23.** 12 **25.** -3 **27.** 32 **29.** -8 **31.** $\frac{1}{2}$ **33.** 4 **35.** 8 **37.** -4 **39.** 4 **41.** -15 **43.** $-\frac{1}{2}$ **45.** 3 **47.** 1 **49.** $\frac{1}{4}$ **51.** -3 **53.** 3 **55.** 2 **57.** $-\frac{3}{2}$ **59.** $-\frac{3}{2}$ **61.** 1 **63.** 1 **65.** -2 **67. a.** $\frac{3}{2}$ **b.** 1 **c.** $-\frac{3}{2}$ **d.** -4 **e.** $\frac{8}{5}$ **69.** 200 tickets **71.** \$1,390.85 per month **73.** 2 **75.** 6 **77.** 3,000 **79.** $3x-11$ **81.** $0.09x+180$ **83.** $-6y+4$ **85.** $4x-11$ **87.** $5x$ **89.** $0.17x$

PROBLEM SET 1.4

1. 3 **3.** -2 **5.** -1 **7.** 2 **9.** -4 **11.** -2 **13.** 0 **15.** 1 **17.** $\frac{1}{2}$ **19.** 7 **21.** 8 **23.** $-\frac{1}{3}$ **25.** $\frac{3}{4}$ **27.** 75 **29.** 2 **31.** 6 **33.** 8 **35.** 0 **37.** $\frac{3}{7}$ **39.** 1 **41.** 1 **43.** -1 **45.** 6 **47.** $\frac{3}{4}$ **49.** 3 **51.** $\frac{3}{4}$ **53.** 8 **55.** 6 **57.** -2 **59.** -2 **61.** 2 **63.** -6 **65.** 2 **67.** 20 **69.** 4,000 **71.** 700 **73.** 11 **75.** 7 **77. a.** $\frac{5}{4}=1.25$ **b.** $\frac{15}{2}=7.5$ **c.** $6x+20$ **d.** 15 **e.** $4x-20$ **f.** $\frac{45}{2}=22.5$ **79.** 14 **81.** -3 **83.** $\frac{1}{4}$ **85.** $\frac{1}{3}$ **87.** $-\frac{3}{2}x+3$

PROBLEM SET 1.5

1. 100 feet **3.** 0 **5.** 2 **7.** 15 **9.** 10 **11.** -2 **13.** 1 **15. a.** 2 **b.** 4 **17. a.** 5 **b.** 18 **19.** $l=\frac{A}{w}$ **21.** $h=\frac{V}{lw}$ **23.** $a=P-b-c$ **25.** $x=3y-1$ **27.** $y=3x+6$ **29.** $y=-\frac{2}{3}x+2$ **31.** $y=-2x-5$ **33.** $y=-\frac{2}{3}x+1$ **35.** $w=\frac{P-2l}{2}$ **37.** $v=\frac{h-16t^2}{t}$ **39.** $h=\frac{A-\pi r^2}{2\pi r}$ **41. a.** $y=\frac{3}{5}x+1$ **b.** $y=\frac{1}{2}x+2$ **c.** $y=4x+3$ **43.** $y=\frac{3}{7}x-3$ **45.** $y=2x+8$ **47.** 60°; 150° **49.** 45°; 135° **51.** 10 **53.** 240 **55.** 25% **57.** 35% **59.** 64 **61.** 2,000 **63.** 100°C; yes **65.** 20°C; yes **67.** $C=\frac{5}{9}(F-32)$ **69.** 4°F over **71.** 7 meters **73.** $\frac{3}{2}$ or 1.5 inches **75.** 132 feet **77.** $\frac{2}{9}$ centimeters **79.** 60% **81.** 26.5% **83.** The sum of 4 and 1 is 5. **85.** The difference of 6 and 2 is 4. **87.** The difference of a number and 5 is -12. **89.** The sum of a number and 3 is four times the difference of that number and 3. **91.** $2(6+3)=18$ **93.** $2(5)+3=13$ **95.** $x+5=13$ **97.** $5(x+7)=30$

PROBLEM SET 1.6

1. 8 **3.** 5 **5.** -1 **7.** 3 and 5 **9.** 6 and 14 **11.** Shelly is 39; Michele is 36 **13.** Evan is 11; Cody is 22 **15.** Barney is 27; Fred is 31 **17.** Lacy is 16; Jack is 32 **19.** Patrick is 18; Pat is 38 **21.** $s=9$ inches **23.** $s=15$ feet **25.** 11 feet, 18 feet, 33 feet **27.** 26 feet, 13 feet, 14 feet **29.** $l=11$ inches; $w=6$ inches **31.** $l=25$ inches; $w=9$ inches

33. $l = 15$ feet; $w = 3$ feet **35.** 9 dimes; 14 quarters **37.** 12 quarters; 27 nickels **39.** 8 nickels; 17 dimes
41. 7 nickels; 10 dimes; 12 quarters **43.** 3 nickels; 9 dimes; 6 quarters **45.** $5x$ **47.** $1.075x$ **49.** $0.09x + 180$ **51.** 6,000 **53.** 30

PROBLEM SET 1.7

1. 5 and 6 **3.** -4 and -5 **5.** 13 and 15 **7.** 52 and 54 **9.** -14 and -16 **11.** 17, 19, and 21 **13.** 42, 44, and 46
15. \$4,000 invested at 8%, \$6,000 invested at 9% **17.** \$700 invested at 10%, \$1,200 invested at 12%
19. \$500 at 8%, \$1,000 at 9%, \$1,500 at 10% **21.** 45°, 45°, 90° **23.** 22.5°, 45°, 112.5° **25.** 53°, 90° **27.** 80°, 60°, 40°
29. 16 adult and 22 children's tickets **31.** 16 minutes **33.** 39 hours **35.** They are in offices 7329 and 7331.
37. Kendra is 8 years old and Marissa is 10 years old. **39.** Jeff **41.** \$10.38 **43.** $l = 12$ meters; $w = 10$ meters **45.** 59°, 60°, 61°
47. \$54.00 **49.** Yes **51.** **a.** 9 **b.** 3 **c.** -9 **d.** -3 **53.** **a.** -8 **b.** 8 **c.** 8 **d.** -8 **55.** -2.3125 **57.** $\frac{10}{3}$

PROBLEM SET 1.8

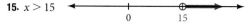

1. $x < 12$ 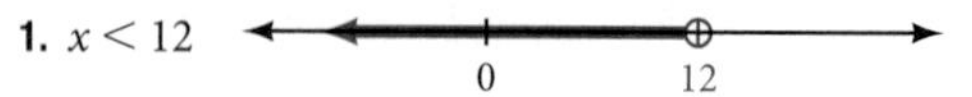
3. $a \le 12$
5. $x > 13$
7. $y \ge 4$
9. $x > 9$
11. $x < 2$
13. $a \le 5$
15. $x > 15$
17. $x < -3$
19. $x \le 6$
21. $x \ge -50$
23. $y < -6$ 
25. $x < 6$
27. $y \ge -5$
29. $x < 3$
31. $x \le 18$
33. $a < -20$
35. $y < 25$
37. $a \le 3$
39. $x \ge \frac{15}{2}$
41. $x < -1$
43. $y \ge -2$
45. $x < -1$
47. $m \le -6$
49. $x \le -5$
51. $y < -\frac{3}{2}x + 3$ **53.** $y < \frac{2}{5}x - 2$ **55.** $y \le \frac{3}{7}x + 3$

57. $y \le \frac{1}{2}x + 1$ **59.** **a.** 3 **b.** 2 **c.** No **d.** $x > 2$ **61.** $x < 3$ **63.** $x \le 3$ **65.** At least 291 **67.** $x < 2$ **69.** $x > -\frac{8}{3}$
71. $x \ge 6$; the width is at least 6 meters. **73.** $x > 6$; the shortest side is even and greater than 6 inches. **75.** $x \ge 2$ **77.** $x < 4$
79. $-1 > x$

PROBLEM SET 1.9

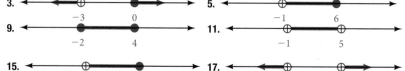

1. **7.** 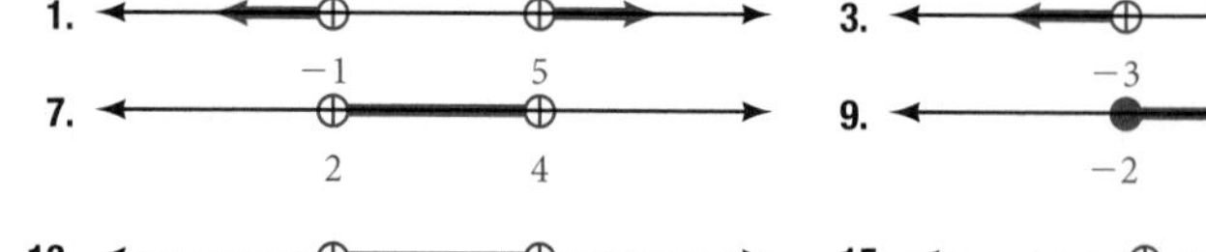
3.
5. **11.** 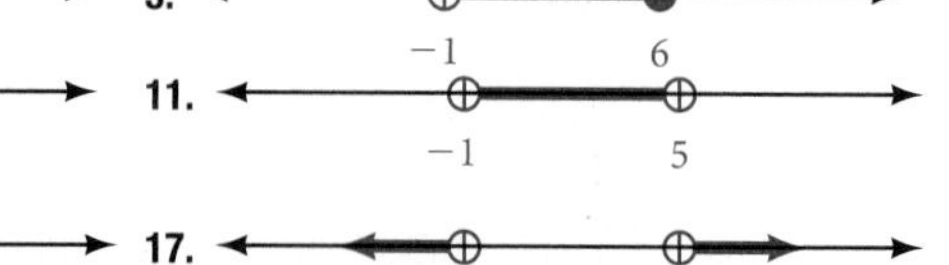
9.
13.
15.
17.

19. 

21. 

23. 

25. 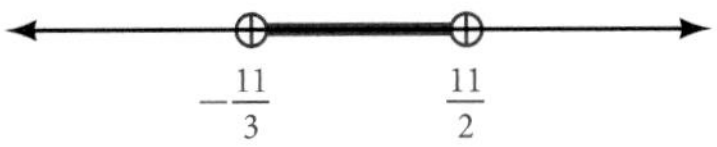

27. 

29. 

31. 

33. 

35. 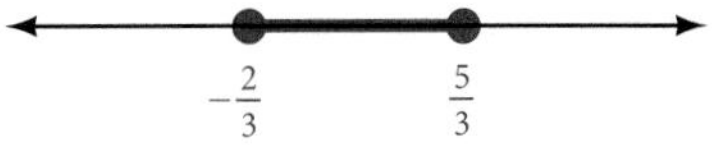

37. 

39. 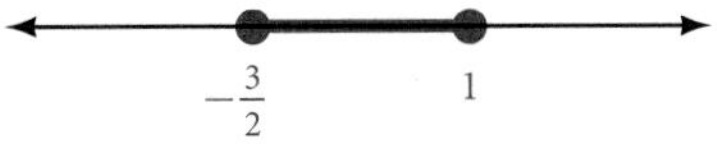

41. $-2 < x < 3$ **43.** $x \le -2$ or $x \ge 3$ **45.** **a.** $2x + x > 10; x + 10 > 2x; 2x + 10 > x$ **b.** $\frac{10}{3} < x < 10$

47. 

49. $4 < x < 5$ **51.** **a.** $20 < P < 30$ **b.** $3 < w < \frac{11}{2}$ **c.** $7 < l \le \frac{19}{2}$

53. 8 **55.** 24 **57.** 25% **59.** 10% **61.** 80 **63.** 400 **65.** -5 **67.** 5 **69.** 7 **71.** 9 **73.** 6 **75.** $2x - 3$ **77.** $-3, 0, 2$

CHAPTER 1 TEST

1. $-y + 1$ **2.** $4x - 1$ **3.** $4 - 2y$ **4.** $x - 22$ **5.** -3 **6.** -4 **7.** **a.**

n	$(n + 2)^2$
1	9
2	16
3	25
4	36

b.

n	$n^2 + 2$
1	3
2	6
3	11
4	18

8. $x = 3$ **9.** $y = -5$ **10.** $x = -3$ **11.** $x = 4$ **12.** $x = 1$ **13.** 55 **14.** $t = -3$ **15.** $x = \frac{10}{4}$ **16.** $x = (.40)(56)$
17. $720 = 0.24x$ **18.** -1 **19.** 8 **20.** $y = 2 - \frac{1}{3x}$ **21.** $a = \frac{x^2 - v^2}{2d}$ **22.** 18, 36 **23.** 20 cm, 55 cm **24.** 6 nickles, 14 dimes
25. \$700, \$1,200 **26.** $x > 10$ 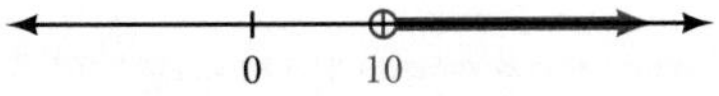

27. $y \ge -4$ 

28. $x > -4$ 

29. $n \le -2$ 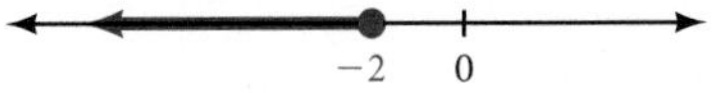

30. $1 > x$ or $x > 3$ 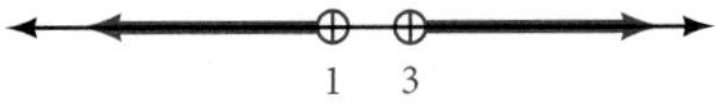

31. $2 \le x \le 8$ 

Chapter 2

PROBLEM SET 2.1

1–17.

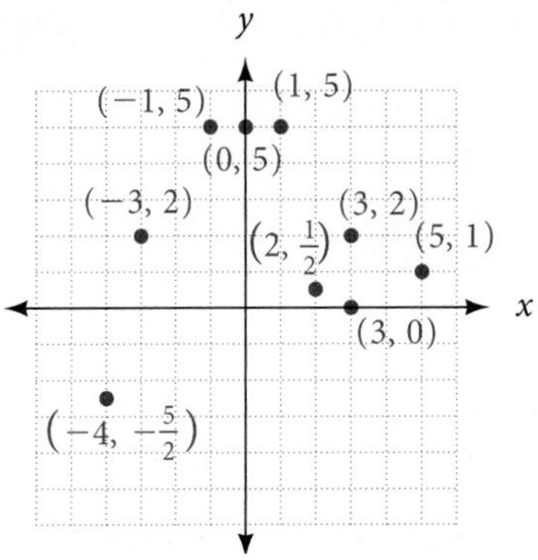

19. $(-4, 4)$ **21.** $(-4, 2)$ **23.** $(-3, 0)$ **25.** $(2, -2)$ **27.** $(-5, -5)$

29. Yes

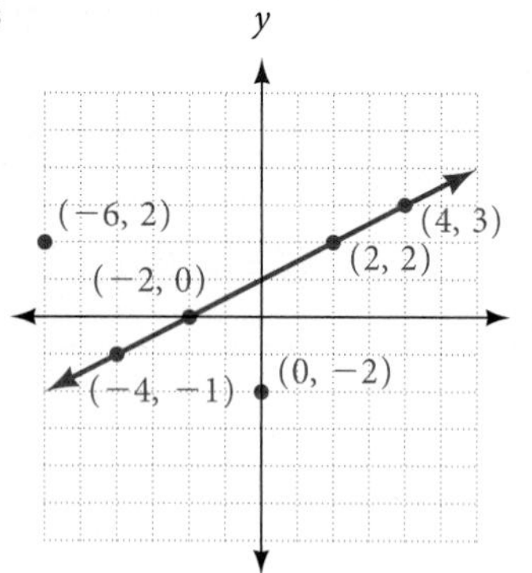

31. No **33.** Yes

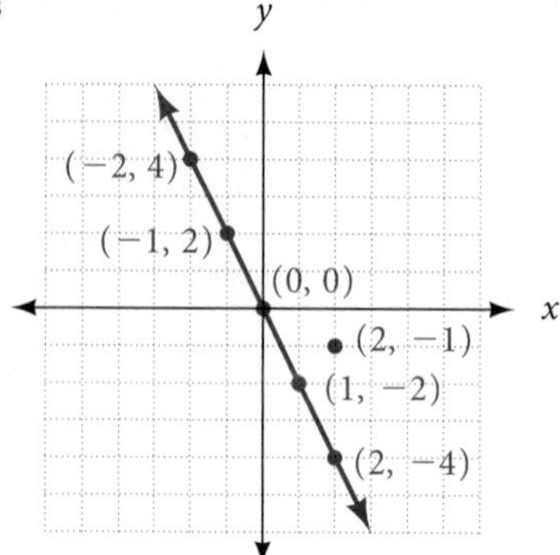

35. No

37. Yes

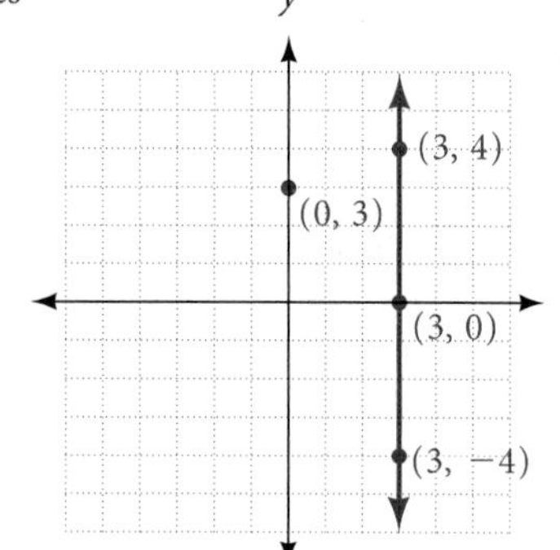

39. No **41.** No

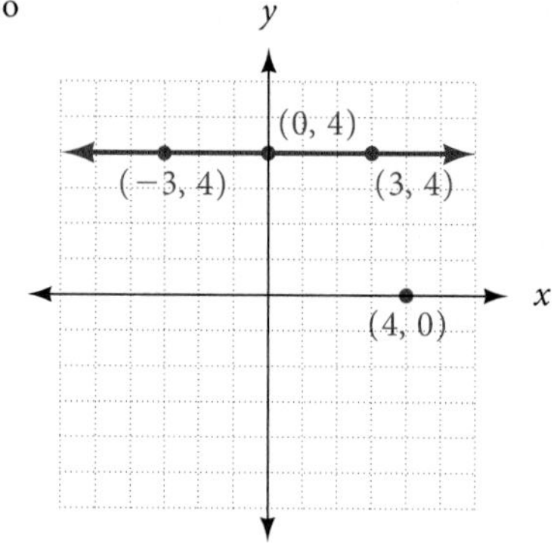

43. No

45. a. (5, 40), (10, 80), (20, 160), Answers may vary **b.** \$320 **c.** 30 hours **d.** No, if she works 35 hours, she should be paid \$280.

47.

Non-Camera Phones (Millions)
300
250
200
150
100
50
0
0 2006 2007 2008 2009 2010
Year

49. (1985, 20.2), (1990, 34.4), (1995, 44.8), (2000, 65.4), (2005, 104)

51. $A = (1, 2), B = (6, 7)$ **53.** $A = (2, 2), B = (2, 5), C = (7, 5)$

55. a. −3 **b.** 6 **c.** 0 **d.** −4 **57. a.** 4 **b.** 2 **c.** −1 **d.** 9

PROBLEM SET 2.2

1. (0, 6), (3,0), (6, −6) **3.** (0, 3), (4, 0), (−4, 6) **5.** (1, 1), $\left(\frac{3}{4}, 0\right)$, (5, 17) **7.** (2, 13), (1, 6), (0, −1) **9.** (−5, 4), (−5, −3), (−5, 0)

11.

x	y
1	3
−3	−9
4	12
6	18

13.

x	y
0	0
$-\frac{1}{2}$	−2
−3	−12
3	12

15.

x	y
2	3
3	2
5	0
9	−4

17.

x	y
2	0
3	2
1	−2
−3	−10

19.

x	y
0	−1
−1	−7
−3	−19
$\frac{3}{2}$	8

21. (0, −2)

23. (1, 5), (0, −2), and (−2, −16) **25.** (1, 6), and (0, 0) **27.** (2, −2) **29.** (3, 0) and (3, −3) **31.** 12 inches

33. a. Yes **b.** No, she should earn \$108 for working 9 hours. **c.** No, she should earn \$84 for working 7 hours. **d.** Yes

35. a. \$375,000 **b.** At the end of 6 years. **c.** No, the crane will be worth \$195,000 after 9 years. **d.** \$600,000

37. −3 **39.** 2 **41.** 0 **43.** $y = -5x + 4$ **45.** $y = \frac{3}{2}x - 3$

Answers to Odd-Numbered Problems

PROBLEM SET 2.3

1. (0, 4), (2, 2), (4, 0)

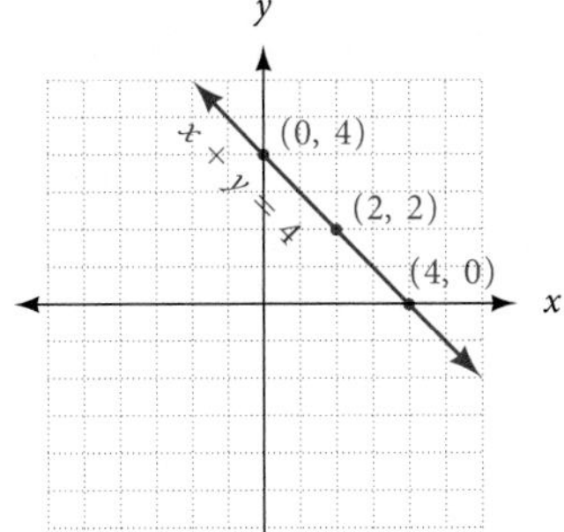

3. (0, 3), (2, 1), (4, −1)

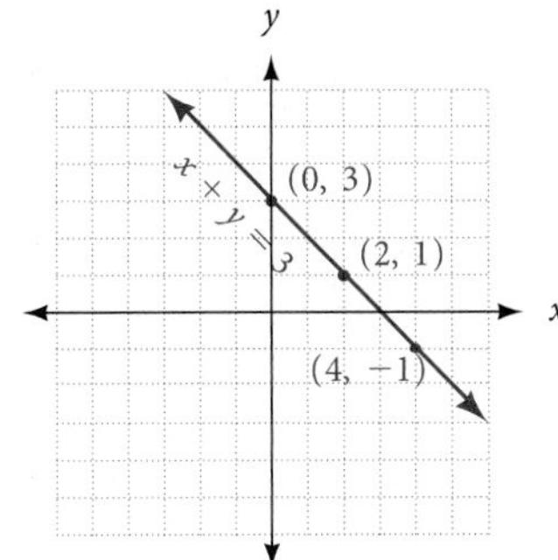

5. (0, 0), (−2, −4), (2, 4)

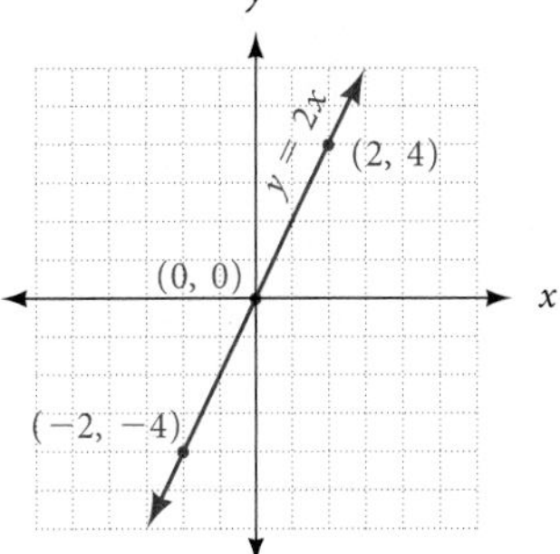

7. (−3, −1), (0, 0), (3, 1)

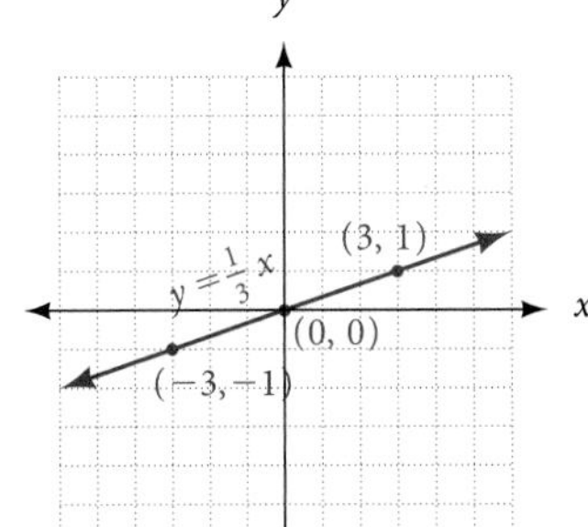

9. (0, 1), (−1, −1), (1, 3)

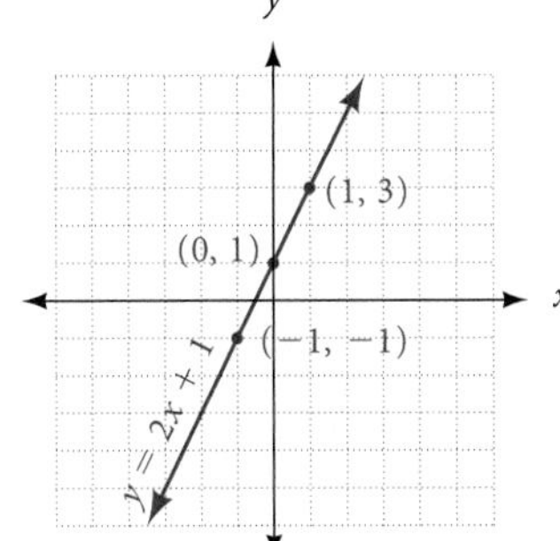

11. (0, 4), (−1, 4), (2, 4)

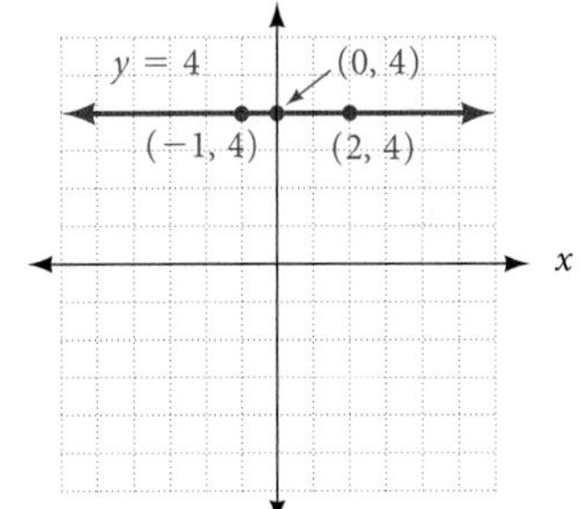

13. (−2, 2), (0, 3), (2, 4)

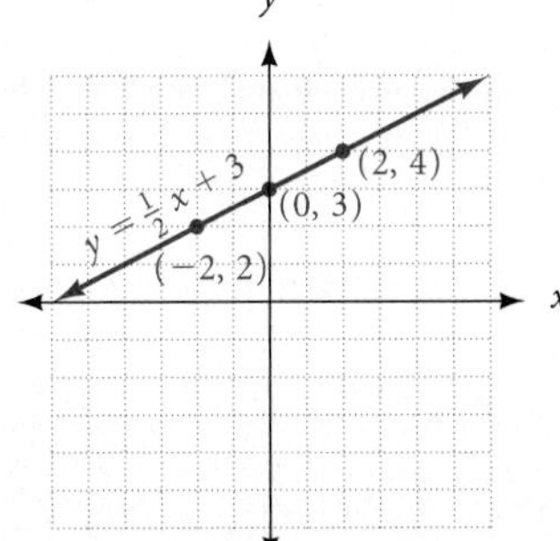

15. (−3, 3), (0, 1), (3, −1)

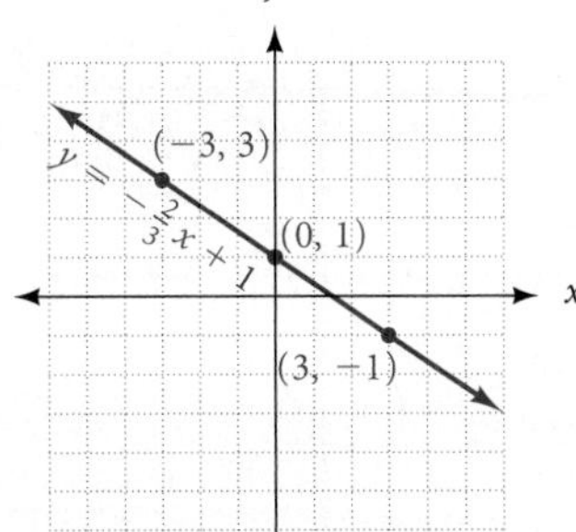

17. (−1, 5), (0, 3), (1, 1)

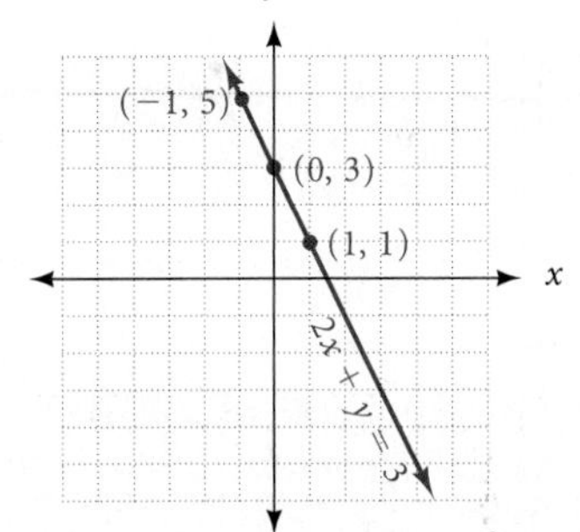

19. (0, 3), (2, 0), (4, −3)

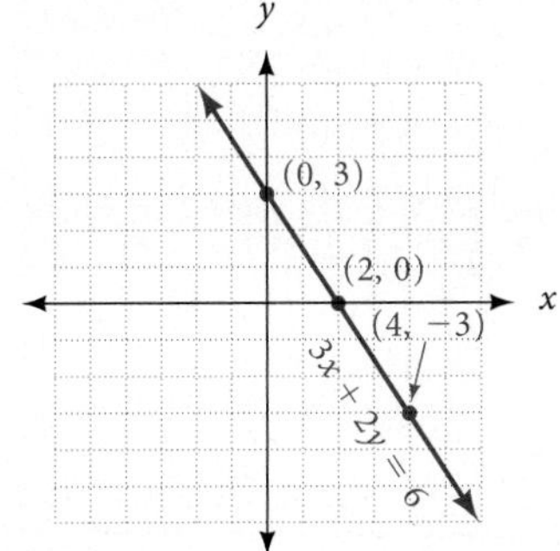

21. (−2, 2), (0, 3), (2, 4)

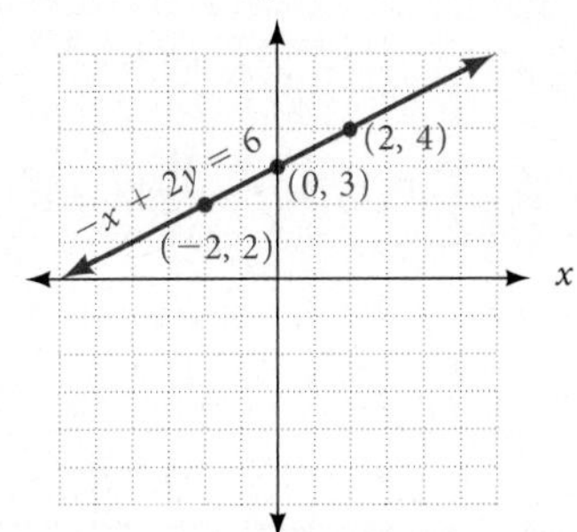

23.
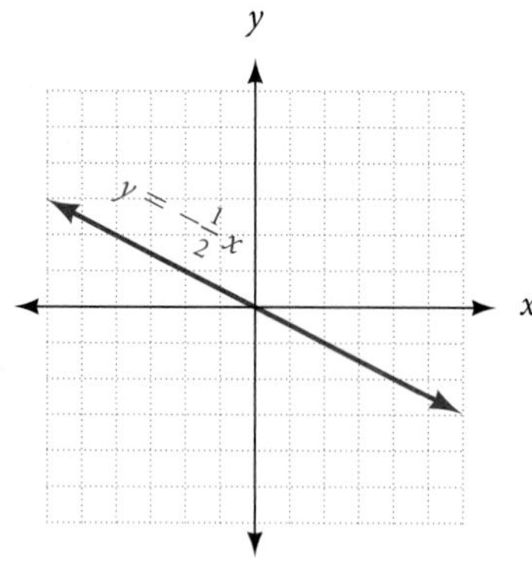

25.
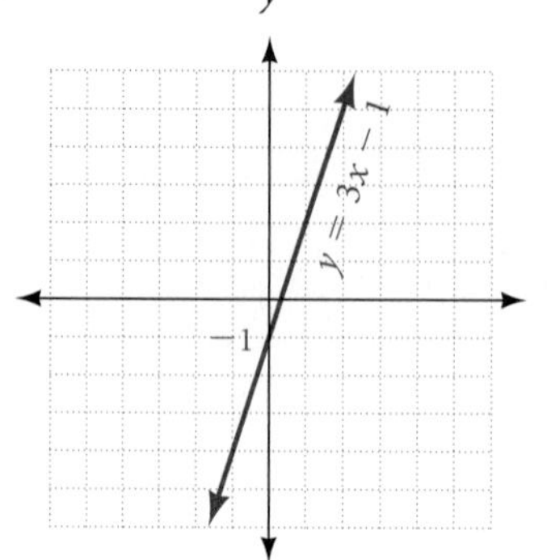

27.
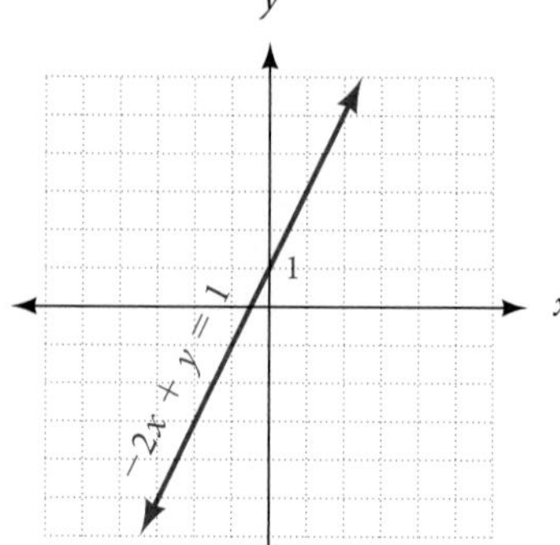

29.
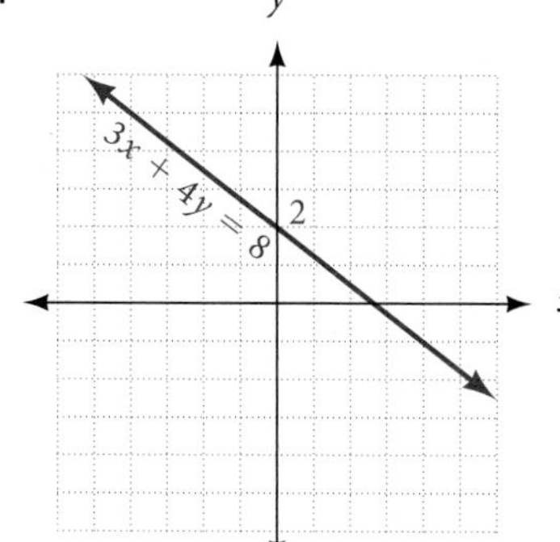

31.
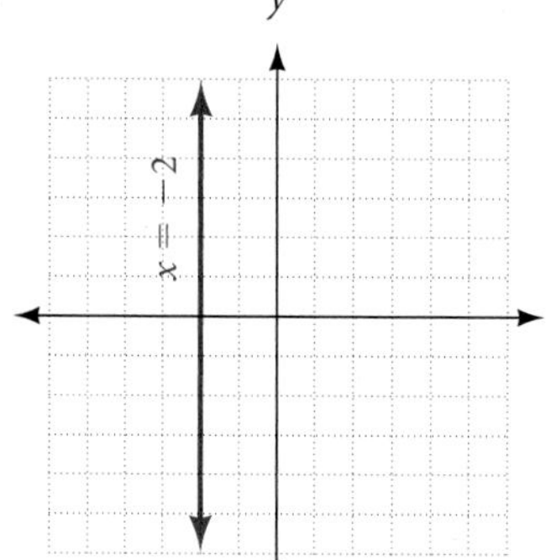

33.
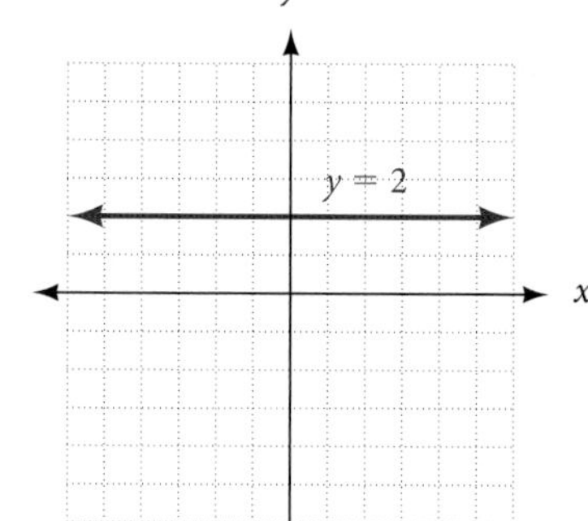

35.
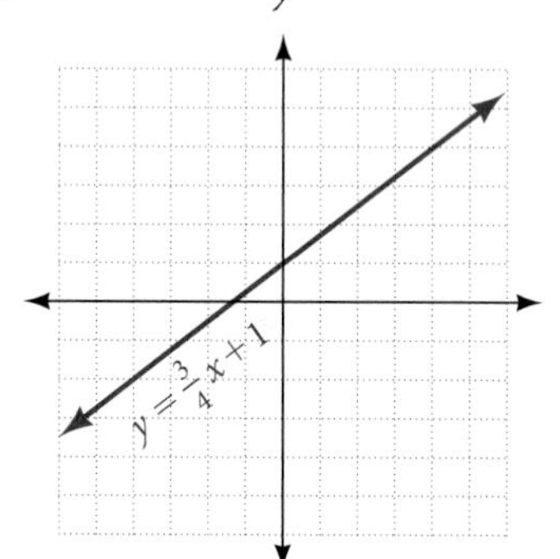

37.
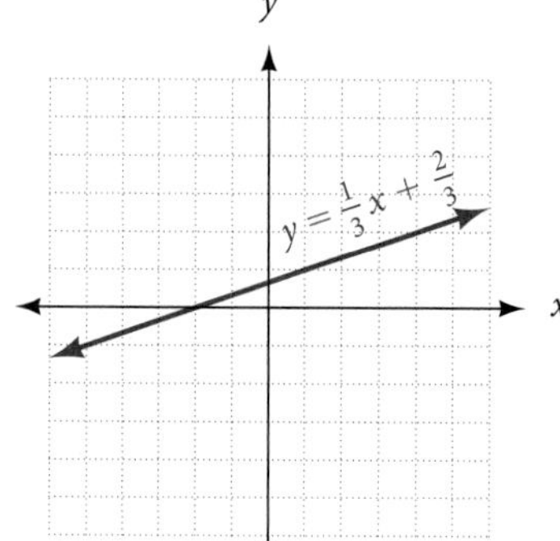

39.
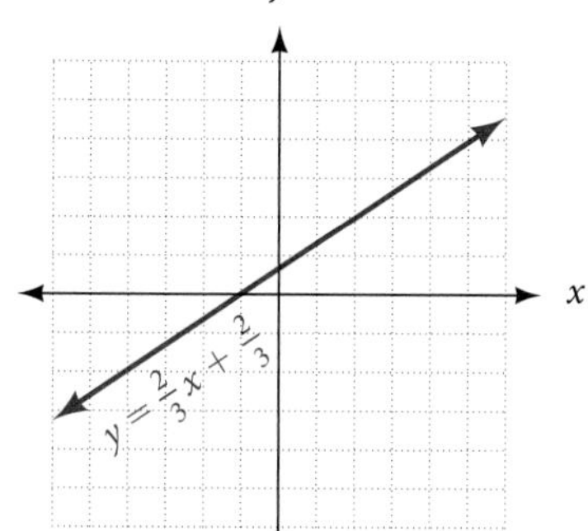

41.

Equation	H, V, and/or O
$x = 3$	V
$y = 3$	H
$y = 3x$	O
$y = 0$	O, H

43.

Equation	H, V, and/or O
$x = -\frac{3}{5}$	V
$y = -\frac{3}{5}$	H
$y = -\frac{3}{5}x$	O
$x = 0$	O, V

45.

x	y
−4	−3
−2	−2
0	−1
2	0
6	2

47. **a.** $\frac{5}{2}$ **b.** 5 **c.** 2 **d.**

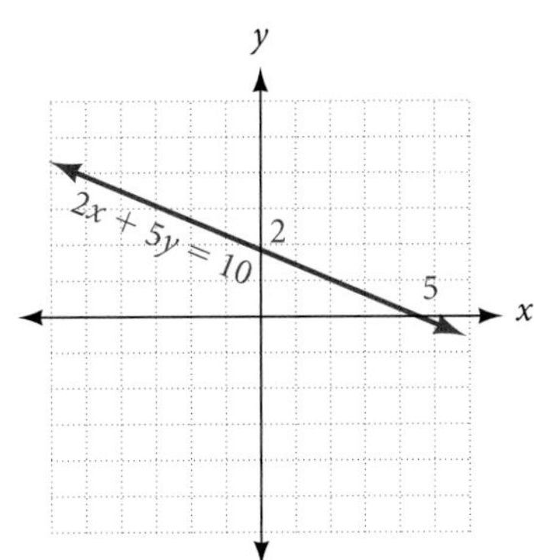

e. $y = -\frac{2}{5}x + 2$

49. **a.** 2 **b.** 3 **51.** **a.** -4 **b.** 2 **53.** **a.** 6 **b.** 2

PROBLEM SET 2.4

1.

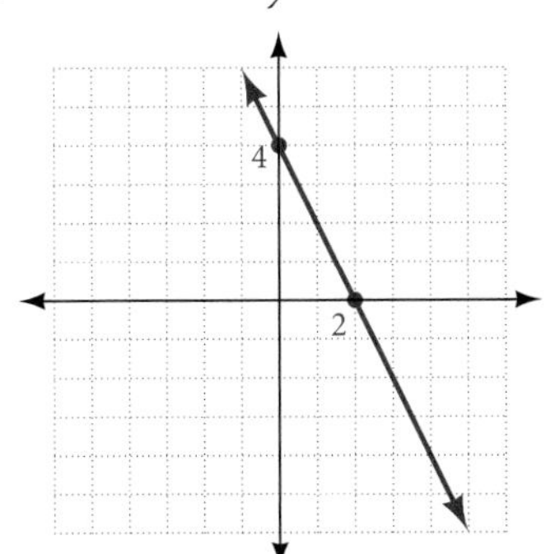

3.

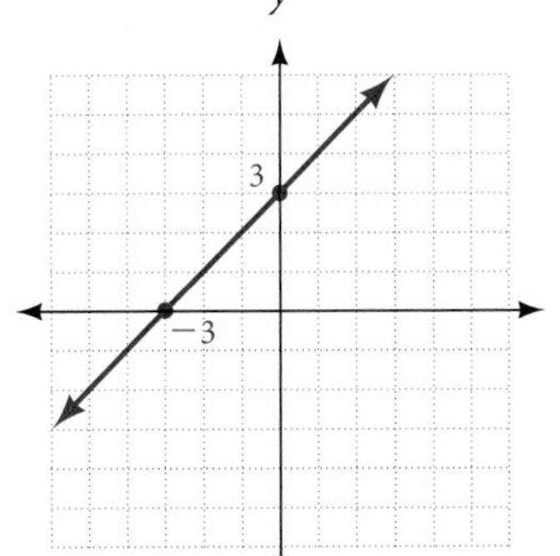

5.

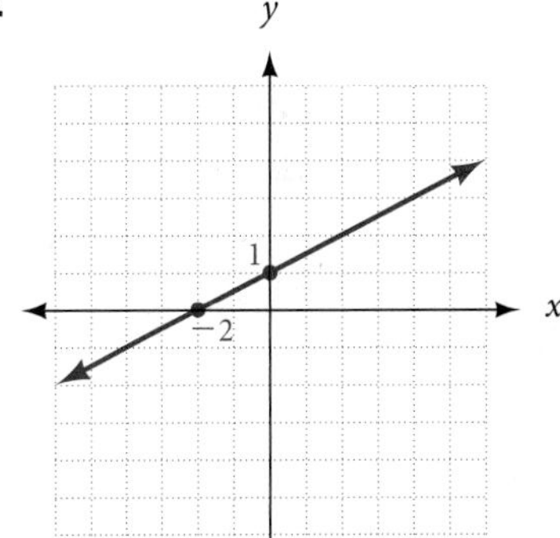

7.

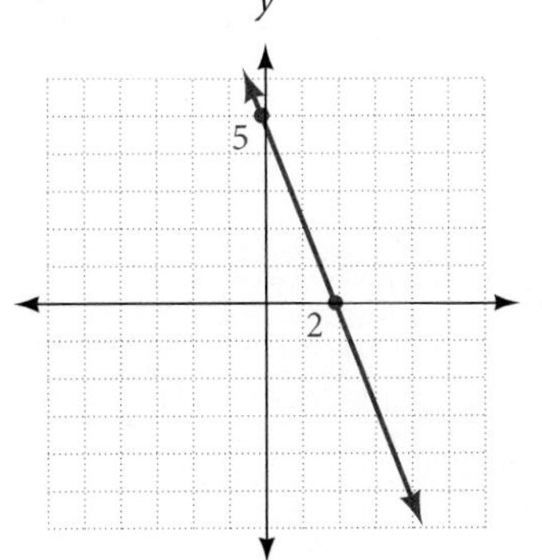

9.

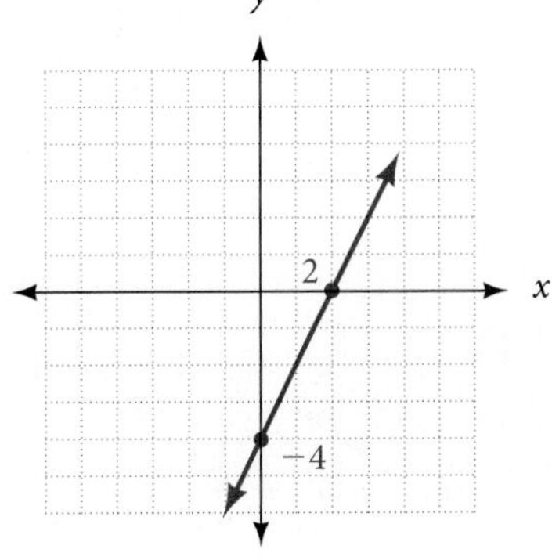

11.

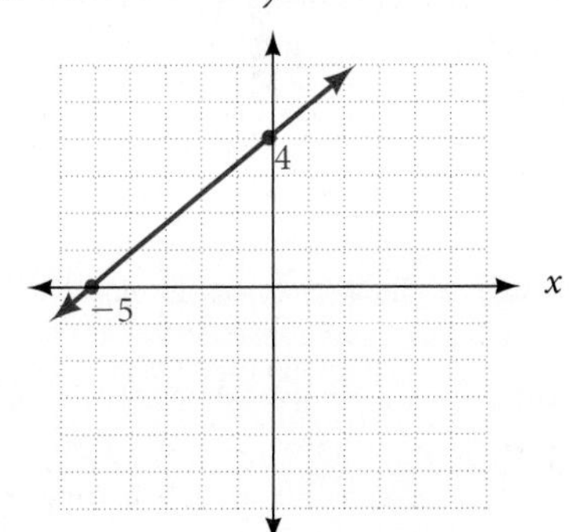

13.

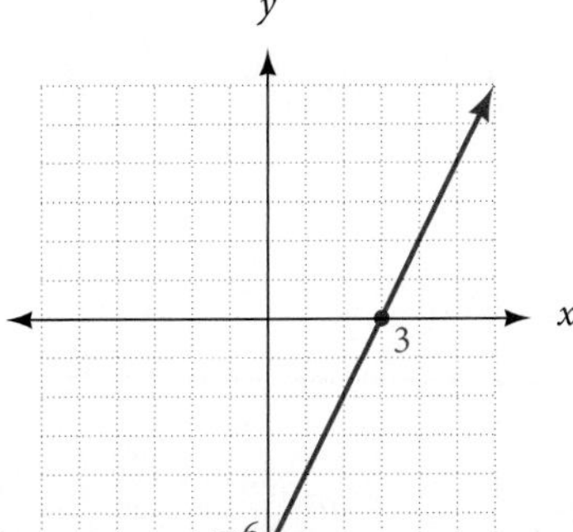

15.

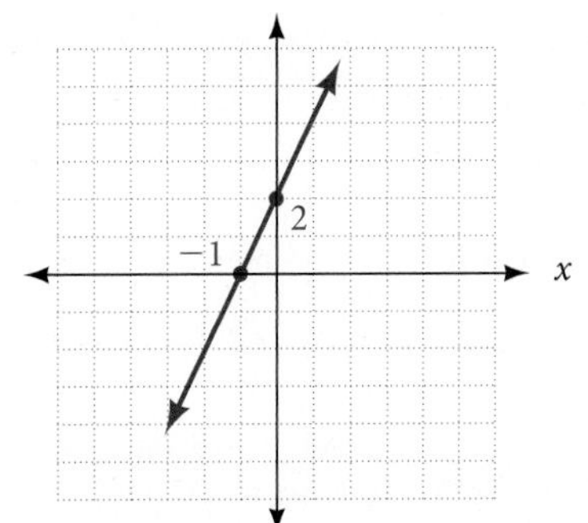

17.

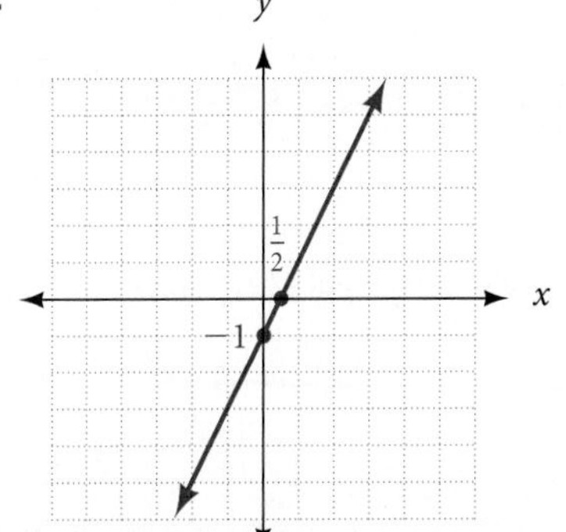

19.

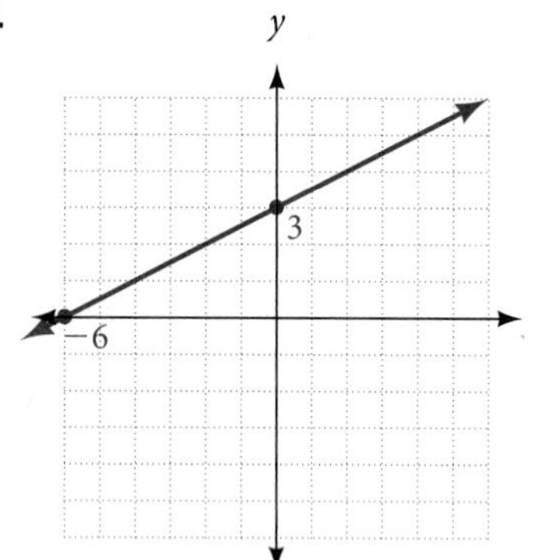

21.

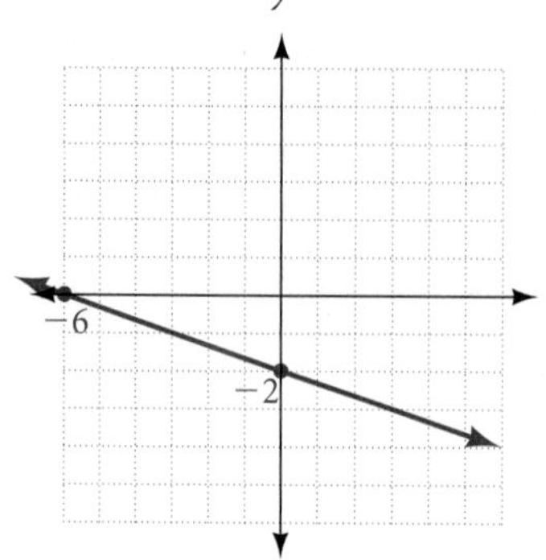

23.

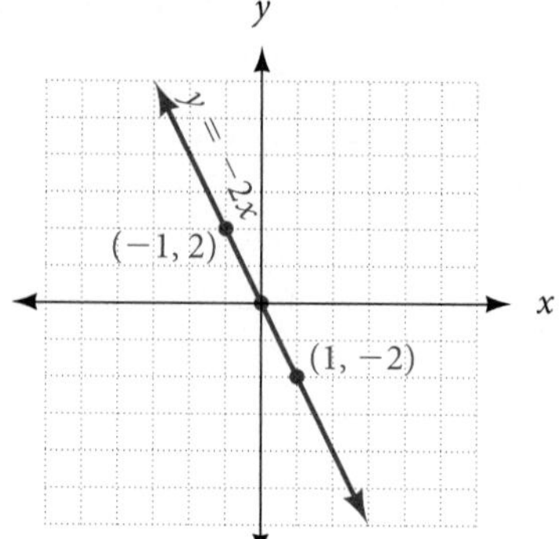

25.

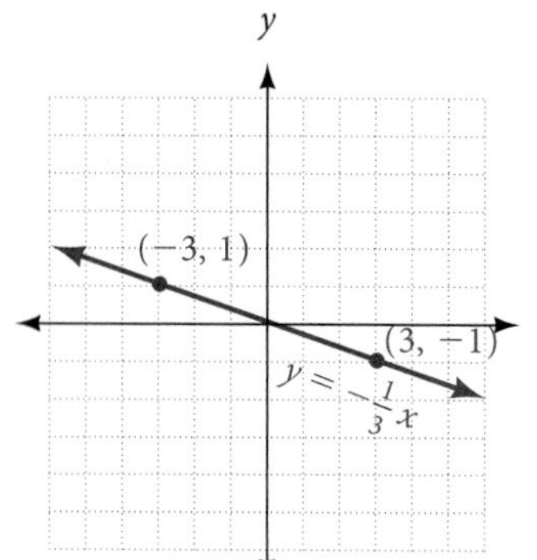

27.

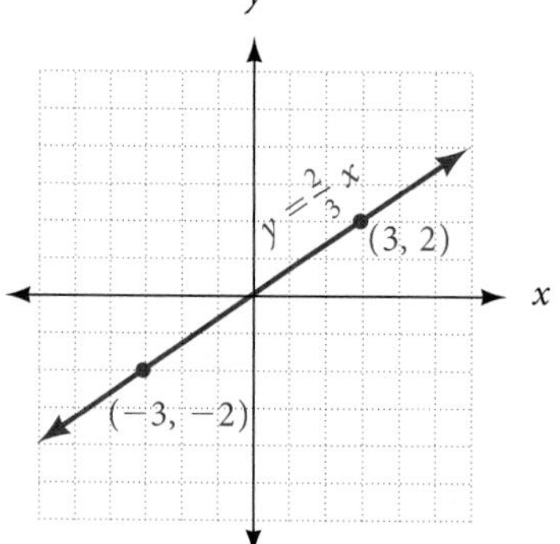

29.

Equation	x-intercept	y-intercept
$3x + 4y = 12$	4	3
$3x + 4y = 4$	$\frac{4}{3}$	1
$3x + 4y = 3$	1	$\frac{3}{4}$
$3x + 4y = 2$	$\frac{2}{3}$	$\frac{1}{2}$

31.

Equation	x-intercept	y-intercept
$x - 3y = 2$	2	$-\frac{2}{3}$
$y = \frac{1}{3}x - \frac{2}{3}$	2	$-\frac{2}{3}$
$x - 3y = 0$	0	0
$y = \frac{1}{3}x$	0	0

33. **a.** 0 **b.** $-\frac{3}{2}$ **c.** 1 **d.**

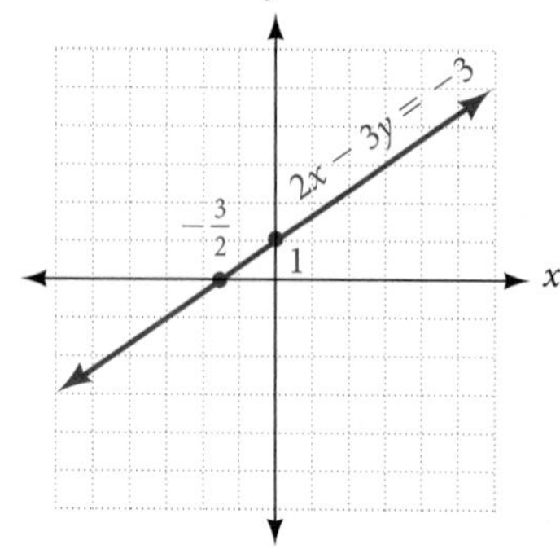

e. $y = \frac{2}{3}x + 1$

35. x-intercept $= 3$; y-intercept $= 5$ **37.** x-intercept $= -1$; y-intercept $= -3$

39.

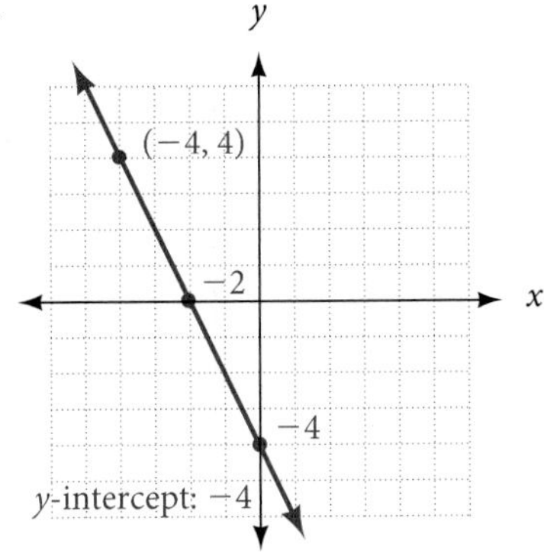

41.

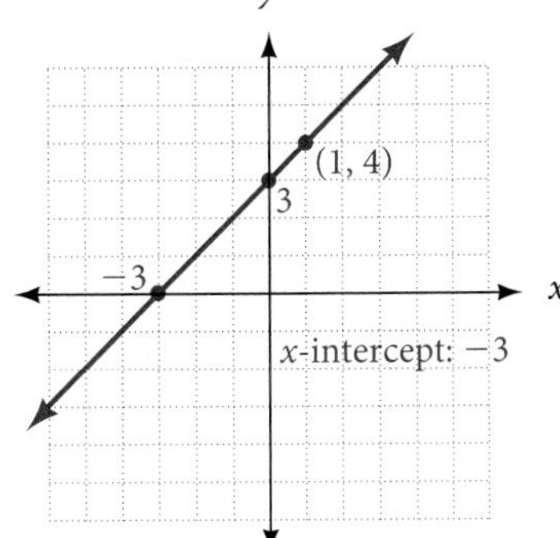

43.

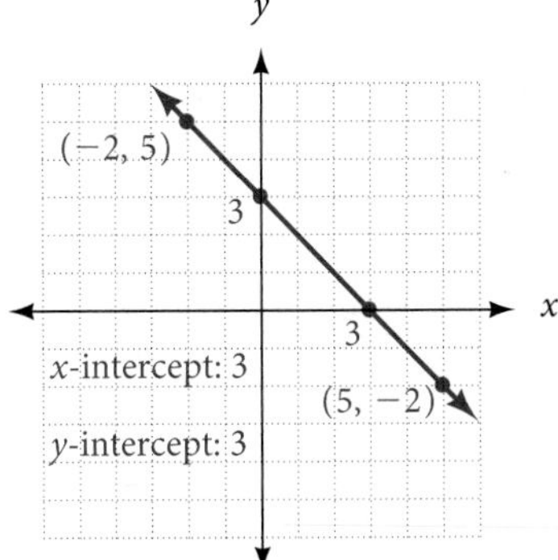

45.

x	y
−2	1
0	−1
−1	0
1	−2

47.

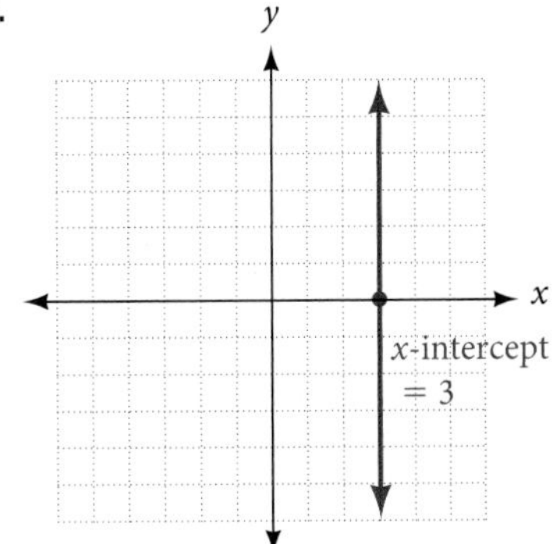

49.

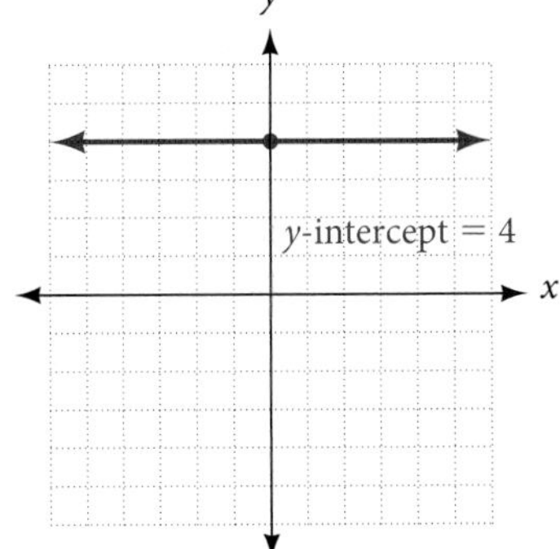

51.

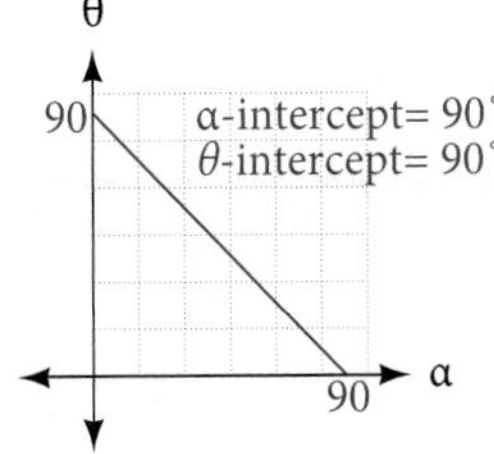

53. **a.** $\frac{3}{2}$ **b.** $\frac{3}{2}$ **55.** **a.** $\frac{3}{2}$ **b.** $\frac{3}{2}$

PROBLEM SET 2.5

1.

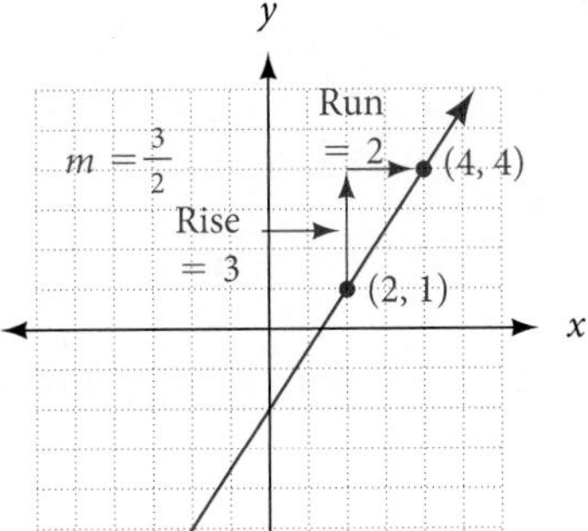

3.

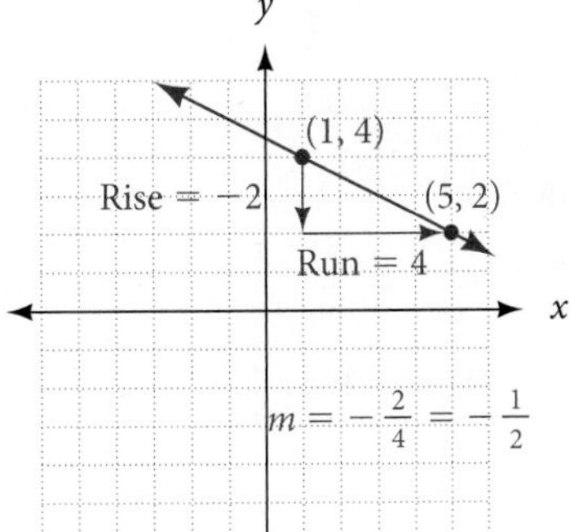

5.

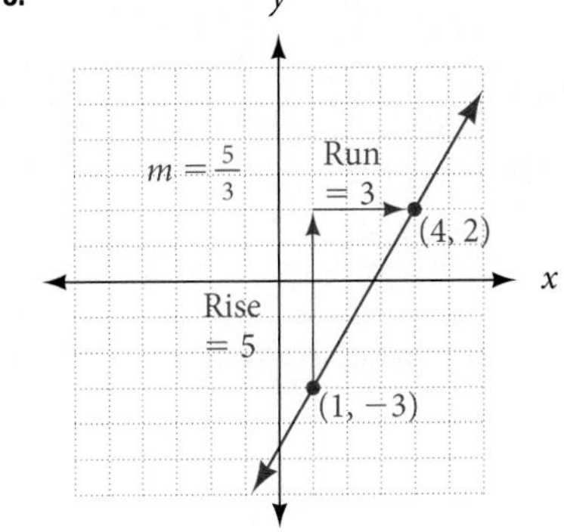

7.

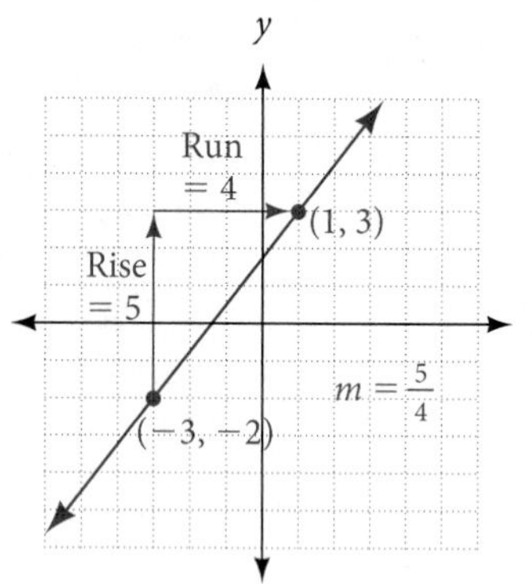

9.

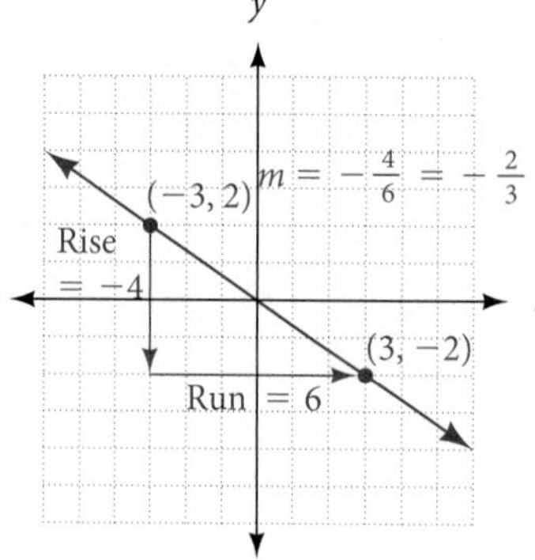

11.

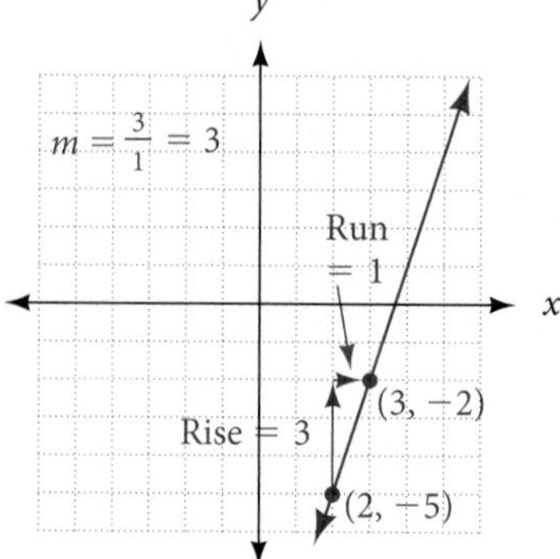

13.

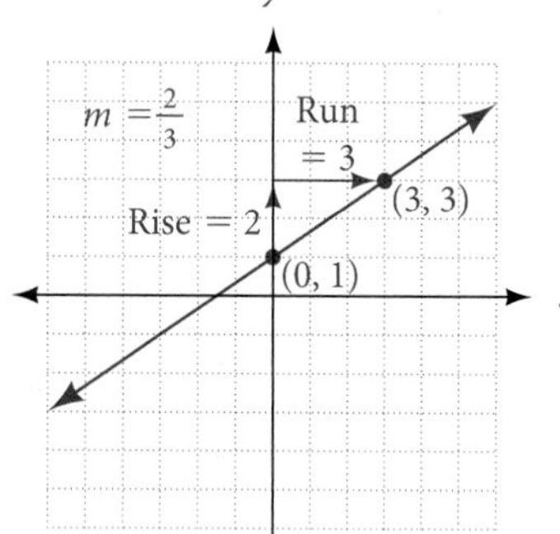

15.

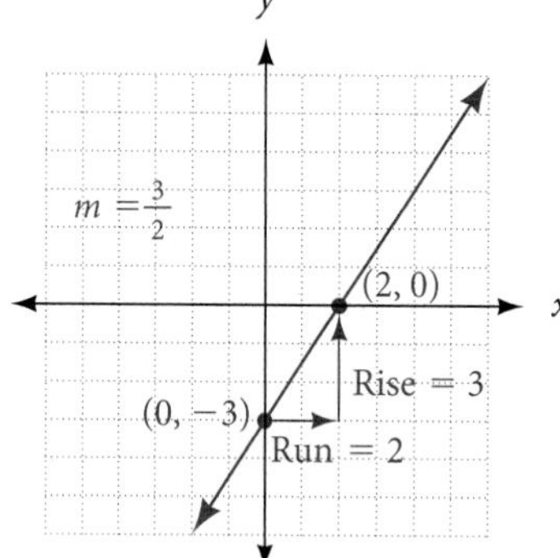

17.

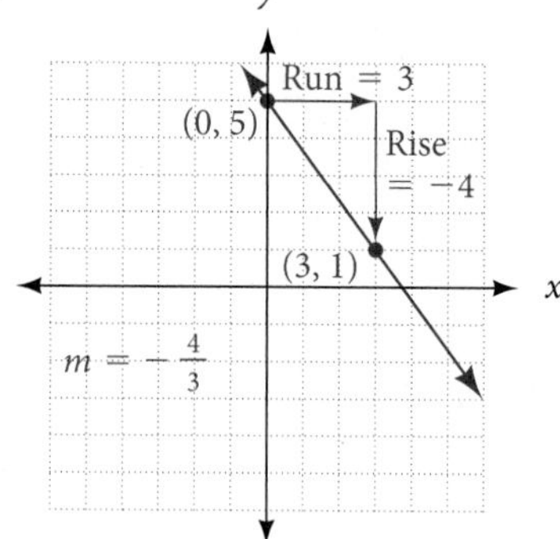

19.

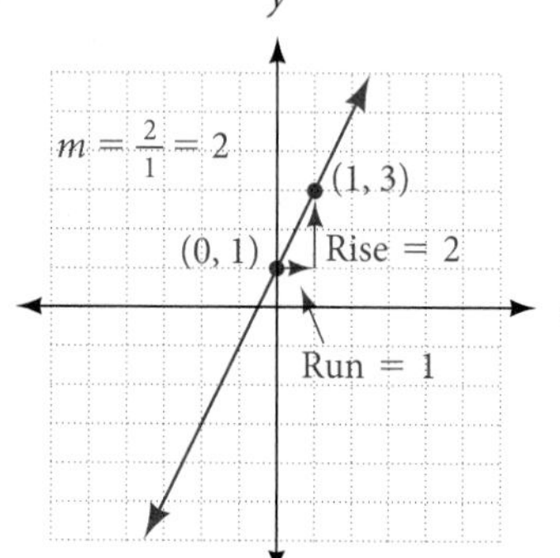

21.

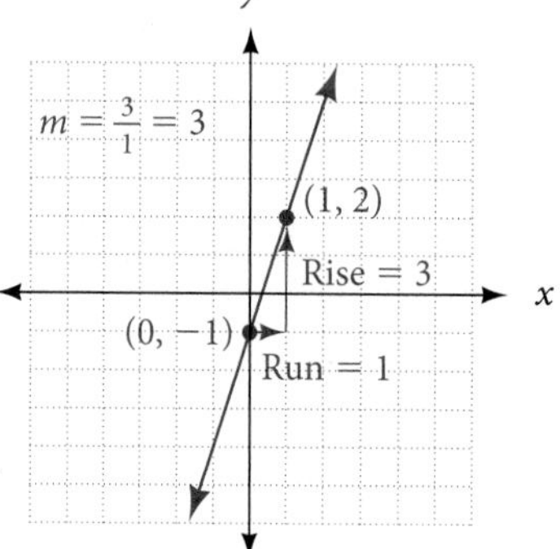

23. Slope = 3; y-intercept = 2 **25.** Slope = 2; y-intercept = −2

27.

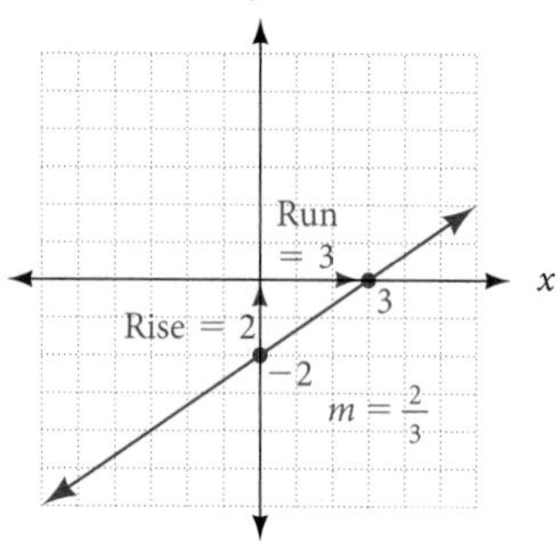

29.

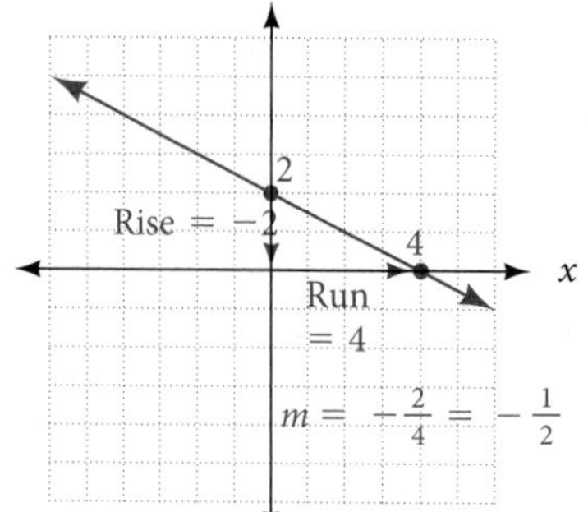

31.

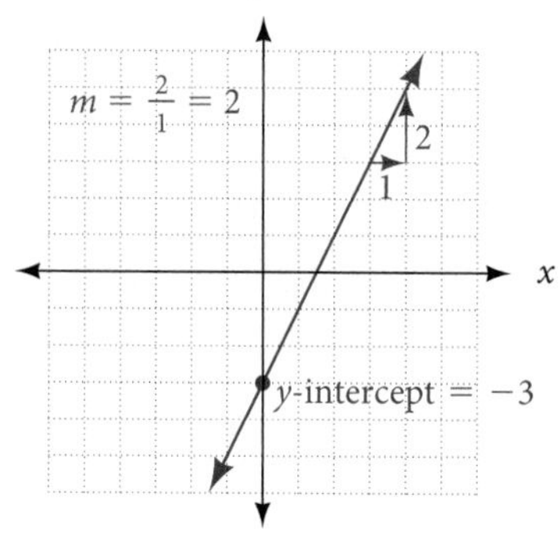

33.

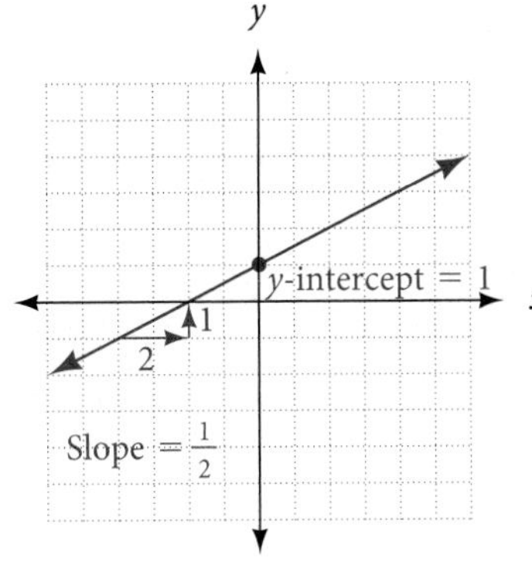

35. 6 **37.**

Equation	Slope
$x = 3$	no slope
$y = 3$	0
$y = 3x$	3

39.

Equation	Slope
$y = -\frac{2}{3}$	0
$x = -\frac{2}{3}$	no slope
$y = -\frac{2}{3}x$	$-\frac{2}{3}$

41. Slopes: *A*, 3.3; *B*, 3.1; *C*, 5.3; *D*, 1.9 **43.** Slopes: *A*, −50; *B*, −75; *C*, −25 **45.** $y = 2x + 4$ **47.** $y = -2x + 3$ **49.** $y = \frac{4}{5}x - 4$ **51.** $y = 2x + 5$ **53.** $y = \frac{2}{3}x - 1$ **55.** $y = -\frac{3}{2}x + 1$

PROBLEM SET 2.6

1. $y = \frac{2}{3}x + 1$ **3.** $y = \frac{3}{2}x - 1$ **5.** $y = -\frac{2}{3}x + 3$ **7.** $y = 2x - 4$

9. $m = 2; b = 4$

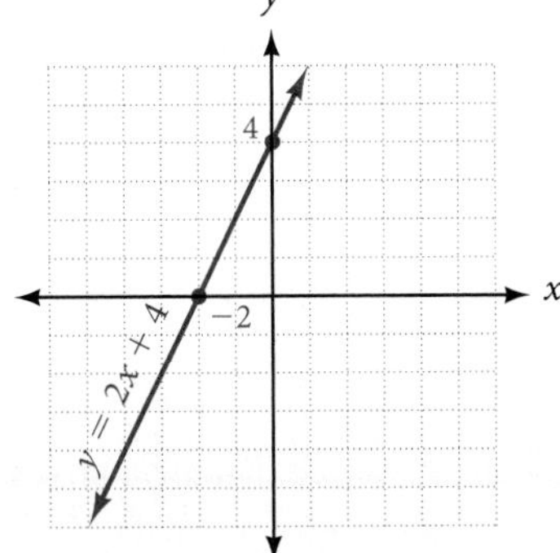

11. $m = -3; b = 3$

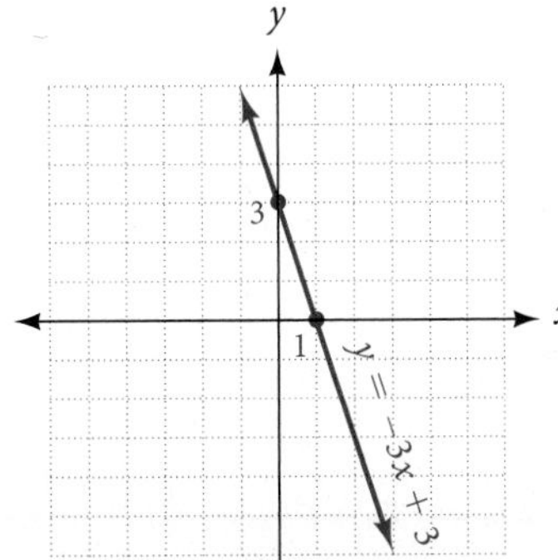

13. $m = -\frac{3}{2}; b = 3$

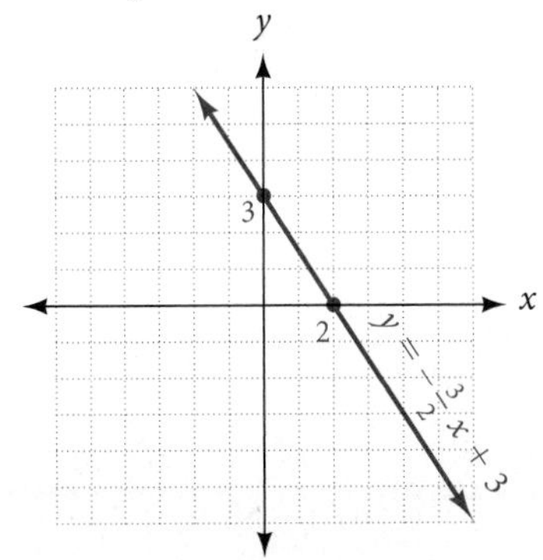

15. $m = \frac{4}{5}; b = -4$

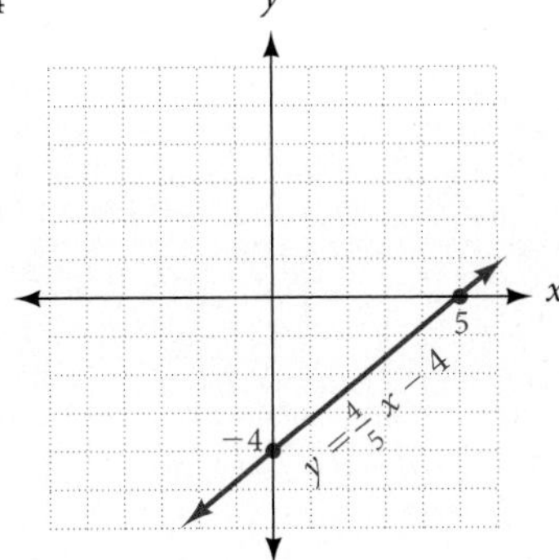

17. $m = -\frac{2}{5}; b = -2$

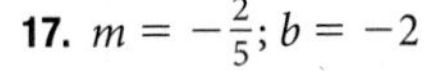

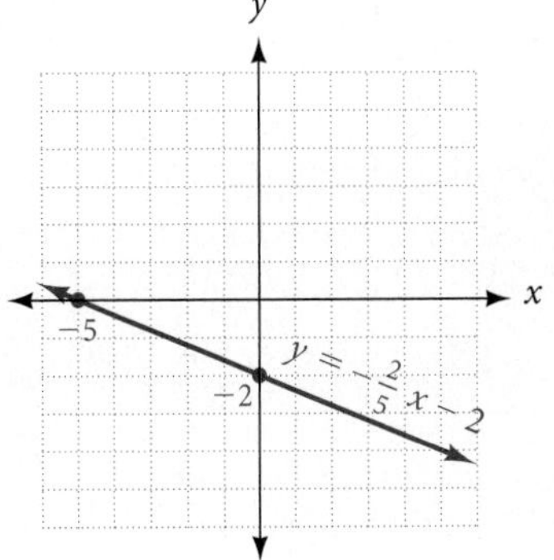

19. $y = 2x - 1$ **21.** $y = -\frac{1}{2}x - 1$ **23.** $y = \frac{3}{2}x - 6$ **25.** $y = -3x + 1$ **27.** $y = x - 2$ **29.** $y = 2x - 3$ **31.** $y = \frac{4}{3}x + 2$ **33.** $y = -\frac{2}{3}x - 3$ **35.** $m = 3, b = 3, y = 3x + 3$ **37.** $m = \frac{1}{4}, b = -1, y = \frac{1}{4}x - 1$

39. **a.** $-\frac{5}{2}$ **b.** $y = 2x + 6$ **c.** 6 **d.** 2 **e.**

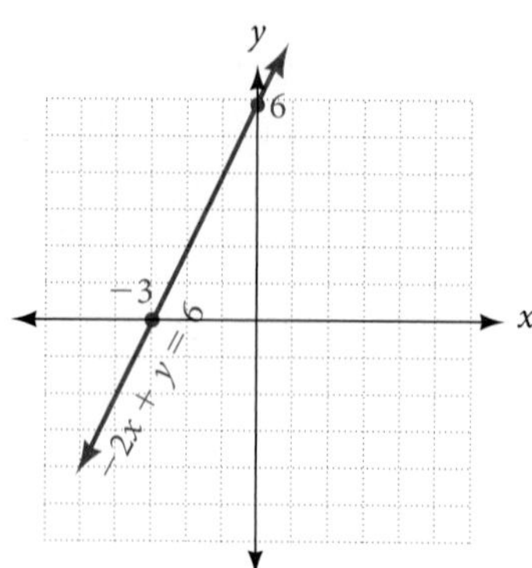

41. $y = -\frac{2}{3}x + 2$ **43.** $y = -\frac{5}{2}x - 5$

45. $x = 3$ **47.** **a.** \$6,000 **b.** 3 years **c.** slope $= -3{,}000$ **d.** \$3,000 **e.** $V = -3{,}000t + 21{,}000$

49.

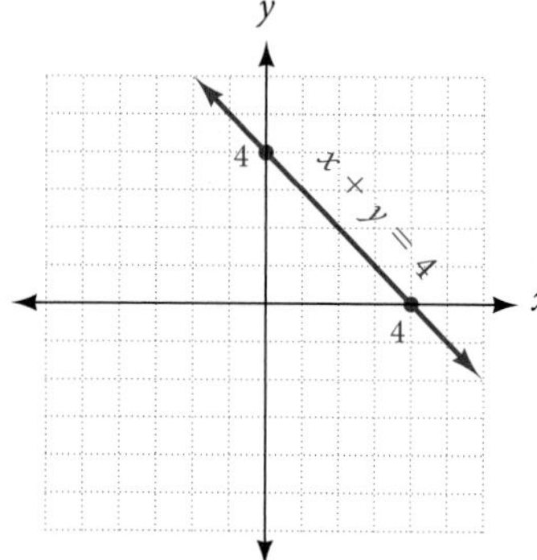

51.

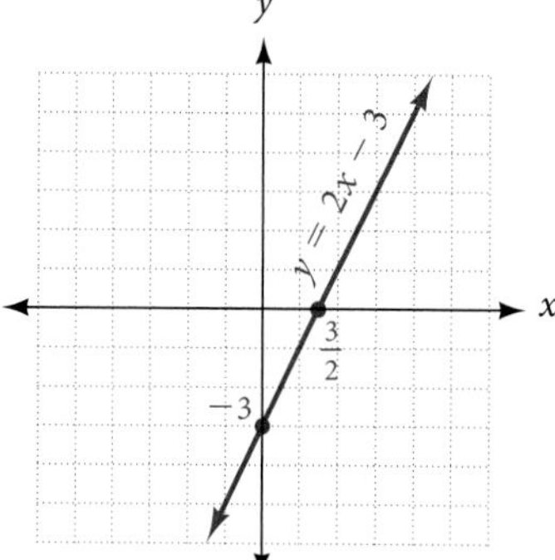

53.

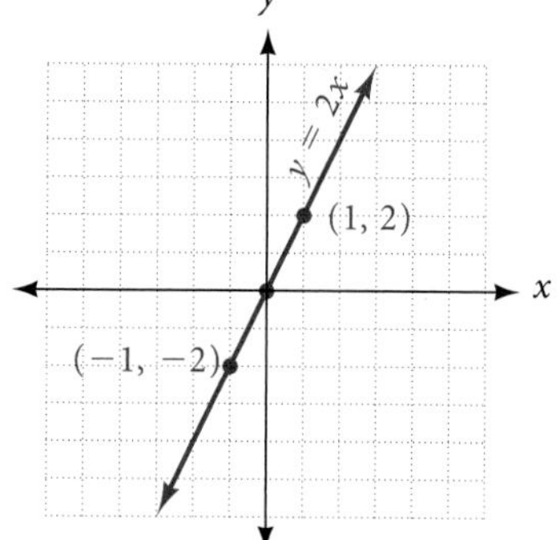

PROBLEM SET 2.7

1.

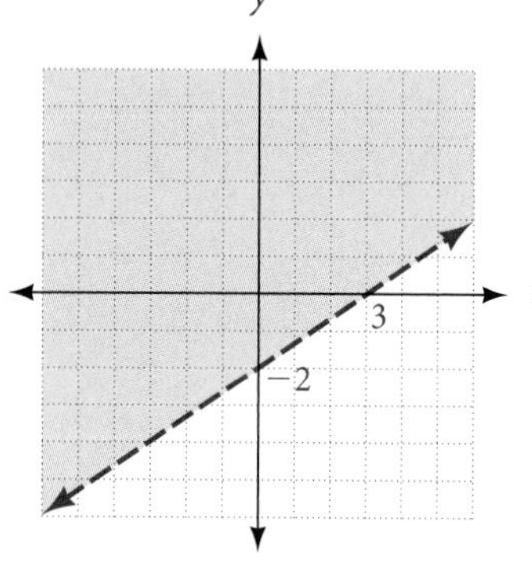

3.

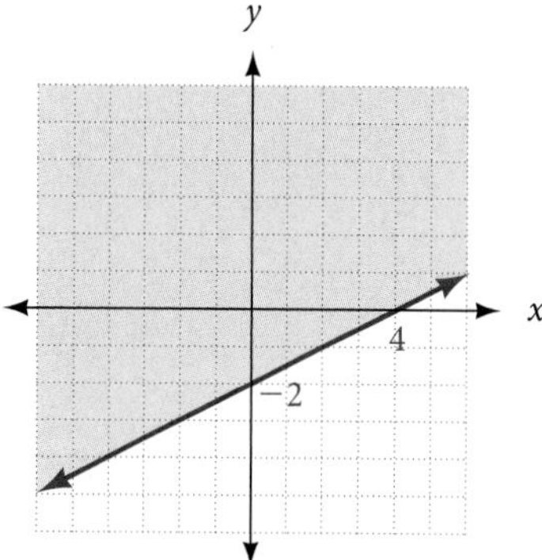

5.

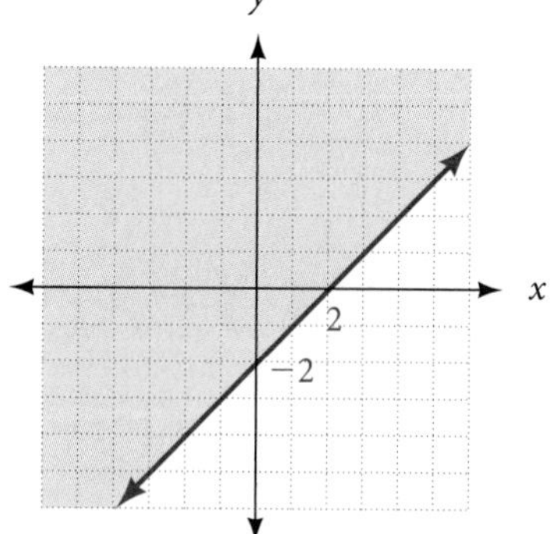

7.

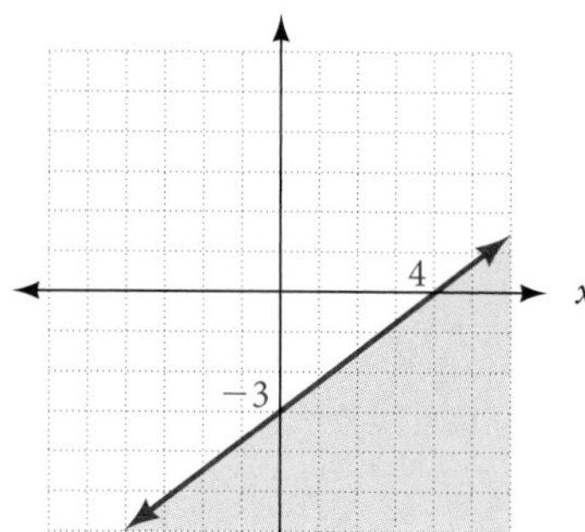

9.

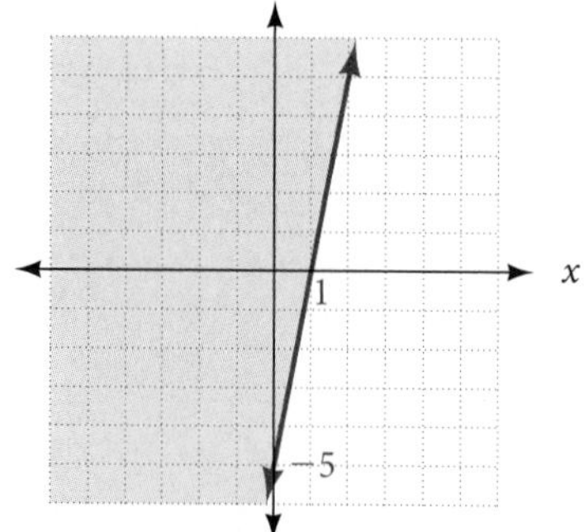

11.

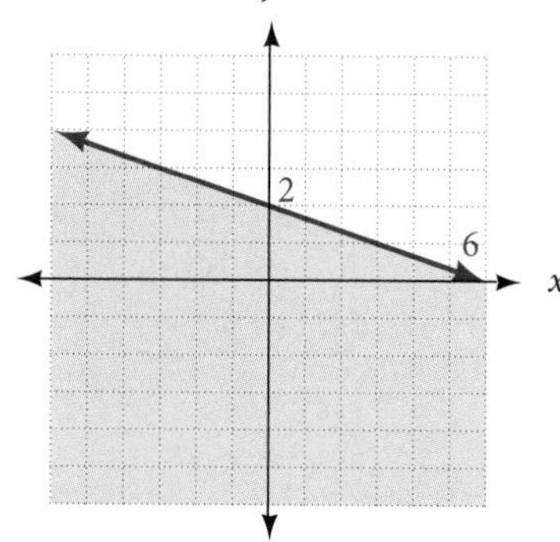

13.
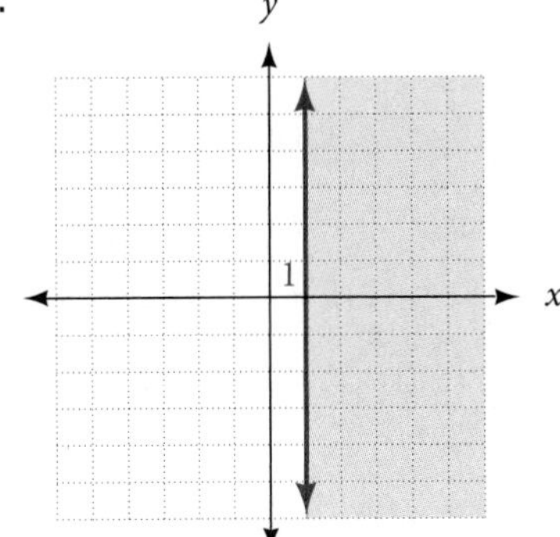

15.
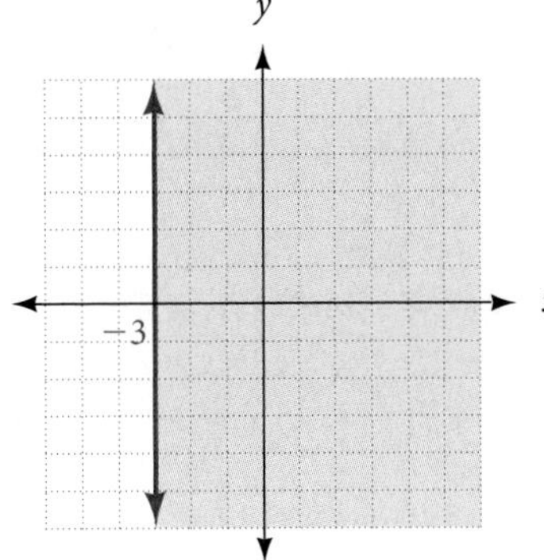

17.
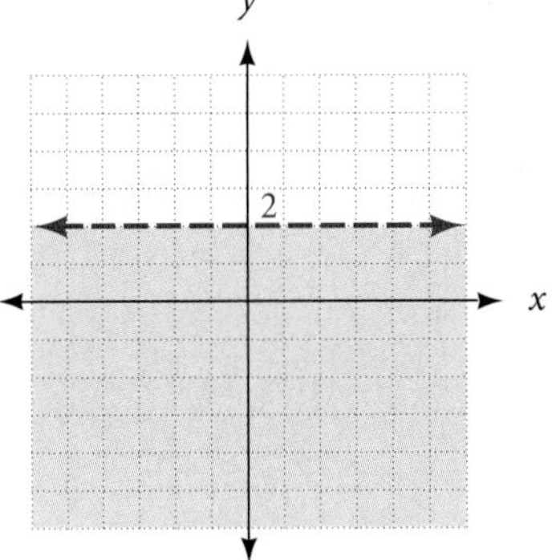

19.
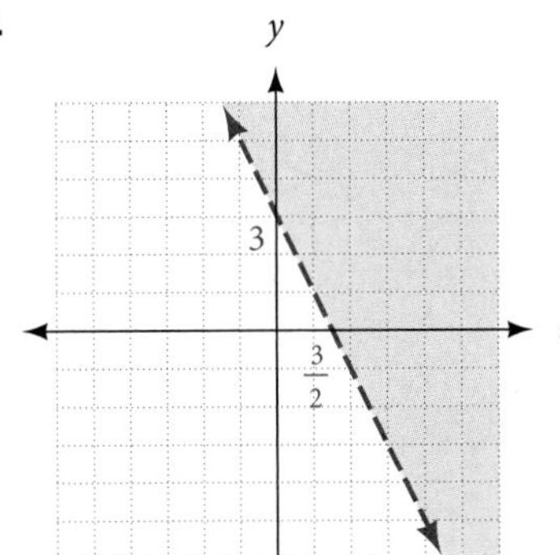

21.
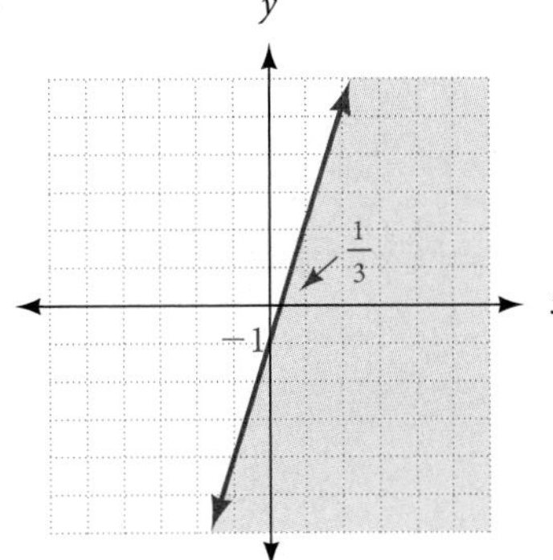

23.
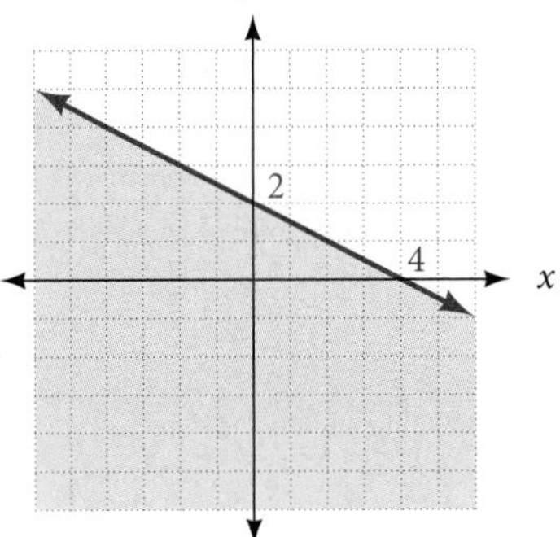

25. **a.** $y < \frac{8}{3}$ **b.** $y > -\frac{8}{3}$ **c.** $y = -\frac{4}{3}x+4$ **d.**

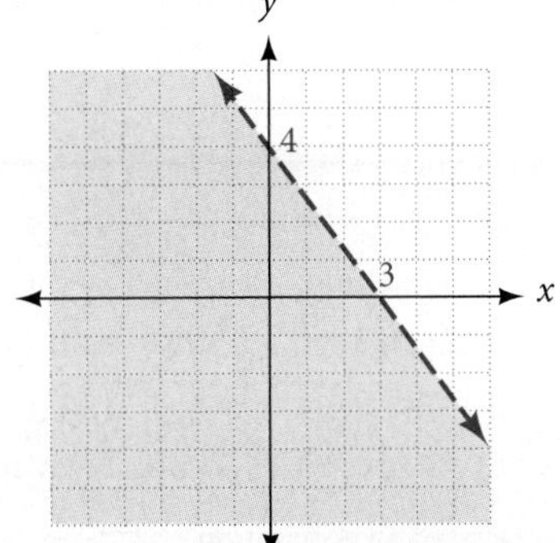

27. **a.** $y = \frac{2}{5}x + 2$ **b.** $y < \frac{2}{5}x + 2$ **c.** $y > \frac{2}{5}x + 2$ **29.** $-6x + 11$ **31.** -8 **33.** -4 **35.** $w = \frac{P - 2l}{2}$

37. (number line: open circle at −1, shaded to the left; marks −1, 0) **39.** $y \geq \frac{3}{2}x - 6$ **41.** Width 2 inches, length 11 inches

CHAPTER 2 TEST

1.-4.

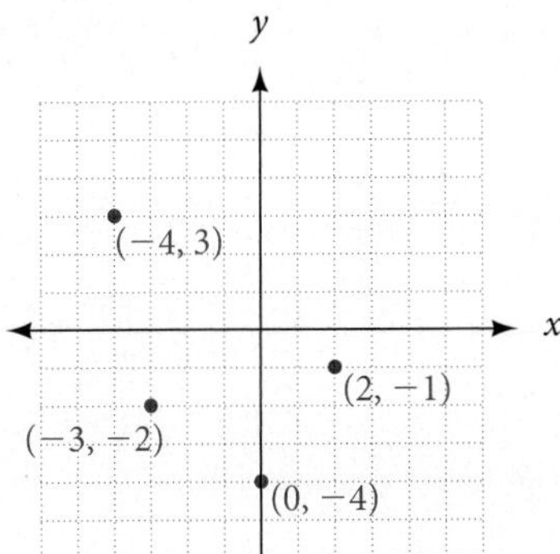

5. $(0, -3), (2, 0), (4, 3), (-2, -6)$ **6.** $(0, 7), (4, -5)$

7.

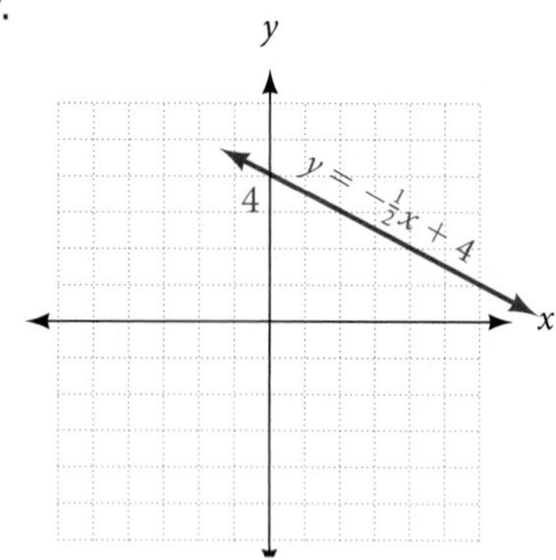

8.

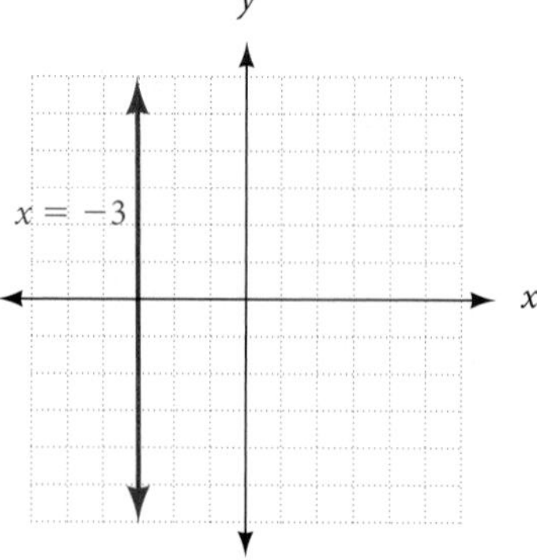

9. (2, 0), (0, −4) **10.** (−4, 0), (0, 6) **11.** No x-intercept. (0, 3) **12.** $-\frac{1}{2}$ **13.** $-\frac{8}{7}$ **14.** −3 **15.** No **16.** 0
17. $y = -\frac{1}{2}x + 3$ **18.** $y = 3x - 5$ **19.** $y = -\frac{2}{3}x - 2$ **20.** $y = -\frac{6}{5}x + \frac{18}{5}$

21.

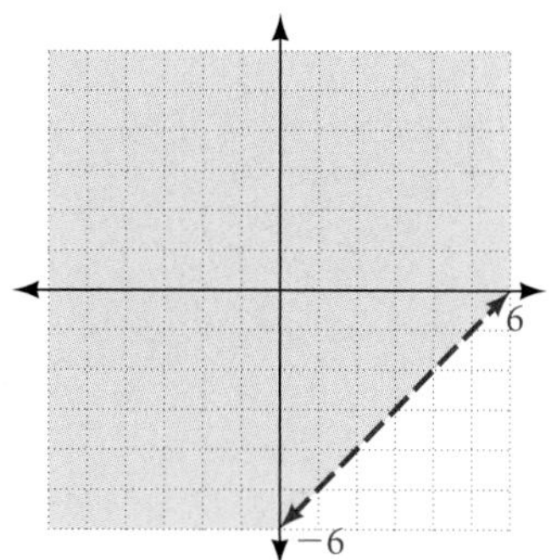

22.

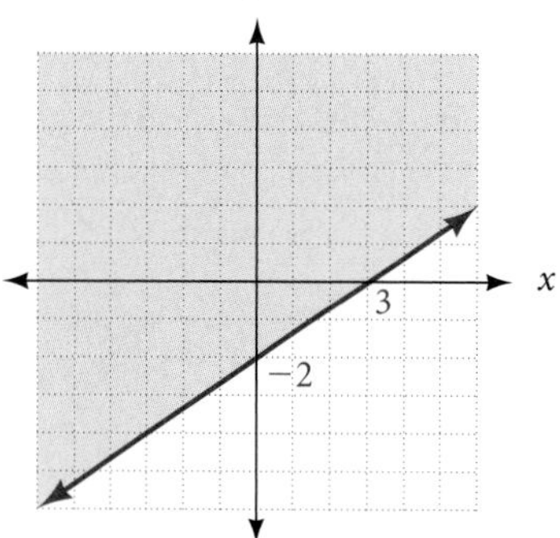

Chapter 3

PROBLEM SET 3.1

1. Domain = {1, 3, 5, 7}, Range = {2, 4, 6, 8}; a function **3.** Domain = {0, 1, 2, 3}, Range = {4, 5, 6}; a function
5. Domain = {a, b, c, d}; Range = {3, 4, 5}; a function **7.** Domain = {a}; Range = {1, 2, 3, 4}; not a function **9.** Yes **11.** No
13. No **15.** Yes **17.** Yes **19.** Domain = $\{x \mid -5 \le x \le 5\}$, Range = $\{y \mid 0 \le y \le 5\}$
21. Domain = $\{x \mid -5 \le x \le 3\}$, Range = $\{y \mid y = 3\}$ **23.** Domain = All real numbers, Range = $\{y \mid y \ge -1\}$, A function
25. Domain = All real numbers, Range = $\{y \mid y \ge 4\}$, A function
27. Domain = $\{x \mid x \ge -1\}$, Range = All real numbers, Not a function
29. Domain = All real numbers, Range = $\{y \mid y \ge 0\}$; a function
31. Domain = $\{x \mid x \ge 0\}$, Range = All real numbers; not a function
33. **a.** $y = 8.5x$ for $10 \le x \le 40$ **b.**

TABLE 4 Weekly Wages

Hours Worked	Function Rule	Gross Pay ($)
x	$y = 8.5x$	y
10	$y = 8.5(10)$	85
20	$y = 8.5(20)$	170
30	$y = 8.5(30)$	255
40	$y = 8.5(40)$	340

c.

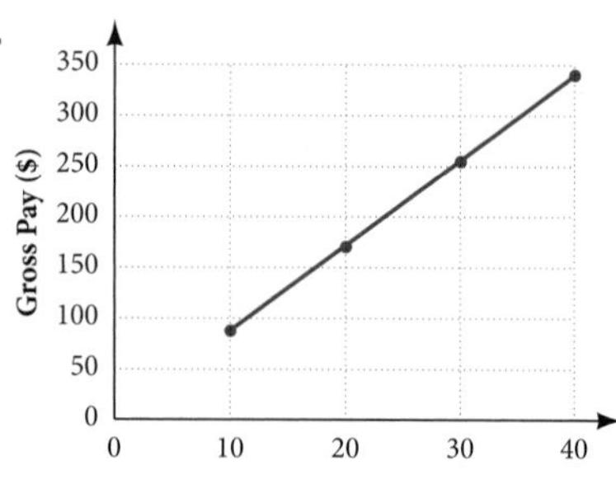

d. Domain = $\{x \mid 10 \le x \le 40\}$; Range = $\{y \mid 85 \le y \le 340\}$ **e.** Minimum = \$85; Maximum = \$340

35. Domain = {2004, 2005, 2006, 2007, 2008, 2009, 2010}, Range = {680, 730, 800, 900, 920, 990, 1030}

37. **a.** III **b.** I **c.** II **d.** IV **39.** 113 **41.** -9 **43.** **a.** 6 **b.** 7.5 **45.** **a.** 27 **b.** 6 **47.** 1 **49.** -3 **51.** $-\frac{6}{5}$ **53.** $-\frac{35}{32}$

PROBLEM SET 3.2

1. -1 **3.** -11 **5.** 2 **7.** 4 **9.** $a^2 + 3a + 4$ **11.** $2a + 7$ **13.** 1 **15.** -9 **17.** 8 **19.** 0 **21.** $3a^2 - 4a + 1$

23. $3a^2 + 8a + 5$ **25.** 4 **27.** 0 **29.** 2 **31.** 24 **33.** -1 **35.** $2x^2 - 19x + 12$ **37.** 99 **39.** $\frac{3}{10}$ **41.** $\frac{2}{5}$ **43.** undefined

45. **a.** $a^2 - 7$ **b.** $a^2 - 6a + 5$ **c.** $x^2 - 2$ **d.** $x^2 + 4x$ **e.** $a^2 + 2ab + b^2 - 4$ **f.** $x^2 + 2xh + h^2 - 4$

47.

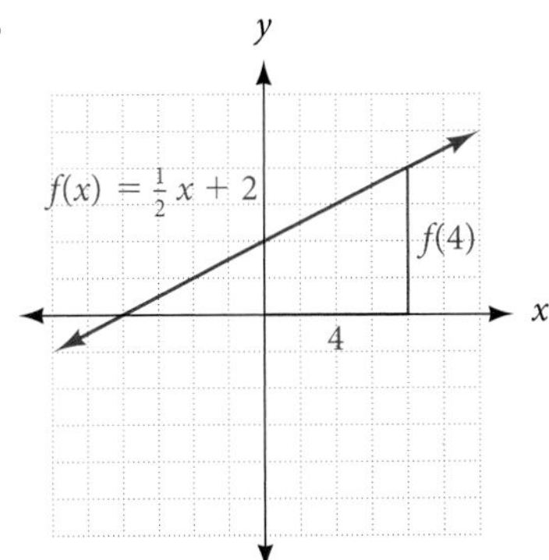

49. $x = 4$ **51.**

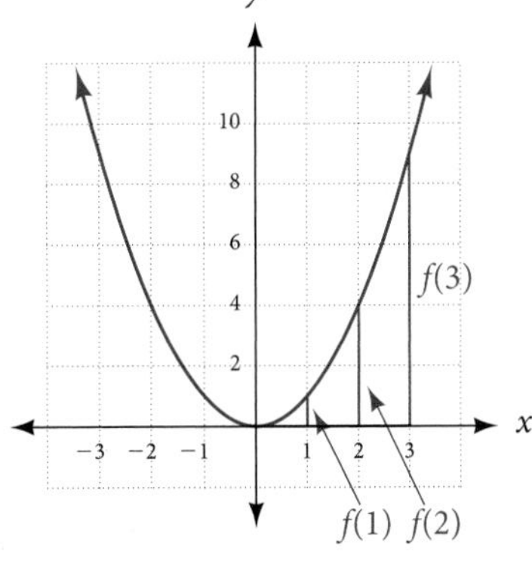

53. $V(3) = 300$, the painting is worth \$300 in 3 years; $V(6) = 600$, the painting is worth \$600 in 6 years.

55. **a.** True **b.** False **c.** True **d.** False **e.** True

57. **a.** \$5,625 **b.** \$1,500 **c.** $\{t \mid 0 \le t \le 5\}$ **d.**

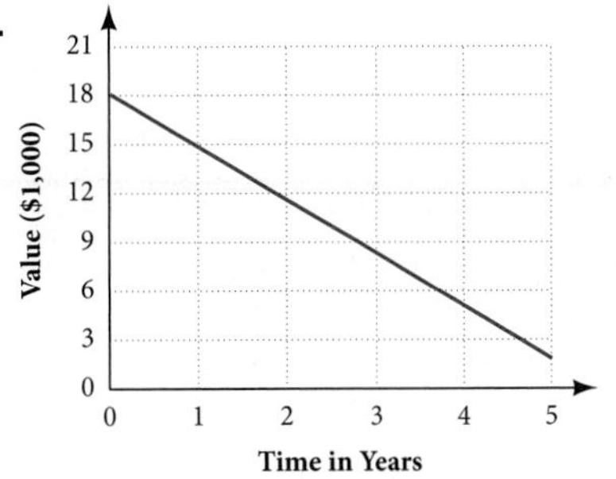

e. $\{V(t) \mid 1{,}500 \le V(t) \le 18{,}000\}$

f. About 2.42 years

59. 196 **61.** 4 **63.** 1.6 **65.** 3 **67.** 2,400

PROBLEM SET 3.3

1. 30 **3.** -6 **5.** 40 **7.** $\frac{81}{5}$ **9.** 64 **11.** 108 **13.** 300 **15.** ± 2 **17.** 1600 **19.** ± 8 **21.** $\frac{50}{3}$ pounds

23. **a.** $T = 4P$ **b.**

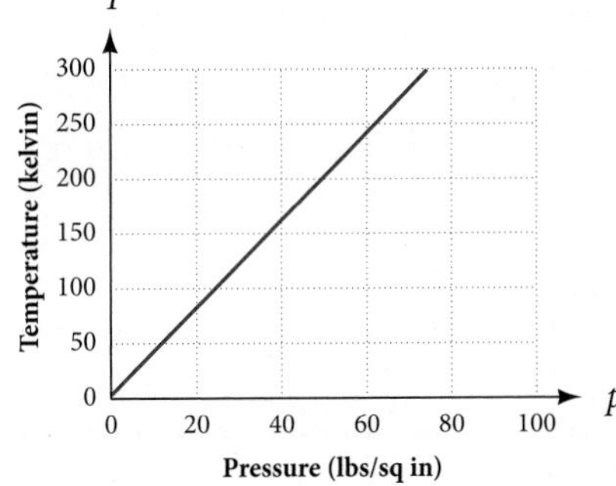

c. 70 pounds per square inch

25. 12 pounds per square inch **27.** **a.** $f = \frac{80}{d}$ **b.**

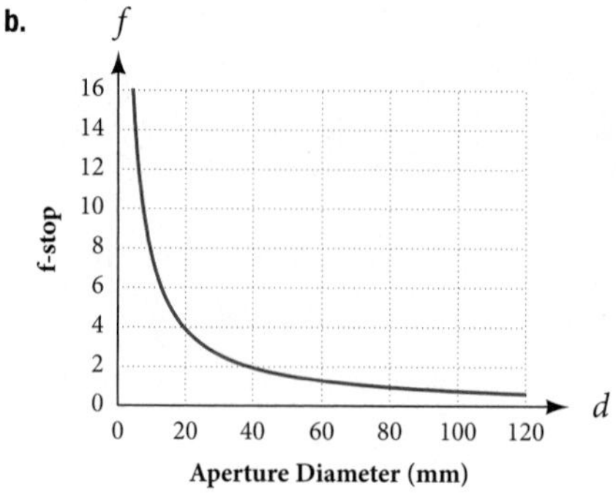

c. An f-stop of 8

29. $\frac{1504}{15}$ square inches **31.** 1.5 ohms **33.** **a.** $P = 0.21\sqrt{L}$ **b.**

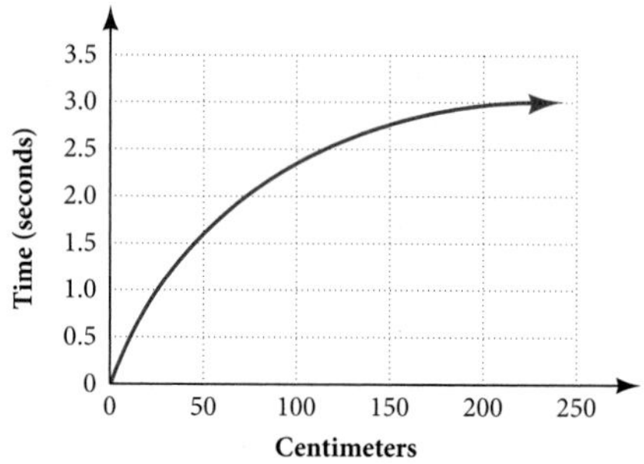

c. 3.15

35. $.6M - 42$ **37.** $16x^3 - 40x^2 + 33x - 9$ **39.** $4x^2 - 3x$ **41.** $6x^2 - 2x - 4$ **43.** 11

PROBLEM SET 3.4

1. $6x + 2$ **3.** $-2x + 8$ **5.** $8x^2 + 14x - 15$ **7.** $\frac{2x+5}{4x-3}$ **9.** $4x - 7$ **11.** $3x^2 - 10x + 8$ **13.** $-2x + 3$ **15.** $3x^2 - 11x + 10$ **17.** $9x^3 - 48x^2 + 85x - 50$ **19.** $x - 2$ **21.** $\frac{1}{x-2}$ **23.** $3x^2 - 7x + 3$ **25.** $6x^2 - 22x + 20$ **27.** 15 **29.** 98 **31.** $\frac{3}{2}$ **33.** 1 **35.** 40 **37.** 147 **39.** **a.** 81 **b.** 29 **c.** $(x + 4)^2$ **d.** $x^2 + 4$ **41.** **a.** -2 **b.** -1 **c.** $16x^2 + 4x - 2$ **d.** $4x^2 + 12x - 1$

43. $(f \circ g)(x) = 5\left[\frac{x+4}{5}\right] - 4 = x + 4 - 4 = x$, $(g \circ f)(x) = \frac{(5x-4)+4}{5} = \frac{5x}{5} = x$

45. **a.** $R(x) = 11.5x - 0.05x^2$ **b.** $C(x) = 2x + 200$ **c.** $P(x) = -0.05x^2 + 9.5x - 200$ **d.** $\overline{C}(x) = 2 + \frac{200}{x}$

47. **a.** $M(x) = 220 - x$ **b.** $M(24) = 196$ **c.** 142 **d.** 135 **e.** 128 **49.** 12 **51.** 28 **53.** $-\frac{7}{4}$ **55.** $w = \frac{P - 2\ell}{2}$

57. $[-6, \infty)$ 

59. $(-\infty, 6)$ 

61. 6, 2 **63.** $\varnothing$

CHAPTER 3 TEST

1. domain = $\{-3, -2\}$, range = $\{0, 1\}$, not a function **2.** domain = all real numbers, range = $\{y \mid y \geq -9\}$, is a function **3.** 11 **4.** -4 **5** 8 **6.** 4 **7.** 18 **8.** $\frac{81}{4}$ **9.** $\frac{2000}{3}$ pounds

Chapter 4

PROBLEM SET 4.1

1.

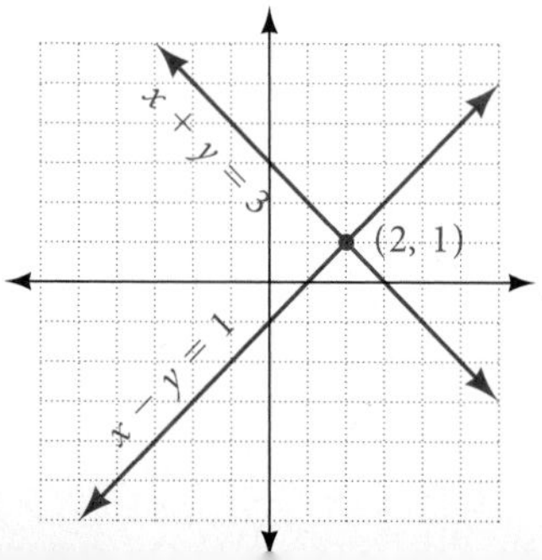

3.

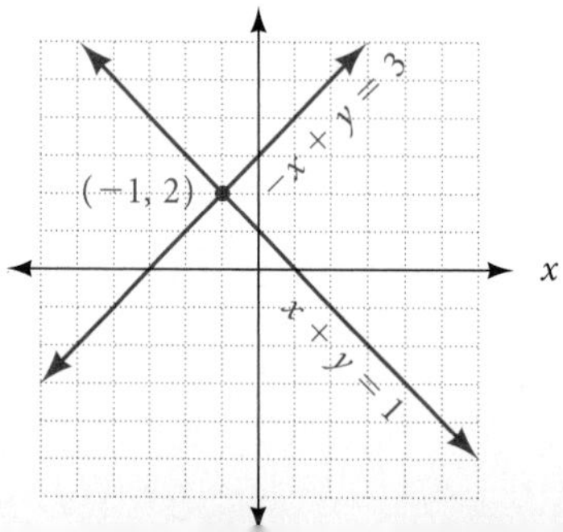

5.

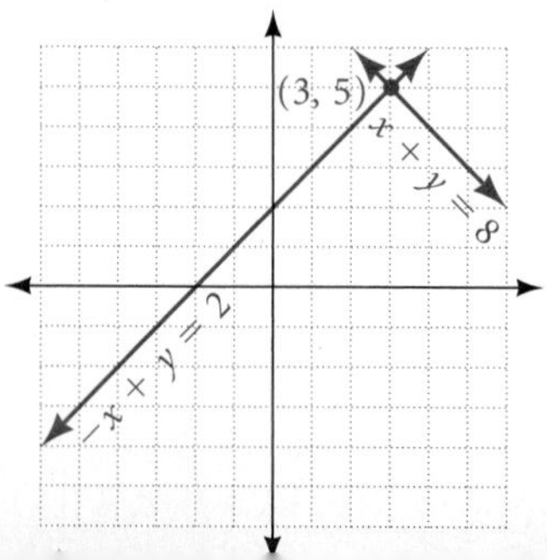

7.

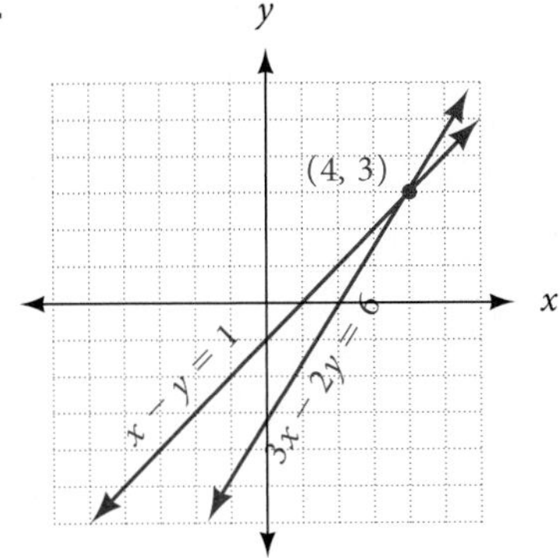

9.

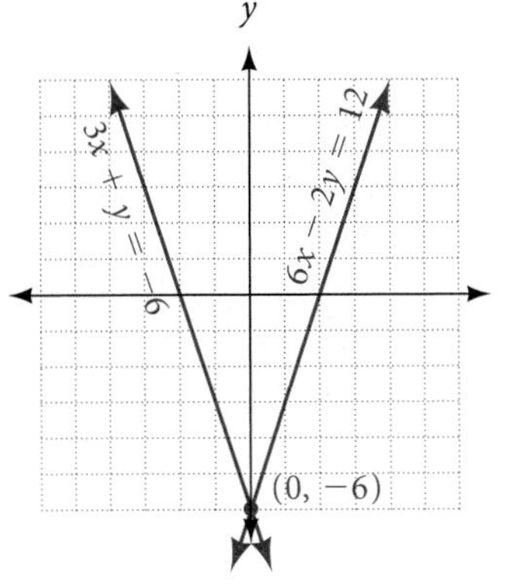

11.

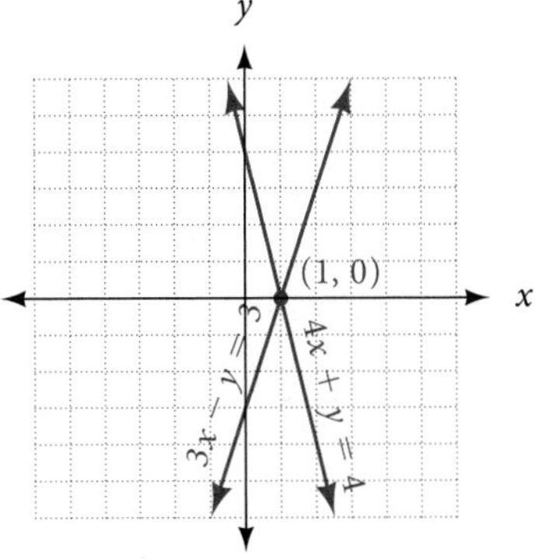

13.

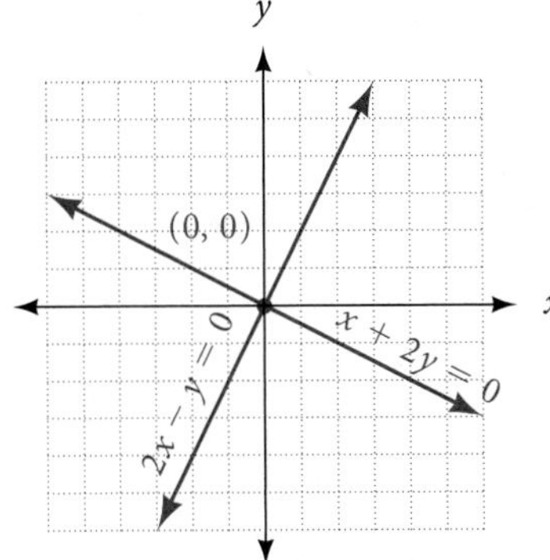

15.

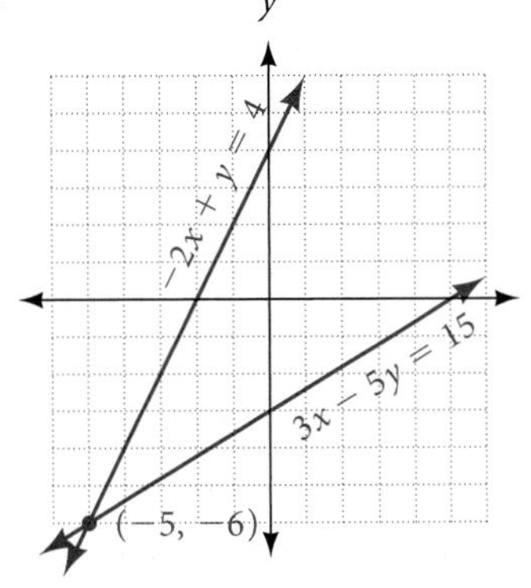

17.

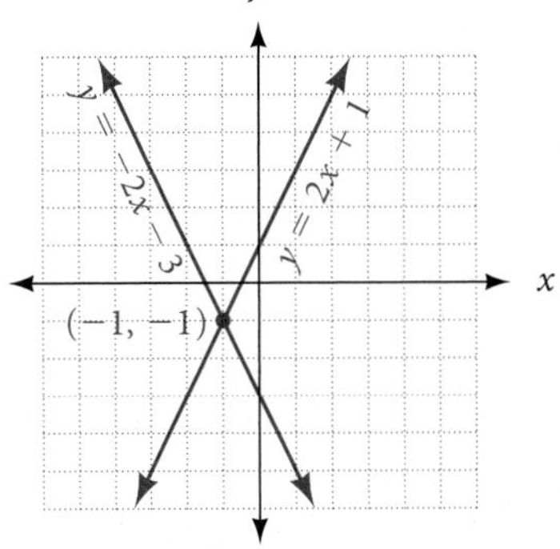

19.

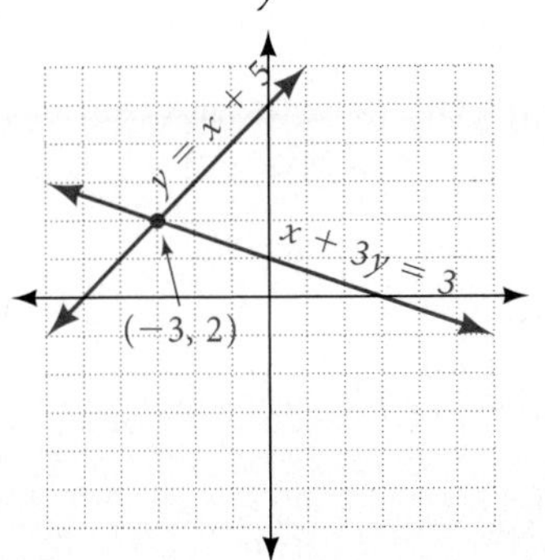

21.

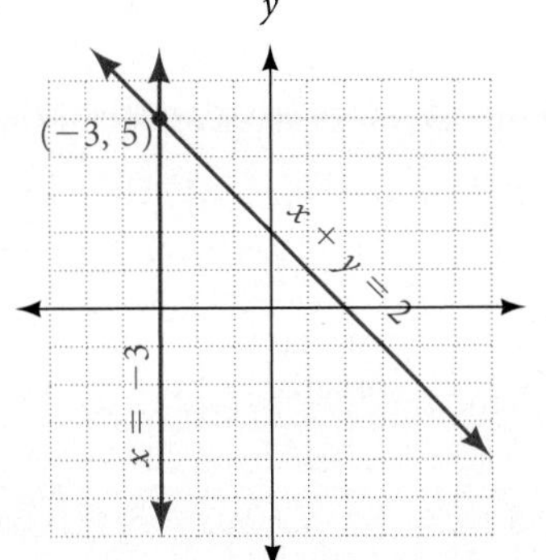

23.

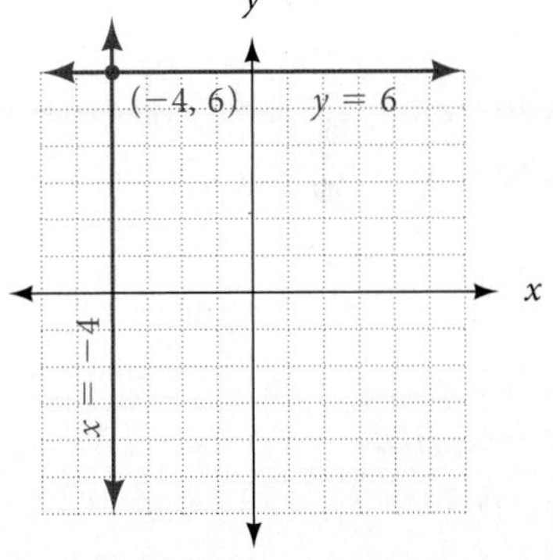

25. ∅

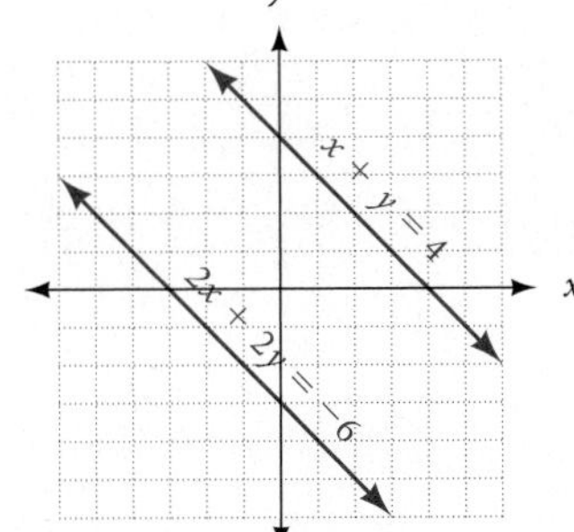

27. Any point on the line

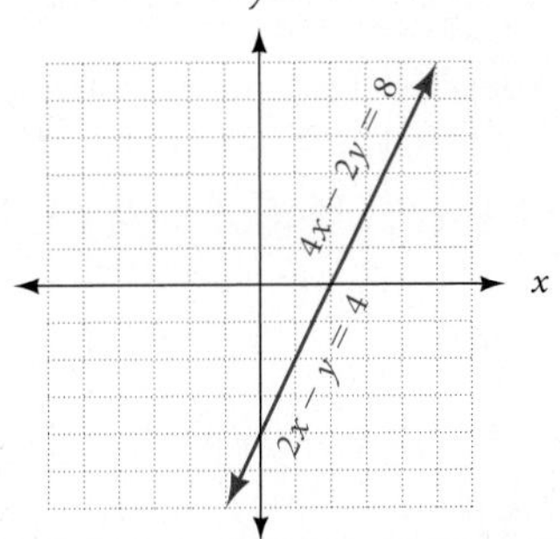

29.

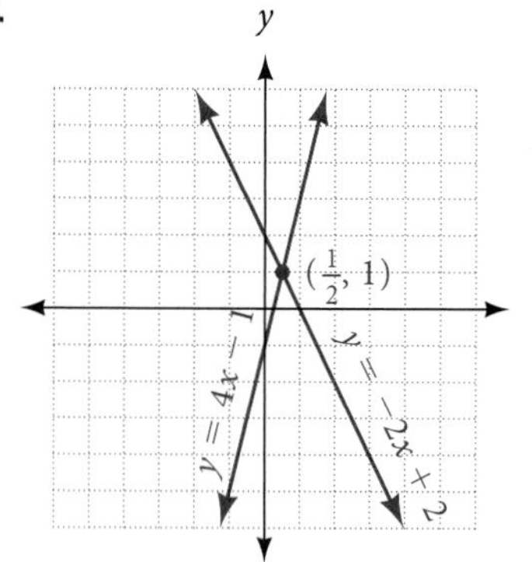

31.

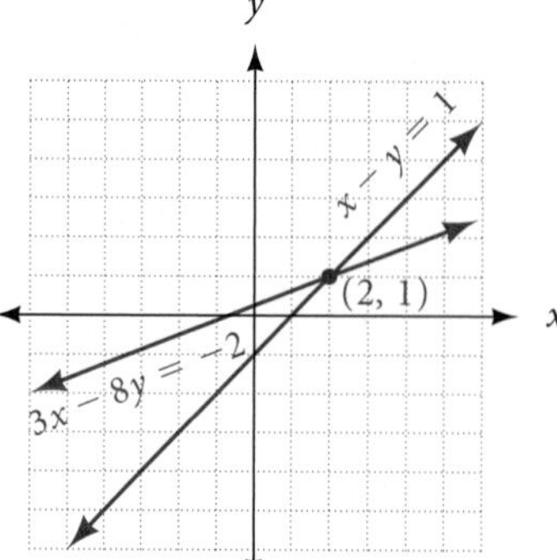

33. **a.** 25 hours **b.** Gigi's **c.** Marcy's

35. $2x$ **37.** $7x$ **39.** $13x$ **41.** $3x - 2y$ **43.** 1 **45.** 0 **47.** -5

PROBLEM SET 4.2

1. (2, 1) **3.** (3, 7) **5.** $(2, -5)$ **7.** $(-1, 0)$ **9.** Lines coincide. **11.** (4, 8) **13.** $\left(\frac{1}{5}, 1\right)$ **15.** (1, 0) **17.** $(-1, -2)$ **19.** $\left(-5, \frac{3}{4}\right)$ **21.** $(-4, 5)$ **23.** $(-3, -10)$ **25.** (3, 2) **27.** $\left(5, \frac{1}{3}\right)$ **29.** $\left(-2, \frac{2}{3}\right)$ **31.** (2, 2) **33.** Lines are parallel. $\varnothing$ **35.** (1, 1) **37.** Lines are parallel. $\varnothing$ **39.** (10, 12) **41.** 1 **43.** 2 **45.** All real numbers. **47.** $x = 3y - 1$ **49.** 1 **51.** 5 **53.** 34.5 **55.** 33.95

PROBLEM SET 4.3

1. (4, 7) **3.** (3, 17) **5.** $\left(\frac{3}{2}, 2\right)$ **7.** (2, 4) **9.** (0, 4) **11.** $(-1, 3)$ **13.** (1, 1) **15.** $(2, -3)$ **17.** $\left(-2, \frac{3}{5}\right)$ **19.** $(-3, 5)$ **21.** Lines are parallel. $\varnothing$ **23.** (3, 1) **25.** $\left(\frac{1}{2}, \frac{3}{4}\right)$ **27.** (2, 6) **29.** (4, 4) **31.** $(5, -2)$ **33.** (18, 10) **35.** Lines coincide. **37.** (10, 12) **39.** **a.** 1,000 miles **b.** Car **c.** Truck **d.** We are only working with positive numbers. **41.** 3 and 23 **43.** 15 and 24 **45.** Length = 23 in.; width = 6 in. **47.** 14 nickels and 10 dimes

PROBLEM SET 4.4

1. 10 and 15 **3.** 3 and 12 **5.** 4 and 9 **7.** 6 and 29 **9.** \$9,000 at 8%, \$11,000 at 6% **11.** \$2,000 at 6%, \$8,000 at 5% **13.** 6 nickels, 8 quarters **15.** 12 dimes, 9 quarters

17. 6 liters of 50% solution, 12 liters of 20% solution **19.** 10 gallons of 10% solution, 20 gallons of 7% solution

	50% Solution	20% Solution	Final Solution
Number of Liters	x	y	18
Liters of Alcohol	0.50x	0.20y	0.30(18)

21. 20 adults, 50 kids **23.** 16 feet wide, 32 feet long **25.** 33 \$5 chips, 12 \$25 chips **27.** 50 at \$11, 100 at \$20 **29.** $\frac{x+2}{x+3}$ **31.** $\frac{2(x+5)}{x-4}$ **33.** $\frac{1}{x-4}$ **35.** $\frac{x+5}{x-3}$ **37.** -2 **39.** 24 hours **41.** 24

CHAPTER 4 TEST

1. (0, 6) **2.**

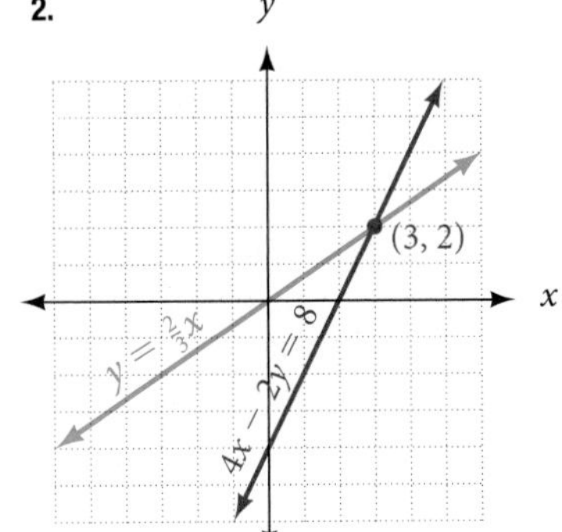

3.

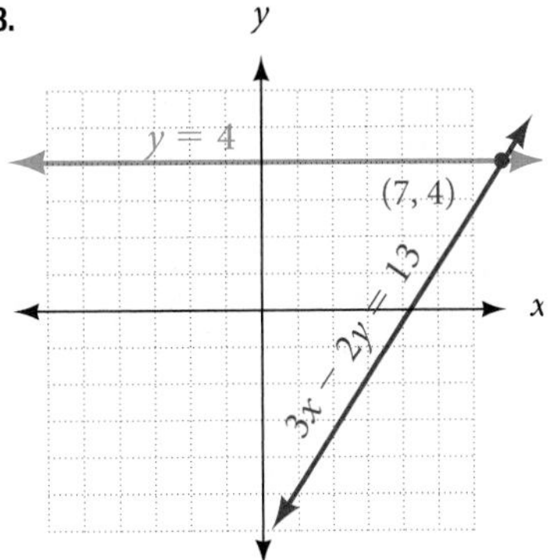

4.

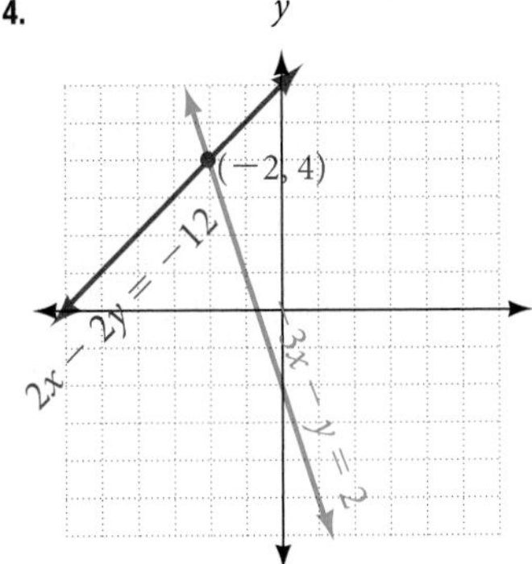

5. (−4, 5) **6.** (1, 2) **7.** (−6, 3) **8.** Lines coincide **9.** (4, 0) **10.** (19, 9) **11.** (−3, 4) **12.** (11, 3) **13.** (13, 2) **14.** 4, 14 **15.** \$1520 at 7%, \$480 at 6% **16.** 8 dimes, 11 nickles **17.** 71 ft × 28 ft

Chapter 5

PROBLEM SET 5.1

1. Base = 4, exponent = 2, 16 **3.** Base = .3, exponent = 2, 0.09 **5.** Base = 4, exponent = 3, 64 **7.** Base = −5, exponent = 2, 25 **9.** Base = 2, exponent = 3, −8 **11.** Base = 3, exponent = 4, 81 **13.** Base = $\frac{2}{3}$, exponent = 2, $\frac{4}{9}$ **15.** Base = $\frac{1}{2}$, exponent = 4, $\frac{1}{16}$

17. a.

Number x	1	2	3	4	5	6	7
Square x^2	1	4	9	16	25	36	49

b. Either *larger* or *greater* will work. **19.** x^9 **21.** y^{30} **23.** 2^{12}

25. x^{28} **27.** x^{10} **29.** 5^{12} **31.** y^9 **33.** 2^{50} **35.** a^{3x} **37.** b^{xy} **39.** $16x^2$ **41.** $32y^5$ **43.** $81x^4$ **45.** $0.25a^2b^2$ **47.** $64x^3y^3z^3$ **49.** $8x^{12}$ **51.** $16a^6$ **53.** x^{14} **55.** a^{11} **57.** $128x^7$ **59.** $432x^{10}$ **61.** $16x^4y^6$ **63.** $\frac{8}{27}a^{12}b^{15}$

65.

Number x	−3	−2	−1	0	1	2	3
Square x^2	9	4	1	0	1	4	9

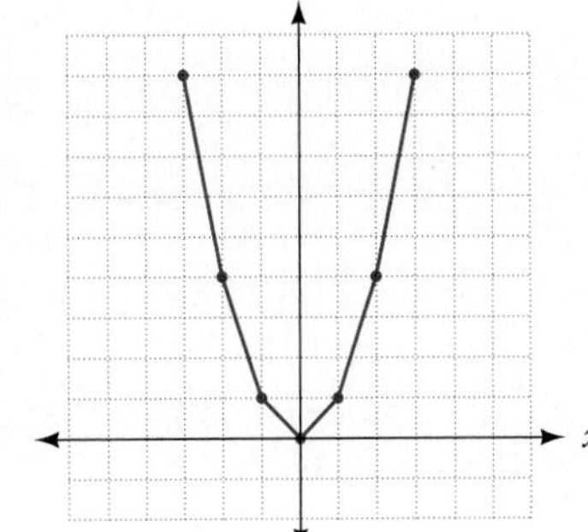

67.

Number x	−2.5	−1.5	−0.5	0	0.5	1.5	2.5
Square x^2	6.25	2.25	0.25	0	0.25	2.25	6.25

69. 4.32×10^4 **71.** 5.7×10^2 **73.** 2.38×10^5 **75.** 2,490 **77.** 352 **79.** 28,000 **81.** 27 inches^3 **83.** 15.6 inches^3 **85.** 36 inches^3 **87.** answers will vary **89.** 6.5×10^8 seconds **91.** \$740,000 **93.** \$180,000 **95.** 219 inches^3 **97.** 182 inches^3 **99.** −3

101. 11 **103.** -5 **105.** 5 **107.** 2 **109.** 6 **111.** 4 **113.** 3

PROBLEM SET 5.2

1. $\frac{1}{9}$ **3.** $\frac{1}{36}$ **5.** $\frac{1}{64}$ **7.** $\frac{1}{125}$ **9.** $\frac{2}{x^3}$ **11.** $\frac{1}{8x^3}$ **13.** $\frac{1}{25y^2}$ **15.** $\frac{1}{100}$ **17.**

Number x	Square x^2	Power of 2 2^x
-3	9	$\frac{1}{8}$
-2	4	$\frac{1}{4}$
-1	1	$\frac{1}{2}$
0	0	1
1	1	2
2	4	4
3	9	8

19. $\frac{1}{25}$ **21.** x^6 **23.** 64 **25.** $8x^3$ **27.** 6^{10} **29.** $\frac{1}{6^{10}}$ **31.** $\frac{1}{2^8}$ **33.** 2^8 **35.** $27x^3$ **37.** $81x^4y^4$ **39.** 1 **41.** $2a^2b$ **43.** $\frac{1}{49y^6}$ **45.** $\frac{1}{x^8}$ **47.** $\frac{1}{y^3}$ **49.** x^2 **51.** a^6 **53.** $\frac{1}{y^9}$ **55.** y^{40} **57.** $\frac{1}{x}$ **59.** x^9 **61.** a^{16} **63.** $\frac{1}{a^4}$

65.

Number x	-3	-2	-1	0	1	2	3
Power of 2 2^x	$\frac{1}{8}$	$\frac{1}{4}$	$\frac{1}{2}$	1	2	4	8

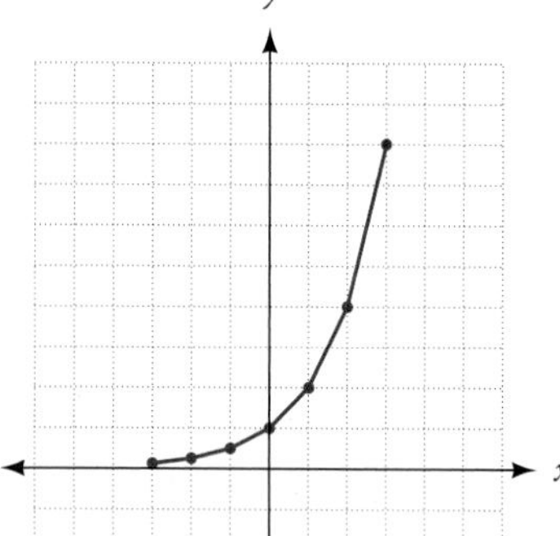

67. 4.8×10^{-3} **69.** 2.5×10^1 **71.** 9×10^{-6} **73.** *shown below*

Expanded Form	Scientific Notation $n \times 10^r$
0.000357	3.57×10^{-4}
0.00357	3.57×10^{-3}
0.0357	3.57×10^{-2}
0.357	3.57×10^{-1}
3.57	3.57×10^0
35.7	3.57×10^1
357	3.57×10^2
3,570	3.57×10^3
35,700	3.57×10^4

75. 0.00423 **77.** 0.00008 **79.** 4.2 **81.** 0.002 **83.** Craven/Bush 2×10^{-3}, Earnhardt/Irvan 5×10^{-3}, Harvick/Gordon 6×10^{-3}, Kahne/Kenseth 1×10^{-2}, Kenseth/Kahne 1×10^{-2}

85. 2.5×10^4 **87.** 2.35×10^5 **89.** 8.2×10^{-4} **91.** 100 inches², 400 inches²; 4 **93.** x^2; $4x^2$; 4

95. 216 inches³; 1,728 inches³; 8 **97.** x^3; $8x^3$; 8 **99.** 13.5 **101.** 8 **103.** 26.52 **105.** 12 **107.** x^8 **109.** x **111.** $\frac{1}{y^2}$ **113.** 340

PROBLEM SET 5.3

1. $12x^7$ **3.** $-16y^{11}$ **5.** $32x^2$ **7.** $200a^6$ **9.** $-24a^3b^3$ **11.** $24x^6y^8$ **13.** $3x$ **15.** $\frac{6}{y^3}$ **17.** $\frac{1}{2a}$ **19.** $-\frac{3a}{b^2}$ **21.** $\frac{x^2}{9z^2}$

23.

a	b	ab	$\frac{a}{b}$	$\frac{b}{a}$
10	$5x$	$50x$	$\frac{2}{x}$	$\frac{x}{2}$
$20x^3$	$6x^2$	$120x^5$	$\frac{10x}{3}$	$\frac{3}{10x}$
$25x^5$	$5x^4$	$125x^9$	$5x$	$\frac{1}{5x}$
$3x^{-2}$	$3x^2$	9	$\frac{1}{x^4}$	x^4
$-2y^4$	$8y^7$	$-16y^{11}$	$-\frac{1}{4y^3}$	$-4y^3$

25. 6×10^8 **27.** 1.75×10^{-1} **29.** 1.21×10^{-6} **31.** 4.2×10^3 **33.** 3×10^{10} **35.** 5×10^{-3} **37.** $8x^2$ **39.** $-11x^5$ **41.** 0 **43.** $4x^3$ **45.** $31ab^2$ **47.**

a	b	ab	a + b
$5x$	$3x$	$15x^2$	$8x$
$4x^2$	$2x^2$	$8x^4$	$6x^2$
$3x^3$	$6x^3$	$18x^6$	$9x^3$
$2x^4$	$-3x^4$	$-6x^8$	$-x^4$
x^5	$7x^5$	$7x^{10}$	$8x^5$

49. $4x^3$ **51.** $\frac{1}{b^2}$ **53.** $\frac{6y^{10}}{x^4}$ **55.** $x^2y + x$ **57.** $x + y$ **59.** $x^2 - 4$ **61.** $x^2 - x - 6$ **63.** $x^2 - 5x$ **65.** $x^2 - 8x$ **67.** 2×10^6

69. 1×10^1 **71.** 4.2×10^{-6} **73.** $9x^3$ **75.** $-20a^2$ **77.** $6x^5y^2$ **79.** -5 **81.** 6 **83.** 76 **85.** $6x^2$ **87.** $2x$ **89.** $-2x - 9$ **91.** 11

PROBLEM SET 5.4

1. Trinomial, 3 **3.** Trinomial, 3 **5.** Binomial, 1 **7.** Binomial, 2 **9.** Monomial, 2 **11.** Monomial, 0 **13.** $5x^2 + 5x + 9$ **15.** $5a^2 - 9a + 7$ **17.** $x^2 + 6x + 8$ **19.** $6x^2 - 13x + 5$ **21.** $x^2 - 9$ **23.** $3y^2 - 11y + 10$ **25.** $6x^3 + 5x^2 - 4x + 3$ **27.** $2x^2 - x + 1$ **29.** $2a^2 - 2a - 2$ **31.** $-\frac{1}{9}x^3 - \frac{2}{3}x^2 - \frac{5}{2}x + \frac{7}{4}$ **33.** $-4y^2 + 15y - 22$ **35.** $x^2 - 33x + 63$ **37.** $8y^2 + 4y + 26$ **39.** $75x^2 - 150x - 75$ **41.** $12x + 2$ **43.** 4 **45.** 56.52 in^3 **47.** 5 **49.** -6 **51.** $-20x^2$ **53.** $-21x$ **55.** $2x$ **57.** $6x - 18$

PROBLEM SET 5.5

1. $6x^2 + 2x$ **3.** $6x^4 - 4x^3 + 2x^2$ **5.** $2a^3b - 2a^2b^2 + 2ab$ **7.** $3y^4 + 9y^3 + 12y^2$ **9.** $8x^5y^2 + 12x^4y^3 + 32x^2y^4$ **11.** $x^2 + 7x + 12$ **13.** $x^2 + 7x + 6$ **15.** $x^2 + 2x + \frac{3}{4}$ **17.** $a^2 + 2a - 15$ **19.** $xy + bx - ay - ab$ **21.** $x^2 - 36$ **23.** $y^2 - \frac{25}{36}$ **25.** $2x^2 - 11x + 12$ **27.** $2a^2 + 3a - 2$ **29.** $6x^2 - 19x + 10$ **31.** $2ax + 8x + 3a + 12$ **33.** $25x^2 - 16$ **35.** $2x^2 + \frac{5}{2}x - \frac{3}{4}$ **37.** $3 - 10a + 8a^2$

39. $(x + 2)(x + 3) = x^2 + 2x + 3x + 6 = x^2 + 5x + 6$

	x	3
x	x^2	$3x$
2	$2x$	6

41. $(x + 1)(2x + 2) = 2x^2 + 4x + 2$

	x	x	2
x	x^2	x^2	$2x$
1	x	x	2

43. $a^3 - 6a^2 + 11a - 6$ **45.** $x^3 + 8$ **47.** $2x^3 + 17x^2 + 26x + 9$ **49.** $5x^4 - 13x^3 + 20x^2 + 7x + 5$ **51.** $2x^4 + x^2 - 15$ **53.** $6a^6 + 15a^4 + 4a^2 + 10$ **55.** $x^3 + 12x^2 + 47x + 60$ **57.** $x^2 - 5x + 8$ **59.** $8x^2 - 6x - 5$ **61.** $x^2 - x - 30$ **63.** $x^2 + 4x - 6$ **65.** $x^2 + 13x$ **67.** $x^2 + 2x - 3$ **69.** $a^2 - 3a + 6$ **71.** $A = x(2x + 5) = 2x^2 + 5x$ **73.** $A = x(x + 1) = x^2 + x$ **75.** 169 **77.** $-10x$ **79.** 0 **81.** 0 **83.** $-12x + 16$ **85.** $x^2 + x - 2$ **87.** $x^2 + 6x + 9$

PROBLEM SET 5.6

1. $x^2 - 4x + 4$ **3.** $a^2 + 6a + 9$ **5.** $x^2 - 10x + 25$ **7.** $a^2 - a + \frac{1}{4}$ **9.** $x^2 + 20x + 100$ **11.** $a^2 + 1.6a + 0.64$ **13.** $4x^2 - 4x + 1$ **15.** $16a^2 + 40a + 25$ **17.** $9x^2 - 12x + 4$ **19.** $9a^2 + 30ab + 25b^2$ **21.** $16x^2 - 40xy + 25y^2$ **23.** $49m^2 + 28mn + 4n^2$ **25.** $36x^2 - 120xy + 100y^2$ **27.** $x^4 + 10x^2 + 25$ **29.** $a^4 + 2a^2 + 1$

31.

x	$(x + 3)^2$	$x^2 + 9$	$x^2 + 6x + 9$
1	16	10	16
2	25	13	25
3	36	18	36
4	49	25	49

33.

a	1	3	3	4
b	1	5	4	5
$(a + b)^2$	4	64	49	81
$a^2 + b^2$	2	34	25	41
$a^2 + ab + b^2$	3	49	37	61
$a^2 + 2ab + b^2$	4	64	49	81

35. $a^2 - 25$ **37.** $y^2 - 1$ **39.** $81 - x^2$ **41.** $4x^2 - 25$ **43.** $16x^2 - \frac{1}{9}$ **45.** $4a^2 - 49$ **47.** $36 - 49x^2$ **49.** $x^4 - 9$ **51.** $a^4 - 16$ **53.** $25y^8 - 64$ **55.** $2x^2 - 34$ **57.** $-12x^2 + 20x + 8$ **59.** $a^2 + 4a + 6$ **61.** $8x^3 + 36x^2 + 54x + 27$ **63.** $(50 - 1)(50 + 1) = 2500 - 1 = 2499$ **65.** Both equal 25. **67.** $x^2 + (x + 1)^2 = 2x^2 + 2x + 1$ **69.** $x^2 + (x + 1)^2 + (x + 2)^2 = 3x^2 + 6x + 5$

71.

	a	b
a	a^2	ab
b	ba	b^2

$a^2 + ab + ba + b^2 = a^2 + 2ab + b^2$ **73.** $2x^2$ **75.** x^2 **77.** $3x$ **79.** $3xy$

PROBLEM SET 5.7

1. $x - 2$ **3.** $3 - 2x^2$ **5.** $5xy - 2y$ **7.** $7x^4 - 6x^3 + 5x^2$ **9.** $10x^4 - 5x^2 + 1$ **11.** $-4a + 2$ **13.** $-8a^4 - 12a^3$ **15.** $-4b - 5a$ **17.** $-6a^2b + 3ab^2 - 7b^3$ **19.** $-\frac{a}{2} - b - \frac{b^2}{2a}$ **21.** $3x + 4y$ **23.** $-y + 3$ **25.** $5y - 4$ **27.** $xy - x^2y^2$ **29.** $-1 + xy$ **31.** $-a + 1$ **33.** $x^2 - 3xy + y^2$ **35.** $2 - 3b + 5b^2$ **37.** $-2xy + 1$ **39.** $xy - \frac{1}{2}$ **41.** $\frac{1}{4x} - \frac{1}{2a} + \frac{3}{4}$ **43.** $\frac{4x^2}{3} + \frac{2}{3x} + \frac{1}{x^2}$ **45.** $3a^{3m} - 9a^m$ **47.** $2x^{4m} - 5x^{2m} + 7$ **49.** $3x^2 - x + 6$ **51.** 4 **53.** $x + 5$ **55.** Both equal 7. **57.** $\frac{3(10) + 8}{2} = 19; 3(10) + 4 = 34$ **59.** $146\frac{20}{27}$ **61.** $2x + 5$ **63.** $x^2 - 3x$ **65.** $2x^3 - 10x^2$ **67.** $-2x$ **69.** 2

PROBLEM SET 5.8

1. $x - 2$ **3.** $a + 4$ **5.** $x - 3$ **7.** $x + 3$ **9.** $a - 5$ **11.** $x + 2 + \frac{2}{x + 3}$ **13.** $a - 2 + \frac{12}{a + 5}$ **15.** $x + 4 + \frac{9}{x - 2}$ **17.** $x + 4 + \frac{-10}{x + 1}$ **19.** $a + 1 + \frac{-1}{a + 2}$ **21.** $x - 3 + \frac{17}{2x + 4}$ **23.** $3a - 2 + \frac{7}{2a + 3}$ **25.** $2a^2 - a - 3$ **27.** $x^2 - x + 5$ **29.** $x^2 + x + 1$ **31.** $x^2 + 2x + 4$ **33.** \$491.17 **35.** \$331.42 **37.** 47 **39.** 14 **41.** 70 **43.** 35 **45.** 5 **47.** 6, 540 **49.** 1, 760 **51.** 20 **53.** 63 **55.** 53

CHAPTER 5 TEST

1. -32 **2.** $\frac{8}{27}$ **3.** $128x^{13}$ **4.** $\frac{1}{16}$ **5.** 1 **6.** x^3 **7.** $\frac{1}{x^3}$ **8.** 4.307×10^{-2} **9.** 7,630,000 **10.** $\frac{y^3z^2}{3x^2}$ **11.** $\frac{a^3b^2}{2}$ **12.** $12x^3$ **13.** 7.5×10^7 **14.** $9x^2 + 5x + 4$ **15.** $2x^2 + 6x - 2$ **16.** $5x - 4$ **17.** 21 **18.** $15x^4 - 6x^3 + 12x^2$ **19.** $x^2 - \frac{1}{12}x - \frac{1}{12}$ **20.** $10x^2 - 3x - 18$ **21.** $x^3 + 64$ **22.** $x^2 - 12x + 36$ **23.** $4a^2 + 16ab + 16b^2$ **24.** $9x^2 - 36$ **25.** $x^4 - 16$ **26.** $3x^2 - 6x + 1$ **27.** $3x - 1 - \frac{5}{3x - 1}$ **28.** $4x + 11 - \frac{38}{x - 4}$ **29.** 32.77 in^3 **30.** $V = w^3$

Chapter 6

PROBLEM SET 6.1

1. $5(3x + 5)$ **3.** $3(2a + 3)$ **5.** $4(x - 2y)$ **7.** $3(x^2 - 2x - 3)$ **9.** $3(a^2 - a - 20)$ **11.** $4(6y^2 - 13y + 6)$ **13.** $x^2(9 - 8x)$ **15.** $13a^2(1 - 2a)$ **17.** $7xy(3x - 4y)$ **19.** $11ab^2(2a - 1)$ **21.** $7x(x^2 + 3x - 4)$ **23.** $11(11y^4 - x^4)$ **25.** $25x^2(4x^2 - 2x + 1)$ **27.** $8(a^2 + 2b^2 + 4c^2)$ **29.** $4ab(a - 4b + 8ab)$ **31.** $11a^2b^2(11a - 2b + 3ab)$ **33.** $12x^2y^3(1 - 6x^3 - 3x^2y)$ **35.** $(x + 3)(y + 5)$ **37.** $(x + 2)(y + 6)$ **39.** $(a - 3)(b + 7)$ **41.** $(a - b)(x + y)$ **43.** $(2x - 5)(a + 3)$ **45.** $(b - 2)(3x - 4)$ **47.** $(x + 2)(x + a)$ **49.** $(x - b)(x - a)$ **51.** $(x + y)(a + b + c)$ **53.** $(3x + 2)(2x + 3)$ **55.** $(10x - 1)(2x + 5)$ **57.** $(4x + 5)(5x + 1)$ **59.** $(x + 2)(x^2 + 3)$ **61.** $(3x - 2)(2x^2 + 5)$ **63.** 6 **65.** $3(4x^2 + 2x + 1)$ **67.** $A = 1{,}000(1 + r)$; \$1120.00 **69.** **a.** $A = 1{,}000{,}000\,(1 + r)$ **b.** 1,300,000 **71.** $x^2 - 5x - 14$ **73.** $x^2 - x - 6$ **75.** $x^3 + 27$ **77.** $2x^3 + 9x^2 - 2x - 3$ **79.** $18x^7 - 12x^6 + 6x^5$ **81.** $x^2 + x + \frac{2}{9}$ **83.** $12x^2 - 10xy - 12y^2$ **85.** $81a^2 - 1$ **87.** $x^2 - 18x + 81$ **89.** $x^3 + 8$

PROBLEM SET 6.2

1. $(x + 3)(x + 4)$ **3.** $(x + 1)(x + 2)$ **5.** $(a + 3)(a + 7)$ **7.** $(x - 2)(x - 5)$ **9.** $(y - 3)(y - 7)$ **11.** $(x - 4)(x + 3)$ **13.** $(y + 4)(y - 3)$ **15.** $(x + 7)(x - 2)$ **17.** $(r - 9)(r + 1)$ **19.** $(x - 6)(x + 5)$ **21.** $(a + 7)(a + 8)$ **23.** $(y + 6)(y - 7)$

25. $(x+6)(x+7)$ **27.** $2(x+1)(x+2)$ **29.** $3(a+4)(a-5)$ **31.** $100(x-2)(x-3)$ **33.** $100(p-5)(p-8)$ **35.** $x^2(x+3)(x-4)$ **37.** $2r(r+5)(r-3)$ **39.** $2y^2(y+1)(y-4)$ **41.** $x^3(x+2)(x+2)$ **43.** $3y^2(y+1)(y-5)$ **45.** $4x^2(x-4)(x-9)$ **47.** $(x+2y)(x+3y)$ **49.** $(x-4y)(x-5y)$ **51.** $(a+4b)(a-2b)$ **53.** $(a-5b)(a-5b)$ **55.** $(a+5b)(a+5b)$ **57.** $(x-6a)(x+8a)$ **59.** $(x+4b)(x-9b)$ **61.** $(x^2-3)(x^2-2)$ **63.** $(x-100)(x+20)$ **65.** $\left(x-\frac{1}{2}\right)\left(x-\frac{1}{2}\right)$ **67.** $(x+0.2)(x+0.4)$ **69.** $x+16$ **71.** $4x^2-x-3$ **73.** $6a^2+13a+2$ **75.** $6a^2+7a+2$ **77.** $6a^2+8a+2$

PROBLEM SET 6.3

1. $(2x+1)(x+3)$ **3.** $(2a-3)(a+1)$ **5.** $(3x+5)(x-1)$ **7.** $(3y+1)(y-5)$ **9.** $(2x+3)(3x+2)$ **11.** $(2x-3y)(2x-3y)$ **13.** $(4y+1)(y-3)$ **15.** $(4x-5)(5x-4)$ **17.** $(10a-b)(2a+5b)$ **19.** $(4x-5)(5x+1)$ **21.** $(6m-1)(2m+3)$ **23.** $(4x+5)(5x+3)$ **25.** $(3a-4b)(4a-3b)$ **27.** $(3x-7y)(x+2y)$ **29.** $(2x+5)(7x-3)$ **31.** $(3x-5)(2x-11)$ **33.** $(5t-19)(3t-2)$ **35.** $2(2x+3)(x-1)$ **37.** $2(4a-3)(3a-4)$ **39.** $x(5x-4)(2x-3)$ **41.** $x^2(3x+2)(2x-5)$ **43.** $2a(5a+2)(a-1)$ **45.** $3x(5x+1)(x-7)$ **47.** $5y(7y+2)(y-2)$ **49.** $a^2(5a+1)(3a-1)$ **51.** $3y(2x-3)(4x+5)$ **53.** $2y(2x-y)(3x-7y)$ **55.** Both equal 25. **57.** $4x^2-9$ **59.** x^4-81 **61.** $h=2(4-t)(1+8t)$

Time t (seconds)	Height h (feet)
0	8
1	54
2	68
3	50
4	0

63. **a.** $V=x\cdot(11-2x)(9-2x)$ **b.** 11 inch × 9 inch

65. x^2-9 **67.** x^2-25 **69.** x^2-49 **71.** x^2-81 **73.** $4x^2-9y^2$ **75.** x^4-16 **77.** x^2+6x+9 **79.** $x^2+10x+25$ **81.** $x^2+14x+49$ **83.** $x^2+18x+81$ **85.** $4x^2+12x+9$ **87.** $16x^2-16xy+4y^2$

PROBLEM SET 6.4

1. $(x+3)(x-3)$ **3.** $(a+6)(a-6)$ **5.** $(x+7)(x-7)$ **7.** $4(a+2)(a-2)$ **9.** Cannot be factored. **11.** $(5x+13)(5x-13)$ **13.** $(3a+4b)(3a-4b)$ **15.** $(3+m)(3-m)$ **17.** $(5+2x)(5-2x)$ **19.** $2(x+3)(x-3)$ **21.** $32(a+2)(a-2)$ **23.** $2y(2x+3)(2x-3)$ **25.** $(a^2+b^2)(a+b)(a-b)$ **27.** $(4m^2+9)(2m+3)(2m-3)$ **29.** $3xy(x+5y)(x-5y)$ **31.** $(x-1)^2$ **33.** $(x+1)^2$ **35.** $(a-5)^2$ **37.** $(y+2)^2$ **39.** $(x-2)^2$ **41.** $(m-6)^2$ **43.** $(2a+3)^2$ **45.** $(7x-1)^2$ **47.** $(3y-5)^2$ **49.** $(x+5y)^2$ **51.** $(3a+b)^2$ **53.** $3(a+3)^2$ **55.** $2(x+5y)^2$ **57.** $5x(x+3y)^2$ **59.** $(x+3+y)(x+3-y)$ **61.** $(x+y+3)(x+y-3)$ **63.** 14 **65.** 25 **67.** **a.** x^2-16 **b.** $(x+4)(x-4)$ **c.**

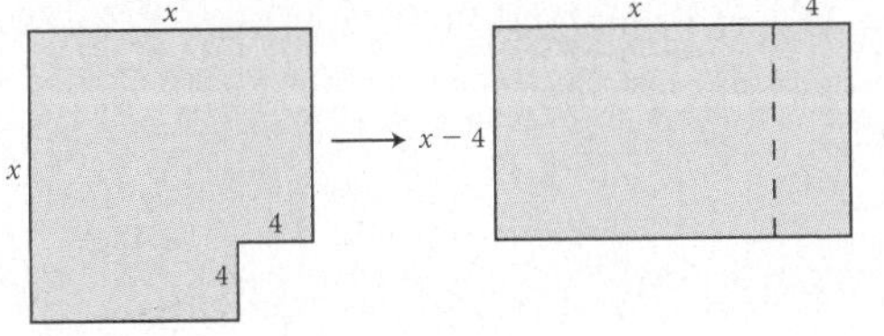

69. $a^2-b^2=(a+b)(a-b)$ **71.** **a.** 1 **b.** 8 **c.** 27 **d.** 64 **e.** 125 **73.** **a.** x^3-x^2+x **b.** x^2-x+1 **c.** x^3+1 **75.** **a.** x^3-2x^2+4x **b.** $2x^2-4x+8$ **c.** x^3+8 **77.** **a.** x^3-3x^2+9x **b.** $3x^2-9x+27$ **c.** x^3+27

PROBLEM SET 6.5

1. $(x-y)(x^2+xy+y^2)$ **3.** $(a+2)(a^2-2a+4)$ **5.** $(3+x)(9-3x+x^2)$ **7.** $(y-1)(y^2+y+1)$ **9.** $(y-4)(y^2+4y+16)$ **11.** $(5h-t)(25h^2+5ht+t^2)$ **13.** $(x-6)(x^2+6x+36)$ **15.** $2(y-3)(y^2+3y+9)$

17. $2(a - 4b)(a^2 + 4ab + 16b^2)$ **19.** $2(x + 6y)(x^2 - 6xy + 36y^2)$ **21.** $10(a - 4b)(a^2 + 4ab + 16b^2)$ **23.** $10(r - 5)(r^2 + 5r + 25)$ **25.** $(4 + 3a)(16 - 12a + 9a^2)$ **27.** $(2x - 3y)(4x^2 + 6xy + 9y^2)$ **29.** $\left(t + \frac{1}{3}\right)\left(t^2 - \frac{1}{3}t + \frac{1}{9}\right)$ **31.** $\left(3x - \frac{1}{3}\right)\left(9x^2 + x + \frac{1}{9}\right)$ **33.** $(4a + 5b)(16a^2 - 20ab + 25b^2)$ **35.** $\left(\frac{1}{2}x - \frac{1}{3}y\right)\left(\frac{1}{4}x^2 + \frac{1}{6}xy + \frac{1}{9}y^2\right)$ **37.** $(a - b)(a^2 + ab + b^2)(a + b)(a^2 - ab + b^2)$ **39.** $(2x - y)(4x^2 + 2xy + y^2)(2x + y)(4x^2 - 2xy + y^2)$ **41.** $(x - 5y)(x^2 + 5xy + 25y^2)(x + 5y)(x^2 - 5xy + 25y^2)$ **43.** $2x^5 - 8x^3$ **45.** $3x^4 - 18x^3 + 27x^2$ **47.** $y^3 + 25y$ **49.** $15a^2 - a - 2$ **51.** $4x^4 - 12x^3 - 40x^2$ **53.** $2ab^5 - 8ab^4 + 2ab^3$

PROBLEM SET 6.6

1. $(x + 9)(x - 9)$ **3.** $(x + 5)(x - 3)$ **5.** $(x + 3)^2$ **7.** $(y - 5)^2$ **9.** $2ab(a^2 + 3a + 1)$ **11.** Cannot be factored. **13.** $3(2a + 5)(2a - 5)$ **15.** $(3x - 2y)^2$ **17.** $4x(x^2 + 4y^2)$ **19.** $2y(y + 5)^2$ **21.** $a^4(a^2 + 4b^2)$ **23.** $(x + 4)(y + 3)$ **25.** $(x^2 + 4)(x + 2)(x - 2)$ **27.** $(x + 2)(y - 5)$ **29.** $5(a + b)^2$ **31.** Cannot be factored. **33.** $3(x + 2y)(x + 3y)$ **35.** $(2x + 19)(x - 2)$ **37.** $100(x - 2)(x - 1)$ **39.** $(x + 8)(x - 8)$ **41.** $(x + a)(x + 3)$ **43.** $a^5(7a + 3)(7a - 3)$ **45.** Cannot be factored. **47.** $a(5a + 1)(5a + 3)$ **49.** $(x + y)(a - b)$ **51.** $3a^2b(4a + 1)(4a - 1)$ **53.** $5x^2(2x + 3)(2x - 3)$ **55.** $(3x + 41y)(x - 2y)$ **57.** $2x^3(2x - 3)(4x - 5)$ **59.** $(2x + 3)(x + a)$ **61.** $(y^2 + 1)(y + 1)(y - 1)$ **63.** $3x^2y^2(2x + 3y)^2$ **65.** 5 **67.** $-\frac{3}{2}$ **69.** $-\frac{3}{4}$

PROBLEM SET 6.7

1. $-2, 1$ **3.** $4, 5$ **5.** $0, -1, 3$ **7.** $-\frac{2}{3}, -\frac{3}{2}$ **9.** $0, -\frac{4}{3}, \frac{4}{3}$ **11.** $0, -\frac{1}{3}, -\frac{3}{5}$ **13.** $-1, -2$ **15.** $4, 5$ **17.** $6, -4$ **19.** $2, 3$ **21.** -3 **23.** $4, -4$ **25.** $\frac{3}{2}, -4$ **27.** $-\frac{2}{3}$ **29.** 5 **31.** $4, -\frac{5}{2}$ **33.** $\frac{5}{3}, -4$ **35.** $\frac{7}{2}, -\frac{7}{2}$ **37.** $0, -6$ **39.** $0, 3$ **41.** $0, 4$ **43.** $0, 5$ **45.** $2, 5$ **47.** $\frac{1}{2}, -\frac{4}{3}$ **49.** $4, -\frac{5}{2}$ **51.** $8, -10$ **53.** $5, 8$ **55.** $6, 8$ **57.** -4 **59.** $5, 8$ **61.** $6, -8$ **63.** $0, -\frac{3}{2}, -4$ **65.** $0, 3, -\frac{5}{2}$ **67.** $0, \frac{1}{2}, -\frac{5}{2}$ **69.** $0, \frac{3}{5}, -\frac{3}{2}$ **71.** $\frac{1}{2}, \frac{3}{2}$ **73.** $-5, 4$ **75.** $-7, -6$ **77.** $-3, -1$ **79.** $2, 3$ **81.** $-15, 10$ **83.** $-5, 3$ **85.** $-3, -2, 2$ **87.** $-4, -1, 4$ **89.** $x(x + 1) = 72$ **91.** $x(x + 2) = 99$ **93.** $x(x + 2) = 5[x + (x + 2)] - 10$ **95.** Bicycle \$75, suit \$15 **97.** House \$2,400, lot \$600

PROBLEM SET 6.8

1. 8, 10 and $-10, -8$ **3.** 9, 11 and $-11, -9$ **5.** 8, 10 and 0, 2 **7.** 8, 6 **9.** 2, 12 and $-\frac{12}{5}, -10$ **11.** 5, 20 and 0, 0 **13.** Width 3 inches, length 4 inches **15.** Base 3 inches **17.** 6 inches and 8 inches **19.** 12 meters **21.** 2 hundred items or 5 hundred items **23.** \$7 or \$10 **25.** **a.** 5 feet **b.** 12 feet **27.** **a.** 25 seconds later **b.**

t	h
0	100
5	1680
10	2460
15	2440
20	1620
25	0

29. $200x^{24}$ **31.** x^7 **33.** 8×10^1 **35.** $10ab^2$ **37.** $6x^4 + 6x^3 - 2x^2$ **39.** $9y^2 - 30y + 25$ **41.** $4a^4 - 49$

CHAPTER 6 TEST

1. $6(x + 3)$ **2.** $4ab(3a - 6 + 2b)$ **3.** $(x - 2b)(x + 3a)$ **4.** $(5y - 4)(3 - x)$ **5.** $(x + 4)(x - 3)$ **6.** $(x - 7)(x + 3)$ **7.** $(x + 5)(x - 5)$ **8.** $(x + 4)(x - 4)$ **9.** Cannot be factored. **10.** $2(3x + 4y)(3y - 4y)$ **11.** $(x^2 - 3)(x + 4)$ **12.** $(x - 3)(x + b)$ **13.** $2(2x - 5)(x + 1)$ **14.** $(4n - 3)(n + 4)$ **15.** $(3c - 2)(4c + 3)$ **16.** $3x(2x - 1)(2x + 3)$ **17.** $(x + 5y)(x^2 - 5xy + 25y^2)$ **18.** $2(3b - 4)(9b^2 + 12b + 16)$ **19.** $5, -3$ **20.** $3, 4$ **21.** $5, -5$ **22.** $-2, 7$ **23.** $5, -6$ **24.** $0, 3, -3$ **25.** $\frac{3}{2}, -4$ **26.** $0, -\frac{5}{3}, 6$ **27.** $6, 12$ **28.** $\{4, 6\}, \{-4, -2\}$ **29.** 4, 13 ft **30.** 6, 8 ft **31.** 3, 2 **32.** \$3, \$5

Answers to Odd-Numbered Problems

Chapter 7

PROBLEM SET 7.1

1. ± 5 **3.** $\pm 3i$ **5.** $\pm\frac{\sqrt{3}}{2}$ **7.** $\pm 2i\sqrt{3}$ **9.** $\pm\frac{3\sqrt{5}}{2}$ **11.** $-2, 3$ **13.** $\frac{-3 \pm 3i}{2}$ **15.** $\frac{-2 \pm 2i\sqrt{2}}{5}$ **17.** $-4 \pm 3i\sqrt{3}$ **19.** $\frac{3 \pm 2i}{2}$ **21.** $36, 6$ **23.** $4, 2$ **25.** $25, 5$ **27.** $\frac{25}{4}, \frac{5}{2}$ **29.** $\frac{49}{4}, \frac{7}{2}$ **31.** $\frac{1}{16}, \frac{1}{4}$ **33.** $\frac{1}{9}, \frac{1}{3}$ **35.** $-6, 2$ **37.** $-3, -9$ **39.** $1 \pm 2i$ **41.** $4 \pm \sqrt{15}$

43. $\frac{5 \pm \sqrt{37}}{2}$ **45.** $1 \pm \sqrt{5}$ **47.** $\frac{4 \pm \sqrt{13}}{3}$ **49.** $\frac{3 \pm i\sqrt{71}}{8}$ **51.** $\frac{-2 \pm \sqrt{7}}{3}$ **53.** $\frac{5 \pm \sqrt{47}}{2}$ **55.** $\frac{5 \pm i\sqrt{19}}{4}$ **57. a.** No **b.** $\pm 3i$ **59. a.** $0, 6$ **b.** $0, 6$ **61. a.** $-7, 5$ **b.** $-7, 5$ **63.** No **65. a.** $\frac{7}{5}$ **b.** 3 **c.** $\frac{7 \pm 2\sqrt{2}}{5}$ **d.** $\frac{71}{5}$ **e.** 3 **67.** $\frac{\sqrt{3}}{2}$ inch, 1 inch **69.** $\sqrt{2}$ inches **71.** 781 feet **73.** 7.3% to the nearest tenth **75.** $20\sqrt{2} \approx 28$ feet **77.** 169 **79.** 49 **81.** $\frac{85}{12}$ **83.** $(3t - 2)(9t^2 + 6t + 4)$

PROBLEM SET 7.2

1. $-3, -2$ **3.** $2 \pm \sqrt{3}$ **5.** $1, 2$ **7.** $\frac{2 \pm i\sqrt{14}}{3}$ **9.** $0, 5$ **11.** $0, -\frac{4}{3}$ **13.** $\frac{3 \pm \sqrt{5}}{4}$ **15.** $-3 \pm \sqrt{17}$ **17.** $\frac{-1 \pm i\sqrt{5}}{2}$ **19.** 1 **21.** $\frac{1 \pm i\sqrt{47}}{6}$ **23.** $4 \pm \sqrt{2}$ **25.** $\frac{1}{2}, 1$ **27.** $-\frac{1}{2}, 3$ **29.** $\frac{-1 \pm i\sqrt{7}}{2}$ **31.** $1 \pm \sqrt{2}$ **33.** $\frac{-3 \pm \sqrt{5}}{2}$ **35.** $3, -5$ **37.** $2, -1 \pm i\sqrt{3}$ **39.** $-\frac{3}{2}, \frac{3 \pm 3i\sqrt{3}}{4}$ **41.** $\frac{1}{5}, \frac{-1 \pm i\sqrt{3}}{10}$ **43.** $0, \frac{-1 \pm i\sqrt{5}}{2}$ **45.** $0, 1 \pm i$ **47.** $0, \frac{-1 \pm i\sqrt{2}}{3}$ **49.** a and b **51. a.** $\frac{5}{3}, 0$ **b.** $\frac{5}{3}, 0$ **53.** No, $2 \pm i\sqrt{3}$ **55.** Yes **57.** 2 seconds **59.** 20 or 60 items **61.** 169 **63.** 0 **65.** ± 12 **67.** $x^2 - x - 6$ **69.** $x^3 - 4x^2 - 3x + 18$

PROBLEM SET 7.3

1. D = 16, two rational **3.** D = 0, one rational **5.** D = 5, two irrational **7.** D = 17, two irrational **9.** D = 36, two rational

11. D = 116, two irrational **13.** ± 10 **15.** ± 12 **17.** 9 **19.** -16 **21.** $\pm 2\sqrt{6}$ **23.** $x^2 - 7x + 10 = 0$ **25.** $t^2 - 3t - 18 = 0$ **27.** $y^3 - 4y^2 - 4y + 16 = 0$ **29.** $2x^2 - 7x + 3 = 0$ **31.** $4t^2 - 9t - 9 = 0$ **33.** $6x^3 - 5x^2 - 54x + 45 = 0$ **35.** $10a^2 - a - 3 = 0$ **37.** $9x^3 - 9x^2 - 4x + 4 = 0$ **39.** $x^4 - 13x^2 + 36 = 0$ **41.** $x^2 - 7 = 0$ **43.** $x^2 + 25 = 0$ **45.** $x^2 - 2x + 2 = 0$ **47.** $x^2 + 4x + 13 = 0$ **49.** $x^3 + 7x^2 - 5x - 75 = 0$ **51.** $x^4 - 18x^2 + 81 = 0$ **53.** $-3, -2, -1$ **55.** $-4, -3, 2$ **57.** $1 \pm i$ **59.** $5, 4 \pm 3i$ **61.** $x^2 + 4x - 5$ **63.** $4a^2 - 30a + 56$ **65.** $32a^2 + 20a - 18$ **67.** $\pm\frac{1}{2}$ **69.** No solution **71.** 1 **73.** $-2, 4$ **75.** $-2, \frac{1}{4}$

PROBLEM SET 7.4

1. $1, 2$ **3.** $-8, -\frac{5}{2}$ **5.** $\pm 3, \pm i\sqrt{3}$ **7.** $\pm 2i, \pm i\sqrt{5}$ **9.** $\frac{7}{2}, 4$ **11.** $-\frac{9}{8}, \frac{1}{2}$ **13.** $\pm\frac{\sqrt{30}}{6}, \pm i$ **15.** $\pm\frac{\sqrt{21}}{3}, \pm\frac{i\sqrt{21}}{3}$ **17.** 4, 25

19. only 25 checks **21.** only $\frac{25}{9}$ checks **23.** 27, 38 **25.** 4, 12 **27.** $t = \frac{v \pm \sqrt{v^2 + 64h}}{32}$ **29.** $x = \frac{-4 \pm 2\sqrt{4 - k}}{k}$ **31.** $x = -y$ **33.** $t = \frac{1 \pm \sqrt{1 + h}}{4}$ **35. a.**

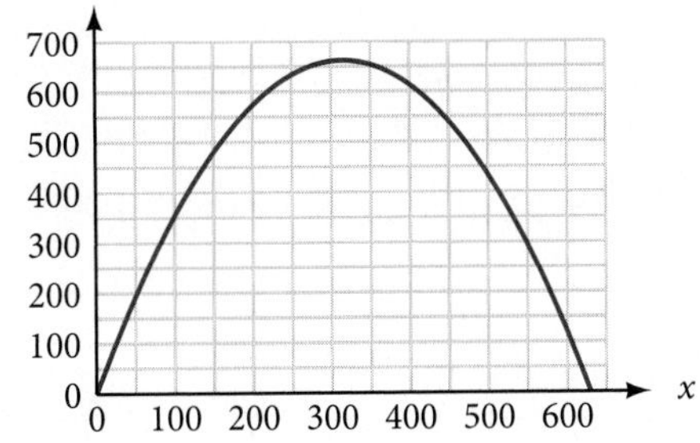

b. 630 ft.

37. **a.** $l + 2w = 160$ **b.** $A = -2w^2 + 160w$ **c.**

w	l	A
50	60	3,000
45	70	3,150
40	80	3,200
35	90	3,150
30	100	3,000

d. 3,200 square yards

39. -2 **41.** 1,322.5 **43.** $-\frac{7}{640}$ **45.** 1, 5 **47.** -3, 1 **49.** $\frac{3}{2} \pm \frac{1}{2}i$ **51.** 9, 3 **53.** 1, 1

PROBLEM SET 7.5

1. x-intercepts $= -3, 1$; vertex $= (-1, -4)$

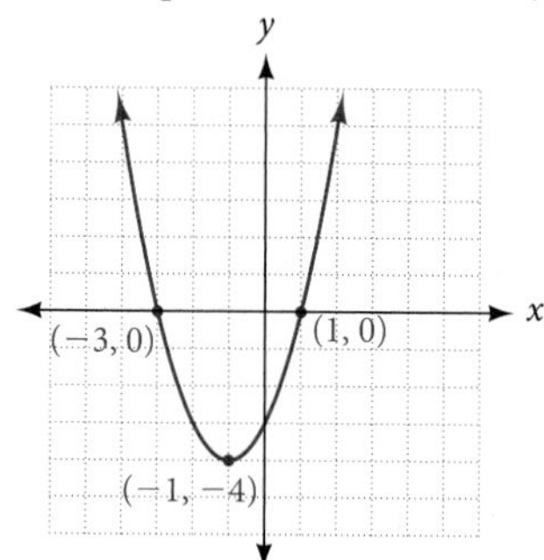

3. x-intercepts $= -5, 1$; vertex $= (-2, 9)$

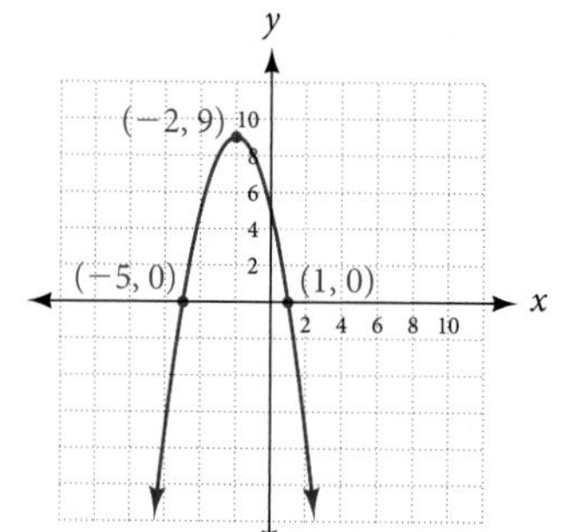

5. x-intercepts $= -1, 1$; vertex $= (0, -1)$

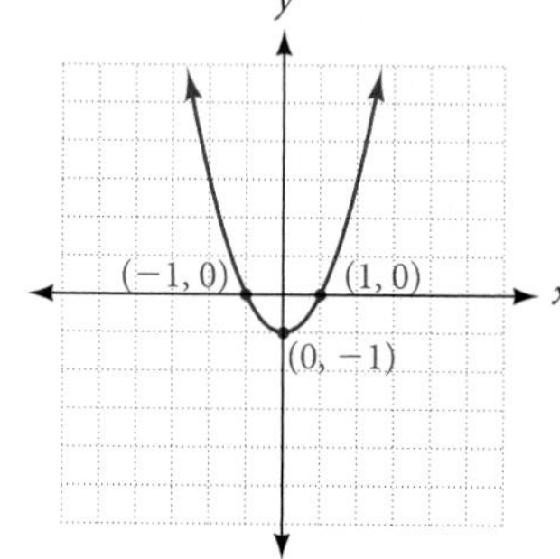

7. x-intercepts $= -3, 3$; vertex $= (0, 9)$

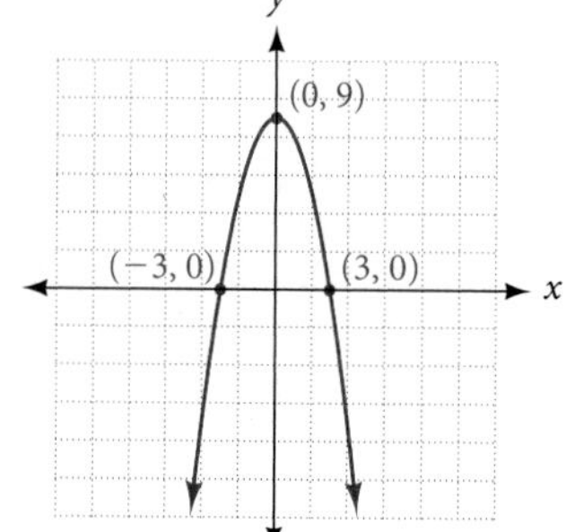

9. x-intercepts $= -1, 3$; vertex $= (1, -8)$

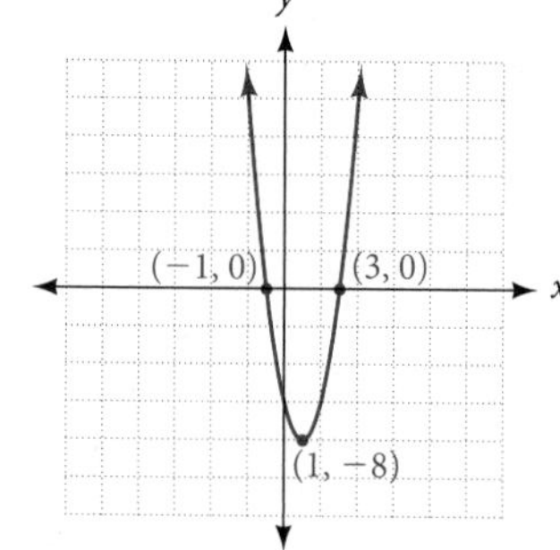

11. x-intercepts $= 1 - \sqrt{5}, 1 + \sqrt{5}$; vertex $= (1, -5)$

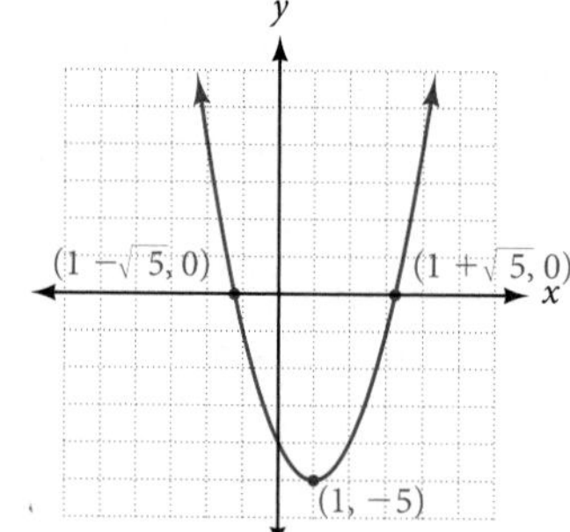

13.

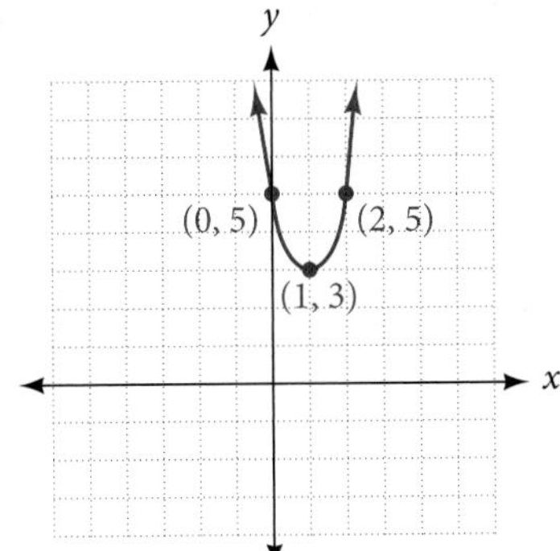

15.

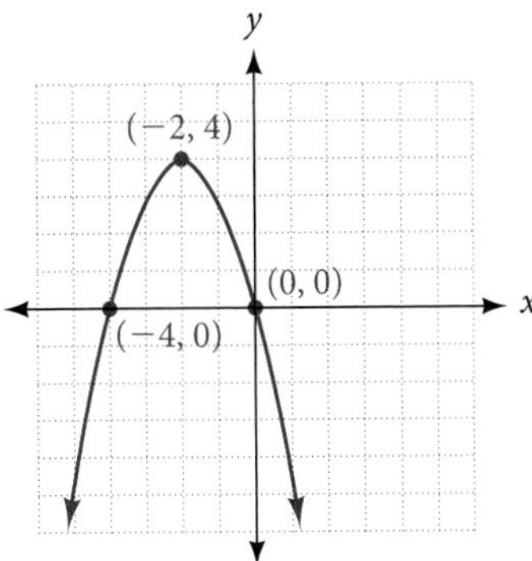

17.

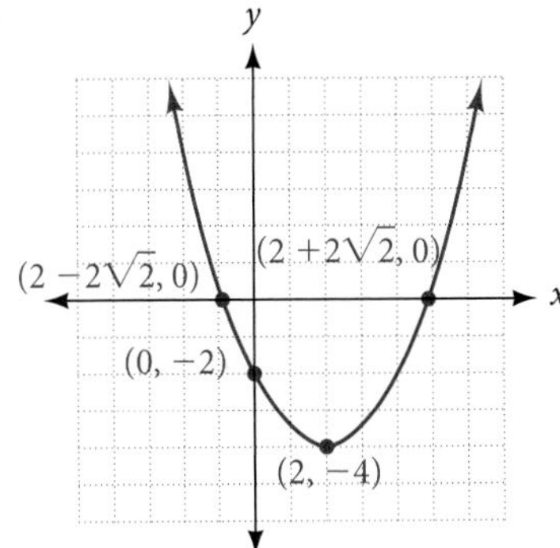

19.

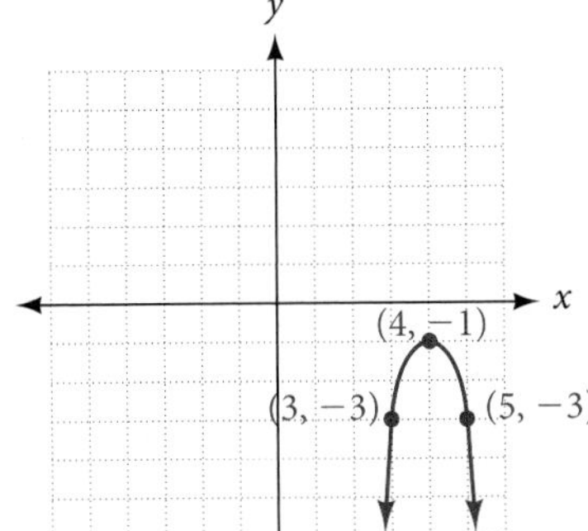

21. vertex = $(2, -8)$

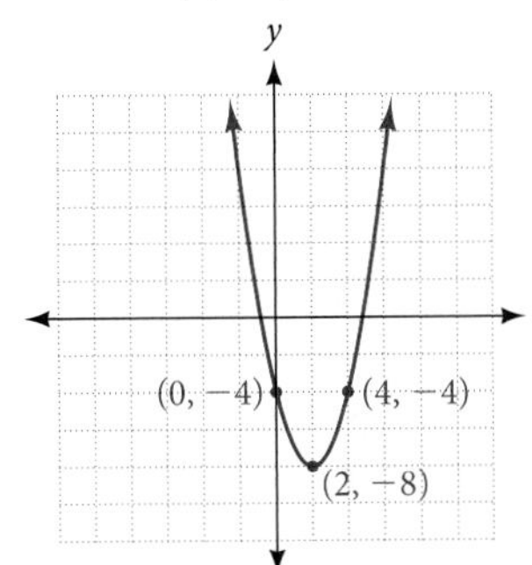

23. vertex = $(1, -4)$

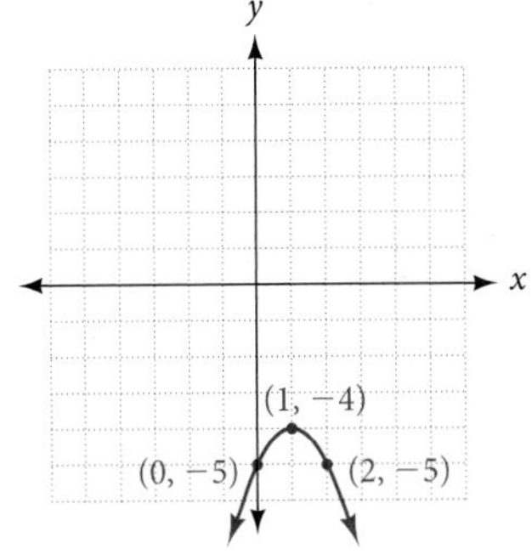

25. vertex = $(0, 1)$

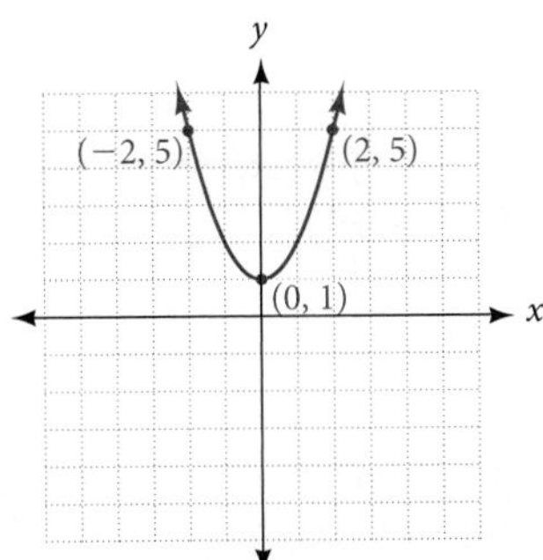

27. vertex = $(0, -3)$

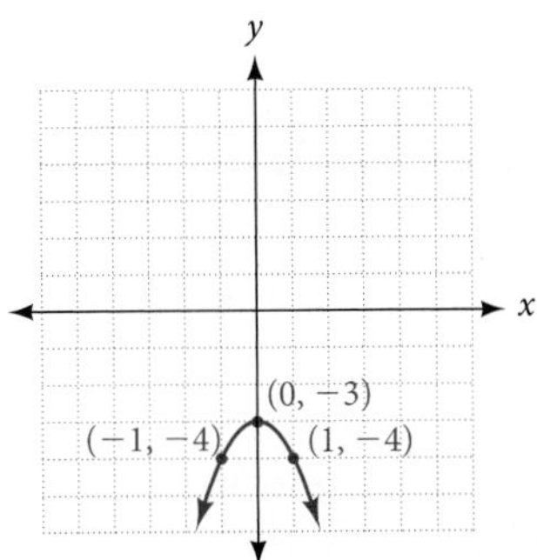

29. vertex = $\left(-\frac{2}{3}, -\frac{1}{3}\right)$

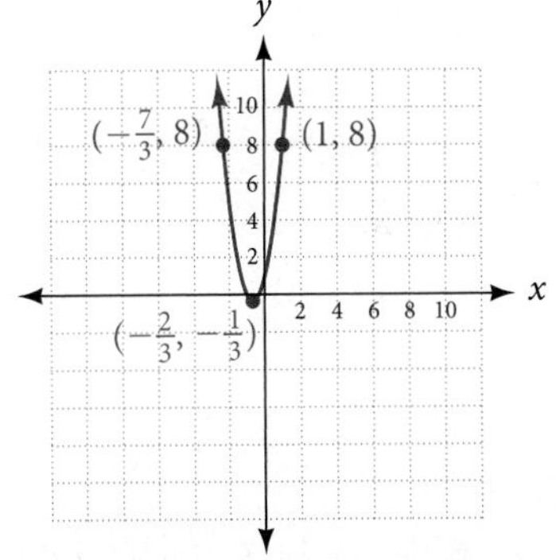

31. $(3, -4)$ lowest 33. $(1, 9)$ highest 35. $(2, 16)$ highest 37. $(-4, 16)$ highest 39. 875 patterns; maximum profit \$731.25

41. The ball is in her hand when $h(t) = 0$, which means $t = 0$ or $t = 2$ seconds. Maximum height is $h(1) = 16$ feet.

43. Maximum $R = \$3{,}600$ when $p = \$6.00$

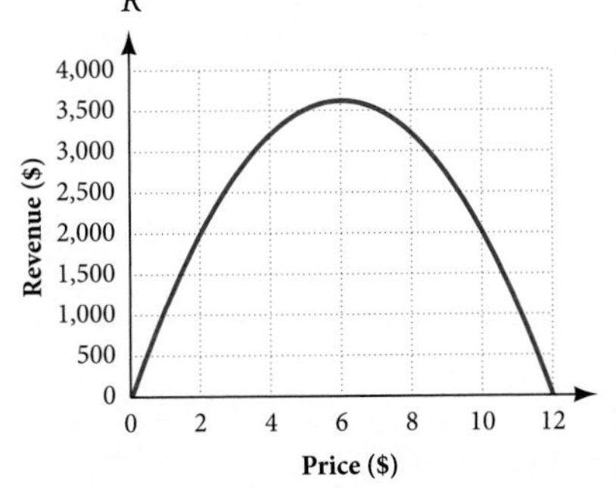

45. Maximum $R = \$7{,}225$ when $p = \$8.50$

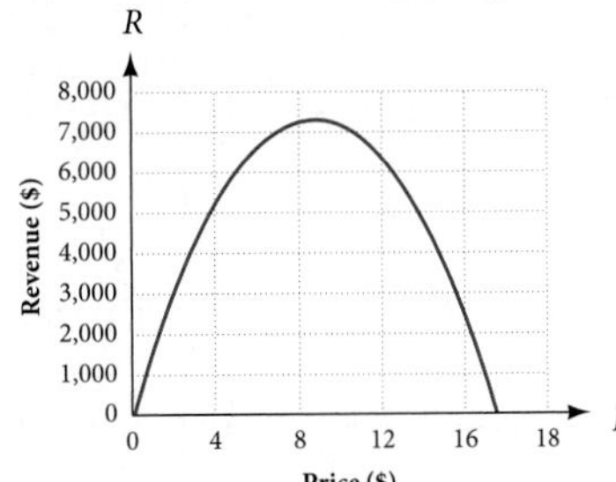

47. $y = -\frac{1}{135}(x - 90)^2 + 60$

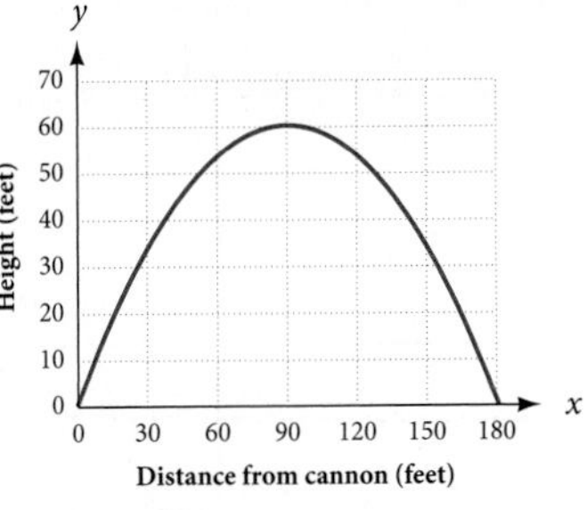

49. $-2, 4$ 51. $-\frac{1}{2}, \frac{2}{3}$ 53. 3

PROBLEM SET 7.6

1. 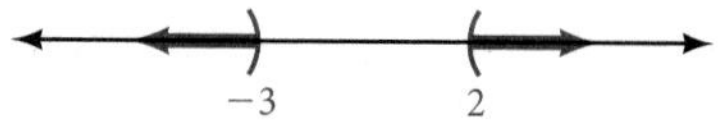

3. 

5. 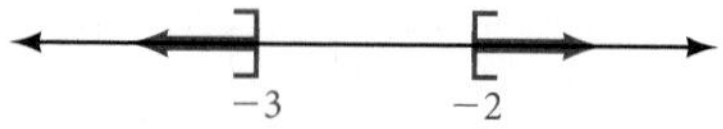

7. 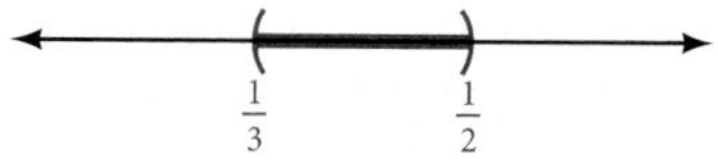

9. 

11. 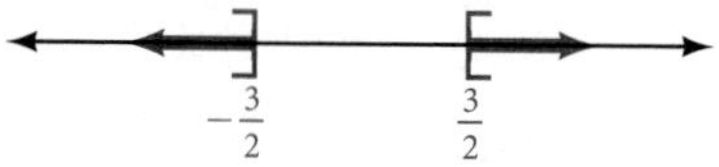

13. 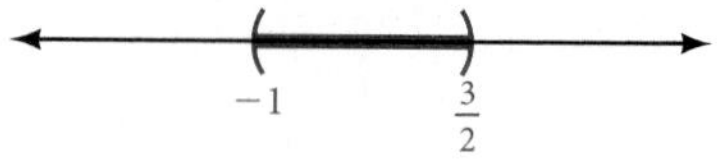

15. All real numbers 

17. ∅ 

19. 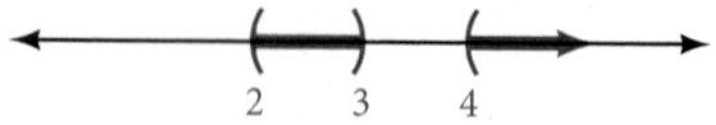

21. 

23. 

25. 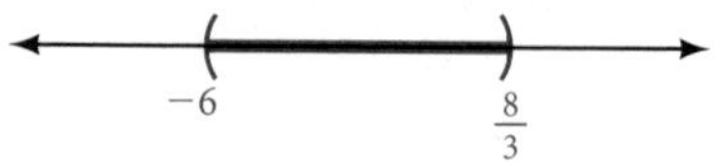

27. 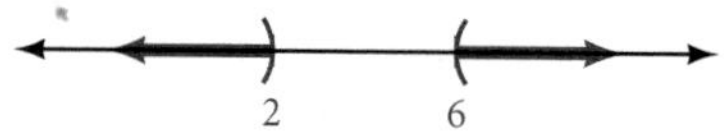

29. 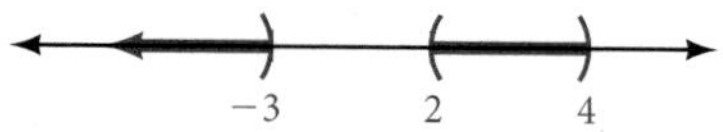

31. 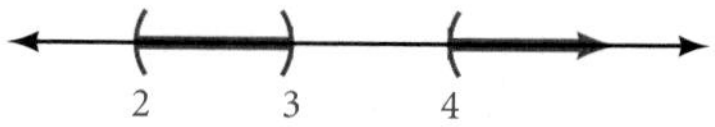

33. 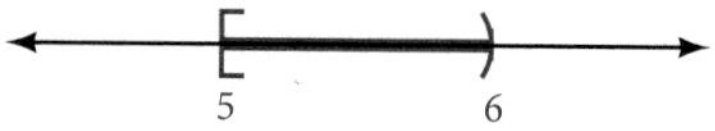

35. **a.** $-2 < x < 2$ **b.** $x < -2$ or $x > 2$ **c.** $x = -2$ or $x = 2$ **37.** **a.** $-2 < x < 5$ **b.** $x < -2$ or $x > 5$ **c.** $x = -2$ or $x = 5$
39. **a.** $x < -1$ or $1 < x < 3$ **b.** $-1 < x < 1$ or $x > 3$ **c.** $x = -1$ or $x = 1$ or $x = 3$ **41.** $x \geq 4$; the width is at least 4 inches
43. $5 \leq p \leq 8$; she should charge at least \$5 but no more than \$8 for each radio **45.** \$300, \$1,800,000 **47.** \$30
49. 1.5625 **51.** 0.6549 **53.** $\frac{2}{3}$ **55.** Possible solutions 1 and 6; only 6 checks; 6 **57.**

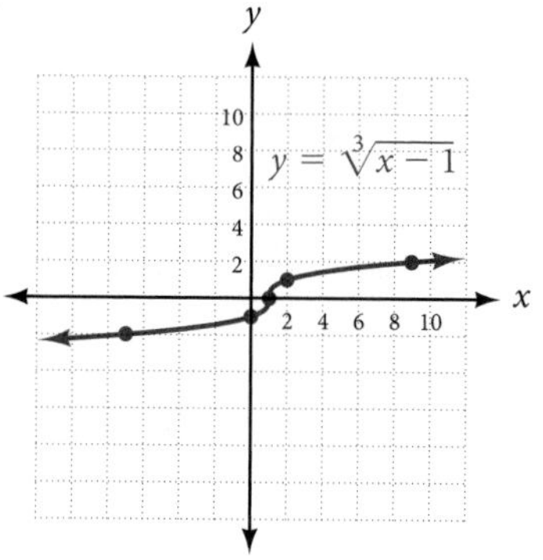

CHAPTER 7 TEST

1. $-\frac{9}{2}, \frac{1}{2}$ **2.** $3 \pm i\sqrt{2}$ **3.** $5 \pm 2i$ **4.** $1 \pm i\sqrt{2}$ **5.** $\frac{5}{2}, \frac{-5 \pm 5i\sqrt{3}}{4}$ **6.** $-1 \pm i\sqrt{5}$ **7.** $r = \pm\frac{\sqrt{A}}{8} - 1$ **8.** $2 \pm \sqrt{2}$
9. $\frac{1}{2}$ or $\frac{3}{2}$ sec **10.** 15 or 100 cups **11.** 9 **12.** $D = 81$; two rational solutions **13.** $3x^2 - 13x - 10 = 0$
14. $x^3 - 7x^2 - 4x + 28 = 0$ **15.** $\pm\sqrt{2}, \pm\frac{1}{2}i$ **16.** $\frac{1}{2}, 1$ **17.** $\frac{1}{4}, 9$ **18.** $t = \frac{7 + \sqrt{49 + 16h}}{16}$

19. vertex: (1, −4)

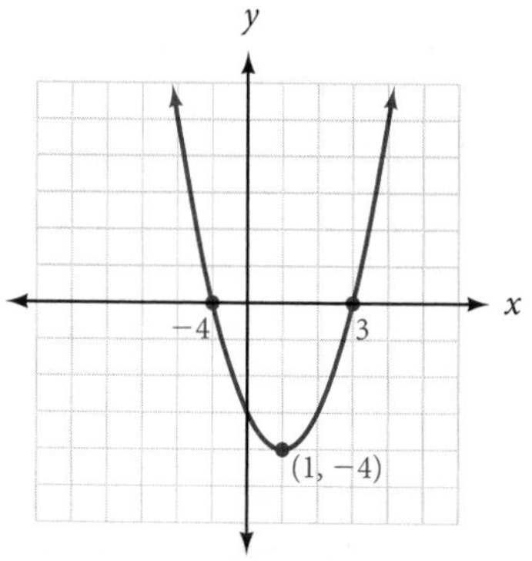

20. vertex: (1, 9)

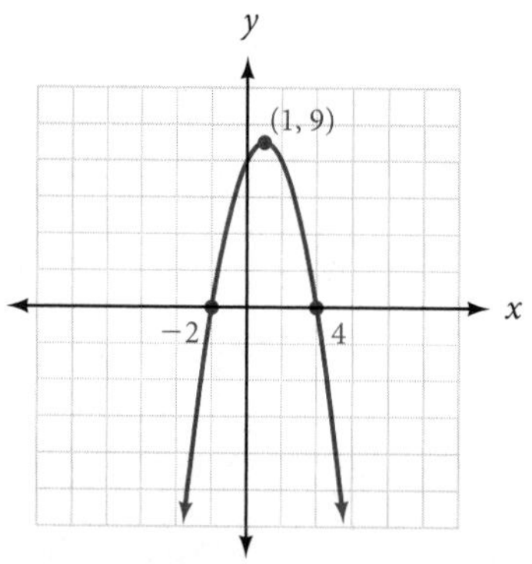

21. $-2 \le x \le 3$ **22.** $x < -3$ or $x > \frac{1}{2}$ **23.** profit = \$900

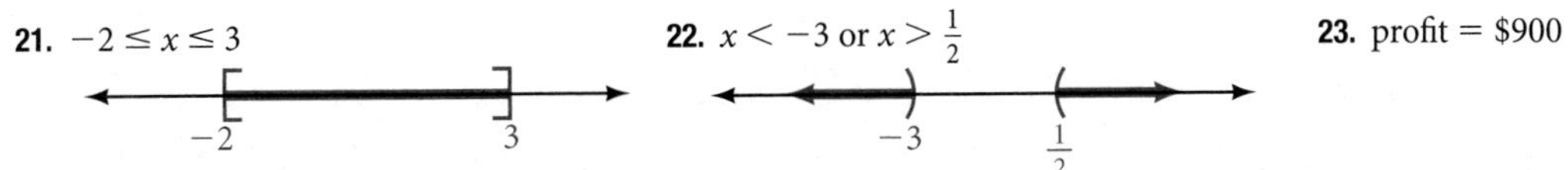

Chapter 8

PROBLEM SET 8.1

1. $g(0) = -3, g(-3) = 0, g(3) = 3, g(-1) = -1, g(1) =$ undefined

3. $h(0) = -3, h(-3) = 3, h(3) = 0, h(-1)$ is undefined, $h(1) = -1$ **5.** $\{x \mid x \neq 1\}$ **7.** $\{x \mid x \neq 2\}$ **9.** $\{t \mid t \neq 4, t \neq -4\}$

11. $\frac{x-4}{6}$ **13.** $(a^2+9)(a+3)$ **15.** $\frac{2y+3}{y+1}$ **17.** $\frac{x-2}{x-1}$ **19.** $\frac{x-3}{x+2}$ **21.** $\frac{x^2-x+1}{x-1}$ **23.** $-\frac{4a}{3}$ **25.** $\frac{b-1}{b+1}$ **27.** $\frac{7x-3}{7x+5}$ **29.** $\frac{4x+3}{4x-3}$ **31.** $\frac{x+5}{2x-7}$ **33.** $\frac{a^2-ab+b^2}{a-b}$ **35.** $\frac{2x-2}{x}$ **37.** $\frac{x+3}{y-4}$ **39.** $x+2$ **41.** $\frac{x^2+2x+4}{x+2}$ **43.** $\frac{4x^2+6x+9}{2x+3}$ **45.** -1

47. $-(y+6)$ **49.** $-\frac{3a+1}{3a-1}$ **51.** 3 **53.** $x+a$ **55.** **a.** 4 **b.** 4 **57.** **a.** 5 **b.** 5 **59.** **a.** $x+a$ **b.** $2x+h$

61. **a.** $x+a$ **b.** $2x+h$ **63.** **a.** $x+a-3$ **b.** $2x+h-3$ **65.** **a.** 2 **b.** -4 **c.** Undefined **d.** 2

67.

Weeks x	Weight (lb) $W(x)$
0	200
1	194
4	184
12	173
24	168

69. $\frac{2}{3}$ **71.** $20x^2y^2$ **73.** $72x^4y^5$ **75.** $(x+2)(x-2)$ **77.** $x^2(x-y)$ **79.** $2(y+1)(y-1)$

PROBLEM SET 8.2

1. $\frac{1}{6}$ **3.** $\frac{9}{4}$ **5.** $\frac{1}{2}$ **7.** $\frac{15y}{x^2}$ **9.** $\frac{b}{a}$ **11.** $\frac{2y^5}{z^3}$ **13.** $\frac{x+3}{x+2}$ **15.** $y+1$ **17.** $\frac{3(x+4)}{x-2}$ **19.** $\frac{y^2}{xy+1}$ **21.** $\frac{x^2+9}{x^2-9}$ **23.** $\frac{1}{4}$ **25.** 1

27. $\frac{(a-2)(a+2)}{a-5}$ **29.** $\frac{9t^2-6t+4}{4t^2-2t+1}$ **31.** $\frac{x+3}{x+4}$ **33.** $\frac{a-b}{5}$ **35.** $\frac{5c-1}{3c-2}$ **37.** $\frac{5a-b}{9a^2+15ab+25b^2}$ **39.** 2 **41.** $x(x-1)(x+1)$

43. $\frac{(a+4b)(a-3b)}{(a-4b)(a+5b)}$ **45.** $\frac{2y-1}{2y-3}$ **47.** $\frac{(y-2)(y+1)}{(y+2)(y-1)}$ **49.** $\frac{x-1}{x+1}$ **51.** $\frac{x-2}{x+3}$ **53.** $\frac{w(y-1)}{w-x}$ **55.** $\frac{(m+2)(x+y)}{(2x+y)^2}$

57. $3x$ **59.** $2(x+5)$ **61.** $x-2$ **63.** $-(y-4)$ or $4-y$ **65.** $(a-5)(a+1)$

67. **a.** $\frac{5}{21}$ **b.** $\frac{5x+3}{25x^2+15x+9}$ **c.** $\frac{5x-3}{25x^2+15x+9}$ **d.** $\frac{5x+3}{5x-3}$ **69.** $\frac{2}{3}$ **71.** $\frac{47}{105}$ **73.** $x-7$ **75.** $(x+1)(x-1)$ **77.** $2(x+5)$

79. $(a-b)(a^2+ab+b^2)$

PROBLEM SET 8.3

1. $\frac{5}{4}$ **3.** $\frac{1}{3}$ **5.** $\frac{41}{24}$ **7.** $\frac{19}{144}$ **9.** $\frac{31}{24}$ **11.** 1 **13.** -1 **15.** $\frac{1}{x+y}$ **17.** 1 **19.** $\frac{a^2+2a-3}{a^3}$ **21.** 1

23. **a.** $\frac{1}{16}$ **b.** $\frac{9}{4}$ **c.** $\frac{13}{24}$ **d.** $\frac{5x+15}{(x-3)^2}$ **e.** $\frac{x+3}{5}$ **f.** $\frac{x-2}{x-3}$ **25.** $\frac{1}{2}$ **27.** $\frac{1}{5}$ **29.** $\frac{x+3}{2(x+1)}$ **31.** $\frac{a-b}{a^2+ab+b^2}$ **33.** $\frac{2y-3}{4y^2+6y+9}$

35. $\frac{2(2x-3)}{(x-3)(x-2)}$ **37.** $\frac{1}{2t-7}$ **39.** $\frac{4}{(a-3)(a+1)}$ **41.** $\frac{-4x^2}{(2x+1)(2x-1)(4x^2+2x+1)}$ **43.** $\frac{2}{(2x+3)(4x+3)}$ **45.** $\frac{a}{(a+4)(a+5)}$

47. $\frac{x+1}{(x-2)(x+3)}$ **49.** $\frac{x-1}{(x+1)(x+2)}$ **51.** $\frac{1}{(x+2)(x+1)}$ **53.** $\frac{1}{(x+2)(x+3)}$ **55.** $\frac{4x+5}{2x+1}$ **57.** $\frac{22-5t}{4-t}$ **59.** $\frac{2x^2+3x-4}{2x+3}$

61. $\frac{2x-3}{2x}$ **63.** $\frac{1}{2}$ **65.** $\frac{3}{x+4}$ **67.** $\frac{(2x+1)(x+5)}{(x-2)(x+1)(x+3)}$ **69.** $\frac{51}{10} = 5.1$ **71.** $x + \frac{4}{x} = \frac{x^2+4}{x}$ **73.** $\frac{1}{x} + \frac{1}{x+1} = \frac{2x+1}{x(x+1)}$

75. $\frac{6}{5}$ **77.** $x+2$ **79.** $3-x$ **81.** $(x+2)(x-2)$

PROBLEM SET 8.4

1. $\frac{9}{8}$ **3.** $\frac{2}{15}$ **5.** $\frac{119}{20}$ **7.** $\frac{1}{x+1}$ **9.** $\frac{a+1}{a-1}$ **11.** $\frac{y-x}{y+x}$ **13.** $\frac{1}{(x+5)(x-2)}$ **15.** $\frac{1}{a^2-a+1}$ **17.** $\frac{x+3}{x+2}$ **19.** $\frac{a+3}{a-2}$ **21.** $\frac{x-3}{x}$

23. $\frac{x+4}{x+2}$ **25.** $\frac{x-3}{x+3}$ **27.** $\frac{a-1}{a+1}$ **29.** $-\frac{x}{3}$ **31.** $\frac{y^2+1}{2y}$ **33.** $\frac{-x^2+x-1}{x-1}$ **35.** $\frac{5}{3}$ **37.** $\frac{2x-1}{2x+3}$ **39.** $-\frac{1}{x(x+h)}$ **41.** $\frac{3c+4a-2b}{5}$

43. $\frac{(t-4)(t+1)}{(t+6)(t-3)}$ **45.** $\frac{(5b-1)(b+5)}{2(2b-11)}$ **47.** $-\frac{3}{2x+14}$ **49.** $2m-9$ **51. a.** $\frac{-4}{ax}$ **b.** $\frac{-1}{(x+1)(a+1)}$ **c.** $-\frac{a+x}{a^2x^2}$

53. a. As v approaches 0, the denominator approaches 1 **b.** $v = \frac{fs}{h} - s$ **55.** $xy - 2x$ **57.** $3x - 18$ **59.** ab

61. $(y+5)(y-5)$ **63.** $x(a+b)$ **65.** 2

PROBLEM SET 8.5

1. $-\frac{35}{3}$ **3.** $-\frac{18}{5}$ **5.** $\frac{36}{11}$ **7.** 2 **9.** 5 **11.** 2 **13.** $-3, 4$ **15.** $1, -\frac{4}{3}$ **17.** Possible solution -1, which does not check; $\varnothing$

19. 5 **21.** $-\frac{1}{2}, \frac{5}{3}$ **23.** $\frac{2}{3}$ **25.** 18 **27.** Possible solution 4, which does not check; $\varnothing$

29. Possible solutions 3 and -4; only -4 checks; -4 **31.** -6 **33.** -5 **35.** $\frac{53}{17}$

37. Possible solutions 1 and 2; only 2 checks; 2 **39.** Possible solution 3, which does not check; $\varnothing$ **41.** $\frac{22}{3}$

43. a. $-\frac{9}{5}, 5$ **b.** $\frac{9}{2}$ **c.** no solution **45. a.** $\frac{1}{3}$ **b.** 3 **c.** 9 **d.** 4 **e.** $\frac{1}{3}, 3$ **47. a.** $\frac{6}{(x-4)(x+3)}$ **b.** $\frac{x-3}{x-4}$ **c.** 5

49. $x = \frac{ab}{a-b}$ **51.** $y = \frac{x-3}{x-1}$ **53.** $y = \frac{1-x}{3x-2}$

55.

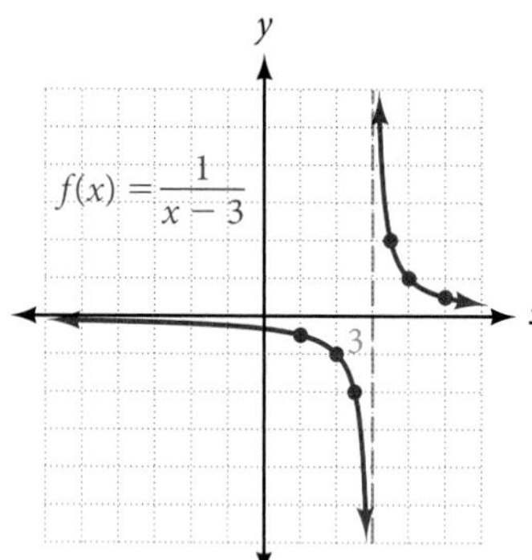

57.

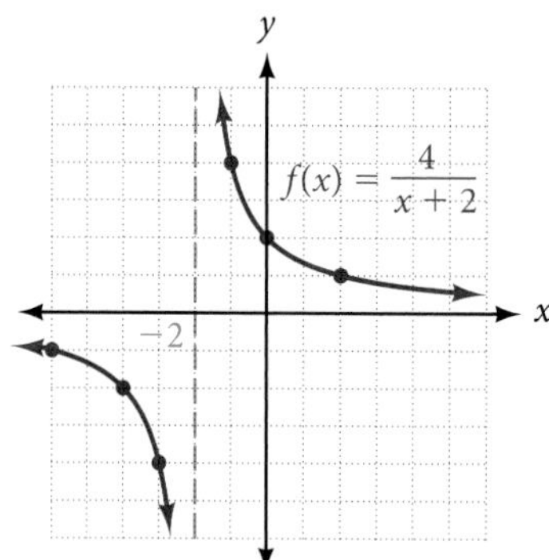

59.

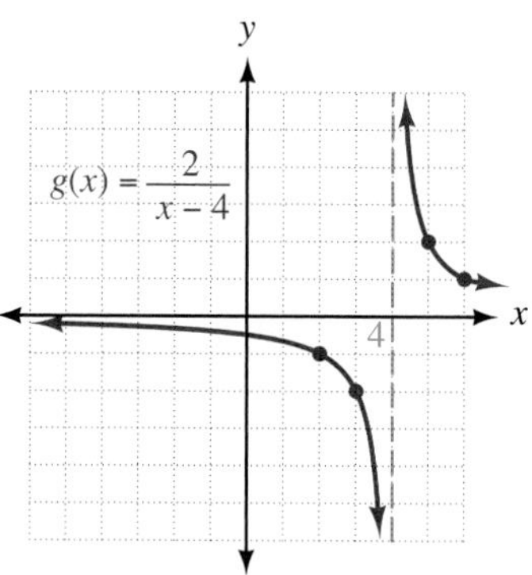

61.

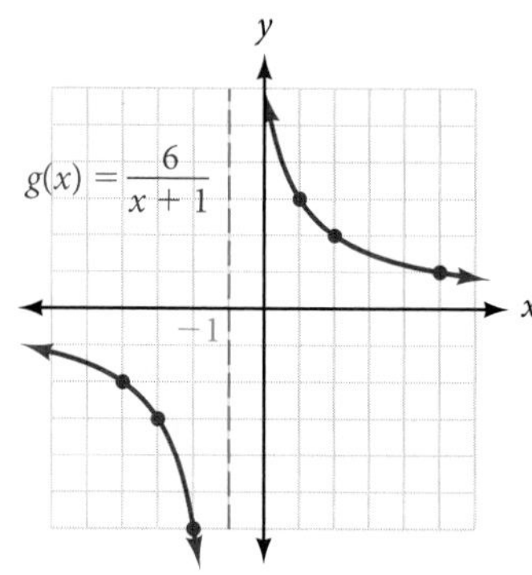

63. $\frac{24}{5}$ feet **65.** 2,358 **67.** 12.3 **69.** 3 **71.** $9, -1$ **73.** 60

PROBLEM SET 8.6

1. $\frac{1}{x} + \frac{1}{3x} = \frac{20}{3}$; $\frac{1}{5}$ and $\frac{3}{5}$ **3.** $x + \frac{1}{x} = \frac{10}{3}$; 3 or $\frac{1}{3}$ **5.** $\frac{1}{x} + \frac{1}{x+1} = \frac{7}{12}$; 3, 4 **7.** $\frac{7+x}{9+x} = \frac{5}{6}$; 3

9. a. and **b.**

	d	r	t
Upstream	1.5	$5 - c$	$\frac{1.5}{5-c}$
Downstream	3	$5 + c$	$\frac{3}{5+c}$

c. They are the same. $\frac{1.5}{5-x} = \frac{3}{5+x}$

d. The speed of the current is 1.7 mph

11. $\frac{8}{x+2} + \frac{8}{x-2} = 3$; 6 mph **13. a.** and **b.**

	d	r	t
Train A	150	$x + 15$	$\frac{150}{x+15}$
Train B	120	x	$\frac{120}{x}$

c. They are the same, $\frac{150}{x+5} = \frac{120}{x}$ **d.** The speed of the train A is 75 mph, train B is 60 mph

15. 540 mph **17.** 54 mph **19.** 16 hours **21.** 15 hours **23.** 5.25 minutes **25.** $10 = \frac{1}{3}\left[\left(x + \frac{2}{3}x\right) + \frac{1}{3}\left(x + \frac{2}{3}x\right)\right]; x = \frac{27}{2}$

27. **a.** 30 grams **b.** 3.25 moles

29.

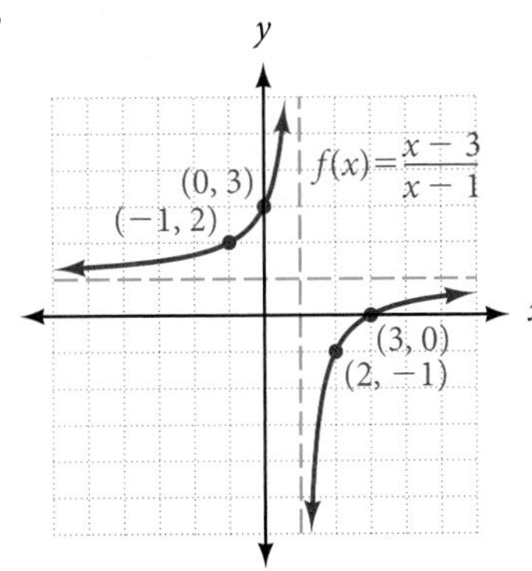

31.

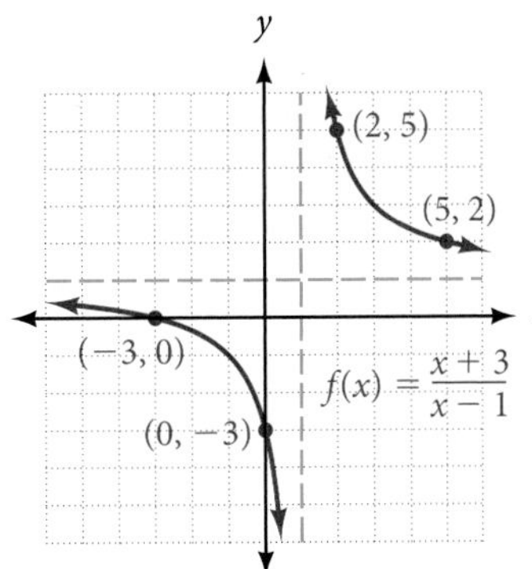

33.

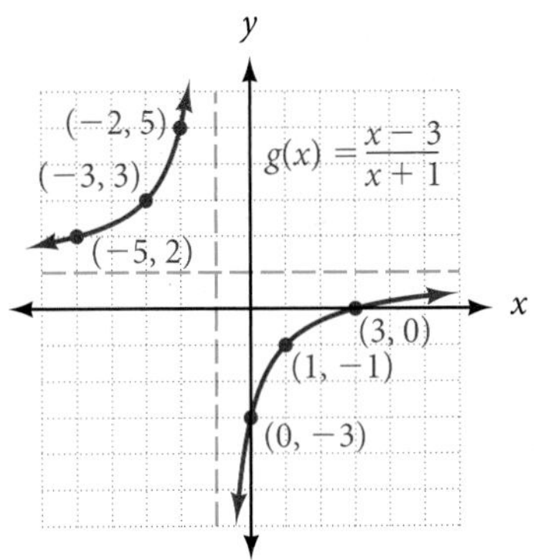

35. 2 **37.** $-2x^2y^2$ **39.** 185.12 **41.** $4x^3 - 8x^2$ **43.** $4x^3 - 6x - 20$ **45.** $-3x + 9$ **47.** $(x + a)(x - a)$ **49.** $(x - 7y)(x + y)$

PROBLEM SET 8.7

1. $2x^2 - 4x + 3$ **3.** $-2x^2 - 3x + 4$ **5.** $2y^2 + \frac{5}{2} - \frac{3}{2y^2}$ **7.** $-\frac{5}{2}x + 4 + \frac{3}{x}$ **9.** $4ab^3 + 6a^2b$ **11.** $-xy + 2y^2 + 3xy^2$

13. $x + 2$ **15.** $a - 3$ **17.** $5x + 6y$ **19.** $x^2 + xy + y^2$ **21.** $(y^2 + 4)(y + 2)$ **23.** $(x + 2)(x + 5)$ **25.** $(2x + 3)(2x - 3)$

27. $x - 7 + \frac{7}{x+2}$ **29.** $2x + 5 + \frac{2}{3x-4}$ **31.** $2x^2 - 5x + 1 + \frac{4}{x+1}$ **33.** $y^2 - 3y - 13$ **35.** $x - 3$ **37.** $3y^2 + 6y + 8 + \frac{37}{2y-4}$

39. $a^3 + 2a^2 + 4a + 6 + \frac{17}{a-2}$ **41.** $y^3 + 2y^2 + 4y + 8$ **43.** $x^2 - 2x + 1$ **45.** $(x + 3)(x + 2)(x + 1)$

47. $(x + 3)(x + 4)(x - 2)$ **49.** yes **51.** same **53.** **a.** $(x - 2)(x^2 - x + 3)$ **b.** $(x - 5)(x^3 - x + 1)$

55. **a.**

x	1	5	10	15	20
C(x)	2.15	2.75	3.50	4.25	5.00

b. $\overline{C}(x) = \frac{2 + 0.15}{x}$ **c.**

x	1	5	10	15	20
$\overline{C}(x)$	2.15	0.55	0.35	0.28	0.25

d. It decreases.
e. $y = C(x)$: domain $= \{x \mid 1 \le x \le 20\}$; range $= \{y \mid 2.15 \le y \le 5.00\}$
$y = \overline{C}(x)$: domain $= \{x \mid 1 \le x \le 20\}$; range $= \{y \mid 0.25 \le y \le 2.15\}$

57. **a.** T(100) = 11.95, T(400) = 32.95, T(500) = 39.95 **b.** $\overline{T}(m) = \frac{4.95 + .07}{m}$ **c.** $\overline{T}(100) = 0.1195$, $\overline{T}(400) = 0.082$, $\overline{T}(500) = 0.0799$

59. $\frac{2}{3a}$ **61.** $(x - 3)(x + 2)$ **63.** 1 **65.** $\frac{3-x}{x+3}$ **67.** no solution

CHAPTER 8 TEST

1. $x + y$ **2.** $\frac{x-1}{x+1}$ **3.** $2(a + 4)$ **4.** $4(a + 3)$ **5.** $x + 3$ **6.** $\frac{38}{105}$ **7.** $\frac{7}{8}$ **8.** $\frac{1}{a-3}$ **9.** $\frac{3(x-1)}{x(x-3)}$ **10.** $\frac{x}{(x+4)(x+5)}$

11. $\frac{x+4}{(x+1)(x+2)}$ **12.** $\frac{3a+8}{3a+10}$ **13.** $\frac{x-3}{x-2}$ **14.** $-\frac{3}{5}$ **15.** no solution (3 does not check) **16.** $\frac{3}{13}$ **17.** −2, 3

18.

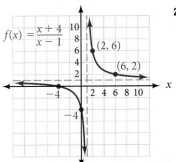

19. −7 **20.** 6 mph **21.** 15 hours **22.** 2.7 miles **23.** 1,012 mph

24. $6x^2 + 3xy - 4y^2$ **25.** $x^2 - 4x - 2 + \frac{8}{2x-1}$

Chapter 9

PROBLEM SET 9.1

1. 12 **3.** Not a real number **5.** -7 **7.** -3 **9.** 2 **11.** Not a real number **13.** 0.2 **15.** 0.2 **17.** $6a^4$ **19.** $3a^4$ **21.** xy^2 **23.** $2x^2y$ **25.** $2a^3b^5$ **27.** 6 **29.** -3 **31.** 2 **33.** -2 **35.** 2 **37.** $\frac{9}{5}$ **39.** $\frac{4}{5}$ **41.** 9 **43.** 125 **45.** 8 **47.** $\frac{1}{3}$ **49.** $\frac{1}{27}$ **51.** $\frac{6}{5}$ **53.** $\frac{8}{27}$ **55.** 7 **57.** $\frac{3}{4}$ **59.** $x^{4/5}$ **61.** a **63.** $\frac{1}{x^{2/5}}$ **65.** $x^{1/6}$ **67.** $x^{9/25}y^{1/2}z^{1/5}$ **69.** $\frac{b^{7/4}}{a^{1/8}}$ **71.** $y^{3/10}$ **73.** $\frac{1}{a^2b^4}$ **75. a.** 5 **b.** 0.5 **c.** 50 **d.** 0.05 **77. a.** $4a^2b^4$ **b.** $2ab^2\sqrt[3]{2ab^2}$ **79.** $(\sqrt{9}+\sqrt{4})^2 \stackrel{?}{=} 9 + 4$; $(3 + 4)^2 \stackrel{?}{=} 13$; $7^2 \stackrel{?}{=} 13$; $49 \neq 13$ **81.** $(a^{1/2})^{1/2} = a^{1/4}$; $a^{1/2 \cdot 1/2} = a^{1/4}$; $a^{1/4} = a^{1/4}$ **83.** 25 mph **85.** 1.618

87. $\frac{13}{8}$. The denominator is the sum of the 2 previous denominators, and the numerator is the sum of the 2 previous numerators.

89. a. 420 picometers **b.** 594 picometers **c.** 5.94×10^{-10} meters **91.** 5 **93.** 6 **95.** $4x^2y$ **97.** $5y$ **99.** 3 **101.** 2 **103.** $2ab$ **105.** 25 **107.** $48x^4y^2$ **109.** $4x^6y^6$

PROBLEM SET 9.2

1. $2\sqrt{2}$ **3.** $7\sqrt{2}$ **5.** $12\sqrt{2}$ **7.** $4\sqrt{5}$ **9.** $4\sqrt{3}$ **11.** $15\sqrt{3}$ **13.** $3\sqrt[3]{2}$ **15.** $4\sqrt[3]{2}$ **17.** $6\sqrt[3]{2}$ **19.** $2\sqrt[5]{2}$ **21.** $3x\sqrt{2x}$ **23.** $2y\sqrt[4]{2y^3}$ **25.** $2xy^2\sqrt{5xy}$ **27.** $4abc^2\sqrt{3b}$ **29.** $2bc\sqrt[3]{6a^2c}$ **31.** $2xy^2\sqrt[5]{2x^3y^2}$ **33.** $3xy^2z\sqrt[5]{x^2}$ **35.** $2\sqrt{3}$

37. $\sqrt{-20}$; not real number **39.** $\frac{\sqrt{11}}{2}$ **41. a.** $\frac{\sqrt{5}}{2}$ **b.** $\frac{2\sqrt{5}}{5}$ **c.** $2 + \sqrt{3}$ **d.** 1

43. a. $2 + \sqrt{3}$ **b.** $-2 + \sqrt{5}$ **c.** $\frac{-2 - 3\sqrt{3}}{6}$ **45.** $\frac{2\sqrt{3}}{3}$ **47.** $\frac{5\sqrt{6}}{6}$ **49.** $\frac{\sqrt{2}}{2}$ **51.** $\frac{\sqrt{5}}{5}$ **53.** $2\sqrt[3]{4}$ **55.** $\frac{2\sqrt[3]{3}}{3}$ **57.** $\frac{\sqrt[4]{24x^2}}{2x}$

59. $\frac{\sqrt[4]{8y^3}}{y}$ **61.** $\frac{\sqrt[3]{36xy^2}}{3y}$ **63.** $\frac{\sqrt[3]{6xy^2}}{3y}$ **65.** $\frac{3x\sqrt{15xy}}{5y}$ **67.** $\frac{5xy\sqrt{6xz}}{2z}$ **69. a.** $\frac{\sqrt{2}}{2}$ **b.** $\frac{\sqrt[3]{4}}{2}$ **c.** $\frac{\sqrt[4]{8}}{2}$ **71.** $5|x|$ 73. $3|xy|$ $\sqrt{3x}$ **75.** $|x - 5|$ **77.** $|2x + 3|$ **79.** $2|a(a + 2)|$ **81.** $2|x|\sqrt{x - 2}$

83. $\sqrt{9 + 16} \stackrel{?}{=} \sqrt{9} + \sqrt{16}$
$\sqrt{25} \stackrel{?}{=} 3 + 4$
$5 \neq 7$

85. $5\sqrt{13}$ feet **87. a.** ≈ 89.4 miles **b.** ≈ 126.5 miles **c.** ≈ 154.9 miles

91. $\sqrt{2}, \sqrt{3}, 2, \sqrt{5}, \sqrt{6}, \sqrt{7}$; $a_{10} = \sqrt{11}$; $a_{100} = \sqrt{101}$ **93.** $7x$ **95.** $27xy^2$ **97.** $\frac{5}{6}x$ **99.** $3\sqrt{2}$ **101.** $5y\sqrt{3xy}$ **103.** $2a\sqrt[3]{ab^2}$

PROBLEM SET 9.3

1. $7\sqrt{5}$ **3.** $-x\sqrt{7}$ **5.** $\sqrt[3]{10}$ **7.** $9\sqrt[5]{6}$ **9.** 0 **11.** $\sqrt{5}$ **13.** $-32\sqrt{2}$ **15.** $-3x\sqrt{2}$ **17.** $-2\sqrt[3]{2}$ **19.** $8x\sqrt[3]{xy^2}$ **21.** $3a^2b\sqrt{3ab}$ **23.** $11ab\sqrt[3]{3a^2b}$ **25.** $10xy\sqrt[4]{3y}$ **27.** $\sqrt{2}$ **29.** $\frac{8\sqrt{5}}{15}$ **31.** $\frac{(x - 1)\sqrt{x}}{x}$ **33.** $\frac{3\sqrt{2}}{2}$ **35.** $\frac{5\sqrt{6}}{6}$ **37.** $\frac{8\sqrt[3]{25}}{5}$ **39.** $\sqrt{12} \approx 3.464$; $2\sqrt{3} \approx 2(1.732) = 3.464$ **41.** $\sqrt{8} + \sqrt{18} \approx 2.828 + 4.243 = 7.071$; $\sqrt{50} \approx 7.071$; $\sqrt{26} \approx 5.099$ **43.** $8\sqrt{2x}$ **45.** 5 **53.** $\sqrt{2}$:1 **55. a.** $\sqrt{2}:1 \approx 1.414:1$ **b.** $5:\sqrt{2}$ **c.** 5:4 **57.** 6 **59.** $4x^2 + 3xy - y^2$ **61.** $x^2 + 6x + 9$ **63.** $x^2 - 4$ **65.** $6\sqrt{2}$ **67.** 6 **69.** $9x$ **71.** $\frac{\sqrt{6}}{2}$

PROBLEM SET 9.4

1. $3\sqrt{2}$ **3.** $10\sqrt{21}$ **5.** 720 **7.** 54 **9.** $\sqrt{6} - 9$ **11.** $24 + 6\sqrt[3]{4}$ **13.** $7 + 2\sqrt{6}$ **15.** $x + 2\sqrt{x} - 15$ **17.** $34 + 20\sqrt{3}$ **19.** $19 + 8\sqrt{3}$ **21.** $x - 6\sqrt{x} + 9$ **23.** $4a - 12\sqrt{ab} + 9b$ **25.** $x + 4\sqrt{x} - 4$ **27.** $x - 6\sqrt{x - 5} + 4$ **29.** 1 **31.** $a - 49$ **33.** $25 - x$ **35.** $x - 8$ **37.** $10 + 6\sqrt{3}$ **39.** $\frac{\sqrt{3} + 1}{2}$ **41.** $\frac{5 - \sqrt{5}}{4}$ **43.** $\frac{x + 3\sqrt{x}}{x - 9}$ **45.** $\frac{10 + 3\sqrt{5}}{11}$ **47.** $\frac{3\sqrt{x} + 3\sqrt{y}}{x - y}$ **49.** $2 + \sqrt{3}$ **51.** $\frac{11 - 4\sqrt{7}}{3}$ **53. a.** $2\sqrt{x}$ **b.** $x - 4$ **c.** $x + 4\sqrt{x} + 4$ **d.** $\frac{x + 4\sqrt{x} + 4}{x - 4}$ **55. a.** 10 **b.** 23 **c.** $27 + 10\sqrt{2}$ **d.** $\frac{27 + 10\sqrt{2}}{23}$ **57. a.** $2\sqrt{2} + \sqrt{6}$ **b.** $2\sqrt{3} + 2$ **c.** $\sqrt{3} + 1$ **d.** $\frac{-1 + \sqrt{3}}{2}$ **59. a.** 1 **b.** -1 **61.** $(\sqrt[3]{2} + \sqrt[3]{3})(\sqrt[3]{4} - \sqrt[3]{6} + \sqrt[3]{9}) = \sqrt[3]{8} - \sqrt[3]{12} + \sqrt[3]{18} + \sqrt[3]{12} - \sqrt[3]{18} + \sqrt[3]{27} = 2 + 3 = 5$ **63.** $10\sqrt{3}$ **65.** $x + 6\sqrt{x} + 9$ **67.** 75 **69.** $\frac{5\sqrt{2}}{4}$ second; $\frac{5}{2}$ second **75.** $t^2 + 10t + 25$ **77.** x **79.** 7 **81.** $-4, -3$ **83.** $-6, -3$ **85.** $-5, -2$ **87.** Yes **89.** No

Answers to Odd-Numbered Problems

PROBLEM SET 9.5

1. 4 **3.** ∅ **5.** 5 **7.** ∅ **9.** $\frac{39}{2}$ **11.** ∅ **13.** 5 **15.** 3 **17.** $-\frac{32}{3}$ **19.** 3, 4 **21.** $-1, -2$ **23.** -1 **25.** ∅ **27.** 7 **29.** 0, 3 **31.** -4 **33.** 8 **35.** 0 **37.** 9 **39.** 0 **41.** 8 **43.** Possible solution 9, which does not check; ∅

45. a. 100 **b.** 40 **c.** ∅ **d.** Possible solutions 5, 8; only 8 checks

47. a. 3 **b.** 9 **c.** 3 **d.** ∅ **e.** 4. **f.** ∅ **g.** Possible solutions 1,4; only 4 checks **49.** $h = 100 - 16t^2$ **51.** $\frac{392}{121} \approx 3.24$ feet

53.

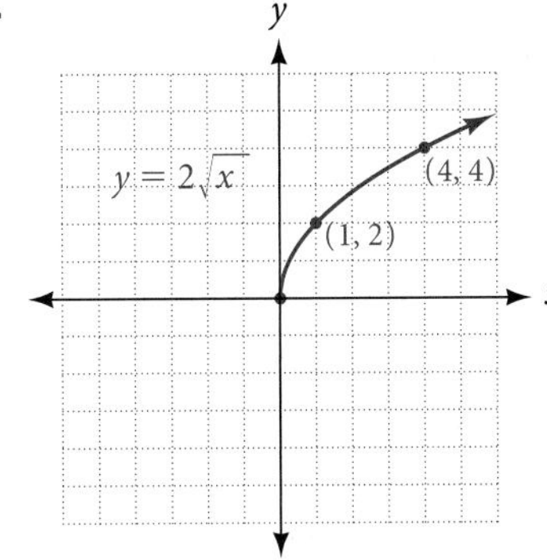

55.

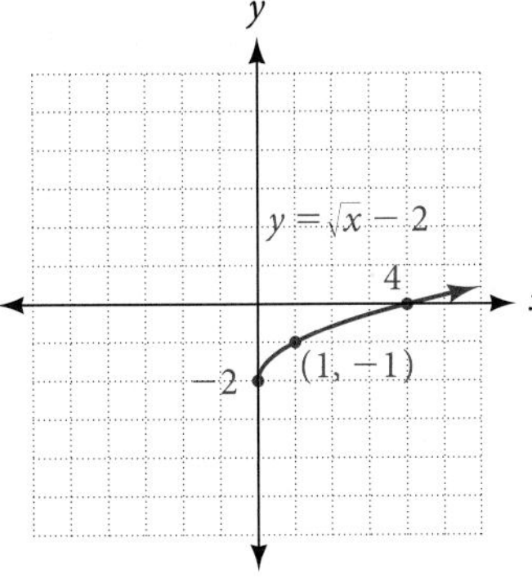

57.

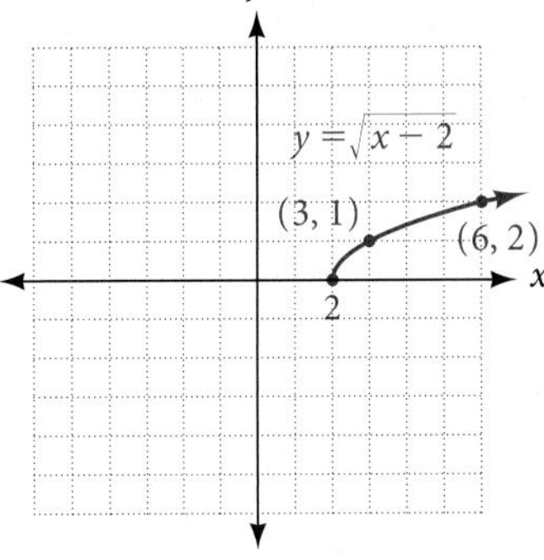

59.

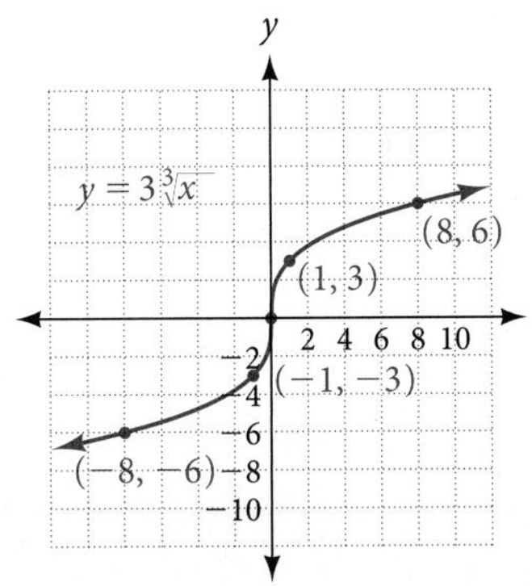

61.

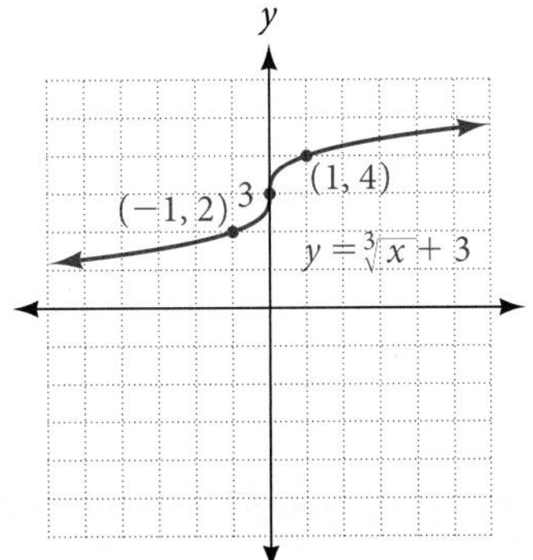

63.

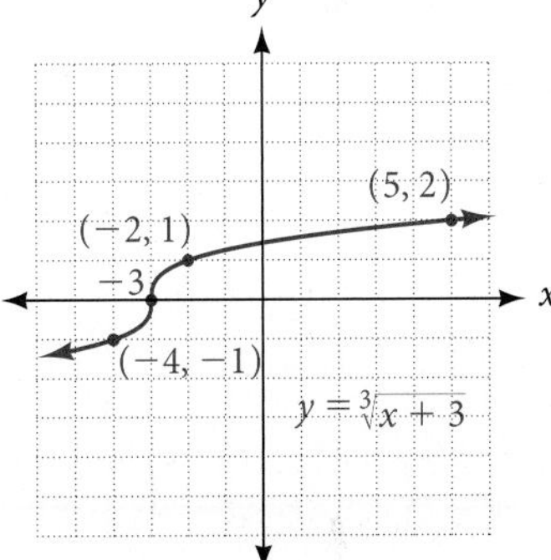

65. 5 **67.** $2\sqrt{3}$ **69.** -1 **71.** 1 **73.** 4 **75.** 2 **77.** $10 - 2x$ **79.** $2 - 3x$ **81.** $6 + 7x - 20x^2$ **83.** $8x - 12x^2$ **85.** $4 + 12x + 9x^2$ **87.** $4 - 9x^2$

PROBLEM SET 9.6

1. $6i$ **3.** $-5i$ **5.** $6i\sqrt{2}$ **7.** $-2i\sqrt{3}$ **9.** 1 **11.** -1 **13.** $-i$ **15.** $x = 3, y = -1$ **17.** $x = -2, y = -\frac{1}{2}$ **19.** $x = -8, y = -5$ **21.** $x = 7, y = \frac{1}{2}$ **23.** $x = \frac{3}{7}, y = \frac{2}{5}$ **25.** $5 + 9i$ **27.** $5 - i$ **29.** $2 - 4i$ **31.** $1 - 6i$ **33.** $2 + 2i$ **35.** $-1 - 7i$ **37.** $6 + 8i$ **39.** $2 - 24i$ **41.** $-15 + 12i$ **43.** $18 + 24i$ **45.** $10 + 11i$ **47.** $21 + 23i$ **49.** $-2 + 2i$ **51.** $2 - 11i$ **53.** $-21 + 20i$ **55.** $-2i$ **57.** $-7 - 24i$ **59.** 5 **61.** 40 **63.** 13 **65.** 164 **67.** $-3 - 2i$ **69.** $-2 + 5i$ **71.** $\frac{8}{13} + \frac{12}{13}i$ **73.** $-\frac{18}{13} - \frac{12}{13}i$ **75.** $-\frac{5}{13} + \frac{12}{13}i$ **77.** $\frac{13}{15} - \frac{2}{5}i$ **79.** $R = -11 - 7i$ ohms **81.** $-\frac{3}{2}$ **83.** $-3, \frac{1}{2}$ **85.** $\frac{5}{4}$ or $\frac{4}{5}$

CHAPTER 9 TEST

1. $\frac{1}{9}$ **2.** $\frac{7}{5}$ **3.** $a^{5/12}$ **4.** $\frac{x^{13/12}}{y}$ **5.** $7x^4y^5$ **6.** $2x^2y^4$ **7.** $2a$ **8.** $x^{n^2-n}y^{1-n^3}$ **9.** $6a^2 - 10a$ **10.** $16a^3 - 40a^{3/2} + 25$ **11.** $(3x^{1/3} - 1)(x^{1/3} + 2)$ **12.** $(3x^{1/3} - 7)(3x^{1/3} + 7)$ **13.** $\frac{x + 4}{x^{1/2}}$ **14.** $\frac{3}{(x^2 - 3)^{1/2}}$ **15.** $5xy^2\sqrt{5xy}$ **16.** $2x^2y^2\sqrt[3]{5xy^2}$ **17.** $\frac{\sqrt{6}}{3}$ **18.** $\frac{2a^2b\sqrt{15bc}}{5c}$ **19.** $-6\sqrt{3}$ **20.** $-3ab\sqrt[3]{3}$ **21.** $x + 3\sqrt{x} - 28$ **22.** $21 - 6\sqrt{6}$ **23.** $\frac{5 + 5\sqrt{3}}{2}$ **24.** $\frac{x - 2\sqrt{2x} + 2}{x - 2}$ **25.** 8 (1 does not check) **26.** -4 **27.** -3

28.

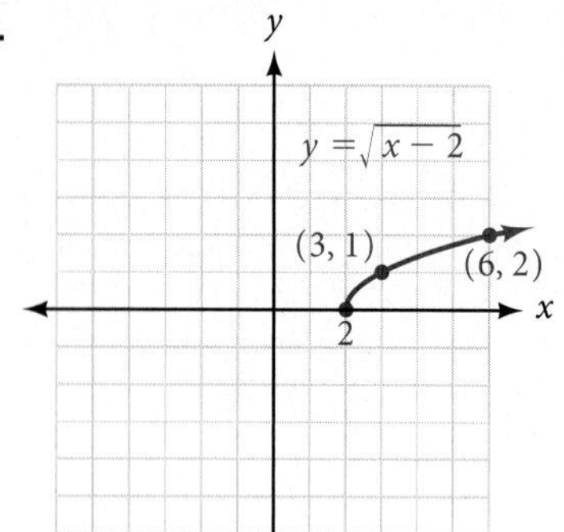

29.

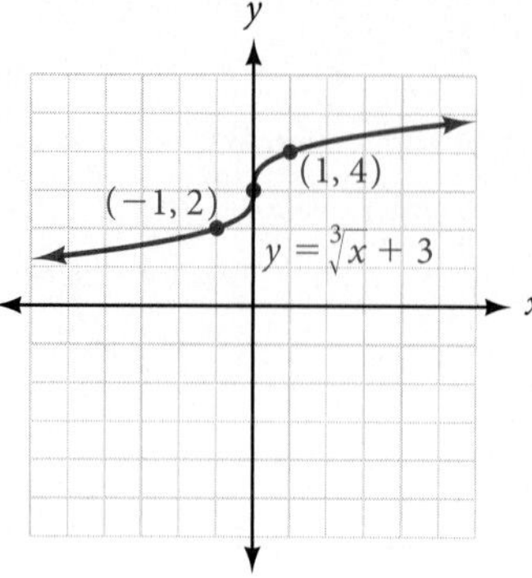

30. $x = \frac{1}{2}, y = 7$ **31.** $6i$ **32.** $17 - 6i$
33. $9 - 40i$ **34.** $-\frac{5}{13} - \frac{12}{13}i$
35. $i^{38} = (i^2)^{19} = (-1)^{19} = -1$

Chapter 10

PROBLEM SET 10.1

1. 1 **3.** 2 **5.** $\frac{1}{27}$ **7.** 13 **9.** $\frac{7}{12}$ **11.** $\frac{3}{16}$ **13.**

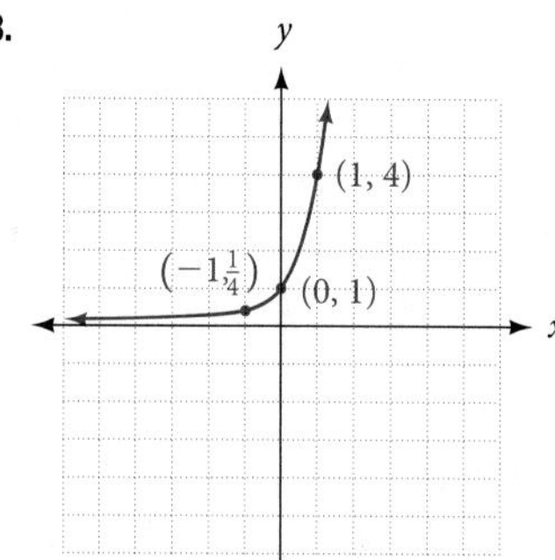

15.

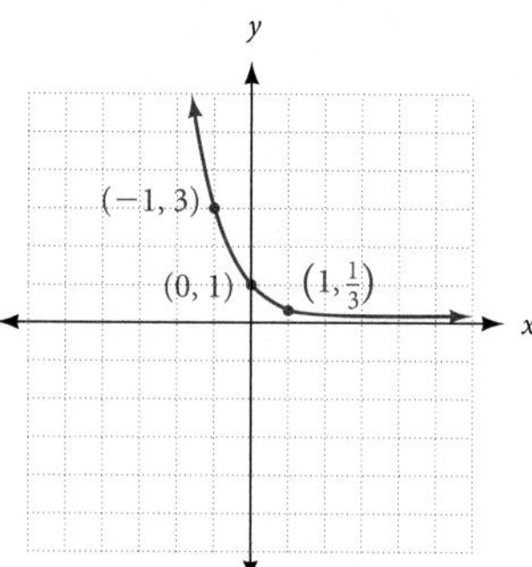

17.

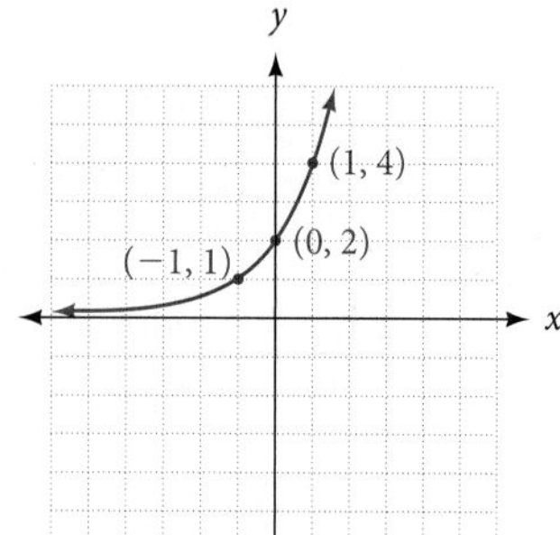

19.

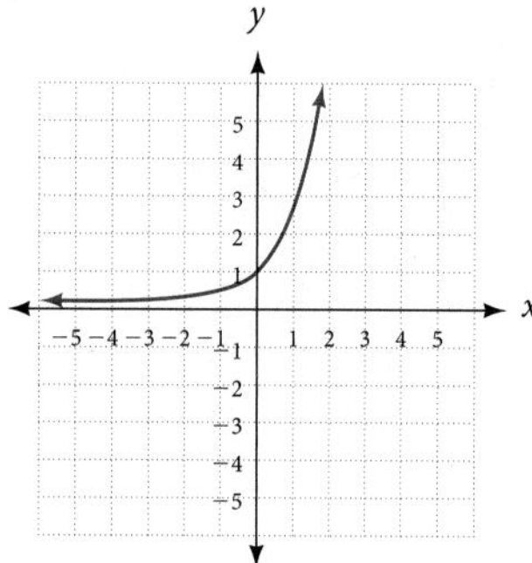

21.

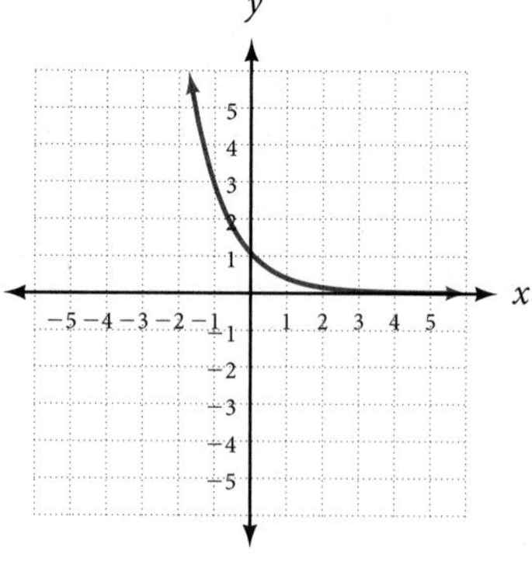

23.

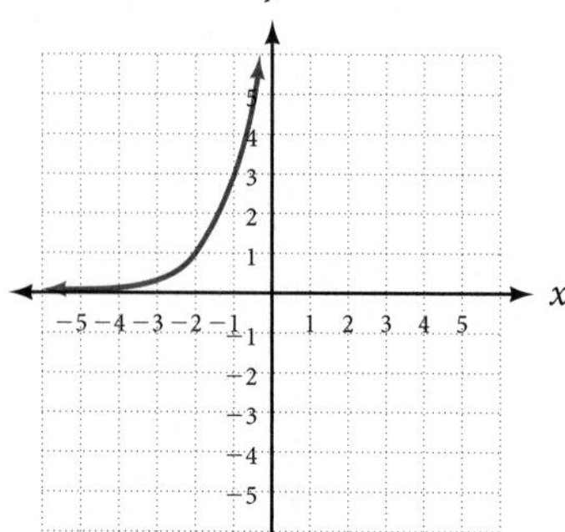

25.

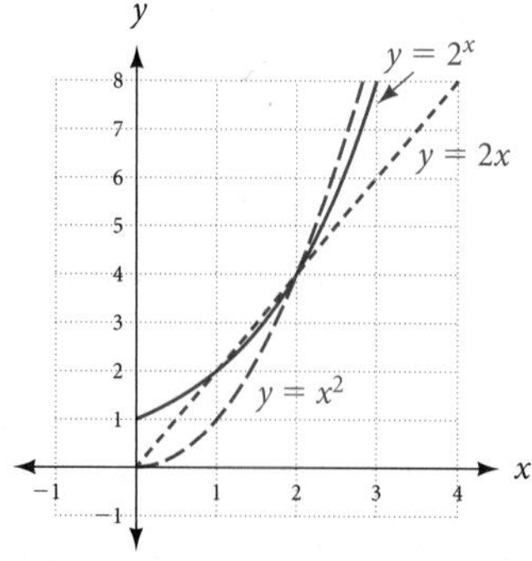

27.

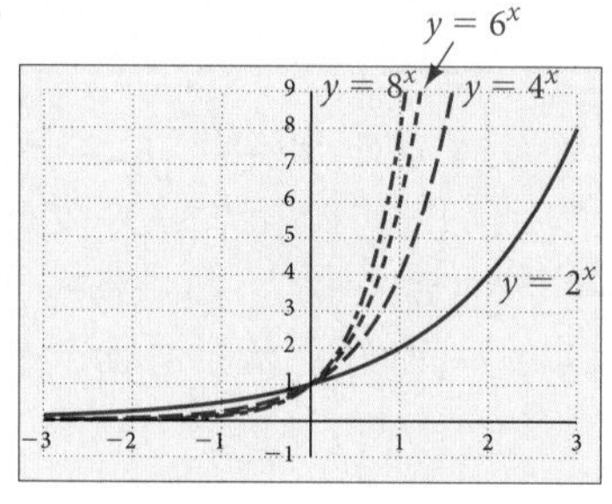

29. $h = 6 \cdot \left(\frac{2}{3}\right)^n$; 5th bounce: $6\left(\frac{2}{3}\right)^5 \approx 0.79$ feet **31.** 4.27 days

33. **a.** $A(t) = 1{,}200\left(1 + \frac{.06}{4}\right)^{4t}$ **b.** \$1,932.39 **c.** About 11.64 years **d.** \$1,939.29

35. **a.** \$129,138.48 **b.** $\{t \mid 0 \le t \le 6\}$ **c.**

d. $\{V(t) \mid 52{,}942.05 \le V(t) \le 450{,}000\}$ **e.** After approximately 4 years and 8 months

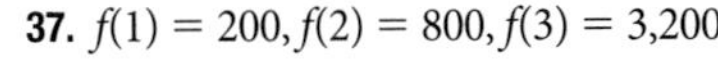

37. $f(1) = 200, f(2) = 800, f(3) = 3{,}200$

39.

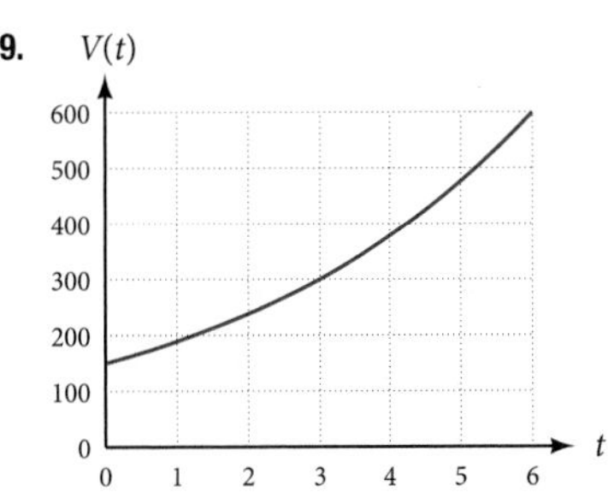

41. **a.** \$0.42 **b.** \$1.00 **c.** \$1.78 **d.** \$17.84

43. 1,258,525 bankruptcies, which is 58,474 less than the actual number. **45.** **a.** 251,437 cells **b.** 12,644 cells **c.** 32 cells

47. $y = \frac{x+3}{2}$ **49.** $y = \pm\sqrt{x+3}$ **51.** $y = \frac{2x-4}{x-1}$ **53.** $y = x^2 + 3$

PROBLEM SET 10.2

1. $f^{-1}(x) = \frac{x+1}{3}$ **3.** $f^{-1}(x) = \sqrt[3]{x}$ **5.** $f^{-1}(x) = \frac{x-3}{x-1}$ **7.** $f^{-1}(x) = 4x + 3$ **9.** $f^{-1}(x) = 2(x+3) = 2x + 6$

11. $f^{-1}(x) = \frac{3}{2}(x+3) = \frac{3}{2}x + \frac{9}{2}$ **13.** $f^{-1}(x) = \sqrt[3]{x+4}$ **15.** $f^{-1}(x) = \frac{x+3}{4-2x}$ **17.** $f^{-1}(x) = \frac{1-x}{3x-2}$

19.

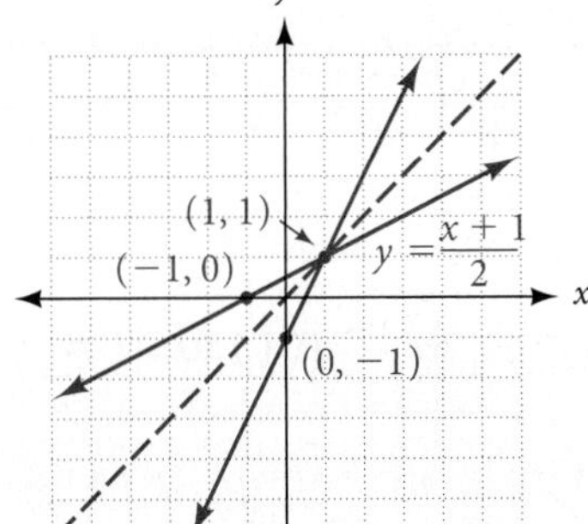

21.

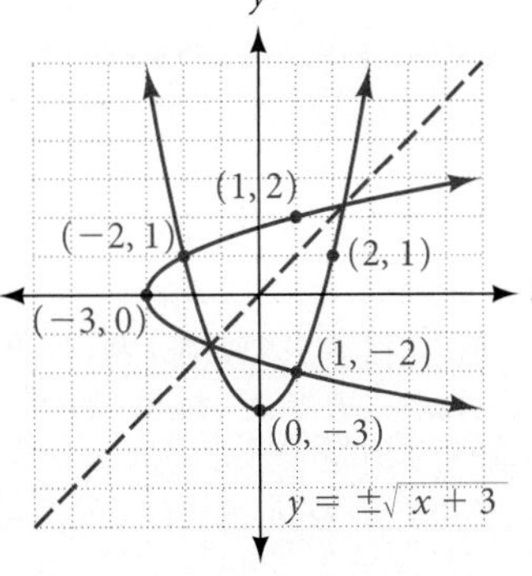

23.

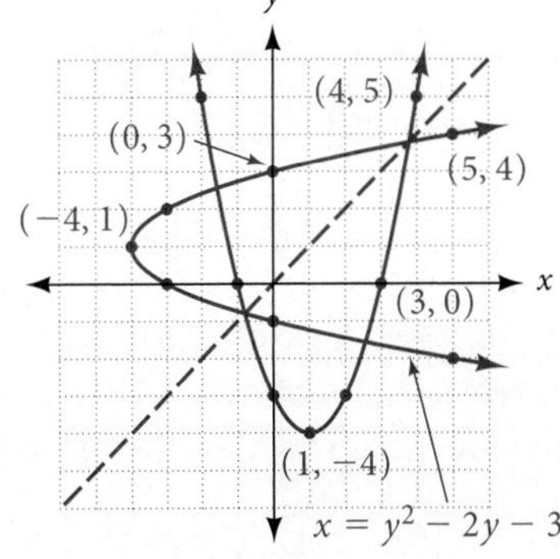

25.

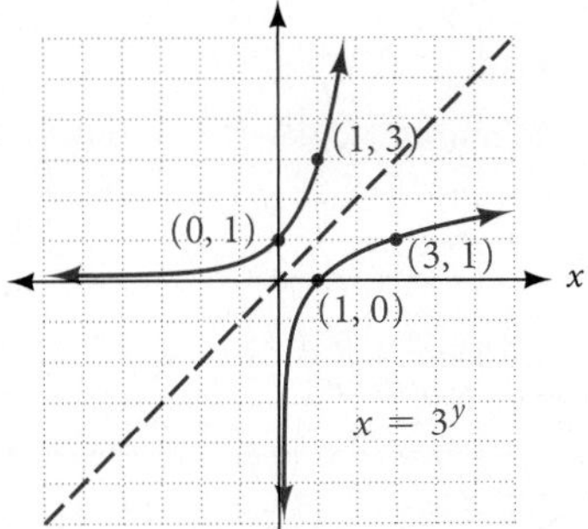

27.

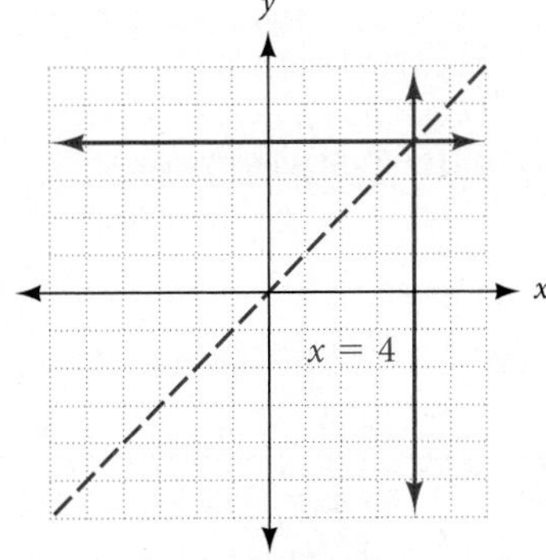

29.

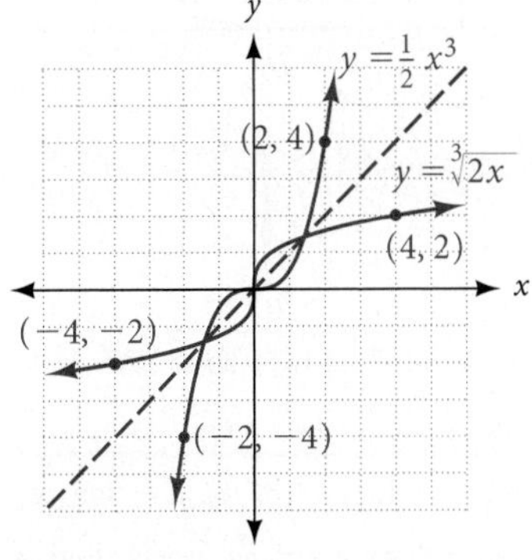

31.

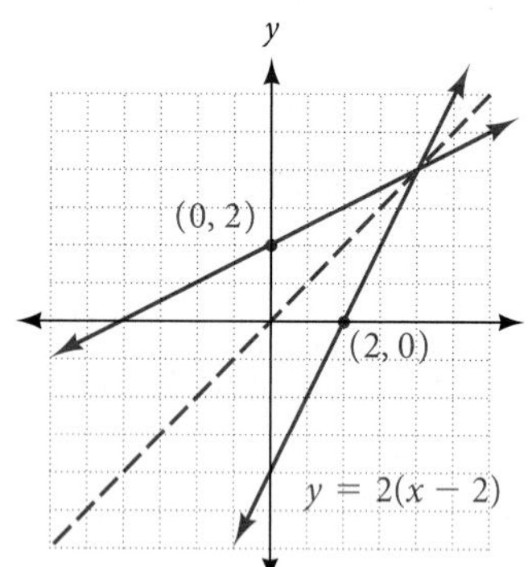

33.

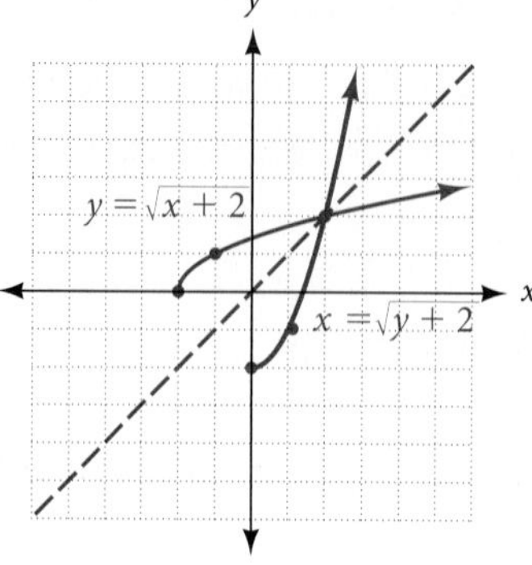

35. **a.** Yes **b.** No **c.** Yes

37. **a.** 4 **b.** $\frac{4}{3}$ **c.** 2 **d.** 2 **39.** $f^{-1}(x) = \frac{1}{x}$ **41.** $f^{-1}(x) = 7(x + 2) = 7x + 14$
43. **a.** -3 **b.** -6 **c.** 2 **d.** 3 **e.** -2 **f.** 3 **g.** inverses **45.** **a.** 489.4 **b.** $s^{-1}(t) = \frac{t - 249.4}{16}$ **c.** 2006
47. **a.** 6629.33 ft/s **b.** $f^{-1}(m) = \frac{15m}{22}$ **c.** 1.36 mph **49.** $\frac{1}{9}$ **51.** $\frac{2}{3}$ **53.** $\sqrt[3]{4}$ **55.** 3 **57.** 4 **59.** 4 **61.** 1

PROBLEM SET 10.3

1. $\log_2 16 = 4$ **3.** $\log_5 125 = 3$ **5.** $\log_{10} 0.01 = -2$ **7.** $\log_2 \frac{1}{32} = -5$ **9.** $\log_{1/2} 8 = -3$ **11.** $\log_3 27 = 3$ **13.** $10^2 = 100$
15. $2^6 = 64$ **17.** $8^0 = 1$ **19.** $10^{-3} = 0.001$ **21.** $6^2 = 36$ **23.** $5^{-2} = \frac{1}{25}$ **25.** 9 **27.** $\frac{1}{125}$ **29.** 4 **31.** $\frac{1}{3}$ **33.** 2 **35.** $\sqrt[3]{5}$
37. 2 **39.** 6 **41.** $\frac{2}{3}$ **43.** $-\frac{1}{2}$ **45.** $\frac{1}{64}$

47.

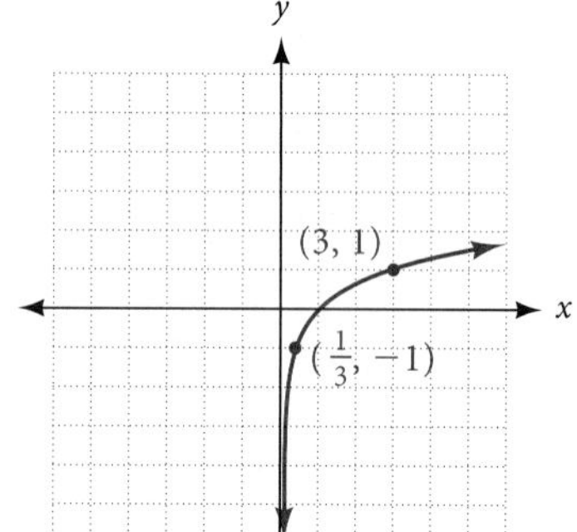

49.

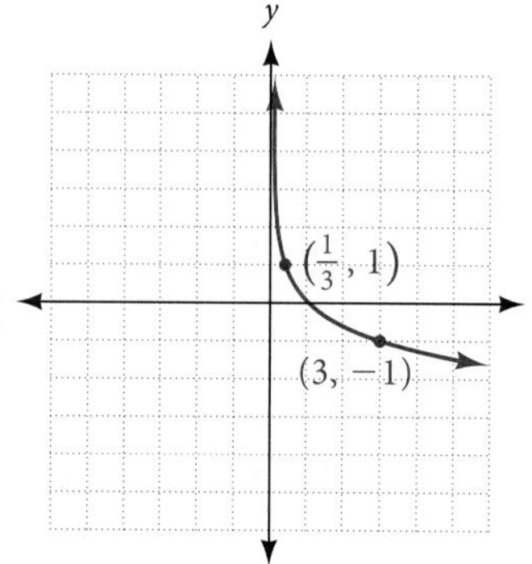

51.

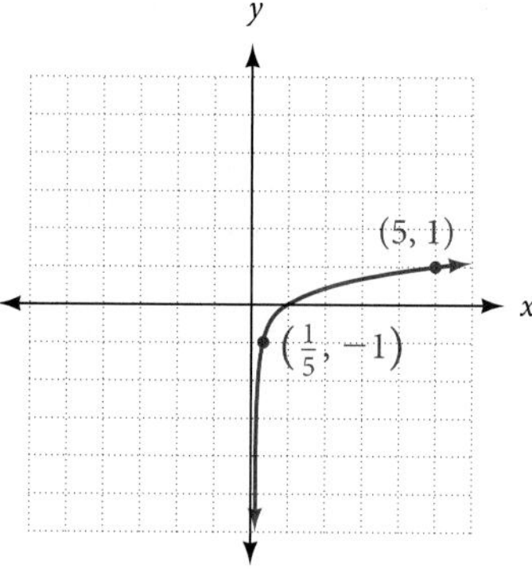

53.

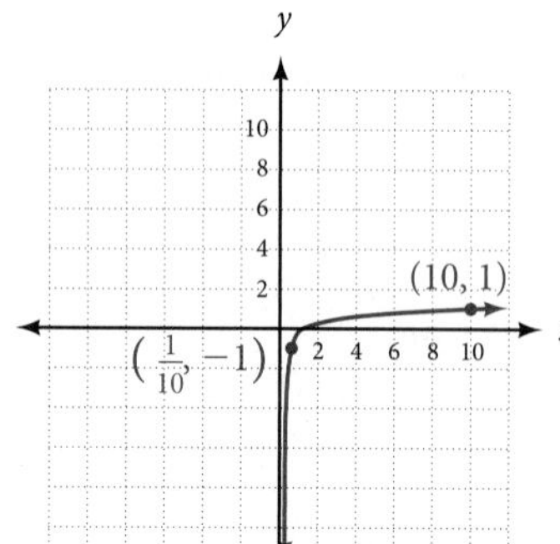

55. $y = 3^x$ **57.** $y = \log_{1/3} x$ **59.** 4 **61.** $\frac{3}{2}$ **63.** 3 **65.** 1 **67.** 0 **69.** 0 **71.** $\frac{1}{2}$
73. $\frac{3}{2}$ **75.** 1 **77.** -2 **79.** 0 **81.** $\frac{1}{2}$

83.

Prefix	Multiplying Factor	$\log_{10}$ (Multiplying Factor)
Nano	0.000 000 001	−9
Micro	0.000 001	−6
Deci	0.1	−1
Giga	1,000,000,000	9
Peta	1,000,000,000,000,000	15

85. 2 **87.** 10^8 times as large **89.** 120 **91.** 4 **93.** $-4, 2$ **95.** $-\frac{11}{8}$ **97.** $2^3 = (x + 2)(x)$ **99.** $3^4 = \frac{x - 2}{x + 1}$

PROBLEM SET 10.4

1. $\log_3 4 + \log_3 x$ **3.** $\log_6 5 - \log_6 x$ **5.** $5 \log_2 y$ **7.** $\frac{1}{3} \log_9 z$ **9.** $2 \log_6 x + 4 \log_6 y$ **11.** $\frac{1}{2} \log_5 x + 4 \log_5 y$
13. $\log_b x + \log_b y - \log_b z$ **15.** $\log_{10} 4 - \log_{10} x - \log_{10} y$ **17.** $2 \log_{10} x + \log_{10} y - \frac{1}{2} \log_{10} z$
19. $3 \log_{10} x + \frac{1}{2} \log_{10} y - 4 \log_{10} z$ **21.** $\frac{2}{3} \log_b x + \frac{1}{3} \log_b y - \frac{4}{3} \log_b z$ **23.** $\frac{2}{3}\log_3 x + \frac{1}{3}\log_3 y - 2\log_3 z$

25. $2\log_a 2 + 5\log_a x - 2\log_a 3 - 2$ **27.** $\log_b xz$ **29.** $\log_3 \frac{x^2}{y^3}$ **31.** $\log_{10} \sqrt{x}\sqrt[3]{y}$ **33.** $\log_2 \frac{x^3\sqrt{y}}{z}$ **35.** $\log_2 \frac{\sqrt{x}}{y^3z^4}$ **37.** $\log_{10} \frac{x^{3/2}}{y^{3/4}z^{4/5}}$ **39.** $\log_5 \frac{\sqrt{x}\cdot\sqrt[3]{y^2}}{z^4}$ **41.** $\log_3 \frac{x-4}{x+4}$ **43.** $\frac{2}{3}$ **45.** 18 **47.** 3 **49.** 3 **51.** 4 **53.** 4 **55.** 1 **57.** 0 **59.** $\frac{3}{2}$ **61.** 27 **63.** $\frac{5}{3}$ **67.** **a.** 1.602 **b.** 2.505 **c.** 3.204 **69.** $\text{pH} = 6.1 + \log_{10}x - \log_{10}y$ **71.** 2.52 **73.** 1 **75.** 1 **77.** 4 **79.** 2.5×10^{-6} **81.** 51

PROBLEM SET 10.5

1. 2.5775 **3.** 1.5775 **5.** 3.5775 **7.** -1.4225 **9.** 4.5775 **11.** 2.7782 **13.** 3.3032 **15.** -2.0128 **17.** -1.5031 **19.** -0.3990 **21.** 759 **23.** 0.00759 **25.** 1,430 **27.** 0.00000447 **29.** 0.0000000918 **31.** 10^{10} **33.** 10^{-10} **35.** 10^{20} **37.** $\frac{1}{100}$ **39.** 1,000 **41.** $\frac{1}{e}$ **43.** 25 **45.** $\frac{1}{8}$ **47.** 1 **49.** 5 **51.** x **53.** 4 **55.** -3 **57.** $\frac{3}{2}$ **59.** $\ln 10 + 3t$ **61.** $\ln A - 2t$ **63.** $2 + 3t\log 1.01$ **65.** $rt + \ln P$ **67.** $3 - \log 4.2$ **69.** 2.7080 **71.** -1.0986 **73.** 2.1972 **75.** 2.7724

77. San Francisco was approx. 2,000 times greater. **79.**

x	$(1 + x)^{1/x}$
1	2
0.5	2.25
0.1	2.5937
0.01	2.7048
0.001	2.7169
0.0001	2.7181
0.00001	2.7183

81. 2009 **83.** Approximately 3.19

85. 1.78×10^{-5} **87.** 3.16×10^5 **89.** 2.00×10^8 **91.**

Location	Date	Magnitude M	Shockwave T
Moresby Island	Jan. 23	4.0	1.00×10^4
Vancouver Island	Apr. 30	5.3	1.99×10^5
Quebec City	June 29	3.2	1.58×10^3
Mould Bay	Nov. 13	5.2	1.58×10^5
St. Lawrence	Dec. 14	3.7	5.01×10^3

SOURCE: *National Resources Canada, National Earthquake Hazards Program.*

93. 12.9% **95.** 5.3% **97.** $\frac{7}{10}$ **99.** 3.1250 **101.** 1.2575 **103.** $t\log 1.05$ **105.** $0.05t$

PROBLEM SET 10.6

1. 1.4650 **3.** 0.6826 **5.** -1.5440 **7.** -0.6477 **9.** -0.3333 **11.** 2 **13.** -0.1845 **15.** 0.1845 **17.** 1.6168 **19.** 2.1131 **21.** -1 **23.** 1.2275 **25.** .3054 **27.** 42.5528 **29.** 6.0147 **31.** 1.333 **33.** 0.75 **35.** 1.3917 **37.** 0.7186 **39.** 2.6356 **41.** 4.1632 **43.** 5.8435 **45.** -1.0642 **47.** 2.3026 **49.** 10.7144 **51.** 11.72 years **53.** 9.25 years **55.** 8.75 years **57.** 18.58 years **59.** 11.55 years **61.** 18.31 years **63.** 11.45 years **65.** October 2018 **67.** 2009 **69.** 1992 **71.** 10.07 years **73.** 2000 **75.** $(-2, -23)$, lowest **77.** $\left(\frac{3}{2}, 9\right)$, highest **79.** 2 seconds, 64 feet

CHAPTER 10 TEST

1.

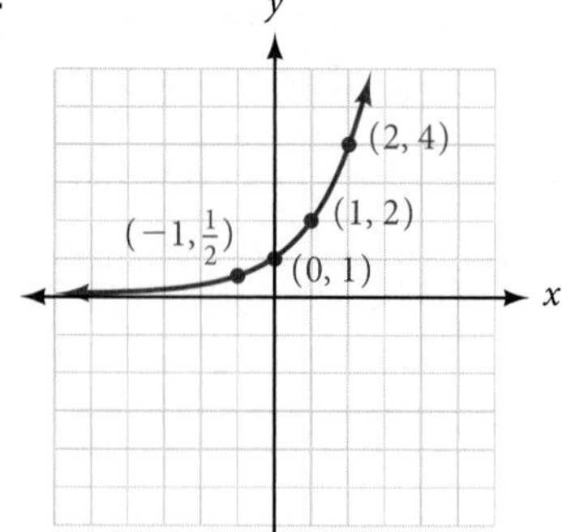

2.

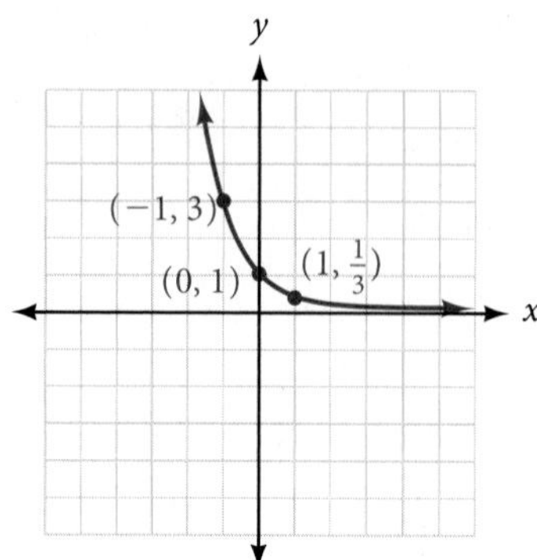

3.

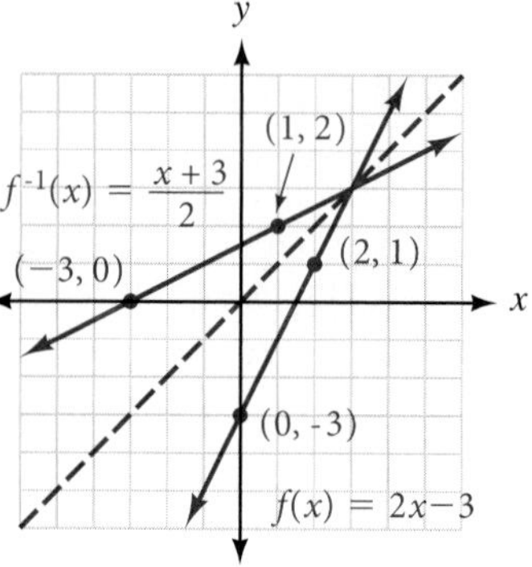

4.

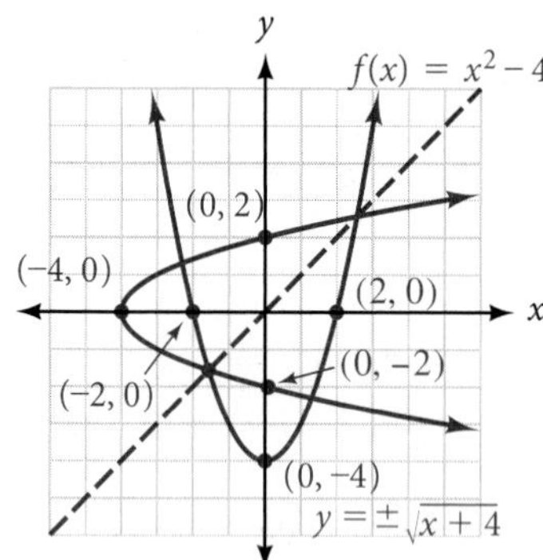

5. 64 **6.** $\sqrt{5}$ **7.**

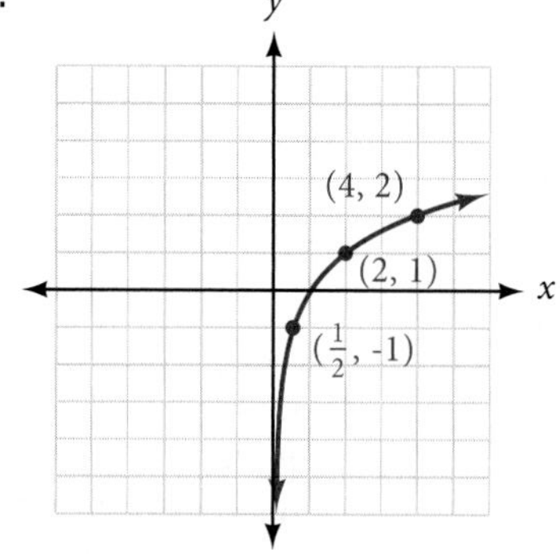

8.

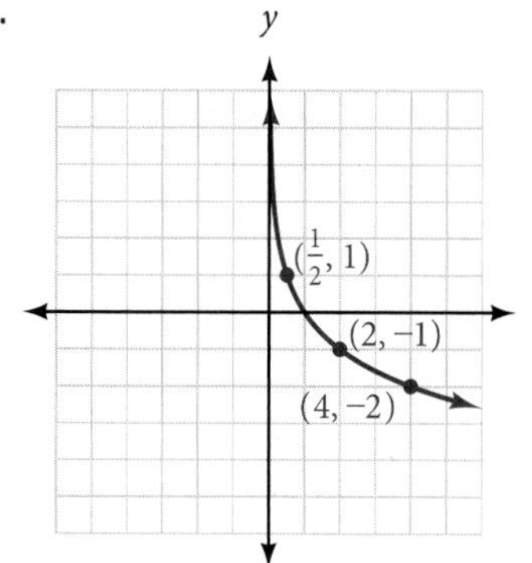

9. $\frac{2}{3}$ **10.** 1.5646 **11.** 4.3692 **12.** −1.9101 **13.** 3.8330 **14.** −3.0748 **15.** $3 + 2\log_2 x - \log_2 y$ **16.** $\frac{1}{2}\log x - 4\log y - \frac{1}{5}\log z$ **17.** $\log_3 \frac{x^2}{\sqrt{y}}$ **18.** $\log_3 \frac{\sqrt[3]{x}}{yz^2}$ **19.** 70,404 **20.** 0.00225 **21.** 1.4650 **22.** $\frac{5}{4}$ **23.** 15 **24.** 8 (−1 does not check) **25.** 6.18 **26.** \$651.56 **27.** 13.87 years **28.** \$7,373

Index

A

Addition
  of polynomials, 267
Addition property
  of equality, 12
  of inequality, 67
Analytic geometry, 103
Antilog, 594
Antilogarithm, 594
Asymptotes, 460, 471
Average cost, 479
Average speed, 167
Axes, 87

B

Binomial, 267
Binomial squares, 279
Boundary, 139

C

Cartesian coordinate system, 103
Change-of-base property, 605
Characteristic, 594
Coefficient, 259
Column method, 274
Common denominator, 441
Common factor, 301
Common logarithm, 593
Complementary angles, 38
Completing the square, 355
Complex conjugates, 547
Complex fraction, 451
Complex number, 543
Composition of functions, 188
Compound inequality, 75
Compound Interest, 560
Concave down, 396
Concave up, 395
Conjugates, 527
Constant of variation, 175
Constant term, 27
Continuously compounded interest, 561
Coordinate system, 89

D

Decibel, 585
Degree of a polynomial, 267
Dependent, 165
Dependent variable, 165
Difference of two squares, 280
Difference quotient, 427
Directly proportional, 175
Direct variation, 175
Discriminant, 377
Division
  of polynomials, 479
  with exponents, 247
Domain, 152, 154
Doubling time, 603

E

Elimination method, 207
Equation
  exponential 603
  linear, 27, 95, 102
  quadratic, 333, 535
Equivalent
  equations, 12
Exponent, 237
  negative integer, 247
  properties of, 239-241
Exponential function, 557
Extraneous solutions, 386

F

Factor
  common, 301
Factoring
  difference of two squares, 319
  by grouping, 303
  trinomials, 309
Family of curves, 560
Foil method, 273
Formulas, 35
Function, 151, 154, 156, 166, 186
Function map, 188
Function notation, 165

G

Golden ratio, 501
Graph
  of linear inequality, 67
  of ordered pairs, 87
  of a straight line, 67
Graphing Parabolas, 393, 399
Greatest common factor, 301

H

Horizontal asymptote, 471
Horizontal Line Test, 569
Hypotenuse, 342

I

Imaginary numbers, 544
Imaginary part, 544
Independent variable, 165
Index, 496
Inequality, 67, 75
Input, 151
Intercepts, 111
Inverse Function, 567
Inversely proportional, 177
Inverse reasoning, 573
Inverse variation, 175

J

Joint variation, 179

L

Least common denominator, 442
Like terms, 3, 261
Line graph, 87
Linear equations
  in one variable, 27
  in two variables, 95
Linear inequality
  in one variable, 67
  in two variables, 139
Linear system, 199
Logarithmic scale, 577
Long division, 289

M

Mantissa, 594
Monomial, 259
  addition of, 261
  division of, 259
  multiplication of, 259
  subtraction of, 261
Multiplication
  of exponents, 239
  of polynomials, 273
Multiplication property
  of equality, 19
  of inequality, 68
Multiplicity, 379

N

Natural logarithm, 596
Negative exponent, 247
Negative Slope, 119
Negative square root, 495
Number *e*, 561
Number *i*, 543
Numerical Coefficient, 259

O

One or more elements, 156
One-to-one function, 569
Ordered pair, 87
Origin, 89
Output, 151

P

Paired data, 87
Parabola, 393
Parallel, 201
Percent problems, 38
Perpendicular, 89
Point-slope form, 133
Polynomial, 267
  addition and subtraction, 268
  division, 285, 289
  long division, 289
  multiplication, 273
Positive Slope, 119
Positive square root, 355, 361
Properties of exponents, 239, 241, 248
Properties of logarithms, 586, 605
Pythagorean theorem, 342

Q

Quadrants, 89
Quadratic equation, 333
Quadratic formula, 367
Quadratic Theorem, 367

R

Radical expression, 496
Radical sign, 496
Radicand, 496
Range, 152
Rational expression, 423
Rational function, 425
Rationalizing the denominator, 509
Real part, 544
Rectangular coordinate system, 89
Recursive, 513
Relation, 156
Rhind Papyrus, 476
Rise, 119
Run, 119

S

Scatter diagram, 87
Section of the coordinate plane, 139
Sequence, 6
Similar radicals, 519
Similar terms, 3, 217
Simplified form, 508
Slope, 119, 120, 133, 427
Slope-intercept form, 129
Solution set, 139
Standard form, 102, 333
Substitution method, 217
Subtraction
  of polynomials, 268
Supplementary angles, 38
Systems of linear equations, 199

T

Term, 3
Trinomial, 267

V

Varies directly, 175
Varies inversely, 178
Varies jointly, 179
Vertex, 394
Vertical asymptote, 460
Vertical Line Test, 157

X

X-axis, 89
X-coordinate, 89
X-intercept, 111

Y

Y-axis, 89
Y-coordinate, 89
Y-intercept, 111

Z

Zero-factor property, 333